Studies in Computational Intelligence

Volume 1077

Series Editor

Janusz Kacprzyk, Polish Academy of Sciences, Warsaw, Poland

The series "Studies in Computational Intelligence" (SCI) publishes new developments and advances in the various areas of computational intelligence—quickly and with a high quality. The intent is to cover the theory, applications, and design methods of computational intelligence, as embedded in the fields of engineering, computer science, physics and life sciences, as well as the methodologies behind them. The series contains monographs, lecture notes and edited volumes in computational intelligence spanning the areas of neural networks, connectionist systems, genetic algorithms, evolutionary computation, artificial intelligence, cellular automata, self-organizing systems, soft computing, fuzzy systems, and hybrid intelligent systems. Of particular value to both the contributors and the readership are the short publication timeframe and the world-wide distribution, which enable both wide and rapid dissemination of research output.

Indexed by SCOPUS, DBLP, WTI Frankfurt eG, zbMATH, SCImago.

All books published in the series are submitted for consideration in Web of Science.

Hocine Cherifi · Rosario Nunzio Mantegna ·
Luis M. Rocha · Chantal Cherifi ·
Salvatore Miccichè
Editors

Complex Networks and Their Applications XI

Proceedings of The Eleventh International Conference on Complex Networks and Their Applications: COMPLEX NETWORKS 2022—Volume 1

Springer

Editors
Hocine Cherifi
University of Burgundy
Dijon, France

Luis M. Rocha
Thomas J. Watson College of Engineering
and Applied Science
Binghamton University
Binghamton, NY, USA

Salvatore Miccichè
Dipartimento di Fisica e Chimica Emilio
Segrè
Università degli Studi Palermo
Palermo, Italy

Rosario Nunzio Mantegna
Dipartimento di Fisica e Chimica Emilio
Segrè
Università degli Studi Palermo
Palermo, Italy

Chantal Cherifi
IUT Lumière—Université Lyon 2
University of Lyon
Bron, France

ISSN 1860-949X ISSN 1860-9503 (electronic)
Studies in Computational Intelligence
ISBN 978-3-031-21126-3 ISBN 978-3-031-21127-0 (eBook)
https://doi.org/10.1007/978-3-031-21127-0

This Springer imprint is published by the registered company Springer Nature Switzerland AG
The registered company address is: Gewerbestrasse 11, 6330 Cham, Switzerland

Organization and Committees

General Chairs

Hocine Cherifi	University of Burgundy, France
Rosario N. Mantegna	Università degli Studi di Palermo, Italy
Luis M. Rocha	Binghamton University, USA

Advisory Board

Jon Crowcroft	University of Cambridge, UK
Raissa D'Souza	University of California, Davis, USA
Eugene Stanley	Boston University, USA
Ben Y. Zhao	University of Chicago, USA

Program Chairs

Chantal Cherifi	University of Lyon, France
Salvatore Miccichè	Università degli Studi di Palermo, Italy

Satellite Chairs

Sabrina Gaito	University of Milan, Italy
Javier Galeano	Universidad Politécnica de Madrid, Spain

Lightning Chairs

Alessandro Rizzo	Politecnico di Torino, Italy
Giancarlo Francesco Ruffo	Università degli Studi di Torino, Italy
Huijuan Wang	TU Delft, Netherlands

Poster Chairs

Manuel Marques-Pita	University Lusófona, Portugal
Michele Tumminello	Università degli Studi di Palermo, Italy
Laura Ricci	Università degli Studi di Pisa, Italy

Publicity Chairs

Christophe Cruz	University of Burgundy, France
Xiangjie Kong	Dalian University of Technology, China

Tutorial Chairs

Luca Maria Aiello	Nokia-Bell Labs, UK
Leto Peel	UC Louvain, Belgium

Special Issues Chair

Sabrina Gaito	University of Milan, Italy

Sponsor Chairs

Roberto Interdonato	CIRAD—UMR TETIS, France
Claudio Fazio	Università degli Studi di Palermo, Italy

Local Committee Chair

Federico Musciotto	Università degli Studi di Palermo, Italy

Local Committee

Jacobo Aguirre	CSIC, Spain
Yuri Antonacci	Università degli Studi di Palermo, Italy
Giosué Lo Bosco	Università degli Studi di Palermo, Italy
Luca Faes	Università degli Studi di Palermo, Italy
Giacomo Fiumara	Università degli Studi di Messina, Italy
Vincenzo Giuseppe Genova	Università degli Studi di Palermo, Italy
Matteo Milazzo	University of Catania, Italy
Pasquale De Meo	Università degli Studi di Messina, Italy
Alessandro Pluchino	Università degli Studi di Catania, Italy
Andrea Rapisarda	Università degli Studi di Catania, Italy

Publication Chair

Matteo Zignani	University of Milan, Italy

Web Chairs

Zakariya Ghalmane	LINEACT, CESI, France
Stephany Rajeh	University of Burgundy, France

Program Committee

Jacobo Aguirre	Centro de Astrobiología (CSIC-INTA), Spain
Amreen Ahmad	Jamia Millia Islamia, India
Masaki Aida	Tokyo Metropolitan University, Japan
Luca Maria Aiello	IT University of Copenhagen, Denmark
Marco Aiello	University of Stuttgart, Germany
Esra Akbas	Georgia State University, USA
Mehmet Aktas	University of Central Oklahoma, USA
Tatsuya Akutsu	Kyoto University, Japan

Reka Albert	The Pennsylvania State University, USA
Laura Maria Alessandretti	ENS Lyon, Italy
Antoine Allard	Université Laval, Canada
Claudio Altafini	Linköping University, Sweden
Lucila G. Alvarez-Zuzek	Fondazione Bruno Kessler, Italy
Luiz G.A. Alves	Northwestern University, USA
Diego Amancio	University of São Paulo, Brazil
G. Ambika	IISER, India
Fred Amblard	IRIT—University Toulouse 1 Capitole, France
Hamed Amini	Georgia State University, USA
Andre Franceschi De Angelis	Unicamp, Brazil
Marco Tulio Angulo	National Autonomous University of Mexico (UNAM), Mexico
Alberto Antonioni	Carlos III University of Madrid, Spain
Nuno Araujo	Universidade de Lisboa, Portugal
Saúl Ares	Centro Nacional de Biotecnología (CNB-CSIC), Spain
Malbor Asllani	University College Dublin, Ireland
Tomaso Aste	University College London, UK
Martin Atzmueller	Osnabrück University, Germany
Konstantin Avrachenkov	INRIA, France
Giacomo Baggio	University of Padova, Italy
Rodolfo Baggio	Bocconi University, Italy
Franco Bagnoli	University of Florence, Italy
Avner Bar-Hen	CNAM, France
Paolo Barucca	University College London, UK
Nikita Basov	St. Petersburg State University, Russia
Gareth Baxter	University of Aveiro, Portugal
Marya Bazzi	University of Oxford, UK
Andras A. Benczur	Hungarian Academy of Sciences, Hungary
Rosa M. Benito	Universidad Politécnica de Madrid, Spain
Kamal Berahmand	Queensland University of technology, Australia
Cyrille Bertelle	Le Havre University, France
Marc Bertin	Université Claude Bernard Lyon 1, France
Ginestra Bianconi	Queen Mary University of London, UK
Christian Bick	University of Exeter, UK
Ofer Biham	The Hebrew University of Jerusalem, Israel
Livio Bioglio	University of Turin, Italy
Hanjo Boekhout	Leiden University, Netherlands
Johan Bollen	Indiana University Bloomington, USA
Anthony Bonato	Toronto Metropolitan University, Canada
Christian Bongiorno	CentraleSupelec, France
Anton Borg	Blekinge Institute of Technology, Sweden
Pierre Borgnat	CNRS, Laboratoire de Physique ENS de Lyon, France

Stefan Bornholdt	Universität Bremen, Germany
Javier Borondo	Upm, Spain
Cécile Bothorel	IMT Atlantique, France
Federico Botta	The University of Exeter, UK
Alexandre Bovet	Mathematical Institute, University of Oxford, UK
Fabian Braesemann	University of Oxford, UK
Dan Braha	NECSI, USA
Ulrik Brandes	ETH Zürich, Switzerland
Rion Brattig Correia	Instituto Gulbenkian de Ciência, Portugal
Markus Brede	University of Southampton, UK
Iulia Martina Bulai	University of Sassari, Italy
Javier M. Buldu	U.R.J.C. and Center for Biomedical Technology, Spain
Raffaella Burioni	Dipartimento di Fisica, Università di Parma, Italy
Noah Burrell	University of Michigan, USA
Fabio Caccioli	University College London, UK
Carmela Calabrese	Aix-Marseille Université, Italy
Paolo Campana	University of Cambridge, UK
M Abdullah Canbaz	University at Albany, SUNY, USA
Carlo Vittorio Cannistraci	TU Dresden, Germany
Vincenza Carchiolo	Università di Catania, Italy
Giona Casiraghi	ETH Zürich, Switzerland
Douglas Castilho	Federal Institute of South of Minas Gerais, Brazil
Costanza Catalano	Central Bank of Italy, Italy
Remy Cazabet	Université Lyon 1, France
Po-An Chen	National Yang Ming Chiao Tung University, Taiwan
Xihui Chen	University of Luxembourg, Luxembourg
Chantal Cherifi	Lyon 2 University, DISP Laboratory, France
Hocine Cherifi	University of Burgundy, France
Peter Chin	Boston University, USA
Matteo Chinazzi	Northeastern University, USA
Matteo Cinelli	Ca' Foscari, University of Venice, Italy
Richard Clegg	Queen Mary University of London, UK
Reuven Cohen	Bar-Ilan University, Israel
Alessio Conte	University of Pisa, Italy
Michele Coscia	IT University of Copenhagen, Denmark
Christophe Crespelle	Université Claude Bernard Lyon 1, France
Regino Criado	Universidad Rey Juan Carlos, Spain
Mihai Cucuringu	University of Oxford, USA
Bhaskar Dasgupta	University of Illinois at Chicago, USA
Joern Davidsen	University of Calgary, Canada
Toby Davies	University College London, UK
Pasquale De Meo	Università degli Studi di Messina, Messina, Italy
Fabrizio De Vico Fallani	Inria—ICM, France
Guillaume Deffuant	Cemagref, France

Charo I. del Genio	Coventry University, UK
Pietro Delellis	University of Naples Federico II, Italy
Karel Devriendt	University of Oxford, UK
Kuntal Dey	Accenture Tech Labs, India
Riccardo Di Clemente	University of Exeter, UK
Matías Di Muro	IFIMAR Institute, Argentina
Marcelo Do Vale Cunha	Federal Institute of Bahia, Brazil, Brazil
Constantine Dovrolis	Georgia Institute of Technology, USA
Ahlem Drif	Sétif 1 University, Algeria
Johan Dubbeldam	Delft University of Technology, Netherlands
Jordi Duch	Universitat Rovira i Virgili, Spain
Nicolas Dugué	LIUM, France
Victor M Eguiluz	IFISC (CSIC-UIB), Spain
Alexandre Evsukoff	COPPE/UFRJ, Brazil
Mauro Faccin	CEPED, Université de Paris, France
Giorgio Fagiolo	Sant'Anna School of Advanced Studies, Italy
Daniel Figueiredo	UFRJ, Brazil
Marco Fiore	IMDEA Networks Institute, Spain
Alessandro Flammini	Indiana University Bloomington, USA
Manuel Foerster	Bielefeld University, Germany
Mattia Frasca	University of Catania, Italy
Angelo Furno	University of Lyon, France
Sabrina Gaito	University of Milan, Italy
Javier Galeano	Universidad Politécnica de Madrid, Spain
Stéphane Galland	UBFC—UTBM, France
Lazaros Gallos	Rutgers University, USA
Riccardo Gallotti	Fondazione Bruno Kessler, Italy
José Manuel Galán	Universidad de Burgos, Spain
Joao Gama	University of Porto, Portugal
Yerali Gandica	CY Cergy Paris Université, France
Jianxi Gao	Rensselaer Polytechnic Institute, USA
Floriana Gargiulo	University of Paris Sorbonne, France
Michael Gastner	Yale-NUS College, Singapore
Alexander Gates	Northeastern University, USA
Vincent Gauthier	Institut Mines Telecom CNRS SAMOVAR, France
Ralucca Gera	Naval Postgraduate School, USA
Raji Ghawi	Technical University of Munich, Germany
Jean-Louis Giavitto	IRCAM, France
Tommaso Gili	IMT, Italy
Silvia Giordano	SUPSI, Switzerland
Pierre-Louis Giscard	Université du Littoral Côte d'Opale, France
David Gleich	Purdue University, USA
Antonia Godoy	University College London, UK
Kwang-Il Goh	Korea University, South Korea
Jesus Gomez-Gardenes	Universidad de Zaragoza, Spain

Antonio Gonzalez-Pardo	Universidad Rey Juan Carlos, Spain
Bruno Gonçalves	Data For Science, Inc, USA
Carlos Gracia-Lázaro	BIFI, Spain
Jean-Loup Guillaume	Université de la Rochelle, France
Mehmet Gunes	Stevens Institute of Technology, USA
Sergio Gómez	Universitat Rovira i Virgili, Spain
Meesoon Ha	Chosun University, South Korea
Jürgen Hackl	University of Liverpool, Switzerland
Aric Hagberg	Los Alamos National Laboratory, USA
Chris Hankin	Imperial College London, UK
Yukio Hayashi	Japan Advanced Institute of Science and Technology, Japan
Mark Heimann	Lawrence Livermore National Laboratory, USA
Torsten Heinrich	Chemnitz University of Technology, Germany
Denis Helic	Graz University of Technology, Austria
Takayuki Hiraoka	Aalto University, Finland
Philipp Hoevel	University College Cork, Ireland
Seok-Hee Hong	University of Sydney, Australia
H. Ulrich Hoppe	University Duisburg-Essen, Germany
Yuichi Ikeda	Kyoto University, Japan
Roberto Interdonato	CIRAD—UMR TETIS, France
Giulia Iori	City University, London, UK
Francesco Iorio	Human Technopole, Italy
Antonio Iovanella	Università degli Studi Internazionali di Roma, Italy
Sarika Jalan	IIT Indore, India
Mahdi Jalili	RMIT University, Australia
Jaroslaw Jankowski	West Pomeranian University of Technology, Poland
Marco Alberto Javarone	Centro Ricerche Enrico Fermi, Italy
Hawoong Jeong	Korea Advanced Institute of Science and Technology, South Korea
Tao Jia	Southwest University, China
Chunheng Jiang	Rensselaer Polytechnic Institute, USA
Jiaojiao Jiang	University of New South Wales, Australia
Di Jin	Amazon, USA
Hang-Hyun Jo	The Catholic University of Korea, South Korea
Ivan Jokić	Delft University of Technology, Netherlands
Marko Jusup	Tokyo Institute of Technology, Japan
Arkadiusz Jędrzejewski	Wrocław University of Science and Technology, Poland
Rushed Kanawati	Université Sorbonne Paris Nord, France
Eytan Katzav	The Hebrew University of Jerusalem, Israel
Mehmet Kaya	Firat University, Turkey
Domokos Kelen	Institute for Computer Science and Control, Hungary
Dror Kenett	Johns Hopkins University, USA

Maria Malek	ETIS, CY Tech, France
Nishant Malik	Rochester Institute of Technology, USA
Fragkiskos Malliaros	Paris-Saclay University, France
Giuseppe Mangioni	University of Catania, Italy
Rosario Nunzio Mantegna	Università degli Studi di Palermo, Italy
Manuel Sebastian Mariani	University of Zurich, Switzerland
Radek Marik	Czech Technical University, Czechia
Daniele Marinazzo	Ghent University, Belgium
Andrea Marino	University of Florence, Italy
Manuel Marques-Pita	Universidade Lusofona, Portugal
Christoph Martin	Hamburg University of Applied Sciences, Germany
Cristina Masoller	Universitat Politècnica de Catalunya, Spain
Rossana Mastrandrea	IMT Institute of Advanced Studies, Italy
Ruth Mateos de Cabo	Universidad CEU San Pablo, Spain
John Matta Southern	Illinois University Edwardsville, USA
Fintan McGee	Luxembourg Institute of Science and Technology, Luxembourg
Matúš Medo	University of Bern, Switzerland
Jose Fernando	Mendes University of Aveiro, Portugal
Humphrey Mensah	Syracuse University, USA
Engelbert Mephu Nguifo	University Clermont Auvergne, France
Anke Meyer-Baese	FSU, USA
Salvatore Miccichè	Università degli Studi di Palermo, Italy
Radosław Michalski	Wrocław University of Science and Technology, Poland
Letizia Milli	University of Pisa, Italy
Boris Mirkin	Higher School of Economics Moscow, UK
Shubhanshu Mishra	University of Illinois at Urbana-Champaign, USA
Marija Mitrović Dankulov	Institute of Physics Belgrade, Serbia
Andrzej Mizera	University of Luxembourg, Luxembourg
Osnat Mokryn	University of Haifa, Israel
Roland Molontay	Budapest University of Technology and Economics, Hungary
Raul Mondragon	Queen Mary University of London, UK
Misael Mongiovì	Consiglio Nazionale delle Ricerche, Italy
Alfredo Morales	MIT Media Lab, USA
Andres Moreira	Universidad Tecnica Federico Santa Maria, Chile
Esteban Moro	Universidad Carlos III de Madrid, Spain
Greg Morrison	University of Houston Department of Physics, USA
Sotiris Moschoyiannis	University of Surrey, UK
Igor Mozetič	Jozef Stefan Institute, Slovenia
Alberto P. Munuzuri	University of Santiago de Compostela, Spain
Masayuki Murata	Osaka University, Japan
Tsuyoshi Murata	Tokyo Institute of Technology, Japan
Katarzyna Musial	University of Technology Sydney, Australia

Muaz Niazi	National University of Science and Technology, Pakistan
Alexandre Nicolas	Université Claude Bernard Lyon 1, France
Peter Niemeyer	Leuphana Universität Lüneburg, Germany
Jordi Nin	ESADE, Universitat Ramon Llull, Spain
Rogier Noldus	Ericsson, Netherlands
El Faouzi Nour-Eddin	Université Gustave Eiffel, France
Masaki Ogura	Osaka University, Japan
Oleh Omelchenko	University of Potsdam, Germany
Andrea Omicini	Università di Bologna, Italy
Gergely Palla	ELTE Institute of Physics, Hungary
Pietro Panzarasa	Queen Mary University of London, UK
Fragkiskos Papadopoulos	Cyprus University of Technology, Cyprus
Symeon Papadopoulos	Information Technologies Institute, Greece
Han Woo Park	YeungNam University, South Korea
Juyong Park	Korea Advanced Institute of Science and Technology, South Korea
Pierre Parrend	EPITA Strasbourg, France
Philip E. Paré	Purdue University, USA
Leto Peel	Maastricht University, Belgium
Tiago Peixoto	Central European University, Germany
Matjaz Perc	University of Maribor, Slovenia
Anthony Perez	Université d'Orléans, France
Lilia Perfeito	Nova School of Business and Economics, Portugal
Giovanni Petri	CENTAI, Italy
Juergen Pfeffer	Technical University of Munich, Germany
Bruno Pinaud	Université de Bordeaux, France
Flavio Pinheiro	NOVA IMS—Universidade NOVA de Lisboa, Portugal
Clara Pizzuti	National Research Council of Italy (CNR), Italy
Pawel Pralat	Toronto Metropolitan University, Canada
Natasa Przulj	University College London, Spain
Oriol Pujol	University of Barcelona, Spain
Rami Puzis	Ben Gurion University of the Negev, Israel
Christian Quadri	University of Milan, Italy
Filippo Radicchi	Indiana University, USA
Andrea Rapisarda	Università di Catania, Italy
Miguel Rebollo	Universitat Politècnica de València, Spain
Juan Restrepo	University of Colorado Boulder, USA
Pedro Ribeiro	University of Porto, Portugal
Laura Ricci	University of Pisa, Italy
Alessandro Rizzo	Politecnico di Torino, Italy
Fernando Rosas	Imperial College London, UK
Giulio Rossetti	KDD Lab ISTI-CNR, Italy
Fabrice Rossi	Université Paris Dauphine PSL, France

Luca Rossi	IT University of Copenhagen, Denmark
Camille Roth	CNRS, Germany
Celine Rozenblat	University of Lausanne Institut de Géographie, Switzerland
Giancarlo Ruffo	Università degli Studi di Torino, Italy
Iraj Saniee	Bell Labs, Nokia, USA
Francisco C. Santos	Universidade de Lisboa, Portugal
Hiroki Sayama	Binghamton University, USA
Nicolas Schabanel	École Normale Supérieure de Lyon, France
Michael Schaub	RWTH Aachen University, Germany
Maximilian Schich	Professor for Cultural Data Analytics, Estonia
Caterina Scoglio	Kansas State University, USA
Hamida Seba	University Lyon1, France
Emre Sefer	Ozyegin University, Turkey
Santiago Segarra	Rice University, USA
Irene Sendiña-Nadal	Rey Juan Carlos University, Spain
Termeh Shafie	Leibniz Institute for the Social Sciences, Germany
Yilun Shang	Northumbria University, UK
Rajesh Sharma	University of Tartu, Estonia
Julian Sienkiewicz	Warsaw University of Technology, Poland
Anurag Singh	National Institute of Technology Delhi, India
Tiratha Raj Singh	Jaypee University of Information technology, India
Rishabh Singhal	Dayalbagh Educational Institute, India
Per Sebastian Skardal	Trinity College, USA
Keith Smith	Nottingham Trent University, UK
Igor Smolyarenko	Brunel University, UK
Zbigniew Smoreda	Orange Labs, France
Annalisa Socievole	National Research Council of Italy (CNR), Italy
Albert Sole	Universitat Oberta de Catalunya, Spain
Sucheta Soundarajan	Syracuse University, USA
Semanur Soyyigit	Kirklareli University, Turkey
Jaya Sreevalsan-Nair	IIIT Bangalore, India
Christoph Stadtfeld	ETH Zurich, Switzerland
Massimo Stella	University of Exeter, UK
Fabian Stephany	University of Oxford, UK
Kashin Sugishita	Tokyo Institute of Technology, Japan
Xiaoqian Sun	Beihang University, China
Xiaoqian Sun	Chinese Academy of Sciences (CAS), China
Bosiljka Tadic	Jozef Stefan Institute, Slovenia
Andrea Tagarelli	DIMES, University of Calabria, Italy
Patrick Taillandier	INRAE/MIAT, France
Kazuhiro Takemoto	Kyushu Institute of Technology, Japan
Frank Takes	Leiden University, Netherlands
Fabien Tarissan	CNRS—ENS Paris-Saclay (ISP), France
Laura Temime	Conservatoire National des Arts et Métiers, France

Claudio Juan Tessone	Universität Zürich, Switzerland
François Théberge	Tutte Institute for Mathematics and Computing, Canada
Michele Tizzoni	ISI Foundation, Italy
Leonardo Torres	Max Planck Institute for Mathematics in the Sciences, Germany
Vincent Antonio Traag	Leiden University, Netherlands
Jan Treur	Vrije Universiteit Amsterdam, Netherlands
Milena Tsvetkova	London School of Economics and Political Science, UK
Liubov Tupikina	Universite de Paris, France
Janos Török	Budapest University of Technology and Economics, Hungary
Stephen Uzzo	New York Hall of Science, USA
Lucas D. Valdez	IFIMAR, Argentina
Pim van der Hoorn	Eindhoven University of Technology, Netherlands
Piet Van Mieghem	Delft University of Technology, Netherlands
Onur Varol	Sabanci University, Turkey
Christian Lyngby Vestergaard	CNRS and Institut Pasteur, France
Johannes Wachs	Central European University, Hungary
Huijuan Wang	Delft University of Technology, Netherlands
Wei Wang	Sichuan University, China
Ingmar Weber	Qatar Computing Research Institute, Qatar
Karoline Wiesner	University of Potsdam, Germany
Gordon Wilfong	Nokia Bell Labs, USA
Mateusz Wilinski	Los Alamos National Laboratory, USA
Richard Wilson	University of York, UK
Dirk Witthaut	Forschungszentrum Jülich, Germany
Jinshan Wu	Beijing Normal University, China
Haoxiang Xia	Dalian University of Technology, China
Fei Xiong	Beijing Jiaotong University, China
Xiaoke Xu	Dalian Minzu University, China
Gitanjali Yadav	University of Cambridge, UK
Gang Yan	Tongji University, China
Xiaoran Yan	Zhejiang Lab, China
Ying Ye	Nanjing University, China
Emilio Zagheni	Max Planck Institute for Demographic Research, Germany
An Zeng	Beijing Normal University, China
Yuexia Zhang	Beijing Information Science and Technology University, China

Zi-Ke Zhang	Hangzhou Normal University, China
Matteo Zignani	University of Milan, Italy
Eugenio Zimeo	University of Sannio, Italy
Lorenzo Zino	University of Groningen, Netherlands
Fabiana Zollo	Ca' Foscari University of Venice, Italy

Preface

The "Università degli Studi di Palermo" in Italy hosts the eleventh edition of the "International Conference on Complex Networks and their Applications" (COMPLEX NETWORKS 2022) from November 8 to November 10, 2022. Every year, COMPLEX NETWORKS brings together researchers from various scientific backgrounds to review the field's current state and formulate new directions. The diversity of the attendees' scientific interests (Finance, Medicine and Neuroscience, Biology and Earth Sciences, Sociology and Politics, Computer Science and Physics, etc.) is a unique opportunity for crossfertilization between fundamental issues and innovative applications.

This edition attracted authors from all over the world, with 313 submissions from 54 countries. Each submission has been peer-reviewed by at least three independent reviewers from the international program committee. The 104 papers included in the proceedings result from this rigorous selection process.

The quality of the contributors is undoubtedly an essential element for a successful edition. The success also goes to the keynote speakers. These leaders and visionaries in their fields present fascinating plenary lectures with big-picture ideas and unique perspectives to help attendees deepen their understanding of scientific challenges. We are delighted to bring together this great line-up of speakers.

- Luís A. Nunes Amaral (Northwestern University, USA)
- Manuel Cebrian (Max Planck Institute for Human Development, Germany)
- Shlomo Havlin (Bar-Ilan University, Israel)
- Giulia Iori (City University, London, UK)
- Melanie Mitchell (Santa Fe Institute, USA)
- Ricard Solé (Universitat Pompeu Fabra, Spain)

Our thanks also goes to the speakers of the traditional tutorial sessions for delivering insightful talks on November 7, 2022.

- Michele COSCIA (IT University of Copenhagen, Denmark)
- Adriana IAMNITCHI (Maastricht University, Netherlands)

The success also relies in the deep involvement of many individuals, institutions, and sponsors.

We sincerely gratify the advisory board members for inspiring the essence of the conference:

Jon Crowcroft (University of Cambridge), Raissa D'Souza (University of California, Davis, USA), Eugene Stanley (Boston University, USA), and Ben Y. Zhao (University of Chicago, USA)

We record our thanks to our fellow members of the organizing committee:

The lightning sessions chairs:

Alessandro Rizzo (Politechnico di Torino, Italy), Giancarlo Francesco Ruffo (Università degli Studi di Torino, Italy), and Huijuan Wang (TU Delft, Netherlands)

The poster sessions chairs:

Manuel Marques Pita (Universidade Lusófona, Portugal), Michele Tumminello (Università degli Studi di Torino, Italy), and Laura Ricci (Università degli Studi di Pisa, Italy)

The tutorial chairs:

Luca Maria Aiello (Nokia-Bell Labs, UK) and Leto Peel (Maastricht University, Netherlands)

The special issue chair:

Sabrina Gaito (University of Milan, Italy)

The publicity chairs:

Christophe Cruz (University of Burgundy, France) and Xiangjie Kong (Zhejiang University of Technology, China)

The sponsor chairs:

Claudio Fazio (Università degli Studi di Palermo, Italy) and Roberto Interdonato (CIRAD—UMR TETIS, Montpellier, France)

The Web chairs:

Zakariya Ghalmane (LINEACT, CESI, France) and Stephany Rajeh (University of Burgundy, France)

Our profound thanks goes to Matteo Zignani (University of Milan, Italy), publication chair, for the tremendous work in managing the submission system and the proceedings publication process.

We would also like to record our appreciation for the work of the local committee chair, Federico Musciotto (Università degli Studi di Palermo, Italy), and all the local committee members, Yuri Antonacci (Università degli Studi di Palermo, Italy), Giosuè Lo Bosco (Università degli Studi di Palermo, Italy), Luca Faes (Università degli Studi di Palermo, Italy), Giacomo Fiumara (Università degli Studi di Messina, Italy), Vincenzo Giuseppe Genova (Università degli Studi di Palermo, Italy), Matteo Milazzo (University of Catania, Italy), Pasquale de Meo (Università degli Studi di Messina, Italy), Alessandro Pluchino (Università degli Studi di Catania, Italy), and Andrea Rapisarda (Università degli Studi di Catania, Italy), for their work in managing the sessions. They intensely participated to the success of this edition.

We would like to express our gratitude to our partner journals: Advances in Complex Systems, Applied Network science, Complex Systems, Entropy, PLOS One, and Social Network Analysis and Mining.

We are thankful to all those who have contributed to the success of this meeting. Sincere thanks to the authors for their creativity.

Finally, we would like to express our most sincere thanks to the program committee members for their considerable efforts in producing high-quality reviews in a minimal time.

These volumes make the most advanced contribution of the international community to the research issues surrounding the fascinating world of complex networks. Their breath, quality, and novelty demonstrate the profound impact of complex networks in understanding our world. We hope you enjoy the papers as much as we enjoyed organizing the conference and putting this collection of articles together.

Hocine Cherifi
Rosario Nunzio Mantegna
Luis M. Rocha
Chantal Cherifi
Salvatore Miccichè

Contents

Information Spreading in Social Media

Cognitive Cascades within Media Ecosystems: Simulating Fragmentation, Selective Exposure and Media Tactics to Investigate Polarization

Nicholas Rabb(✉) and Lenore Cowen

Tufts University, Medford, MA 02144, USA
nicholas.rabb@tufts.edu

Abstract. This work introduces a simple extension to the recent Cognitive Cascades model of Rabb et al. with modeling of multiple media agents, to begin to investigate how the media ecosystem might influence the spread of beliefs (such as beliefs around COVID-19 vaccination). We perform some initial simulations to see how parameters modeling audience fragmentation, selective exposure, and responsiveness of media agents to the beliefs of their subscribers influence polarization. We find that media ecosystem fragmentation and echo-chambers may not in themselves be as polarizing as initially postulated, in the absence of outside fixed media messages that are polarizing.

Keywords: Computational social science · Network science · Cognitive modeling

1 Introduction

The media ecosystem is defined by Benkler, Faris, and Roberts as "the outlets and influencers who form networks, the structure of networks, and the flow of information in networks," and includes social media, blogs, major news channels, and newspapers [1]. It describes which media organizations are serving information to media consumers, and how they interface with each other. The concept evokes an array of media providers who have influence on information distribution and the subsequent opinion formation of a public.

Media studies, with the advent of social media and the internet, have increasingly focused on analyzing the effect of *audience fragmentation* (the partition of media consumers into many small audiences) [2] and *selective exposure* (the tendency of media consumers to consume messages in a systematically biased manner diverging from the composition of available messages) [3–7] on media consumers. Social media, by making more media sources available and algorithmically serving bias-confirming content, is argued facilitate selective exposure and subsequent fragmentation (the *high-choice news avoidance* thesis [5]), leading to political *polarization* [8–12].

H. Cherifi et al. (Eds.): COMPLEX NETWORKS 2016 2022, SCI 1077, pp. 3–15, 2023.
https://doi.org/10.1007/978-3-031-21127-0_1

This line of thinking—that high media availability combined with selective exposure leads to fragmentation and echo-chambers, which then lead to population polarization—is what we call the *high-choice polarization thesis*. The purpose of this paper is to study a simple example of this thesis, and whether it does, indeed, lead to a polarized public. To do this, we extend a simple, but cognitively- and socially-grounded, model of public opinion formation that we developed in [13].

We find, at least in our simple model, that polarization trends are quite robust to changes in audience fragmentation and selective exposure, contradicting the thesis that the high-choice news avoidance drives polarization in the current media ecosystem. Even when we simulate high fragmentation, strong selective exposure, and the presence of echo-chambers, most conditions fail to polarize. We observe a noticeable effect of selective belief in messages on polarization, but a striking difference in polarization trends when comparing models where media sources continuously broadcast a fixed belief to models where media sources instead try to appeal to a median or mean belief of their subscribers—*even when subscribers are assumed to only subscribe to news sources that are in concordance with their initial beliefs*. The difference in messaging tactics is further marked by whether or not the media ecosystem is more polarized than the population. This begins to hint that the media ecosystem fragmentation and echo-chambers may not in themselves be as polarizing as initially postulated, in the absence of outside fixed media messages that are polarizing.

Our current models have limitations (as discussed in Sect. 6). In particular, we have not yet modeled how and when individuals switch which media sources they subscribe to over time (as the messages and/or their beliefs evolve), nor have we created algorithms for determining which messages to broadcast that might also take into account desires to gain subscribers, or at the least, not lose any. Moreover, the models are not yet empirically validated. However, these models already present a framework in which to study a rich and subtle set of parameters affecting polarization in order to test theories about causes of the polarization of political opinion.

2 Background

2.1 Media Ecosystem

Media scholars describe the media ecosystem interacting with media consumers in two ways: through top-down and bottom-up processes. Top-down processes communicate a discursive structure of society and ideology relayed by media organizations and political elites [2,14]. Traditional media is argued to be a top-down, one way communication regime where information is selected, curated, and distributed by professional organizations with certain norms and motivations [1,15]. Media organizations are driven by the motivations and goals of those who run them, and may differ in tactics, but are generally interested in maximizing their audience [1,14].

Bottom-up processes involve the formation of opinion through interaction with media that involves *motivated cognition* on the part of individuals and groups [1,14]. Individuals have social and psychological motives that drive their belief or rejection of certain information. These may involve socioeconomic position, identity, or other psychological traits [14,16].

Two major developments in the media ecology that have sparked new study by researchers are the widespread adoption of social media and the perceived rise of misinformation and disinformation. These developments have led to many studying the phenomena of *audience fragmentation* (the partition of media consumers into many small audiences) [2] and *selective exposure* (the tendency of media consumers to consume messages in a systematically biased manner diverging from the composition of available messages [3–6]). Social media has been argued to increase fragmentation and the ability to selective expose oneself to bias-confirming media [7].

This all has led many media researchers to argue that our difficult political moment, where *polarization* of political opinion is frequently reported [8–12], is being caused by fragmentation, selective exposure, and its resultant *echo-chambers*. Though countervailing evidence exists against fragmentation [1,2], the presence of echo-chambers [3,4], or disinformation being primarily caused by them [1,16–20], the causal link stands to be further investigated.

2.2 Opinion Diffusion Models

Computational models of opinion formation can be used to understand the interplay of complicated mechanisms governing messaging adoption and spreading. Opinion diffusion models are often based on what are called social contagion models [21]. We focus, in this work, on a form of cognitive contagion model; the name we give to models which give simulated, networked agents rules governing their adoption of beliefs that are grounded in some sort of cognitive or psychological process. Several models in this category have been used to try to understand complicated social phenomena such as polarization [22–25], and contagion governed by cognitive dissonance [13,26].

Our previous work introduced the *cognitive cascade* class of models that combined a simple cognitive model with a concept in network science called network cascades [27]. As stories spread through a network like social media, they are said to "cascade" as they are shared. Our cognitive cascade model captured this cascading behavior by modeling institutions, message-passing behavior, and individual cognitive models in each of the networked agents [13]. This model was formed around describing the spread of misinformation, and lent insights and a simple framework to develop on top of to further investigate the phenomenon.

3 Static Media Ecosystem Model

Our primary objective in this study was to test the high-choice polarization thesis within our cognitive cascade model [13]. To do so, we extended this model to include more institutional agents serving as media organizations, and diversified their messaging tactics.

3.1 Review of Cognitive Cascade Model

The cognitive cascade model operates on a graph $G = (V, E)$ of *citizen agents* $v \in V$ who are connected in a network. Each citizen in the graph holds beliefs $B \subseteq \mathcal{B}$ from the universe of possible propositions to believe in $\mathcal{B}$, that each span the range $[0, 1]$, which, without loss of generality, we represent discretely as $B = \{0 <= b <= 6\}$ where B_i is some proposition (e.g. "COVID is real"), and $b = 0$ represents strong disbelief in B_i while $b = 6$ represents strong belief. The model allows for agents to hold beliefs over many propositions, though in [13], we primarily focused on just one.

Our cognitive cascade model captured the cascading nature of messages studied in network science by introducing *institutional* agents into the model, and modeling contagion of *messages* they share – originating from institutions and cascading through the network. Institutional agents $i \in I$ connect to citizens based on their b_i value and an ϵ parameter. If $|b_u - b_i| \leq \epsilon$, then agent u will *subscribe* to institution i. An institution's subscribers $S \subseteq V$ are thus the citizens who first receive messages, $M_i \ : \ t \rightarrow (m_0, m_1, ..., m_j)$, $m_j \in \mathcal{M}, j \geq 0$, that the institution broadcasts at each time step t. The set of all possible messages, $\mathcal{M}$ spans all permutations of modeled beliefs B_i. For simplicity, our work in [13] mainly focused on *one* institutional agent broadcasting messages to subscribers – a simplification that allowed us to begin a preliminary analysis of the model dynamics.

Messages that originate with institutions, and encode beliefs from B, are either believed or rejected by citizen agents based on a function $\beta(b_u, b_m)$ that yields a probability of believing the message m. If any citizen believes a message, it will update its value of b to match b_m, and then send the message to its neighbors, who engage in the same process. Otherwise, the citizen agent does not share the message further. The function β can be tuned to capture different cognitive phenomena that govern belief or rejection of information. In [13], we tuned this function to capture dissonance and exposure phenomena (commonly cited as being at play in the belief of misinformation narratives [17,28–32]) in a *defensive cognitive contagion* (DCC) function, described in Eq. 1.

$$\beta(b_{u,(t+1)}, b_v) = \frac{1}{1 + e^{\alpha(|b_{u,t} - b_v| - \gamma)}} \tag{1}$$

In [13], we tested these dynamics on several random graph topologies: the Erdős-Rényi random graph [33], Watts-Strogatz small world network [34], Barabási-Albert preferential attachment network [35], and the multiplicative attribute graph [36].

3.2 Extension to Model the Media Ecosystem

A Diverse Array of Media Sources To model diverse static media ecosystems, we increased the number of institutional agents and gave each institutional agent i a randomly drawn belief b_i that represents their stance on the proposition B. In our model, the set of institutional beliefs, $\mathcal{I}_i$, for agent i can be any probability distribution over B_i; in our experiments below we model uniform, normal, and polarized initial distributions to simplistically model commonly chosen prior belief distributions (normal and uniform) and those described in media studies literature (polarized).

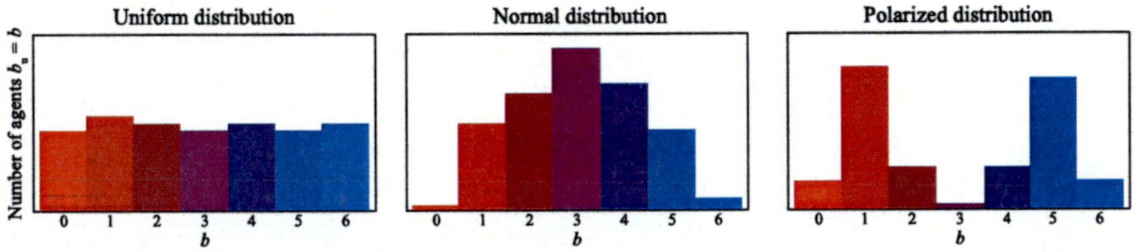

Fig. 1. An example of initial distributions over B (from left to right: uniform $\mathcal{U}$, normal $\mathcal{N}$, polarized $\mathcal{P}$)

Because we added more institutional agents, citizen agents connected with many more message sources based on their ϵ parameter—sources which expose them to different messages. While this model has citizens that subscribe to multiple different media sources, this is a *static* media model because we do not (yet) allow for subscribers to *change* which media sources they subscribe to in a dynamic fashion, in response to different media messages, as their beliefs evolve. This reconnection to different media sources is something that is hypothesized to happen under the high-choice news availability thesis [5], but will be left to future work.

Media Messaging Patterns In our model, institutional agents utilize different *messaging patterns*, $\varphi : S_{i,t} \rightarrow m, m \in \mathcal{M}$. This function φ determines which message, m to broadcast to subscribers of institution i at time t.

4 Experiments

To test the high-choice polarization thesis, we chose certain model parameters to serve as corollaries of audience fragmentation and selective exposure, and varied these with different initial conditions and media tactics to determine what configurations led to opinion polarization. These provided the basis for simulation experiments that were developed and run with NetLogo 6.1 [37] and Python 3.8 scripts that interfaced with the simulation. Experiments were designed and run with NetLogo's BehaviorSpace extension. Source code and replication instructions are available publicly on GitHub at https://github.com/ricknabb/cog-cascades-media-ecosystem. Results were analyzed using Python 3.8, and analysis code is also available on GitHub.

Our experiments aggregate simulation results across 5 trials per parameter combination (with 486 combinations across parameters in Tables 1 and 2) and analyze the means of the measures over time. Additionally, each unique parameter combination was tested on two random graphs (5 trials on each). The trends we report are observed across all such results.

4.1 Institutional Belief Distributions

We considered three types of distributions for the belief functions of the Institutional agents: namely, we allowed the institutional beliefs $\mathcal{I}$ to be uniformly distributed, normally (truncated) distributed $\mathcal{N}(\mu_i, \sigma_i, a, b)$ (where a and b are the lower and upper limits), or polarized where the distribution draws from two truncated normal distributions $\mathcal{P} = \mathcal{N}_l(1, 1, 0, 6) \cup \mathcal{N}_u(5, 1, 0, 6)$.

4.2 Media Messaging Patterns

We modeled the messaging patterns in our experiments off of simple heuristics and tactics observed in media study. One pattern, *broadcast*, makes institutional agents send a message encoding their belief b_i at each time step ($\varphi(S_{i,t}) = b_i$)). The heuristic patterns allow institutional agents to assess the *mean* or *median* belief of their subscribers at time step t and send a message encoding that belief (e.g. *mean*: $\varphi(S_{i,t}) = \sum b_s/|S_{i,t}|, s \in S_{i,t}$. The former messaging pattern may represent media organizations that are attempting to push a certain view [1,16], and the latter, organizations that try to measure audience metrics to maximize the reception of their messages [2].

4.3 Graph topologies and homophily

We chose to run simulations on a graph of citizen (N = 100) and media ($|I| = 20$) agents, where citizens are connected by the Barabási-Albert [35] preferential attachment method that can be tuned to be more or less homophilic with a scalar parameter h_G. This graph topology mimics essential aspects of social networks, and is thus a good choice for a social phenomenon.

This scalar, h_G can be used in the agent connection process that yields the probability of connection p between agent u and v. In conjunction with the Barabási-Albert process, while u is being added to $G = (V, E)$, h_G modifies the connection probability as follows:

$$p((u, v) \in E) = \frac{h_G * k_u}{\sum_v k_v} \tag{2}$$

If homophily in the graph is desired, h_G could be calculated with a function $h(b_u, b_v)$, a scalar between 0 and 1 based on the similarity of agent beliefs. Setting

$h_G = 1$ would yield a graph with no added homophily. For our homophilic graphs, we set $h(b_u, b_v)$ as a linear combination of agent beliefs:

$$h(b_u, b_v) = \frac{1}{1 + (|b_u - b_v|)} \tag{3}$$

4.4 Parameters of Fragmentation and Selective Exposure

In our experiments, we varied certain model parameters we viewed as related to fragmentation and selective exposure to find conditions that led to polarization. Additionally, we varied citizens' initial belief distribution, media's initial belief distribution, and media messaging patterns to determine if results of those fragmentation and exposure parameters were robust across different initial conditions. Variations in parameters can be found in Table 1.

Table 1. Experimental parameter values for those chosen to represent audience fragmentation and selective exposure.

Phenomenon	Parameter	Possible values
Fragmentation (subscriber threshold)	ϵ	0, 1, 2
Selective exposure (dissonance threshold)	γ term in β (Eq. 1)	0, 1, 2
Selective exposure (homophily)	h_G (Eq. 3)	1, $h(b_u, b_v)$

The model parameter we chose to tune audience fragmentation was the ϵ parameter, representing the threshold that agents use to subscribe to different media. A lower value of ϵ will mean that agents subscribe to media that is increasingly the same as what they already believe. Lower values thus lead to an initial condition in the system that represents a fragmented audience.

To tune selective exposure, we varied the translation parameter γ in the DCC function governing belief update, and in the homophily of the graph itself through the parameter h_G. Cognitive dissonance reduction is argued to underlie selective exposure, so our previous work can be utilized [8]. The DCC translation parameter represents where the fuzzy threshold lies given a belief difference. A lower value (e.g. 1) means that agents will increasingly only believe messages that are close to their prior belief (e.g. a distance of 1 away). Moreover, increasing the homophily in the graph structure increases the presence of echo-chamber-like structures in the network – neighborhoods of agents who already agree with each other.

Changing the initial conditions of the experiment allowed testing the robustness of these parameters potentially leading to polarization. We varied institutional messaging patterns φ as described above in Sect. 3 and initial media and citizen belief distribution, described in Table 2.

Table 2. Experimental parameter values for those chosen to vary initial model conditions: the media messaging tactic φ, initial media distribution $\mathcal{I}$, and initial citizen belief distribution $\mathcal{C}$.

Parameter	Possible values
φ	Broadcast, appeal (mean, median)
$\mathcal{I}$	Uniform, normal, polarized
$\mathcal{C}$	Uniform, normal, polarized

4.5 Measures of Polarization

We measured polarization using a measure developed by [38], but modified for our model. Polarization was measured via the equation:

$$P(G) = \overline{\boldsymbol{b}}^T \overline{\boldsymbol{b}}, \tag{4}$$

where $\boldsymbol{b}$ is a vector of all beliefs b_u for $u \in V$ (and $\boldsymbol{b}_u$ is $\frac{b_u}{max(B)}$ to normalize to a $[0, 1]$ scale as in [38]) and $\overline{\boldsymbol{b}}$ is the mean-centered vector of beliefs. This measure and the citizen belief distribution was recorded at each time step and their means across simulation runs were calculated.

To measure trends in polarization over time, we fit linear regression lines to the means of polarization results over simulation trials. By using the slope (β) and intercept (α) of those lines, we were able to partition polarization results into several sets. If results were *polarized*, the slope of the regression fit to them was positive above a certain threshold, or the slope was near zero but the intercept was sufficiently high ($\beta \geq 0.01$, or $-0.01 \leq \beta \leq 0.01$ and $\alpha \geq 8.5$). Results were *nonpolarized* or *failed to polarize* if they had a sufficiently negative slope or had a near zero slope and a lower intercept below a threshold ($\beta \leq -0.01$, or $-0.01 \leq \beta \leq 0.01$ and $\alpha < 8.5$). Note that nonpolarized results also include those which *depolarized* the population ($\beta \leq -0.01$). A slope threshold of 0.01 was chosen to include even results that somewhat polarized over time. An intercept threshold of 8.5 (half of the maximum polarization value) was used to include results with near-zero slopes but which started and remained polarized or nonpolarized.

When partitioning results into polarized and nonpolarized results, we measured how often, out of five simulations, trials for a given parameter combination yielded the same polarization result as the mean across trials. We found that 5/5 trials matched the mean classification in 62% of our results, 4/5 in 20% of results, 3/5 in 13%, 2/5 in 4% and 1/5 in 1%. This gave us confidence that classifying the means of the trials would be representative.

5 Results

5.1 Fragmentation and Exposure do not Cause Polarization

We observed that polarization occurred in the population regardless of γ, h_G, and ϵ – the parameters for selective exposure and fragmentation, respectively.

This runs contrary to arguments in favor of these phenomena being the cause of polarization [3–5,7,8].

Table 3. Percentage of polarized/nonpolarized results (over the 36 experiments in each cell) broken down by selective exposure (γ and h_G) and fragmentation (ϵ).

h_G	1			$h(b_u, b_v)$		
$\epsilon = 2$	19/81	17/83	9/91	15/85	15/85	9/91
$\epsilon = 1$	15/85	11/89	11/89	20/80	17/83	9/91
$\epsilon = 0$	13/87	15/85	6/94	22/78	13/87	7/93
γ	0	1	2	0	1	2

Table 3 shows what percentage of simulations polarized, holding constant combinations of ϵ, γ, and h_G. Each cell contains the percent of simulation results that polarized, keeping those parameter values constant while letting others vary. The results show little effect from changes in ϵ and h_G, but a noticeable effect as γ changes: as it increases (agents are more likely to believe messages dissimilar to their prior belief), polarization decreases. Logistic regression on ϵ, γ, and h_G to predict the polarization outcome confirmed these observations. Regression results showed only γ significantly contributing to polarization outcomes (one unit increase decreased odds of polarization by 69%, 95% CI [54%, 86%], $p = 0.001$), with ϵ and h_G failing to reach significance ($p_\epsilon = 0.644$, $p_{h_G} = 0.509$).

5.2 Most Appeals that Adjust to Subscriber Beliefs Fail to Polarize

Our model found that in cases where media used tactics of appealing to subscribers, population beliefs were far less likely to polarize. By instead splitting the data by φ and γ, we found that polarization results differ significantly based on φ. The percent of results that polarized, keeping constant combinations of φ and γ, are shown in Table 4.

Table 4. Percentage of polarized/nonpolarized results (over the 108 experiments in each cell) broken down by media tactic (φ) and selective exposure from cognitive dissonance reduction (γ).

φ	Broadcast			Appeal mean			Appeal median		
	49/51	42/58	26/74	2/98	0/100	0/100	1/99	2/98	0/100
γ	0	1	2	0	1	2	0	1	2

While γ has an effect on polarization, φ has a stronger one. The broadcast tactic comprised almost all the polarized results, while the mean and median appeal tactics almost never polarized. This further challenges intuition, which imagines that appealing to subscribers leads to polarization. Notably, these results include cases where the media ecosystem and citizen population are *both* polarized, and where fragmentation and selective exposure are at their highest.

Focusing on results that did polarize, where $\varphi = broadcast$, broken down by the initial institutional and citizen belief distributions, reveals more about their effect on polarization. This is displayed in Table 5.

Table 5. Percentage of polarized/nonpolarized results when $\varphi = broadcast$ (over the 12 experiments in each cell) broken down by selective exposure from dissonance (γ), initial citizen distribution ($\mathcal{C}$) and initial institutional distribution ($\mathcal{I}$).

$\varphi = Broadcast$									
γ	0			1			2		
$\mathcal{C} = \mathcal{P}$	58/42	0/100	75/25	42/58	0/100	67/33	17/83	0/100	67/33
$\mathcal{C} = \mathcal{N}$	75/25	0/100	100/0	67/33	0/100	92/8	33/67	0/100	42/58
$\mathcal{C} = \mathcal{U}$	75/25	0/100	58/42	42/58	0/100	67/33	17/83	0/100	58/42
$\mathcal{I}$	$\mathcal{U}$	$\mathcal{N}$	$\mathcal{P}$	$\mathcal{U}$	$\mathcal{N}$	$\mathcal{P}$	$\mathcal{U}$	$\mathcal{N}$	$\mathcal{P}$

When the media tactic is broadcasting, the population polarizes only when $\mathcal{I}$ is uniform or polarized. Particularly, there is more polarization when $\mathcal{I}$ is more polarized than $\mathcal{C}$—as polarized distributions yield a higher polarization value by Eq. 4 than uniform distributions, and uniform more than normal. Logistic regression on ϵ, γ, h_G, and $\mathcal{C}$—on a subset of data where $\varphi = broadcast$ and $\mathcal{I} = \mathcal{U}$—confirmed a significant effect of γ (one unit increase decreased odds of polarization by 34%, 95% CI [20%, 59%], $p < 0.001$) and nearly significant effect of $\mathcal{C} = \mathcal{P}$ (decreased odds of polarization by 39%, 95% CI [14%, 91%], $p = 0.076$). Another regression on the same variables, but data where $\varphi = broadcast$ and $\mathcal{I} = \mathcal{P}$ showed only a nearly significant effect of γ (one unit increase decreased odds of polarization by 57%; 95% CI [33%, 97%], $p = 0.039$).

6 Discussion

The results that our static media ecosystem model produce are notable because they contradict popular theoretical assumptions made when reasoning about the state of political polarization in a high-choice media ecosystem. They show that media tactics combined with initial media and citizen belief distributions have the strongest effect on polarization, with selective exposure regarding belief in messages having a less striking, but significant effect. This disrupts the typical causal argument of the high-choice polarization thesis: that high availability of media, combined with selective exposure, leads to echo-chambers and fragmentation, which leads to polarization [3–5,7,8]. Our models rather hint at reasons to further explore the effect of polarized media and stubbornness to update beliefs on political polarization.

While this model yields interesting results, it is surely a strong simplification of a complex social system. Most notably, our static media model does not allow breaking and forming subscriber edges, and is thus limited. In particular, it will be important to test if the same dynamics are observed in a more general model; the high-choice news avoidance thesis [5] involves media consumers switching sources if they fail to sufficiently satisfy consumer desires. Adding this and other features like effects on belief based on trust in different sources, or affective polarization, remains important future work, where we also then intend to try to validate portions of these extended models against existing empirical data.

Acknowledgement. NR thanks NSF-NRT 2021874 for support, and LC and NR thank NSF 1934553 for additional support.

References

1. Benkler, Y., Faris, R., Roberts, H.: Manipulation, Disinformation, and Radicalization in American Politics. Oxford University Press, Network propaganda (2018)
2. Webster, J.G., Ksiazek, T.B.: The dynamics of audience fragmentation: Public attention in an age of digital media. J. Commun. **62**(1), 39–56 (2012)
3. Cardenal, A.S., Aguilar-Paredes, C., Galais, C., Pérez-Montoro, M.: Digital technologies and selective exposure: how choice and filter bubbles shape news media exposure. Int. J. Press Polit. **24**(4), 465–486 (2019)
4. Messing, S., Westwood, S.J.: Selective exposure in the age of social media: endorsements trump partisan source affiliation when selecting news online. Commun. Res. **41**(8), 1042–1063 (2014)
5. Karlsen, R., Beyer, A., Kari, S.-J.: Do high-choice media environments facilitate news avoidance? A longitudinal study 1997–2016. J. Broadcast. Electron. Med. **64**(5), 794–814 (2020)
6. Knobloch-Westerwick, S.: Choice and Preference in Media Use: Advances in Selective Exposure Theory and Research. Routledge (2014)
7. Arendt, F., Northup, T., Camaj, L.: Selective exposure and news media brands: Implicit and explicit attitudes as predictors of news choice. Med. Psychol. **22**(3), 526–543 (2019)
8. Iyengar, S., Hahn, K.S.: Red media, blue media: evidence of ideological selectivity in media use. J. Commun. **59**(1), 19–39 (2009)
9. Clinton, J., Cohen, J., Lapinski, J.S., Trussler, M.: Partisan pandemic: How partisanship and public health concerns affect individuals' social distancing during covid-19. Available at SSRN 3633934 (2020)
10. Bakshy, E., Messing, S., Adamic, L.A.: Exposure to ideologically diverse news and opinion on facebook. Science **348**(6239), 1130–1132 (2015)
11. Conover, M.D., Ratkiewicz, J., Francisco, M., Gonçalves, B., Menczer, F., Flammini, A.: Political polarization on twitter. In: Fifth international AAAI Conference on Weblogs and Social Media (2011)
12. Gramlich, J.: 20 striking findings from 2020. Technical report (2020)

13. Rabb, N., Cowen, L., de Ruiter, J.P., Scheutz, M.: Cognitive cascades: How to model (and potentially counter) the spread of fake news. PLoS ONE (2022)
14. Jost, J.T., Federico, C.M., Napier, J.L.: Political ideology: its structure, functions, and elective affinities. Ann. Rev. Psychol. **60**, 307–337 (2009)
15. Shehata, A., Strömbäck, J.: Learning political news from social media: network media logic and current affairs news learning in a high-choice media environment. Commun. Res. **48**(1), 125–147 (2021)
16. Marwick, A.E.: Why do people share fake news? A sociotechnical model of media effects. Georgetown Law Technol. Rev. **2**(2), 474–512 (2018)
17. Stanley, J..: How Propaganda Works. Princeton University Press, In How Propaganda Works (2015)
18. Koltai, K., Moran, R.E., Grasso, I.: Addressing the root of vaccine hesitancy during the covid-19 pandemic. XRDS: Crossroads ACM Magaz. Stud. **28**(2), 34–38 (2022)
19. Kuo, R..: Marwick, Alice: Critical disinformation studies: history, power, and politics. Harvard Kennedy School Misinform. Rev. **2**(4), 1–11 (2021)
20. Wendy, H., Kyong, C.: Beyond verification: algorithmic authenticity and polarizing trust. Beyond Disinformation (2021)
21. Christakis, N.A., Fowler, J.H.: Social contagion theory: examining dynamic social networks and human behavior. Stat. Med. **32**(4):556–577 (2013)
22. Dandekar, P., Goel, A., Lee, D.T.: Biased assimilation, homophily, and the dynamics of polarization. Proc. Nat. Acad. Sci. **110**(15), 5791–5796 (2013)
23. DellaPosta, D., Shi, Y., Macy, M.: Why do liberals drink lattes? Am. J. Sociol. **120**(5), 1473–1511 (2015)
24. Goldberg, A., Stein, S.K.: Beyond social contagion: associative diffusion and the emergence of cultural variation. Am. Sociol. Rev. **83**(5), 897–932 (2018)
25. Sikder, O., Smith, R.E., Vivo, P., Livan, G.: A minimalistic model of bias, polarization and misinformation in social networks. Sci. Rep. **10**(1), 1–11 (2020)
26. Li, K., Liang, H., Kou, G., Dong, Y.: Opinion dynamics model based on the cognitive dissonance an agent-based simulation. Inform. Fus. **56**, 1–14 (2020)
27. Del Vicario, M.., Bessi, A., Zollo, F., Petroni, F., Scala, A., Caldarelli, G., Eugene Stanley, H, Quattrociocchi, W.: The spreading of misinformation online. Proc. Nat. Acad. Sci. **113**(3), 554–559 (2016)
28. Wiley, C. :Mindf*ck: Cambridge Analytica and the Plot to Break America. Random House/Penguin Random House LLC (2019)
29. Van Bavel, J.J., Pereira, A.: The partisan brain: an identity-based model of political belief. Trends Cogn. Sci. **22**(3), 213–224 (2018)
30. Swire-Thompson, B., DeGutis, J., Lazer, D.: Measurement and Design Considerations, Searching for the Backfire Effect (2020)
31. Festinger, L.: A Theory of Cognitive Dissonance, vol. 2. Stanford University Press (1957)
32. Porot, N., Mandelbaum, E.: The Science of Belief: A Progress Report. Wiley Interdisciplinary Reviews: Cognitive Science, pp. e1539 (2020)
33. Erdős, Paul: Rényi, Alfréd: On the evolution of random graphs. Publ. Math. Inst. Hung. Acad. Sci **5**(1), 17–60 (1960)
34. Watts, D.J., Strogatz, S.H.: Collective dynamics of 'small-world' networks. Nature **393**(6684), 440–442 (1998)
35. Barabási, A., and Albert, R.: Emergence of scaling in random networks. Science **286**(5439), 509–512 (1999)
36. Kim, M., Leskovec, J.: Modeling social networks with node attributes using the multiplicative attribute graph model (2011). arXiv preprint http://arxiv.org/abs/1106.5053arXiv:1106.5053

37. Wilensky, U.: Netlogo itself. http://ccl.northwestern.edu/netlogo/ (1999). Accessed: 14 December 2020
38. Musco, C., Musco, C., Tsourakakis, C.E.: Minimizing polarization and disagreement in social networks. In: Proceedings of the 2018 World Wide Web Conference, pp. 369–378 (2018)

Check for updates

Properties of Reddit News Topical Interactions

Gaël Poux-Médard(✉), Julien Velcin, and Sabine Loudcher

Université de Lyon, Lyon 2, ERIC UR 3083, Bron 69676, France
gael.poux-medard@univ-lyon2.fr
https://gaelpouxmedard.github.io

Abstract. Most models of information diffusion online rely on the assumption that pieces of information spread independently from each other. However, several works pointed out the necessity of investigating the role of interactions in real-world processes, and highlighted possible difficulties in doing so: interactions are sparse and brief. As an answer, recent advances developed models to account for interactions in underlying publication dynamics. In this article, we propose to extend and apply one such model to determine whether interactions between news headlines on Reddit play a significant role in their underlying publication mechanisms. After conducting an in-depth case study on 100,000 news headline from 2019, we retrieve state-of-the-art conclusions about interactions and conclude that they play a minor role in this dataset.

Keywords: Dirichlet Hawkes process · Topic modeling · Interaction · Information spread

1 Introduction

As the volume of data available on the Internet keeps on growing exponentially, so does the need for efficient data-processing tools and methods. In particular, the user-generated content produced on online social platforms provides a detailed snapshot of the world population's thoughts and interests. This kind of data can be used for many different applications (e.g. in marketing, opinion mining, fake news control, summary generation, etc.). However, we need a fine understanding of the underlying data-generation mechanisms at stake to refine the results of these possible applications.

Early works on information spread consider that a user publishing a piece of information (or meme) did so due to an earlier exposure to this same meme [11]. Users are represented as a network on which edges pieces of information flow independently from each other. This model has seen several refinements, that consider additional information about nodes and edges to model the way information spreads. Nodes participate the spread when exposed a certain number of times [15,17], have a different influence on their neighbours depending on their position in the network [12], can publish exogenous memes (e.g. that were not

H. Cherifi et al. (Eds.): COMPLEX NETWORKS 2016 2022, SCI 1077, pp. 16–28, 2023.
https://doi.org/10.1007/978-3-031-21127-0_2

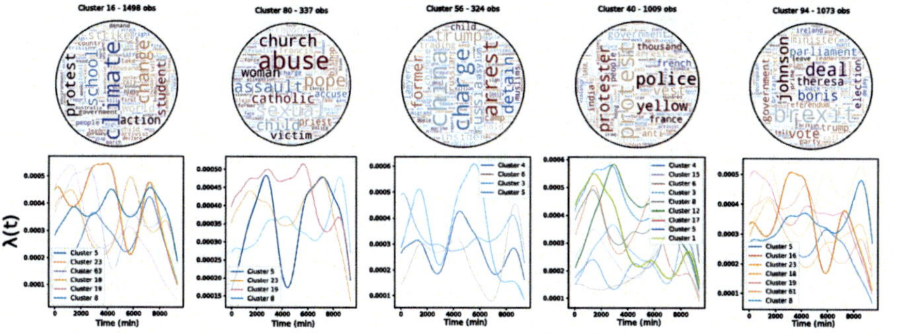

Fig. 1. An output of the proposed approach—(Top) A set of inferred topics along with the vocabulary of their documents. (Bottom) Instantaneous probability for one observation of a topic to trigger other observations in other topics.

flowing on the network beforehand) [10,16], etc. Edges between nodes represent the likeliness of a meme flowing from one node to the other; later works consider edges whose intensity depends on the meme considered [1,6], on the temporal distance to the last exposure [8,9], etc.

However, a core assumption of most of these models and extensions is that memes spread independently from each other. It has been highlighted in some occasions that modeling the interaction between pieces of information might bring interesting insights in the way data gets generated online [3,7,15]. For instance, we expect memes about two politically opposed candidates to have some interaction (the publication of one would inhibit the publication chances of the other one); memes raising climate change awareness might be more likely to spread when coexisting with memes about ecologic disasters, etc.

In this work, we conduct an in-depth study of such interaction mechanisms on a large-scale Reddit dataset—gathered from 9 news subreddits over 2019. We propose to use a model that answers the challenges of interaction modeling raised in recent works on interactions—interactions between memes are likely sparse and brief. This model, the Multivariate Powered Dirichlet-Hawkes Process (MPDHP), groups textual memes into topics. A possible output is represented in Fig. 1. Topics interact in pairs to yield a probability for a new meme about a given topic to get published, in a continuous-time setting. The output of MPDHP is a time-dependent topical interaction network. Once this network has been inferred, we can analyse the properties of topical interactions. We conduct experiments using several iterations of MPDHP (accounting for different timescales and topical modeling), so that we get an exhaustive panorama of interactions at stake. Overall, our conclusions hint that interactions play a minor role on news subreddits; most information indeed spreads independently in this context. From a broader perspective, we introduce a methodology to assess the role of temporal topical interactions textual datasets.

2 Background

Modeling interactions In this section, we briefly review previous works on interaction modeling in spreading processes. To the best of our knowledge, the first work tackling the problem proposes a SIR-based model that comprises an interaction term β between two co-existing viruses—in their case, the adoption of either Firefox or Chrome web browser [3]. Interaction is modeled as a hyper-parameter, that has to be tuned manually to retrieve global shares of each virus over time. Building on this work, [15] later proposed to model interaction between memes at the agent level: given a user is exposed to tweet A at time t_A, what is the probability to retweet tweet B at a time t_B later. This model attempts to learn the interaction parameter β from the data instead of fine-tuning it. Their conclusion is that interactions can have a large overall effect of the data-generation processes (here retweeting mechanisms), but that most interactions have little influence. Later works tackling the problem from a similar perspective on several real-world datasets find similar results: significant interactions between clusters of memes are sparse [19,21]. This highlight the need to cluster memes so that it becomes possible to retrieve meaningful interaction terms.

Some other works tackled the problem of pair-interaction modeling from a temporal perspective. Simple considerations show that interactions cannot remain constant over time; a user that read a meme 5 min earlier and another meme 5 days earlier is likely to be much more influenced the first one than the second one. Interactions are likely to be brief. This temporal dependence is explicitly modeled in [18], where the interaction strength is shown to typically decay exponentially, which correlated the findings of [4], and of [15] to a certain extent. It highlights the need to consider time to relevantly model interactions. Thus the need for a model able to handle large piles of data, performing topic inference, allowing interactions between these topics, and accounting for time.

Dirichlet-Hawkes processes The Dirichlet-Hawkes process [5] seem to qualify for these requirements. In particular, it has been extended as the Powered Dirichlet-Hawkes process [20] to allows for extended modeling flexibility—to which extend should we favor the textual information over the temporal information. However, both these models do not account for pair-interaction between the inferred topics. Instead, topics can only trigger new observations from themselves. In this work, we consider a Multivariate extension of the Powered Dirichlet-Hawkes process (MPDHP). This extension boils down to substituting the Hawkes process described in [5,20] by a multivariate Hawkes process [13], so that topics can influence the probability of new observations to belong to either other topics.

The principle of the Dirichlet-Hawkes approaches relies on Bayesian inference. An online language model accounts for the textual content of documents (Fig. 1-top). This model, typically a Dirichlet-Multinomial bag of words as we will use later, expresses the likelihood for a new document to belong to any existing topic. This language model is coupled to a Dirichlet-Hawkes prior, that can be expressed as a sequential process. The prior assigns to a new observation a prior probability to belong to either topic based on its temporal interaction with earlier publications (Fig. 1-middle). In the remaining of this section, we present the modified formulation of the base models [5,20] so that it accounts for topical pair-interactions.

We first rewrite and detail the expression of the Dirichlet-Hawkes process as introduced in [5,20]:

$$\begin{aligned}
&P(C_i = c|n_i, N_{<i,c}, t_i, \lambda(t_i), \mathcal{H}, \theta_0, \lambda_0, r) \\
&\quad \propto \underbrace{P(n_i|C_i = c, N_{<i,c}, \theta_0)}_{\substack{\text{Textual likelihood}\\ \text{(Dirichlet-Multinomial))}}} \underbrace{P(C_i = c|t_i, \lambda(t_i), \mathcal{H}, \lambda_0, r)}_{\substack{\text{Temporal prior}\\ \text{(MPDHP)}}} \\
&\quad = \frac{\Gamma(N_c + \theta_0)}{\Gamma(N_c + n_i + \theta_0)} \prod_v \frac{\Gamma(N_{c,v} + n_{i,v} + \theta_{0,v})}{\Gamma(N_{c,v} + \theta_0)} \begin{cases} \frac{\lambda_c^r(t_i)}{\lambda_0 + \sum\limits_{c'} \lambda_{c'}^r(t_i)} \text{if c} \leq \text{K} \\ \frac{\lambda_0}{\lambda_0 + \sum\limits_{c'} \lambda_{c'}^r(t_i)} \text{ if c} = \text{K} + 1 \end{cases}
\end{aligned} \tag{1}$$

where C_i represents the cluster chosen by the ith document, c is the random variable accounting for this allocation, n_i is a vector that whose vth entry represents the count of word v in document i, $N_{<i,c}$ is the vector whose entry v represents the total count of word v in all documents up to i that belong to cluster c, t_i the arrival time of document i, $\lambda(t_i)$ the vector of intensity functions at time t_i whose cth entry corresponds to cluster c, and $\mathcal{H}$ the history of all previous documents that appeared before time t_i. The three last symbols are hyperparameters: θ_0 is the concentration parameter of the Dirichlet-Multinomial language model [5,22], λ_0 is the concentration parameter of the MPDHP temporal prior [5], and r controls the extent to which we rely of either textual of temporal information [20]. Data is processed sequentially using a Sequential Monte Carlo (SMC) algorithm similar to [5,14,20]. For each new observation, we get from Eq. 1 the posterior probability that it belongs to either of K existing clusters, or to start a cluster of its own ($K + 1$). The SMC algorithm accounts for several allocation hypotheses at once, and discard the most unlikely ones every other iteration.

Now, the extension of the Dirichlet-Hawkes process to the multivariate case boils down to giving a new definition for the vector $\lambda(t)$. We express it as:

$$\lambda_c(t) = \sum_{t_i^{c'} < t} \boldsymbol{\alpha}_{c,c'}^T \cdot \boldsymbol{\kappa}(t - t_i^{c'}) \quad \text{where} \quad \kappa_l(\Delta t) = \frac{1}{\sqrt{2\pi\sigma_l^2}} e^{-\frac{(\Delta t - \mu_l)^2}{2\sigma_l^2}} \tag{2}$$

where $\boldsymbol{\alpha}_{c,c'}$ is a vector of L parameters to infer, and $\boldsymbol{\kappa}(t - t_i^{c'})$ is a vector of L given temporal kernel functions depending only on the time difference between two events. We consider $\boldsymbol{\kappa}(\Delta t)$ to be a Gaussian RBF kernel with fixed mean $\boldsymbol{\mu}$ and deviation $\boldsymbol{\sigma}$, which allows us to model a range of different intensity functions. Each parameter $\alpha_{c,c',l}$ accounts for the influence of c' on c according to the lth entry of the temporal kernel. The dot product of α with κ yields the intensity function vector λ, which represents the topical interactions' temporal adjacency matrix (represented in an alternative way in Fig. 1-middle).

3 Experimental Setup

Dataset The dataset used in this study has been gathered from the Pushshift Reddit repository [2], which contains archives of the entirety of Reddit posts and comments up to June 2021. For each Reddit post, we can retrieve the subreddit it came from, the title of the publication, its publication date and its score (number of upvotes minus number of downvotes).

We choose to only consider popular English news subreddits. Namely, we select only posts from 2019 published on the following subreddits: inthenews, neutralnews, news, nottheonion, offbeat, open news, qualitynews, truenews, worldnews. This leaves us with a corpus of 867,328 headlines, which makes a total of 1,111,955 words drawn from a vocabulary of size 36,284.

Finally, we discard uninformative words and documents from the dataset. Explicitly, we remove the stopwords, punctuation signs, web addresses, words whose length is lesser than 4 characters, and words that appear less than 3 times in the whole dataset. Then, we remove publications that carry lesser textual or temporal information. Firstly, we choose not to consider the publications that have a popularity lesser than 20—meaning that they received less than 20 positive votes more than negative votes. We make this choice so that we consider publications that are visible enough to have any influence on the data generation

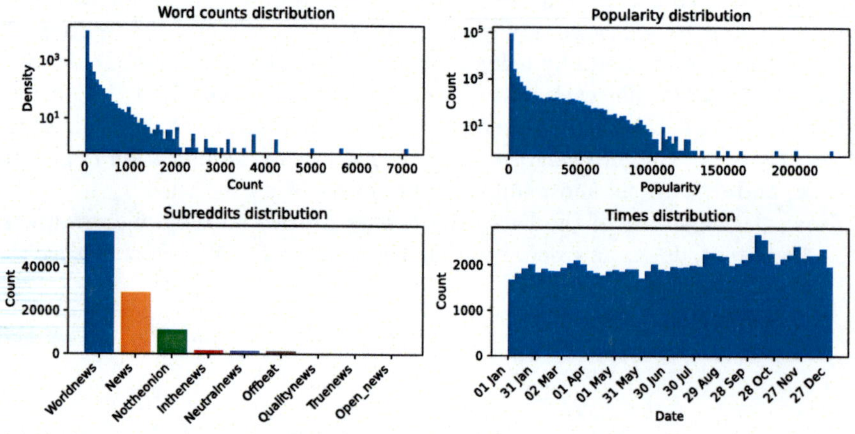

Fig. 2. Characteristics of the News dataset—For ~100,000 headlines and ~13,000 different words: (Top-Left) Distribution of the words count. (Top-Right) Distribution of headlines popularity. (Bottom-Left) Distribution of headlines over subreddits. (Bottom-Right) Distribution of headlines over time.

process. Secondly, we remove publications that comprise less than 3 words. The semantic information so-carried is expected to be poor and is not considered in our analysis.

After curating the dataset in the way described above, we are left with 102,045 news headlines (one-eighth of the original data), which makes a total of 875,334 tokens (named entities, verbs, numbers, etc.) drawn from a vocabulary of size 13,241 (one-third of the original vocabulary). The characteristics of this dataset are shown in Fig. 2.

Temporal kernel We run our experiments using three different RBF kernels, which account for publication dynamics at three different timescales: minute, hour, and day. The "**Minute**" RBF kernel is made of Gaussian functions centered at the following times: $[0, 10, 20, 30, 40, 05, 60, 70, 80]$ minutes; each entry shares a same standard deviation σ of 5 minutes, and $\lambda_0 = 0.01$. The "**Hour**" RBF kernel has Gaussians centered around $[0, 2, 4, 6, 8]$ hours, with a standard deviation σ of 1 hour, and $\lambda_0 = 0.001$. The "**Day**" RBF kernel is centered around $[0, 1, 2, 3, 4, 5, 6]$ days, with a standard deviation σ of 0.5 days, and $\lambda_0 = 0.0001$. For each of these kernels, we set the concentration parameter λ_0 so that it reaches roughly the value of one Gaussian function evaluated at 2σ. It means that an event which is 2σ away from the center of the Gaussian kernel of a single observation has 50% chances of getting associated with this Gaussian kernel entry, and 50% chances of opening a new cluster.

Hyper-parameters We consider two values for the concentration parameter of the language model: $\theta_0 = 0.001$ and $\theta_0 = 0.01$. The choice of this range is standard in the literature [5] and supported by our own observations. A larger value of θ_0 makes the inferred clusters cover a broader range of document types, whereas a small value makes the inferred clusters more specific to a topic.

The value of r is chosen to be either 0 (no use of the temporal information), 0.5, 1 and 1.5. The larger r, the more the inference relies of the temporal dynamics instead of the textual content of the documents.

The SMC algorithm described in [5,20] is run using 8 particles and 100,000 samples used to estimate the matrix of parameters α.

4 Results

4.1 Overview of the Experiments

In Table 1, we represent the main characteristics of each run in terms of number of inferred clusters K, the average cluster population $\langle N \rangle$ (where $\langle \cdot \rangle$ denotes the average), the average normalized entropy of the vocabulary of the top 20 clusters $S_{text}^{(20)}$, the average normalized entropy of the subreddits partition of the most populated 20 clusters $S_{sub}^{(20)}$. The normalized entropy is bounded between 0 and 1. It is defined so that a low entropy $S_{text}^{(20)}$ (resp. $S_{sub}^{(20)}$) means that

Table 1. Results for each experiment—We characterize each run in terms of clusters (number, size, textual entropy, subreddits entropy). The standard deviation on the last digits is given between parenthesis.

$\kappa(t)$	θ_0	r	K	$< N >$	$S^{(20)}_{text}$	$S^{(20)}_{sub}$
Minute	**0.01**	0.0	16150	6	0.744 (68)	0.400 (36)
		0.5	8498	12	0.796 (55)	0.441 (24)
		1.0	5069	20	0.790 (49)	0.476 (53)
		1.5	2730	37	0.808 (43)	0.490 (52)
	0.001	0.0	46304	2	0.475 (48)	0.239 (56)
		0.5	37277	3	0.485 (40)	0.256 (60)
		1.0	26858	4	0.501 (37)	0.266 (72)
		1.5	19275	5	0.506 (39)	0.280 (58)
Hour	**0.01**	0.0	3792	27	0.798 (52)	0.474 (106)
		0.5	1735	59	0.791 (45)	0.469 (77)
		1.0	825	124	0.803 (47)	0.484 (75)
		1.5	426	240	0.795 (37)	0.489 (69)
	0.001	0.0	18012	6	0.760 (86)	0.397 (63)
		0.5	11923	9	0.784 (72)	0.425 (38)
		1.0	4837	21	0.821 (50)	0.497 (30)
		1.5	2368	43	0.814 (42)	0.481 (91)
Day	**0.01**	0.0	609	168	0.713 (34)	0.413 (103)
		0.5	326	313	0.728 (33)	0.429 (98)
		1.0	172	593	0.743 (31)	0.461 (94)
		1.5	96	1063	0.755 (36)	0.464 (86)
	0.001	0.0	4349	23	0.705 (49)	0.396 (103)
		0.5	2654	38	0.721 (59)	0.404 (104)
		1.0	1399	73	0.734 (57)	0.431 (106)
		1.5	764	134	0.741 (54)	0.442 (98)

the top 20 clusters contain documents that are concentrated around a reduced set of words (resp. of subreddits); conversely, a large entropy means that these top 20 clusters do not account for documents concentrated around a specific vocabulary (resp. set of subreddits). We can make several observations from Table 1:

- The number of inferred clusters decreases with r, and their average population increases.
- The number of clusters grows large for the "Minute" kernel. This is because the short time range considered does not allow for clusters to last in time. A cluster that does not replicate within 1h30 is forgotten.

- We recover the fact that textual clusters have a lower entropy for small values of r [20]; this is because their creation is based more on textual coherence than on temporal coherence.
- The subreddit entropy seems to increase as r grows. A possible interpretation is that favouring the temporal information for cluster creation results in larger clusters (see $\langle N \rangle$). They would be too large to account for subreddit-specific dynamics. However, the entropy remains lower than the entropy of the distribution Fig. 2-top-left, equal to 0.51.

4.2 Quantifying Interactions

Effective interaction We introduce the parameters we are going to use in follow-up analyses. The output of MPDHP consists of a list of clusters comprising timestamped bags of words—news headlines. Between each pair of clusters, MPDHP inferred a temporal influence function $\lambda(t)$, that represents the probability for one cluster to trigger publications from another. Therefore, our model yields an adjacency matrix $A \in \mathbb{R}^{K \times K \times L}$, where K is the number of clusters and L the size of the RBF kernel $\boldsymbol{\kappa}(t)$. One entry $a_{i,j,l}$ represents the strength of the influence of j in i due to the lth entry of $\boldsymbol{\kappa}(t)$.

However, we must consider the effective number of interactions, that is to which extent the intensity function has effectively had a role in the inference. To do so, we simply consider a weight matrix $W \in \mathbb{R}^{K \times K \times L}$, whose entries $w_{i,j,l}$ are the average of the intensity of i above λ_0 due to j from the kernel entry l for all observations. Explicitly:

$$w_{i,j,l} = \frac{1}{|\mathcal{H}_i|} \sum_{t_i \in \mathcal{H}_i} \sum_{t_j < t_i} \max(a_{i,j,l}\kappa_l(t_i - t_j) - \lambda_0, 0) \tag{3}$$

The notations are the same as in Eqs. 1 and 2. Note that we retract λ_0 from the intensity term, because it is considered as a background probability for a publication to happen. Therefore, W can also be interpreted as the instantaneous increase in probability due to interactions.

Interactions strength In Table 2, we investigate the effective impact of interactions in the dataset. We consider the following metrics:

- $\langle A \rangle$: the average value of the whole adjacency matrix; to which extent topics interact with each other according to MPDHP.
- $\langle W \rangle$: the average value of the effective interactions; the extent to which the interactions (encoded in A) effectively happen in the dataset.
- $\langle A \rangle_W$: the average of the inferred interaction matrix A weighted by the effective interactions W. In this case, W can be interpreted as our confidence in the corresponding entries of A.
- $\frac{\langle W^{intra} \rangle}{\langle W^{extra} \rangle}$: ratio of the intra-cluster effective interactions with the extra-cluster effective interactions; how much clusters self-interact versus how much they interact with other ones.

Table 2. Interaction strength—Overall, interaction between clusters is weak. The standard deviation on the last digits are given between parenthesis. The large standard deviations suggest that there is a large variety of interacting behaviours. Interactions tend to happen within a cluster (self-interactions).

$\kappa(t)$	θ_0	r	$\langle A\rangle$ (10^{-3})	$\langle W\rangle$ (10^{-5})	$\langle A\rangle_W$ (10^{-3})	$\frac{\langle W^{intra}\rangle}{\langle W^{extra}\rangle}$
Minute	**0.01**	0.5	49 (21)	342 (889)	66 (17)	1.8 (62)
		1.0	48 (20)	478 (1124)	60 (17)	1.4 (43)
		1.5	48 (20)	746 (1901)	60 (17)	1.0 (33)
	0.001	0.5	50 (22)	316 (882)	66 (17)	3.1 (138)
		1.0	50 (21)	279 (752)	67 (16)	2.6 (105)
		1.5	50 (22)	268 (665)	67 (16)	2.3 (84)
Hour	**0.01**	0.5	49 (18)	389 (843)	56 (17)	0.5 (13)
		1.0	49 (18)	478 (1187)	56 (17)	0.6 (15)
		1.5	48 (17)	471 (789)	52 (15)	0.7 (13)
	0.001	0.5	50 (21)	110 (398)	61 (17)	1.7 (67)
		1.0	50 (18)	133 (506)	57 (17)	1.4 (60)
		1.5	49 (17)	183 (554)	55 (17)	1.1 (37)
Day	**0.01**	0.5	49 (18)	41 (97)	55 (17)	1.2 (34)
		1.0	49 (19)	63 (131)	54 (17)	1.2 (31)
		1.5	49 (19)	91 (187)	53 (18)	1.2 (31)
	0.001	0.5	50 (20)	18 (90)	60 (19)	1.1 (59)
		1.0	50 (19)	23 (101)	58 (17)	1.0 (50)
		1.5	50 (19)	37 (111)	56 (18)	1.0 (36)

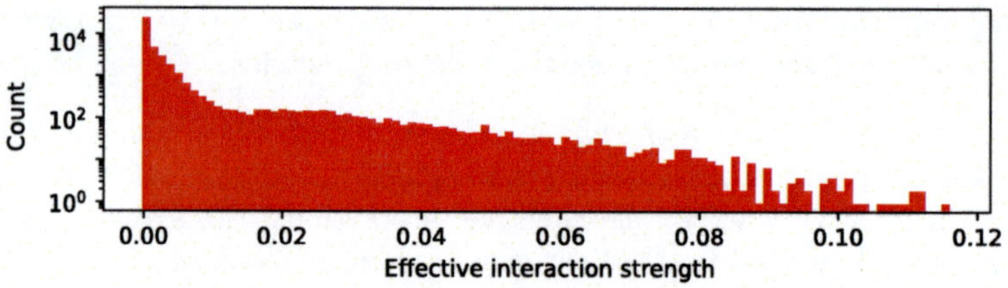

Fig. 3. Distribution of interaction strength—A large effective interaction strength means that the publication of a new document can be significantly explained by the presence of previously published news. We see that most interactions are weak.

The main conclusion of the results Table 2 is that most interactions are weak. The average value of A tells us that the average value of the inferred parameters is around 0.05, which is few given entries of A are bounded between 0 and 1. The metric $\langle W \rangle$ tells us that on all events, the interaction between clusters rose the probability of publication by 0.1–1% on average. We can also note that the values of $\langle W \rangle$ are of the same order of magnitude as λ_0 (0.01 for the "Minute" kernel, 0.001 for "Hour", and 0.0001 for "Day"). We can interpret this as the probability for a new document belonging to a cluster or being from a new cluster is roughly the same from a temporal perspective. The metric $\langle A \rangle_W$ tells us that when weighting the average of A with the effective interaction, the values of A are slightly higher than 0.05; we can now be confident in this value, given it has been inferred on a statistically significant number of observations. Still, only some interactions seem to be significant, which correlates with [15,19]. Finally, the last metric $\frac{\langle W^{intra} \rangle}{\langle W^{extra} \rangle}$ finds that most effective interactions take place more within the same cluster; clusters tend to self-replicate. Further studies of this phenomena would involve an extension of the MPDHP that considers Nested Dirichlet Processes instead of Powered Dirichlet Processes. Clusters would then be broken down into smaller ones, whose interactions could be analyzed.

Another major observation from Table 2 is that standard deviations of effective interactions are large: it hints that some of interactions may play a more significant role in the dataset. In Fig. 3, we plot the distribution of effective interactions for one specific run ("Hour" kernel, $\theta_0 = 0.01$, $r = 1$; we recover the same trend in all other experiments). The results of this figure are similar to the ones of previous studies [15,19].

Interactions range Finally, in Table 3, we investigate the range of effective interactions. We compute the effective interaction for each entry of the temporal kernel individually and average it over all existing clusters. Importantly, the effective interaction of κ_1 is smaller than others. This is induced by our kernel choice: because κ_1 is centered around $t = 0$, half of the associated Gaussian function accounts negative time differences, which never happens by design. Therefore, only other kernels entries can contribute on both sides of their mean.

We see in Table 2 that influence tends to decrease over time for all the kernels considered. Overall, the interaction between documents still plays a marginal role. We did not plot the standard deviation for visualization purposes, but they are similar as in Table 2; *most* interactions do not play a significant role in the publication of subsequent documents over time. Overall, the increase in probability for a new document to belong to a cluster due to interactions is within 0.1–1%.

5 Conclusion

In this work, we conducted extensive experiments on a real-world large-scale dataset from Reddit. We conducted 24 different experiments, each accounting

Table 3. Interaction range—All the values for effective interaction are given in ten-thousandth (10^{-5}). Influence tends to decrease over time for all kernels.

$\kappa(t)$	θ_0	r	κ_1	κ_2	κ_3	κ_4	κ_5	κ_6	κ_7	κ_8	κ_9
			0 m	10 m	20 m	30 m	40 m	50 m	60 m	70 m	80 m
Minute (m)	**0.01**	0.5	133	421	407	451	428	403	395	345	121
		1	198	591	580	607	532	575	521	507	224
		1.5	308	937	893	914	955	840	808	810	304
	0.001	0.5	218	509	457	424	371	340	313	178	52
		1	142	435	396	388	343	327	272	187	45
		1.5	104	388	366	353	326	333	290	215	61
			0 h	2 h	4 h	6 h	8 h	–	–	–	–
Hour (h)	**0.01**	0.5	247	430	502	456	324				
		1	329	538	549	542	451				
		1.5	229	615	532	526	411				
	0.001	0.5	62	149	119	137	92				
		1	77	164	172	149	111				
		1.5	104	244	223	197	156				
			0 d	1 d	2 d	3 d	4 d	5 d	6 d	–	–
Day (d)	**0.01**	0.5	22	45	47	46	47	47	37		
		1	35	71	72	68	68	70	61		
		1.5	51	100	101	105	98	105	82		
	0.001	0.5	9	20	21	21	21	21	17		
		1	12	26	24	27	28	26	22		
		1.5	11	41	42	41	41	41	35		

for a given combination of parameters, that determine the timescale considered ($\kappa(\Delta t)$), the sparsity of the language modeling (θ_0), and the extent on which we rely on text or time during the inference (r).

Our experiments hint that interactions do not play a significant role in this dataset. We proposed several ways to assess the role of interactions in the dataset. In particular, we introduced the notion of *effective interaction* as a way to evaluate how confident we are in MPDHP's output. On this basis, we analysed the importance of interactions in general, as well as from a temporal perspective. We recovered the conclusions of prior works: interactions are sparse and decay over time. By looking at the global effective interaction average, we conclude that interactions play a minor role this dataset. Overall, they only increase the instantaneous probability for a new observation to appear by 1%. Even the most extreme interactions seem to only increase this probability by 12% top.

However, despite intending our study as exhaustive, there is room for improvement in interaction modelling using MPDHP. In particular, there are two biases that we could not explore in this work. Firstly, the parameter λ_0 has been set according to a heuristic (so that a new cluster is opened with fifty percent chances when we are 95% sure that it does not match the existing one). Its direct inference would robustify the approach. Nevertheless, this inclusion sounds challenging: λ_0 does not account for individual events realizations, but for Hawkes processes starts, whose inference is not a trivial extension. Another improvement would be to allow clusters to passively replicate—without the need for an interaction. We expect that this would boil down to adding a time-independent kernel entry to $\boldsymbol{\kappa}$. However, other questions may arise from such modification: when to consider a cluster as extinct given a non-fading kernel? How should this kernel relate to the temporal concentration parameter λ_0? We believe such improvements would make MPDHP more robust and interpretable, and find applications beyond interaction modelling.

References

1. Barbieri, N., Manco, G., Ritacco, E.: Survival factorization on diffusion networks. In: Machine Learning and Knowledge Discovery in Databases, pp. 684–700 (2017)
2. Baumgartner, J., Zannettou, S., Keegan, B., Squire, M., Blackburn, J.: The pushshift reddit dataset. Proc. Int. AAAI Conf. Web Social Media **14**(1), 830–839 (2020)
3. Beutel, A., Prakash, B., Rosenfeld, R., Faloutsos, C.: Interacting viruses in networks: can both survive? Proc. ACM SIGKDD (2012)
4. Cao, J., Sun, W.: Sequential choice bandits: learning with marketing fatigue. AAAI-19 (2019)
5. Du, N., Farajtabar, M., Ahmed, A., Smola, A., Song, L.: Dirichlet-Hawkes processes with applications to clustering continuous-time document streams. In: 21th ACM SIGKDD Conference on Knowledge Discovery and Data Mining (2015)
6. Du, N., Song, L., Woo, H., Zha, H.: Uncover topic-sensitive information diffusion networks. Proc. Sixteenth Int. Conf. Artif. Intell. Stat. **31**, 229–237 (2013)
7. Gomez-Rodriguez, M.: Structure and Dynamics of Diffusion Networks. Ph.D. thesis (2013)

8. Gomez-Rodriguez, M., Balduzzi, D., Schölkopf, B.: Uncovering the temporal dynamics of diffusion networks. ICML, pp. 561–568 (2011)
9. Gomez-Rodriguez, M., Leskovec, J., Schölkopf, B.: Structure and dynamics of information pathways in online media. WSDM (2013)
10. He, X., Rekatsinas, T., Foulds, J.R., Getoor, L., Liu, Y.: Hawkestopic: A joint model for network inference and topic modeling from text-based cascades. ICML (2015)
11. Kempe, D., Kleinberg, J., Tardos, E.: Maximizing the spread of influence through a social network. In: Proceedings of the ACM SIGKDD International Conference on Knowledge Discovery and Data Mining, pp. 137–146 (2003)
12. Larremore, D., Carpenter, M., Ott, E., Restrepo, J.: Statistical properties of avalanches in networks. Phys. Rev. E **85** (2012)
13. Liniger, T.J.: Multivariate Hawkes processes. Ph.D. thesis (ETH Zurich) (2009)
14. Mavroforakis, C., Valera, I., Gomez-Rodriguez, M.: Modeling the dynamics of learning activity on the web. In: Proceedings of the 26th International Conference on World Wide Web, pp. 1421–1430 (2017)
15. Myers, S.A., Leskovec, J.: Clash of the contagions: Cooperation and competition in information diffusion. In: 2012 IEEE 12th International Conference on Data Mining, pp. 539–548 (2012)
16. Myers, S.A., Zhu, C., Leskovec, J.: Information diffusion and external influence in networks. In: Proceedings of the 18th ACM SIGKDD International Conference on Knowledge Discovery and Data Mining, pp. 33–41 (2012)
17. Niemczura, B., Gliwa, B., Zygmunt, A.: Linear Threshold Behavioral Model for the Spread of Influence in Recommendation Services, pp. 98–105 (2015)
18. Poux-Médard, G., Velcin, J., Loudcher, S.: Information interaction profile of choice adoption. In: Machine Learning and Knowledge Discovery in Databases (ECML-PKDD). Research Track, pp. 103–118 (2021)
19. Poux-Médard, G., Velcin, J., Loudcher, S.: Information interactions in outcome prediction: quantification and interpretation using stochastic block models. In: Fifteenth ACM Conference on Recommender Systems (RecSys), pp. 199–208 (2021)
20. Poux-Médard, G., Velcin, J., Loudcher, S.: Powered hawkes-dirichlet process: challenging textual clustering using a flexible temporal prior. In: 2021 IEEE International Conference on Data Mining (ICDM), pp. 509–518 (2021)
21. Poux-Médard, G.: Interactions in information spread. In: WWW 2022: Companion Proceedings of the Web Conference 2022, pp 313–317 (2022). https://doi.org/10.1145/3487553.3524190
22. Yin, J., Chao, D., Liu, Z., Zhang, W., Yu, X., Wang, J.: Model-based clustering of short text streams. In: Proceedings of the 24th ACM SIGKDD International Conference on Knowledge Discovery and Data Mining, pp. 2634–2642 (2018)

Will You Take the Knee? Italian Twitter Echo Chambers' Genesis During EURO 2020

Chiara Buongiovanni[1], Roswita Candusso[1], Giacomo Cerretini[1], Diego Febbe[2], Virginia Morini[3,4(✉)], and Giulio Rossetti[4]

[1] Department of Philology, Literature and Linguistics, University of Pisa, Pisa, Italy
c.buongiovanni@studenti.unipi.it, r.candusso@studenti.unipi.it, g.cerretini2@studenti.unipi.it
[2] Department of Physics, University of Pisa, Pisa, Italy
d.febbe@studenti.unipi.it
[3] Department of Computer Science, University of Pisa, Pisa, Italy
virginia.morini@phd.unipi.it,virginia.morini@isti.cnr.it
[4] KDD Laboratory, ISTI, National Research Council, Pisa, Italy
giulio.rossetti@isti.cnr.it

Abstract. Echo chambers can be described as situations in which individuals encounter and interact only with viewpoints that confirm their own, thus moving, as a group, to more polarized and extreme positions. Recent literature mainly focuses on characterizing such entities via static observations, thus disregarding their temporal dimension. In this work, distancing from such a trend, we study, at multiple topological levels, echo chambers genesis related to the social discussions that took place in Italy during the EURO 2020 Championship. Our analysis focuses on a well-defined topic (i.e., BLM/racism) discussed on Twitter during a perfect temporally bound (sporting) event. Such characteristics allow us to track the rise and evolution of echo chambers in time, thus relating their existence to specific episodes.

Keywords: Echo chambers · Online communication networks · Black lives matter

1 Introduction

In the social media era, a widely debated issue is whether the algorithmic bias emerging from digital platforms reinforces the need for confirmation of each individual (i.e., confirmation bias), fostering the radicalization of opinions and the emergence of echo chambers (henceforth ECs) [7,16,18]. Commonly speaking, an echo chamber can be considered as a closed system, insulated from rebuttal, in which beliefs are amplified and polarized by communication repetition. Unfortunately, there is concern that ECs might lead to several alarming episodes [15] such as hate speech, misinformation, and minority discrimination. Indeed,

H. Cherifi et al. (Eds.): COMPLEX NETWORKS 2016 2022, SCI 1077, pp. 29–40, 2023.
https://doi.org/10.1007/978-3-031-21127-0_3

since debates, campaigns, and movements taking place on online platforms also resonate in the physical world, their effects should not be relegated only to the virtual realm. Accordingly, detecting and characterizing ECs is of utmost importance since it is the first step toward deploying actionable strategies to mitigate its effects.

For such reasons, a large body of scientific works [2,5,9,10,12,14,17] has addressed the issue of echo chambers detection, often focusing on social media online discussions around highly divisive topics. However, the ill-posedness of ECs definition, along with the absence of standard strategies to support their identification, has often led to conflicting and hard to generalize experimental results [7]. Giving a brief overview, quantitative analytical methods proposed in such studies can be classified into two different families: *content-based* and *network-based.* The former subset relies on the assumption that polarized environments are detectable by looking at the leaning of content shared or consumed by a user and analyzing its sentiment on the controversy, regardless of its interactions with others. For instance, [1] explores the US debate between Liberals and Conservatives on Facebook and Twitter, looking for partisan users—i.e., those sharing articles conforming to their political beliefs—while [2] also considers users' exposure to crosscutting contents from the news feed or friends. On the other hand, the latter subset mainly focuses on finding clustered topologies in users' interactions rather than on their content homophily. Dealing with it, the authors of [11], first define the conversational network of Facebook users discussing the 2014 Thai election and then partition it into well-knit communities from a topological point of view. Nevertheless, hybrid methodologies to detect ECs—e.g., taking into account both users' ideology as well as their interactions with each other—also exist, as in [3] where the authors study online communications to understand when they resemble ECs, collecting several million tweets concerning twelve political and non-political issues. Authors infer users' ideology relying on their *follow* to popular controversial accounts then define their interaction network via *retweet.* Similarly, [8], first estimates users' leaning on political controversy based on the media slant that they share and consume and thus defines the debate network through the *follow* relationship. Indeed, one of the limitations of existing studies is that they primarily focus on open discussions not bounded to a specific time window in which users' opinions—when observed—are already formed. Accordingly, ECs are often extracted from years-long online discussions centered around ongoing, evolving, and recurring hot topics (e.g., immigration, war...), thus making impossible to investigate when and how users find themselves trapped in ECs. Moreover, such studies often discard the temporal dimension collapsing the observed behaviors into a single, timeless snapshot describing the studied phenomena as a whole.

Here we propose a longitudinal analysis of ECs related to a controversial event: namely, the Twitter discussion around "Taking the knee" that emerged in Italy during the EURO 2020 football championship. The specificity of the selected phenomenon—i.e., being temporal bounded—allowed us to not only identify and characterize ECs at different topological scales (macro, meso, and micro level) but also to dynamically track and evaluate their formation process in time—from the beginning to the conclusion of a specific framing context—thus relating them to those events that linchpin the online discussion.

The rest of the paper is organized as follows. In Sect. 2, we overview the societal issues that arose around the EURO 2020. Section 3 discusses how we manipulate Twitter data in order to assess echo chambers' existence and evolution over time. Then, in Sect. 4, we evaluate the presence of echo chambers in the debate, both considering their evolution in time and differences in scale. Finally, Sect. 5 concludes the paper with a discussion of results and directions for future work.

2 EURO 2020: Beyond the Sportive Event

The event that we want to explore from an echo chamber point of view concerns the 2020 UEFA European Football Championship (EURO 2020), i.e., a sporting event soon became a public theatre to debate racial issues. It all started when some teams participating in the championship decided to take the knee to show support for the Black Lives Matter (BLM) movement[1] and, as a consequence, the offline and online debate was monopolized about whether the players should or should not show their support for the movement.

The act of taking the knee has its origins with Martin Luther King, who knelt on one knee, praying, after many arrests occurred during a peaceful protest in Selma, Alabama in 1964. The gesture finds its way into the sports fields thanks to Colin Kaepernick, a football player, who in 2016 knelt during the performance of the American anthem, as a protest against racial discrimination suffered by black people.[2] In the context of EURO 2020, the gesture of taking the knee in support of the BLM movement started spreading thanks to the Belgium national football team who decided to kneel just before the kick-off of the matches. Such a stance was imitated by some national teams (e.g., Wales, England) and not supported by others (e.g., Hungary, Russia, Holland). In between those two opposite sides stands the Italian Football Federation (FIGC[3]) position that did not give clear support to the movement but they left the players free to behave as they thought it was better. Accordingly, during their third match (i.e., Italy-Wales), five Italian players took the knee, while the others remained standing. This event triggered a heated discussion on Twitter where people took two sides: the players standing were either labeled as racist or on the contrary, the bastion of free-thinking against the "dictatorship of politically correct". The hashtag #iononmiinginocchio[4] went immediately trending. Finally, the Italian players

[1] BLM is a social movement that protests against violence committed by the US police, and more generally against racism towards black people. It was born in 2013, and reached its peak in 2020 when George Floyd was killed by the policeman Derek Chauvin, triggering violent protests all around the US and being known worldwide.

[2] The Week, https://bit.ly/3GE3p1m, last visited: 1/06/2022

[3] https://www.figc.it/en/home/.

[4] In English "I don't take the knee".

stated their final decision just minutes before their 4th match (round of 16, Italy-Austria): they would take the knee only if the players of the other team did the same, to sustain the opponents' choices but not being supportive of the BLM movement itself.[5]

3 The Online Debate: Will You Take the Knee?

Given the high polarization of opinions around EURO 2020 in Italy, we focus our analysis on the Italian scenario by assessing whether and how echo chambers are born, strengthened, and evolved during the seven football matches. We decided to select Twitter as a data source for such a purpose since the debate about taking the knee during the football event started spreading from the beginning. In this section, we discuss how we leverage Twitter data to infer user stances on the debate as well as define the interaction network between users. The data and the code used for this study are available on Github.[6]

Dataset. Our Twitter data collection covers roughly one month—starting on June 10, the eve of the EURO 2020 opening match played by Italy and Turkey, and ending on July 13, two days after the final match, Italy-England. The conversations it encompasses gravitate towards a predefined set of hashtags we used to filter our collection pipeline, all referring to Italy's played matches, to the competition in general, and to taking the knee. We collect a total of 38,908 tweets made by 16,235 different users.

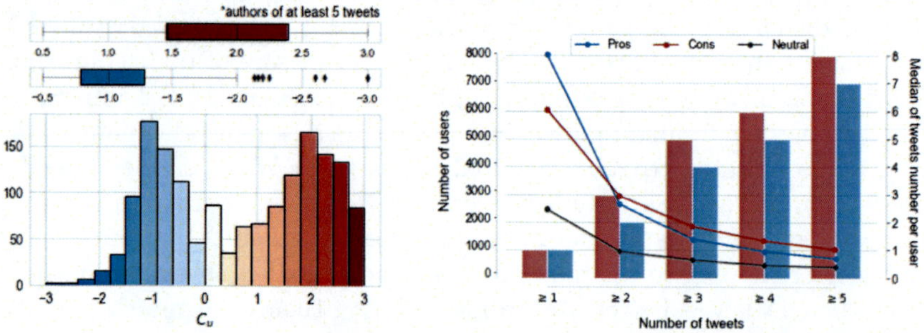

(a) (b)

Fig. 1. Users Profiling. (a) Bottom: distribution of users who wrote at least 5 tweets (to ensure that the average activity of these users covers most of the events) with respect to the opinion C_u. Top: boxplots of the distribution of non-neutral opinions, from 0.5 to the extreme limit (±3). (b) The line graph (left axes) shows the number of pros, against, and neutral users who posted at least 1 tweet (≥ 1), at least 2 tweets (≥ 2), and so on. The bar graph (right axes) shows the median of the number of tweets posted by users.

[5] La Repubblica, https://bit.ly/38VmRKN, last visited: 1/06/2022.
[6] GitHub: t.ly/-c1p.

Opinions Estimate. To classify the users' opinions, we chose a hashtag-oriented approach [6,19]. We classified, by performing a manual annotation, 2304 hashtags used in the dataset. We associate every hashtag with a numerical value. After several attempts leading to similar results, we chose the following values: ± 3 if the hashtag express a clear position, cons (+), e.g. **#iononmiinginocchio**, or pro (-), e.g. **#iomiinginocchio**,[7] on taking the knee during EURO 2020; ± 1 if the hashtag is close to the faction cons (+), e.g. **#noblm**, or pro (-), e.g. **#BlackLivesMatter**, even though not directly related to EURO 2020; 0 for the neutral and/or not relevant hashtag, e.g. **#vacciniamoci**.[8] Only 15.2% of the classified hashtags are not neutral: 180 support taking the knee, and 170 are against it. Among them, only 14 and 40 hashtags explicitly refer to kneeling, respectively for the supporters and the opponents. For every tweet, we set its value of classification C_t by computing the average value of the classification of non-neutral hashtag C_h in it. For every user u we found its classification C_u by averaging the classification values of their tweets. Looking at the distribution of users' opinions C_u in Fig. 1a, we observe the typical distribution of polarized issues with a neat prevalence of extreme values and a small number of users having a neutral position. Further, the boxplots also highlight the asymmetry with respect to $C_u = 0$ of this distribution: for $C_u \geq 0.5$ users have on average more extreme opinions, while for $C_u \leq -0.5$ there are more moderate values.

Further, we can obtain additional insights by looking at the users in the dataset. We find 7949 in favor of kneeling ($C_u \leq -0.5$) and 5970 against it ($C_u \geq 0.5$). Although the proponents outnumber the opponents by about 2000 users, the scenario changes when considering the users who posted more than one tweet during the event: the numerical difference not only decreases but overturns. The opponents slightly exceed those in favor (Fig. 1b), suggesting that the proponents might be less involved with a milder opinion regarding the kneeling act in itself (Fig. 1a). This result confirms what has already emerged in the hashtag classification operations: opponents are more likely to go explicitly against taking the knee, while proponents, rather than supporting the kneeling itself, put emphasis on the ethical and moral reasons behind it. Indeed, while those against kneeling used the hashtag **#iononmiinginocchio** more than 14,000 times, the counterpart tweeted the hashtag **#iomiinginocchio** less than 2000 times, preferring the hashtag **#blacklivesmatter** instead. In addition to the main hashtags, we notice a consistent use of the hashtag **#razzismo**[9] on both sides, highlighting the ethical implication of the issue. Further, the hashtags used by supporters are mostly slogans against racism and Nazi-fascist dictatorships, while the opponents' ones often refer to a wide range of current political and social issues (e.g., immigration, LGBTQ+ rights, Euroscepticism).

Static Network modeling. We built a weighted undirected graph, where the nodes are the users, and the edges represent their interactions (i.e., retweet, mention, quote, and reply). We chose to distinguish between "active" (or wilful)

[7] In English: "I take the knee".

[8] In English: "let's take the vaccine".

[9] In English: "racism".

interaction ($w = 1$), i.e. retweets or quote, and "passive" (not wilful) interaction ($w = 0.5$). The network is composed by $N = 15,378$ nodes and $L = 36,496$ edges. The degree distribution suggests a power law description $p(k) = CK^{-\gamma}$ and through a fit algorithm in the regime region, we obtain $\gamma = 2.12 \pm 0.04$, meaning that a scale-free regime well describes our network. Measuring different kinds of centrality gives us a complete picture of its most influential nodes. Table 1 shows nodes with the highest score for various centralities. Among them, we can distinguish two groups of people. The first is composed by quite popular accounts, mostly journalist, very productive on Twitter (e.g. `@PBerizzi`, `@Giorgiolaporta`); the second one by prominent figures, who did not join the debate directly on Twitter, although they explicitly took sides on other media, such as television or newspapers (e.g. `@FedericoRampini`, `@EnricoLetta`).

Table 1. Static Network. Top 5 nodes for different centrality scores.

Degree		Eigenvector		Closeness		Betweeness	
PBerizzi	1179	fratotolo2	0.291	JFSebastian146	0.402	PBerizzi	0.125
Giorgiolaporta	1074	FedericoRampini	0.264	fratotolo2	0.368	JFSebastian146	0.122
fratotolo2	855	Giorgiolaporta	0.242	PBerizzi	0.367	Giorgiolaporta	0.083
lefrasidiosho	793	Gianmar26145917	0.235	Vivo Azzurro	0.365	lefrasidiosho	0.076
FedericoRampini	788	Vivo Azzurro	0.195	EnricoLetta	0.361	Vivo Azzurro	0.066

Temporal Network modeling. Finally, to better highlight the evolution of the online discussion, we broke down such flat network into seven temporal-bounded snapshots, each corresponding to one of the matches played by Italy. Such modeling allowed us to longitudinally estimate and discuss ECs, as will emerge in the forthcoming section. As shown in Table 2, the network considerably grows between G3 and G4, increasing both the number of nodes and links.

Table 2. Network Snapshots. For each snapshot, match phase, reference and result, whether people knelt (N: no, Y: yes, P: partially), temporal coverage, and the number of nodes and edges.

Network	Phase	Match	Result	Kneeling	From	To	$\|V\|$	$\|E\|$
G_1	Group A	Turkey–Italy	0-3	N	10/06	14/06	78	88
G_2	Group A	Italy–Switzerland	3-0	N	15/06	19/06	102	114
G_3	Group A	Italy–Wales	1-0	P	20/06	24/06	3374	4994
G_4	Round of 16	Italy–Austria	2-1	N	25/06	29/06	14,332	32,610
G_5	Quarter-finals	Belgium–Italy	1-2	Y	30/06	04/07	15,113	35,711
G_6	Semi-finals	Italy–Spain	1-1 (4-2p)	N	05/07	09/07	15,174	35,936
G_7	Final	Italy–England	1-1 (3-2p)	Y	10/07	12/07	15,378	36,496

4 Echo Chambers: From Global to Local

To provide a comprehensive view of the online discussion under analysis, we measure and characterize echo chambers by focusing on different topological levels. In particular, we discuss them by observing patterns emerging at the *macro* (network-wide), *meso* (community-wide) and *micro* (node) level. While the first method can be exploited to identify well-separated echo chambers at an aggregated level, the meso-scale approach allows us to identify multiple ECs by taking into account differences within certain areas of the network, e.g., there can be more than one EC with the same ideology. The third method, instead, outputs the level of echo each individual node is subject to.

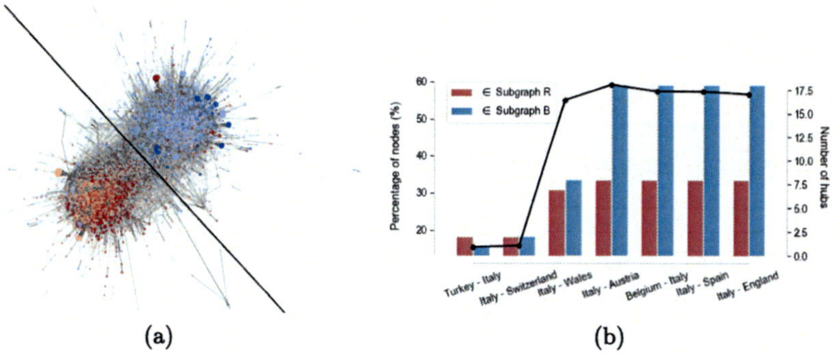

Fig. 2. Macro-scale ECs. a Visualization of the time-aggregated representation of our network, with spatial separation of the two echo chambers. **b** The line graph (left axes) shows the percentage of nodes in the network connected to hubs. The bar graph (right axes) shows the number of hubs present at different time intervals in echo chambers R (in red) and B (in blue).

Macro-scale. To qualitatively assess the presence of macro-scale echo chambers, we visualize the time-flattened network with the Force Atlas 2 graphical layout. In Fig. 2a, we can quickly identify the line that best divides the two echo chambers. Accordingly, we generate two subgraphs, one with the nodes and links of the section below the line (subgraph R), and one with those above it (subgraph B). The number of links that crosses the two subgraphs is 1889, just 5.22% of all links in the original network. Subgraph B turns out to be the largest (8352 nodes and 17,205 links vs. 6514 nodes and 17,088 links) although sparser (0.0004 vs. 0.0008) than subgraph R, which also shows higher values for the average degree (4.11 vs. 5.24) and transitivity value (0.007 vs. 0.012). Subgraph R hosts mostly users opposed to kneeling (in red, 68% of all its nodes), with an average opinion of 1.29. On the contrary, in subgraph B, 75% of the nodes have a favorable opinion regarding the issue discussed. Notice that subgraph R hosts a higher percentage of users with the opposite opinions (12% vs. 9.66%) and neutral or unclassifiable ones (19.65% vs. 14.9%) than subgraph B, which is more homogeneous. The bar graph in Fig. 2b shows another difference between the two ECs: subgraph B contains most of the hubs of the entire network (21 out of 31), however, hubs in subgraph R have a much higher average degree (646 vs. 434). Therefore, hubs in both subgraphs have been the linchpin of the debate. This is supported by the fact that their entrance into the network, between G_3 and G_4, correspond to the growth peak of the network, as shown in Table 2. More specifically, during this short time window, almost 80% of the entirety of the nodes join the network, of which about 60% through direct links with the hubs (line graph in Fig. 2b). Among the causes of the growth in G_3 we find Enrico Letta (`@EnricoLetta`) former leader of the Democratic Party ended up at the center of the controversy for the pro-kneeling statements released on television on June 21.[10] Looking at the biggest peak in G_4 we find Federico Rampini (`@FedericoRampini`) journalist and essayist, who in an interview on June 25 expressed his concerns about taking the knee,[11] Giorgio Chiellini (`@chiellini`) criticized for his gaffe about Nazism on June 26,[12] as well as Roberto Saviano (`@robertosaviano`) famous writer[13] and Le frasi di Osho (`@lefrasidiosho`), Twitter account of a popular satirical Facebook page.[14] All of these public interventions ignited the debate, encouraging the polarization of users' opinion, hence setting the stage for the ECs.

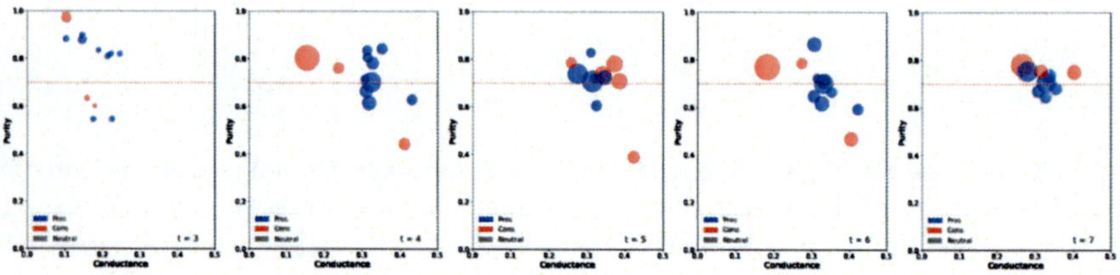

(a) G_3 (b) G_4 (c) G_5 (d) G_6 (e) G_7

Fig. 3. Meso-scale ECs. The scatter plots display the Conductance (x-axis) and Purity (y-axis) scores for the biggest detected communities. Circles represent Louvain communities, where red denotes communities populated by most opponents users, blue proponents ones. The red line marks the Purity threshold (0.7). Due to the early stages of the event, communities extracted in G_1 and G_2 are too small to be meaningful, therefore we omitted the related visualization.

Meso-scale. To refine the macro-scale analysis, we then focus on assessing the presence of echo chambers-like meso-scale topologies (e.g., well-separated communities composed of like-minded users)—and on studying if/how and at what

[10] Otto e mezzo on LA7, https://bit.ly/3tgM0q6, last visited: 5/06/2022.
[11] Stasera Italia on Rete4, https://bit.ly/3x7H1Jw at min: 15, last visited: 5/06/2022.
[12] Il Giornale, https://bit.ly/3zaf2eH, last visited: 2/06/2022.
[13] Twitter, https://t.co/wBqvDwuNde, 26/06/2021.
[14] Twitter, https://t.co/4mBKxgIlpU, 26/06/2021.

moment users happen to gather in that form. To do so, we cluster each temporal network snapshot independently using the Louvain algorithm [4]. Once identified the network communities, we evaluate their echo chamberness using Conductance (i.e., the fraction of total edge volume that points outside the community) and Purity (i.e., the product of the frequencies of the most frequent labels carried by nodes of a community), following the approach introduced in [13]. To such an extent, we label each user as: Pros (proponents) ($C_u \leq -0.5$), Cons (opponents) ($C_u \geq 0.5$), Neutral ($-0.5 < C_u < 0.5$). We set the Conductance threshold value at 0.5 to ensure that more than half of the total edges in the community remain within its boundaries. For Purity, we set a threshold equal to 0.7 to make sure that most of the users in a community share the same opinion.

In Fig. 3 we show the communities evaluation process over time. The first thing we notice is the network growth, and therefore the number of the communities along with their size, around the Italy-Wales match, between the second and the fourth time frame. This match (Fig. 3b) is the one in which only five Italian players took the knee, following the Welsh players' example. The press was finally talking about it, and the discussion on Twitter went on and on. From that moment, the network grows steadily until the Italy-Austria match (Fig. 3d) and remains stable in the subsequent matches. In each scatter plot, we can classify as echo chambers the communities that lie above the dashed red line. The colors used to portray the opinion are blue for the users that support the kneeling act, red for those who are against it, and grey for the neutral ones. In the first and the second time frames, the network is too small to be meaningful from a meso-scale perspective, so, excluding those two, the overall scenario is almost the same: there are lots of communities classifiable as echo chambers, more of each of them for proponents and opponents; on average, 70% of the users are confined in echo chambers; in almost every time-frame the supporters outnumber the opponents, while neutral users appear only at $t = 4$ (G_4) and they do not exceed the 5%. Moreover, by globally looking longitudinally at the various communities identified, we can observe that the only big-sized ones with a lower score of Conductance are those prevalently composed by opponents. Indeed, as we can see in Fig. 1a, the opponents are more extreme in expressing their thoughts: their opinion being further from 0 than the one of those in favor. Such behavior could mean that users with a highly polarized opinion are more likely to be trapped in well-defined echo chambers.

Micro-scale. Finally, to understand if, individually, the observed users are embedded in local echo chambers, we compare their opinions with their neighbors' one. To such an extent, we assume that user-centric chambers are present if/when nodes tend to be connected primarily to peers sharing a similar opinion. Therefore, to measure such a tendency, we define for every node u the average neighbors' opinion as $C_{N(u)}$. Figure 4 shows the correlation between a user's opinion C_u and the position of his nearest neighbors $C_{N(u)}$ in each snapshot. Initially, the users spread out over a very large area (Pearson's coefficient $\rho \simeq 0.49$), but as time passes, the tendency of the density to spread out diago-

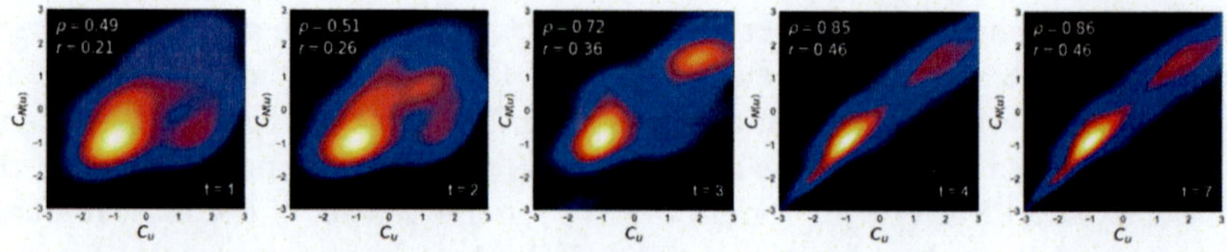

(a) G_1 (b) G_2 (c) G_3 (d) G_4 (e) G_7

Fig. 4. Micro-scale ECs. Contour map for average opinion of neighbors $C_{N(u)}$ against the average opinion of a user C_u. The colors represent the density of users: the lighter it is, the greater the number of users. Each plot refers to a temporal network snapshot . G_5 and G_6 are omitted due to their negligible changes w.r.t. G_7. ρ is the Pearson's coefficient. r is the Assortative Mixing.

nally gradually becomes more pronounced until it grows considerably ($\rho = 0.72$) and become statistically significant ($\rho = 0.85$, $p - value \simeq 0$) at the G_4 in Fig. 4d, i.e., between Italy-Wales and Italy-Austria, after which the situation remains almost unchanged. We observe the same trend by looking at networks assortativity (r) that is always greater than 0 and increases around G_3, stabilizing itself from G_4 at $r = 0.46$, meaning that networks are quite assortative. Such results unveil the presence of micro-scale echo chambers: users who express an opinion in favor/against taking the knee have a higher probability of interacting with peers sharing the same opinion.

5 Conclusions

In this work, we proposed a longitudinal analysis of the emergence of echo chambers, witnessing their evolution from the very moment they are born to their stabilization. The event we explored concerns EURO 2020 and the Italian Twitter debate born around the players taking the knee, or not, in support of the BLM movement.

Our time-aware analysis unfolds across three different topological scales (macro, meso, micro) and highlights a consistent behavior of the system as a whole as well as when its components are taken independently. At a macro-scale analysis, we identified two well-separated ECs. Next, at the meso-scale level, we observed how ECs started appearing around G_3 and strengthened in G_4—a result also confirmed by our node-level analysis that highlighted how users' opinions become considerably correlated to their neighbors' one, starting from G_4, the time window when all the hubs entered the network.

We aim to extend the proposed longitudinal analytical framework to enhance the understanding of ECs formation and, at the same time, to open the discussion on the degree of predictability of EC-like phenomena in online debates.

Acknowledgments. This work was supported by the scheme 'INFRAIA-01-2018-2019: Research and Innovation action', Grant Agreement No. 871042 'SoBigData++:

European Integrated Infrastructure for Social Mining and Big Data Analytics and by the HumaneAI-Net project', Grant Agreement No. 952026.

References

1. An, J., Quercia, D., Crowcroft, J.: Partisan sharing: Facebook evidence and societal consequences. In: Proceedings of the Second ACM Conference on Online Social Networks, pp. 13–24 (2014)
2. Bakshy, E., Messing, S., Adamic, L.A.: Exposure to ideologically diverse news and opinion on Facebook. Science **348**(6239), 1130–1132 (2015)
3. Barberá, P., Jost, J.T., Nagler, J., Tucker, J.A., Bonneau, R.: Tweeting from left to right: is online political communication more than an echo chamber? Psychol. Sci. **26**(10), 1531–1542 (2015)
4. Blondel, V.D., Guillaume, J.L., Lambiotte, R., Lefebvre, E.: Fast unfolding of communities in large networks. J. Stat. Mech. Theor. Exper. **2008**(10), P10008 (2008)
5. Cinelli, M., Morales, G.D.F., Galeazzi, A., Quattrociocchi, W., Starnini, M.: The echo chamber effect on social media. Proc. Nat. Acad. Sci. **118**(9) (2021)
6. Cota, W., Ferreira, S.C., Pastor-Satorras, R., Starnini, M.: Quantifying echo chamber effects in information spreading over political communication networks. EPJ Data Sci. **8**(1), 1–13 (2019)
7. Dubois, E., Blank, G.: The echo chamber is overstated: the moderating effect of political interest and diverse media. Inf. Commun. Soc. **21**(5), 729–745 (2018)
8. Garimella, K., De Francisci Morales, G., Gionis, A., Mathioudakis, M.: Political discourse on social media: echo chambers, gatekeepers, and the price of bipartisanship. In: Proceedings of the 2018 World Wide Web Conference, pp. 913–922 (2018)
9. Ge, Y., Zhao, S., Zhou, H., Pei, C., Sun, F., Ou, W., Zhang, Y.: Understanding echo chambers in e-commerce recommender systems. In: Proceedings of the 43rd International ACM SIGIR Conference on Research and Development in Information Retrieval, pp. 2261–2270 (2020)
10. Gilbert, E., Bergstrom, T., Karahalios, K.: Blogs are echo chambers: blogs are echo chambers. In: 2009 42nd Hawaii International Conference on System Sciences, pp. 1–10. IEEE (2009)
11. Grömping, M.:'Echo chambers' partisan Facebook groups during the 2014 Thai election. Asia Pacific Media Educator **24**(1), 39–59 (2014)
12. Hilbert, M., Ahmed, S., Cho, J., Liu, B., Luu, J.: Communicating with algorithms: a transfer entropy analysis of emotions-based escapes from online echo chambers. Commun. Methods Measures **12**(4), 260–275 (2018)
13. Morini, V., Pollacci, L., Rossetti, G.: Toward a standard approach for echo chamber detection: reddit case study. Appl. Sci. **11**(12), 5390 (2021)
14. Nguyen, T.T., Hui, P.M., Harper, F.M., Terveen, L., Konstan, J.A.: Exploring the filter bubble: the effect of using recommender systems on content diversity. In: Proceedings of the 23rd International Conference on World Wide Web, pp. 677–686 (2014)
15. O'Connor, C., Weatherall, J.O.: The Misinformation Age: How False Beliefs Spread. Yale University Press (2019)
16. Pariser, E.: The Filter Bubble: What the Internet is Hiding From You. Penguin, UK (2011)

17. Quattrociocchi, W., Scala, A., Sunstein, C.R.: Echo chambers on Facebook. Available at SSRN 2795110 (2016)
18. Sunstein, C.R.: Republic.Com 2.0. Princeton University Press, USA (2007)
19. Williams, H.T., McMurray, J.R., Kurz, T., Lambert, F.H.: Network analysis reveals open forums and echo chambers in social media discussions of climate change. Glob. Environ. Change **32**, 126–138 (2015)

A Simple Model of Knowledge Scaffolding

Franco Bagnoli[1,2(✉)] and Guido de Bonfioli Cavalcabo[1]

[1] Department of Physics and Astronomy and CSDC, University of Florence, via G. Sansone 1, 50019 Sesto Fiorentino, Italy
franco.bagnoli@unifi.it, guido.debonfiolicavalcabo@stud.unifi.it
[2] INFN, Sezione di Firenze, Italy

Abstract. We introduce a simple model of knowledge scaffolding, simulating the process of building a corpus of knowledge based on logic derivations starting from a set of "axioms". The starting idea around which we developed the model is that each new contribution, still not present in the corpus of knowledge, can be accepted only if it is based on a given number of items already belonging to the corpus. When a new item is acquired by the corpus we impose a limit to the maximum growth of knowledge for every step that we call the "jump" in knowledge. We analyze the growth with time of the corpus and the maximum knowledge and analyzing the results of our simulations we managed to show that they both follow a power law. Another result is that the number of "holes" in the knowledge corpus always remains limited. Using an approach based on a death-birth Markov process we were able to derive some analytical approximation of it.

Keywords: Learner model · Knowledge organization · Scaffolding model · Knowledge modelling · Knowledge visualization

1 Introduction

Whether knowledge accumulation is equivalent to scientific progress [3,4] or not [9] or whether this latter concept is broader than either theories suggest [8], we can assume that an increasing amount of knowledge is fundamental for the advancement of science, as actually reported in many fields [5,7,11]. Moreover, accumulation of information and knowledge is as much important to groups as it is for individuals, and, In general, knowledge accumulates over separate yet related learning episodes [2].

In all aspects of scientific disciplines based on deduction, such as most of mathematics and many parts of physics, among others, the derivation of higher-level achievements depends on previous knowledge. In these cases, new pieces of knowledge that are accepted and inserted into the existing corpus are based on previous results. Therefore it is not correct to define this process by the word "accumulation", while "scaffolding" is best suited, in our opinion.

H. Cherifi et al. (Eds.): COMPLEX NETWORKS 2016 2022, SCI 1077, pp. 41–49, 2023.
https://doi.org/10.1007/978-3-031-21127-0_4

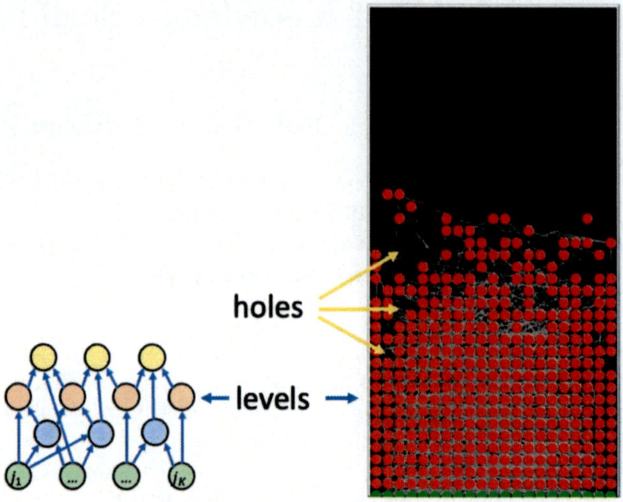

Fig. 1. The fundamental knowledge scaffolding model. (left) Knowledge bits are represented as nodes of a network, where different colors represent different levels, and nodes at a certain level only depend on a certain number of nodes at lower levels. Green (basic) nodes represent axioms. (right) Observing the filling of the network (here with fixed width and with fixed number of dependencies), one can detect holes that are filled after the appearance of nodes at higher levels.

A graphical representation of this process is given by an oriented or growing network [1] in which the single bits of knowledge are the nodes, and the links represent connection between the new items and its prerequisites, i.e., the elements of the existing corpus needed to prove the new result.

As shown in Fig. 1, this feed-forward structure can be seen as a layered network in which an item at any level depends on other items at lower levels. In this way one can put into evidence the structures that group together all simultaneous and independent derivations.

In this model it is easy to identify the "knowledge holes", which are missing items at levels which are lower than the highest one. Notice that holes are evident in a static snapshot of the fixed-width version (Fig. 1-right), but can be identified also in the variable-width model by observing the temporal filling of the structure.

The "filling" of knowledge holes have been studied for instance in world learning [10]. In this case, however, authors only had at their dispositions a static snapshot (of an arbitrary semantic network) so they had to resort to the analysis of the connectivity (robustness) of the network to identify the location of possible missing items. However, it is not granted that the possible missing items will appear, for instance, in the example provided in the cited article [10], the missing item, identified by the words "large" and "yellow", was a school-bus, but school-buses are not always yellow in every country.

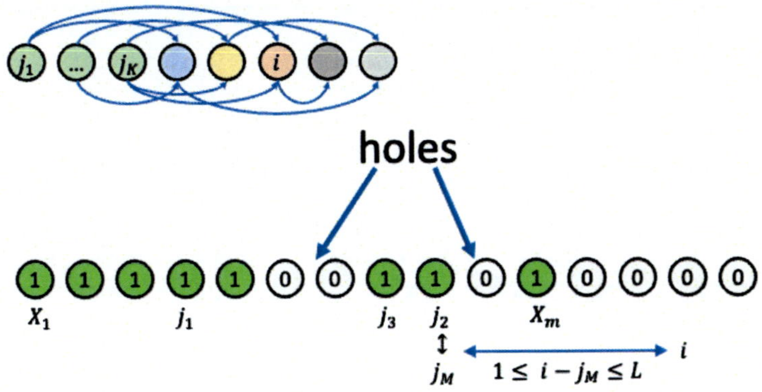

Fig. 2. The linear scaffolding model. In this version there are no well-defined levels. Known items are denoted by ones in a linear array of elements (X), and unknown ones by zeros. The "highest" known item is at position m. New items (i) become known ($X_i = 1$) if all random prerequisites j_k are known. The maximum knowledge jump L limits the choice of i so that $1 \leq i - j_M \leq L$, where j_M is the highest of prerequisites j_k.

In the following we shall use a simpler, unstructured layout representing a linear progress, illustrated in Fig. 2. This is of course a simplification of the previous process, since even the existence of yet unknown pieces of knowledge is in many cases not knowable: while in the layered structure one can add a variable number of intermediate nodes, in the linear version this is not possible. However, the analysis in this case is easier, and in a practical case it may correspond to an "ex-post" analysis of the temporal development of a knowledge space, for which the final number of items, their dependencies and the temporal sequence of the filling are known.

In summary, in this paper we want to investigate such a model of *linear* knowledge scaffolding, that despite its simplicity shows interesting behavior and can be approximated in an analytic way.

The model is presented in a detailed way in the following section, numerical and analytical results are reported in Sects. 2 and 3, respectively. Conclusions are drawn in the last section.

2 The Model

The knowledge space is represented by an array X_i of N items, which can be known ($X_i = 1$) or unknown ($x_i = 0$). We denote by

$$c(t) = \sum_i X_i$$

the size of the knowledge at a certain time.

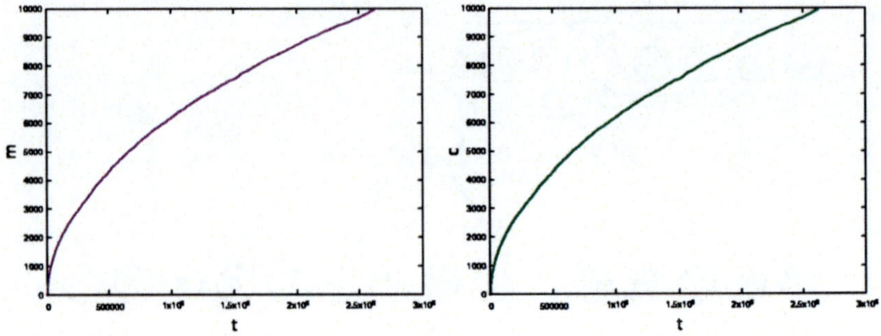

(a) (b)

Fig. 3. Evolution of knowledge corpus and maximum knowledge with $N = 10000$, $K = 4$ and $L = 5$. **a** Maximum knowledge (m) versus time. **b** Knowledge corpus (c) versus time.

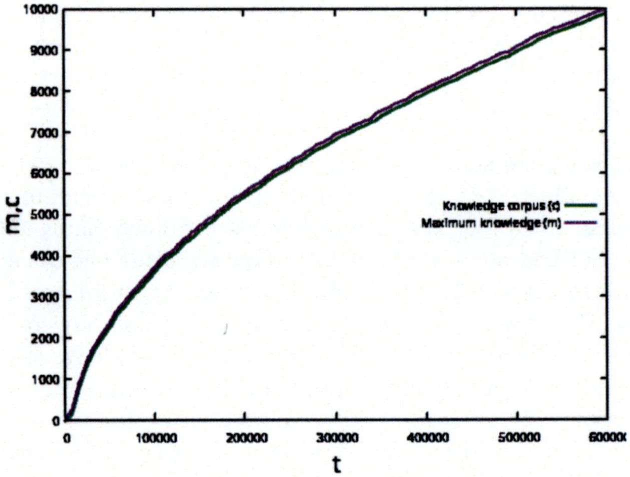

Fig. 4. Comparison between the evolution of knowledge corpus (c) and maximum knowledge (m) versus time for $N = 10000$, $K = 4$ and $L = 20$.

The linear scaffolding process can be modeled by a random proposal of *theorems*, each of which depends on a certain number, say K, of prerequisites, and furnishes a *higher* contribution. We assume that there is a limit to *jumps* in knowledge, and therefore the newly proposed item cannot be at a distance greater than L from the highest prerequisite (Fig. 2).

Let us denote by m the index of the highest known item in the corpus, i.e., $X_m = 1$ and $X_j = 0 \forall j > m$. The corpus can contain holes, i.e., $X_j = 0$ with $j < m$.

A theorem cannot be based on prerequisites higher than m, but could be rejected because it is based on holes in the present corpus. Redundant theorems (i.e., theorems providing known items) are not considered even if they are valid.

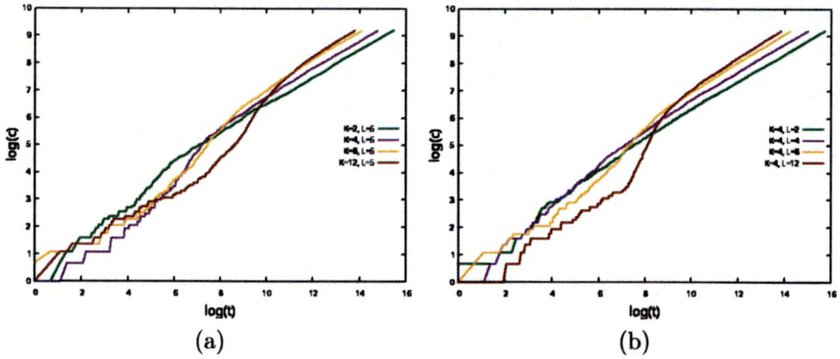

Fig. 5. Logarithmic plot of the knowledge corpus c for $N = 10{,}000$. **a** Fixed $L = 5$ and five values of K. **b** Fixed $K = 4$ and five values of L.

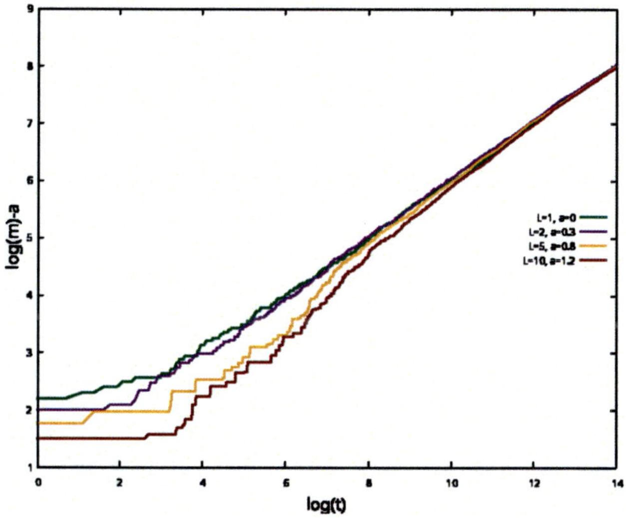

Fig. 6. Logarithmic plot of knowledge corpus c for $N = 10000$ and $K = 4$. It is possible to numerically rescale the function with different L to obtain an asymptotic convergence. The same results could be obtained by varying K.

The corpus filling proceeds by randomly choosing K items, $j_1, \dots, j_k, \dots, j_K$, with $j_k \le m$, as the prerequisite for the new theorem. It may happen that two or more of the j_k correspond to the same item, since they are chosen at random in the interval $0, \dots, m$.

We denote by j_M the largest values of the selected j_k, and we extract a random integer i such that

$$i \in (j_M + 1, \dots, j_M + 1 + L)$$

as a candidate for the new piece of knowledge.

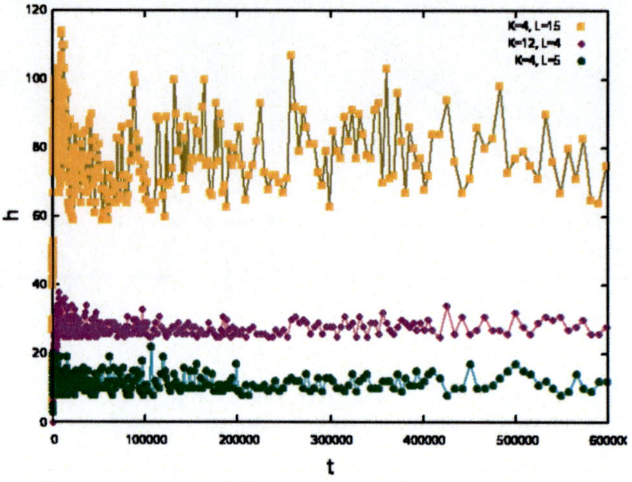

Fig. 7. Time evolution of the number of holes (h) for $N = 10000$ and different values of K and L.

If all the prerequisites are known (i.e., $X_{j_k} = 1 \quad \forall k$) and the piece of knowledge X_i is not already known (i.e., $X_i = 0$), then this derivation is added to the corpus ($X_i = 1$ and c is incremented by one), and if $i > m$ then $m = i$. In any case, the time t is incremented by one.

At the beginning we start with a knowledge vector of zeros, except the $2K$ smallest locations, that represent the axioms from which the knowledge structure is built.

The choice of the number of axioms is arbitrary, and in our case $2K$ was chosen so that the first theorem is not forced to use all present axioms (since each theorem is based on K prerequisites), but the number of starting axioms does not influence the evolution of the corpus.

For $L > 1$ the knowledge corpus may contain holes, i.e., locations ℓ with $\ell < m$ and $X_\ell = 0$. We denote by $h(t)$ the number of holes in the corpus at time t.

We repeat the previous steps until the maximum knowledge m is equal to N.

3 Numerical Results

A typical evolution for the maximum knowledge $m(t)$ and corpus size $c(t)$ are reported in Fig. 3. As one can see in Fig. 4, both values grows with the same trend regardless of the value of K and L, separated by a gap that remains finite after an initial growth. This gap is related to the number of holes in the knowledge corpus.

By plotting $c(t)$ in a log-log scale, as shown in Fig. 5, we can notice that, regardless of the values chosen for L and K, it is asymptotically approximated by an power law

$$m(t) = \mu(K, L)t^{\alpha}$$

and

$$c(t) = \gamma(K, L)t^{\alpha},$$

with $\alpha = 1/2$.

Looking at Fig. 5, we can notice that the transient depends on K and L: increasing these values the curve becomes less linear in the left part of the graph. But the asymptotic trend is independent of K and L, as shown in Fig. 6.

By investigating the number of holes $h(t)$, one sees that the value remains quite constant and that this effect is present regardless of the value of L and K, as reported in Fig. 7. Indeed, h increases at first, and the value around which the number of holes oscillates increases with L and, in a less evident way, also with of K, but eventually, for every K and L, the the filling of inner holes takes place, in correspondence with a slowing growth of $c(t)$.

4 Markov Birth-Death Approach

Let us consider the simplest case possible: $K = L = 1$, for which there is no hole in the corpus, since every piece of knowledge depends on the immediately previous one. In this case $c = m$.

For each time step, either the knowledge increases by one, or stays constant. Therefore it is an example of a birth-death Markov Chain, with no deaths [6].

Let us denote by $P(y, t)$ the probability that at time t the corpus or maximum knowledge is y. The evolution of P is given by

$$P(y, t + 1) = P(y, t)\left(\frac{y - 1}{y}\right) + P(y - 1, t)\left(\frac{1}{y - 1}\right), \tag{1}$$

where there are two processes: the knowledge increases by one if the j_1 prerequisite is equal to $y - 1$ (with probability $1/(y - 1)$, or stays constant and equal to y if j_1 is one of the other $y - 1$ possibilities.

The average knowledge $\bar{y}(t)$ at time t is defined as

$$\bar{y}(t) = \sum_{y} yP(y, t). \tag{2}$$

The Markov process starts from the condition $P(y, 0) = \delta_{y,2K}$.

The time and space continuous approximation of the Markov equation (1) (valid far from the initial conditions, for $y \gg 2K$ and $t \gg 0$) is:

$$\frac{\partial P}{\partial t} = -\frac{1}{y}\frac{\partial P}{\partial y}, \tag{3}$$

which implies that P is a function of $y^2 - 2t$.

Therefore, in this approximation

$$\bar{y}(t) \propto \sqrt{2t} \tag{4}$$

for large t.

Indeed, this is the case, as shown in Fig. 8.

The approximation Eq (1) has been obtained in the case $L = 1$, $K = 1$, for which there is no hole. It is possible to generalize it to the case $L > 1$, imposing the absence of holes, as

$$P(y, t+1) = \frac{1}{Z(t)} \sum_{k=0}^{L} P(y-k-1, t) \left(\frac{L-k+1}{L(y-k-1)} \right),$$

where $y - k - 1 > 0$ and Z is a normalization constant, such that $\sum_y P(y, t) = 1$. This generalization is justified by the fact that even for $L > 1$ the number of holes stay limited. This approximation reproduces the power-law growth of the corpus.

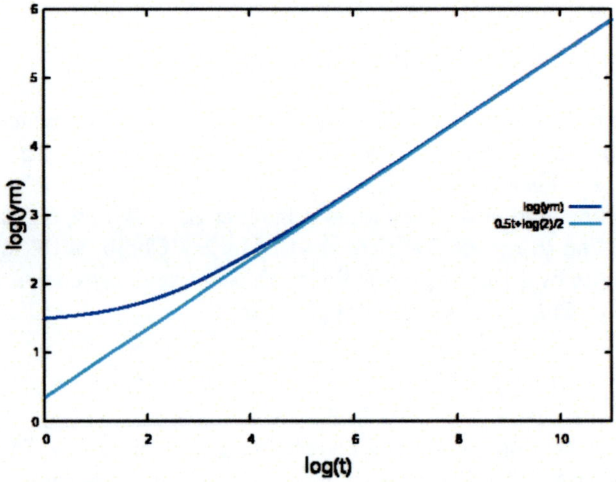

Fig. 8. Comparison of the $\bar{y}(t)$ (ym in the graph) with the function $0.5 \cdot t + \frac{\log(2)}{2}$, with $N = 10000$ and the total time of the simulation $T = 400000$. As shown in Eq. (4) there is a linear proportionality between time and $\bar{y}(t)$.

5 Conclusions

We have presented and analyzed a simple model of knowledge scaffolding, approximated by the growth and filling of a knowledge vector. The idea is that of providing a model corresponding to an ex-post analysis of the temporal development of a knowledge space.

The elementary step of our model is the proposal of a new theorem depending on at most K prerequisites and providing a knowledge jump of at most L steps.

This move is accepted and inserted in the corpus if the prerequisite does already belong to the known corpus and the result does not.

The size of the corpus, i.e., the number of known items and the maximum knowledge grows in time following a power law with exponent 1/2, regardless of the number of input items K and jump L.

An analytical approximation, based on a death-birth Markov process is proposed, reproducing the power law.

We believe that this very simple and approximate model can be used as a basis for a qualitative description of more complex systems.

References

1. Barabási, A.L.: Network science: Luck or reason. Nature **489**(7417), 507–508 (2012). https://doi.org/10.1038/nature11486
2. Bauer, P.J.: We know more than we ever learned: Processes involved in accumulation of world knowledge. Child Dev. Perspect. **15**(4), 220–227 (2021). https://doi.org/10.1111/cdep.12430
3. Bird, A.: What is scientific progress? Noûs **41**(1), 64–89 (2007). https://doi.org/10.1111/j.1468-0068.2007.00638.x
4. Bird, A.: Scientific progress as accumulation of knowledge: a reply to rowbottom. Stud. History Philos. Sci. Part A **39**(2), 279–281 (2008). https://doi.org/10.1016/j.shpsa.2008.03.019
5. Cong, L.W., Xie, D., Zhang, L.: Knowledge accumulation, privacy, and growth in a data economy. Manage. Sci. **67**(10), 6480–6492 (2021). https://doi.org/10.1287/mnsc.2021.3986
6. Karlin, S.: A First Course in Stochastic Processes. Academic Press (2014)
7. Kuo, C.I., Wu, C.H., Lin, B.W.: Gaining from scientific knowledge: the role of knowledge accumulation and knowledge combination. R&D Manage. **49**(2), 252–263 (2018). https://doi.org/10.1111/radm.12322
8. Mizrahi, M.: What is scientific progress? lessons from scientific practice. J. General Philos. Sci. **44**(2), 375–390 (2013). https://doi.org/10.1007/s10838-013-9229-1
9. Rowbottom, D.P.: N-rays and the semantic view of scientific progress. Stud. History Philos. Sci. Part A **39**(2), 277–278 (2008). https://doi.org/10.1016/j.shpsa.2008.03.010
10. Sizemore, A.E., Karuza, E.A., Giusti, C., Bassett, D.S.: Knowledge gaps in the early growth of semantic feature networks. Nat. Human Behaviour **2**(9), 682–692 (2018). https://doi.org/10.1038/s41562-018-0422-4
11. Wagener, T., Gleeson, T., Coxon, G., Hartmann, A., Howden, N., Pianosi, F., Rahman, M., Rosolem, R., Stein, L., Woods, R.: On doing hydrology with dragons: realizing the value of perceptual models and knowledge accumulation. WIREs Water **8**(6) (2021). https://doi.org/10.1002/wat2.1550

Using Knowledge Graphs to Detect Partisanship in Online Political Discourse

Ari Decter-Frain[1] and Vlad Barash[2(✉)]

[1] Cornell University, Ithaca, NY 14850, USA
agd75@cornell.edu
[2] Graphika Technologies Inc., New York, NY 10020, USA
vlad.barash@graphika.com

Abstract. Existing methods for detecting partisanship and polarization on social media focus on either linguistic or network aspects of online communication, and tend to study a single platform. We explore the possibility of using knowledge graph embeddings to detect and analyze partisanship in online discourse. Knowledge graphs can potentially combine linguistic and network information across multiple platforms to enable more accurate discovery of a political dimension in online space. We train embeddings on heterogeneous graphs with different combinations of information text, network, single- and multi-platform information. Building on previous work, we develop a semi-supervised approach for uncovering a political dimension in the embedding space from a handful of labelled observations, and show that this method enables more accurate differentiation between liberal and conservative Twitter accounts. These results indicate that knowledge graphs can potentially be useful tools for analyzing online discourse.

Keywords: Polarization · Knowledge graphs · Graph embeddings

1 Introduction

In recent years researchers have sought to quantify online polarization via a number of different methods, using various measures of partisanship and polarization. In general, these methods focus either on the content produced by users [8,11,15,20] or the networks formed by their interactions [1,13,19]. Fewer papers consider how focusing on each of these components may change our conclusions about polarization, or how results may generalize across platforms.

In this paper, we conduct a first exploration into the potential for knowledge graphs to enhance our understanding of political partisanship and polarization in online discourse. We fit knowledge graph embeddings (KGEs) via the library dgl-ke [21]. KGEs are embeddings trained on graphs with heterogeneous nodes (called entities) and edges (called relations). They encode information about network position while incorporating heterogeneous sources of information [17]. KGEs can potentially yield more detailed summaries of political discourse than

H. Cherifi et al. (Eds.): COMPLEX NETWORKS 2016 2022, SCI 1077, pp. 50–61, 2023.
https://doi.org/10.1007/978-3-031-21127-0_5

graph embeddings based on retweets, or text embeddings based on content. Furthermore, they enable the inclusion of entities from multiple social media platforms into the same network, and support analysis of how discourses play out across the social media landscape as well as how they differ between platforms.

We begin from the following question: Can unsupervised methods detect a partisanship dimension in KGEs trained on online political discourse? Furthermore, how does detection of this dimension improve as we incorporate different types of information into the graph? In addressing this question, we contribute to an ongoing line of research developing methods for quantifying controversy, partisanship, and polarization in social media discourse.

We conduct our analyses using data collected surrounding two prominent online discourses. The first pertains to disinformation about the COVID-19 pandemic (COV) and the second is about the veracity of climate change (CLM). We chose these discourses for two reasons: first, both are major topics of public discussion that generate a high volume of engagement online across multiple platforms. Chen et al. [5] found approximately 500,000 Twitter posts tagged with climate-change related hashtags in a period of just twenty days (Jan 7 2020–Jan 27 2020), while Treen et al. [18] found nearly 300,000 posts and comments . Similarly, Cinelli et al. [4] find 1,300,000 posts related to COVID-19 across multiple Gab.ai, Instagram, Reddit, Twitter, and YouTube between Jan 1 2020–Feb 14 2020, early in the pandemic. Second, COV and CLM span a wide variety of sociocultural contexts and platforms; both discourses are characterized by a high degree of political polarization [14,18], but in neither case is the polarization confined to a specific sociocultural or political environment. We hope that our choice of discourses will allow the results of our analysis to generalize to a broad range of polarized topics.

2 Related Work

Several efforts have been made to characterize the political landscape on online social media platforms. Conover and colleagues found that a basic clustering algorithm applied to retweet and mention graphs in a political sample of Twitter data yields two large clusters [7]. Barbera developed a latent-space model for quantifying online partisanship in retweet networks [1,2]. Interian and Ribeiro developed an empirical model of polarization that builds up from the homophily of each individual's ego network [13]. Morales et al. [16] use a network-based approach to demonstrate how online spaces reflect the polarization in physical space. Waller and Anderson performed community detection across a full historical dataset of Reddit posts, and developed a procedure for detecting social dimensions, including partisanship in embedding space [19]. Below, we implement a variation on this procedure to measure partisanship.

Several papers focus on partisan content. Gentzkow and colleagues model polarization as the proportion of users whose partisanship can be correctly classified using the distribution of tokens from their posts [11]. Mehova et al. [15] use a combination of crowdsourcing and dictionary-based methods to characterize bias and emotion in controversial news discourse. Demsky and colleagues combine lexical and embedding-based measures to analyze discourse surrounding mass shooting events [8]. Yan et al. [20] carry out a similar embedding exercise and validate their embedding-based measure of polarization through a stance detection task.

Some consider ways of interpreting both text and network features as indicative of polarization. Conover et al. [7] find that word frequencies differ between polarized network clusters. Chin and colleagues use both approaches to study polarization in a multiparty context [6]. Garimela and Weber follow polarization longitudinally using text- and network-based measures [10]. Garimela et al. [9] consider a method for quantifying online controversy that uses information on shared content to de-noise a retweet graph.

To our knowledge, no previous work has attempted to quantify polarization by incorporating both content and network information into a heterogeneous graph and embedding both types of information.

2.1 Data

We collected data pertaining to each of the two high profile narratives using a set of key terms to search across platforms. We used different methods for collecting data from different platforms, as follows: for Twitter, we used the public Twitter API to collect posts authored active users who posted tweets with at least one of a set of key terms, described below. Due to Twitter's API limitations, our Twitter data collection was limited to a 30-day period. For Reddit, we used the Pushshift service [3] to collect any post with one or more of the key terms since 2011. Finally, for Telegram we constructed our own collection service to collect posts within 30 days from the collection time from public channels related to CLM or COV respectively, as identified by subject matter expert analysts who collaborated with us on this project.

For CLM, we did not discover any major public Telegram channels, so we collected data from Twitter and Reddit. The key terms for CLM are listed in the Appendix. This resulted in 799 Reddit posts between February 22, 2011 and March 15, 2022; and 155,403 Twitter posts between July 5, 2021 and August 4, 2021. For COV, we collected data from Twitter, Reddit, and Telegram. The key terms for COV are listed in the Appendix. This resulted in 1793 Reddit posts between May 25, 2011 and August 18, 2021; 3889 Telegram posts between July 15, 2021 and August 15, 2021; and 405,507 Twitter posts between May 31, 2021 and June 30, 2021.

3 Methods

3.1 Knowledge Graph Generation

We extracted knowledge graph relations from the data. Knowledge graphs consist of two basic objects: entities, which represent users who author social media posts, the context where that post appeared (e.g. a subreddit in Reddit), and the informational content (e.g. hashtags, n-grams, URLs) present in the post text or metadata; and relations, which represent directed connections between the entities. For example, if user **@user** writes a post with hashtag **#hashtag**, we can add the entities **@user** and **#hashtag** and the relation (**@user**, **#hashtag**, "hashtag") to the knowledge graph, where the relation is represented as an ordered triple with the "head" (the source entity of the relation), the "tail" (the target entity of the relation), and the "type" (a label for the type of the relation). Posts on different platforms sometimes, but not always, imply different relation types. For example, Twitter has a unique relation type "retweet," but Reddit, Telegram, and Twitter all share the relation type "n-gram" (representing a user using a specific n-gram in the text of their post). We distinguish between content-relations between the author entity of a post and informational content within the same post; and network-relations between the author entity and another author entity or a context entity. We list the full set of relation types we parse from the posts described above in the Appendix. To increase graph density, we duplicated each relation to include its reverse.

Figure 1 shows a sample knowledge graph schematic using the example above. For clarity of visualization, not all the reverse relations are shown.

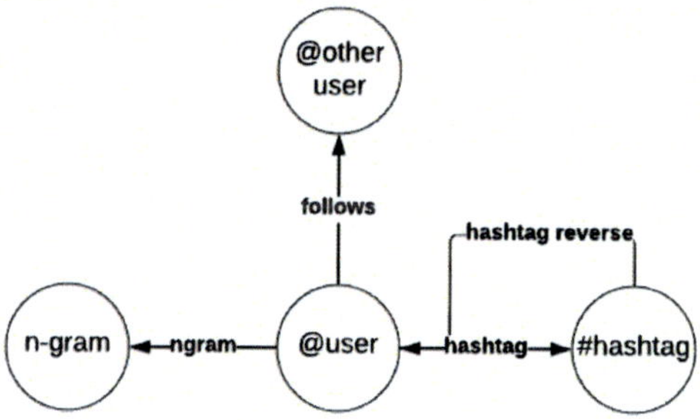

Fig. 1. Example of the relational structure of a heterogeneous knowledge graph used in this study.

As a final step, we performed K-core reduction [22] on the graphs because sparse graphs tend to inhibit convergence in dgl-ke. Beginning from $K = 10$, we computed the estimated memory requirement for training embeddings on the graph and incremented K until the estimated memory requirement fell beneath our available 32GB memory limit. We used an equation for estimating the memory burden of an embedding in a github issue [12]. Table 1 presents the breakdown of entities included in each final graph, along with the total number of relations.

Table 1 summarizes the contents of each graph, including the number of each type of entity, and the total number of relations between entities. Comparing the third and fourth column for each graph shows the effect on the total graph size from adding additional platform data. Adding data from more platforms does add more entities to the graph, particularly accounts. However, the data we collected from other platforms generally adds few additional relations.

Table 1. Contents of each graph for which we trained embeddings. Entities are generated for each entity, and entities are linked by various relations. Min. degree describes the value used for K-core reduction to ensure each graph is as large as possible while still enabling training within memory constraints.

	Covid disinformation				Climate change Denial			
	Twitter			All	Twitter			All
	Network	Content	All	All	Network	Content	All	All
Min. degree	10	10	20	20	10	10	10	10
N. relations	106,307	1,382,892	1,289,267	1,370,592	317,751	350,360	668,165	686,029
Account	1872	2363	2573	3538	16,926	257	16,926	17,159
Post	3279	–	1397	1397	229	–	229	229
Unigram	–	20,886	11,706	11,957	–	8,310	8309	8485
Bigram	–	8	–	–	–	–	–	2
Named entity	–	2808	1359	1395	–	1018	1018	1041
Hashtag	–	831	424	420	–	138	138	136
Media	–	225	95	95	–	3	3	3
URL	–	585	339	339	–	231	231	231

3.2 Training Embeddings

To train embeddings, we first split the arcs (triplets indicating the origin, relation type, destination of a particular relation) into training, test, and validation sets using a 90/5/5 split. We used the training and validation sets to train each embedding. We trained the embeddings using the dgl-ke library, using a fixed set of hyperparameters manually tuned to obtain adequate performance during evaluation [21]. We used the ComplEx model type because it is both flexible and efficient. We trained models for 600,000 iterations with a batch size of 1024 and a negative sample size of 512. We trained 512D embeddings using a learning rate equal to 0.5 and a regularization coefficient equal to 2.00E-6.

We evaluated the embeddings using the standard approach, by assessing their link prediction accuracy on unseen data. Link prediction asks the trained model to differentiate between a real triplet in the data (origin entity, relation, target entity), and a large number (256) of fake triplets. For each of the real and fake triplets, the model produces a probability that the triplet is real. These probabilities are ranked, and performance is measured based on where in the ranking the real triplet ends up [21]. Strong link prediction performance indicates that embeddings have encoded adequate information to identify and differentiate entities in the graph.

Table 2 presents the evaluation metrics for each trained embedding. For all models, we obtained satisfactory performance on the link prediction evaluation, which indicates the resulting embeddings have encoded adequate information about the structure of the underlying graphs. Figure 2 summarizes the data collection, knowledge graph generation, embedding, and evaluation pipeline we describe above.

Table 2. Evaluation metrics for each trained embedding. Evaluation involves generating 512 random negative samples for each real arc, then having the model rank all 513 of these arcs in order of their probability of belonging in the graph. HITS@1 indicates the share of the test set for which the model assigns the most probability to the real arc. HITS@3 and HITS@10 are the share for which it is in the top 3 and 10, respectively. MRR is mean reciprocal rank, defined as $MRR = \frac{1}{Q}\sum_{i=1}^{Q}\frac{1}{rank_i}$, where Q is the number of test cases, and $rank_i$ is the rank assigned to the real triplet among the negative samples. All evaluations take place on a 5% test set unseen during training.

	COVID-19 disinformation				Climate change Denial			
	Twitter			All	Twitter			All
	Network	Content	All	All	Network	Content	All	All
Test size	21,262	276,579	257,854	274,119	63,551	70,072	133,633	137,206
MRR	0.960	0.722	0.689	0.680	0.833	0.884	0.898	0.895
HITS@1	0.951	0.613	0.577	0.567	0.763	0.853	0.866	0.862
HITS@3	0.963	0.798	0.762	0.752	0.898	0.923	0.923	0.920
HITS@10	0.977	0.924	0.905	0.899	0.929	0.948	0.948	0.948

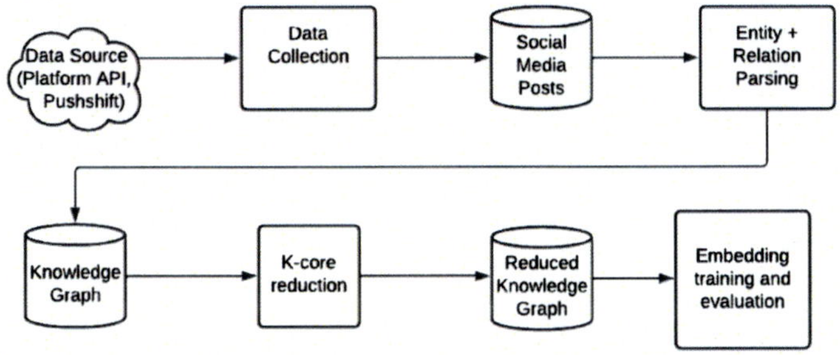

Fig. 2. Architecture diagram for data collection and knowledge graph generation.

3.3 Detecting Partisanship

To detect partisanship in knowledge graph embeddings, we apply a semi-supervised approach that builds on previous work by Waller and Anderson [19]. Our approach relies upon a small number of seed entities to construct a vector in the embedding space that represents a given social dimension. First, for each graph, we select five pairs of entities that differ in partisanship but are otherwise similar. To select seeds, we first collected Twitter IDs for all media sources in the allsides.com media bias list[1] We joined these Twitter IDs to the entities in the graph to obtain a subset of each graph with ground-truth partisan labels. Next we computed cosine similarity between all pairs of 'left' and 'right' leaning entities and selected the five most similar pairs. We end up with five pairs per graph. Each pair contains one 'left' and one 'right' labelled entity that are otherwise relatively similar (as indicated by their proximity in the embedding space).

Next, we take the vector difference between the normalized embeddings for each pair of seeds, and average across the five pairs. The result is a 512 dimension vector that theoretically represents political partisanship in the embedding space. Once we have this vector, we can compute a partisan score for every other entity in the graph as the dot product between their normalized embeddings and the partisanship vector. Waller and Anderson show this is equivalent to taking each entity's average similarity to the left of the partisanship vector, minus its average similarity to the right [19].

3.4 Evaluation

If our measure of partisanship accurately reflects the political partisanship of the entities in the graph, we should be able to use this measure to correctly differentiate between entities with ground-truth partisan labels. We collect a set of labelled Twitter accounts by combining two sources. First, we use the allsides.com dataset described above. Second, we collect Twitter IDs for all current members of US Congress. We recoded the labels from these two sources into three categories: Liberal, Conservative, and Center. Next, we then joined these labels into each graph. For the COV network-only and CLM content-only graphs, we obtained fewer than 50 labelled cases. Thus, we omit evaluation on these graphs. For the other six graphs, we obtained between 56 and 285 labelled cases. Using the labelled subsets of each graph, we fit logistic regressions to predict ground-truth partisanship using the partisanship scores we constructed as a predictor.

We compare performance against two baselines. First, we construct a random model where each entity's partisanship is randomly drawn from a uniform distribution between zero and one. Second, we try using the first principal component of the embedding space as a measure of partisanship.

4 Results

Figure 3 presents two-dimensional representations of the entity embeddings obtained via principal components analysis (PCA). Each point represents an entity, with its type given by its shape, and its inferred partisanship given by its

[1] Allsides.com uses independent expert panels to assign partisan bias ratings to news outlets and journalists. https://www.allsides.com/media-bias/ratings.

color. For several of the graphs, some degree of polarization appears to emerge along the principal components. The principal components also appear to reflect differences between entity types. For instance, text and account entities appear to cluster somewhat separately in the graphs where the two are most heavily combined.

Interestingly, some of the text clusters appear to have strong partisan alignments. In particular, substantial clusters of blue (i.e. left-leaning) words emerge in the graphs with network and content information. These words are more 'similar' to the liberal seed accounts than the conservative seed accounts. This suggests that certain words, phrases, and entities tend to be used by accounts that are more similar to the liberal seeds. The knowledge graph embedding encodes this heterogeneous graphical information simultaneously.

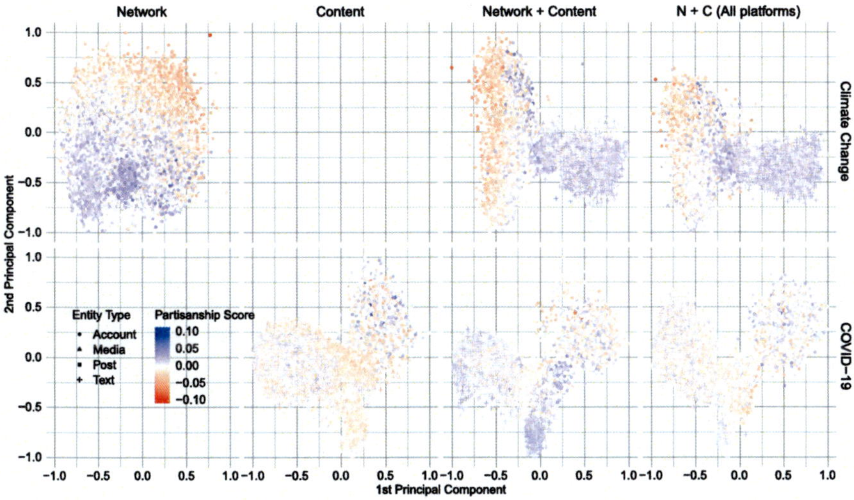

Fig. 3. Scatterplots of two-dimensional embedding space after performing knowledge graph embeddings on heterogeneous representations of online discourse. The 512 dimension embeddings are reduced to two dimensions through principal components analysis. The two omitted graphs lacked sufficient labelled data to construct/evaluate a measure of political partisanship. 'Media' includes URL, hashtag, and image entities. 'Text' includes unigrams, bigrams, and named entities.

Finally, Table 3 presents the results of our evaluation exercise. The seeded approach to measuring partisanship improves upon random and PCA baselines in every case we could measure. Improvements are largest for CLM, which may be due to the superior fit of the embeddings over these graphs. It is also notable that models with more information do not consistently outperform simpler representations of the discourse. For COV, content graphs appear better suited to detecting partisanship than more complex graphs containing content and network information. For CLM, combining network and content information yields marginally better performance than a graph with only network information.

Table 3. In-sample performance metrics for multinomial logistic regression models to trained to distinguish between liberal, conservative, and centrist media outlets using unsupervised measures of partisanship. 'Seeded' refers to the seed-based partisanship score. COVID-19 network and Climate Change content are omitted because these graphs lacked a sufficient number of labelled media organizations to reliably estimate performance.

		Accuracy			F1 score		
		Random	PCA	Seeded	Random	PCA	Seeded
COVID-19							
	Network	–	–	–	–	–	–
	Content	0.482	0.518	0.589	0.363	0.392	0.445
	Network + content	0.435	0.468	0.484	0.334	0.357	0.371
	N + C (All platforms)	0.484	0.419	0.484	0.370	0.321	0.370
Climate change							
	Network	0.589	0.589	0.716	0.247	0.247	0.536
	Content	–	–	–	–	–	–
	Network + content	0.589	0.600	0.740	0.247	0.289	0.563
	N + C (all platforms)	0.589	0.632	0.723	0.247	0.379	0.526

5 Discussion

In this paper we made a first attempt to apply knowledge graph embeddings to the task of learning users' partisan affiliation via unsupervised or semi-supervised methods. We trained embeddings using graphs with different combinations of content and network information, and estimated the partisanship of every node using a seed-based procedure. These estimates were substantially better than random guessing and using the first principal component of the embeddings as a proxy for partisanship.

The data upon which we relied for our analyses have important limitations that may have factored into our results. First, several of our sampled datasets contained few accounts for which we could find ground-truth partisan labels. In two cases, this prevented us from generating and evaluating partisanship measures. Our approach of collecting data from search terms contrasts with many studies that begin from a set of labelled accounts and sample outward. This latter approach is more common and guarantees ample data for evaluation, but also risks making the problem easier by building a graph around the entities of greatest interest. Our results suggest that such a sampling strategy may lead to overly optimistic conclusions about our ability to detect partisanship.

Another challenge we ran into pertains to the magnitudes of the datasets sampled from different platforms. Although we collected data from Twitter, Telegram, and Reddit, the samples we obtained from Twitter were much larger than the other platforms. This may partly reflect the reality that Twitter contains the most discourse about the topics under study, and may also partly reflect differences in the particularities of their APIs. Here, the small sizes of the Telegram and Reddit datasets prevented us from looking at them individually, and we could only include them as additions to the Twitter network. Future work will need to more carefully develop a framework for integrating multiplatform data into a knowledge graph to maximize information gain while accounting for platform and API idiosyncrasies.

Methods for unsupervised and semi-supervised detection of social dimensions in online discourse remain important for applied and theoretical research. This paper represents a first exploration of how knowledge graphs and knowledge graph embeddings can improve performance on this task. While our study has several important limitations, we believe it points to a promising area of research that will provide deeper, more generalizable insights into political polarization. Polarization in today's online discourse is cross-platform as well as highly dependent on the socio-cultural context of the discourse—approaches that rely on pure text or pure network analysis are likely to draw the wrong conclusions about the mechanisms behind a polarized environment, and therefore, about the likely trajectory of its evolution.

Future research should explore solutions to the challenges we identified. This includes more systematically evaluating the effects of partisan imbalance, developing methods for incorporating multiple platforms of different sizes, and understanding when an embedding has been trained well enough to identify latent social dimensions in the data.

6 Appendix

6.1 Key Terms

For CLM, we used the following key terms: #ClimateEmegency, #EnviroScam, #ClimateAction, #climatefraud, #GreenNewDeal, #climatescare, #ActOnClimate, #NoClimateCrisis, #climatestrike, #NoClimateEmergency, #GlobalGaslighting, #climatehoax, #GlobalCooling, #ClimateReality, #climatechangehoax, #climatechangescam, #ClimateHoaxers, #ClimateScam, #GlobalWhining, #ItsCalledWeather.

For COV, we used the following key terms: #BillGatesBioTerrorist, #BillGatesVaccine, #vaxxed, #mandatoryvaccination, #forcedvaccination, #populationcontrol, #depopulation, #vaccineawareness, #BillGatesVaccine, #novaccinemandates, #medicaltyranny, #vaccineinjury, #learntherisk, #vacc, #medicalfreedom, #maskless, #uniteforfreedom, #onemillionplus, #wewillALLbethere, #vaers, #stopnewnormal, #wedonotcomply, #nomasksinclass, #novaccineforme, #destinationdepop, #nurembergtrials, #nomoremedicaltyranny, #fuckcovidjab, #vaxkseen, #nuremberg2, #ScreenB4Vaccine, #NaturalImmunity, #vaccineskill, #HumanExperiment, #UNMASKOURCHILDREN, #CrimesAgainstHumanity, #FauciLiedPeopleDied, #Agenda21.

6.2 Relation Types

See Table 4.

Table 4. Relation types by platform.

Platform	Head	Tail	Relation type
Reddit	User	Named entity	Named entity
Reddit	User	n-gram	ngram
Reddit	User	Subreddit	Subreddit
Reddit	User	URL	URL
Telegram	User	channel	Forward
Telegram	User	Named entity	Named entity
Telegram	User	n-gram	ngram
Telegram	User	User	Reply
Telegram	User	URL	URL
Twitter	User	User	Follow
Twitter	User	Hashtag	Hashtag
Twitter	User	User	Mention
Twitter	User	Named entity	Named entity
Twitter	User	n-gram	ngram
Twitter	User	User	Quote-tweet
Twitter	User	User	Reply
Twitter	User	User	Retweet
Twitter	User	URL	URL

References

1. Barberá, P.: Birds of the same feather tweet together: Bayesian ideal point estimation using twitter data. Polit. Anal. **23**(1), 76–91 (2015). https://doi.org/10.1093/pan/mpu011
2. Barberá, P., Jost, J.T., Nagler, J., Tucker, J.A., Bonneau, R.: Tweeting from left to right: is online political communication more than an echo chamber? Psychol. Sci. **26**(10), 1531–1542 (2015). https://doi.org/10.1177/0956797615594620
3. Baumgartner, J., Zannettou, S., Keegan, B., Squire, M., Blackburn, J.: The Pushshift reddit dataset. In: Proceedings of the Fourteenth International AAAI Conference on Web and Social Media (ICWSM 2020)
4. Cinelli, M., Quattrociocchi, W., Galeazzi, A. et al.: The COVID-19 social media infodemic. Sci. Rep. **10**, 16598 (2020). https://doi.org/10.1038/s41598-020-73510-5
5. Chen, C., Shi, W., Yang, J., Fu, H.: Social bots' role in climate change discussion on twitter: measuring standpoints, topics, and interaction strategies. Adv. Clim. Change Res. **12**(6), 913–923 (2021). https://doi.org/10.1016/j.accre.2021.09.011

6. Chin, A., Coimbra Vieira, C., Kim, J.: Evaluating digital polarization in multiparty systems: evidence from the German Bundestag. In: 14th ACM Web Science Conference 2022, Barcelona Spain, Jun 2022, pp. 296–301. https://doi.org/10.1145/3501247.3531547
7. Conover, M.D., Ratkiewicz, J., Francisco, M., Goncalves, B., Flammini, A., Menczer, F.: Political Polarization on Twitter, p. 8
8. Demszky, D., et al.: Analyzing polarization in social media: method and application to tweets on 21 Mass shootings. In: Proceedings of the 2019 Conference of the North, Minneapolis, Minnesota, 2019, pp. 2970–3005. https://doi.org/10.18653/v1/N19-1304
9. Garimella, K., Morales, G.D.F., Gionis, A., Mathioudakis, M.: Quantifying controversy in social media. arXiv:1507.05224 [cs], Sep 2017, Accessed Apr 12, 2022 [online]. Available http://arxiv.org/abs/1507.05224
10. Garimella , K., Weber, I.: A Long-Term Analysis of Polarization on Twitter, p. 5
11. Gentzkow, M., Shapiro, J.M., Taddy, M.: Measuring group differences in high-dimensional choices: method and application to congressional speech. ECTA **87**(4), 1307–1340 (2019). https://doi.org/10.3982/ECTA16566
12. GitHub Issue on dgl-ke. Bus error (core dumped). https://github.com/awslabs/dgl-ke/issues/174. Accessed on 07/11/2022
13. Interian, R., Ribeiro, C.C.: An empirical investigation of network polarization. Appl. Math. Comput. **339**, 651–662 (2018). https://doi.org/10.1016/j.amc.2018.07.066
14. Jiang, J., Chen, E., Lerman, K., and Ferrar, E.: Political polarization drives online conversations about COVID-19 in the United States. Human Behav. Emerg. Technol. **2** (2020). https://doi.org/10.1002/hbe2.202
15. Mejova, Y., Zhang, A.X., Diakopoulos, N., Castillo, C.: Controversy and sentiment in online news. arXiv:1409.8152 [cs], Sep 2014, Accessed Apr 12, 2022 [online]. Available http://arxiv.org/abs/1409.8152
16. Morales, A.J., Dong, X., Bar-Yam, Y., Pentland, A.: Segregation and polarization in urban areas. R. Soc. Open Sci. **6**(10). https://doi.org/10.1098/rsos.190573
17. Nickel, M., Murphy, K., Tresp, V., Gabrilovich, E.: A review of relational machine learning for knowledge graphs. Proc. IEEE **104**(1), 1 (2016). https://doi.org/10.1109/JPROC.2015.2483592
18. Treen, K., Williams, H., O'Neill, S., Coan, T.G.: Discussion of climate change on reddit: polarized discourse or deliberative debate? Environ. Commun. (2022). https://doi.org/10.1080/17524032.2022.2050776
19. Waller, I., Anderson, A.: Quantifying social organization and political polarization in online platforms. Nature **600**(7888), 264–268 (2021). https://doi.org/10.1038/s41586-021-04167-x
20. Yan, M., Wen, X., Lin, Y.-R., Deng, L.: Quantifying content polarization on Twitter. In: 2017 IEEE 3rd International Conference on Collaboration and Internet Computing (CIC), San Jose, CA, Oct 2017, pp. 299–308. https://doi.org/10.1109/CIC.2017.00047
21. Zheng, D., et al.: DGL-KE: Training Knowledge Graph Embeddings at Scale. Apr 18, 2020. Accessed May 27, 2022 [online]. Available http://arxiv.org/abs/2004.08532
22. Seidman, S.B.: Network structure and minimum degree. Soc. Netw. **5**(3), 269–287 (1983). https://doi.org/10.1016/0378-8733(83)90028-X

The `wisdom_of_crowds`: An Efficient, Philosophically-Validated, Social Epistemological Network Profiling Toolkit

Colin Klein[1], Marc Cheong[2(✉)], Marinus Ferreira[3], Emily Sullivan[4], and Mark Alfano[3]

[1] The Australian National University, Canberra, ACT, Australia
colin.klein@anu.edu.au
[2] University of Melbourne, Melbourne, VIC, Australia
marc.cheong@unimelb.edu.au
[3] Macquarie University, Sydney, NSW, Australia
[4] Eindhoven University, Eindhoven, The Netherlands

Abstract. The epistemic position of an agent often depends on their position in a larger network of other agents who provide them with information. In general, agents are better off if they have diverse and independent sources. Sullivan et al. [19] developed a method for quantitatively characterizing the epistemic position of individuals in a network that takes into account both diversity and independence and presented a proof-of-concept, closed-source implementation on a small graph derived from Twitter data [19]. This paper reports on an open-source re-implementation of their algorithm in Python, optimized to be usable on much larger networks. In addition to the algorithm and package, we also show the ability to scale up our package to large synthetic social network graph profiling, and finally demonstrate its utility in analyzing real-world empirical evidence of 'echo chambers' on online social media, as well as interdisciplinary diversity in an academic communications network.

Keywords: Social epistemology · Python · Social network analysis · Testimonial networks

1 Introduction

Most of what we know we know because we learned about it from other people.[1] *Social epistemology* is the subfield of philosophy that studies how knowledge and justification depend on the testimony of others transmitted through social networks [7]. A focus on networks has been influential because it allows philosophers to connect their concerns to the substantial body of empirical and simulation work on real-world networks and their graph-theoretic properties. Sullivan et al.

[1] Our acknowledgments to Ehsan Abedin, Ignacio Ojea Quintana, and Ritsaart Reimann for their kind feedback on this paper.

H. Cherifi et al. (Eds.): COMPLEX NETWORKS 2016 2022, SCI 1077, pp. 62–73, 2023.
https://doi.org/10.1007/978-3-031-21127-0_6

[19] presented a method for quantitatively characterizing the epistemic position of individuals in a network. Broadly speaking, individuals are in a better epistemic position if they are receiving information from *diverse* and *independent* sources, with the more diversity and independence the better. This method was based on, amongst others, the *Wisdom of Crowds* hypothesis, that the aggregated judgements of many individuals can systematically be more accurate than the judgements of those individuals taken singly [20]. They then operationalized these two concepts in a way that allowed them to provide an interesting profile of a small 185-member Twitter community [19]. That work relied on a bespoke, closed-source codebase. As it was built as a proof-of-concept, it was also not optimized in ways that naturally scaled to larger networks. This made it difficult to apply the technique to other datasets, such as networks from other social media sites, or networks created from artificial social simulation algorithms, e.g. Laputa [16]: all of which are of interest to both philosophers and computer scientists alike. That work's codebase had an emphasis on the generation of Java-based visualizations—using a combination of several platforms and toolchains—which does not lend itself to convenient large-scale network analysis due to performance limitations.

To make this tool more widely available to researchers, we therefore present **`wisdom_of_crowds`**, a complete ground-up re-implementation in Python of the core Sullivan et al. [19] concepts. The code is optimized to deal with larger networks. It also includes some standardized helper functions to allow for coordinating results between research groups and data scientists. We have made the code for this package open source, under the *GNU General Public License 3.0*[2], on GitHub (https://github.com/cvklein/wisdom-of-crowds/). It has also been accepted to the Python Package Index (PyPI, at http://pypi.org/project/wisdom-of-crowds), and is available for any user to install via `pip`.[3]

As much as possible, we have relied on open-source Python packages such as `networkx` [17] and `matplotlib` [9] as they have been rigorously tested and are freely available for auditing and peer review. For good practices in verification and validation, the `pytest` [10] library is used to provide a unit-testing framework. Having created this new implementation, we sought out to deploy it in our investigation of contemporary social network data, to provide a data-driven perspective to complement existing conceptual and theoretical work. This paper thus aims to present three key findings:

- the core concepts of the `wisdom_of_crowds` package, including compatibility with (and optimization for) contemporary network-related datasets and packages in Python. This includes improvements to the base [19] algorithm, by clearly defining the bounds (and justifications for) parameters used, as well as suggested extensions including the *h*-measure derived from [8].
- application of `wisdom_of_crowds` on simulated large networks; to investigate its feasibility/performance.

[2] Full license terms can be found at https://www.gnu.org/licenses/gpl-3.0.en.html.

[3] It can be installed by the Python community via: `pip install wisdom_of_crowds`.

- application of `wisdom_of_crowds` on actual real-world networks—Twitter discourse about the Black Lives Matter (BLM) movement between January and July 2020 centering around the May 25th 2020 murder of George Floyd [2]; and the *email-Eu-core* network on communications patterns in "a large European research institution" [11,12,23]—to corroborate theoretical findings on modern-day social networks centered on current phenomena.

2 Background and Methods

2.1 Core Concepts

Epistemology is a branch of philosophy which, simply put, "is concerned with how people should go about the business of trying to determine what is true" [18]. *Social epistemology* concerns the testimony of others embedded in social contexts [7]; in contrast with 'individual' epistemology which concerns how an individual conducts reasoning, abstracted away from their "social environment" [18].

In recent years, social epistemologists have moved away from considering dyadic relationships between individuals to consider the ways in which social epistemic *networks* shape the information we receive [4,15]. Consider an epistemic network G where nodes are epistemic agents and edges represent the relationship of receiving information via testimony. '*Testimony*' is used broadly in social epistemology for any way in which one source delivers information to another, and includes speech, writing, and other forms of media. All things being equal, a node is better off receiving information from more and more diverse nodes. However, testimony is often transmitted in chains, and this transmission need carry only the content of the information, not (meta-)information about the original source or the intermediate links. This complicates the position of any individual who is trying to learn from multiple sources. For example, a piece of gossip heard from two people seems more reliable than from one, but that reliability is undermined if both heard it from the same person [3].

2.2 Sullivan et al. (2020)'s Operationalizations

Defining the *m, k* observer: Following [19], we say that a node n is an m, k-observer just in case it receives information from a set of at least k different nodes which are pairwise at least m steps away from one another, when considered on the subgraph of G that does not contain n. If G is directed, then candidate sources must be at least m steps away in both directions. The removal of n from consideration in the case of distances is necessary for directed graphs, as otherwise all sources to n are trivially at most 2 steps apart; we carry over that requirement to undirected graphs as well.

In this work, we assume $1 \leq m \leq 5$, as most real life networks have length 6 paths between most arbitrarily chosen nodes [13]. We bound $2 \leq k \leq 5$, because a node with a single source is in a very poor epistemic position with respect to diversity of input. Note that it is a consequence of the definition that if n is an m, k-observer, it is also both an $m-1, k$-observer and an $m, k-1$-observer (assuming $m-1$ and $k-1$, respectively, are permissible values).

Given this definition, the core concepts in [19] are defined as follows.

***S(n)*, independence of sources.** $S(n)$ gives a measure of the independence of sources to node n. Consider the set s of possible pairs (m, k) for which n is an m, k-observer. Then define

$$S(n) = \begin{cases} 0 & \text{if } s = \emptyset \\ \max\{mk: (m, k) \in s\} & \text{otherwise} \end{cases} \tag{1}$$

In other words, $S(n)$ is just the largest mk such that the n is an m, k-observer. If S has 0 or 1 nodes as sources, they are considered as being in an epistemically bad position, and so $S(n) = 0$. Note that given this definition, possible S values do not increase smoothly. Given the bounds set out above, $S(n) \in \{0, 2, 3, 4, 5, 6, 8, 9, 10, 12, 15, 16, 20, 25\}$.

***D(n)*, diversity of sources.** $D(n)$ measures the diversity of the sources that contribute information to n. Let each node i be associated with a set a_i of epistemically-relevant attributes. These might be group affiliations, topics of interest, scientific approaches, political leanings, and so on. Let s be the set of n's sources. Then define

$$D(n) = \mid \bigcup\{a_i : i \in s\} \mid \tag{2}$$

That is, D gives the number of distinct *types* of information that feed into n.

***π(n)*, epistemic position.** Finally, as the epistemic position of a node is a function of both the diversity and independence of sources, we define

$$\pi(n) = S(n)D(n) \tag{3}$$

2.3 Caveats and Considerations

There are a few notes to make about the implications of the Sullivan et al. [19] core concepts in real social networks.

Regarding m, k-observers, while higher rankings are better and all the nodes with a specific value of m and k are members of an equivalence class, the framework does not posit which of two m, k-observers is better positioned if one has a higher m value and the other a higher k value—for instance, whether 2, 3-observers are better- or worse-placed than 3, 2-observers. The framework thus does not provide a total order but instead provides a collection of *partial* orders. Intuitively, for instance, a 1, 2-observer is worse placed than a 2, 3-observer despite having neither an m or a k value in common, because a 2, 3-observer is better placed than a 2, 2-observer when considering k values, which in turn is better placed than a 1, 2-observer when comparing m values. Thus, a node's S value does provide a way to more easily compare the epistemic position of nodes, by combining their m and k values into a single value.

The choice of multiplying m and k values together to come to a single measure is ultimately arbitrary. The S value contains less information than the m and k values does, because it fails to preserve the difference between being a $2, 3$-observer and being a $3, 2$-observer (the former has fewer independent sources, but these have a greater degree of independence from each other than in the latter's). Simultaneously, it posits that a node which is a $3, 3$-observer is determinately better placed than either a $4, 2$-observer or a $2, 4$-observer, but worse than a $5, 2$-observer or a $2, 5$-observer.

In short, while a node's S value is one measure of the independence of its sources, it is a measure that throws away information and makes finer-grained distinctions than may be warranted in some cases. This was the impetus for us introducing a further measure, the `h_measure(n)` [8] which returns the highest h such that n is a h, h-observer, as discussed below.

2.4 Our Re-implementation

The core of the `wisdom_of_crowds` is a class `Crowd`.[4] `Crowd` is initialized with a *NetworkX* graph (encapsulating the social network's edges and nodes), and provides various functions to calculate the metrics defined above.

Calculating whether a node is an m, k-observer combines multiple shortest-path problems with clique-finding problems. Naïve approches to the latter have complexity $O(n^k)$ [22]. Given that we are considering unweighted paths, the shortest-path problem has a reasonably efficient linear-time solution, but the requirement to remove n from the distance calculations also means that network shortest paths cannot simply be calculated at the outset. In the worst case scenario, they must be recalculated for each pair of sources for each node.

Hence this is a computationally difficult problem to brute-force. In practice, efficient caching and testing of seen paths plus greedy k-clique algorithms means that worst-case performance can often be avoided for realistic networks: we *memoize* (i.e., cache) intermediate path values, trading-off space for a reduced processing time. As the envisioned use case for large graphs is for one-shot batch processing, our code allows for easy multithreading or multiprocessing (e.g. using *multiprocessing* or *concurrent.futures*) allowing it to attain substantial performance gains, in conjunction with the memoization mechanism.

The m, k-observer functionality is the basis for calculating D, S, and π. D is calculated on node attributes, and users can specify the appropriate key for the attribute. If a single attribute is supplied, D is calculated using the singleton set containing that attribute.

In addition to the standard measures, we also introduce an improvement: the `h_measure()` of a node n as the largest x such that n is an x, x-observer; comparable to the standard definition of Hirsch's h-index in citation practices [8]. As suggested by [19], being a 3,3-observer is the minimal secure epistemic

[4] Due to space constraints, we only discuss our substantial contributions within this paper. The full documentation for all implemented methods is available at https://github.com/cvklein/wisdom-of-crowds/blob/main/docs/wisdom_of_crowds.py.md.

position, and the use of a single non-multiplicative standard may be useful for some cases.

Finally, the package includes two helper functions to allow for comparable reporting and display across different users. `iteratively_prune_graph()` takes a *NetworkX* graph, removes small-degree nodes, small-weight edges, and takes the largest connected component in what remains, iterating this process until the graph is stable. The thresholds are parameterized; the default is for $indegree + outdegree \leq 1$ and no edge culling, as per [19]. In the spirit of scientific reproducibility, `make_sullivanplot()` makes a summary figure of a whole network in the style of [19, refer to their Fig. 7]. It can produce standalone plots or return a subplot in a specified `matplotlib` axis.

3 Experimental Results

3.1 Efficiency Tests

Firstly, to benchmark the ability of `wisdom_of_crowds` on different magnitudes of nodes, typical of modern-day datasets, we sought to batch calculate S for different magnitudes of node sizes ($|N|$), given various probabilities of edge connections.

Figure 1 shows the efficiency of batch calculating S for each node of a `Crowd` on random graphs with varying parameters for probability of edge connection. Random directed graphs were generated using the `networkx` generator `fast_gnp_random_graph()` with the parameters indicated in the figure. Timing was done using the python `timeit` package over a single iteration.

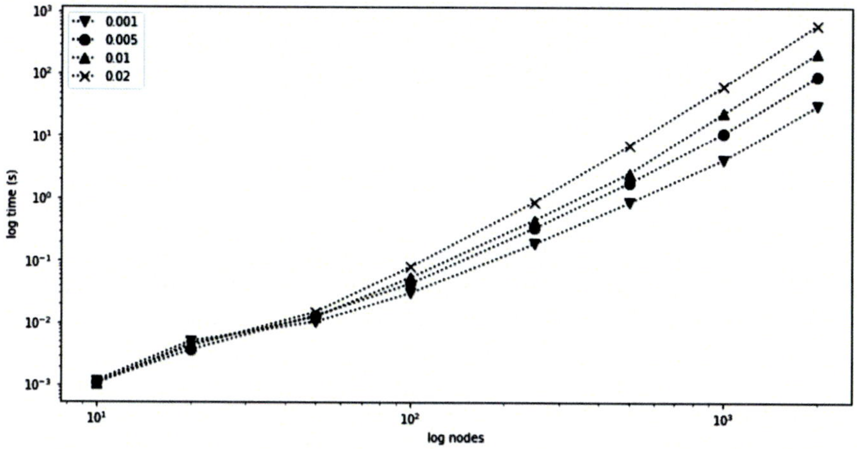

Fig. 1. Timing curve for random graphs by number of nodes. Different markers represent different connection probabilities for nodes.

As the log-linear plot in Fig. 1 shows, there is a roughly exponential relationship between the number of nodes and runtime, with the exponent a function of the number of edges. This suggests that the efficiency of our code approaches what would be expected given the fundamental complexity of the clique-finding problem. Note that the exponential growth means that the boundary between computationally tractable and intractable graphs can be relatively tight. Judicious pruning often makes a difference.

Further, note that this was achieved in standard Python operating conditions, i.e. in the absence of any multiprocessing/multithreading support (see also Sect. 2.4). Performance gains will be attained for batch calculations on high-CPU (16-or-more CPU cores) systems.

3.2 Application on Social Network Data: BLM on Twitter

Secondly, as part of our investigation of real-world phenomena, in line with [19], we replicate their findings on a real-world Twitter retweet network to examine the information-sharing dynamics during the Black Lives Matter movement. An earlier version of this dataset was used by [19] who were able to examine a network of 185 nodes. For further information on the dataset[5] used, see [2]. For brevity, the key details are summarised here.

A *retweet network* [2,19] was generated: i.e., a weighted directed network where nodes are authors and the weight of an edge from node u to node v represents the number of times that user v retweeted user u. We took the largest connected component of this graph as the starting point for cluster-analysis [6,21]. Following [2], We found first-level clusters using Modularity Vertex Partitioning, preserving clusters with more than 10% of the original nodes. This gave 4 clusters, covering 83% of the graph. Next, we manually inspected the 100 most-influential nodes within each group, characterizing the communities as Activists, Center-Left Democrats, Republicans, and a set of "Boosters" who mainly amplified the content of the first two groups. Topic models were fit using *scikit-learn*'s non-negative matrix factorization (NMF), fit on a *tf-idf* representation of the Twitter text (post-sanitization) with `min_df=0.05`. Finally, we used `iteratively_prune_graph()` with a node and weight threshold of 3, which resulted in a tractable subgraph with $|N| = 16249$ nodes and $|E| = 145246$ edges. This subgraph had very little representation from the 'booster' group, so they were omitted from further analysis. By comparison with [19], our analysis is an increase in the number of nodes by a magnitude of $\sim 100\times$. Batch processing took about 6.25 h on a 2017 desktop iMac.

Figure 2 plots S, D, and π for the current experiment. For the left side of 2, D was calculated using the cluster identity of the node. For the right side, D was calculated using the `argmax` of the fitted and normalized W matrix for the topic model. This gives the topic that is most distinctive of each user's tweets.

[5] We queried the Twitter Streaming API with a series of Black Lives Matter (BLM)-related keywords, hashtags, and short expressions in a window between January and July 2020.The dataset comprised $\sim$ 4.6 M original tweets between January 13th and July 18th and $\sim$ 94.5 M retweets from January 18th to July 23rd; these tweets were produced by $\sim$ 2.0 M distinct authors.

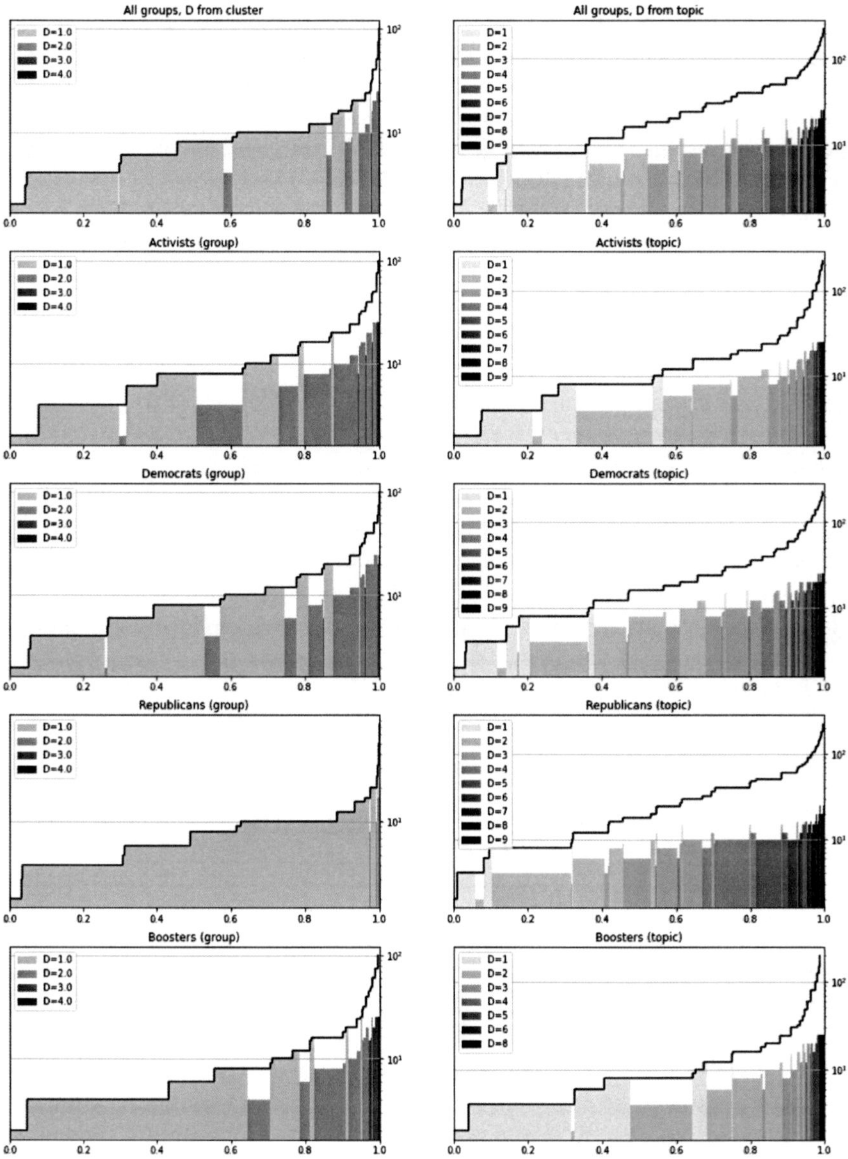

Fig. 2. Profile plots for entire network and subgroups looking at clusters (left) and topics (right). X axis is proportion of total, Y axis shows both S (height of bars) and π (black line), plotted on a log scale.

We examine both the network as a whole and three identified subclusters in the graph. The left half of Fig. 2 shows S, D, and π for the network, where D is calculated via the subcluster identity of sources. The right half of Fig. 2 recalculates D and π based on a 9-topic NMF topic model of aggregated tweets (compared with the 3-topic model of [19]). The results illustrate the utility of profiling networks using our toolkit. On the left, one can see that Republicans appear to be in the worst epistemic position in terms of the other subgroups with which they interact: they have a generally low D, suggesting that they tend to listen mostly to in-group members. However, they have a relatively high S and therefore a π comparable to other subgroups. Compare this to the topic-wise graph, in which Republicans have a relatively high diversity for *topics*, one at least as good as other groups. The activist group shows something of the inverse pattern. That is, they show a more varied range of S and D values for group-group interactions, but a comparatively lighter graph with fewer topics for the broad span.

These results might suggest that both groups are part of 'echo chambers' [1,14], but in different ways: the right tends to be a monoculture socially but a polyculture topically, with a converse pattern on the left. Finally, we note that all subgroups, in both domains of measurement, tend to have an $S < 10$ for more than half the population. This replicates the observations of [19], in which most participants end up in a comparatively poor epistemic position. However, most groups tend to contain at least a small sub-population which is well-connected and often with a relatively high D. We note that this is especially the case with our 'Booster' group, a small subset who seemed to be mainly concerned with amplifying and relaying the content of other groups.

3.3 Application on Communication Network Data: *email-Eu-core*

Finally, we apply our `wisdom_of_crowds` analysis onto the *email-Eu-core* network, an existing dataset curated by [12,23] on the Stanford Large Network Dataset Collection [11]. The rationale behind this is to apply our social epistemological lens to analyze an existing network which is hitherto (to our knowledge) never been examined using the tools we have at hand.

Briefly, *email-Eu-core* consists of "email data from a large European research institution" [11], represented as a digraph where an edge $(u,\ v)$ exists "if [researcher] u sent [researcher] v at least one email". The beauty of *email-Eu-core* lies in the fact that it only considers internal communications in the institution, ignoring any noise resulting from possible non-academic communication originating from/in reply to outside actors; and that the ground-truth membership of each node has already been established, i.e., as belonging to any "one of 42 departments at the research institute" [11].

Compared to more modern social networks, *email-Eu-core* has a comparatively small $|N| = 1005$ and $|E| = 25571$. However, again, we note that this is already $\sim 10\times$ the magnitude as compared to the hitherto state-of-the-art [19]. The total running time on such a dataset is comparatively trivial (less than 5 min). As we do not have e-mail text associated with the original nodes in this dataset, we sought only to analyze the 'overall epistemic picture' of the network, as a whole. In the same vein as Sect. 3.2, Fig. 3 illustrates the profile plot for *email-Eu-core.*

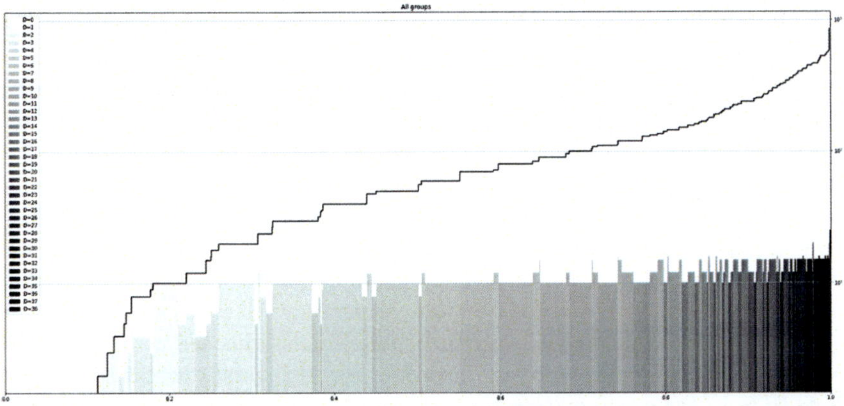

Fig. 3. Profile plots for the entire *email-Eu-core* network [11,12,23]. The X axis is proportion of total, Y axis shows both S (height of bars) and π (black line), plotted on a log scale.

Again, the utility of our approach is evident. Given the results, we observe that the distribution of S is fairly consistent for all the nodes (researchers) in the research institution's academic network. However, the progressively darker bars illustrate that researchers who have connections with a more diverse amount of departments (thereby maximizing D) can vastly optimize their epistemic position. Roughly, about $\sim$ 40% of researchers—i.e., the right-most data points in Fig. 3—have a π of about 100 or more, despite having roughly the *same* S values. To our knowledge, this is the first time empirical social epistemological analyses in the spirit of [19] have been conducted on such email networks.

4 Discussion and Conclusion

Our results show that it is possible to replicate the methodology used by [19] to larger networks, and that insights about the relative epistemic positions of different communities within a network can be drawn from plotting these parameters. As our package and its dependencies are all open source, this makes it possible for researchers in a range of fields (including philosophy, psychology, sociology, anthropology, communications, and network science) both to conduct new research and to reanalyze networks that they have previously studied.

Our work is unique as there are hitherto no comparable tools that analyze properties of extant networks, excepting [19]. Existing (social) network metrics—such as PageRank, eigenvector centrality, and in- and out-degree—are premised on the *connectedness* and topology of nodes, rather than their *epistemic* qualities. We anticipate that future research will expand the types of social networks under study. Other sources from online social media such as Facebook, Reddit, and YouTube all seem to be viable candidates for study. Considering offline epistemic networks would be especially valuable, such as those found in the landmark Bernard-Kilworth-Sailer (*BKS*) analyses of social networks [5], as their structure may be interestingly different from the structures found online. `wisdom_of_crowds` also allows us to conduct research on epistemic network simulations, created with tools such as Laputa [16], which can quickly scale up to thousands of nodes. We expect that studies of friend networks, organizational networks in industry and the military, networks of sources used by journalists, criminal cartel networks, and academic citation networks would prove valuable.

Moving beyond that, it would be interesting to study networks with more than one type of testimonial edge (e.g., public communications versus private ones). One intriguing hypothesis is that these may differ in structure even if they contain the same nodes, and that individuals who are central in public networks but peripheral in private networks (or vice versa) would tend to play unique roles in the social epistemology of those networks. For instance, someone who is privately in communication with a very large number of others but not publicly visible is in a position to exert influence because the others may assume that they have a much better epistemic position than they actually do. We also envision future work in measuring *epistemic power* by reversing the direction of the edges (i.e., from hearer to testifier).

The exploratory profiling made possible by our tool reveals patterns of epistemic isolation and interaction across real-world networks, and suggests possibilities for more specific analyses. By providing it to the community at large, we hope to facilitate further modeling of epistemic networks across a variety of domains.

Funding

Work on this paper was supported by ARC Grant DP190101507 (to C.K. and M.A.) and by Templeton Grant 61378 (to M.A. and M.C.). This research was supported by use of the Nectar Research Cloud and by Melbourne Research Cloud (University of Melbourne) (to M.C.). The Nectar Research Cloud is a collaborative Australian research platform supported by the NCRIS-funded Australian Research Data Commons (ARDC).

References

1. Alfano, M., Carter, J.A., Cheong, M.: Technological seduction and self-radicalization. J. Am. Philos. Assoc. **4**(3), 298–322 (2018)
2. Alfano, M., Reimann, R., Quintana, I. O., Chan, A., Cheong, M., Klein, C.: The affiliative use of emoji and hashtags in the Black Lives Matter movement on Twitter. In: Social Science Computer Review. Accepted for publication (forthcoming) (2022)

3. Alfano, M., Robinson, B.: Gossip as a burdened virtue. Ethical Theor. Moral Practice **20**(3), 473–487 (2017)
4. Alfano, M., Sullivan, E.: Humility in social networks. In: The Routledge Handbook of Philosophy of Humility, pp. 484–494. Routledge (2020)
5. Bernard, H.R., Killworth, P.D., Sailer, L.: Informant accuracy in social network data IV: a comparison of clique-level structure in behavioral and cognitive network data. Soc. Netw. **2**(3), 191–218 (1979)
6. Csardi, G., Nepusz, T.: The iGraph software package for complex network research. InterJournal Complex Syst. 1695 (2005)
7. Goldman, A.I.: Knowledge in a Social World. Oxford University Press (1999)
8. Hirsch, J.E.: An index to quantify an individual's scientific research output. Proc. Nat. Acad. Sci. **102**(46), 16569–16572 (2005)
9. Hunter, J.D.: Matplotlib: a 2D graphics environment. Comput. Sci. Eng. **9**(3), 90–95 (2007)
10. Krekel, H., Oliveira, B., Pfannschmidt, R., Bruynooghe, F., Laugher, B., Bruhin, F.: pytest 6.2.5 (2004)
11. Leskovec, J. (n.d.).: Snap: Network datasets: email-eu-core Network. https://snap.stanford.edu/data/email-Eu-core.html
12. Leskovec, J., Kleinberg, J., Faloutsos, C.: Graph evolution: densification and shrinking diameters. ACM Trans. Knowl. Discov. Data **1**(1), 2-es (2007)
13. Milgram, S.: The small world problem. Psychol. Today **2**(1), 60–67 (1967)
14. Nguyen, C.T.: Echo chambers and epistemic bubbles. Episteme **17**(2), 141–161 (2020)
15. O'Connor, C., Weatherall, J.O.: The Misinformation Age. Yale University Press (2019)
16. Olsson, E.J.: A simulation approach to veritistic social epistemology. Episteme **8**(2), 127–143 (2011)
17. Schult, D.A.: Exploring network structure, dynamics, and function using network. In: In Proceedings of the 7th Python in Science Conference (SciPy). Citeseer (2008)
18. Steup, M., Neta, R.: Epistemology. In: Zalta, E.N. (ed.) The Stanford Encyclopedia of Philosophy (Fall 2020 Edition) (2020). https://plato.stanford.edu/archives/fall2020/entries/epistemology/
19. Sullivan, E., Sondag, M., Rutter, I., Meulemans, W., Cunningham, S., Speckmann, B., Alfano, M.: Vulnerability in social epistemic networks. Int. J. Philos. Stud. **28**(5), 731–753 (2020)
20. Surowiecki, J.: The Wisdom of Crowds. Abacus, London (2005)
21. Traag, V., Waltman, L., van Eck, N.J.: From Louvain to Leiden: Guaranteeing well-connected communities. Sci. Rep. **9**, 5233 (2019)
22. Vassilevska, V.: Efficient algorithms for clique problems. Inf. Process. Lett. **109**(4), 254–257 (2009)
23. Yin, H., Benson, A.R., Leskovec, J., Gleich, D.F.: Local higher-order graph clustering. In: Proceedings of the 23rd ACM SIGKDD International Conference on Knowledge Discovery and Data Mining, KDD '17, pp. 555–564. Association for Computing Machinery, New York, NY, USA (2017)

Opening up Echo Chambers via Optimal Content Recommendation

Antoine Vendeville[1,2(✉)], Anastasios Giovanidis[3], Effrosyni Papanastasiou[3], and Benjamin Guedj[1,2]

[1] University College London, Department of Computer Science, Centre for Doctoral Training in Cybersecurity and Centre for Artificial Intelligence, London, UK
a.vendeville@ucl.ac.uk, b.guedj@ucl.ac.uk
[2] Inria, Lille—Nord Europe Research Centre and Inria London, Lille, France
[3] Sorbonne University, CNRS, LIP6, 75005 Paris, France
anastasios.giovanidis@lip6.fr,
effrosyni.papanastasiou@etu.sorbonne-universite.fr
http://antoinevendeville.github.io

Abstract. Online social platforms have become central in the political debate. In this context, the existence of echo chambers is a problem of primary relevance. These clusters of like-minded individuals tend to reinforce prior beliefs, elicit animosity towards others and aggravate the spread of misinformation. We study this phenomenon on a Twitter dataset related to the 2017 French presidential elections and propose a method to tackle it with content recommendations. We use a quadratic program to find optimal recommendations that maximise the diversity of content users are exposed to, while still accounting for their preferences. Our method relies on a theoretical model that can sufficiently describe how content flows through the platform. We show that the model provides good approximations of empirical measures and demonstrate the effectiveness of the optimisation algorithm at mitigating the echo chamber effect on this dataset, even with limited budget for recommendations.

Keywords: Social networks · Echo chambers · Recommendations · Elections · Optimisation

1 Introduction

The advent of Online Social Platforms (OSPs) in the last decade has irremediably changed our societies by allowing us to communicate on an unprecedented scale. This has not come without drawbacks, and concerns are rising about potentially nefarious consequences of OSPs [9]. In particular, the emergence of so-called *echo chambers* have sparked a growing interest in the scientific community. These clusters of like-minded users foster a continuous reinforcement of prior beliefs and strongly reject opposing ideas, thus hindering democratic debate and providing a fertile breeding ground for extremism and conspiracy theories. Occurrences of this phenomenon have been observed in online discussions surrounding various

H. Cherifi et al. (Eds.): COMPLEX NETWORKS 2016 2022, SCI 1077, pp. 74–85, 2023.
https://doi.org/10.1007/978-3-031-21127-0_7

political and controversial topics [1,12,22], although the amount of users they impact and the extent to which they do so is still up to debate [3,4].

Multiple factors have been advanced as explanations for this phenomenon, such as homophily, confirmation bias, negativity or information overload [10,15]. These natural biases are exacerbated by the personalisation algorithms used by the platforms: to sort through the enormous mass of information constantly flowing online, OSPs carefully filter, select and rank only the most relevant content to show to their users, to maximise their engagement. These algorithms tend to hide under the rug anything that support different views, leading to the entrapment of users into their own personalised *filter bubble* [17]. This in turn fosters the polarisation of opinions, animosity towards others with different views and the formation of echo chambers [12,21].

The whole personalisation algorithm used by an OSP is often referred to as a recommendation algorithm. The term may be misleading as these algorithms often mix content the user specifically asked for (by *following* a person or a page) with other items suggested by the platform ('you may be interested in...'). In addition these algorithms may also include filtering or ranking of content. In this paper we use the term *recommendation* specifically to refer to the platform inserting carefully curated content into the newsfeed of users. This mechanism aside we hypothesise that users are presented items as they are posted by their *leaders*—i.e. those whom they follow, and do not make any other assumption on the personalisation policy of the platform.

We introduce a novel method for OSPs to mitigate the echo chamber effect via content recommendation. We start by discussing related works in Sect. 2. In Sect. 3 we present the dataset [6] and exhibit the existence of echo chambers therein. To counter it we propose to increase the diversity of content that users are exposed to. We account for the long-term impact of recommendations by extending in Sect. 4 an existing model [8] that describes how content spreads throughout the network. Section 5 is then devoted to specifying our mitigation method as a budgeted quadratic program with linear constraints that relies on the model equations. We apply it on the dataset and present the results in Sect. 6. We conclude in Sect. 7.

2 Related Literature

With the advent of OSPs, the impact of algorithmic personalisation on social structures and opinions has attracted an unwavering interest in the scientific community. Many works have studied the consequences of recommending either social connections or content on echo chambers and opinion polarisation, with a variety of metrics proposed to measure these effects.

Rewiring connections based on similarity, be it of opinion or common acquaintances, tend to foster the emergence of echo chambers [2,5,21]. Notably, Cinus et al. [2] propose a framework to study the impact of recommending either social connections or content on any opinion dynamics model. Presenting too much congenial information to users often leads to a feedback loop as both opinions and recommendations get more extreme [20]. Perra and Rocha [18] find that the presence of triangles in the graph is a key driver of the emergence of echo chambers under various content recommendation policies. However note that under certain conditions, the OSP algorithm may actually reduce polarisation [2,5,19,21].

To prevent the formation of echo chambers, Musco et al. [16] propose a method to dynamically nudge friendship weights while preserving relevance of content shown to the users. Garimella et al. [7] search for friendship recommendations with high chances of being accepted. It may also be useful to inject some randomness into the recommender system [20]. Finally, some empirical methods exhibit good performance [13,23].

In this paper we quantify the impact of recommendations by the diversity of content users are exposed to. To compute it we extend a previous work [8] that describes the flow of content throughout an OSP. Analogously to Matakos et al. [14] we seek to maximise the diversity of information presented to users, but we also account for the dynamical flow of content through the network.

3 Problem Statement

First we present the dataset and propose a measure to quantify the diversity of information users are exposed to. This allows us to unveil the existence of echo chambers and we introduce our paradigm for countering the phenomenon.

3.1 Dataset

For the purposes of this work we find it crucial to work with real-world OSP data through which we can quantify the existence of echo chambers. We use the Twitter dataset from [6] relating to the 2017 French presidential campaign. It includes 2,414,584 tweets and 7,763,931 retweets from 22,853 Twitter profiles ('users' hereafter) discussing the election between November 2016 and May 2017. Users have been manually annotated by experts with political affiliations describing support for up to two of the main competing parties:

FI France Insoumise, far-left party (candidate Jean-Luc Mélenchon),
PS Parti Socialiste, left-wing party (candidate Benoit Hamon),
EM En Marche, centre party (candidate Emmanuel Macron),
LR Les Républicains, right-wing party (candidate François Fillon),
FN Front National, far-right party (candidate Marine Le Pen).

For some users the affiliation is unknown and we remove those from our study. Amongst remaining users, a small percentage are affiliated to two different parties (e.g. "PS/EM"). In addition we crawled the followers graph, which was not part of the original dataset. We removed users who did not tweet nor retweeted anything and restricted ourselves to the largest strongly connected component of the followers graph. Finally some profiles were not available anymore and we end up with an anonymised dataset $\mathcal{D}$ that features N = 8,277 users and $E = 975,168$ edges. The share of the considered population for each affiliation is shown in Fig. 1 (top left).

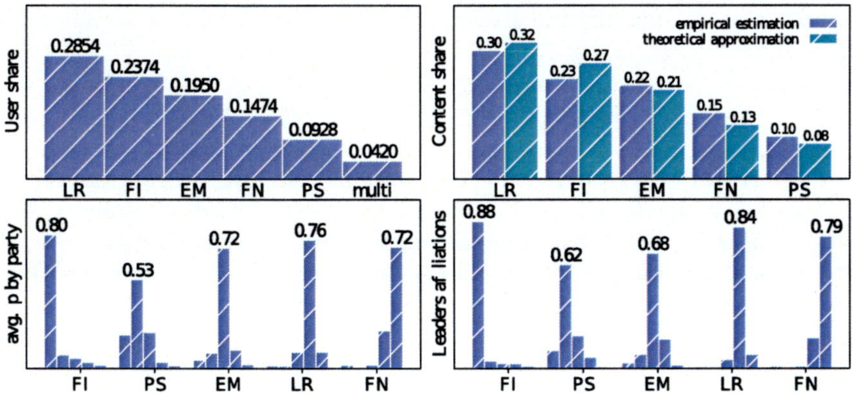

Fig. 1. **Top left**: user share for each affiliation. **Top right**: content share for each affiliation. **Bottom left**: distribution of content on the newsfeeds. **Bottom right**: distribution of leaders' affiliations by party. Each bar represents a party, in the same order as the x-axis.

3.2 Echo Chambers

Assume each user is endowed with a *newsfeed* that contains tweets propagated by their *leaders*—i.e. other users they follow. Tweets are labelled by the party their creator supports. We define $p_s^{(n)}$ as the average proportion of posts labelled with party s, thereafter s-posts, on the newsfeed of n. The vector $p^{(n)}$ thus describes the distribution of political leanings in the posts user n is exposed to. To quantify the echo chamber effect we look at $echo(n)$, the entry of $p^{(n)}$ corresponding to the party supported by n. The higher this value, the larger share of content presented to n comes from supporters of the same party.

We evaluate $p_s^{(n)}$ empirically for every user n and each party s via the procedure detailed in the Appendix. As seen in Fig. 1 (bottom left), the exposure to congenial content varies between 53% (PS) and 80% (FI). The average *echo* is 0.73, indicating a strong echo chamber effect overall. Notably, about 30% of users have $echo > 0.9$, meaning that almost a third of the online population is barely confronted with opposite ideas. About 5% of users are even trapped in *perfect* echo chambers, as they have $echo = 1$.

These results should not come as too surprising since the user graph exhibits strong homophily, as seen in Fig. 1 (bottom right). Between 62% (PS) and 88% (FI) of leaders support the same party, hence congenial information is heavily represented in the newsfeeds.

3.3 Promoting Content Diversity

In order to reduce the echo chamber effect, we seek to increase the diversity of content presented to users on their newsfeeds. In addition we will detail exactly how to do so in practice via personalised recommendations from the platform administrator. To quantify the diversity of the newsfeed of user n we rely on the following metric:

$$\Phi_n(p) \stackrel{\text{def}}{=} \frac{S}{S-1}\sum_{s=1}^{S} p_s^{(n)}(1 - p_s^{(n)}). \tag{1}$$

Here $S = 5$ is the number of parties. The constant in front of the sum ensures that Φ_n ranges in $[0, 1]$. A value of 0 indicates that the newsfeed of n only contains posts referring to a single party, describing a perfect echo chamber. On the other hand when $\Phi_n = 1$ all parties are equally represented on the newsfeed with the same average proportion of $1/S$, meaning a maximally balanced information diet.

Let $\bar{\Phi} \stackrel{\text{def}}{=} \sum_n \Phi_n(p)/N$ denote the average diversity of newsfeed over the whole platform. We are looking to solve an optimisation problem of the form

$$\underset{x,p}{\text{argmax}}\ \bar{\Phi} \tag{P0}$$

where x are personalised recommendation policies that describe what type of content should be inserted by the platform into the newsfeeds and when. The dependency of $\bar{\Phi}$ on x is detailed in Sect. 5. Importantly the impact of inserting an item into a newsfeed is not limited to an immediate change therein, but may also include a broader effect on the whole network as the concerned user can share it to their followers, who may share it further, and so on. The spreading behaviour of the individuals on the platform thus affects the results of the recommendation policies. To model this effect, in the next section we extend an existing model [8] that describes the flow of content throughout the newsfeeds. This allows us to quantify the long-term impact of recommendations on the platform as a whole.

Finally note that we could have chosen to measure diversity via the entropy $-\sum_s p_s^{(n)} \log p_s^{(n)}$ of the newsfeed of n. Its use is ubiquitous and its maximum lies in the same place as Φ_n. However the latter has the nice property of being quadratic in $p^{(n)}$ which allows for considerably more efficient optimisation.

4 Theoretical Model

The model developed in [8] describes the diffusion of content throughout an OSP. Consider N users who repeatedly create new content (*selfposts*) or spread content created by others (*reposts*). The newsfeed of a user contains posts propagated by their leaders. User n is endowed with two exponential clocks[1] of respective parameters $\lambda^{(n)}$ and $\mu^{(n)}$. Whenever the first one rings they create a new selfpost, while the second one prompts them to visit their newsfeed and select an item to repost amongst all content available there. All newsfeeds are of finite size and we make the assumptions that *(i)* the selection of an item to repost when visiting the newsfeed is made uniformly at random, and *(ii)* when a newsfeed is full any new entry will evict an older one chosen uniformly at random.[2]

[1] Differently distributed times induce similar equilibrium behaviour [8].
[2] Other selection and eviction policies induce similar equilibrium behaviour [8].

The system evolves towards a unique equilibrium characterised by the average state of the newsfeeds, which is solution to a linear system [8]. From there we derive a metric to quantify the overall influence of a user, and compare its values against empirical estimates. The contribution of the present paper related to the original model is twofold. First we introduce opinions by labelling posts with the political party they support. The model is now able to describe opinion dynamics by taking into account not only the topology of the graph but also the activity of its users. We evaluate its performance on the dataset by comparing its predictions with empirical estimates. Second we modify the equilibrium equations to account for the insertion of recommended content into the newsfeeds. This in turn allows us to formulate our method for countering the echo chamber effect as a budgeted quadratic program with linear constraints.

4.1 Introducing Opinions

We assume that user n produces selfposts supporting party s at rate $\lambda_s^{(n)}$, so that $\sum_{s\in\mathcal{S}} \lambda_s^{(n)} = \lambda^{(n)}$. In other words a proportion $\lambda_s^{(n)}/\lambda^{(n)}$ of all selfposts from n that refer to party s. Accounting for preferences when reposting introduces theoretical difficulties so we leave the mathematical analysis of that case for future research.[3] Let $\mathcal{L}^{(n)}$ be the set of all leaders of user n. The average proportion $p_s^{(n)}$ of s-posts on the newsfeed of n at equilibrium can be computed via Thm. 1.

Theorem 1 (Balance of opinions on newsfeeds). *For any party s the average proportions $p_s^{(1)}, \dots, p_s^{(N)}$ of s-posts on each newsfeed at equilibrium are solution of the following linear system:*

$$p_s^{(n)} \sum_{k\in\mathcal{L}^{(n)}} (\lambda^{(k)} + \mu^{(k)}) = \sum_{k\in\mathcal{L}^{(n)}} (\lambda_s^{(k)} + \mu^{(k)} p_s^{(k)}), \quad n = 1, \dots, N. \tag{2}$$

Assuming the user graph is strongly connected and at least one user has $\lambda > 0$, the system has a unique solution.

This is a direct reformulation of [8, Eq. (8)] with posts no more labeled by their creator but by a political opinion. It is a balance equation that equates the input and output rates of s-posts on the newsfeed of n induced by the activity of its leaders. On the left-hand side is the rate at which s-posts are evicted at random from the newsfeed to be replaced by fresher content. On the right-hand side is the rate at which s-posts propagated by the leaders of n enter the newsfeed. It decomposes in the selfposting rates of leaders of n about party s, and the reposting rates of content s from leaders of n, through their newsfeeds. Note that $p_s^{(n)}$ is the probability to uniformly select a post about opinion s, in the steady state. Solving Eq. (2) relies on matrix inversion which is not tractable for large systems, so we use a power iteration algorithm as described in [8].

[3] Certain contexts might actually benefit from this simplification. For example mentions and replies on Twitter may sometimes represent support and sometimes hostility towards the initial content [22].

4.2 Comparison with Empirical Results

The accuracy of the original model when compared with empirical estimates of p was demonstrated in [8] on two real-life datasets. For the present extension we compute theoretical values of p as given by Eq. (2) and compare them with empirical evaluations made in Sect. 3.2. We estimate $\lambda^{(n)}$ (*resp.* $\mu^{(n)}$) as the total number of tweets (*resp.* retweets) posted by n divided by the duration covered by the dataset. To label posts we make the simplifying assumption that all tweets created by a user are labelled as the user's affiliation. In practice people may post about other parties, and we leave to future research the estimation of more precise labels based on richer features.

We display in Fig. 2 the values of $p_s^{(n)}$ and $echo(n)$ from both the empirical evaluation and the theoretical model. The difference between both is 0.093 on average, and overall tendencies are respected as the Pearson correlation coefficients are close to 1. Moreover the ranking of parties based on their overall share of content, defined for party s as the average of p_s over all users, is the same as for empirical estimations (Fig. 1, top right). Note that it is also identical to the ranking of parties according to the number of users affiliated to them (Fig. 1, top left).

However the model is too moderate as it tends to overestimate the small values of $p_s^{(n)}$ and underestimate the high values. The average *echo* is 0.55 instead of 0.73 evaluated on the data, indicating a lower echo chamber effect. This is not surprising due to the random reposting policy: in the model users repost without distinction for the label of the posts, which is not the case in practice as people are more inclined to interact with congenial content—about 89% of retweets in the dataset are between supporters of the same party.

Hence the model captures the echo chamber effect as induced by the graph topology, the differences in posting preferences and the unequal posting and

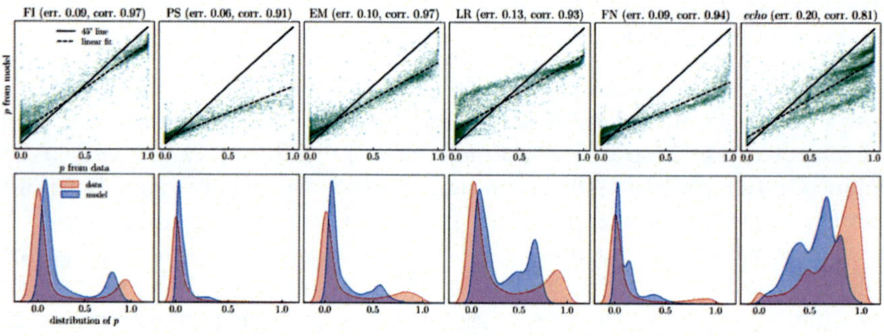

Fig. 2. Comparison between p estimated from data and p given by the model. Average errors ('err.') and Pearson correlations ('corr.') between estimations and model are indicated above. **Top**: scatter plots. **Bottom**: distributions.

reposting activity. But it does not include the reposting preferences, which as seen in these plots play also an important role in the precise prediction of content in the newsfeeds. Future research shall take this preferential reposting behaviour into account in order to obtain more accurate results.

5 Maximising Content Diversity

We are now able to give a precise formulation of problem (P0). Assume that the system can be acted upon by the platform administrator via personalised recommendations, that consist in selecting posts to insert into the newsfeeds of others. Let $x_s^{(n)}$ be the rate at which an s-post recommendation is inserted into the newfeed of user n this way. We are looking for the values of x that maximise the overall diversity $\bar{\Phi}$. For the sake of equity we would like all newsfeeds at equilibrium to contain on average the same proportion $B < 1$ of recommended content. Users with very active leaders and thus fast-changing newsfeeds should get new recommendations more often than those with quieter leaders. Formally we require for any user n:

$$\sum_s x_s^{(n)} = \frac{B}{1-B} \sum_{k \in \mathcal{L}^{(n)}} \left(\lambda^{(k)} + \mu^{(k)}\right). \tag{3}$$

This ensures that a proportion B of all content arriving on the newsfeed of n is a recommendation. The recommender system can be seen as if each user n has an artificial leader controlled by the platform, who creates s-posts at rates $x_s^{(n)}$ and these immediately appear on the newsfeed of user n, as recommendations. Hence, the steady-state of the newsfeeds is now given for all n and s by

$$p_s^{(n)} \left(\sum_s x_s^{(n)} + \sum_{k \in \mathcal{L}^{(n)}} (\lambda^{(k)} + \mu^{(k)}) \right) = x_s^{(n)} + \sum_{k \in \mathcal{L}^{(n)}} \left(\lambda_s^{(k)} + \mu^{(k)} p_s^{(k)}\right). \tag{4}$$

This is straightforward from (2). Inserting (3) in (4) then yields for all n and s,

$$\frac{p_s^{(n)}}{1-B} \sum_{k \in \mathcal{L}^{(n)}} \left(\lambda^{(k)} + \mu^{(k)}\right) = x_s^{(n)} + \sum_{k \in \mathcal{L}^{(n)}} \left(\lambda_s^{(k)} + \mu^{(k)} p_s^{(k)}\right). \tag{5}$$

Such values $p_s^{(n)}$ exist and are unique, as proved in the Appendix. We are now able to formulate our optimisation problem.

Theorem 2 (Diversity maximisation). *The optimal recommendation rates that maximise the average diversity of content on the newsfeeds under budget B* *can be computed via the following quadratic program with linear constraints.*

$$\begin{aligned}
&\underset{x,p}{\text{argmax}} \quad \bar{\Phi} && \text{(P)}\\
&\text{s.t.} \quad \frac{p_s^{(n)}}{1-B} \sum_{k \in \mathcal{L}^{(n)}} (\lambda^{(k)} + \mu^{(k)}) = x_s^{(n)} + \sum_{k \in \mathcal{L}^{(n)}} (\lambda_s^{(k)} + \mu^{(k)} p_s^{(k)}), \quad \forall n, s,\\
&\sum_s x_s^{(n)} = \frac{B}{1-B} \sum_{k \in \mathcal{L}^{(n)}} (\lambda^{(k)} + \mu^{(k)}), \quad \forall n,\\
&x_s^{(n)}, p_s^{(n)} \geq 0 \quad \forall n, s.
\end{aligned}$$

With $N = 8,277$ nodes and five categories, problem (P) has 82,770 variables and 49,662 linear constraints. Note the presence of p in the optimisation variables, due to its values being dependent on the recommendations x.

6 Results

We solve (P) for various values of the budget B and present the results in Fig. 3. Experiments are run on a virtual machine with 40 vCPUs and 256GB RAM. For the solution of the optimization problem, we configure a Gurobi solver with the barrier algorithm. The runtime is less than 10 min.

Our method is able to drastically improve the diversity of newsfeeds Φ_n and the echo chamber intensities *echo* (top left and middle plots). As B increases Φ_n converges towards 1 and *echo*(n) towards 0.2 which is $1/S$. An ideally diverse newsfeed has a p vector with all entries equal to $1/S$. The higher the budget the better the results, but small values of B already yield significant improvements. Indeed with just $B = 0.02$, meaning users encounter about 2% of curated recommended content on average, the average newsfeed diversity $\bar{\Phi}$ is up to 0.77 from 0.38 initially and the average *echo* is down to 0.50 from 0.73. In the top right plot we show for each value of B the relative change in $\bar{\Phi}$ and average *echo* per budget unit. Interestingly, the curves are decreasing, showing that higher budgets have diminishing returns.

The bottom left and middle plot underline how important the diffusion aspect captured by the model is. If we assume that recommended content is never propagated by users, the states of the newsfeeds are given by $p^{(n)} = (1-B)p^{(n)} + Bx^{(n)}/\|x^{(n)}\|_1$ where p are initial estimates made before optimisation. In that case (bottom left and middle plots), the distributions of Φ and *echo* keep their initial shape and their means improve way less.

Finally in the bottom right plot we show the overall share of recommended content that supports each party, that is for each party s we plot $\sum_{n=1}^{N} x_s(n)/N$ while the budget varies. These shares, while rather spread at first, converge towards 0.2 as the budget increases. Indeed when $B \to 1$ the newsfeeds only contain curated recommended content so that maximal diversity is achieved with a balanced representation of all parties. Finally note that the more a party was initially represented in the newsfeeds (Fig. 1, top right), the least they get recommended.

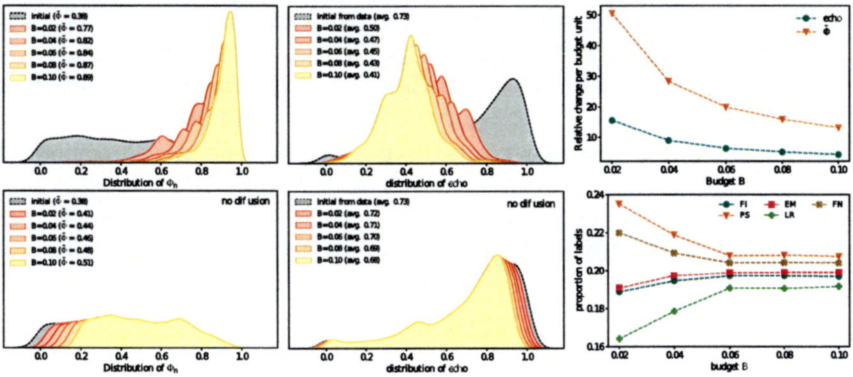

Fig. 3. Optimisation results for various budgets. **Top**: Distribution of newsfeed diversity (left), echo chambers intensity (middle), and relative impact of each budget unit on $\bar{\Phi}$ and *echo* (right). **Bottom**: results without diffusion of recommended content (left and middle), and distribution of labels in recommended posts (right). Initial distributions correspond to empirical estimates before optimisation.

7 Conclusion

The mitigation of the echo chamber effect through recommendations even with small budgets emphasizes the critical role and responsibility of the social platforms themselves. For the sake of space constraint, we leave for a follow up paper the task of improving the model's accuracy with preferential reposting behaviour. To avoid backfire effects and to preserve their user base, platforms may wish to not feed individuals with too much cross-cutting content. This can be achieved by penalising the change in newsfeeds in the optimisation objective. Moreover because certain parties benefit more from recommendations than others, we may want to enforce equality in the share of recommendations dedicated to each party. Finally we leave for future work the exciting task to consider filtering the content entering a newsfeed or recommending users to one another.

Acknowledgements. The authors have no competing interests to declare. This project was funded by the UK EPSRC grant EP/S022503/1 that supports the Centre for Doctoral Training in Cybersecurity delivered by UCL's Departments of Computer Science, Security and Crime Science, and Science, Technology, Engineering and Public Policy. The research of A.G. and E.P. is supported by the French National Agency of Research (ANR) through the FairEngine project under Grant ANR-19-CE25-0011.

Appendix

Empirical evaluation of *p*

First we label each tweet by the affiliation(s) of its original creator, or by '?' if the affiliation is unknown. Because the size of the newsfeeds does not matter according to the model [8], we assume for simplicity that all users have newsfeeds of size 1. We do not know the initial content of each newsfeed and assume they all contain a post of unknown origin, labelled '?'. As soon as a user tweets or retweets something, the post is inserted into the newsfeeds of its followers, evicting any previous post that was there. Finally to obtain $p_s^{(n)}$ we compute for each user n and label s the proportion of time their newsfeed contained a post labelled s. The original affiliation of most tweets is unknown and for each user we disregard periods during which the newsfeed contained a post labelled '?'.

Existence and Unicity of p With Recommendations

Let us write equation (5) in matrix form: $p_s = \mathbf{A}p_s + \mathbf{b}$. As long as the spectral radius $\rho(\mathbf{A})$ of $\mathbf{A}$ is stricly less than 1, the system has a unique solution $p_s = (\mathbf{I} - \mathbf{A})^{-1}\mathbf{b}$. The entries of $\mathbf{A}$ are given by

$$a_{ij} = (1 - B)\frac{\mu^{(j)}}{\sum_{k\in\mathcal{L}^{(i)}} \lambda^{(k)} + \mu^{(k)}}\mathbf{1}_{j\in\mathcal{L}^{(i)}} \tag{6}$$

and because $B > 0$ it holds that $\sum_j a_{ij} < 1$ for any row i. But from [11, Thm. 8.1.22] it we have $\rho(\mathbf{A}) \leq \max_i \sum_j a_{ij}$ and thus p_s exists and is unique.

Availability of code

The code for the optimisation problem is available at https://github.com/AntoineVendeville/Opening-up-echo-chambers.

References

1. Cinelli, M., De Francisci Morales, G., Galeazzi, A., Quattrociocchi, W., Starnini, M.: The echo chamber effect on social media. Proc. Natl. Acad. Sci. **118**(9), e2023301118 (2021)
2. Cinus, F., Minici, M., Monti, C., Bonchi, F.: The effect of people recommenders on echo chambers and polarization. arXiv:2112.00626 [physics] (Dec 2021)
3. De Francisci Morales, G., Monti, C., Starnini, M.: No echo in the chambers of political interactions on Reddit. Sci. Rep. **11**(1), 2818 (2021)
4. Dubois, E., Blank, G.: The echo chamber is overstated: the moderating effect of political interest and diverse media. Inf. Commun. Soc. **21**(5), 729–745 (2018)
5. Ferraz de Arruda, H., Maciel Cardoso, F., Ferraz de Arruda, G., R. Hernández, A., da Fontoura Costa, L., Moreno, Y.: Modelling how social network algorithms can influence opinion polarization. Inf. Sci. **588**, 265–278 (2022)
6. Fraisier, O., Cabanac, G., Pitarch, Y., Besançon, R., Boughanem, M.: #Élysée2017fr: the 2017 French presidential campaign on Twitter. In: Proceedings of the 12th International AAAI Conference on Web and Social Media (2018). Dataset available at https://zenodo.org/record/5535333#.Yp974OxBxH0
7. Garimella, K., De Francisci Morales, G., Gionis, A., Mathioudakis, M.: Reducing controversy by connecting opposing views. In: Proceedings of the Tenth ACM International Conference on Web Search and Data Mining, pp. 81–90. WSDM '17, Association for Computing Machinery, New York, NY, USA (Feb 2017)

8. Giovanidis, A., Baynat, B., Magnien, C., Vendeville, A.: Ranking online social users by their influence. IEEE/ACM Trans. Netw. **29**(5), 2198–2214 (2021)
9. Haidt, J.: Why the past 10 years of American life have been uniquely stupid. The Atlantic (2022). https://www.theatlantic.com/magazine/archive/2022/05/social-media-democracy-trust-babel/629369/. Accessed on June 6, 2022
10. Hills, T.T.: The dark side of information proliferation. Perspect. Psychol. Sci. **14**(3), 323–330 (2019)
11. Horn, R.A., Johnson, C.R.: Matrix Analysis. Cambridge University Press (1990)
12. Kirdemir, B., Agarwal, N.: Exploring bias and information bubbles in YouTube's video recommendation networks. Complex Netw. Appl. **X**, 166–177 (2022)
13. Liao, Q.V., Fu, W.T.: Can you hear me now? Mitigating the echo chamber effect by source position indicators. In: CSCW'14, pp. 184–196 (Feb 2014)
14. Matakos, A., Tu, S., Gionis, A.: Tell me something my friends do not know: diversity maximization in social networks. Knowl. Inf. Syst. **62**(9), 3697–3726 (2020)
15. McPherson, M., Smith-Lovin, L., Cook, J.M.: Birds of a feather: homophily in social networks. Annu. Rev. Sociol. **27**(1), 415–444 (2001)
16. Musco, C., Musco, C., Tsourakakis, C.E.: Minimizing polarization and disagreement in social networks. In: Proceedings of the 2018 World Wide Web Conference, pp. 369–378. WWW'18, International World Wide Web Conferences Steering Committee, Republic and Canton of Geneva, CHE (Apr 2018)
17. Pariser, E.: The Filter Bubble: What the Internet Is Hiding from You. The Penguin Group (2011)
18. Perra, N., Rocha, L.E.C.: Modelling opinion dynamics in the age of algorithmic personalisation. Sci. Rep. **9**(1), 7261 (2019)
19. Ramaciotti Morales, P., Cointet, J.P.: Auditing the effect of social network recommendations on polarization in geometrical ideological spaces. In: RecSys'21: 15th ACM Conference on Recommender Systems. Amsterdam, Netherlands (Sep 2021)
20. Rossi, W.S., Polderman, J.W., Frasca, P.: The closed loop between opinion formation and personalised recommendations. IEEE Trans. Control Netw. Syst. 1 (2021)
21. Santos, F.P., Lelkes, Y., Levin, S.A.: Link recommendation algorithms and dynamics of polarization in online social networks. Proc. Natl. Acad. Sci. USA **118**(50), e2102141118 (2021)
22. Williams, H.T.P., McMurray, J.R., Kurz, T., Hugo Lambert, F.: Network analysis reveals open forums and echo chambers in social media discussions of climate change. Glob. Environ. Change **32**, 126–138 (May 2015)
23. Yang, Q., Qureshi, K., Zaman, T.: Mitigating the backfire effect using pacing and leading. Complex Netw. Appl. **X, 1016**, 156–165 (2022)

Change My Mind: Data Driven Estimate of Open-Mindedness from Political Discussions

Valentina Pansanella[1], Virginia Morini[2(✉)], Tiziano Squartini[3], and Giulio Rossetti[4]

[1] Scuola Normale Superiore, Pisa, Italy
valentina.pansanella@sns.it
[2] Department of Computer Science, University of Pisa, Pisa, Italy
virginia.morini@phd.unipi.it
[3] IMT School for Advanced Studies, Lucca, Italy
tiziano.squartini@imtlucca.it
[4] KDD Laboratory, ISTI, National Research Council, Pisa, Italy
giulio.rossetti@isti.cnr.it

Abstract. One of the main dimensions characterizing the unfolding of opinion formation processes in social debates is the degree of open-mindedness of the involved population. Opinion dynamic modeling studies have tried to capture such a peculiar expression of individuals' personalities and relate it to emerging phenomena like polarization, radicalization, and ideology fragmentation. However, one of their major limitations lies in the strong assumptions they make on the initial distribution of such characteristics, often fixed so as to satisfy a normality hypothesis. Here we propose a data-driven methodology to estimate users' open-mindedness from online discussion data. Our analysis—focused on the political discussion taking place on Reddit during the first two years of the Trump presidency—unveils the existence of statistically diverse distributions of open-mindedness in annotated sub-populations (i.e., Republicans, Democrats, and Moderates/Neutrals). Moreover, such distributions appear to be stable across time and generated by individual users' behaviors that remain consistent and underdispersed.

Keywords: Online debates · Open-Mindedness · Political polarization · Opinion dynamics

1 Introduction

One of the most debated and analyzed phenomena on online social networks is the tendency to observe political polarization [1,5,19,20], i.e., the divergence of political attitudes to ideological extremes not aiming at reaching any form of synthesis. Indeed, social media blurred the boundaries of communication and democratized content diffusion. However, the fact that we can potentially engage

H. Cherifi et al. (Eds.): COMPLEX NETWORKS 2016 2022, SCI 1077, pp. 86–97, 2023.
https://doi.org/10.1007/978-3-031-21127-0_8

with a lot of different information does not mean that we can engage with and be influenced by opposing or even mildly different stances [25].

Furthermore, humans are far from being perfectly rational individuals and are affected by a series of cognitive biases[1] in the process of forming their belief system. Among these, it is worth mentioning confirmation bias [14,24], the human tendency to ignore content that counters their prior beliefs by (a) choosing to interact with like-minded individuals [17] (b) ignoring interactions with the "opposite faction" [8]. Unfortunately, by connecting with like-minded individuals and ignoring contrasting views, users can easily compromise their open-mindedness [35], exacerbating and polarizing their views even more, and, in the end, giving rise to the formation and persistence of echo chambers [32].

One of the most exploited approaches for understanding the effects of different kinds of biases on public opinions—especially political opinions—is through mathematical models of opinion formation [29], where parameters incorporate psychological factors (e.g., cognitive biases) affecting individual opinion evolution. Despite having some advantages, opinion dynamics models lack empirical validation [26]. However, thanks to the advent of the Internet—and with the rise of social media—an increasing part of human interactions leave a massive digital footprint that can be exploited to study the dynamics of opinion formation and diffusion. Following such reasoning, since different models or different parameter values can predict different, even opposite, effects of biases on opinions [16], there is a crucial need for empirical works to study and quantify socio-psychological and external drivers of the dynamics.

In this work, we move in this direction by studying how being politically polarized affects users' open-mindedness over time. In detail, we focus on a twenty-months discussion on `r/politics` (i.e., the largest political subreddit on Reddit), computing the probability of changing one's political leaning over time and estimating the distribution of open-mindedness within the user population and its temporal evolution. Firstly, this analysis aims to understand how open-mindedness is distributed within the population and over time. A second aim is to understand if there is a substantial difference in the level of open-mindedness in relation to the political leaning, i.e., if more extreme individuals are more close-minded than moderate ones or vice versa. Our results suggest that the different subpopulations' open-mindedness distributions are stable in time and statistically different: Moderates/Neutrals and Republicans showing more nuanced close-mindedness patterns w.r.t. Democrats. Finally, our analysis also underlines how Reddit users tend to be consistent in their open-mindedness attitude across time, showing, on average, low degrees of variance and dispersion.

The rest of the paper is organized as follows. In Sect. 2, we discuss the literature involving open-mindedness modeling both from an opinion dynamics and data-driven point of view. Section 3 discusses how we leverage Reddit data in order to model opinion evolution over time. Then, in Sect. 4, we propose an algorithm to estimate users' open-mindedness and we longitudinally study its evolution in three well-characterized sub-populations. Finally, Sect. 5 concludes

[1] Cognitive biases cheat sheet: shorturl.at/afjpM, last visited June 2022.

the paper with a brief summary of the main results and discusses limitations and directions for future work.

2 Related Work

In the process of opinion formation cognitive biases play a role because people usually feel discomfort when encountering divergent opinions, thus tending to select information that confirm their pre-existing beliefs [24]. One of the first attempts at understanding the effects of such kind of cognitive bias is the model by Deffuant, and Weisbuch [8]. In this case, the bias is introduced with a single parameter, namely "confidence bound", which indicates that agents only influence each other if their opinion distance is below a certain threshold. The main result of this model is that if the population is "close-minded", polarization arises. So it appears that cognitive biases may be enough to justify opinion polarization, even without considering additional factors, such as the network structure or the information environment. However, the effects of open-mindedness alone and in interplay with other factors have been extensively studied through opinion dynamics models, extending the baseline [8] with additional agent-level characteristics [15], considering multi-dimensional opinions [7] or adding repulsive behaviors [4]. Despite conclusions from these models appear to be realistic and seem to explain why polarization arises and resists in many social networks, there is a need for empirical works unveiling the level of open-mindedness within polarizing topics. Indeed, to the best of our knowledge, there are still no works leveraging the vast amount of big data traces left from online social networks to tackle such a task. With respect to modelling works, simulation results have been compared to election data [34], survey data [30], social experiments data [31], or social media data [36] in order to manually tune parameters or learn them with automated approaches [13,18,28,33]. Different data-driven works [1,10–12,23,27] have shown how the polarization of opinions around political issues led the people to enter in closed polarized systems insulated from rebuttal, i.e., echo chambers. Interestingly, authors in [9] found that echo chambers do not arise around topics not politically polarized (e.g., TV shows, food), thus suggesting that the echo chamber phenomenon and users' open-mindedness might be strictly related to the discussed topic.

3 Analyzing the Political Debate Online

Data Collection. For the purpose of this work, we built a dataset of political discussions on Reddit. Reddit is a popular social platform that allows users to post content to individual forums called subreddits, each dedicated to a specific topic. Such a categorized structure makes it easy to find users involved in specific debates. To assess the open-mindedness of users over time, we decide to select a quite controversial domain, i.e., Politics. Among the thousands of subreddits talking about politics we choose `r/politics`,[2] since it is the largest political sub-

[2] https://www.reddit.com/r/politics/ "`r/politics` is for news and discussion about US politics."

reddit and further, it is not aligned with a specific ideology but rather is visited regularly by users having different political beliefs. Notice that, as highlighted in the subreddit description[1], `r/politics` mainly refers to political discussion in the US. Thanks to the Pushshift API [3], we collected all posts and comments shared on the subreddit from May 2018 to December 2019, i.e., about one year and a half of Donald Trump's presidency—covering all discussions of 1,089795 users. As shown in Fig. 1a, the numbers of users who participated in the debate tend to increase over time, and further, more than 60% of users are stable across contiguous months, meaning that they continue posting or commenting at least for two consecutive months. The code and the data used for this work are available in a dedicated GitHub repository.[3]

Ideology Estimate. To assess if the open-mindedness of users evolves over time, we have to establish the ideology of users in different time periods. Since we are dealing with users debating political issues in the US, we try to categorize them with respect to the US two-party system: *Republican* and *Democratic.*

For such a purpose, we model the task of predicting users' political alignment as a text classification problem. In other words, we leverage users' posts to measure their degree of alignment with Republican and Democrat ideologies. To accomplish this task, we leverage an LSTM model that we trained on Reddit US political texts in our previous works [21,22]. In detail, to train such a model we defined a ground truth composed of Reddit posts belonging to subreddits known to be either Pro-Trump or Anti-Trump (i.e., `r/The_Donald` for the first group and `r/Fuckthealtright` and `r/EnoughTrumpSpam` for the second). Accordingly, we modeled the text classification task as a binary problem. During model selection, we perform a 3-fold Cross-Validation trying different hyper-parameters configurations and obtaining the best performances on the validation set (i.e., average accuracy of 82.9%) using GloVe word embeddings and 128 LSTM units.[4] Consequently, we apply the model to the `r/politics` dataset in order to infer posts leaning for all the population. Notice that we apply it separately for each month in order to assess users' ideology stability over time. For each textual content, we obtain model predictions ranging from 0 to 1 (i.e., the model confidence), where 1 means that the post aligns itself with Republican (specifically Pro-Trump) ideologies while 0 with Democrats (specifically Anti-Trump) ones. Then, for each user u who participated in the debate in each month m we compute his *leaning score*, $L_{u,m}$, as the average value of his monthly posts leaning as follows:

$$L_{u,m} = \frac{\sum_{p \in P_{u,m}} \text{Prediction\_Score}(p)}{|P_{u,m}|} \tag{1}$$

where $P_{u,m}$ is the set of posts shared by a user u in each month m. Since we are interested in assessing if users with polarized opinions tend to move to more

[3] https://github.com/ValentinaPansanella/OpenMindednessReddit.git.

[4] For further details on the model selection and evaluation steps, the reader should refer to our previous works [21,22].

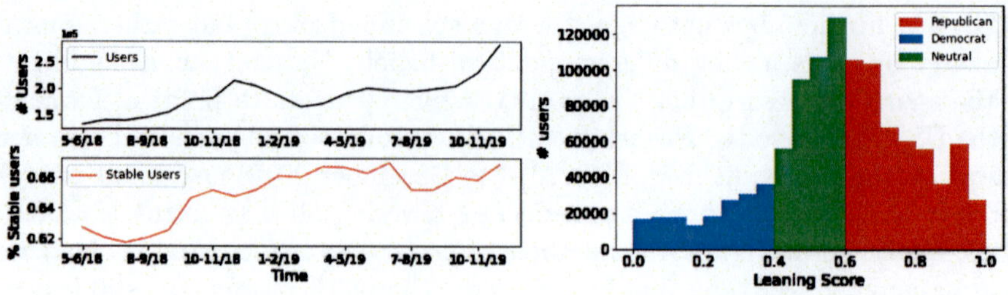

(a) (b)

Fig. 1. **a** Top: for each month, the number of users who participated in the debate. Bottom: For each month, the percentage of users that are stable across contiguous months. **b** Authors' leaning distribution in the whole time period.

moderate positions, we discretized such leanings into three intervals: *Democrat* if $L_{u,m} \leq 0.4$; *Republican* if $L_{u,m} \geq 0.6$; while *Neutral* if $0.4 < L_{u,m} < 0.6$. By adding a third label, we make sure to capture users with highly polarized ideologies. Figure 1b shows the authors' leaning score distribution for the whole time period obtained by averaging users' monthly scores. Such a distribution confirms what observed in a recent work [6] that focus the analysis on the r/politics subreddit too: Republican users (530,909) outnumber Democrat ones (185,256) and Neutrals users (373,630) show a tendency towards republicans beliefs.

Network Definition. Given the users for which we inferred their ideology, we define their interaction networks for each month to take into account the evolution of leanings in time. The resulting networks have Reddit users as node sets, V, and as edges the set of pair $(u, v) \in V$ for which a reply of u to a v's post or a comment exists. We set each edge weight to represent the total number of comments exchanged between two users. Also, we label users (i.e., nodes) with their discretized *leaning score* $L_{u,m}$ (i.e., Republican, Neutral and Democrat). In Table 1, we provide the main averaged network statistics across the 20 considered months.

Table 1. Network statistics averaged across the 20 considered months: number of users N, divided in Republican N_R, Democrat N_D and Neutral N_N, number of edges E, network average degree $\langle k \rangle$, and network assortativity r with respect to the political leaning.

N	N_R	N_D	N_N	E	$\langle k \rangle$	r
183,296	88,285	36,385	58,626	1,271,080	6.97	0.020

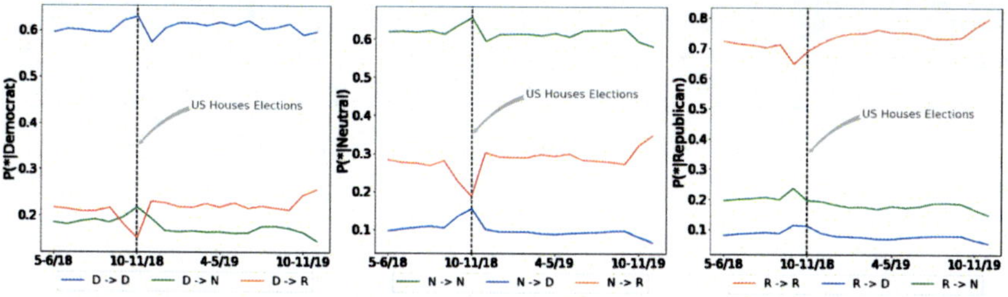

Fig. 2. For each ideology, users' transition probabilities over contiguous months.

4 Political Leaning: Stability and Open-Mindedness

Ideology Stability over Time. As a preliminary analysis, we try to understand how users' ideologies evolve. In detail, we are interested in exploring if users are stable and consistent with the same ideology or instead tend to change opinions according to specific events. We model such an issue in terms of transition probabilities for such a purpose. In other words, for each user, we compute his probability p_{ij} to move from *state* i to *state* j over contiguous months. Notice that, in this scenario, *state* stands for the user's ideology (i.e., Republican, Neutral or Democrat).

In Fig. 2 we show, for each ideology, the probability of users to change or remain in their states (i.e., opinions) over contiguous months. At a first glance, we can observe a similar behavior for all the political ideologies: users tend to be rooted in their position over months, with a probability to remain in their state ranging around 0.60 for Democrat and Neutral leanings and around 0.70 for Republican. Moreover, Republican users—when changing their state—foster a more neutral position instead of moving to Democrat beliefs (differently from Democrats). These attitudes reflect the extreme polarization that characterizes the US political debate in the Trump Era [2]. However, for all the considered ideologies, we also notice an evident fluctuation of opinions in favor of Democrats and Neutrals around November 2018, precisely when (i.e., 6 November) the Democratic Party won control of the US House from the Republican Party. Indeed, this was a decisive moment in which Democrats won the seats needed to take the House after capturing districts where President Trump was unpopular.

Estimating Open-mindedness. In order to understand the levels of open-mindedness involved in the process of changing one's political leaning, we started by assuming a simple process of opinion evolution at the individual level, based on a very well known model of opinion formation [8]. In opinion dynamics models, agents update their opinions after interacting with their neighbors according to simple mathematical rules. For example, in [8] agents average their opinion with the opinion of their interacting peer, which is randomly chosen from the pool of their neighbors, if and only if their opinion distance is below a certain threshold representing the open-mindedness of the population. The hypothesis that open-mindedness is a characteristic trait of an entire population and not

Algorithm 1 Confidence bound estimation algorithm. G_t = Weighted undirected interaction network at time t; V_t = set of nodes at time t; E_t = set of weighted edges at time t; $x_u(t)$ = opinion of agent u at time t; $d_{u,v} = |x_u(t) - x_v(t)|$ = opinion distance between $u, v \in V$ at time t; $\widehat{CB}$ = estimated confidence bounds.

1: **if** $u \in V_{t+1}$ **then**
2: Procedure to estimate $\widehat{x_u}(t+1)$ and $\widehat{CB}_u$
3: $N_{u,t} = \{v|(u,v) \in E_t\}$; $|N_{u,t}| = n$
4: $X_{N_{u,t}}[1...n]$ = Array of opinions of nodes $v \in N_{u,t}$
5: **if** $N_{u,t} \neq \emptyset$ **then**
6: Sort $X_{N_{u,t}}[1...n]$ *by* $d_{u,v}$ in ascending order.
7: $\widehat{X_u}(t+1)[0...n]$ array of estimated opinions
8: $\widehat{X_u}(t+1)[0] = x_u(t)$
9: $E = [0...n]$ array of estimation errors
10: $E[0] = 1.0$
11: **for** i=1; i=n; i++ **do**
12: $x_v = X_{N_{u,t}}[i]$
13: $\widehat{X_u}(t+1)[i] = \frac{\widehat{X_u}(t+1)[i-1]+x_v}{2}$
14: $E[i] = |\widehat{X_u}(t+1)[i] - x_u(t+1)|$
15: $min_e = E[n]$
16: **for** i=n; i=0; i- **do**
17: $e = E[i]$
18: **if** $e \leq min_e$ **then**
19: $min_e = e$
20: $j = i$
21: **end if**
22: **end for**
23: **end for**
24: $\widehat{x_u}(t+1) = \hat{X}_u(t+1)[j]$
25: $\widehat{CB} = |x_u(t) - X_{N_{v,t}}[j]|$
26: **end if**
27: **end if**

a characteristic that varies from individual to individual is strong and probably unrealistic. For this reason, in the present work, we assume a Deffuant-like process of opinion update (i.e., users averaging their opinions with the opinions of their interacting partners in a pairwise fashion) and provide a data-driven time-aware estimate of individual-level open-mindedness. To estimate users' tendency to be influenced by their neighbors, we developed a simple approach (see Algorithm 1 for implementation details) that takes as input the weighted interaction network at time t and the opinions of the agents at time t and time $t+1$. In the estimation procedure, we select each node u for which we have both opinions $x_u(t)$ and $x_u(t+1)$ (Algorithm 1 line 1) so that we can estimate how much the interactions that happened in that time step influenced the opinion change and therefore obtaining an estimate for the level of bounded confidence. After selecting u, we order all the interacting partners (the neighbours of the node in the snapshot network) from the closer to the further by the opinion distance absolute value $d_{u,v}(t) = |x_u(t) - x_i(t)|$ (Algorithm 1 line 6). Then we compute an estimate $\widehat{x_u}(t+1)$ by iteratively averaging the new estimate with the interacting neighbours (Algorithm 1 line 13). The final estimated value $\widehat{x_u}(t+1)$ is the one that minimizes the error with respect to the real value $x_u(t+1)$ (Algorithm 1 lines 15–22). Finally, we compute the confidence bound as the distance in absolute value with the neighbor that represents the point of minimum in the estimation errors sequence (Algorithm 1 line 25). With the proposed approach, we can compute an estimate only for the subset of users present in two consecutive observations (i.e., months) and have at least one link (i.e., interaction) in the snapshot graph.

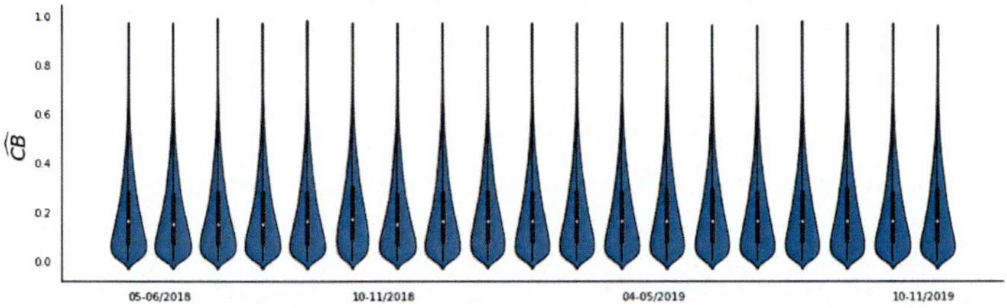

Fig. 3. Estimated open-mindedness $\widehat{CB}$ distribution from May-2018 to December-2019. Distributions of the estimated confidence bound $\widehat{CB}$ over the whole time period. All distributions are positively skewed and constant over time.

Figure. 3 underlines that the distribution of the estimated open-mindedness $\widehat{CB}$ is stable during the considered time period in the analyzed discussion. Moreover, it also highlights that the majority of the analyzed users is "close-minded", i.e., their confidence bound is $\widehat{CB} \leq 0.2$, which is considered to be a sufficient condition for the population to become polarized in the long term, according to [8]. This means that most of the users participating in this discussion can hardly be influenced by neighbors holding distant opinions, even if they interact with these users during the considered time period, like in the case of this discussion, where the network has a low assortativity with respect to the political leaning (see network assortativity in Table 1). However, we can also see that the distributions at each time step have a very high variance, allowing the presence of individuals having a level of open-mindedness close to 1.0—indicating that some of the users can also be influenced by people holding very different opinions and changing their expressed political leaning accordingly. The distribution of the estimated confidence bound for the opinions is less skewed between October and November 2018, i.e., around the US House of Representatives elections. Such behavior confirms the data-driven analysis based on transition probabilities performed in Sect. 3: in this time window, Republicans, which normally have a highly skewed distribution, seemed to be somehow more open-minded, and their average confidence bound is higher. In Fig. 4 we reported the estimated distribution for Democrat, Neutral, and Republican users (we took the orientation at time t and estimated the confidence bound $\widehat{CB}$ between time t and t+1). Only three months are present in this figure, i.e., the three months around the US House election, but conclusions still hold for the other months considered in this work.

We can see from Fig. 4 that there are differences in the distribution of open-mindedness when we consider political orientation. Both neutral and republican distributions are positively skewed, and the distribution has a long right tail. Also, the Democrats distribution has a right tail, but much less than the others. From our analysis, it appears that Neutral individuals are also the most close-minded, while Democrats have a wider range of confidence bound levels. Their distribution is not as skewed as the others, and many users have a very high level of bounded confidence, i.e., they change their opinion significantly over contiguous time periods under the influence of their neighbors. Republicans, like Neutrals, have a positively skewed confidence bound distribution, even with a higher average confidence bound. We performed a 2-sample KS-test comparing the distributions of the estimated confidence bound for each political leaning (e.g., Democrat vs. Republican $\widehat{CB}$) within each time step obtaining a p-value ≈ 0.0, supporting the conclusions that distributions are different for the three political leanings. Finally, while we can say that population-level open-mindedness is reasonably constant over time, i.e., we have the same mix of open-minded and close-minded individuals participating in the discussion, we do not have information about how variable open-mindedness may be at the individual level. In this analysis, each user may have a different value of open-mindedness at each time step, making not only the overall distribution heterogeneous but also the distribution at the individual-level. To understand how much individual bounded confidence may vary, we computed the standard deviation of our estimate for each observed user—Fig. 5a. Reddit users' open-mindedness tendencies appear stable in time, showing a characteristic low standard deviation, σ. Such a result is also confirmed by the distribution of the Fano dispersion index (Fig. 5b)—i.e., the ratio between variance σ^2 and mean value, μ, of the estimated individual open-mindedness scores. The observed Fano values, prevalently distributed in the range $0 < \frac{\sigma^2}{\mu} < 1$, identify under-dispersed behaviors, thus expressing consistent patterns of stability.

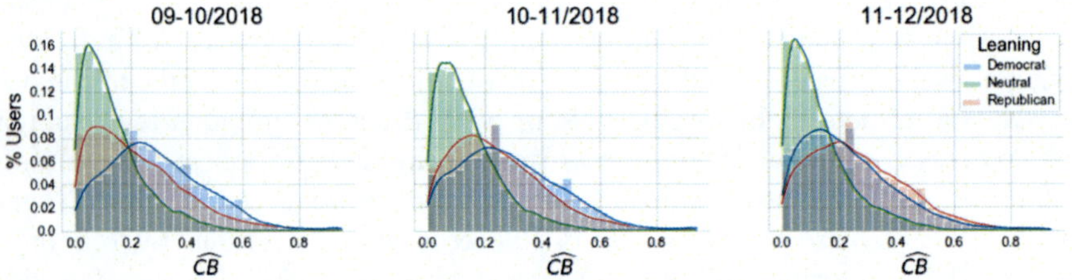

Fig. 4. Estimated open-mindedness distribution in the period September—December 2018. Distributions of the estimated confidence bound $\widehat{CB}$ for the different political leanings: Democrat (blue), Neutral (green), and Republican (red) from September to December 2018.

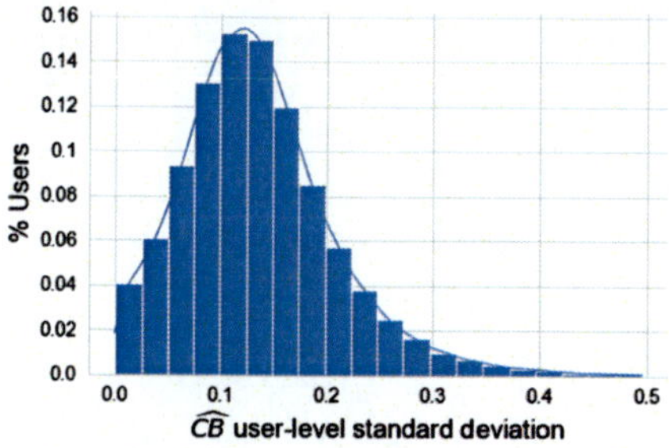

(a)

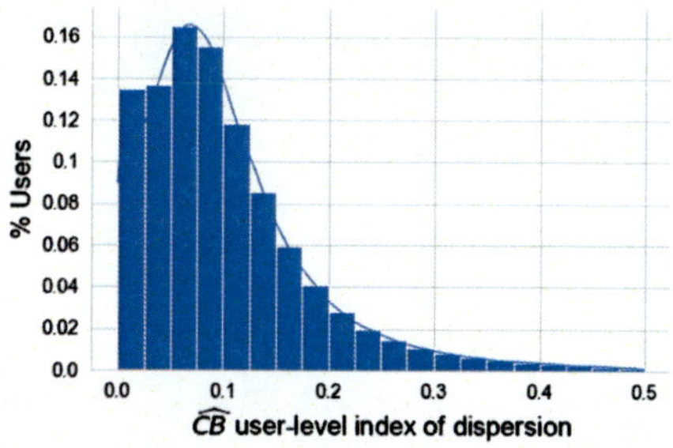

(b)

Fig. 5. Users' open-mindedness stability analysis. **a** Distribution of individuals' open-mindedness standard deviations; **b** distribution of individuals' open-mindedness dispersion indexes (variance over mean value).

5 Conclusions

In this work, we introduced a methodology to estimate individuals' open-mindedness from online discussion data. When applied to study the political debate that took place on Reddit during the first two years of the Trump presidency, our proxy unveiled the existence of characteristic distributions for well-defined sub-populations: Moderates/Neutrals and Republicans being more close-minded on average than Democrats. The proposed longitudinal analysis also unveiled that the observed Reddit users tend to maintain a stable behavior for what concerns their open-mindedness, exhibiting low variance and underdispersion.

Indeed the current study, like all empirical ones, suffers from limitations. In particular, it leverages a data-driven estimate of the political leaning that can be subject to errors and cannot be fully validated on ground truth external data. Moreover, although Reddit users tend to be particularly involved in political discussions, the population variability in time and the sparsity of observation data do not allow to estimate the open-mindedness of less active individuals. For each month, we were not able to estimate the open-mindedness of $40 - 50\%$ of nodes since we have no information on the opinion at time $t + 1$ or the user does not have neighbors on the network at time t.

As future works, we plan to enhance the proposed estimation procedure, allowing for asymmetric open-mindedness. Additionally, we will investigate the interplay of open-mindedness and known polarization phenomena (e.g., the presence of echo chambers) in order to better characterize the role of different individuals in their emergence.

Acknowledgments. This work was supported by the scheme 'INFRAIA-01-2018-2019: Research and Innovation action', Grant Agreement n. 871042 'SoBigData++: European Integrated Infrastructure for Social Mining and Big Data Analytics, by the CHIST-ERA grant CHIST-ERA-19-XAI-010, by MUR (grant No. not yet available), FWF (grant No. I 5205), EPSRC (grant No. EP/V055712/1), NCN (grant No. 2020/02/Y/ST6/00064), ETAg (grant No. SLTAT21096), BNSF (grant No. o КП-06-ДОО2/5)

References

1. Bakshy, E., Messing, S., Adamic, L.A.: Exposure to ideologically diverse news and opinion on Facebook. Science **348**(6239), 1130–1132 (2015)
2. Bartels, L.M.: Partisanship in the trump era. J. Polit. **80**(4), 1483–1494 (2018)
3. Baumgartner, J., Zannettou, S., Keegan, B., Squire, M., Blackburn, J.: The pushshift reddit dataset. In: Proceedings of the International AAAI Conference on Web and Social Media, vol. 14, pp. 830–839 (2020)
4. Chen, X., Tsaparas, P., Lijffijt, J., Bie, T.D.: Opinion dynamics with backfire effect and biased assimilation. PLoS One **16** (2021)
5. Conover, M., Ratkiewicz, J., Francisco, M., Gonçalves, B., Menczer, F., Flammini, A.: Political polarization on twitter. In: Proceedings of the International AAAI Conference on Web and Social Media, vol. 5, pp. 89–96 (2011)
6. De Francisci Morales, G., Monti, C., Starnini, M.: No echo in the chambers of political interactions on reddit. Sci. Rep. **11**(1), 1–12 (2021)
7. Deffuant, G., Carletti, T., Huet, S.: The leviathan model: absolute dominance, generalised distrust, small worlds and other patterns emerging from combining vanity with opinion propagation. J. Artif. Soc. Soc. Simul. **16** (2013)
8. Deffuant, G., Neau, D., Amblard, F., Weisbuch, G.: Mixing beliefs among interacting agents. Adv. Complex Syst. **3**, 87–98 (2000)
9. Garimella, K., De Francisci Morales, G., Gionis, A., Mathioudakis, M.: Political discourse on social media: echo chambers, gatekeepers, and the price of bipartisanship. In: Proceedings of the 2018 World Wide Web Conference, pp. 913–922 (2018)
10. Ge, Y., Zhao, S., Zhou, H., Pei, C., Sun, F., Ou, W., Zhang, Y.: Understanding echo chambers in e-commerce recommender systems. In: Proceedings of the 43rd International ACM SIGIR Conference on Research and Development in Information Retrieval, pp. 2261–2270 (2020)
11. Gilbert, E., Bergstrom, T., Karahalios, K.: Blogs are echo chambers: blogs are echo chambers. In: 2009 42nd Hawaii International Conference on System Sciences, pp. 1–10. IEEE (2009)
12. Hilbert, M., Ahmed, S., Cho, J., Liu, B., Luu, J.: Communicating with algorithms: a transfer entropy analysis of emotions-based escapes from online echo chambers. Commun. Methods Measures **12**(4), 260–275 (2018)
13. Kulkarni, B., Agarwal, S., De, A., Bhattacharya, S., Ganguly, N.: Slant+: A nonlinear model for opinion dynamics in social networks. In: 2017 IEEE International Conference on Data Mining (ICDM), pp. 931–936. IEEE (2017)
14. Kunda, Z.: The case for motivated reasoning. Psychol. Bull. **108**(3), 480 (1990)
15. Lorenz, J.: Heterogeneous bounds of confidence: meet, discuss and find consensus! Complexity **15**(4), 43–52 (2010)
16. Maes, M., Bischofberger, L.: Will the personalization of online social networks foster opinion polarization? Available at SSRN 2553436 (2015)
17. Mcpherson, S.-L.: Birds of a feather: homophily in social networks. Ann. Rev. Sociol. **27**, 415 (2001)

18. Monti, C., De Francisci Morales, G., Bonchi, F.: Learning opinion dynamics from social traces. In: Proceedings of the 26th ACM SIGKDD International Conference on Knowledge Discovery & Data Mining, pp. 764–773 (2020)
19. Morales, A.J., Borondo, J., Losada, J.C., Benito, R.M.: Measuring political polarization: Twitter shows the two sides of Venezuela. Chaos Interdisc. J. Nonl. Sci. **25**(3), 033114 (2015)
20. Morini, V., Pollacci, L., Rossetti, G.: Capturing political polarization of reddit submissions in the trump era. In: SEBD, pp. 80–87 (2020)
21. Morini, V., Pollacci, L., Rossetti, G.: Toward a standard approach for echo chamber detection: reddit case study. Appl. Sci **11**(12), 5390 (2021)
22. Morini, V., Rossetti, G., Pollacci, L.: Capturing political polarization of reddit submissions in the trump era. In: Italian Symposium on Advanced Database Systems (2020)
23. Nguyen, T.T., Hui, P.M., Harper, F.M., Terveen, L., Konstan, J.A.: Exploring the filter bubble: the effect of using recommender systems on content diversity. In: Proceedings of the 23rd International Conference on World Wide Web, pp. 677–686 (2014)
24. Nickerson, R.S.: Confirmation bias: a ubiquitous phenomenon in many guises. Rev. General Psychol. **2**(2), 175–220 (1998)
25. Pariser, E.: The Filter Bubble: What the Internet is Hiding From You. Penguin, UK (2011)
26. Peralta, A.F., Kertész, J., Iñiguez, G.: Opinion Dynamics in Social Networks: From Models to Data (2022)
27. Quattrociocchi, W., Scala, A., Sunstein, C.R.: Echo chambers on Facebook. Available at SSRN 2795110 (2016)
28. Sichani, O.A., Jalili, M.: Inference of hidden social power through opinion formation in complex networks. IEEE Trans. Netw. Sci. Eng. **4**(3), 154–164 (2017)
29. Sîrbu, A., Loreto, V., Servedio, V.D., Tria, F.: Opinion dynamics: models, extensions and external effects. In: Participatory Sensing, Opinions and Collective Awareness, pp. 363–401. Springer (2017)
30. Sobkowicz, P.: Quantitative agent based model of opinion dynamics: polish elections of 2015. PLoS One **11** (2016)
31. Stewart, A.J., Mosleh, M., Diakonova, M., Arechar, A.A., Rand, D.G., Plotkin, J.: Information gerrymandering and undemocratic decisions. Nature **573**, 117–121 (2019)
32. Sunstein, C.R.: Republic.Com 2.0. Princeton University Press, USA (2007)
33. Valensise, C.M., Cinelli, M., Quattrociocchi, W.: The dynamics of online polarization (2022)
34. Vendeville, A., Guedj, B., Zhou, S.: Forecasting elections results via the voter model with stubborn nodes. Appl. Netw. Sci. **6**, 1–13 (2021)
35. Verducci, S.: Critical thinking and open-mindedness in polarized times. Encounters Theo. History Educ. **20**(1), 6–23 (2019)
36. Xiong, F., Liu, Y.: Opinion formation on social media: an empirical approach. Chaos **24**(1), 013130 (2014)

The Effects of Message Sorting in the Diffusion of Information in Online Social Media

Diego F. M. Oliveira[1,2,3,4](✉), Kevin S. Chan[5], and Peter J. Mucha[6]

[1] Statistical and Applied Mathematical Sciences Institute, Durhan, NC, USA
diegofregolente@gmail.com
[2] Department of Mathematics, University of North Carolina, Chapel Hill, NC, USA
[3] Social Data Science Center, University of Maryland, College Park, MD, USA
[4] College of Information Studies, University of Maryland, College Park, MD, USA
[5] U.S Army Research Laboratory, 2800 Powder Mill Rd, Adelphi, MD, USA
[6] Department of Mathematics, Dartmouth College, Hanover, NH, USA

Abstract. In this work, we propose an agent-based model to study the effects of message sorting on the diffusion of low- and high-quality information in online social networks. We investigate the case in which each piece of information has a numerical proxy representing its quality, and the higher the quality, the greater are the chances of being transmitted further in the network. The model allows us to study how sorting information in the agent's attention list according to their quality, node's influence and popularity affect the overall system's quality, diversity and discriminative power. We compare the three scenarios with a baseline model where the information is organized in a first-in first-out manner. Our results indicate that such an approach intensifies the exposure of high-quality information increasing the overall system's quality while preserving its diversity. However, it significantly decreases the system's discriminative power.

Keywords: Networks · Competition · Limited attention · Information load

1 Introduction

The introduction of online social media platforms such as Twitter and Facebook have completely changed the ways the modern civilization consumes and shares information. Due to the low cost of information production and broadcasting, we are exposed to hundreds if not thousands of messages, or memes [1], every day which exceeds by far our capacity of content consumption [2], and each piece of information must compete for our limited attention. As a result, only a tiny fraction of the information created ends up going viral, while the vast majority will simply never be re-transmitted and quickly forgotten. If from one side online social media can facilitate the interaction between people from different

H. Cherifi et al. (Eds.): COMPLEX NETWORKS 2016 2022, SCI 1077, pp. 98–107, 2023.
https://doi.org/10.1007/978-3-031-21127-0_9

parts of the globe, they also provide the perfect ecosystem for the spreading of low-quality information such as fake news and misinformation (i.e., information that is misleading or inaccurate) that can be very harmful to our society. Examples of misuses of such platforms can be seen during events like terrorist attacks, natural disasters and even to negatively affect our economy [3–6]. Such an interesting behaviour and real life implications caught the attention of the scientific community and the field of information diffusion experienced a significant growth and every year new models are proposed to better understand the relationship between humans and the new universe of online social media [7–12].

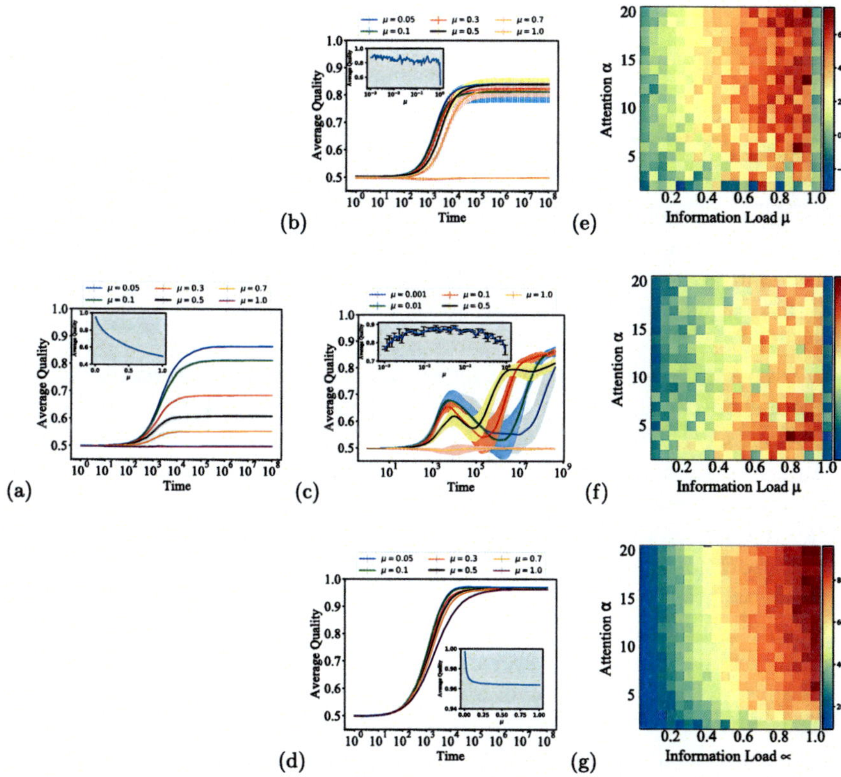

Fig. 1. Behaviour of the average system's quality as a function of time for **a** the baseline model and the model with sorted attention list according to **b** the meme's popularity, **c** node's influence and **d** meme's quality. The insets show the behaviour of the average system's quality at steady state for different values of the information load μ and with $\alpha = 14$. **e–g** Phase diagrams for different values of attention α and different values of the information load μ of the improvement in percent in the average system's quality at the steady state for the three models with respect to the baseline model. The error bars in (a–d) represent the standard deviation obtained from ten different simulations.

Traditionally, models of information diffusion are based on tools borrowed from theoretical epidemiology where susceptible agents became infected by interactions with infected agents and, in spite of their simplicity, they were able to reproduce several empirical observations [13–15]. For example, Weng [16], Gleeson [17] and Notarmuzi [18] have shown that a very simple model of information diffusion can produce a flat-tailed distribution for the popularity of a given meme. Such kinds of behaviors were commonly observed in a variety of systems that include citations [19–21], hashtags and URLs on Twitter [22], videos on YouTube [23], among many others [24–27]. Qiu et al. proposed a diffusion model that considers the user's limited attention and the quality of the information being transmitted. The authors showed that there exists a tradeoff between discriminative power and diversity. However, in realistic conditions, the model predicts that high-quality information has little advantage over low-quality information [28]. Simultaneously, Sreenivasan et al. [29] proposed a model of information cascades on feed-based networks also considering the finite attention, innovations and message diffusion. In such a case, the authors estimated the branching factor associated with the cascade process for different attention spans and different forwarding probabilities. They demonstrated that beyond a certain attention span, cascades tend to become viral. Ciampaglia et al. [30] proposed a model in which memes are selected based on their popularity or quality and the authors found that popularity bias hinders average quality when users are capable of exploring many items, as well as when they only consider very few top items due to scarce attention. They also found that an intermediate regime exists in which some popularity bias is good in distinguishing high-quality information, but too much can harm the system. More recently, Oliveira et al. investigated the impact of influential nodes on the spreading of information. The authors showed that a meme's quality does not guarantee virality, but there is a strong correlation between the meme's success and the influence of the agent who introduced it. Additionally, when trust is introduced into the model and the agents can decide whether or not to accept a meme, the authors observed that both lifetime and popularity distributions have broad power-law tails indicating that only a few memes spread virally through the population reproducing perfectly the broad distributions obtained from empirical data [31]. When considering situations where agents with heterogeneous criteria of quality, Cisneros-Velarde et al. proposed a simple method for enhancing the spread of high-quality information. Their results consist of strategically re-sorting the information feeds of users that share low-quality information. Under different settings of types of users, the authors showed that this policy has the best performance on homogeneous agents with a good criterion of what constitutes "good information". Moreover, they found that even in the case where agents are either purely malicious or have an opposite criterion of what constitutes high-quality information, the policy greatly reduces the spread of low quality information [32].

In situations in which quality is not easily quantified, other metrics—such as ratings, number of views, likes, number of downloads, etc.—can be used to enhance the exposure of certain content to people. In principle, such approaches would allow high-quality information to prevail. However, this popularity-based approach can create bias since the systems can be easily manipulated by social bots, for example [12]. Another disadvantage was highlighted by Sunstein [33,34] and Pariser [35]. The authors argued that the reliance on personalization and social media can lead people to being exposed to a narrow set of point of views [36] and one's existing beliefs would be reinforced because they are locked inside so-called filter bubbles or echo chambers, which prevent the users from engaging with ideas different from their own. Such selective exposure could facilitate confirmation bias [37] and possibly create a fertile ground for polarization and misinformed opinions [38,39]. Although several other works [40–43] have been done trying to address to the crucial importance for the problem of competition for attention, there is still a lack of a better understanding of how memes behave in on-line social networks. In this work, we investigate how the adoption of different messages sorting mechanisms on the users' news feed will affect the system's quality, diversity and discriminative power. We assume that each piece of information carries a numerical proxy representing its quality, interestingness or truthfulness. We anticipate that by sorting the memes, will increase the exposure of high-quality information, therefore, increasing the overall system's quality. However, it is still unknown how it will affect other characteristics of the systems such as diversity of information and discriminative power.

2 The Model and Numerical Results

The model consists of a network with N agents, each of them equipped with a finite memory (or attention) containing α memes. Each meme is equipped with a numerical value drawn from an uniform distribution representing its quality or truthfulness. Furthermore, new memes are continuously introduced into the system in an exogenous way. The rate at which this happens determines the amount of diversity in the system in the sense that the higher information load μ, the higher the diversity and as a consequence, the harsher the competition. We assume that at time $t = t_0$ the system is in a state of higher diversity where each node has α unique memes. At every time step a node i is selected at random and with probability μ it introduces a new meme into the system by adding it to its attention list and sharing it with all its neighbors. On the other hand, with probability $1 - \mu$ the selected node chooses a meme from its attention list and, then, it transmits it to all of its neighbors. In this scenario, the exposed agent will accept the meme only if it is not already in his/her attention list. Furthermore, the probability that an agent selects a specific meme m from its list to transmit is proportional to the meme's quality $f(m)$. For example, if node i has a set of memes $m(i)$, the probability of meme m_k being selected is

$$P_i(k) = \frac{f(m_k)}{\sum_{j=1}^{\alpha} f_i(m_j)}, \tag{1}$$

therefore, the higher the quality, the higher the chances of being transmitted. Once all neighbors receive the meme, the memes at the bottom of the nodes'

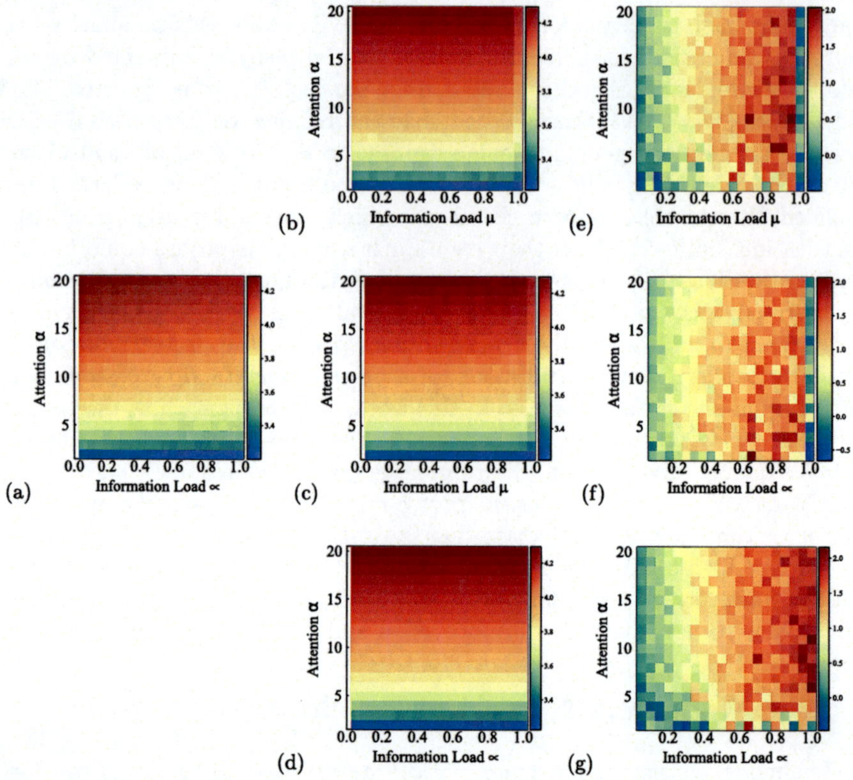

Fig. 2. The Diversity H (color scale bar) as a function of intensity of information load and attention for **a** the baseline model and the model with a sorted attention list according to **b** the meme's popularity, **c** node's influence and **d** meme's quality. **e–g** Difference in percentage between the baseline model and the models with different sorting mechanisms. The results indicate that sorting preserves the overall system's diversity.

attention lists are removed or forgotten to make space for the incoming message. Here, our goal is to understand how sorting the memes according to their quality, popularity and based on the source's influence (or degree) will affect the overall system's quality, diversity and discriminative power. We will compare these three scenarios with a baseline model where memes are organized in a first-in-first-out manner. The results were obtained from a scale-free network with $N = 1000$ and average degree $\overline{k} = 20$ for each of the scenarios considered. We run each simulation until the system reaches a steady state and once in such a state [44], we follow 4×10^6 memes for each combination of the control parameters considered from the moment they were first introduced into the model until they completely disappear from the network recording their quality, popularity, lifetime as well as some of the characteristics of the source of information. Here, we define popularity as the number of times a given meme is selected to be

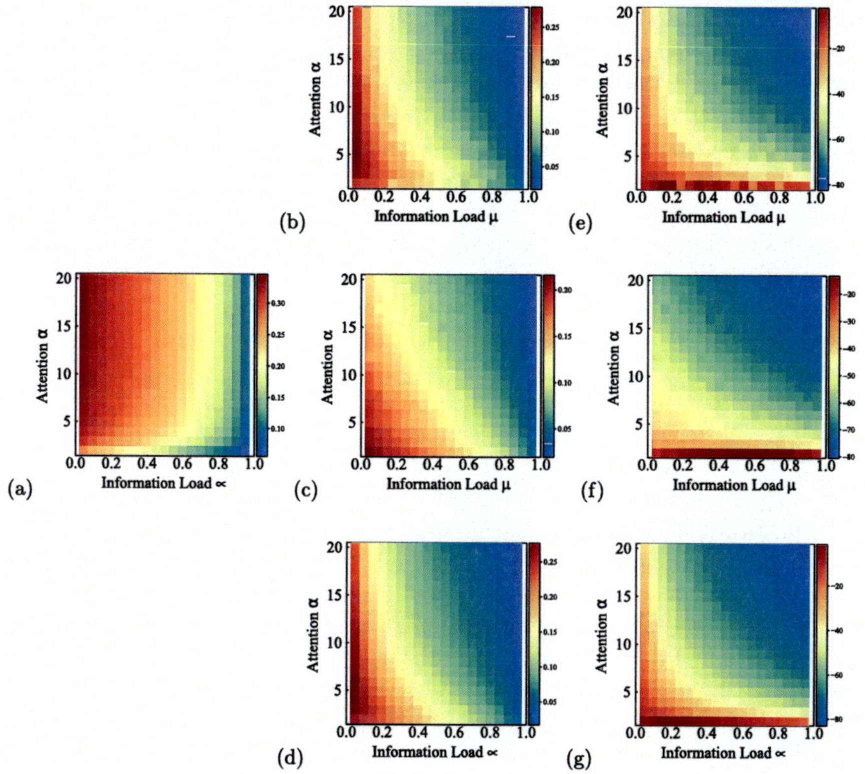

Fig. 3. The Kendall Tau (color scale bar) as a function of intensity of information load and attention for the **a** baseline model and the model with memes sorted according to **b** the meme's popularity, **c** node's influence and **d** meme's quality. **e–g** Difference in percentage between the baseline model and the models with sorted attention list. In this case, the introduction of sorting hinders the system's discriminative power with differences between models as high as 82.5%.

transmitted, lifetime is the time passed between the meme's creation and its extinction and the influence or degree of the node that generated the meme. In all situations, the selection depends purely on the meme's quality, however, in principle, sorting the user's attention list will enhance the exposure of high-quality information. In order to verify if our assumption is correct, Fig. 1 shows the behavior of the system's average quality as a function of time for different values of information load μ for (a) baseline model and the models with sorted attention list according to (b) meme's popularity (c) node's influence and (d) meme's quality. As one can see in Fig. 1 (a–d), at time $t = 0$, the average system's quality is 0.5 since all memes are drawn at random from a uniform distribution between 0 and 1. However, as the competition starts to take place, the average quality starts to increase in a way that highly depends on the information load μ and the sorting mechanism. The inset shows the dependence of the system's average quality $\overline{Q}$ on μ at the steady state. Observe that as $\mu \to 1$, $\overline{Q} \to 0.5$ which is equivalent to the case where there is no diffusion, only innovation with a new meme being introduced at every time step. The only exception is Fig. 1 (d). In such a case, the system is able to converge to an optimal state with only high quality information even when $\mu = 1$. Finally, Fig. 1 (e–g) shows the behavior of the attention α as a function of the information load μ. Color represents the difference in percent between the system's quality at the steady state of the baseline and the model with a sorted attention list. Observe that the overall system's quality improves significantly in all cases, reaching as high as 96% improvement for some combination of the control parameters. Confirming our hypotheses that by sorting the memes will enhance the exposure of high-quality information and as a consequence increase the overall system's quality.

Next, we investigate how the diversity of information and the system's discriminative power are affected by sorting mechanisms. To measure the amount of diversity in the system at the steady state, we start from the entropy $H = -\sum_m P(m) \log P(m)$ where $P(m)$ is the portion of attention received by meme m, i.e., the fraction of messages with m across all of the user feeds. The sum runs over all memes present at a given time and is averaged over a long period after stationarity has been achieved. Figure 2 (a–d) shows the behavior of the diversity (system's entropy) for (a) the baseline model and the models with sorted attention list according to the (b) meme's popularity (c) node's influence and (c) meme's quality for different values of α and μ. Observe that the information load does not significantly affect the system's diversity, however, as expected, it increases as the user's attention increases. Furthermore, Fig. 2 (e–g) shows the difference in percent between the baseline and the models with a sorted attention list. We observe that sorting does not affect the diversity of information in any significant way in any of these cases. On the other hand, as we will show next, it does considerably decrease the system's ability to distinguish between memes.

To measure the system's discriminative power, we employ the Kendall rank correlation [45] between popularity and quality, which is computed by ranking memes according to the two criteria and then counting the number of meme pairs for which the two rankings are concordant or discordant, properly accounting for ties. The extreme case $\tau = 1$ indicates a perfect correlation between quality and popularity and fitter memes are more likely to go viral. On the other hand, if $\tau = -1$, the two rankings are completely discordant. Figure 3(a-d) shows in color the Kendall correlation rank for the four models considered for different values of α and μ. We observed that in general the rank correlation decreases as the information load increases and a comparison between the models review that the introduction of the sorting mechanisms hinders the system's discriminative power with differences between models being as high as 82.5% as shown in Fig. 2(e-g).

3 Conclusions

We have considered an agent-based model to study the effects of message sorting on the diffusion of low and high-quality information in online social networks. We have considered three scenarios where memes are sorted according to their popularity, the influence of the node that posted the meme or the meme's quality. In order to understand the effects of such changes on the overall system's quality, diversity and discriminative power, we have compared these three situations with a baseline model where memes are organized in a first-in first-out manner. The results indicate that such approaches intensify the exposure of high-quality information, increasing the overall system's quality while preserving its diversity. On the other hand, a more significant change was observed when considering the system's discriminative power τ. In all cases, namely, the baseline model and the model with sorted attention lists, τ decreases as $\mu \rightarrow 1$, however when comparing them, sorting hinders the system's discriminative power with differences as high as 82.5% for some combination of the control parameters.

Acknowledgment. The research was supported partially by NSF-DMS Grant 1929298 and ARL through ARO Grant W911NF-16-1-0524. The views and conclusions contained in this document are those of the authors and should not be interpreted as representing the official policies, either expressed or implied, of the Army Research Laboratory or the U.S. Government. The U.S. Government is authorized to reproduce and distribute reprints for Government purposes notwithstanding any copyright notation here on.

References

1. Dawkins, R.: The Selfish Gene, no. 199. Oxford University Press (2006)
2. Dunbar, R.: The social brain hypothesis. Brain **9**(10), 178–190 (1998)
3. Starbird, K., Maddock, J., Orand, M., Achterman, P., Mason, R.M.: Rumors, false flags, and digital vigilantes: misinformation on Twitter after the 2013 Boston marathon bombing. In: IConference 2014 Proceedings (2014)
4. Mendoza, M., Poblete, B., Castillo, C.: Twitter under crisis: can we trust what we RT? In: Proceedings of the First Workshop on Social Media Analytics, pp. 71–79. ACM (2010)
5. Gupta, A., Lamba, H., Kumaraguru, P., Joshi, A.: Faking Sandy: characterizing and identifying fake images on Twitter during hurricane Sandy. In: Proceedings of the 22nd International Conference on World Wide Web, pp. 729–736. ACM (2013)
6. Domm, P.: False rumor of explosion at white house causes stocks to briefly plunge; AP confirms its Twitter feed was hacked. http://cnb.cx/2thGiq5, 2013. Online. Accessed 05-Oct-2020
7. Rosen, L.D., Carrier, L.M., Cheever, N.A.: Facebook and texting made me do it: media-induced task-switching while studying. Comput. Human Behav. **29**(3), 948–958 (2013)
8. Junco, R.: Too much face and not enough books: the relationship between multiple indices of Facebook use and academic performance. Comput. Human Behav. **28**(1), 187–198 (2012)

9. World Economic Forum: Digital Wildfires in a Hyperconnected World. http://bit.ly/2sY6djW, 2013. Online. Accessed 30-Jan-2019
10. Tambuscio, M., Oliveira, D.F., Ciampaglia, G.L., Ruffo, G.: Network segregation in a model of misinformation and fact-checking. J. Comput. Social Sci. **1**(2), 261–275 (2018)
11. Lazer, D.M., Baum, M.A., Benkler, Y., Berinsky, A.J., Greenhill, K.M., Menczer, F., Metzger, M.J., Nyhan, B., Pennycook, G., Rothschild, D., et al.: The science of fake news. Science **359**(6380), 1094–1096 (2018)
12. Ferrara, E., Varol, O., Davis, C., Menczer, F., Flammini, A.: The rise of social bots. Commun. ACM **59**(7), 96–104 (2016)
13. Newman, M.: Networks: An Introduction. Oxford University Press (2010)
14. Easley, D., Kleinberg, J.: Networks, Crowds, and Markets: Reasoning About a Highly Connected World. Cambridge University Press (2010)
15. Goffman, W., Newill, V.: Generalization of epidemic theory: an application to the transmission of ideas. Nature **204**(4955), 225–228 (1964)
16. Weng, L., Flammini, A., Vespignani, A., Menczer, F.: Competition among memes in a world with limited attention. Sci. Rep. **2** (2012)
17. Gleeson, J.P., Ward, J.A., O'Sullivan, K.P., Lee, W.T.: Competition-induced criticality in a model of meme popularity. Phys. Rev. Lett. **112**(4), 048701 (2014)
18. Notarmuzi, D., Castellano, C.: Analytical study of quality-biased competition dynamics for memes in social media. EPL (Europhys. Lett.) **122**(2), 28002 (2018)
19. Stringer, M.J., Sales-Pardo, M., Amaral, L.A.N.: Statistical validation of a global model for the distribution of the ultimate number of citations accrued by papers published in a scientific journal. J. Am. Soc. Inf. Sci. Technol. **61**(7), 1377–1385 (2010)
20. Wang, D., Song, C., Barabási, A.-L.: Quantifying long-term scientific impact. Science **342**(6154), 127–132 (2013)
21. Penner, O., Pan, R.K., Petersen, A.M., Kaski, K., Fortunato, S.: On the predictability of future impact in science. Sci. Rep. **3** (2013)
22. Lerman, K., Ghosh, R., Surachawala, T.: Social contagion: an empirical study of information spread on Digg and Twitter follower graphs. arXiv:1202.3162 (2012)
23. Crane, R., Sornette, D.: Robust dynamic classes revealed by measuring the response function of a social system. Proc. Nat. Acad. Sci. **105**(41), 15649–15653 (2008)
24. Leskovec, J., Adamic, L.A., Huberman, B.A.: The dynamics of viral marketing. ACM Trans. Web (TWEB) **1**(1), 5 (2007)
25. Aral, S., Walker, D.: Creating social contagion through viral product design: a randomized trial of peer influence in networks. Manage. Sci. **57**(9), 1623–1639 (2011)
26. Rogers, E.M.: Diffusion of Innovations. Simon and Schuster (2010)
27. Jamali, S.: Comment mining, popularity prediction, and social network analysis. PhD thesis, George Mason University (2010)
28. Qiu, X., Oliveira, D.F., Shirazi, A.S., Flammini, A., Menczer, F.: Limited individual attention and online virality of low-quality information. Nat. Human Behav. **1**(7), s41562-017 (2017)
29. Sreenivasan, S., Chan, K.S., Swami, A., Korniss, G., Szymanski, B.: Information cascades in feed-based networks of users with limited attention. IEEE Trans. Netw. Sci. Eng. (2017)
30. Ciampaglia, G.L., Nematzadeh, A., Menczer, F., Flammini, A.: How algorithmic popularity bias hinders or promotes quality. Sci. Rep. **8**(1), 15951 (2018)

31. Oliveira, D.F., Chan, K.S.: The effects of trust and influence on the spreading of low and high quality information. Phys. A Stat. Mech. Appl. **525**, 657–663 (2019)
32. Cisneros-Velarde, P., Oliveira, D.F., Chan, K.S.: Spread and control of misinformation with heterogeneous agents. In: International Workshop on Complex Networks, pp. 75–83 (2019)
33. Sunstein, C.R.: The law of group polarization. J. Polit. Philos. **10**(2), 175–195 (2002)
34. Sunstien, C.R.: Republic 2.0. Princeton University Press, Princeton (2009)
35. Pariser, E.: The filter bubble: How the new personalized web is changing what we read and how we think. Penguin (2011)
36. Nikolov, D., Oliveira, D.F., Flammini, A., Menczer, F.: Measuring online social bubbles. PeerJ Comput. Sci. **1**, e38 (2015)
37. Baron, J.: Thinking and Deciding. Cambridge University Press (2000)
38. Nyhan, B., Reifler, J.: When corrections fail: the persistence of political misperceptions. Polit. Behav. **32**(2), 303–330 (2010)
39. Stanovich, K.E., West, R.F., Toplak, M.E.: Myside bias, rational thinking, and intelligence. Curr. Directions Psychol. Sci. **22**(4), 259–264 (2013)
40. Ratkiewicz, J., Fortunato, S., Flammini, A., Menczer, F., Vespignani, A.: Characterizing and modeling the dynamics of online popularity. Phys. Rev. letters **105**(15), 158701 (2010)
41. Lerman, K., Ghosh, R.: Information contagion: an empirical study of the spread of news on Digg and Twitter social networks. ICWSM-Int. Conf. Weblogs Social Media **10**, 90–97 (2010)
42. González-Bailón, S., Borge-Holthoefer, J., Rivero, A., Moreno, Y.: The dynamics of protest recruitment through an online network. Sci. Rep. **1**, 197 (2011)
43. Baños, R.A., Borge-Holthoefer, J., Moreno, Y.: The role of hidden influentials in the diffusion of online information cascades. EPJ Data Sci. **2**(1), 6 (2013)
44. Oliveira, D.F., Chan, K.S., Leonel, E.D.: Scaling invariance in a social network with limited attention and innovation. Phys. Lett. A **382**(47), 3376–3380 (2018)
45. Kendall, M.G.: A new measure of rank correlation. Biometrika **30**(1/2), 81–93 (1938)

Gradual Network Sparsification and Georeferencing for Location-Aware Event Detection in Microblogging Services

Diaoulé Diallo(✉) and Tobias Hecking

German Aerospace Center (DLR) Institute for Software Technology, Linder Höhe, Cologne 51147, Germany
diaoule.diallo@dlr.de

Abstract. Event detection in microblogging services such as Twitter has become a challenging research topic within the fields of social network analysis and natural language processing. Many works focus on the identification of general events with event types ranging from political news and soccer games to entertainment. However, in application contexts like crisis management, traffic planning, or monitoring people's mobility during pandemic scenarios, there is a high need for detecting localisable physical events. To address this need, this paper introduces an extension of an existing event detection framework by combining machine learning-based geo-localisation of tweets and network analysis to reveal events from Twitter distributed in time and space. Gradual network sparsification is introduced to improve the detection events of different granularity and to derive a hierarchical event structure. Results show that the proposed method is able to detect meaningful events including their geo-locations. This constitutes a step towards using social media data to inform, for example, traffic demand models, inform about infection risks in certain places, or the identification of points of interest.

Keywords: Event detection · Social media analysis · Twitter analysis · Data science · Georeferencing · Network clustering

1 Introduction

Microblogging has become an important medium to exchange information, messages, opinions, and upcoming events. Private individuals as well as politicians, celebrities and in general people with great influence use the platform to disseminate information. Recognizing that rich and user-generated content is of great utility to identify events happening in time and space, the challenging discipline of event detection has emerged. As one of the largest microblogging services, which also offers an easily accessible API,[1] Twitter has become a common data

[1] https://developer.twitter.com/en/docs/twitter-api.

H. Cherifi et al. (Eds.): COMPLEX NETWORKS 2016 2022, SCI 1077, pp. 108–120, 2023.
https://doi.org/10.1007/978-3-031-21127-0_10

source for event detection in social media content. Current Twitter event detection methods differ not only in terms of the underlying methods, but also in terms of the type of events to be detected [1–3]. The obtaining of tweets from a specific country for the purpose of event location detection is usually connected with the necessity of a geo-tagged Twitter stream. The Twitter API can only provide geo-tagged tweets when a request is limited by a geographical area. As noted in previous work [4–6], a common issue is that only about 1–2% of all tweets are geo-tagged with exact coordinates. Therefore, a geo-tagged stream might reduce the overall quality of the detected events or even change the kind of detected events in general. With the qualitative improvements of geo-place name extractors and generally named entity recognition models in different languages, the possibilities to localize events have improved. Besides the focus on the localisation of events, the exhaustive collection of events of different granularity as well as hierarchical structures that can reveal sub-events are of great importance. To advance event detection towards serious applicability in real-world use cases we present the following contributions:

- We adapt an existing state-of-the-art event detection approach to find more events in the same time window using gradual sparsification of a co-occurence network of named entities, hashtags, and geo-locations. With these modification also the dependence on external information sources (in this case Wikipedia) is reduced.
- We analyse the resulting events in terms of the type, granularity and hierarchical structure revealing an higher diversity and sub-events.
- We investigate how well events can be localized using geo-information from tweet texts and the application of a microblog-specific place name extractor.

In the following, Sect. 2 presents related work in the area of event detection and local event detection. Subsequently, Sect. 3 presents the adaption of an existing event detection algorithm as well as the introduction of event localisation. We present our results in Sect. 4 and provide a brief conclusion in Sect. 5.

2 Related Work

There have been many attempts to define events in social media [1,2,7,8]. Following these explanations, events in online streams can be defined as the occurrence of topics and entities of substantial volume during a certain period.

We can distinguish between event detection using external information and event detection without any prior knowledge. Thus, in the literature two problem formulation exists, namely specified event detection and unspecified event detection [1,2]. The former uses external knowledge about the type of event and can thus, for example, selectively collect tweets that contain specific keywords. The more important and also more challenging discipline is that of unspecified event detection. In that case, events are detected without the use of further information about the event.

Event detection techniques include but are not limited to graph theory [9–11], burst detection in time series [12–14], and clustering [12,15,16], using different text representations, tweet meta data, hashtags and entities.

SEDTWik [12] combines a burstiness score with clustering and Wikipedia in order to detect unspecified events in an unsupervised manner. The algorithm takes into account not only the tweet texts but also various meta data and can be easily adapted for different time windows. It is still referred to SEDTWik as state-of-the-art for comparison [17,18]. As the algorithm is ideally suited to be extended in the direction of real-world event detection including event localisation and sub-events, it serves as the basis in this work. We provide a detailed description of SEDTWik in Sect. 3.

Many event detection techniques which attempt to detect event locations use geo-tagged tweets exclusively [19–22]. However, since geo-tagged tweets only account for 1% of the data [23], the quality of the resulting events will suffer. Consequently geo-tagged twitter streams are not suited as the only source for detecting and localising events. Recent changes in the Twitter policies [4] further support the claim that local event detection techniques need to take account of named entity recognition (NER) models to identify place names in microblogs used for location inference. Unakard et al. [6] use the reported user location as well as location entities extracted with the Stanford Named Entity Recognizer [24]. More recently location extraction has further improved with named entity recognition models (NER) specifically created for detecting place names in microblog entries [25].

3 Method

The basic intuition behind SEDTWik [12] is to use tweet texts, hashtags, as well as user mentions and other metadata to extract entities (or segments) that occur more often than expected in a certain period and relate them based on similarity scores. Events are then identified by clustering the resulting entity network and ranked using auxiliary information from Wikipedia. In this work an adaptation of SEDTWik is presented that aims to increase the number of meaningful events detected without requiring the time-consuming computation of news values from Wikipedia by using a gradual network sparsification algorithm. Furthermore, different NER approaches are used to extract geo-locations from tweet texts and associate events with locations. These basic steps are described below.

Tweet segmentation In the tweet segmentation step, tweet texts are preprocessed[2] and segments are extracted. Segments are defined as all uni- and multigrams where at least one word exists as Wikipedia title. Since hashtags and user mentions are commonly used to thematically label tweets, these are always considered and not depending being matched with a Wikipedia title. Hashtags are split for uppercase letters so that hashtags i.e. *#SpaceShuttleEndeavour* result in *space shuttle endeavour*.

[2] https://github.com/kevalmorabia97/pyTweetCleaner.

In addition to the original SEDTWik approach the extracted segments are associated with geo-locations based on co-occurrence as a prerequisite to location-aware events detection. To do so, tweets are searched for locations using GazPNE2 [25] and spaCy [26]. GazPNE2 combines global gazetteers, deep learning and pretrained transformer models to extract geo place names from microblogs. Since the model achieves state of the art results on a variety of twitter datasets, it is perfectly suited for our case. In addition we use the well-known NER model provided by spaCy [26]. For a segment s, we store all locations L_s that occur together with segment s in a tweet, as well as the associated location frequencies.

Bursty segment extraction Segments occurring more often than expected in a time window and belong to tweets that gain attention (measured by retweets and follower count) are considered as "bursty". For details on calculating the burstiness score we refer to [12]. In total the top $K = \sqrt{N_w}$ bursty segments are further processed, with N_w being the total number of tweets in time window w.

Event identification The identified bursty segments and their similarity scores constitute a weighted network G. In SEDTWik this network is reduced to a mutual $k-$nearest neighbour network by retaining only edges (s_i, s_j) if s_i belongs to the $k-$nearest neighbours of s_j and vice versa. The components of the resulting disconnected network are considered as event candidates. The events are ranked using Wikipedia "keyphraseness" values [27]. Keyphraseness is calculated as the probability that the entity appears as anchor text in Wikipedia articles containing the entity [12,27]. The "newsworthiness" of an event candidate is calculated from the keyphraseness of contained entities and their similarities. By discarding event candidates with newsworthiness below a threshold T, clusters of generic terms that emerge randomly are filtered out.

While the approach described above works well for identifying major events on Twitter, it has several shortcomings that are addressed by our alternative method using *gradual network sparsification.*

In the original approach segments are only considered if they are in any $k-$nearest neighbour list of another segment. However, since the similarity value could be high for several segments, important information is dropped here. On the right-hand side of Fig. 1 one can see an example where a segment (dark node) is removed from the graph although is has high similarities to other segments that may be a candidate for an event. Contrarily, parts of the similarity graph with overall low similarity scores can accidentally be detected as an event candidate as it can be seen in the left example in Fig. 1.

A less restrictive method for event identification that does not delete nodes based on a fixed parameter k regardless of whether there are highly similar but not mutually top-k neighbours is presented in the following. Starting with G where initially almost all nodes are connected to each other by edges weighted based on their similarity of their endpoints. An initial value for an edge weight threshold $t_{sim}^{(1)}$ is chosen. All edges with weight below $t_{sim}^{(1)}$ are deleted. Next, the similarity threshold $t_{sim}^{(i+1)} = t_{sim}^{(i)} + \Delta_{t_{sim}}$ is increased by a defined value $\Delta_{t_{sim}}$, and the edge deletion procedure is repeated. After each edge removal step all

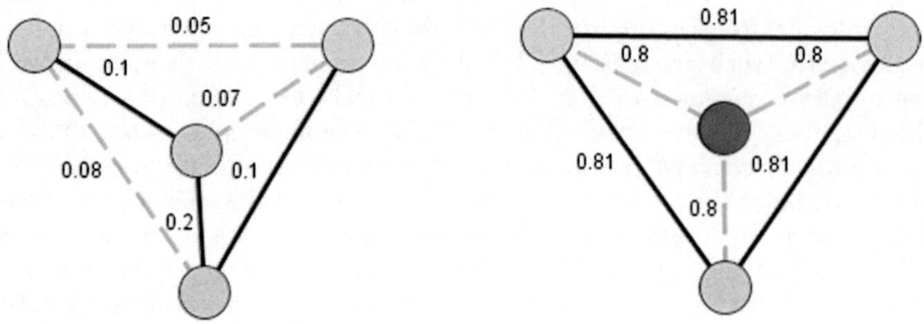

Fig. 1. Dashed edges do not belong to the mutual k-nearest neighbour graph. Left: Segment cluster with overall low similarity scores detected as event candidate. Right: The darker node will be removed from the graph although it probably belongs to an event candidate. For simplicity, k is set to 2 in this example.

isolated nodes are removed from the graph as well. In each iteration connected components of $size_{cc}$ nodes for which $size_{cc}^{min} \leq size_{cc} \leq size_{cc}^{max}$ are stored as events. This gradual network sparsification is repeated until the graph contains no more nodes. At that point, all nodes were either removed because they had no neighbors left or because they were assigned to an event. The algorithm is summarized in pseudo-code below:

Algorithm 1 Gradual network sparsification

```
t_sim^(1) ← 0.1
while G is not empty do
    for e in G do
        if sim_e ≤ t_sim then
            - Remove e from G
        end if
    end for
    for cc in connected_components(G) do
        if size_cc^min ≤ size_cc ≤ size_cc^max then
            - Store cc in event list
            - Remove nodes and edges of cc
        end if
    end for
    t_sim^(i+1) = t_sim^(i) + Δ_t_sim
end while
```

As the basic SEDTWik algorithm tends to cluster similar segments such as *wan na hear, wan na talk, wan na know, know takes, wa na see*[3] and *good day, today day, today good day, good morning, early morning*, it is reasonable

[3] Some of these segments result from the resolving of hashtags by capital letters.

to apply some method to detect and remove such *star clusters*.[4] Therefore, as a final step, sequence matching is applied, where we evaluate the term-based similarity of the different segments in a segment cluster. A segment cluster with the segments *this day, today day, good day* has a term-based similarity of 0.5, since each segment shares half of the containing terms with every other segment from the segment cluster. Previous experiments have shown that a term-based threshold of 0.3 detects most star events without removing significant segment clusters. The overall approach does not suffer from the issues of event detection based on a k-nearest neighbours graph depicted in Fig. 1. It is also possible to flexibly define how large or small segment cluster should be, and thus, allows for tuning granularity. Gradual network sparsification allows for subdivide larger segment clusters into smaller segment clusters representing sub-events. Another advantage over the original SEDTWik approach is that event identification does not rely on the costly calculation newsworthiness values from Wikipedia. Our method has certain similarities to the Girvan-Newman algorithm [28] for community in networks where edges are iteratively removed and re-evaluated based on their edge-betweenness centrality.

Event localisation The presented approach can generally detect all types of events. However, since we are interested in localisable real-world events the previously determined co-occurences of tweet segments and identified place names are especially considered. In large datasets almost all segments co-occur with some locations just by chance. However, in the case of localisable events we observe that the defining segments co-occur way more often with locations where the event takes place, while the frequency of other locations is almost uniformly distributed. Consequently, localisable events can be classified using the Gini coefficient of the association strength (number of co-occurences) of tweet segments that define an event and location names. Since the Gini coefficient is a measure how skewed a distribution is the value is high when one or a few place names stand out and almost 0 for non-localisable events.

4 Results

In the following, the results for event detection using gradual network sparsification are compared with the original version of the SEDTWik algorithm.

4.1 Experimental Setting

Dataset and Metrics. For our experiments, we used the same dataset as the authors of SEDTWik. They used seven days (October 11, 2012 to October 17, 2012) from the event detection dataset *Event2012* created by McMinn et al. [29]. Regarding Twitter data, there are many difficulties to face as for example in the creation, dissemination and comparability of Twitter datasets (for more detailed information see [3,4,30,31]). Hettiarachchi et al. [17] showed that only

[4] As the connected components often look like stars with one central node.

65.8% of the provided tweet ids from the *Event2012* dataset could be received. Nevertheless, the sample from the *Event2012* dataset used here is sufficient to evaluate the improvements of the adapted algorithm. The majority of all publications create their own tweet collections, which makes comparisons generally difficult [1].

According to the authors' description, the streaming API was used without further filtering and then non-English tweets and spam were removed [29]. As noted by SEDTWik's authors, the dataset lacks an exhaustive event list associated with the tweets [12]. Since there are no predefined labels, we evaluated all resulting events manually. This means that for each event consisting of a set of segments and a tweet summary, it is verified whether it is an actual event via web search. Only the events that were detected on the correct day are counted as valid events. For example, if the release of an album is detected as an event within the tweets of 15 October 2012, but the album was already released on 14 October 2012, this event will not be counted as a correctly detected event.

The number of detected valid events will serve as the main metric as in [12]. The authors also use a precision and a DERate as metrics. However, since a valid event is defined differently in this work (see evaluation subsection below) and due to subjective components in the verification of the events, we limit the evaluation metric to the number of detected actual events.

Parameter Setting As long as t_{sim} is initiated with a low value and $\Delta_{t_{sim}}$ is chosen sufficiently small in the network sparsification step, no events are lost. We found that setting an initial t_{sim}^1 of 0.05 and $\Delta_{t_{sim}}$ to 0.01 is small enough while the computation time is acceptable. Setting $\Delta_{t_{sim}}$ too high could cause several segments to lose their edges within one step, causing them to be discarded instead of being recognized as an event. *lower_bound* is set to 3 and *upper_bound* to 10.

4.2 Effect on Gradual Network Sparsification on Event Detection

In the following, the variation of the SEDTWik approach including gradual network sparsification described in Sect. 3 is compared to the original formulation of the method in terms of the number and nature of detected events. For a fair comparison the events found by SEDTWik and reported by [12] (available here[5]) were evaluated. Taking into account that (1) no events are valid that were detected for the wrong day and that (2) simple discussions in tweets about past events are also not valid (for example, last weekend's football game), we have to reduce the number of detected events reported by SEDTWik from 79 to 71. Both SEDTWik and our approach detect the *breast cancer awareness month*, which we exclude based on the criteria described in Sect. 4.1.

In total, we were able to detect 90 actual events that were manually verified through a web search, which is 21.1% more than the events detected by SEDTWik. This can be attributed to the applied gradual networks sparsification algorithm for event detection described in Sect. 3, that is less restrictive

[5] https://github.com/kevalmorabia97/SEDTWik-Event-Detection-from-Tweets.

regarding which tweet segments are considered to belong to an event and does not need to filter events based on their newsworthiness score calculated using Wikipedia as external knowledge base.

For a more detailed comparison Table 1 shows the number of events detected per day for each of the two methods. It can be seen that the proposed method never detects fewer actual (verifiable) events on each day than SEDTWik. Surprisingly although our approach yields more events not all SEDTWik events are included in the result set. One explanation can be that gradual network sparsification tends to discard loosely coupled subgraphs of the similarity graph (i.e. segments that make up an event but having low overall similarity) as shown on the left side of Fig. 1. While this property is helpful to filter in many cases there are also sometimes events that are described by very general segments that do not have a strong temporal correlation. As mentioned in Sect. 4.1, we will not focus on precision, recall or similar measures, which depend highly on subjective interpretation. However, approximately 70% of all detected events could be verified as actual events following the definition of Sect. 4.1. Considering the precision values published by Morabia et al. [12], our approach shows a slightly increased false positive rate. Nevertheless, it must be emphasized that a comparability is not given for the reasons already explained above.

Table 1. Comparison of number of detected events per date.

	Detected actual events		
Date	SEDTWik	Our approach	In common
2012/10/11	13	15	7
2012/10/12	8	8	7
2012/10/13	9	9	1
2012/10/14	10	13	7
2012/10/15	10	12	8
2012/10/16	14	19	5
2012/10/17	8	14	4

Table 2 shows some examples of actual events that could not be detected by SEDTWik but by our approach. As can be seen, the types of events range from announcements, current political world events, content about famous people, and sports to daily news and election reminders. The type of events discovered by SEDTWik are to some degree comparable, nevertheless the majority of them are sports-related events. The events detected by our approach are more diverse and not as biased towards sport events. This can be explained by the fact that SEDTWik discards events whose newsworthiness is below a threshold, to exclude non significant segment clusters. The drawback of this approach is that at the same time meaningful segment clusters with low newsworthiness are also removed. Also, our observation does not confirm that events with high newsworthiness are necessarily more important events. In contrast to SEDTWik, we remove such segment clusters using the term similarity of the segments in one segment cluster, as explained in Sect. 3.

Table 2. Some events which were not detected by SEDTWik but our approach.

Date	Event Info
2012/10/11	– Samsung confirms Galaxy S III Mini
2012/10/12	– NASA space shuttle Endeavour begins 12-mile trip by road to new home – President Obama's campaign said Springsteen will appear along Clinton a pro-Obama rally
2012/10/13	– Jonathan Ross show with David Walliams, Ed Sheeran and more
2012/10/14	– Start of new seaseon Walking Dead
2012/10/15	– Man seriously hurt after being stabbed near Candlestick Park after San Francisco 49er game – Patriots versus Seahawks game
2012/10/16	– Last day to register vote in Maryland – Microsoft announces new surface details and starts pre-order
2012/10/17	– Police Taser blind man mistaking his white stick for a samurai – Google throws open doors to its top-secret data center

4.3 Localising Events

We categorised the detected verified events as localisable or non-localisable using the Gini coefficient as explained in Sect. 3. Out of 90 actual events, 32 could be associated with a specific location. Since it is noticeable that the granularity of the locations varies greatly and to a certain degree also depends on the nature of the event, it is difficult to make an absolute statement about the quality of the location extraction. However, to make a simple evaluation of the location determination procedures, we examined the two locations with the highest frequencies for each localisable event to see if they match the actual event location. In 28 of the 32 cases, the list of the two top locations contains at least one correct event location. The identified event locations are at state level 7 times, at state level 2 times, at city level 18 times, and at place level once. For most sport events the event locations can be found on city level. The extracted locations are greatly influenced by the team names, which in almost all cases already have locations in their names. Events like the *vice presidential debate* and *Ella Enderson and James Arthur performing at X Factor* could only be located on country-level. *Dancing with the Stars: Paula Abdul Guest Star Edition* is an example for a localisable event which was incorrectly assigned by our approach. However, as the focus of TV shows is often not on the shooting location it is questionable in how far these kinds of events are localisable. We observe that the granularity of the detected locations indicate the 'spatial extent' of the event. For example, the detected location for the *presidential debate* was *America* while the sports event involving the *New York Yankees* was associated with *new york*. In these cases, the detected locations rather.

4.4 Hierachical Sub-event Structures

A closer look at the events reveals that they partly overlap thematically, but show a different focus. For example for the vice presidential debate on 2012/10/12 as well as for the presidential debate on 2012/10/17, sub-events emerged. Certain debate topics can be found in different segment clusters which indicates the potential to generate hierarchical sub-event structures by using gradual network sparsification. Topics around the presidential debate and therefore the different detected events were ranging from taxes over parenthood to gun violence (see Fig. 2). Figure 2 shows the state of the segment network during gradual network sparsification. As we can see, the cluster around *energy independence* has just been identified as an event as it becomes an isolated connected component. In the step before, the segments of that connected component were still part of the larger segment network.

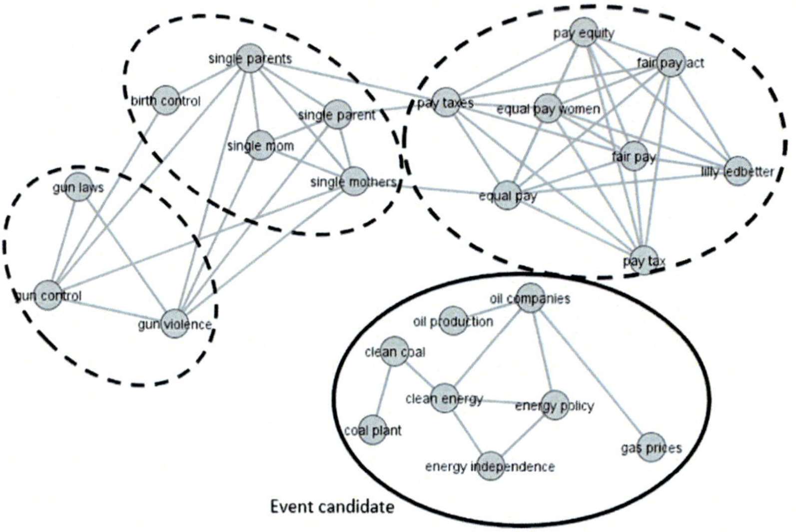

Fig. 2. An example of an intermediate step of the gradual network sparsification algorithm. Aspects of the presidential debate are first related but become event candidates after a sufficient number of iterations.

5 Conclusion

This paper described a method for localisable event detection in Twitter streams. It adapted the state-of-the-art method proposed by Morabia et al. [12] by incorporating place name extraction as well as a novel gradual network sparsification algorithm that reduces a co-occurrence network of tweet segments to meaningful components representing events. It could be shown that this approach leads to the detection of a significantly higher number of verifiable events along with physical locations.

In future works it will be investigated how to tune the parameters of the event detection algorithm according to spatial granularity (i.e. city, state, country level). Further improvements can also be made regarding the precision of the results since some detected events could not be verified.

References

1. Saeed, Z., Abbasi, R.A., Maqbool, O., Sadaf, A., Razzak, I., Daud, A., et al.: What's happening around the world? a survey and framework on event detection techniques on twitter. J. Grid Comput. **17**(2), 279–312 (2019)
2. Hasan, M., Orgun, M.A., Schwitter, R.: A survey on real-time event detection from the twitter data stream. J. Inf. Sci. **44**(4), 443–463 (2018)
3. Weiler, A., Grossniklaus, M., Scholl, M.H.: Survey and experimental analysis of event detection techniques for twitter. Comput. J. **60**(3), 329–346 (2017)
4. Kruspe, A., Häberle, M., Hoffmann, E. J., Rode-Hasinger, S., Abdulahhad, K., Zhu, X. X.: Changes in Twitter Geolocations: Insights and Suggestions for Future Usage. arXiv:2108.12251 (2021)
5. Choi, D., Park, S., Ham, D., Lim, H., Bok, K., Yoo, J.: Local event detection scheme by analyzing relevant documents in social networks. Appl. Sci. **11**(2) (2021)
6. Unankard, S., Li, X., Sharaf, M.A.: Emerging event detection in social networks with location sensitivity. In: World Wide Web, pp. 1393–1417 (2015)
7. Becker, H., Naaman, M., Graano, L.; Beyond trending topics: Real-world event identification on twitter. In: AAAI Conference on Web and Social Media, vol. 5, no. 1 (2011)
8. Dou, W., Wang, X., Ribarsky, W., Zhou, M.: Event detection in social media data. In: IEEE VisWeek Workshop on Interactive Visual Text Analytics-task Driven Analytics of Social Media Content, pp. 971–980 (2012)
9. Sayyadi, H., Hurst, M., Maykov, A.; Event detection and tracking in social streams. In: AAAI Conference on Web and Social Media, vol. 3, no. 1, pp. 311–314 (2009)
10. Schinas, M., Papadopoulos, S., Petkos, G., Kompatsiaris, Y., Mitkas, P. A.: Multimodal graph-based event detection and summarization in social media streams. In: 23rd ACM International Conference on Multimedia, pp. 189–192 (2015)
11. Edouard, A., Cabrio, E., Tonelli, S., Le Thanh, N.: Graph-based event extraction from twitter. In: Recent Advances in Natural Language Processing (2017)
12. Morabia, K., Murthy, N., Malapati, A., Samant, S.: SEDTWik: Segmentation-based event detection from tweets using Wikipedia. In: Conference of the North American Chapter of the Association for Computational Linguistics: Student Research Workshop, pp. 77–85 (2019)

13. Xie, W., Zhu, F., Jiang, J., Lim, E. P., Wang, K.: Topicsketch: Real-time bursty topic detection from twitter. IEEE Trans. Knowl. Data Eng. **28**(8), 2216–2229 (2016)
14. Corney, D., Martin, C., Göker, A.: Spot the ball: Detecting sports events on Twitter. In: European Conference on Information Retrieval, pp. 449–454 (2014)
15. Hasan, M., Orgun, M.A., Schwitter, R.: Real-time event detection from the Twitter data stream using the TwitterNews+ framework. Inf. Process. Manage. **56**(3), 1146–1165 (2019)
16. Nguyen, S., Ngo, B., Vo, C., Cao, T.: Hot topic detection on twitter data streams with incremental clustering using named entities and central centroids. In: IEEE-RIVF International Conference on Computing and Communication Technologies, pp. 1–6 (2019)
17. Hettiarachchi, H., Adedoyin-Olowe, M., Bhogal, J., Gaber, M.M.: Embed2Detect: temporally clustered embedded words for event detection in social media. Mach. Learn. **111**(1), 49–87 (2022)
18. Ahmad, F., Abbasi, A., Kitchens, B., Adjeroh, D.A., Zeng, D.: Deep learning for adverse event detection from web search. IEEE Trans. Knowl. Data Eng. **34**(06), 2681–2695 (2022)
19. Zhang, C., Lei, D., Yuan, Q., Zhuang, H., Kaplan, L., Wang, S., Han, J.: GeoBurst+ effective and real-time local event detection in geo-tagged tweet streams. ACM Trans. Intell. Syst. Technol. **9**(3), 1–24 (2018)
20. He, J., Liu, Y., Jia, Y.: EventGraph based events detection in social media. In: International Conference of Pioneering Computer Scientists, Engineers and Educators, pp. 150–160 (2018)
21. George, Y., Karunasekera, S., Harwood, A., Li, K. H.: Spatio-temporal event detection using poisson model and quad-tree on geo-tagged social media. In: IEEE International Conference on Big Data, pp. 2247–2256 (2019)
22. Wei, H., Zhou, H., Sankaranarayanan, J., Sengupta, S., Samet, H.: Delle: detecting latest local events from geo-tagged tweets. In: 3rd ACM International Workshop on Analytics for Local Events and News, pp. 1–10 (2019)
23. Abdelhaq, H., Sengstock, C., Gertz, M.: Eventweet: online localized event detection from twitter. VLDB Endow. **6**(12), 1326–1329 (2013)
24. Jurafsky, D., Martin, J.H.: Speech and Language Processing: An Introduction to Natural Language Processing, Computational Linguistics, and Speech Recognition (2000)
25. Hu, X., Zhou, Z., Sun, Y., Kersten, J., Klan, F., Fan, H., Wiegmann, M.: GazPNE2: a general place name extractor for microblogs fusing gazetteers and pretrained transformer models. IEEE Internet Things J. (2022)
26. Honnibal, M., Montani, I., Van Landeghem, S., Boyd, A.: spaCy: Industrial-Strength Natural Language Processing in Python (2020). https://doi.org/10.5281/zenodo.1212303
27. Li, C., Sun, A.,Datta, A.,. Twevent: segment-based event detection from tweets. In: 21st ACM International Conference on Information and Knowledge Management, pp. 155–164 (2012)
28. Girvan, M., Newman, M.E.: Community structure in social and biological networks. PNAS **99**(12), 7821–7826 (2002)
29. McMinn, A. J., Moshfeghi, Y., Jose, J. M.: Building a large-scale corpus for evaluating event detection on twitter. In: Proceedings of the 22nd ACM International Conference on Information & Knowledge Management, pp. 409–418 (2013)

30. Tromble, R., Storz, A., Stockmann, D.: We don't know what we don't know: when and how the use of Twitter's public APIs biases scientific inference. Available at SSRN 3079927 (2017)
31. Campan, A., Atnafu, T., Truta, T.M., Nolan, J.: Is data collection through twitter streaming API useful for academic research? In: 2018 IEEE International Conference on Big Data, pp. 3638–3643 (2018)

Manipulation During the French Presidential Campaign: Coordinated Inauthentic Behaviors and Astroturfing Analysis on Text and Images

Victor Chomel[1,2](✉), Maziyar Panahi[1], and David Chavalarias[1,2]

[1] Centre National de la Recherche Scientifique (CNRS), Institut des Systèmes Complexes Paris Ile-de-France (ISC-PIF), Paris, France
victor.chomel@iscpif.fr

[2] Ecole des Hautes Etudes en Sciences Sociales (EHESS), Paris, France

Abstract. In April 2022, the French presidential election took place, and social media played a prominent role in it. By analyzing more than 150 million interactions on French Twitter, this study aims to provide evidence of coordinated behaviors from political parties. We find that extreme parties left and right, appear with a particular internal structure compared to moderate parties. Moreover, by examining similar patterns in community structures but also in duplicated tweets, we unveil online astroturfing strategies of the main parties online, and in particular the extreme right.

Keywords: Computational social science · Misinformation · Disinformation · Astroturfing

1 Introduction

It is believed that social media exerts an increasing influence on the outcome of elections [3]. Indeed, the recent 2022 French presidential election anticipated social media to play a crucial role in shaping election results. During the political campaign season, political parties across the spectrum sought to impose their platform in the public discourse and mobilize their constituencies online in a coordinated, strategic manner. Our study proposes and validates a method to measure the scale of these online information campaigns by probing the coordination patterns and structure of online communities. For instance, in order to reach a larger audience, political campaigns may use coordination techniques that are considered *disinformation* [18]. One such case is posting numerous identical media instead of retweeting. Contrary to a retweet that explicitly references another user, this method masks the real coordination behind these tweets and can artificially enlarge the audience. These *astroturfing* methods can be detected and measured for each community. Astroturfing can be seen as an attempt to give the false impression that a message or organization enjoys widespread grassroots

H. Cherifi et al. (Eds.): COMPLEX NETWORKS 2016 2022, SCI 1077, pp. 121–134, 2023.
https://doi.org/10.1007/978-3-031-21127-0_11

support of the community when little such support exists. Due to astroturfing, some communities consistently push their narratives into trending topics for the entire country. With only a few hundred accounts, political parties manage daily to reach out to hundreds of thousands of accounts outside of their original community.

Related Work Misinformation and disinformation have gained momentum in public discourse during the last few years. A broad definition of misinformation is incorrect or misleading information. It is differentiated from disinformation, which is deliberately deceptive. Information spread regarding COVID-19 has accelerated both the production of disinformation and misinformation as well as their detection [9]. There is a growing literature that seeks to understand the roots of misinformation and disinformation, including evaluating their volume [11], determining fake news ecosystems [1], detecting misleading content [7], characterizing astroturfing [17], viewing activism as advertising [5], or disinformation as distraction [16]. The worldwide prevalence of astroturfing and coordinated inauthentic behaviors [14,24], especially in a political context [8], demonstrates a need to analyze the last French presidential election in this framework.

Notably, methods have been developed to uncover coordinated disinformation networks based on Twitter data [22]. These astroturfing campaigns [19] have sometimes been seen as related to foreign countries trying to undermine the trust in the political system [2] or promoting their authoritarian views [21]. Most of them are now cross-platform, duplicating content and creating narratives before coming to a large audience [13,20]. These coordination patterns can come from several potential coordination mechanisms such as several accounts run by the same person, a shared channel with instructions on how/when to share, or semi-organic behavior by users that rapidly amplifies target accounts.

Many papers analyze bot activity to answer the astroturfing problem [6,25,26]. However, analyzing disinformation using bot detection is being increasingly called into question [23]. Though bots participate in astroturfing campaigns, they do not represent the entirety of astroturfing activity [15]. These information campaigns are largely planned and relayed by true accounts and even accounts with an already substantial audience.

Most of these studies focus only on one factor such as time coordination [4], structure, or identical content. In this study, we propose to gather these solutions to show that, across different modalities (text, image, video), they all point in the same direction– that is, to the presence of coordinated information campaigns by online political communities.

1.1 Data

The study period starts in October 1st, 2021 and ends on the day of the second round of the French presidential election, April 24th, 2022, covering the French presidential electoral campaign. The data has been collected with the Twitter Public Stream API using the same methodology than Gaumont et al. [10]. Most of top French political accounts (Governments, Members of parliament, Members of European parliament, Senators, Mayors of big cities, etc.) from all political parties are followed with the API. There are more than 2000 selected accounts. Moreover, thousands of elections keywords such as names of candidates and parties are tracked. If a selected account tweets, retweets or quotes, of if it is retweeted, mentioned, quoted, the interaction is collected. If any message (tweet, retweet or quote) has a keyword in it, it is collected. About 160 million interactions (tweets, retweets, and quotes) have been collected over this period.

1.2 French Political Landscape

Political communities are defined based on the retweet graph. A retweet graph is a graph with users as nodes and retweets without quotes as edges. We compute the retweet graph over the whole period (from October 2021 to April 2022), then perform a clustering, using the Louvain algorithm, to which we assign political labels. In addition to the previous political communities, we define a "media community" in our clustering, consisting of users who primarily engage with major media outlets. Communities size and average degrees are reported in Table 1. In previous work, this method has been tested and had an averaged precision above 90%, so with few false positives [10]. During these seven months, the set of political communities remained stable and thus their internal structures can be studied.

Table 1. Main political communities with number of users and the average number of retweets per user

Political community with their leaders	Number of nodes	Average number of retweets per user over the whole period
Alt. Right—Dupont-Aignan	39,330	27.4
Ext. Left—Mélenchon	39,098	22.4
Center—Macron	33,306	19.2
Ext. RightZemmour	30,783	45.4
Right—Pécresse	11,341	18.7
Ext. Right—Le Pen	9060	21.2
Left—Hidalgo	5770	8.3
Green—Jadot	5582	9.4
Ext. Left—Roussel	3634	10.3
Media	22,848	4.1
Other	45,808	3.4

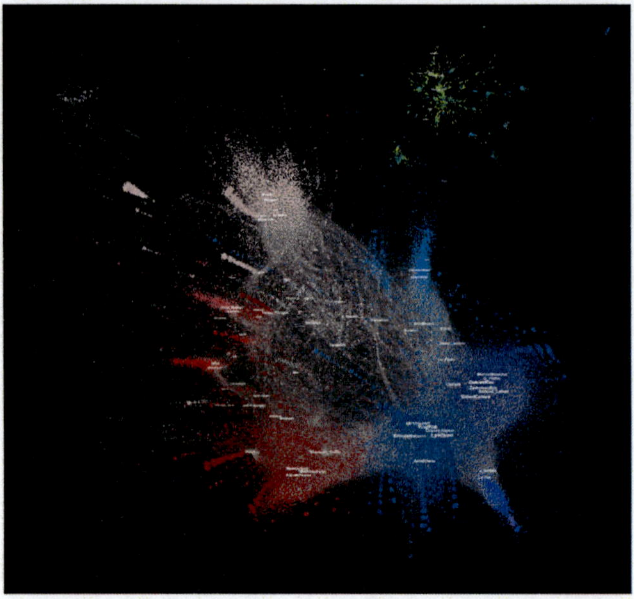

Fig. 1. Retweet graph of the French political debate—March 2022. Nodes are located with ForceAtlas2 in Gephi [12]. Ukraine community, on the top right in yellow/green, gathers users from Ukraine mentioning and retweeting Emmanuel Macron to support the war effort.

The French political landscape (Fig. 1) is composed of 9 online political communities, with their candidates in italics, divided into three main blocks:

- Left: Parti socialiste—*Hidalgo* (Left party, pink on the left of Fig. 1), EELV—*Jadot* (Green party, red on the left), Parti communiste français—*Roussel* (Extreme Left, red on the bottom left) and France Insoumise—*Mélenchon* (Extreme Left, red on the bottom)
- Center: République En Marche—*Macron* (Government, Light pink on the top)
- Right: Les Républicains—*Pécresse* (Right, light blue), Rassemblement National—*Le Pen* (Extreme Right, blue on the bottom right), Reconquête—*Zemmour* (Extreme Right, dark blue on the right) and an alt-right community including several parties (dark blue on the bottom).

2 Astroturfing Criterion and Definition of Communities

Intra-community Dynamics We first compare the extent to which each community engages its users. To do so, we measure the online social mobility of each community using a method based on Markov chains [27]. Each community's

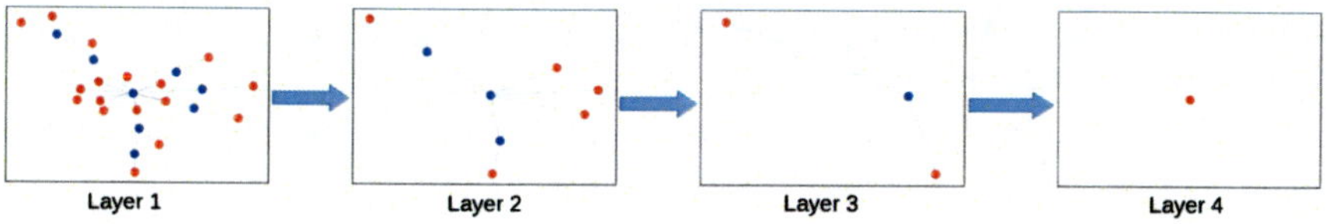

Fig. 2. Successive layers on a toy example. In blue are the remaining nodes for the next step. In red are the nodes in the layer.

users are categorized into layers of increasing importance. This is done by iteratively removing nodes with the lowest degree (Fig. 2). As this layer distribution is recomputed each two weeks, the layer of each node can be computed for all time steps. Then, the probability of transitioning between layers at a given time step may be computed for a given community. If a node leaves a community, it goes to layer 0. For each community, the transition matrices are then averaged over all time steps, from October 2021 to April 2022.

The probability matrix of moving from layer i to layer j may be formulated as the transition matrix of a Markov chain, with layers as the state. This allows us to capture the internal dynamics of communities, the extent to which users remain in a given community, and the extent to which users participate during the political campaign. This is summarized by the stationary distribution of the Markov chain.

Communities which experience high artificial tweet volume and coordinated campaigns get a particular type of signature in this method. To measure these coordinations, several other indicators are used such as co-retweet graphs, identical tweets, or identical images.

Co-retweet Graphs A co-retweet graph $\mathcal{G}_T$ is a graph where nodes are user accounts. Two nodes are linked if the two accounts retweet the same tweet within T seconds from each other, that is to say, they *co-retweet* each other. To capture changes in the network on an informative timescale T, we use small values of T because we are looking for coordination. The lower the T, the more there is coordination. The weight of an edge is defined as the number of co-retweets between two users that occur within a time difference of T. A co-retweet graph is an aggregation over a longer period, typically a week. As we seek to find clusters of nodes that are strongly connected– knowing that they all participate in the same coordinated information campaign– we excise all spurious or weak connections. The probability of two users retweeting the same content within a small amount of time is really low, it rarely happens twice, spurious connections are thus defined by edge weight ≤ 2.

Identical Tweets Campaign During the presidential campaign, a strategy used to give the impression of volume was to post numerous exactly identical messages without retweeting the original poster, which we will refer to as an *information campaign*. Users coordinated on private discussion but also on other platforms such as 4chan to publish the same content at the same time [29]. Another strategy is to have a trending hashtag although the tweets may originate from a very small group of coordinated users. In what follows, we will consider all coordinated campaigns with least five different users.

Coordinated Identical Images and Video Campaign Posted images and videos need not be exactly identical to be considered part of the same information campaign. For example, images may be cropped or otherwise slightly edited. Therefore, we expand our criterion for image "sameness" to include these altered images. We collect images and videos attached to tweets with at least one retweet in order to gather "relevant" content, or content that has been shared at least once. For the sake of simplicity, only the first frame of videos is kept and used as image. Due to the volume of the data, only the last two weeks before the presidential elections (28/03/22–10/04/22) and the two weeks between the two rounds of voting (11/04/22–24/04/22) are collected. About 100,000 images and videos are collected for each period.

The media are embedded as 512 dimensional vectors using a pre-trained ResNet18 from img2vec. For each image, cosine similarities with all other images are computed. Only almost identical images (similarity above 0.95) are kept. To understand information campaign coordination, we order images by their number of tweets (and not retweets) to see how many times they have been posted. For each image, we collect the posters and the users who retweet the image. The 100 most tweeted and retweeted images and videos were checked by hand to validate the method. On this small sample, the method has a precision of 1, meaning that all videos and images labeled as identical or near identical are correctly labeled (no false positives). Accuracy or recall requires to count false negatives which is extremely hard due to the size of the dataset.

3 Results

3.1 Intra-community Dynamics

Stationary distributions of user engagement for seven political blocs and the media can be found in Fig. 3. Here, the user engagement is defined as the ability of a party to mobilize its users and push them towards a higher online activity. The higher the layer, the more engaged users are. By taking a look at the values at $x = 0$, or "user attrition", one can see that the extreme parties have low attrition, thus more stable communities. This is confirmed by the percentage of supporters (value at $x = 1$). The only two anomalous cases (the right and

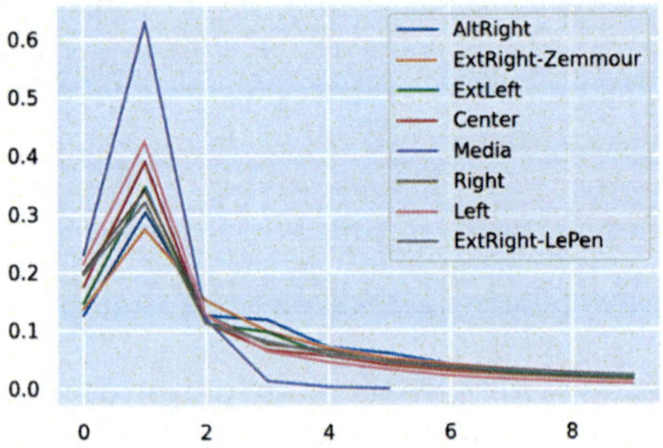

Fig. 3. Stationary probability of being in each layer by community.

Table 2. Level of political engagement quantified by the mean layer of the stationary distribution by community, ordered descending from left to right.

Community	ExtRight Zemmour	AltRight	ExtLeft	ExtRight Le Pen	Center	Right	Left	Media
Mean layer of stationary distribution	3.44	3.29	3.15	2.93	2.67	2.35	1.67	0.93

the extreme-right - Le Pen) reflect the fact that a substantial part of their supporters left to join Zemmour, reflecting higher attrition. As expected, the media community is mainly composed of users that retweet occasionally, leading to a high attrition and a high percentage of supporters.

The mean of the distribution can also be a good proxy to characterize the engagement (Table 2). The mean engagement of extreme parties is higher, while on the other hand, that of centrist parties and the media is lower.

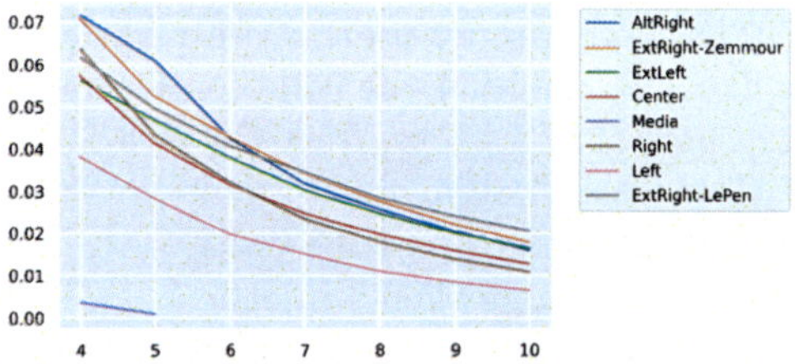

Fig. 4. Stationary probability of being in each layer by community. Zoom on high layers.

3.2 Co-retweet Graphs

We compute co-retweet graphs from the last part of the political campaign using several time differences T between retweets.

We quantify the level of coordination in online communities by computing a *coordination ratio*, defined as the number of nodes in the co-retweet graph divided by the number of nodes in the retweet graph by community. This value is plotted as a function of time between retweets (Fig. 5). This ratio captures coordination as it represents the share of the community engaged in coordination (measured by co-retweets), correcting the community size effect.

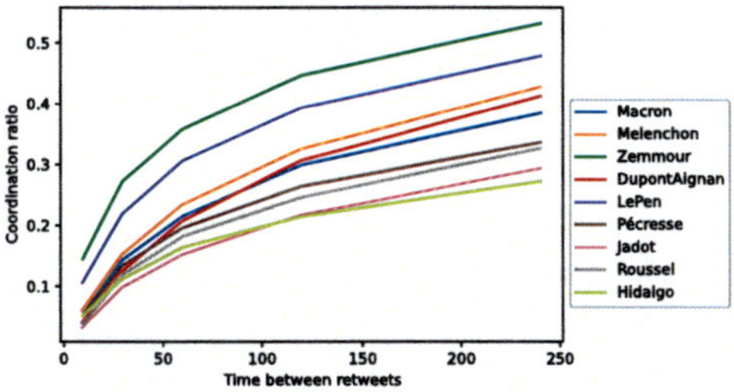

Fig. 5. Coordination ratio against the time between retweets

We find that, mirroring our results on levels of political engagement, extreme communities tend to have higher coordination ratios, with the main exception being the Alt-Right community. For lower timescales such as 30 or 60 s, the Alt-Right community appears less coordinated than the Center. This may be explained by the fact that it is not a well organised community but mainly a gathering of people coming for different reasons (vaccine criticism, conspiracy theories, sanitary pass opponents, etc.).

However, this ratio does not take into account the volume of the activity of the community. A large part of the network could be involved in a low-intensity coordinated campaign for example. The average degree in the co-retweet graph per user against the time between retweets can be found in Fig. 6.

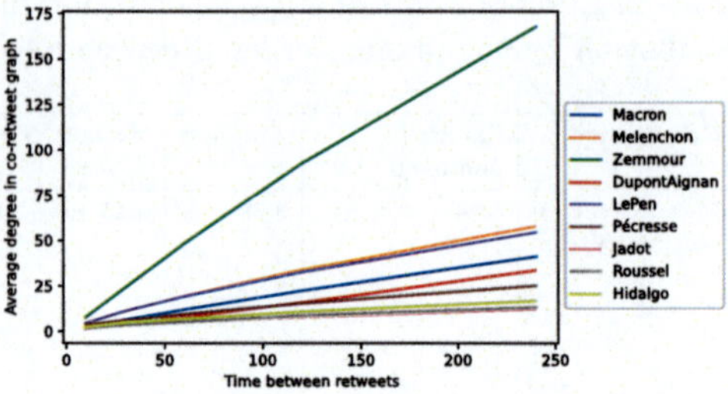

Fig. 6. Average degree in co-retweet graph by community against time between retweets

Based on these first indications and on previous work [24], it appears that the right value to detect the coordination (for example in the case of the Alt-Right) is at $T = 60$ s. The value is large enough to let users retweet but low enough to exhibit coordination. We fix $T = 60$ s and plot the coordination ratio and average degree, computed over snapshots of two weeks, from October 2021 to April 2022.

The coordination ratio and the average degree over time track political campaign events in each community (Figs. 7 and 8). For example, we see the main political parties appear to get increasingly coordinated with the incoming deadline of the first round ($t = 23$), then witness a dip in coordination ratio for several communities explained by their loss in the first round. Compare to the Center, headed by Macron, and the Extreme Right with Le Pen, whose coordination continue to increase to week 25. Moreover, the only primary election that occurred during the period was the Right party primary located in the beginning at the sixth week.

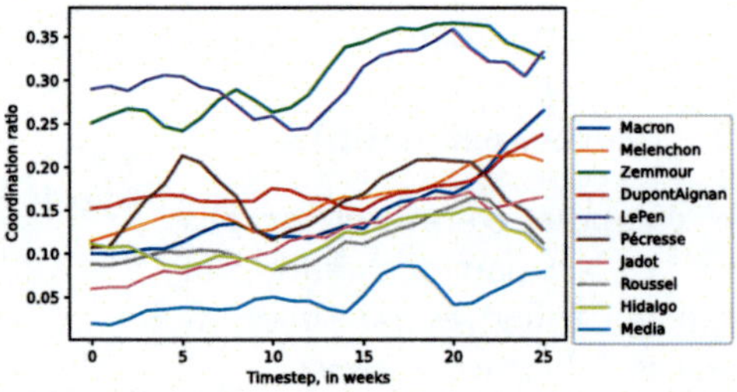

Fig. 7. Moving average (window = 3 time steps) of the coordination ratio as a function of timescale T in seconds (from Oct. 2021 to April 2022).

3.3 Identical Tweets

Most of the information campaigns are coming from abroad (Tigray war, Russo-Ukrainian war, etc.) but their influence on French public discourse is limited, as they do not come from users who are well-rooted in political communities. To measure the impact of campaigns,

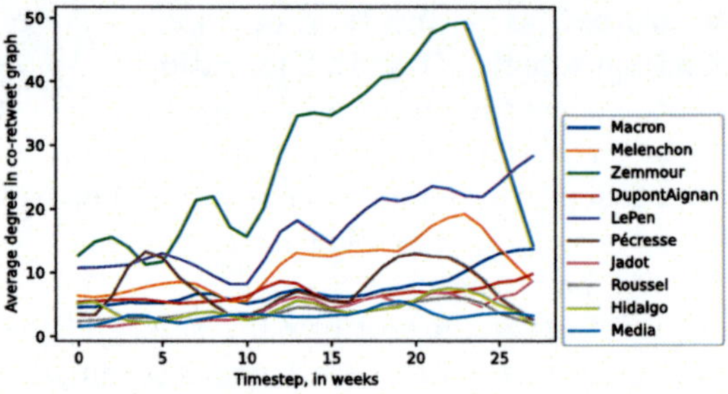

Fig. 8. Moving average (window = 3 time steps) of the average degree in the co-retweet graph against time steps (form Oct. 2021 to April 2022).

we thus rank users according to the number of campaigns they participate in. The political colors of each of them are computed based on the results of the clustering from the retweet graph. A limit of this analysis is that a user who, on purpose, only tweets and never retweets (for example a bot), would not be politically labeled but it can clearly support a candidate.

We remark that the extreme right community of Zemmour is over-represented in top users: all the top 100 users belong to this community. Among the top 500 users, 257 of them come from this community and 213 are not politically affiliated. The extreme right community behind Eric Zemmour highly push content and hashtags on a daily basis. Indeed, their intention was to break into trending topics as much as possible to reach users outside of their community.

Users posting same tweets can be seen as a graph with users as nodes and edges between users having shared at least a campaign. Two of such graphs have been computed with two thresholds on edge weights. Two users are linked if they share at least a defined number of campaigns. Many values have been tested that can be gathered in two cases. A low value (typically 2 or 3) leads to many users with weak ties. A higher value (typically 5 or higher) is useful to detect the core of users that always act together. The presented values are 2 and 5.

As one may expect, the resulting graph is a union of many small connected components. Most of them are related to an above mentioned political community or to a special issue such as the war in Ukraine.

Interestingly, we have found a large connected component (more than 1000 users) that doesn't belong to a unique political community, in the lower threshold settings. It gathers accounts that campaigns in favor of ecological issues and try to push users towards petitions or crowdfunding. By changing the threshold of the degrees from 2 to 5, the community disappears meaning that it gathers a lot of users participating to low intensity campaigns.

To verify if accounts in this community are labeled as bots, the Botometer [28] has been used. This tool gives a score between 0 and 1 to each account. The bot score distribution on this large component has a bimodal pattern as reported in the original paper but with slightly more bots. About one third of the accounts scores above 0.7 (probably bots) and half of them below 0.3 (not detected as bots). As expected, many bots belong to this community but surprisingly, about half of the accounts seems to, at least partially, be run by humans.

3.4 Identical Images

Images and videos are key vectors during a political campaign to communicate campaign goals, or push narratives. Media that are part of the campaign: meetings, posters or images of the candidates, are widely posted and shared intra-community. However, other images which, e.g. criticise the government, are shared extra-community. We compare who creates, repeatedly posts, and retweets such content and examine the communities of origin and dissemination. Images can be found in Appendix 4.

When examining information campaigns, e.g. the relationship between Putin and Le Pen, it appears that most posting users come from the center community

Table 3. Several main image-narratives developed during the campaign, ordered by the number of tweets.

Period	Information campaign	Tweets	Retweets
Two weeks before the first round	Campaign Zemmour	967	30,158
	Videos Zemmour	640	7223
	Macron McKinsey	582	5787
	State Violence	361	7621
	Putin with Lepen	314	5667
	Ecology from Macron	476	4410
Between the two rounds	Putin with Le Pen	433	14,645
	State violence	360	2566
	Putin with Macron	254	20,906
	Anti-elite images	71	11,963

but with retweets from non-affiliated nodes. Indeed, among the retweets, more than 60% of the retweets come from another community than center. This shows that the political center manages to take root in public opinion.

After the first round, the two remaining parties were the center of Macron and the extreme right of Le Pen. To be elected, both candidates attempted to garner the vote of the extreme left (around 20% of votes in the first round). To do so, Macron defended his performance on environmental issues during his last presidential term. Even if they posted hundreds of messages (Table 3), the candidate did not receive the expected response. Among the more than 4000 retweets, more than 95% of them circulated within the original community.

On the other hand, the extreme right launched a narrative around images of Macron with Putin as an answer to information campaign concerning Putin and Le Pen. Even though this narrative has been pushed by the extreme right parties, more than half of the retweets are outside of extreme right parties.

4 Discussion

Our examination of internal community dynamics over time reveals a complex picture of online political discourse. We detect and quantify coordination strategies used by communities (especially the extreme right), such as posting numerous identical tweets to lend the community artificial volume, strategies which have contributed to a never-before-seen level of *astroturfing* in French politics. To compare, when the same method for identical messages has been applied to the 2017 presidential election, only one tweet was identically tweeted more than 1000 times (compared to 26 in 2022), 2975 tweets were tweeted more than 5 times (26489 in 2022) and 1101 users participated to campaigns with at least 5 tweets (6507 in 2022).

Coordination, as it has been defined in this paper, has some limits. For example, several accounts tweeting the same image because it's coming from another platform can happen. Co-retweeting popular content isn't rare. To avoid these pitfalls, the accounts and the contents appearing the more "coordinated" according to the metrics have been manually checked. Almost all of them denoted an intention to disinform in some ways (fake accounts, repeatedly posting and deleting contents, extremely high activity, etc.). Another limit of this approach is that a counterfactual network without coordination can only be theoretical. The question of the expected level of coordination of an uncoordinated political community is open.

If election outcomes do not seem to reflect a candidate's online influence, especially for the extreme-right candidate Zemmour, it has pushed the boundaries of the extreme right sphere due to visibility given by inflated media coverage.

Acknowledgement. We would like to express our very great appreciation to Emily S. Cheng and Jeanne Bruneau–Bongard for their valuable and constructive suggestions during the writing of this research work. We would also like to thank reviewers for taking the time and effort necessary to review the manuscript and appreciate their valuable comments and suggestions for the discussion.

Appendix

Fig. 9. Example of repeatedly posted images within two narratives : the proximity of Marine Le Pen and Putin and the violence of the French police, especially during the Yellow Jackets movement

Fig. 10. Most posted images during the two weeks before the first round (left), between the two rounds (right). The first one is a manipulation of the Macron political campaign poster, with a caption about the McKinsey scandal. The second one is part of Macron's campaign against Le Pen.

References

1. Allen, J., Howland, B., Mobius, M., Rothschild, D., Watts, D.J.: Evaluating the fake news problem at the scale of the information ecosystem. Sci. Adv. **6**(14), eaay3539 (2020). https://www.science.org/doi/abs/10.1126/sciadv.aay3539
2. Bail, C.A., Guay, B., Maloney, E., Combs, A., Hillygus, D.S., Merhout, F., Freelon, D., Volfovsky, A.: Assessing the Russian internet research agency impact on the political attitudes and behaviors of American twitter users in late 2017. Proc. Nat. Acad. Sci. **117**(1), 243–250 (2020). https://www.pnas.org/doi/abs/10.1073/pnas.1906420116
3. Bright, J., Hale, S., Ganesh, B., Bulovsky, A., Margetts, H., Howard, P.: Does campaigning on social media make a difference? evidence from candidate use of twitter during the 2015 and 2017 U.K. elections. Commun. Res. **47**(7), 988–1009 (2020). https://doi.org/10.1177/0093650219872394
4. Cinelli, M., Cresci, S., Quattrociocchi, W., Tesconi, M., Zola, P.: Coordinated inauthentic behavior and information spreading on twitter. In: Decision Support Systems, p. 113819 (2022). https://doi.org/10.1016/j.dss.2022.113819
5. Enos, R.D., Hersh, E.D.: Party activists as campaign advertisers: the ground campaign as a principal-agent problem. Am. Polit. Sci. Rev. **109**(2), 252–278 (2015). https://doi.org/10.1017/s0003055415000064
6. Ferrara, E., Varol, O., Davis, C., Menczer, F., Flammini, A.: The rise of social bots. Commun. ACM **59**(7), 96–104 (2016). https://doi.org/10.1145/2818717
7. Ferreira, R.R.: Liquid disinformation tactics: overcoming social media countermeasures through misleading content. J. Pract. 1–22 (2021). https://doi.org/10.1080/17512786.2021.1914707

8. Freelon, D., Wells, C.: Disinformation as political communication. Polit. Commun. **37**(2), 145–156 (2020). https://doi.org/10.1080/10584609.2020.1723755
9. Gallotti, R., Valle, F., Castaldo, N., Sacco, P., Domenico, M.D.: Assessing the risks of 'infodemics' in response to COVID-19 epidemics. Nature Human Behav. **4**(12), 1285–1293 (2020). https://doi.org/10.1038/s41562-020-00994-6
10. Gaumont, N., Panahi, M., Chavalarias, D.: Reconstruction of the socio-semantic dynamics of political activist twitter networks-method and application to the 2017 French presidential election. PLOS ONE **13**(9), 1–38 (2018). https://doi.org/10.1371/journal.pone.0201879
11. Guess, A.M., Nyhan, B., Reifler, J.: Exposure to untrustworthy websites in the 2016 US election. Nature Human Beh. **4**(5), 472–480 (2020). https://doi.org/10.1038/s41562-020-0833-x
12. Jacomy, M., Venturini, T., Heymann, S., Bastian, M.: Forceatlas2, a continuous graph layout algorithm for handy network visualization designed for the gephi software. PLOS ONE (2014). https://doi.org/10.1371/journal.pone.0098679
13. K., S., Wilson, T.: Cross-platform disinformation campaigns: lessons learned and next steps. Harvard Kennedy School (HKS) Misinf. Rev. (2020). https://doi.org/10.37016/mr-2020-002
14. Keller, F., Schoch, D., Stier, S., Yang, J.: How to manipulate social media: analyzing political astroturfing using ground truth data from South Korea. Proc. Int. AAAI Conf. Web Social Media **11**(1), 564–567 (2017). https://ojs.aaai.org/index.php/ICWSM/article/view/14941
15. Keller, F.B., Schoch, D., Stier, S., Yang, J.: Political astroturfing on twitter: how to coordinate a disinformation campaign. Polit. Commun. **37**(2), 256–280 (2020). https://doi.org/10.1080/10584609.2019.1661888
16. King, G., Pan, J., Roberts, M.E.: How the Chinese government fabricates social media posts for strategic distraction, not engaged argument. Am. Polit. Sci. Rev. **111**(3), 484–501 (2017). https://doi.org/10.1017/s0003055417000144
17. Kovic, M., Rauchfleisch, A., Sele, M., Caspar, C.: Digital astroturfing in politics: definition, typology, and countermeasures. Stud. Commun. Sci. **18**(1) (2018). https://doi.org/10.24434/j.scoms.2018.01.005
18. Lazer, D.M.J., Baum, M.A., Benkler, Y., Berinsky, A.J., Greenhill, K.M., Menczer, F., Metzger, M.J., Nyhan, B., Pennycook, G., Rothschild, D., Schudson, M., Sloman, S.A., Sunstein, C.R., Thorson, E.A., Watts, D.J., Zittrain, J.L.: The science of fake news. Science **359**(6380), 1094–1096 (2018). https://www.science.org/doi/abs/10.1126/science.aao2998
19. Linvill, D.L., Warren, P.L.: Troll factories: manufacturing specialized disinformation on twitter. Polit. Commun. **37**(4), 447–467 (2020). https://doi.org/10.1080/10584609.2020.1718257
20. Lukito, J.: Coordinating a multi-platform disinformation campaign: internet research agency activity on three U.S. social media platforms, 2015–2017. Polit. Commun. **37**(2), 238–255 (2020). https://doi.org/10.1080/10584609.2019.1661889
21. Miller, B.A.P.: Automatic detection of comment propaganda in Chinese media. SSRN Electron. J. (2015). https://doi.org/10.2139/ssrn.2738325
22. Pacheco, D., Hui, P., Torres-Lugo, C., Truong, B.T., Flammini, A., Menczer, F.: Uncovering coordinated networks on social media. CoRR abs/2001.05658 (2020). https://arxiv.org/abs/2001.05658
23. Rauchfleisch, A., Kaiser, J.: The false positive problem of automatic bot detection in social science research. PLOS ONE **15**(10), 1–20 (2020). https://doi.org/10.1371/journal.pone.0241045

24. Schoch, D., Keller, F.B., Stier, S., Yang, J.: Coordination patterns reveal online political astroturfing across the world. Sci. Rep. **12**(1) (2022). https://doi.org/10.1038/s41598-022-08404-9
25. Stella, M., Ferrara, E., Domenico, M.D.: Bots increase exposure to negative and inflammatory content in online social systems. Proc. Nat. Acad. Sci. **115**(49), 12435–12440 (2018). https://www.pnas.org/doi/abs/10.1073/pnas.1803470115
26. Stukal, D., Sanovich, S., Bonneau, R., Tucker, J.A.: Detecting bots on Russian political twitter. Big Data **5**(4), 310–324 (2017). https://doi.org/10.1089/big.2017.0038
27. Chomel, V., Panahi, M., Chavalarias, D.: A Markov chain based method to model the dynamical properties of misinformation networks—application to climate change on twitter. In: Proceedings of Disinformation in Open Online Media: Second Multidisciplinary International Symposium (2022)
28. Yang, K.C., Ferrara, E., Menczer, F.: Botometer 101: social bot practicum for computational social scientists (2022). https://arxiv.org/abs/2201.01608
29. Zannettou, S., Caulfield, T., De Cristofaro, E., Sirivianos, M., Stringhini, G., Blackburn, J.: Disinformation warfare: understanding state-sponsored trolls on twitter and their influence on the web. In: Companion Proceedings of The 2019 World Wide Web Conference, pp. 218–226. WWW '19, Association for Computing Machinery, New York, NY, USA (2019). https://doi.org/10.1145/3308560.3316495

Modeling Human Behavior

Lexical Networks Constructed to Correspond to Students' Short Written Responses: A Quantum Semantic Approach

Ismo T. Koponen[1(✉)], Ilona Södervik[2], and Maija Nousiainen[1]

[1] Department of Physics, University of Helsinki, Helsinki, Finland
ismo.koponen@helsinki.fi

[2] Centre for University Teaching and Learning (HYPE), University of Helsinki, Helsinki, Finland

Abstract. A simple method to construct lexical networks (lexicons) of how students use scientific terms in written texts is introduced. The method is based on a recently introduced quantum semantics generalization of a word-pair co-occurrence. Quantum semantics allows entangled co-occurrence, thus allowing to model the effect of subjective bias on weighting the importance of word co-occurrence. Using such a generalized word-pair co-occurrence counting, we construct students' lexicons of scientific (life-science) terms they use in their written responses to questions concerning food chains in life-science contexts. The method allows us to construct ensembles of lexicons that probabilistically simulate the variability of individual lexicons. The re-analyses of the written reports show that while sets of top-ranking terms contain nearly the same terms irrespective of details of the method used to count co-occurrences, the relative rankings of some key-terms may be different in quantum semantic analysis.

Keywords: Lexical networks · Quantum semantics · Education

1 Introduction

Science learning requires from students adoption of specific type of scientific language, which has its own vocabulary, semantics and syntax; the language of science, as some researchers call it (see e.g., [1]. Learning the language of science, its terms and their correct use is central in the process of learning science and scientific thinking [2,3] as well as in building scientific claims and communicating scientific knowledge [4]. To learn the language of science, students must learn a kind of new lexicon (i.e. lexical network) consisting of specific words, terms

H. Cherifi et al. (Eds.): COMPLEX NETWORKS 2016 2022, SCI 1077, pp. 137–149, 2023.
https://doi.org/10.1007/978-3-031-21127-0_12

and concepts and understand, for example, new phenomena by using that new lexicon. Consequently, to study how language of science is learned, constructing lexicons of students' use of scientific vocabulary and terms in a written text is an important problem in many settings of educational research [5–7].

In this study, we introduce a simple method to investigate how students use scientific terms in written texts, i.e. how they master the basic vocabulary of scientific language. Such a method is needed for three reasons. First, students' written answers are often too short and sparse to allow individual responses being analysed, in which case only aggregated answers yield to analysis. Second, we need an approach that allows us to estimate the extent of individual variations in lexical networks. Third, in educational applications simple enough methods for text analysis are needed because advanced methods of text analysis [8–11] require expertise and are thus not likely to be adopted.

The method introduced here is designed to meet these three demands. The method is based on the recently introduced quantum semantic approach allowing entanglement -kind interdependence of word-pairs as a generalization of word-pair co-occurrence counting, and thus, allows to model a kind of subjective bias in word co-occurrence [12–14]. Such a subjective bias refers to a possibility that different readers or writers of the text may lay a different emphasis on co-occurrence of words in a sentence regarding meaning of words (for a more detailed discussion, see refs. [12–14]). Due to the possibility of taking into account subjective bias we are able to simulate probabilistically the range of individual variations of lexicons corresponding short texts. Consequently, this study contributes to developing a simple method to investigate students' lexicons of language of science for educational applications.

To demonstrate the viability of the method we use as an example first-year life science students' written answers to tasks related to the role of photosynthesizing plants in the ecosystem from the viewpoints of food chains. The same topic was investigated in our previous study [5], where we utilized more conventional network methods [7] to analyze the students' responses but only in the form of an aggregated lexicon. In the previous study [5], the analysis was hampered in many ways by the sparsity of texts given as responses; each text contained only few key-words and sentences were too short to allow detailed analysis to take sentence structure into account and perform co-occurrence counts (in short sentences, co-occurrences are too rare). Therefore, it was possible to analyze only an aggregated lexicon, aggregating about 100 individual texts. Aggregation, however, always creates artefactual connections, and it is difficult to estimate how severely such artefacts bias inferences about the significance of key-terms.

The approach presented here is still based on aggregation of lexicons, but now utilizing quantum semantics [12–14], which now allows us to to estimate variability in linking key-terms in individual lexicons, owing to possibility of entanglement as a model of subjective bias. Consequently, we can then construct ensembles of lexicons that probabilistically simulate the variability of individual lexicons. The re-analysis of the written reports studied previously [5] show that while sets of top-ranking terms contain nearly the same terms irrespective of details of the method to count co-occurrences, the relative rankings of some key-terms may be different in quantum semantic analysis.

2 Methods and Materials

The data analyzed here comes from our previous study [5], where the written texts analyzed were short answers by first-year university students' of life sciences to questions regarding photosynthesis and its role in food chains in ecosystems, a topic which is of major importance in studying life sciences. Data were collected using open-ended questions that required application of basic conceptual knowledge. The tasks and the instruction were: explain the role of plants in the ecosystem from the food chain's point of view. The tasks entailed an understanding of basic biological phenomena, the most important being photosynthesis. The participants were 150 first-year university-level life science students. The answers to the questions were short and restricted to six lines. Of all answers 100 were long enough for analysis. For this study, no new data were collected; only the data from a previous study [5], made available in fully anonymized, transcribed and lemmatized form is used here.

Other details of the task, data collections, and issues related to it of the educational aspects and uses of the topic are reported elsewhere [5] and are not of further interest here. In what follows, we focus on developing and discussing the methods of analysis that can be used in analyzing sparse texts in educational settings.

2.1 Word Co-occurrence Counts and Concurrence

The first step in transforming the written texts into semantic networks, in the form of lexicons of terms, consisted of splitting the sentences into clauses, in order of their appearance, and after that, recognizing the nouns and within them, term-like words. Note that sentence structure was preserved, instead of using fixed lengths of text sequences as is often done in co-occurrence counting. Term-like words that appeared only once were discarded, and from the remaining set, about 70 were chosen for closer attention. The set of potentially interesting key-words were first selected as basis of their frequency of appearance and second, on basis of their importance for the topic in question. Words and terms deemed to be irrelevant or too commonplace of being further interest we discarded (for details, see [5]). Consequently, the selection process is quite simple and misses many finer points, but contains basic and robust elements commonly identified in analyzing speech and writing.

The second step consisted of finding the co-occurrence statistics for the 70 selected terms. To do so, we utilized quantum semantics approach [12–14], which can be considered a generalization of contingency based analysis of co-occurrence and in particular, as related Yule's Y-factor as a measure of contingency [15,16] (see also Appendix A).

To obtain concurrence Q (degree of entanglement) between words A and B in a text, we count four different frequencies of co-occurrence: the frequency n_{11} that A and B both occur in a given block of clauses at least once, n_{00} that neither A and B occurs, n_{10} that A occurs at least once but B does not occur, and n_{01} that A does not occur but B occurs at least once. With these frequencies, the concurrence Q as a measure of entanglement is given as [12,13].

$$Q = Q_0\sqrt{(1-\Theta R)}\,,\quad 0 \le Q \le 1\,, \tag{1}$$

where $-1 \le \Theta \le 1$ is the (compound) phase factor (see Appendix A for details and derivation), is taken here as a free parameter to account for a kind of subjective bias' [12–14]. In that, the subjective bias refers to various subjective ways to read meanings into the co-occurrence of terms A and B, i.e. to emphasize importance of co-occurrence differently [12–14]. The prefactor $Q_0 = 2\sqrt{n_{01}\,n_{10} + n_{00}\,n_{11}})$ corresponds to concurrence obtained with $\Theta = 0$ and takes into account the marginals of conditional probabilities $n_{01}\,n_{10}$ and $n_{00}\,n_{11}$ (note that rarely occurring pairs are here of no interest). The factor $R = 2\sqrt{n_{11}\,n_{00}\,n_{10}\,n_{01}}/(n_{11}\,n_{00} + n_{10}\,n_{01})$ is the ratio of the geometric mean of frequencies $n_{11}\,n_{00}$ and $n_{10}\,n_{01}$ to their arithmetic mean. The entanglement is now possible only in cases $R \neq 0$.

2.2 Constructing Lexicons

The concurrence $Q(i,j) = Q(j,i)$ of all pairs of terms i and j is obtained by using Eq. (1) for all pairs of terms of interest in students' responses (aggregated). The pairwise values $Q(i,j) = Q(j,i)$ are then used to form the weighted (symmetric) adjacency matrix Q with elements $[\mathrm{Q}]_{ij} = Q(i,j)$ describing the connectivity of terms in the lexicon.

Aggregated lexicon is then used as to generate individual lexicons. The existence and concurrence of links between the terms in an individual lexicon is modelled using parameter Θ as a random variable, to create three cohorts of interest: cohorts corresponding to low $(0.5 \le \Theta \le 1.0)$, high $(-1.0 \le \Theta \le -0.5)$ and averaged values $(-1 \le \Theta \le 1)$, respectively, to create ensembles of possible model lexicons. The model lexicons contain thus the same terms (nodes) as the individual texts but connections are predicted on the basis of aggregated lexicon by using random factors Θ. These ensembles provide then information of bounds of variation allowed in individual lexicons even if co-occurrence frequencies n_{01}, n_{10}, n_{00} and n_{11} are fixed.

2.3 Finding Key-Terms

A correlation matrix describing how nodes are correlated can be introduced by utilizing an exponential adjacency kernel (matrix exponential transformation) [17] to define a correlation matrix $\mathbf{G}$ of the form

$$\mathbf{G} = \mathbf{D}^{-\frac{1}{2}}\ \exp[\beta\mathbf{Q}]\ \mathbf{D}^{-\frac{1}{2}}\,, \tag{2}$$

where $\mathbf{D} = \mathrm{Diag}[\exp[\beta\mathbf{Q}]]$ is a diagonal matrix. Here, correlations refer to"positions" of nodes in the network, position meaning how through different paths of different lengths nodes can be reached; the more easily the nodes can be reached from each the larger their correlation (for a detailed discussion, see ref. [17]).

The elements of $[\mathbf{G}]_{ij}$ of the correlation matrix can be shown to be directly related to the covariance of values of nodes i and j in the network [17], providing a way to find the key-nodes based on correlations. Towards this end, we define a correlation centrality as an off-diagonal sum

$$\Gamma_i = \sum_{j \neq i} [\mathbf{G}]_{ij} , \tag{3}$$

which closely resembles communicability centrality [18,19]. The correlation centrality is used here as a basis to define key-terms and their rankings.

2.4 Similarity Comparison

The similarity of lexicons can also be operationalized based on correlation centrality Γ_i. Forming for lexicons L and L' centrality vectors $\bar{\Gamma}$ and $\bar{\Gamma}'$ consisting of centralities Γ_i and Γ_i' of nodes, respectively, we can define the similarity S_0 as a dot-product [20,21]

$$S_0[\mathrm{L}||\mathrm{L}'] = \bar{\Gamma} \cdot \bar{\Gamma}' , \tag{4}$$

which projects the centrality vectors to each other. For comparison we use normalized similarity (called cosine-similarity) given by $S_N[\mathrm{L}||\mathrm{L}'] = S_0[\mathrm{L}||\mathrm{L}']/(|\bar{\Gamma}|\,|\bar{\Gamma}'|)$, with range limited to $0 \leq S_N \leq 1$. The similarity S_0 takes into account the size of the lexicons, while S_N does not depend explicitly on size; high similarity lexicons have high values of both similarity measures.

3 Results

The frequency of occurrences of words and terms in a text is the simplest thinkable measure characterizing the lexicons. An obvious step further is to perform co-occurrence counts of word pairs, to be used as a basis to construct lexicons, where co-occurrence correlations are represented as links between pair of words. Such a lexical network of connections constructed as an aggregate of connection in all sentences in 100 written sparse texts is shown in Fig. 1. The size of the nodes is proportional to correlation centrality as defined by Eq. (3). The network shown is for a moderate entanglement with $\Theta = -0.5$. Symbols denoting terms are explained in Table 1.

In the aggregated lexicon shown in Fig. 1, the six most important terms are: "plant" (P), "food chain" (Fc), "energy" (E), "producer" (p), "organism" (O) and "nourishment" (N) (for other terms, see Table 1). Different choices for factor

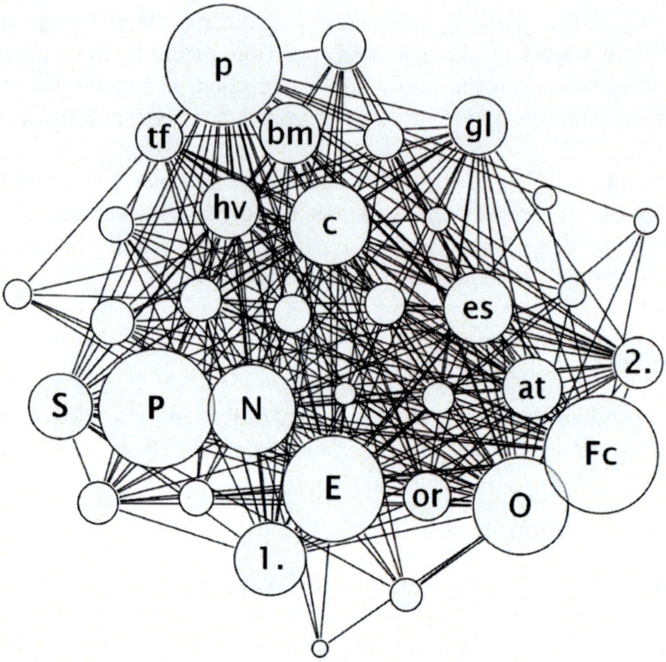

Fig. 1. The aggregated lexicon (lexical network) of key-terms in students' responses to questions about the role of plants in the ecosystem from the viewpoint of the food chain. The symbols denoting key-terms are explained in Table 1. The links that exceed a threshold-value of 0.30 are shown. The size of the nodes correspond to their correlation centrality as defined in Eq. (3).

Θ produce different entanglements, and thus different rankings, but five of the six top terms are always present in the top-five cohort.

The texts analyzed here are sparse, with quite short sentences (mostly statements), so that formation of lexicons corresponding to an individual text is in most cases impossible. However, the aggregated lexicons shown in Fig. 1 now provide information on how probable a given connection is in the whole set of individual texts. Note, however, that only a fraction of the terms might be present in individual texts. To predict the existence of connections between the terms in an individual lexicon, we use random phases, distributed into three cohorts corresponding to low values of concurrence (Low=L, with $0.5 \leq \Theta \leq 1.0$, high values of concurrence (High=H, with $-1.0 \leq \Theta \leq -0.5$, and total range (Total=T, with $-1 \leq \Theta \leq 1$, to estimate the individual variations in strength of links. Only links exceeding the threshold value Q^{*}=0.30 are included in final lexicons. It should be noted that the results remain nearly intact if threshold values lower than 0.30 are used because most below that threshold have very low values $Q < 0.2$, and thus, do not affect the correlation centrality in any essential way.

For the constructed individual lexicons, the importance of nodes and their rankings is based on correlation centrality as defined in Eq. (3). Lexicons then appear quite different, although it will be seen that many of them have about six to eight common terms, which have a high correlation centrality (i.e. key terms). Figure 2 shows eight lexicons that occur quite frequently as similar to other lexicons. It is obvious that it is difficult to compare the lexicons by visual inspection or to figure out how similar or different they are, or how key-terms appear in the set of 100 analyzed here. Nevertheless, Figs. 1 and 2 provide an overall idea of the appearance of the individual lexicons.

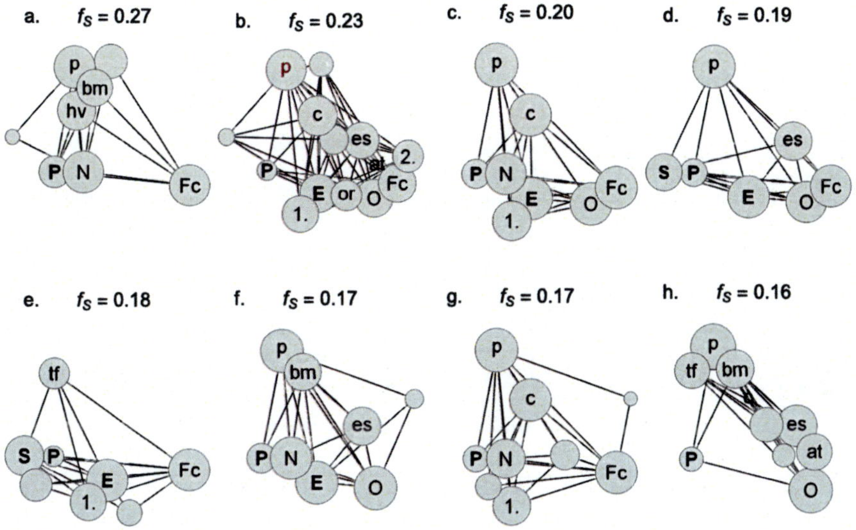

Fig. 2. Eight lexicons (from a to h) that appear most often in high similarity pairs, exceeding threshold $S^* = 0.55$. The frequency f_S of their appearance in similarity pairs is denoted.

To get an idea of importance of key-terms in the individual lexicons, we form rankings on the basis of correlation centrality Γ_k of a given term k in each individual lexicon, and then, find out how often the terms appear in the top-five, -ten and -15 cohorts of key-terms. Obviously, if a certain term appears in the top-five cohort in many individual lexicons, it is a key-term that deserves attention; the same applies, to a more moderate degree, to those terms that appear in the top-ten and -15 cohorts most often. Table 1 shows the rankings of key concepts, according to the different measures used as the basis for rankings.

The simplest measure is the frequency of occurrence (f) based ranking R_f. On the basis of the frequency of their appearance, the highest ranking concepts are, "plant" (occurs 208 times in 100 lexicons, corresponding to relative frequency f=1.0), "food chain" (f=0.45), "energy" (f=0.44), "producer" (f= 0.38) and "organism" (f= 0.28). These concepts are relevant in explaining the role of

Table 1. The key-terms and their rankings R_X appearing in students' responses. The subscripts X denote rankings based on: frequencies (f); top-15 terms for total range of phases $-1 \le \Theta \le 1$ as a stable reference value (r) and three cases of top-5 terms, which correspond to cohorts corresponding to total range (T), low quantile (L) and high quantile (H), as explained in the main text. Acronyms of terms in column A are as in Figs. 1 and 2

Term	A	R_f		R_r		R_T	R_L	R_H	Term	R_f	R_r	R_T	R_L	R_H
Plant	P	1	1.0	1	0.96	7	3	16	Decomposer	18	18	18	17	17
Food chain	Fc	2	0.45	3	0.70	2	2	2	Oxygen	19	22	20	23	22
Energy	E	3	0.44	4	0.56	3	6	3	Inorganic	20	23	23	20	25
Producer	P	4	0.38	2	0.71	1	1	1	Solar energy	21	20	16	14	20
Organism	O	5	0.28	5	0.47	4	5	4	Sugar	22	25	24	27	21
Nourishment	N	6	0.23	6	0.40	6	7	5	Photosynthesis	23	19	21	26	15
Consumer	C	7	0.18	7	0.34	5	4	6	Carbon dioxide	24	29	29	29	27
Ecosystem	Es	8	0.12	9	0.26	9	9	7	Carbon	25	–	–	–	–
Autotrophic	At	9	0.12	12	0.22	11	12	12	Nutrition	26	–	–	–	–
The first	1	10	0.11	8	0.27	8	8	8	Heterotrophic	27	24	19	15	23
Glucose	Gl	11	0.11	14	0.19	13	16	11	Food source	28	32	–	–	–
Biomass	Bm	12	0.10	11	0.22	14	19	9	Animal	29	26	27	21	–
Sun	S	13	0.10	10	0.24	12	11	13	Carnivore	30	–	–	–	–
Herbivore	Hv	14	0.09	13	0.20	10	10	10	Life	31	34	–	–	–
Organic	Or	15	0.08	17	0.14	17	18	14	Energy source	32	–	–	–	–
The second	2	16	0.07	15	0.15	15	13	19	Primary prod	33	27	–	–	–
Trophy level	Tf	17	0.06	16	0.14	22	24	18	Predator	–	31	28	28	-

plants as primary producers in the biosphere. In addition, concepts referring to systems, like "ecosystem", "the first" and "the second" (referring to levels in food chain) are common in the texts. However, for example, the occurrence of the concept of "photosynthesis" is relatively rare. Furthermore, concepts such as "the Sun" or "light energy", that refer to the origin of the energy, received less attention than expected.

The rankings of terms in Table 1 as based on values of correlation centrality as defined in Eq. (3) and calculated on basis of concurrence provide a picture somewhat different from frequency based rankings. The lowest value of concurrence Q obtained for $\Theta = 1$ is roughly proportional to classical contingency as measured by Yule's Y-factor multiplied by a factor Q_0 (see Appendix A) corresponding to correlations that are less affected by entanglement. The maximal co-occurrence with maximal entanglement is obtained with $\Theta = -1$. However, now we do not have enough information of individual lexicons, and thus, we generate ensembles of possible individual lexicons by letting parameter Θ vary. These ensembles of lexicons provide bounds for variance of diversity of connections but preserve the average frequency counts of corresponding to aggregated lexicons. With the ensemble of possible individual lexicons, we can estimate the ranking of terms based on correlation centrality by counting how often a given concept appears among the cohorts of top-five, -ten or -15 terms within the ensembles Low (L), High (H) and Total (T). Rankings based on these ensembles are summarized in Table 1 for top-5 cohort denoted by R_X, with $X \in T, H, L$ and for top-15 cohort for ensemble T, which is a reference (denoted by subscript r) most closely related to frequency-based counting (i.e. containing low correlations due to co-occurrence) but selective about the top (15) key terms .

From the results in Table 1, we see that some key-terms have different rankings from the frequency-based rankings. For example, in ensemble H, the highest ranking term is now "producer" and the term "plant" has a much reduced ranking, down to R_{H}=16 from $R_{\mathrm{f}} = 1$. As expected, however, among an extensive enough cohort of top-15 terms ensemble, the term "plant" has rank 1 as in frequency-based ranking. This shows that while the term "plant" appears quite often, its entangled connections to other terms are weak, i.e. it appears in trivial ways in many sentences (e.g. as parts of lists, not in more complicated correlative connections). Such a conclusion is in agreement with results obtained previously, where "plant" is often connected to auxiliary other terms. Therefore, in the current analysis, with increased weight on entanglement, its importance and rankings are diminished. Due to entanglement based correlations, "producer" is in the topmost ranked term in the top-5 ensemble. This can be interpreted as "producer" being a kind of cohesion bringing term, connecting several other terms in a network through different sentences, not only through unstructured statements or lists. Some other terms, like "photosynthesis", "biomass", and "herbivore" also have increased rankings in comparison to frequency-based rankings. However, many terms retain their rankings as they appear in frequency-based analysis.

Finally, it is interesting to compare the similarity of lexicons. Figure 3 shows the distribution (in the form of a box-whisker plot) of normalized similarities S_N plotted against unnormalized similarities corresponding to values $S_0 = 1, 2, 3, \ldots, 7$ averaged (binned) over a range ±0.5. The values of S_0 correspond roughly to a number of important nodes, which is at most about 6–7 in the most extensive individual lexicons. As seen from the box-whisker plots, values S_N and S_0 correlate, but S_0 is a better indicator for comparisons of similarities, because S_N has wide bounds of variation for most values of S_0. The box-whisker plots show cases obtained for cohorts Low (L), High (H) and Total (T). The corresponding histograms for probability density distributions p are shown in the lower row, in Fig. 3d–f. It is seen that Θ has a clear and significant effect on distributions of similarity values, high values of Q corresponding to substantial increased similarities. These results can be compared with Fig. 2 displaying eight examples of lexicons which appear most often in pairs of high similarity lexicons.

In the results shown in Fig. 3 the high values of Q (i.e. corresponding to increased correlations due to entanglement) are responsible of increased similarity of lexicons for cohort High (H) as seen in Fig 3c and in Fig 3f, where peaks in histogram of similarity distributions is due to such terms. This means that the terms with large values for correlation centrality are globally the most important, overall coherence-producing terms for all lexicons, and thus, the most important terms for learning language of science. In particular, term "producer"

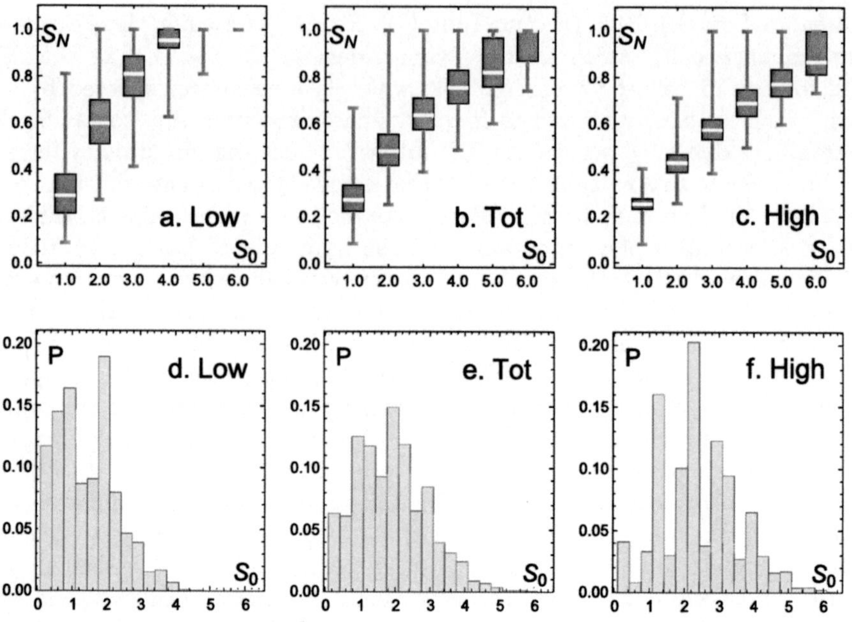

Fig. 3. Distributions of cosine- similarity values S_N and unnormalized similarity values S_0 for all lexicon pairs. The upper row shows a box-whisker plots of distribution of (normalized) cosine-similarities S_N when unnormalized values S_N, fall in the range $k - 0.5 < S_0 < k + 0.5$, with $k = 1, 2, 3, \ldots, 6$. Note that 99 % of values of S_0 are less than seven. Distributions are shown for cases corresponding to lowest (Low) and highest (High) quantiles of phase factors Θ, to be compared with values obtained for total range (Tot) of phase factors. The distributions are shown as standard box-whisker plots. Corresponding histograms of distributions of S_0 are shown in the lower row in the form of probability density p

attains then attains the most important role instead of "plant", which appears as the most important concept in frequency-based analysis as well as in traditional analysis ignoring entanglement (compare also results reported in ref. [5]). It is of importance that quantum sematics provides now a completely different picture of the importance of this ter. On the other hand, results obtained for low values of Q, where entanglement reduces correlations, do not differ significantly from frequency-based rankings nor from cases corresponding to total range of values (Tot), which minimize all correlations.

4 Discussion and Conclusions

The quantum semantics approach to constructing lexical networks (lexicons) that correspond to short written texts, with sparse co-occurrence of key terms, was applied to a sample of 100 students' texts about a topic in life -science (specifically, role of plants in food chains). The method allowed us to use entanglement to model the effects of individual subjective bias in weighting the importance of co-occurrence of key terms in the text and thus, to use aggregated lexicons as a starting point to generate model ensembles of individual lexicons. Previously, with the same sample of texts, such analysis of individual lexicons was not possible due to the shortness of the texts, when co-occurrences of words in a given text are rare.

The results are mostly in line with previous analysis as well as with basic occurrence frequency counts. Nevertheless, the generalized co-occurrence counts and lexicons based on them showed some interesting and important differences. The key-terms that appear to be most central on the basis of their frequency of appearance turn out to not always be central in co-occurrence analysis. When the co-occurrence correlations correspond to maximal entanglement, the changes in rankings of key-terms can be significant. For example, the most frequent term ("plant") drops to a lower-ranking position and another term, not so frequently occurring ("producer") but more connected (entangled) in the lexical network takes the highest ranking position. The cohorts of top-five, -10 and -15 terms remain nearly the same but within the cohorts, changes in rankings may take place. The key-terms are identified as terms, which attain high values of concurrence and which are shared by many lexicons (i.e. lexicons having high similarity). Such terms are supposedly also the most important ones for learning the language of science.

The method introduced here is meant to be simple enough for practical applications in educational research, where advanced and complicated methods of text analysis are not likely to be widely adopted. However, it is useful to be able to go beyond frequency-counts and estimate co-occurrence of word-pairs and to construct lexical networks. The case study presented here suggests that the generalized method introduced here is suitable for such a purpose. The ability to estimate such variations is useful for educational applications, as a first step to give an idea of how the appearance of terms in a text may be related to different ways to read the texts.

Appendix A

In quantum semantics [12], word co-occurrence is described as a two-state event: words A and B may both occur at least once in a block of text, only one of them might occur (at least once), or neither A or B occurs. Occurrence is described by tag 1 while non-occurrence is described by 0. Co-occurrence can be then coded as pairs $|00\rangle$ (neither A nor B occur), $|01\rangle$ (A does not occur but B occurs), $|10\rangle$ (A occurs but not B), $|11\rangle$ (A and B both occur). These are the basic building blocks of the qubit states of two two-state systems [12,13]. The frequencies of each of them, found by counting co-occurrence, are denoted by n_{00}, n_{01}, n_{10} and n_{11}, respectively. We can construct a superposition of the two qubit states, corresponding to all potential ways to build up the word meaning as it is realized in different sentences [12]

$$|\Psi\rangle = c_{00}|00\rangle + c_{01}|01\rangle + c_{10}|10\rangle + c_{11}|11\rangle \tag{A.1}$$

where amplitudes $c_{ij} = \sqrt{n_{ij}}\, e^{i\phi ij}$ are complex numbers related to real valued frequency n_{ij} with normalization $\Sigma n_{ij} = \Sigma|c_{ij}|^2 = 1$ and also including phase factors ϕ_{ij}. The phase factors ϕ_{ij} can be taken as parameters that describe various subjective ways to read meanings into the co-occurrence of terms A and B, i.e. to emphasize the meaning of connection differently [12–14]. The way to model subjective bias through entanglement is a first step to overcome a situation, where co-occurrence frequency is completely fixed and thus, needs to be associated with only one possible way to contribute on the meaning of words through their occurrences in sentences.

Factorization of state $|\Psi\rangle$ in Eq. (A.1) into independent marginal states of one qubits is not possible in general; the state is said to be entangled. Different entanglements then correspond to different subjective, biased readings of the meaning of co-occurrence. The degree of entanglement can be quantified as a difference between the factorized state and the entangled state, called a concurrence Q. Here, we omit the details of the derivation (lengthy but straightforward) of concurrence, to be found elsewhere [12] and state the final result

$$Q = 2|c_{01}c_{10} - c_{00}c_{11}|, \quad 0 \leq Q \leq 1 \tag{A.2}$$

Substituting in Eq. (A.3) the amplitudes in Eq. (A.2) we finally get [12]

$$Q = 2\sqrt{n_{01}n_{10} + n_{00}n_{11}}\sqrt{(1 - \Theta\frac{\sqrt{n_{01}n_{10}n_{00}n_{11}}}{(n_{01}n_{10} + n_{00}n_{11})/2}}, \tag{A.3}$$

where $\Theta = \cos(\phi_{01} + \phi_{10} - \phi_{00} - \phi_{11})$ is the compound phase factor, taken here as the free parameter to account for (unknown) phase factors. The concurrence in Eq. A.2, using abbreviations Q_0 and R for the prefactor and fractional term, leads then to expression$Q = Q_0\sqrt{(1 - \Theta R)}$ of Eq. (1).

An analogy of concurrence to a so-called mean square contingency factor and Yule-factors [15,16] is interesting to note [12]. The mean square contingency (closely related to the so-called odds -ratio) is proportional to a factor $n_{11}n_{00} - n_{01}n_{01}$, to be compared with (A.2). The factor in Eq. A.2, without phase factors ϕ, would yield a factor $|\sqrt{n_{11}n_{00}} - \sqrt{n_{01}n_{01}}|$ reminiscent of Yule's Y-factor [16]. The main difference with concurrence Q and different forms of such contingency measures is that contingency measures are made independent of marginal distributions of co-occurrences, but here, it is important to preserve the total probability (rarely occurring co-occurrences are of no interest), taken into account by factor Q_0. Of most importance here is the notion that all measures of contingency as well as the concurrence Q take a value of zero when all clauses are randomly mixed [15,16] (see also ref. [12]), in which case associations between occurrence of words A and B disappear. The same condition also leads to $Q = 0$ and thus, to the disappearance of entanglement.

References

1. Fang, Z.: The language demands of science reading in middle school. Int. J. Sci. Educ. **28**, 491–520 (2006)
2. Ford, A., Peat, F.D.: The role of language in science. Found. Phys. **18**, 1233–1242 (1988)
3. Lemke, J.: Talking science: language. Learning and values. Language and educational processes. Ablex Publishing Corporation, Norwood, NJ (1990)
4. Yore, L.D., Hand, B., Goldman, S.R., Hilderbrand, G.M., Osborne, J.F., Treagust, D.F., Wallace, C.S.: New directions in language and science education research. Read. Res. Quart. **39**, 347–352 (2004)
5. Södervik, I., Nousiainen, M., Koponen, I.T.: First-year life science students' understanding of the role of plants in the ecosystem—a concept network analysis. Educ. Sci. **11**, 369 (2021)
6. Yun, E., Park, Y.: Extraction of scientific semantic networks from science textbooks and comparison with science teachers' spoken language by text network analysis. Int. J. Sci. Educ. **40**, 2118–2136 (2018)
7. Nousiainen, M., Koponen, I.T.: Pre-service teachers' declarative knowledge of wave-particle dualism of electrons and photons: finding lexicons by using network analysis. Educ. Sci. **10**, 76 (2020)
8. Christianson, N.H., Sizemore, B.A., Bassett, D.S.: Architecture and evolution of semantic networks in mathematics texts. Proc. R. Soc. A **476**, 20190741 (2020)
9. Chai, L.R., Zhou, D., Bassett, D.S.: Evolution of semantic networks in biomedical texts. J. Complex Netw. **8**, cnz023 (2020)
10. Ribeiro, E., Teixeira, A.S., Ribeiro, R., de Matos, D.M.: Semantic frame induction through the detection of communities of verbs and their arguments. Appl. Netw. Sci. **5**, 69 (2020)
11. Medeuov, D., Roth, C., Puzyreva, K., Basov, N.: Appraising discrepancies and similarities in semantic networks using concept-centered subnetworks. Appl. Netw. Sci. **6**, 66 (2021)
12. Surov, I.A., Semenenko, E., Platonov, A.V., Bessmertny, I.A., Galofaro, F., Toffano, Z., Khrennikov, A.Y., Alodjants, A.P.: Quantum semantics of text perception. Sci. Rep. **11**, 4193 (2021)
13. Surov, I.A.: Quantum cognitive triad: semantic geometry of context representation. Found. Sci. **26**, 947–975 (2021)
14. Galofaro, F., Toffano, Z., Doan, B.-L.: A quantum-based semiotic model for textual semantics. Kybernetes **47**, 307–320 (2018)
15. Johnson, R.A., Wichern, D.W.: Applied Multivariate Statistical Analysis. Pearson, New Jersey (2008)
16. Bonett, D.G., Price, R.M.: Statistical inference for generalized yule coefficients in 2x2 contingency tables. Soc. Meth. Res. **35**, 429–446 (2007)
17. Estrada, E.: Informational cost and networks navigability. Appl. Math. Comp. **397**, 125914 (2021)
18. Estrada, E.: The Structure of Complex Networks: Theory and Applications. Oxford University Press, Oxford (2012)
19. Estrada, E., Hatano, N., Benzi, M.: The physics of communicability in complex networks. Phys. Rep. **514**, 89–119 (2012)
20. Han, J., Kamber, M., Pei, J.: Data Mining: Concepts and Techniques, 3rd edn. Elsevier, Amsterdam (2012)
21. Newman, M.: Networks, 2nd edn. Oxford University Press, Oxford (2018)

Attributed Stream-Hypernetwork Analysis: Homophilic Behaviors in Pairwise and Group Political Discussions on Reddit

Andrea Failla[1], Salvatore Citraro[1(✉)], and Giulio Rossetti[2]

[1] Computer Science department, University of Pisa, Pisa, Italy
a.failla@studenti.unipi.it, salvatore.citraro@phd.unipi.it
[2] KDD Laboratory, ISTI, National Research Council, Pisa, Italy
giulio.rossetti@isti.cnr.it

Abstract. Complex networks are solid models to describe human behavior. However, most analyses employing them are bounded to observations made on dyadic connectivity, whereas complex human dynamics involve higher-order relations as well. In the last few years, hypergraph models are rising as promising tools to better understand the behavior of social groups. Yet even such higher-order representations ignore the importance of the rich attributes carried by the nodes. In this work we introduce ASH, an Attributed Stream-Hypernetwork framework to model higher-order temporal networks with attributes on nodes. We leverage ASH to study pairwise and group political discussions on the well-known Reddit platform. Our analysis unveils different patterns while looking at either a pairwise or a higher-order structure for the same phenomena. In particular, we find out that Reddit users tend to surround themselves by like-minded peers with respect to their political leaning when online discussions are proxied by pairwise interactions; conversely, such a tendency significantly decreases when considering nodes embedded in higher-order contexts - that often describe heterophilic discussions.

Keywords: High-order networks · Dynamic networks · Feature-rich networks · Data analysis

1 Introduction

Several human behaviors can be described through the lens of complex network analysis. From homophily [20] to the time-evolving nature of social groups [14], complex networks can shed lights on the laws governing nodes' wiring patterns and their evolution as time goes by as well. However, the intrinsic nature of network connectivity can not explicitly go beyond dyadic patterns, i.e., coupling between pairs of nodes. Such pairwise constraints must be taken into account when aiming to investigate the behavior of social groups and their complex

H. Cherifi et al. (Eds.): COMPLEX NETWORKS 2016 2022, SCI 1077, pp. 150–161, 2023.
https://doi.org/10.1007/978-3-031-21127-0_13

dynamics. Hypernetwork science is the new, cutting-edge line of research that aims to address higher-order structures for representing and analyzing complex systems [1,3,27]. In this work we consider such higher-order structures involved in dynamic and attribute-aware environments as well. This way we propose to study a phenomenon along three main dimensions: topology, dynamic features and node attributes, plus any preferred combination of these. To this purpose we introduce *ASH*, an *A*ttributed *S*tream-*H*ypernetwork model to represent higher-order temporal networks with attributive information on nodes. To test ASH's potentiality, we infer the attribute-rich higher-order temporal structure of group political discussions on the well-known Reddit platform. In particular, we focus on data collected from the debate between Trump supporters and anti-Trump citizens as described in [18] along three main topics: gun legalization, minorities discrimination and general debates on the US political sphere. We aim to explore to what extent users are homogeneously embedded in group discussions when modeling the higher-order structure, where *homogeneity* is computed as the fraction of users within the higher-order neighborhood of a target node that share the same attribute value of this latter one. Then, we aim to compare these results to the ones obtained by mining the pairwise network only.

The rest of the work is organized as follows. Section 2 sums up the principal literature on the three main complex network contexts surrounding this work, namely dynamic, node-attributed, and higher-order networks. Section 3 introduces our Attributed Stream-Hypernetwork model. Section 4 discusses our main results on the selected datasets. Finally, Sect. 5 concludes the work and discusses promising lines of research left open for the future.

2 Related Work

We provide in the following an overview of the main enriched network models used in this work. First, we discuss dynamic and node-attributed network representations [16]; then, we sum up the emerging contributions about higher-order representations for complex systems.

Dynamics of networks. The temporal dimension is often available from network data representing human behaviors. However, choosing a proper representation for modeling interaction dynamics is not a straightforward task. Friendships are persistent over time, whereas face-to-face interactions involve a certain duration, and e-mails or financial transactions are even instantaneous. Different temporal semantics impose different modeling strategies [14,24]. We can identify dynamic network models by selection of different criteria: (i) stability, e.g., the snapshot sequence from a time-window aggregation strategy [6,23]; (ii) duration, e.g., the interval graph model [14]; and (iii) immediacy, e.g., the stream graph [17], where a dynamic network comes as a stream of both temporal nodes and interactions. Stream graphs, in particular, can easily extend and generalize classic centrality measures on static graphs [26], and multi-layer structure as well [21]. Among the most interesting and cutting-edge tasks on dynamic networks we can mention community detection [24], link prediction [13], and mixing pattern estimation [10].

Networks with metadata. Similarly to temporal information, metadata or attributes describing nodes' properties are often available from network data. Metadata-enriched models can support augmented mining and analyses about the relationships between the structural and the attributive information in complex systems. Most of the works focusing on the interplay between structure and attributes aim to study the correlation between the two dimensions, searching for bridges between tabular and networked data [28]. Node attributes can be fruitfully used for improving the community detection task, where both tight connectivity and label-homogeneity within communities need to be guaranteed [8]. They can be also leveraged for estimating heterogeneous mixing patterns in complex networks [22,25]. The distribution of metadata surrounding a single node, e.g., the distribution of features within the node's ego-network, is particularly useful in the node classification and in the link prediction tasks [4].

Higher-order networks. Although traditional network science mostly addressed analyses on pairwise interactions, many network dynamics can be better modelled by higher-order representations involving relations between groups of nodes. As an emerging line of research [3,27], the expressive power of such representations is yet largely unexplored. The interest in the physics of higher-order interactions is growing [2], and it has been mainly explored in the context of diffusion analysis, e.g., studying social contagion with simplicial complexes [15], in time-varying settings as well [7]. Higher-order structures varying in time are an important and emerging trend of research [5]. In [12], three distinct types of hypergraph neighborhoods are introduced: star, radial and contracted ego-networks. The star ego-network of a node u is the set of all the hyperedges that include u, whereas the other two include also the hyperedges whose nodes within are all u's neighbors (radial) or further extends u's neighborhood to the hyperedges that interact with nodes outside of u's ego-network (contracted). The differences between the three ego-networks is interesting when studying them from a temporal perspective: some hyperedges included in the radial and contracted neighborhoods may have begun a time before the target node u joined the context [12]. In most of such analyses, the higher-order structure of static/dynamic networks is addressed by investigating datasets originally designed for graph-based analysis, thus one of the most intriguing future challenges is the inference of statistically significant higher-order interactions from complex systems [19]. Moreover, higher-order-based techniques are emerging to generalize well-known graph-based techniques or to conservatively shift to them, as in the case of s-line graph analysis for hypergraph models [1].

3 Attributed Stream-Hypergraphs

To study dynamic high-order social interactions, simply borrowing results from the existing literature is not enough. Hypergraphs and simplicial complexes, to name the most used high-order representation frameworks [3,27], have both strengths and weaknesses. None of them has been adequately defined in the presence of evolving topologies. Indeed, their applicability to online social environments needs to be carefully analyzed to understand if the constraint they come with aligns with the semantics expressed by social interaction networks. Moreover, individuals embedded in a social system can often be characterized by multiple features—*profiles* that contextualize some of the key properties playing a role for social interactions (e.g., nationality, gender, age...). To start filling the existing gap in high-order dynamic and feature-rich modeling of social systems, here we propose the framework of Attributed Stream Hypergraphs.

Definition 1. *[Attributed Stream Hypergraph (ASH)]* Let $\mathcal{S} = (T, V, W, E, L)$ be a stream hypergraph, where:

- $T = [\mathrm{A}, \Omega]$ is the set of discrete time instants, with A and Ω the initial and final instants;
- V is the set of the nodes of the temporally flattened hypergraph;
- $W \subseteq T \times V$ is the set of temporal nodes;
- $E \subseteq T \times V^n$ is the set of temporal hyperedges such that $(t, N) \in E$ implies that $N \subseteq V$ and $\forall u_i \in N, (t, u_i) \in W$;
- L is the set of temporal node attributes such that $L(t, u)$ with $(t, u) \in W$ and $t \in T$, identifies the set of categorical values associated to u at time t.

ASHs are a conservative extension of well known modeling frameworks, namely hypergraphs [3] and stream graphs [17]. For instance, temporal nodes and temporal edges are a peculiarity of the stream graph representation, where the nature of nodes and edges is analyzed with respect to the time they appear in the temporal stream. A node is a temporal entity that can be present or absent at a certain time in the stream, so that a node is said to be*equal to 1*—i.e., it is represented as a whole quantity as in a classic static graph—only if it is present all the time in the stream. Similarly, an edge accounts for 1 only if it is present all the time in the stream. From both hypergraphs and stream graphs ASH inherits several analytical peculiarities and, from their union, it is able to provide novel insights that the original models are not able to unveil independently. Temporal hyperedges indeed are an example of such novelty, where the temporal presence of an interaction is accounted for higher-order groups and not only pairwise interactions. Moreover, integrating time evolving node attributes, it allows to study not only how individuals' characteristics change (e.g., opinions, political leaning) but also how such changes relate/affect the topological structure surrounding them.

In the following we provide a first example of how ASH can be used to study real world phenomena. Our analysis will focus on some aspects of the three dimensions modeled by ASH: high-order topology, node semantics and time.

4 Experiments

Sociopolitical Reddit data. We focus on data collected from the debate between Trump supporters and anti-Trump citizens during the first two and half years of Donald Trump's presidency, covering a period between January 2017 and July 2019. Data collection, users' ideology inference and network construction are properly described in the reference paper [18] (thus identifying three users' families: protrump, antitrump and neutral). Leveraging the original temporal network,[1] we infer the hypergraph structure by means of all the maximal cliques. As in the original analysis about this dataset, [18], we consider a time-window of six months when analyzing system interactions' dynamics. Average statistics for the pairwise graphs are shown in Table 1.

The three main topics considered are the followings:

- *Gun Control*: this topic is identified by collecting lists of subreddits that either support gun legalization or are against it;
- *Minorities Discrimination*: identified by considering groups that promote gender/racial/sexual equality and groups showing more conservative attitudes;
- *Political Sphere*: identified by covering different US political ideologies such as Republicans, Democrats, Liberals, and Populists.

Table 1. Network statistics (averaged across semesters). size of the network in terms of nodes and edges, number of users with a pro-trump, anti-trump or neutral leaning score.

Topic	# Nodes	# Edges	# Pro-trump	# Anti-trump	# Neutral
Gun Control	4991	15,298	3346	1645	–
Minorities Discrimination	5540	12,605	3318	2222	–
Political Sphere	4509	7079	1280	2395	834

Among these topics, the first two ones are controversial/polarizing sociopolitical issues, whereas the latter one is intended to offer a more broad spectrum of the US political debate.

Analytical setting. We set a four-fold framework to analyze ideology homogeneity from different perspectives as in the following:

i. we promote an analysis on dyadic interactions, and we measure how much users are homogeneously embedded in their pairwise ego-networks;
ii. we shift the focus from individual users to groups, and we measure the homophily of such groups/contexts modelled as hyperedges;
iii. we return to individual users, adopting the single user point of view while measuring how much an user is embedded in the contexts/hyperedges where he/she participates;
iv. we introduce a time-aware analysis to track stability or variations in ideology homogeneity.

[1] Original network data available at https://github.com/virgiiim/EC_Reddit_Case Study.

As a preliminary question we aim to explore whether different behaviours emerge among individual users (i) and groups (ii): can higher-order interactions capture patterns which are invisible to dyadic interactions? Then, we aim to understand the role of single users in the several contexts where they participate (iii), as a meeting point between the two previous points: can higher-order neighborhoods capture different patterns that graph ego-networks cannot? Finally, the focus on interactions' dynamics (iv) would allow us to track stable or mutable patterns as time goes by.

It should be noted that computations in (i) and (iii) are different from (ii). In (i) and (iii) we use a measure of homogeneity to estimate target nodes' similarity within nodes own contexts. The focus is on single nodes and we aim to measure the homogeneity of the contexts with respect to the political leaning of a specific target node. In detail, the homogeneity within the context of a target node with respect to the node's own ideology is computed as the relative frequency of the attribute value carried by the node within the context that embeds it:

$$Homogeneity(u) = \frac{|\{v|v \in c \wedge l_u = l_v\}|}{|c|}, \tag{1}$$

where u is the target node with l_u as attribute value, c is the context where u is embedded, and l_v is the attribute value of a node v that belongs to c. In the pairwise network, c is the first-order neighborhood, namely the set of nodes adjacent to u. In the hypergraph, c is the set of hyperedges of u's star ego-network. Since a hyperedge includes several nodes, we must associate to the context expressed by the hyperedge a characteristic value of such a context; coherently with the choice to measure the homogeneity with respect to a target node neighborhood, we set as the characteristic value of a hyperedge the most frequent attribute value of the nodes participating in it.

Finally, the homogeneity of the hypergraph is defined as the average of all its nodes' homogeneities:

$$Homogeneity = \frac{1}{|V|} \sum_{u \in V} Homogeneity(u). \tag{2}$$

Conversely, in (ii) the focus is on hyperedges' homogeneities. Thus, we use a particular instantiation of the homogeneity measure, namely the purity [11], that computes the relative frequency of the most frequent attribute value within the hyperedge:

$$Purity(e) = \frac{\max_{l \in L}(\sum_{v \in e} l_v)}{|e|}, \tag{3}$$

where e is the target hyperedge, and v is a node participating in it with l_v as attribute value. This way we only capture the characteristic behaviour of a target hyperedge.

Finally, the purity of the hypergraph is defined as the average of all its hyperedges' purities:

$$Purity = \frac{1}{|E|} \sum_{e \in E} Purity(e). \tag{4}$$

Results

Pairwise ego-networks reveal both homophilic and heterophilic users' preferences. Fig. 1 outlines graph ego-networks' homogeneities in the three topics considered. Results are aggregated over the semesters. The analysis on pairwise interactions captures both homophilic and heterophilic patterns. In *GunControl*, there emerges heterogeneity in antitrump and protrump users, where the latter ones appear to be more likely to engage with similar users. Conversely, an overall strong homophilic distribution emerges in *Minority*, whose homogeneity values for both antitrump and protrump nodes mostly gather around the upper bound. A strong difference is visible in *Politics*, where protrump and neutral users are less homogeneous than antitrump ones. All these observations are coherent with the analyses performed on the original data paper [18], where in *Minority* and *Politics* it is more likely to observe echo-chambers—oriented towards a protrump political leaning in *Minority*, and antitrump in *Politics*, while *GunControl* discussions are less polarized.

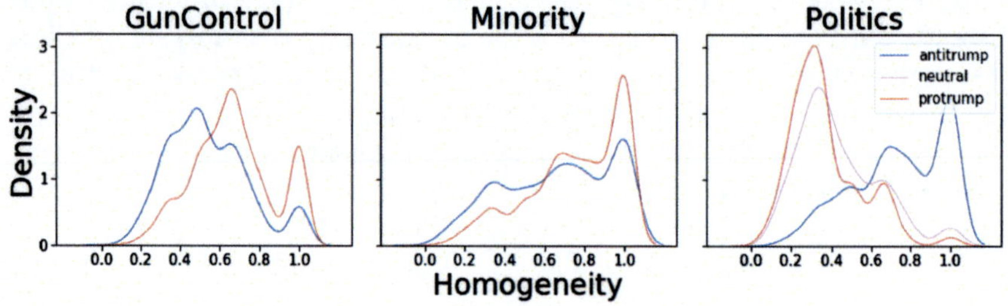

Fig. 1. KDE distributions of pairwise ego-networks' homogeneity among the three different Reddit communities.

Hyperedges' purity emphasizes homogeneous discussions.
Figure 2(a) gives a global overview of ideology homogeneity within the contexts/hyperedges in the three topics considered. Results are aggregated over the semesters. The bimodal distributions observed in all the categories reflect similarly to graph ego-networks what already emerged on the same data while identifying echo chambers at the meso-scale community level [18]: *GunControl does not present strongly polarized communities [i.e., echo chambers] among different semesters* (p. 12) [18] as well as it seems that only a bunch of contexts present quite perfect purity (Fig. 2(a), leftmost); *[in Minority] on average, more*

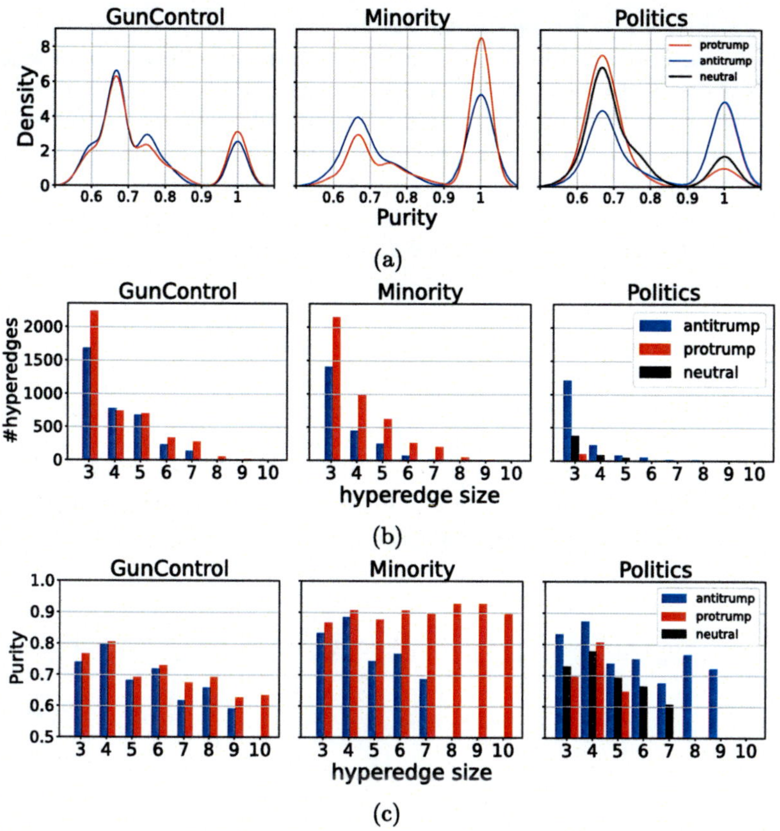

Fig. 2. KDE distributions of hyperedges' purity (a), number of pure hyperedges (b) and average purity (c) in function of hyperedge size among the three different Reddit communities.

than half of total users are trapped in echo chambers (p. 12) [18], and hyperedge purities show a quite similar pattern as well, with a tendency of protrump users to form more homogeneous groups (Fig. 2 (a), center); also *Politics* presents high homogeneity contexts, where antitrump users are more likely to form homogeneous groups (Fig. 2(a), rightmost). Nonetheless, this analysis leveraging higher-order contexts reveals more heterogeneity in the dataset and emphasizes different behaviours among users, even when focusing on the purity only. Moreover, all these patterns are observed with respect to the hyperedge size as well. Figure 2(b, c) highlight, respectively, the number (b) and the average purity (c) of pure-groups in function of the group size. For instance, in *Minority* we observe that only protrump pure discussions involve groups with more than 7 participants, and that they are quite pure, 0.9. The same does not happen in *GunControl*, while in *Politics* the biggest contexts involve antitrump users only but with a lower purity than the one of protrump users in *Minority*.

Users are involved in heterogeneous debates.
As can be observed in Fig. 3, the topics show diversified behaviours when the analysis shifts to star egos, which describe the hypernetworks at the meso-scale level while maintaining local information about individual nodes. Indeed, there is no more trace of the bimodal patterns observed in Fig. 2(a), with the sole exceptions of antitrump nodes in *Minority* and neutral ones in *Politics*. The key insight, however, relates to the heterogeneity of user debates. While engaging in relatively homogeneous contexts (Fig. 2), it seems that users find themselves in rather mixed collections of debates. That is to say, although homophilic behaviour is highlighted in most debates (i.e., hyperedges), the set of contexts a node is involved in (i.e., its star) is generally diversified w.r.t. ideology/political leaning. This is especially true in *GunControl*, where protrump users appear to engage in a more heterogeneous set of debates than their counterparts, as opposed to what was noted in Fig. 2(a). The same holds for *Politics*, which displays a peak in heterogeneous protrump stars while the antitrump ones show more homophilic behavior. *Minority*, instead, still shows strong homogeneity traits for both antitrump and protrump users, thus confirming previous observations.

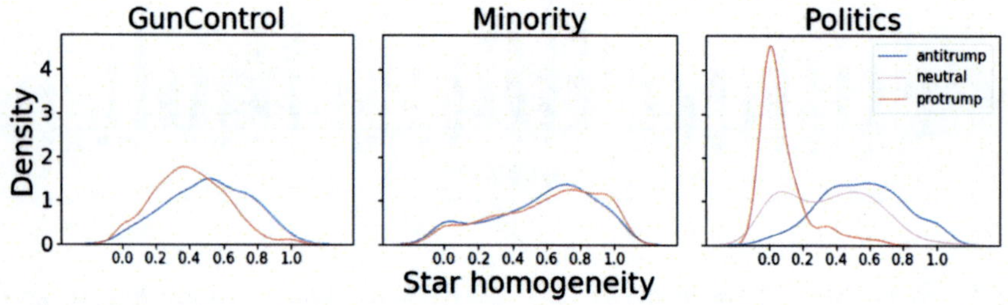

Fig. 3. KDE distributions of hypergraph star ego-networks' homogeneity among the three different Reddit communities.

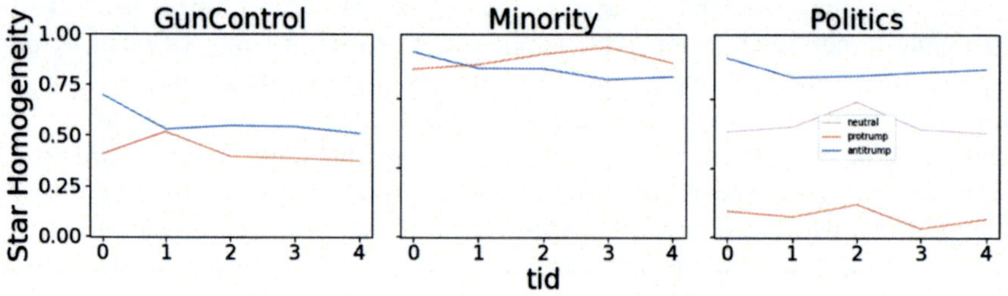

Fig. 4. Average hypergraph star ego-networks' homogeneity over time among the three different Reddit communities.

Interactions' dynamics: users' preferences tend to be consistent in time.
As far as the temporal dimension is concerned, a certain degree of consistency w.r.t. debates homogeneity/heterogeneity can be observed. As a matter of fact, the average star homogeneity outlines almost-flat trends (Fig. 4), indicating minor variations. Here, *GunControl* and *Minority* reveal near-constant heterogeneity/homogeneity for both political alignments; lastly, *Politics* displays only a small bump during the third semester concerning protrump and neutral users.

5 Discussion and Conclusions

In this work we described and applied ASH, our new framework to represent streams of higher-order interactions enriched with node attributes as Attributed Stream-Hypernetworks. This way a complex phenomenon can be described along several dimensions, namely topology, node-features, and interaction dynamics, plus any combination of these as preferred by the analyst of a specific domain [16]. We applied our framework to study group political discussions on Reddit, analyzing to what extent politically oriented users are homogeneously embedded in group discussions with respect to group's representative political leaning. Even if not mainly related to the identification/analysis of echo chambers, this work gave us another perspective to interpret the homophilic patterns emerging from controversial political debates on social platforms. Our results can support the claim of users aggregation in similar/homophilic contexts/hyperedges, but this result may not necessarily involve segregation [9]. In fact, while applying our framework to the Reddit communities, we observed strong homophilic behaviors among groups/hyperedges with respect to users' political leaning. However, while focusing on the preferences of single nodes, namely on how much a target node is homogeneously embedded with respect to the representative political leaning of the groups/hyperedges it belongs to, we mostly observe a relevant decrease in nodes' homophilic behaviors. As a consequence, we observe that users prefer participating in contexts whose representative leaning is different than the target node's own label, although hyperedges are strongly homophilic per se. Interestingly, this pattern can not be observed when looking at the pairwise ego-networks only. Finally, regarding the interaction dynamics, we mainly observe stability and consistency over time with respect to users' political leaning. In future works, we plan to work with more datasets that can better involve changes along the temporal dimension. Also, we plan to define domain-specific measures based on a solid characterization of node profiles, leveraging them to better analyze higher-order structure dynamics. Lastly, we plan to formalize more consistently the ASH framework, as already done for recent generalizations of dynamic networks as stream graphs [17], and we plan to focus on the constraints that stream hypergraphs could eventually raise, such as the issues of under/over fitting social data or the robustness of the measures to missing data.

Acknowledgments. This work is supported by the European Union—Horizon 2020 Program under the scheme "INFRAIA-01-2018-2019—Integrating Activities for Advanced Communities", Grant Agreement n. 871042, "SoBigData++: European Integrated Infrastructure for Social Mining and Big Data Analytics" (http://www.sobigdata.eu).

References

1. Aksoy, S.G., Joslyn, C., Marrero, C.O., Praggastis, B., Purvine, E.: Hypernetwork science via high-order hypergraph walks. EPJ Data Sci. **9**(1), 16 (2020)
2. Battiston, F., Amico, E., Barrat, A., Bianconi, G., Ferraz de Arruda, G., Franceschiello, B., Iacopini, I., Kéfi, S., Latora, V., Moreno, Y., et al.: The physics of higher-order interactions in complex systems. Nature Phys. **17**(10), 1093–1098 (2021)
3. Battiston, F., Cencetti, G., Iacopini, I., Latora, V., Lucas, M., Patania, A., Young, J.G., Petri, G.: Networks beyond pairwise interactions: structure and dynamics. Phys. Rep. **874**, 1–92 (2020)
4. Bhagat, S., Cormode, G., Muthukrishnan, S.: Node classification in social networks. In: Social Network Data Analytics, pp. 115–148. Springer (2011)
5. Cencetti, G., Battiston, F., Lepri, B., Karsai, M.: Temporal properties of higher-order interactions in social networks. Sci. Rep. **11**(1), 1–10 (2021)
6. Chiappori, A., Cazabet, R.: Quantitative evaluation of snapshot graphs for the analysis of temporal networks. In: International Conference on Complex Networks and Their Applications, pp. 566–577. Springer (2021)
7. Chowdhary, S., Kumar, A., Cencetti, G., Iacopini, I., Battiston, F.: Simplicial contagion in temporal higher-order networks. J. Phys. Complex. **2**(3), 035019 (2021)
8. Chunaev, P.: Community detection in node-attributed social networks: a survey. Comput. Sci. Rev. **37**, 100286 (2020)
9. Cinelli, M., Morales, G.D.F., Galeazzi, A., Quattrociocchi, W., Starnini, M.: The echo chamber effect on social media. Proc. Nat. Acad. Sci. **118**(9) (2021)
10. Citraro, S., Milli, L., Cazabet, R., Rossetti, G.: {*Delta*}-conformity: multi-scale node assortativity in feature-rich stream graphs (2021). arXiv:2111.15534
11. Citraro, S., Rossetti, G.: Identifying and exploiting homogeneous communities in labeled networks. Appl. Netw. Sci. **5**(1), 1–20 (2020)
12. Comrie, C., Kleinberg, J.: Hypergraph ego-networks and their temporal evolution. In: 2021 IEEE International Conference on Data Mining (ICDM), pp. 91–100. IEEE (2021)
13. Divakaran, A., Mohan, A.: Temporal link prediction: a survey. New Gen. Comput. **38**(1), 213–258 (2020)
14. Holme, P., Saramäki, J.: Temporal networks. Phys. Rep. **519**(3), 97–125 (2012)
15. Iacopini, I., Petri, G., Barrat, A., Latora, V.: Simplicial models of social contagion. Nature Commun. **10**(1), 1–9 (2019)
16. Interdonato, R., Atzmueller, M., Gaito, S., Kanawati, R., Largeron, C., Sala, A.: Feature-rich networks: going beyond complex network topologies. App. Netw. Sci. **4**(1), 1–13 (2019)
17. Latapy, M., Viard, T., Magnien, C.: Stream graphs and link streams for the modeling of interactions over time. Social Netw. Anal. Mining **8**(1), 1–29 (2018)
18. Morini, V., Pollacci, L., Rossetti, G.: Toward a standard approach for echo chamber detection: reddit case study. Appl. Sci. **11**(12), 5390 (2021)

19. Musciotto, F., Battiston, F., Mantegna, R.N.: Detecting informative higher-order interactions in statistically validated hypergraphs. Commun. Phys. **4**(1), 1–9 (2021)
20. Newman, M.E.: Mixing patterns in networks. Phys. Rev. E **67**(2), 026126 (2003)
21. Parmentier, P., Viard, T., Renoust, B., Baffier, J.F.: Introducing multilayer stream graphs and layer centralities. In: International Conference on Complex Networks and Their Applications, pp. 684–696. Springer (2019)
22. Peel, L., Delvenne, J.C., Lambiotte, R.: Multiscale mixing patterns in networks. Proc. Nat. Acad. Sci. **115**(16), 4057–4062 (2018)
23. Ribeiro, B., Perra, N., Baronchelli, A.: Quantifying the effect of temporal resolution on time-varying networks. Sci. Rep. **3**(1), 1–5 (2013)
24. Rossetti, G., Cazabet, R.: Community discovery in dynamic networks: a survey. ACM Comput. Surv. (CSUR) **51**(2), 1–37 (2018)
25. Rossetti, G., Citraro, S., Milli, L.: Conformity: a path-aware homophily measure for node-attributed networks. IEEE Intel. Syst. **36**(1), 25–34 (2021)
26. Simard, F., Magnien, C., Latapy, M.: Computing betweenness centrality in link streams (2021). arXiv:2102.06543
27. Torres, L., Blevins, A.S., Bassett, D., Eliassi-Rad, T.: The why, how, and when of representations for complex systems. SIAM Rev. **63**(3), 435–485 (2021)
28. Zanin, M., Papo, D., Sousa, P.A., Menasalvas, E., Nicchi, A., Kubik, E., Boccaletti, S.: Combining complex networks and data mining: why and how. Phys. Rep. **635**, 1–44 (2016)

Individual Fairness for Social Media Influencers

Stefania Ionescu(✉), Nicolò Pagan, and Anikó Hannák

Social Computing Group, University of Zürich, Zürich 8050, Switzerland
ionescu@ifi.uzh.ch, pagan@ifi.uzh.ch, hannak@ifi.uzh.ch

Abstract. Nowadays, many social media platforms are centered around content creators (CC). On these platforms, the tie formation process depends on two factors: (a) the exposure of users to CCs (decided by, e.g., a recommender system), and (b) the following decision-making process of users. Recent research studies underlined the importance of content quality by showing that under exploratory recommendation strategies, the network eventually converges to a state where the higher the quality of the CC, the higher their expected number of followers. In this paper, we extend prior work by (a) looking beyond averages to assess the fairness of the process and (b) investigating the importance of exploratory recommendations for achieving fair outcomes. Using an analytical approach, we show that non-exploratory recommendations converge fast but usually lead to unfair outcomes. Moreover, even with exploration, we are only guaranteed fair outcomes for the highest (and lowest) quality CCs.

Keywords: Social network formation · Individual fairness · Markov chains

1 Introduction

The past couple of decades brought a steep increase in the impact social media platforms have on our lives, e.g., by shaping the information we receive [1] and the opinions we form [2]. During this time, platforms previously designed to only connect real-life friends slowly encouraged users to follow strangers based on their content. Today, platforms such as YouTube, Twitter, Instagram, and TikTok are heavily centred around User Generated Content (UGC) and use recommender systems (RSs) to facilitate the exploration of content. In response to this change, some users specialize in creating even semi-professional content that can attract more and more followers to the point that they can make revenue based on their audience. Similarly to the labour market [3], it is then natural to expect that these online platforms guarantee fairness for the content creators (CCs) in a way that equally good CCs should be rewarded similarly in terms of visibility and audience, and ultimately of income. Since the network formation process is often heavily mediated by the RSs, it is thus appropriate to ask whether they produce fair outcomes for the CCs.

H. Cherifi et al. (Eds.): COMPLEX NETWORKS 2016 2022, SCI 1077, pp. 162–175, 2023.
https://doi.org/10.1007/978-3-031-21127-0_14

To first understand the structure and the properties of these social media platforms, [4] proposed a simple model in which (a) each CC has an intrinsic and objective quality, (b) in each round, users receive a recommendation for a CC (which could be drawn from the uniform distribution, or from a preferential attachment (PA) process), and (c) users follow the recommended CC if this CC has a higher quality than all the user's current followees. By simulating this model, Pagan et al. [4] showed that the expected number of followers (in-degree) of the CCs follows a Zipf's law [5]. In particular, the expected rankings of CCs given by their quality and by the number of followers are the same.

While the findings of [4] suggest UGC-centered platforms are fair for the CCs, some empirical evidence suggests differently. In cultural markets, for example, it is often difficult to anticipate success [6]. The experimental study of [7] revealed this is at least partially caused by the social influence. More precisely, as the amount of information users receive about the prior choices of others increases, the predictability of the items' popularity decreases. This unpredictability can be interpreted as a lack of fairness guarantees on some UGC-centered platforms.

We believe this apparent incongruity in previous work is mainly due to two model limitations of [4]. First, the analysis is restricted to two exploratory recommendation processes (i.e., processes which take risky actions in order to uncover better options and are thus limited in the amount of social influence they exert). However, in practice, not all RSs are as such. Recent literature in RSs argued why and how we should encourage diversity by finding the right balance between exploration and exploitation [8–11]. In practice, even pairs of popular items might not be jointly accessible to users, i.e., if a user is recommended and follows one of the two, they will not be recommended the other [12]. This puts real-world RSs in stark contrast with PA and uniform random (UR) recommendations where every user can be recommended any CC. Second, the authors only focus on the *expected* number of followers *at convergence* (i.e., after all users were eventually recommended the best CC). However, such ex-ante fairness does not imply ex-post fairness (i.e., even if in expectation CCs receive a number of followers proportional to their quality, many of the actual outcomes that could materialize are unfair) [13]. Moreover, as also noted by the authors [4], there might be long times to convergence. This means that even if a fair outcome would eventually be reached, this might not happen within reasonable time.

This paper aims to address these two challenges by bridging between the network science, RS, and algorithmic fairness communities. More precisely, the current work (a) defines both ex-ante and ex-post fairness metrics for CCs, (b) extends the model with extreme PA (a non-exploratory RS, inspired by the popularity RS[14], which only recommends the most followed CCs), (c) uses Markov Chains to theoretically study the network formation process and its fairness under extreme PA, and (d) compares the network formation processes and their fairness under extreme PA with the ones under PA and UR recommendations (either by referring to prior work, or by novel analytical and numerical results).

2 Related Work

Networks. After the pioneering work on the random graph model [15], the complex networks community started developing simple yet insightful mechanisms that explain the formation of social networks (e.g., the small-world network model [16], and the preferential attachment model -PA- [17]). In PA, newborns form connections to existing nodes with a probability proportional to the degree. This rich-get-richer phenomenon successfully reproduces the idea that popular users experience higher visibility, which in turns brings them more popularity. On the other hand, it gives little emphasis on the socio-economic microscopic foundations that explain why individuals make certain connections. Focusing on this alternative approach, one line of research in sociology (Stochastic Actor Oriented Models [18]) and one in economics (strategic network formation models [19]) take an utilitarian perspective: agents build their connections to maximize some benefit e.g., their network centrality. The quality-based model of Pagan et al. [4] combines these approaches by using a UR or PA-based RSs and a utilitarian following decision-making function for users. To enhance our understanding on the coupling between RS and human network behavior, we add a non-exploratory RS and investigate the fairness of the resulting outcomes.

Fairness. Researchers are not only concerned with the average performances of processes, but also with the equity of these processes in impacting individuals. This is reflected by the extensive work on developing fairness measures as well as a methodology to choose the most suitable one depending on the application domain [20–22]. From these various fairness metrics, we focus on *individual fairness* which assesses the degree similarly qualified individuals receive similarly quality outcomes (see [23] for an overview of its importance and apparent incompatibility with other fairness metrics). An important phenomenon is the *timing effect*, according to which it is not enough to specify a welfare function but also when this should be measured (ex-ante or ex-post) [13]. Building on these, we define and investigate both the ex-ante and ex-post individual fairness for CCs.

Recommender Systems. Recently, the RS community argued for the importance of looking beyond accuracy [8] in RS-evaluation. Design-wise, there is an ongoing struggle to develop diverse RSs [9–11] which, moreover, ensure that any two items could be recommended jointly to users [12]. This perhaps explains why the popularity-based algorithms implemented within the RS community [14,24] differ from PA [4,17] by not allowing for a complete exploration of alternatives. Our motivation to analyse extreme PA is grounded in this prior work, although our goal is to understand the network-formation process and its fairness. We also depart from the experimental study of [7] as we (a) formally define fairness metrics, (b) use theoretical tools to understand *why* social influence decreases fairness, and (c) also look at network-specific metrics (the time to convergence).

3 Model

As mentioned in the previous two sections, our work is based on the quality-based model proposed by [4] and briefly presented below. As any model, this is based on some simplifying assumptions, for which we refer the user back to the original paper. Note that our notation occasionally differs from [4] in order to allow for future extensions. The next two subsections contain our advancements of the model: (a) a RS-based formulation of the network formation process (plus the definition of the new RS, extreme PA), and (b) the fairness definitions.

We assume users can be partitioned into $n \geq 2$ content creators (CCs) and m (non-CC) users, with $m \gg n$. We refer to CCs as $CC_1, CC_2, \ldots, CC_n$ and assume they are ordered strictly by an existing objective quality. That is, CC_1 is the absolute best CC, CC_2 the next best CC, and so on. Every user can follow any content creator thus leading to a bipartite, unweighted, and directed network.[1] We represent this network by its adjacency matrix $A \in \{0,1\}^{m \times n}$, such that (s.t.) $a_{u,i}$ is 1 if user u follows CC_i, and 0 otherwise.

The network formation is a sequential dynamic process where (a) the network is initially empty, and (b) at each timestep, each user is recommended a CC which they can follow or not. Formally, let A^t capture the state of the network at time t. By assumption (a), $a^0_{u,i} = 0$ for each user $u \in \overline{m}$ and CC $i \in \overline{n}$.[2] The next subsection presents how recommendations are made and what are their effects.

3.1 Recommendation Function

We view recommendations as functions which map users to items, i.e., the recommendations at time t are given by $R^t : \overline{m} \to \overline{n}$.[3] The recommender system is the algorithm which produces such a recommender function based on the current state of the network. In this paper, we look at three such algorithms which we list below from the most to the least exploratory one:

- *Uniform Random (UR).* Each user can be recommended any CC with an equal probability, i.e., $R^t_{\text{UR}}(u) \sim \mathcal{U}(\overline{n})$.
- *Preferential Attachment (PA).* Each content creator can be recommended with a chance proportional to their current number of followers. That is, if

[1] Although, in principle, on influencer-centered platforms, e.g., YouTube, any user could create content, reports show that only a small number of users do so [25]. This also means that in practice CCs can also follow other CCs. However, since they are relatively few compared to non-CCs, we can opt for simplicity and ignore such edges. Our results can be generalised though when allowing for such edges and thus not restricting the network to a bipartite structure. A more detailed discussion on this issue can be found in [4].

[2] Throughout the paper we use $\overline{k}$ to denote the set of non-zero natural numbers that are at most equal to k, i.e. $\overline{k} := \{1, 2, \ldots k\}$.

[3] Note that under full generality each user receives a *list* of recommendations. However, allowing for a single recommendation per user makes our model comparable with [4].

$a_{\cdot,i} := \sum_{u \in \overline{m}} a_{u,i}$ is the number of followers of CC_i, then:

$$\mathbb{P}(R^t_{\text{PA}}(u) = i) = \frac{a^t_{\cdot,i} + 1}{\sum_{j \in \overline{n}} (a^t_{\cdot,j} + 1)}.$$

- *ExtremePA*. An extreme version of preferential attachment which recommends only (one of) the most popular items:

$$\mathbb{P}(R^t_{\text{ext}}(u) = i) = \begin{cases} 0 & \text{, if } a^t_{\cdot,i} < \max_j a^t_{\cdot,j}, \\ 1/|\{i, \text{ s.t. } i \in \arg\max_j a^t_{\cdot,j}\}| & \text{, otherwise.} \end{cases}$$

After each user u receives their associated recommendation $R^t(u)$ they decide to follow $R^t(u)$ iff $R^t(u)$ is better (perhaps vacuously) than any followee of u:

$$a^{t+1}_{u,i} = \begin{cases} 1 & \text{, if } R^t(u) = i \text{ and } i < j \ \forall j \text{ s.t. } a^t_{u,j} = 1 \\ a^t_{u,i} & \text{, otherwise.} \end{cases}$$

3.2 Metrics of Interest

Since the network at time $t+1$ only depends on the network at time t, we can view this network formation process as a Markov Chain (MC). More details will follow in Sect. 4. With this interpretation, we first want to investigate (a) whether this is an absorbing MC (i.e., whether there exist some networks - absorbing states - which will eventually be reached and which will not change no matter how many timesteps proceed), and, if so, (b) the expected number of steps until absorption (i.e., until reaching such a state).

However, as mentioned in the introduction, we are also interested in *the fairness* of the overall process, both ex-ante and ex-post. Thus we define:

- *Ex-post individual fairness for CCs.* We say that a network A is *(individually) fair* if the number of followers of CCs have the same ranking as the CCs, i.e., if $a_{\cdot,1} \geq a_{\cdot,2} \geq \cdots \geq a_{\cdot,n}$. We also say that A is *(individually) CC_i-fair* if CC_i is one of the top i ranked CCs, i.e. if $|\{j : a_{\cdot,j} > a_{\cdot,i}\}| < i$. In particular, this means that a network is fair iff it is CC_i-fair for all $i \in \overline{n}$.[4]
- *Ex-ante individual fairness for CCs.* When this is an absorbing MC, we can also look at the ex-ante fairness of the network formation process. More precisely, we say that a process is *ex-ante (individually) fair* if the expected number of followers of CCs at absorption is decreasing, i.e., if $\mathbb{E}[a^\infty_{\cdot,1}] \geq \mathbb{E}[a^\infty_{\cdot,2}] \geq \cdots \geq \mathbb{E}[a^\infty_{\cdot,n}]$. Similarly, we say that a process is *ex-ante (individually) CC_i-fair* if $|\{j : \mathbb{E}[a^\infty_{\cdot,i}] < \mathbb{E}[a^\infty_{\cdot,j}]\}| < i$.

[4] Note that this is a weak version of the fairness definition. Alternatively, we can say a network A is fair if $a_{\cdot,1} > a_{\cdot,2} > \cdots > a_{\cdot,n}$. However, as we will see later in the results section, as the number of users goes to infinity the chance of achieving equality goes to zero. So, in the limit, the two definitions are equivalent.

4 Results

In our results section we investigate the metrics of interest introduced in Sect. 3.2. As such, Sect. 4.1 shows how this process can be viewed as a MC and what are the absorbing states. Section 4.2 looks at the expected time to absorption. Finally, in Sect. 4.3, we build on the prior two and investigate fairness. Each of these subsections starts with (a) a summary paragraph which gives an overview of the results, and (b) a take-away paragraph which discusses the relevance of these results. The reminder of each subsection is used for the results themselves. In the interest of space, in a few of the later proofs we omit details and only provide the proof's outline and intuition.

4.1 An Absorbing Markov Chain

Summary. This subsection starts by proving that our process is an absorbing MC (see Theorem 1). During the proof we also define the transition matrix. Figure 1 illustrates this process for the small case of two CCs and two (non-CC) users. The remaining results characterize the absorbing states. First, Theorem 2 shows that under exploratory recommendation processes (i.e., PA or UR) the absorbing states are the ones where each user follows the best CC. Second, Theorem 3 shows that, under ExtremePA, a state reachable from the **0** state is absorbing if (a) there is a unique CC with the maximum number number of followers and (b) no new user would like to follow this CC. For the latter, we use Lemma 1, which states that any state reachable from **0** has, for each CC_i with the maximum number of followers, a user who does not follow any CC better than CC_i.

Take-away. The results of this subsection build a representation for the process which facilitates its understanding. Importantly, by describing the transition matrix we see the differences between the three RSs: (a) ExtremePA has more absorbing states, (b) out of these, the states where CC_1 is the most followed CC (i.e., the CC_1-fair ones) are the absorbing states under the exploratory RSs, (c) PA situates itself between ExtremePA and UR in terms of exploration. By the latter we mean that, when compared to UR, PA could remain for longer times in transient states (which are fairness-wise similar to the states which are absorbing only under ExtremePA).

Theorem 1. *$(A^t)_{t \geq 0}$ is an absorbing MC (for all three RSs).*

Proof. First, $(A^t)_{t \geq 0}$ is a MC, as A^{t+1} depends only on the network configuration at the previous step. More precisely, it is a MC where: (a) the state space is $\{0,1\}^{m \times n}$, (b) the initial distribution is λ where $\lambda_A = 1$ iff A is the zero matrix, and (c) the transition matrix is given by $p_{B,C} := \mathbb{P}(A^{t+1} = C | A^t = B)$, where:

- from the zero matrix we can only transit to a matrix where each user follows exactly one CC, i.e.: $p_{\mathbf{0},C} = \begin{cases} 1/n^m & \text{, if } c_{u,\cdot} = 1 \ \forall \ u \in \overline{m} \\ 0 & \text{, otherwise} \end{cases}$;

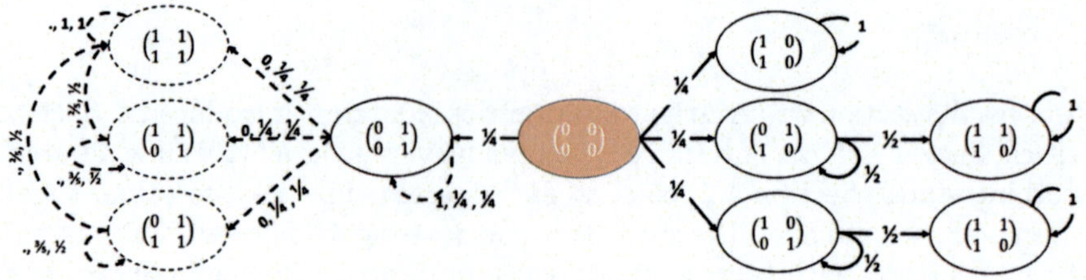

Fig. 1. An example of the MC representation for $n = m = 2$. The coloured node is the starting state. We use (a) full edges for transitions that are the same under the three RSs, and (b) dotted edges for transitions that differ (labeled with the respective probabilities for (1) ExtremePA, (2) PA, (3) UR). Dots replace probabilities when the starting state is not reachable from **0** under ExtremePA.

- from any other matrix we can only transit to a new matrix where each user either (a) follows the same CCs as before, or (b) follows exactly one more CC which is better than the best CC they followed so far; the probabilities of such transitions depend on the recommendation process.

Moreover, from the shape of the transition matrix it follows that this process is an absorbing MC. To see this, note that $p_{B,C}$ is non-zero iff $C = B$ or there is some user u who follows one more CC, i.e. $c_{u,.} > b_{u,.}$. So, any state B is either absorbing, or can transit to a state of a strictly higher sum of elements. Since the sum of elements of B is bounded by $m \cdot n$, such a sequence of transitions must be finite. Thus, in this latter case, an absorbing state will be eventually reached through a sequence of transitions. □

Theorem 2. *Under PA and UR recommendations, a state B is absorbing iff all users follow the best CC, i.e. iff $b_{u,1} = 1$ for all $u \in \overline{m}$.*

Proof. ($\Rightarrow$) Assume there exist some user $u \in \overline{m}$ s.t. $b_{u,1} = 0$. Under PA and UR recommendations there is always a non-zero chance u is recommended CC_1. If this happens, then u follows CC_1. So we can transit to a state $C \neq B$ where $c_{u,1} = 1$ with a non-zero probability. Hence, B is not absorbing.

($\Leftarrow$) The reverse is straightforward. If $b_{u,1} = 1$, then no recommendation will change the followees of u. This holds for all users u, so B is absorbing. □

Lemma 1. *For any state B reachable from **0** and any $i \in \overline{n}$ s.t. $b_{.,i} = \max_j b_{.,j}$ there exists some user $u \in \overline{m}$ s.t. $b_{u,j} = 0$ for all $j < i$.*

Proof. We use an inductive argument. The claim is obviously true for any state directly reachable from **0** (i.e., any network achievable in one timestep), as, in such states, each user follows exactly one CC. Next, we assume the claim is true for some state B. From this state, we either remain in B, or we transit to a state C where a subset of these CCs have a higher maximum number of followers. As each CC_i in this subset increased their number of followers they needed to be recommended to a user u that preferred them to any CC they previously followed. As such, CC_i is the best CC followed by u in C, i.e., $i = \min\{j : c_{u,j} = 1\}$. □

Theorem 3. *Under Extreme PA, all absorbing states B reachable from* **0** *are such that (a) there exists some* $i \in \overline{n}$ *s.t.* $b_{\cdot,i} > b_{\cdot,j}$ *for all* $j \in \overline{n} - \{i\}$*, and (b) for all* $u \in \overline{m}$ *there exists some* $j \leq i$ *s.t.* $b_{u,j} = 1$*. Moreover, every state which satisfies (a) and (b) is absorbing.*

Proof. For the first part, let B be an absorbing state reachable from **0**. If (a) is not the case then there are at least two CCs, say CC_i and CC_j with $i < j$, that have the maximum number of followers. By Lemma 1, there exists some user u who does not follow CC_i yet. Therefore, if everybody is recommended CC_i, only the number of followers of CC_i will increase by at least one (as u will follow them). There is a non-zero probability of this happening, and, in such a case, a new state is reached. Therefore, B is not an absorbing state if (a) does not hold. Next, if (a) holds but (b) does not then in the next round (when the unique most followed CC_i is recommended to everybody) new users will follow CC_i. So, again, if (a) holds but (b) does not then B cannot be absorbing.

The second part is straightforward. If (a) holds then only the unique CC with the highest number of followers can be recommended in the next round. Since (b) holds, no new user will follow this CC. So, there is no chance of transiting to a new state, i.e. the current state is absorbing. □

4.2 Expected Time to Absorption

Summary. Theorem 4 shows that as the number of users goes to infinity, the expected time to absorption under ExtremePA is less than 2. Two preliminary results are needed to prove this statement: (a) the chance of having ties in the number of followers after the first round of recommendations (Lemma 2) and (b) the chance that in the second round no one new would follow the most followed CC_i (if $i \neq n$, see Lemma 3) both go to 0 as the number of users goes to infinity.[5] These results are used for the annotations in Fig. 2, which summarizes the process and provides the intuition for the proof of the theorem.

Take-away. This puts ExtremePA in sharp contrast with the exploratory RSs, as prior work indicates that, under the UR and PA recommendation scenarios, the convergence time increases logarithmically in the number of CCs and linearly (or sub-linearly) in the number of users (see Figs. 2 and 7b of [4]). Importantly, it indicates that, while for ExtremePA it could be sufficient to analyse the fairness in the absorbing states alone, this might not be the case for PA and UR: as these RSs could lead, in practice, to long convergence times the fairness in the transient states should also be evaluated. This is particularly true for PA, which, as discussed earlier, could remain in unfair states for longer than UR.

Lemma 2. *For any* $i \neq j \in \overline{n}$, $\mathbb{P}(a^1_{\cdot,i} = a^1_{\cdot,j}) \to 0$ *as* $m \to \infty$.

[5] Note that the two preliminary results are independent on the RS.

Proof. First, let Y_u be the following random variables depending on the recommendations in the first round:

$$Y_u = \begin{cases} 0 & \text{, if user } u \text{ is recommended some } CC_k \text{ with } k \neq i, j; \\ 1 & \text{, if user } u \text{ is recommended } CC_i; \\ -1 & \text{, if user } u \text{ is recommended } CC_j. \end{cases}$$

Since the first recommendations are uniform random, $(Y_u)_{u \in \overline{m}}$ are i.i.d. (with $\mathbb{P}(Y_u = 0) = (n-2)/n$ and $\mathbb{P}(Y_u = c) = 1/n$ when $c = \pm 1$). Hence, $\mathbb{E}[Y_u] = 0$ and $\text{Var}(Y_u) = \mathbb{E}[Y_u^2] = 2/n$. By the central limit theorem it follows that,

$$\frac{\sum_{u \in \overline{m}} Y_u - m \cdot 0}{2/n \cdot \sqrt{m}} \to^{\mathcal{D}} \mathcal{N}(0, 1).$$

In particular, this implies that $\lim_{m \to \infty} \mathbb{P}(a^1_{\cdot,i} = a^1_{\cdot,j}) = 0$. □

Lemma 3. *For any* $i \in \overline{n-1}$, $\mathbb{P}((a^1_{\cdot,i} > a^1_{\cdot,j}\ \forall j \in \overline{n}) \wedge (\forall u \in \overline{m},\ \exists j \in \overline{i}$ *s.t.* $a^1_{u,j} = 1)) \to 0$ *as* $m \to \infty$.

Proof. This follows as the latter part of the conjunction corresponds to the scenario of all users being recommended one of the top i CCs in the first round. So, $\mathbb{P}((a^1_{\cdot,i} > a^1_{\cdot,j}\ \forall j \in \overline{n}) \wedge (\forall u \in \overline{m},\ \exists j \in \overline{i}\ a^1_{u,j} = 1)) \leq \mathbb{P}(\forall u \in \overline{m},\ \exists j \in \overline{i}\ a^1_{u,j} = 1) = \left(\frac{i}{n}\right)^m \to 0$ as $m \to \infty$. □

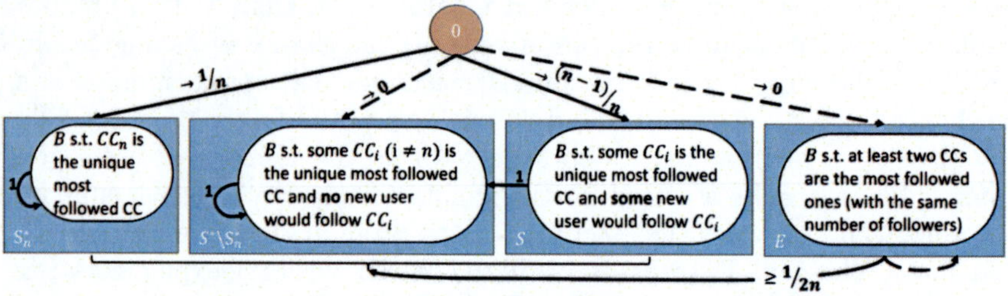

Fig. 2. A summary of the MC for ExtremePA as $m \to \infty$. Rectangles represent sets of states. Transition probabilities are annotated in the limit. Dotted lines correspond to transitions who are negligible as $m \to \infty$.

Theorem 4. *Under ExtremePA, the expected time to absorption goes to* $2 - 1/n$ *as* m *goes to infinity.*

Proof. Based on Theorems 1 and 3, we can group states in subsets, as shown in Fig. 2. Let μ_B be the expected time from the state B to absorption. Then $\mu_{B: B \in S^*} = 0$ (as all states in S^* are absorbing) and $\mu_{B: B \in S} = 1$ (as all states in S lead in one timestep to a state in S^*). Therefore,

$$\mu_{\mathbf{0}} = 1 \cdot p_{\mathbf{0}, S_n^*} + 1 \cdot p_{\mathbf{0}, S - S_n^*} + 2 \cdot p_{\mathbf{0}, S} + \sum_{B \in E} (1 + \mu_B) \cdot p_{\mathbf{0}, B}$$

We can use Lemmas 2 and 3 to find the transition probabilities between the sets of states as $m \to \infty$: (a) $p_{\mathbf{0},S_n^*} \to \frac{1}{n}$, (b) $p_{\mathbf{0},S^*-S_n^*} \to 0$, (c) $p_{\mathbf{0},S} \to \frac{n-1}{n}$, and (d) $p_{\mathbf{0},E} \to 0$. Since, in addition, $\sum_{B \in E} p_{\mathbf{0},B} \cdot (1+\mu_B) \leq p_{\mathbf{0},E} \cdot (1+\mu_{B^*})$ (where $B^* = \arg\max_{B \in E} \mu_B$) and $\mu_{B^*} \leq c$ (c constant),[6] the result follows. □

4.3 Fairness for Content Creators

Summary. Building on previous results, we show that the exploratory RSs are both ex-ante and ex-post CC_1-fair (Corollary 1), while ExtremePA is only ex-ante fair (Theorem 5). In fact, Corollary 2 shows the probability of achieving a CC_1-fair absorbing state under ExtremePA goes to $1/n$ as the number of users goes to infinity, while Corollary 3 shows that the probability of achieving a fair outcome for all CCs goes to $1/n!$. Finally, Fig. 3 depicts the probability of achieving a CC_i fair outcome for each CC under any of the three RSs.

Take-away. This analysis carries several important messages. First, it shows that exploration is key in achieving CC_1-fair outcomes. Second, although ExtremePA is ex-ante fair it is rarely ex-post fair, thus underlying the importance of looking beyond the number of followers in expectation. This finding is congruent with the experimental results of [7],[7] thus it confirms that real-life RSs share important similarities with ExtremePA. Third, our numerical analysis reveals that although exploration always leads to a CC_1-fair outcome, the fairness for the other CCs is not guaranteed. However, it suggests that exploratory RSs distribute fairness more homogeneously compared to ExtremePA.

Corollary 1. *Under UR and PA recommendations, $(A^t)_t$ is both ex-post and ex-ante CC_1-fair.*

Proof. This is an immediate consequence of Theorem 2. Since B is absorbing iff $b_{\cdot,1} = m$, all absorbing states are CC_1 fair (i.e., we have ex-post fairness) and $\mathbb{E}[b_{\cdot,1}] = m \geq \mathbb{E}[b_{\cdot,i}]$ (i.e., ex-ante CC_1-fairness). □

Corollary 2. *Under ExtremePA, the probability the outcome is ex-post CC_1-fair goes to $1/n$ as $m \to \infty$.*

Proof. The final outcome (a) is always CC_1-fair if CC_1 is the unique most followed CC after the first round, and (b) is CC_1-fair only if CC_1 is one of the CCs with a maximum number of followers after the first round. So, using the notation from Fig. 2: $p_{\mathbf{0},S_1 \cup S_1^*} \leq \mathbb{P}(CC_1 - \text{fair}) \leq p_{\mathbf{0},S_1 \cup S_1^*} + p_{\mathbf{0},E}$. The claim follows, since, as shown previously, $p_{\mathbf{0},S_1 \cup S_1^*} \to 1/n$ and $p_{\mathbf{0},E} \to 0$. □

Corollary 3. *Under ExtremePA, the probability the outcome is ex-post fair goes to $1/n!$ as $m \to \infty$.*

---

[6] This can be easily shown for $c = 4n/(n-1)$ by using Lemma 1 to prove that $p_{B^*,S \cup S^*} \geq 1/2n$ (although better bounds can be obtained).

[7] "Although, on average, quality is positively related to success, songs of any given quality can experience a wide range of outcomes" (from [7], p. 855).

Proof. By Lemma 2, when $m \to \infty$ the probability of achieving a tie in the number of followers after the first round goes to zero. After ignoring ties, only outcomes with $a^1_{\cdot,1} > \cdots > a^1_{\cdot,n}$ lead to a fair outcome. By symmetry, all strict orderings of $(a^1_{\cdot,i})_i$ have an equal probability. Since there are $n!$ such orderings, $\mathbb{P}(a^1_{\cdot,1} > \cdots > a^1_{\cdot,n}) \to 1/n!$ as $m \to \infty$. The conclusion follows. □

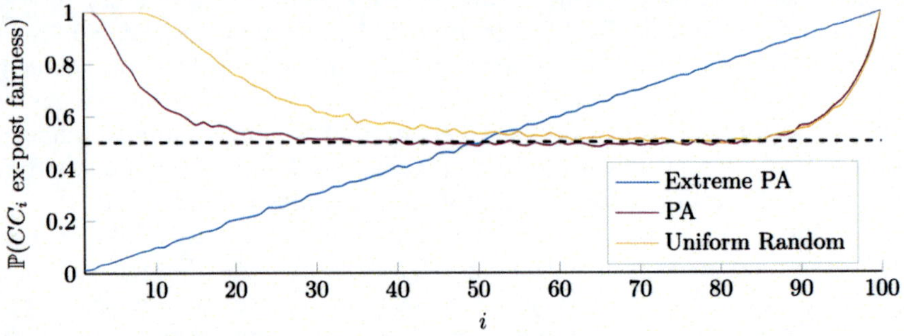

Fig. 3. Sampling probability of ex-post CC_i-fairness, under the three RSs: ExtremePA (blue), PA (purple), and UR (orange). For each RS, we run 10,000 simulations until convergence with $n = 100$ and m =10,000. The dashed line denotes a reference value of 0.5. Fairness under ExtremePA increases from best- to lowest-quality CCs. PA and UR achieve higher fairness for the best CCs.

The behavior depicted in Fig. 3 is consistent with the theoretical results on ex-post fairness. In particular, while UR and PA guarantee CC_1-fairness at convergence, ExtremePA only achieves it with a low probability (about 1%). More precisely, the probability of ex-post CC_i-fairness grows linearly from $1/n$ to 1 with the quality index i. In contrast, UR and PA generally achieve more (less) fair outcomes for the top (lowest) half of CCs. Between the two, the most exploratory recommendation strategy (UR) is more fair for most high-quality CCs, while PA is fair only for a small number of top (and bottom) quality CCs.

Theorem 5. *Under ExtremePA, the final outcome is ex-ante fair.*

Proof. The proof is based on the following observation: when CC_i becomes the most followed CC, they will be eventually followed by all CCs who did not follow a better-quality CC before. For example, if CC_2 is the most followed after the first round, all users (except those who were recommended CC_1) will follow CC_2 after round 2. Differently, if CC_1 was the most followed, then everybody will end up following CC_1 next, while if CC_3 was the most followed, then everybody except those who originally followed CC_1 or CC_2 will follow CC_3. This intuitively leads to CC_2 having more followers in expectation than CC_3 and fewer than CC_1.

For simplicity, we will only formalize this intuition for $n = 2$:

$$\mathbb{E}[a^{\infty}_{\cdot,1}] = \sum_{k=0}^{m} \mathbb{P}(a^{1}_{\cdot,1} = k) \cdot \mathbb{E}[a^{\infty}_{\cdot,1} | a^{1}_{\cdot,1} = k] = \sum_{k=0}^{\lceil \frac{m-1}{2} \rceil} \frac{\binom{k}{m}}{n^m} \cdot k + \sum_{k=\lceil \frac{m-1}{2} \rceil + 1}^{m} \frac{\binom{k}{m}}{n^m} \cdot m.$$

As the sum is larger than $\mathbb{E}[a^{\infty}_{\cdot,2}] = \sum_{k=0}^{m} \frac{\binom{k}{m}}{n^m} \cdot k$, ExtremePA is ex-ante fair. □

5 Conclusion

In this work we analyzed the role recommendations play in UGC-based social networks with respect to the individual fairness for CCs. To do so, we (a) extended prior models with a RS-framework, (b) introduced a non-exploratory RS, and (c) defined ex-ante and ex-post measures of fairness. Our results showed that the network formation process is an absorbing Markov Chain with different absorbing states, expected times, and fairness guarantees depending on the RS. In particular, the expected absorbing time under ExtremePA is bounded by 2, i.e., much faster than the ones under UR or PA (which are linearly or sublinearly increasing in m). Furthermore, while all studied RSs guarantee ex-ante fairness, ex-post fairness is rarely attainable: without exploration (i.e., under ExtremePA), ex-post fairness is achieved with probability $1/n!$, and even with exploration (i.e., under UR or PA), ex-post fairness is guaranteed only for the best (and lowest) CCs. In essence, exploration in RSs trades faster absorption times for higher probabilities of achieving fair outcomes for the best CCs.

A key feature of our model is its simplicity. On the one hand, some level of model simplicity is both inevitable for a feasible theoretical analysis and useful for its clarity. On the other hand, the assumptions about both users and the RS are not always fulfilled by real-world systems. For example, here we reduced both the interests of users and the performance of CCs to one ordinal dimension: quality. Subsequent iterations of this model, which include multiple and different types of attributes, would automatically go beyond a uniform preference of users. Therefore, they could investigate the effect of personalized RSs (e.g., collaborative filtering), look beyond individual fairness (e.g., group fairness), and capture the evolution of users (e.g., changes in the taste of users). In short, UGC-centered platforms are complex sociotechnical systems, and our work is just one piece of this elaborate puzzle. Although there are still many unanswered questions, we believe this mathematical analysis carries one cornerstone message: if we want a fair community, one which rewards its creators based on the quality of their work, then we must encourage exploration.

Acknowledgements. The authors gratefully acknowledge financial support from the University of Zürich and the SNSF under the NCCR Automation grant (Grant Number 180545). We also thank the anonymous reviewers for their valuable feedback.

References

1. Bakshy, E., Rosenn, I., Marlow, C., Adamic, L.: The role of social networks in information diffusion. In: Proceedings of the 21st international conference on World Wide Web, pp. 519–528 (2012)
2. Hall, W., Tinati, Ramine, Jennings, W.: From Brexit to Trump: social media's role in democracy. Computer **51**(1), 18–27 (2018)
3. Stacy Adams, J.: Towards an understanding of inequity. J Abnormal Social Psychol. **67**(5), 422 (1963)
4. Pagan, N., Mei, W., Li, C., Dörfler, F.: A meritocratic network formation model for the rise of social media influencers. Nature Commun. **12**(1), 1–12 (2021)
5. Zipf, G.K.: Human behavior and the principle of least effort: an introduction to human ecology. Ravenio Books (2016)
6. Vany, A.D.: Hollywood economics: how extreme uncertainty shapes the film industry. Routledge (2003)
7. Salganik, M.J., Dodds, P.S., Watts, D.J.: Experimental study of inequality and unpredictability in an artificial cultural market. Science **311**(5762), 854–856 (2006)
8. McNee, S.M., Riedl, J., Konstan, J.A.: Being accurate is not enough: how accuracy metrics have hurt recommender systems. In: *CHI'06 Extended Abstracts on Human Factors in Computing Systems*, pp. 1097–1101 (2006)
9. Kunaver, M., Požrl, T.: Diversity in recommender systems-a survey. Knowl. Syst. **123**, 154–162 (2017)
10. Helberger, N., Karppinen, Kari, D'Acunto, L.: Exposure diversity as a design principle for recommender systems. Inf. Commun. Soc. **21**(2), 191–207 (2018)
11. Gravino, P., Monechi, B., Loreto, V.: Towards novelty-driven recommender systems. Comptes Rendus Physique **20**(4), 371–379 (2019)
12. Guo, W., Krauth, K.,Jordan, M., Garg, N.: The stereotyping problem in collaboratively filtered recommender systems. In: Equity and Access in Algorithms, Mechanisms, and Optimization, pp. 1–10. ACM (2021)
13. Myerson, R.B.: Utilitarianism, egalitarianism, and the timing effect in social choice problems. Econometrica J. Econ. Soc. 883–897 (1981)
14. Chaney, A.G.B., Stewart, B.M., Engelhardt, B.E.: How algorithmic confounding in recommendation systems increases homogeneity and decreases utility. In: Proceedings of the 12th ACM Conference on Recommender Systems, pp. 224–232 (2018)
15. Erdös, P., Rényi, Alfréd: On random graphs, i. Publicationes Mathematicae (Debrecen) **6**, 290–297 (1959)
16. Watts, D.J., Strogatz, S.H.: Collective dynamics of "small-world" networks. Nature **393**(6684), 440 (1998)
17. Barabási, A., Albert, R.: Emergence of scaling in random networks. Science **286**(5439), 509–512 (1999)
18. Snijders, T.A.B.: Stochastic actor-oriented models for network change. J. Math. Sociol. **21**(1–2), 149–172 (1996)
19. Jackson, M.O.: Social and Economic Networks. Princeton University Press (2010)
20. Verma, S., Rubin, J.: Fairness definitions explained. In: 2018 IEEE/ACM International Workshop on Software Fairness (Fairware), pp. 1–7. IEEE (2018)
21. Garg, P., Villasenor, J., Foggo, V.: Fairness metrics: a comparative analysis. In: 2020 IEEE International Conference on Big Data (Big Data), pp. 3662–3666. IEEE (2020)
22. Mitchell, S., Potash, E., Barocas, S., D'Amour, A., Lum, K.: Algorithmic fairness: choices, assumptions, and definitions. Annual Rev. Stat. Appl. **8**, 141–163 (2021)

23. Binns, R.: On the apparent conflict between individual and group fairness. In: Proceedings of the 2020 Conference on Fairness, Accountability, and Transparency, pp. 514–524 (2020)
24. Lucherini, E., Sun, M., Winecoff, A., Narayanan, A.: T-recs: a simulation tool to study the societal impact of recommender systems (2021). arXiv:2107.08959
25. Nielsen, J.: Participation inequality: lurkers vs. contributors in internet communities. Jakob Nielsen's Alertbox **107**, 108 (2006)

Multidimensional Online American Politics: Mining Emergent Social Cleavages in Social Graphs

Pedro Ramaciotti Morales(✉)

médialab, Sciences Po & LPI, Université Paris Cité, Paris, France
pedro.ramaciottimorales@sciencespo.fr

Abstract. Dysfunctions in online social networks (e.g., echo chambers or filter bubbles) are studied by characterizing the opinion of users, for example, as Democrat- or Republican-leaning, or in continuous scales ranging from most liberal to most conservative. Recent studies have stressed the need for studying these phenomena in complex social networks in additional dimensions of social cleavage, including anti-elite polarization and attitudes towards changing cultural issues. The study of social networks in high-dimensional opinion spaces remains challenging in settings such as that of the US, both because of the dominance of a principal liberal-conservative cleavage, and because two-party political systems structure preferences of users and the tools to measure them. This article builds on embedding of social graphs in multi-dimensional ideological spaces and NLP methods to identify additional cleavage dimensions linked to cultural, policy, social, and ideological groups and preferences. Using Twitter social graph data I infer the political stance of nearly 2 million users connected to the political debate in the US for several issue dimensions of public debate. The proposed method shows that it is possible to identify several dimensions structuring social graphs, non-aligned to liberal-conservative divides and related to new emergent social cleavages. These results also shed a new light on ideological scaling methods gaining attention in many disciplines, allowing to identify and test the nature of spatial dimensions mined on social graphs.

Keywords: Social graphs · Graph embedding · Network homophily · Ideological scaling · Ideologies · Polarization

1 Introduction

The study of socio-political dysfunctions or disorders unfolding in digital social media and social networks [7] has raised to prominence in the past decade, including studies of algorithmic bias [3], extremism [20], or echo chambers [5]. These studies hinge on ontologies for the political positions or stances of online

H. Cherifi et al. (Eds.): COMPLEX NETWORKS 2016 2022, SCI 1077, pp. 176–189, 2023.
https://doi.org/10.1007/978-3-031-21127-0_15

users. Bakshy et al. [3], for example, classified users and content on Facebook as Democrat- or Republican-leaning to analyze cross-cutting recommendations, and Barbera et al. [5] positioned Twitter users on liberal-to-conservative continuous scales to investigate the so-called echo chambers. These *categorical* (e.g., user classification) or *geometrical* (e.g., opinion scales) have been leveraged in several studies. In many settings, including that of the US, categorical approaches are limited to binary classifications, while geometrical ones are often limited to one-dimensional opinion scales. This stems from the fact that the US is in essence a two-party system, and that the US has undergone several decades of *issue alignment* [22]: the alignment of issue positions on several matters, resulting in highly-correlated views on gun control, abortion, racial issues, among several others [18]. Methods for embedding social networks in higher dimensional opinion spaces (with dimensions standing for opinion indicators for different issues of social cleavage) have been developed only recently. These methods allow to tackle, for example, several European socio-political settings, known to require higher *dimensionality* [2] to account for observed social choice data, from roll call voting [10] to online social network activity [28]. Accordingly, recent methods have proposed to embed online social graphs in empirical geometrical spaces of several dimensions, where dimensions stand for indicators of opinions on immigration, income redistribution, or ideological stances such as left- and right-wing positions [27]. In settings such as that of the US and that display a main dominant dimension of social cleavage, there is comparatively less development on methods to mine additional dimensions. In this context, *cleavages* refers to historically-determined political, social or cultural dimensions dividing society in the sense that individuals into groups that structure political debate [15]. The study of online social networks in higher opinion dimensions calls nonetheless for renewed efforts, as recent research highlights the importance of new additional cleavages, in particular relating to anti-elite sentiment and populism, as driving forces in many relevant political and informational phenomena in the US [32].

This article builds on recent ideological scaling [14] and aforementioned graph embedding methods for spatializing social graphs in multi-dimensional ideological spaces [27]. Exploiting graph embedding and NLP methods, it proposes a method to mine cleavage dimensions linked to cultural, policy, social, and ideological groups and preferences in social graphs. The method is applied to Twitter social graph data of nearly 2 million users strongly connected to the online political debate in the US. The proposed method shows that several cleavage dimensions traditionally considered in social network analysis (e.g., conservatism, gun control, patriotism, religion) are indeed strongly related, as most studies find, but further allows to quantitatively measure the relative alignment of these issues. More importantly, the method identifies and computes positions of large numbers of users in emerging, non-aligned dimensions identified in political science research as emerging cleavages, and that several works in social science have hypothesized and put forward in recent years. These results extend previous research works showing that it is possible to spatialize users in the US in several space dimensions that are not highly correlated. These results also shed light on the ubiquitous practice of analyzing social networks with ideological scaling methods, by showing ways to assess its validity and improving existing methods. While traditional ideological scaling methods yield a scale that is highly correlated with Democrat-Republican cleavages (resulting from an issue alignment process), party cleavage in specifically can be more accurately determined among many other aligned cleavages.

2 Related Work

Binary categorical classification of social media and network users counts numerous works, from exploitation of self-reporting and surveys [3] to sophisticated methods using neural networks on heterogeneous graphs [33]. More related to this article, ideological spatialization of relational behavioral data can be traced back at least to the Nominal Three-Step Estimation (NOMINATE) method [21]. The NOMINATE method is used to position parliamentarians and bills in one-dimensional liberal-conservative scales. It assumes (1) the existence of unobservable geometrical ideological parameters or positions, and (2) an underlying homophilic generative process (i.e., parliamentarians vote to approve bills that are ideologically *close* to them in the unobservable space), and uses roll call data in Bayesian inference of the unobservable parameters. This process, called *ideological scaling*, has been used in numerous settings involving social choice: court rulings, campaign donations, parliamentary vote, and social network activity among others [14]. In its most wide-spread form, the homophilic probabilistic generative process is modeled as [4]:

$$P\left(i \leftarrow j|\alpha, \beta, \boldsymbol{\phi}_i, \boldsymbol{\phi}_j\right) = \text{logit}^{-1}\left(\alpha - \beta\|\boldsymbol{\phi}_i - \boldsymbol{\phi}_j\|^2\right), \quad (1)$$

where the probability of observing user j interacting with user i (i.e., $i \leftarrow j$) depends on position and scale parameters α and β, and, most importantly, on the distance between the unobservable positions $\boldsymbol{\phi}_i$ and $\boldsymbol{\phi}_j$ of users i and j. Social choice data (i.e., pairs $i \leftarrow j$) can then be used to infer positions $\boldsymbol{\phi}_i$ for all users i. A large number of works assumes that one dimension is enough to retrieve the main social cleavage in the US, implicitly assuming it is the liberal-conservative one, and uses social network choice data (e.g., Facebook likes [8] or Twitter following link [4]) to compute the position of users in this dimension. These works often rely on ex post validation using text cues to argue that the latent dimension indeed represents the political positions of users. Multi-dimensional inference for $\boldsymbol{\phi}_i$ can be achieved in a computationally-tractable manner with Correspondence Analysis [12], as it has been shown to approximate the inference of unobservable parameters of (1), both theoretically [16] and empirically [5].

Because (1) depends on unobservable parameters $\boldsymbol{\phi}$ through pairwise distances, their inference is invariant to isometric transformations. In particular rotation transformations mean that retrieved dimensions cannot be assured to be aligned with social cleavages that might be structuring social choice. In European settings, using the position of referential users such as politicians of known political parties, and party positions in reference issue spaces (provided, e.g., by political polls or surveys), it has been shown how inferred dimensions display different levels of alignment with issues of public political debate [28]. This means that, in general, it cannot be assured that a single-dimensional ideological scaling model will yield a political opinion scale completely aligned with some presumed main left-right or liberal-conservative cleavage. Using the position of several political parties, this fact has been leveraged in embedding large numbers of users in multi-dimensional space where dimensions stand for identifiable and separate political issues, not requiring ex post interpretation or validation [27]. Because the US is in essence a two-party system, and because of the difficulty in obtaining reference users that have both (1) known positions ϕ in latent space and (2) positions known through political surveys (in particular for issues that might not be highly correlated with liberal-conservative cleavages), similar methods have proved to be more challenging in this setting. This article thus proposes a method for constructing such referential positions for large numbers of users, relating to hypothesized cleavages, based on text descriptions written by users in their online profiles on social network sites. Using both, multidimensional scaling and embedding based on (1), and NLP methods for constructing groups of referential users, this article proposes a method to mine several spatial opinion dimensions, and most importantly, emergent opinions that are not highly-correlated with liberal-conservative cleavages.

3 Social Network Data

Following multidimensional ideological scaling works in Europe [28] and in the US [5], we select a bipartite sub-graph of the Twitter social graph. To capture online social choices that might be revealing of several social and political preferences we take as reference users members of the US Congress. We manually annotate the Twitter accounts of 550 members of the 116th United States Congress (looking for verified accounts corresponding to each congressperson), and collected their 17 952 824 followers (collection performed using Twitter's API in October 27th, 2020, see the Acknowledgements section for privacy-compliance information and references). To minimize the probability of followers being *bots*, we follow criteria adopted by several studies [23, 25, 28, 29] and further identified followers with more than 25 followers (7 325 940), and users that have posted more at least 100 tweets (7 471 365). See [4] for further details behind the rationale for these parameters. To identify users that are strongly connected to political debate and that follow spatial preference models we identify followers that follow at least three members of congress (3 846 925) [4]. We select the 1 821 272 unique followers that satisfy all three conditions.

To establish reference points in latent space, we collect the text self-descriptions made by users in their Twitter profiles (also on October 27th, 2020). Out of 1 821 272 users, 1 442 716 had written any text entry in their Twitter profiles. This collection, performed in the days leading to the 2020 US Presidential Election allows us to investigate cleavages in candidate preferences. The strategy presented in this article consists in finding space dimensions capable of dichotomizing pairs of groups of users identified with keywords used in their profiles, and revealing of social cleavages. We follow several studies to identify several possible cleavages of interest, including [6,21,27,32]: party, candidate, racial cleavages, cleavages in regional politics (urban vs rural, or state vs federal positions), religious cleavages, cleavages relating to gun control, the long-standing issue of communism in the US, cleavages related to liberal "life-styles" [1] (e.g., homosexuality, feminism), welfare, military, patriotism, globalization, conspirationism and mistrust (including mistrust of the establishment and experts), and cleavages around attitudes towards elites. Table 1 summarizes proposed issues of partition according to the surveyed literature, with the keywords identified for the classification of users in binary partitions. User classification relies on a keyword-based strategy of minimalist choice: keywords do not intend to portrait the diverse forms of expression of position on each issue, but rather to identify groups of users with low ambiguity of position. In addition to keywords, we rely on sentiment analysis of profile text to distinguish positive and negative mentions of keywords (using a pre-trained BERT base model for uncased words [11]), assigning to each profile text a sentiment from 1 (very negative) to 5 (very positive). We label text profiles as negative [−] if sentiment is equal to 1, and as positive [+] if sentiment is equal to 5. In Table 1 we also distinguish users whose profiles are not negative [$\not-$]. This is needed, for example, to identify users that might use the word "republican" in their profiles, but in order to utter critique (e.g., "I hate republicans!").

4 Homophily Network Embedding

To identify dimensions that might be revealing of cleavages, we first produce a multi-dimensional space embedding in which these dimensions might emerge as directions. For this, we take the bipartite social subgraph of the 550 members of congress and their 1 821 272 followers to produce an homophily embedding using Correspondence Analysis on the adjacency matrix to compute values ϕ of (1) following the procedure in [28, Section IV]. In this multi-dimensional space, dimensions δ_i ($i = 1, 2, ...$) are ranked according to the information they contain about choices represented in the bipartite social graph, as measured by the inertia. Figure 1 shows the inertia of each dimension and their relative gain, showing that at most the three first dimensions are relatively more informative than the rest. Figure 1 also shows the embedding positions of both, congressional members and followers, and the marginal density on these first three dimensions, estimated for the purposes of visualization with kernel density estimation. We can compute party positions as the mean position of congressional members. As expected, the first—most explicative—dimension, δ_1, stands as a good candidate of dimension for attitudes towards parties (or related cleavages, such as liberal-conservative). However, the question remains whether isometric transformations can improve the distinction made by any dimension between Democrat- and Republican-leaning followers. In particular, we do not know if a rotation might improve the ability of a classifier to distinguish between Democrat and Republicans on the first dimension. We know that δ_1 stands for a latent cleavage, and

Table 1. Summary of the proposed issue partitions of users into minimal groups for mining spatial cleavage dimensions capable of classifying them. For each issue we identify two disjoint groups defined by queries of the Twitter profile text descriptions, including keywords (case insensitive, here all written in lowercase), and sentiment: positive [+], negative [−], and non-negative [$\not-$].

Partition	Values	Criteria: keywords and sentiment in profile text	Users
Party	Republican	"republican(s)" AND [$\not-$]	7 895
	Democrat	"democrat(s)" AND [$\not-$]	14 481
Candidate	Trump sup.	("Trump" OR "MAGA") AND [+]	50 159
	Biden sup.	("Biden" OR "Build Back Better") AND [+]	10 229
Black Lives Matter	Pro-BLM	("BLM" OR "black lives matter"] AND [+]	13 350
	Anti-BLM	("BLM" OR "black lives matter") AND [−]	12 017
Black/BlueLivesMatter	BlackLivesMatter	("black lives matter" OR "blm") AND [$\not-$]	16 690
	BlueLivesMatter	"blue lives matter AND [$\not-$]	240
Rural/Urban	Rural	" rural" AND [$\not-$]	567
	Urban	" urban" AND [$\not-$]	1 724
State/Federal	State	"state" AND [$\not-$]	18 369
	Federal	"federal" AND [$\not-$]	1 233
Religion	Christian	"christian" AND [$\not-$]	22 735
	Atheist	"atheist" AND [$\not-$]	1081
Guns	Pro-Gun	"pro gun" OR "gunner" OR ("guns" AND [$\not-$])	605
	Anti-Gun	"anti gun" OR ("gun laws" OR "gun control" AND [$\not-$])	496
Communism	Anti-Communism	("communist" OR "communism") AND [−]	1 160
	Pro-Communism	("communist" OR "communism") AND [+]	448
Liberal LifeStyle	Pro-Lib. LifeStyle	("gay" OR "feminist" OR "lgbt" OR "feminism") AND [+]	9 659
	Anti-Lib. LifeStyle	("gay" OR "feminist" OR "lgbt" OR "feminism") AND [−]	9 921
Welfare	Libertarian	"libertarian" AND [$\not-$]	3 045
	Welfare	"welfare" AND [$\not-$]	681
Police	Pro-Police	"police" AND [+]	1 811
	Anti-Police	("police" AND [−]) OR "defund the police" or "fuck the police"	1 686
Military	Pro-Military	("army" OR "navy" OR "air force" OR "military") AND [+]	17 721
	Anti-Military	("army" OR "navy" OR "air force" OR "military") AND [−]	8 187
Patriotism	Patriot	"patriot"	27 907
	Anti-imperialist	"imperialism"	120
Local/Global focus	Local focus	"local" AND [$\not-$]	5 532
	Global focus	"global" AND [$\not-$]	6 283
DeepState/Rationalist	DeepState	"deep state"	486
	Rationalist	"rationalist"	111
Anti-elitism	Elite	"ceo" OR "founder" OR "representative" OR "expert" OR "phd" OR "professor"	47 032
	Anti-Elite	"people" OR "elite" OR "bankers" OR "politicians" OR "millionaires" OR "experts"	38 086

we know that it is highly aligned with Democrat-Republican cleavages, but we do not know if it is the best spatial direction for distinguishing these two groups. More broadly, it is not trivial to attribute an inductive meaning to what δ_2 and δ_3 might stand for, or to any other space direction for that matter.

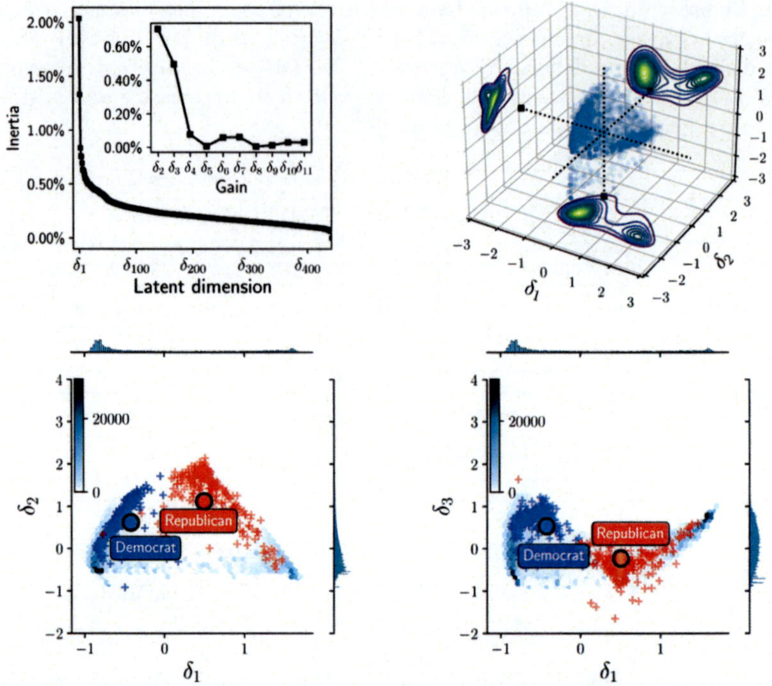

Fig. 1. Multi-dimensional homophily embedding of the collected Twitter network. Dimensions ranked by inertia, and relative gain of including each dimension (top left). Scatter plot and estimated density of the conditional probabilities for the position of users in first three dimensions (top right). Density of followers and positions of members of congress colored by party, and party positions as mean of groups shown in first three dimensions (bottom).

5 Mining Cleavage Dimensions

To extract cleavage dimensions as directions in this latent ideological space, we leverage our proposed set of binary labels from Sect. 3. We first illustrate this on party cleavages. Among users our users, 7895 use the word "republican" and 14 481 the word "democrat" in their profile without negative sentiment. To measure the degree to which δ_1, δ_2 or δ_3 might be good candidate directions for distinguishing these two groups, we fit a logistic regression model on each dimension. With this model, we can compute users that might be rightly or wrongly classified with it, and thus compute a precision, recall and F1-score metric. Figure 2 shows these values and the distribution of these two groups along δ_1, δ_2 and δ_3, showing that δ_1 indeed is the only dimension among the three to produce a meaningful distinction. Alternatively, instead of using a given dimension, we can fit a multivariate logistic regression model, and identify the direction perpendicular to the decision hyper-plane boundary. In the case of our three-dimensional model, the decision boundary will be a plane and the direction a three-dimensional vector (see in Fig. 2). This discovered direction separating these two groups of users is well aligned with δ_1, but it does not produce an improvement in the F1-score. Still, ideological scaling cannot rely on the *a priori* assumption that this will always be the case as it is standard practice in many works in several disciplines, especially in light of research suggesting a decline in left-right cleavages structuring social choice [13,28,32].

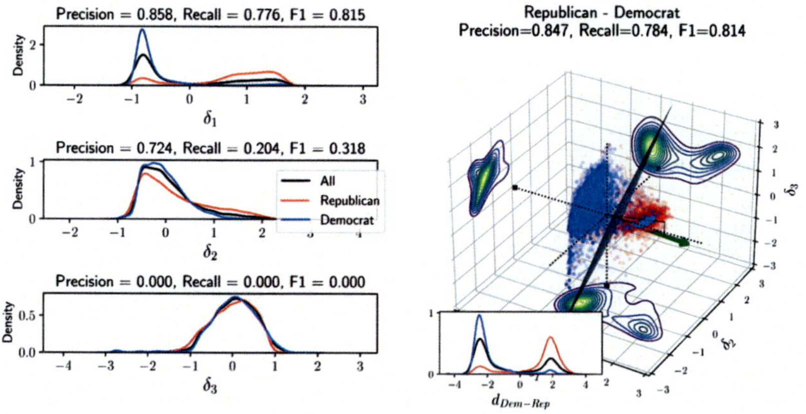

Fig. 2. Distribution Republican- or Democrat-leaning according to their Twitter text profile description, their distribution along the first 3 latent space dimensions, and the accuracy of logistic regression models fitted on each dimension (left). Conditional distributions and positions of labeled users in three-dimensions, and distributions along the direction perpendicular to the boundary of a multivariate logistic regression (right).

This method can be further used to test all proposed binary groups of Table 1, hypothesized to represent potentially relevant cleavages. We fit a multivariate logistic regression model for each pair of groups, and measure the classification accuracy of the model, reported in Table 2, highlighting the cases with F1-score accuracy equal or greater to 0.6. As some pairs are highly imbalanced (e.g., for religious cleavages), we systematically sub-sample the majority group with a Near-Miss strategy [17]. Figure 3 shows a selection of groups of pairs of labeled users, with the decision boundary and discovered orthogonal direction of the fitted multivariate decision model. This selection highlights the different qualities in the accuracy of the multivariate logistic regression classifier, corresponding to different strengths of cleavages for the pairs in each labeled group, under the assumption that the chosen criteria identify a relevant group of users.

The identification of different plausible dimensions of social cleavage presents us with several interesting possibilities. First, the coordinate of each user along these new identified directions can be used to disentangle issue alignment: e.g., we can produce separate and explicit dimensions for party cleavages and attitudes towards gun control. In contrast, δ_1 (usually reported in works using ideological

Table 2. Groups of pairs of labeled users (according to criteria of Table 1), naming of the mined dimension perpendicular to the decision boundary of a multivariate logistic regression classification model, and the accuracy of the fitted model.

Label 1	Label 2	Dimension	Precision	Recall	F1-score
Democrat	Republican	$d_{Dem-Rep}$	0.814	0.823	**0.819**
Trump sup.	Biden sup.	$d_{Trump-Biden}$	0.944	0.874	**0.908**
Pro-BLM	Anti-BLM	d_{BLM}	0.488	0.205	0.289
BlackLivesMatter	BlueLivesMatter	d_{Police}	0.687	0.925	**0.789**
Urban	Rural	d_{Urban}	0.529	0.577	0.552
State	Federal	d_{St-Fed}	0.626	0.536	0.578
Christian	Atheist	$d_{Christian}$	0.706	0.559	0.624
Pro-Gun	Anti-Gun	$d_{Pro-Gun}$	0.713	0.544	**0.617**
Pro-Communism	Anti-Communism	$d_{Comm.}$	0.647	0.650	**0.648**
Anti-Liberal LifeStyle	Pro-Liberal LifeStyle	$d_{LifeStyle}$	0.537	0.867	**0.664**
Libertarian	Welfare	$d_{Welfare}$	0.695	0.720	**0.707**
Pro-Police	Anti-Police	d_{Police}	0.617	0.624	**0.620**
Pro-Military	Anti-Military	$d_{Military}$	0.599	0.474	0.529
Patriot	Anti-imperialist	$d_{Patriot}$	0.709	0.692	**0.700**
Global focus	Local focus	d_{Local}	0.547	0.676	**0.605**
DeepState	Rationalist	$d_{DeepState}$	0.569	0.667	**0.614**
Elite	Anti-Elite	$d_{AntiElite}$	0.549	0.439	0.488

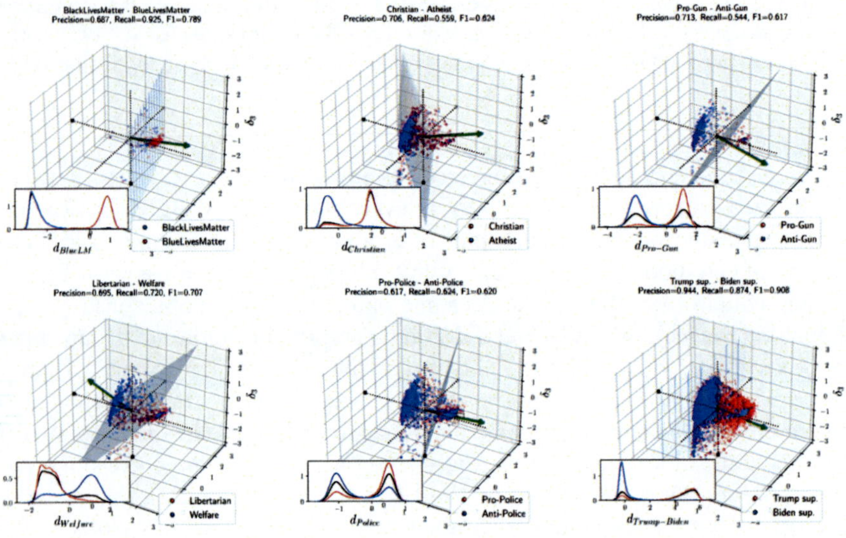

Fig. 3. Selection of groups of pairs of labeled users, with the decision boundary and orthogonal direction of the fitted multivariate decision model.

scaling) is a proxy for party cleavages, but also for other positions on correlated issues (e.g., racial or religious issues, see Fig. 3) in non-explicit ways. Similarly, by inspecting the alignment between different cleavage dimensions we can identify and quantify this issue polarization. Figure 4 shows the mined cleavage dimensions (i.e., with F1-score ≥ 0.6) and their pairwise angular distance, with metrical clusters computed with Un-weighted Pair Group Method with Arithmetic (UPGMA) mean [31]. The mined cleavages can be organized in four *ideologies* (in the sense of issue alignment): (1) a dominant ideology comprising party, candidate, and other stances correlated with δ_1, (2) an ideology separating people defining themselves using the words "local" and "global", (3) an ideology separating those defining themselves using the words "welfare" and "libertarian", and (4) an ideology separating those with positive and negative mentions of issues relating to sexual diversity and feminism, and the use of the word "communism". Four directions cannot be perfectly orthogonal in three-dimensions, but any two cleavages belonging to two different identified ideological groups will display enough angular distance, so as to not be considered as highly correlated. Being able to disentangle issues in separate dimensions, allows us to conduct different investigations against the map positions of actors in now identifiable axes. Because we can also measure the position of reference users (politicians) in mined cleavage dimensions, we can investigate intra-party diversity on separate issues: e.g., of support for their presidential candidate, or attitudes towards welfare, religious diversity, or diversity of views on racial issues. Figure 5 shows, for example, that Republicans are more heterogeneous in their support for Donald Trump than the Democrats in their support for Joseph Biden, both the members of congress (in crosses in Fig. 5) and the followers (density shown in light blue in Fig. 5).

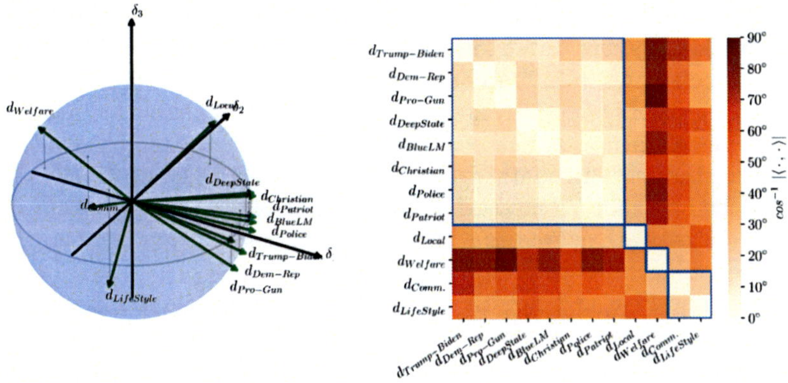

Fig. 4. Mined cleavage dimensions (left) can be organized in four *ideologies* (in the sense of issue polarization), shown in four groups in blue in the angular distance matrix (right).

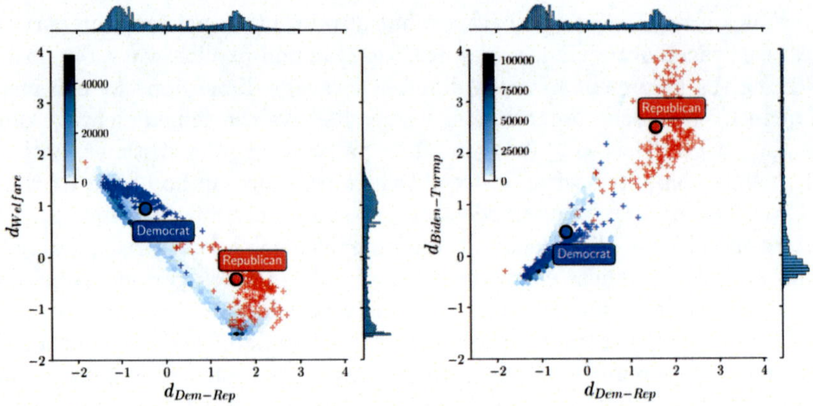

Fig. 5. Density of Twitter users and positions of members of congress along mined dimensions. The distribution of members of congress shows the intra-party diversity of stances towards presidential candidates and welfare.

6 Discussion and Conclusions

This article argued for the importance of re-examining the assumptions implicitly leveraged by studies that use ideological scaling, showing that the dominant latent dimensions cannot be assumed to be the direction that is most aligned with traditional Democrat-Republican cleavages in the US. The degree of alignment between latent dimension and different cleavages can and must be determined and measured. This article further presented a way of mining dimensions of social cleavage with explicit meaning using both network embedding and NLP methods. Furthermore, using this combination of methods, this article analyzed the case of a political Twitter friend network in the US, identifying the main dominant cleavage, but also additional ones hypothesized as relevant by recent studies in social sciences [32]. In particular, four ideologies, or bundled groups of cleavage dimensions were identified. These groups of dimensions are not highly aligned among them, and represent new cleavage dimensions that can be used in further studies. This method also offers the possibility of developing new applications for explicitly measuring issue polarization as the alignment of bundled social cleavages, as well as a method for projecting large numbers of users onto space dimensions with explicit meaning in terms of the issues to which it measures positive and negative views. This new possibility, it was argued, opens interesting paths for research, which was illustrated with a brief example. By measuring positions of Democrat and Republican congressional members on both, party cleavage and candidate cleavage dimensions with data collected in the days leading to the 2020 US presidential election, this article showed that, when compared with Democrats, it may be proved that Republicans display higher heterogeneity in their support for their candidate. Beyond this example, many others could leverage these results and methods. In particular, having multidimensional distributions of political attitudes could be leveraged in the study of social mobilization by analyzing the determinants and characteristics of political movements (see for example [10,24,26]). Additionally, by leveraging information consumption practices and media diets, attitudinal positions could be attributed to news media articles and outlets, allowing for the study of diversity, or lack thereof, in information consumption patterns [19,30]. This, in turn, presents interesting possibilities for large-scale analysis of wide news and informational ecosystems [9].

Acknowledgments. This work has been been funded by the "European Polarisation Observatory" (EPO) of the CIVICA Consortium, and by the French National Agency for Research (ANR) under grants ANR-19-CE38-0006 "Geometry of Public Issues" (GOPI) and ANR-18-IDEX-0001 "IdEx Université de Paris". Data declared the 19 March 2020 and 15 July 2021 at the registry of data processing at the *Fondation Nationale de Sciences Politiques* (Sciences Po) in accordance with General Data Protection Regulation 2016/679 (GDPR) and Twitter policy. For further details and the respective legal notice, please visit https://medialab.sciencespo.fr/en/activities/epo/.

References

1. Bakker, R., Hooghe, L., Jolly, S., Marks, G., Polk, J., Rovny, J., Steenbergen, M., Vachudova, M.A.: 2019 Chapel Hill expert survey. Chapel Hill (2020). www.chesdata.eu
2. Bakker, R., Jolly, S., Polk, J.: Complexity in the European party space: exploring dimensionality with experts. Eur. Union Polit. **13**(2), 219–245 (2012)
3. Bakshy, E., Messing, S., Adamic, L.A.: Exposure to ideologically diverse news and opinion on Facebook. Science **348**(6239), 1130–1132 (2015)
4. Barberá, P.: Birds of the same feather tweet together: Bayesian ideal point estimation using twitter data. Polit. Anal. (2015)
5. Barberá, P., Jost, J.T., Nagler, J., Tucker, J.A., Bonneau, R.: Tweeting from left to right: is online political communication more than an echo chamber? Psychol. Sci. **26**(10), 1531–1542 (2015)
6. Baumann, F., Lorenz-Spreen, P., Sokolov, I.M., Starnini, M.: Modeling echo chambers and polarization dynamics in social networks. Phys. Rev. Lett. **124**(4), 048301 (2020)
7. Benkler, Y., Faris, R., Roberts, H.: Network propaganda: manipulation, disinformation, and radicalization in American politics. Oxford University Press (2018)
8. Bond, R., Messing, S.: Quantifying social media's political space: estimating ideology from publicly revealed preferences on Facebook. Am. Polit. Sci. Rev. **19** (2015)
9. Cointet, J.P., Cardon, D., Mogoutov, A., Ooghe-Tabanou, B., Plique, G., Ramaciotti Morales, P.: Uncovering the structure of the French media ecosystem (2021). arXiv:2107.12073
10. Cointet, J.P., Ramaciotti Morales, P., Cardon, D., Froio, C., Mogoutov, A., Ooghe-Tabanou, B., Plique, G.: What colours are the yellow vests? an ideological scaling of facebook groups. Statistique et Société (2021)
11. Devlin, J., Chang, M.W., Lee, K., Toutanova, K.: Bert: pre-training of deep bidirectional transformers for language understanding (2018). arXiv:1810.04805
12. Greenacre, M.: Correspondence analysis in practice (2017)

13. Grossman, E., Sauger, N.: Economic internationalization and the decline of the left–right dimension. Party Politics (2019)
14. Imai, K., Lo, J., Olmsted, J.: Fast estimation of ideal points with massive data. Am. Polit. Sci. Rev. **110**(4), 631–656 (2016)
15. Lipset, S.M., Rokkan, S., et al.: Cleavage structures, party systems, and voter alignments: an introduction (1967)
16. Lowe, W.: Understanding wordscores. Political Analysis (2008)
17. Mani, I., Zhang, I.: knn approach to unbalanced data distributions: a case study involving information extraction. In: Proceedings of Workshop on Learning from Imbalanced Datasets, vol. 126, pp. 1–7. ICML (2003)
18. Mason, L.: "i disrespectfully agree": the differential effects of partisan sorting on social and issue polarization. Am. J. Polit. Sci. **59**(1), 128–145 (2015)
19. Morales, P.R., Lamarche-Perrin, R., Fournier-S'Niehotta, R., Poulain, R., Tabourier, L., Tarissan, F.: Measuring diversity in heterogeneous information networks. Theoret. Comput. Sci. **859**, 80–115 (2021)
20. O'Callaghan, D., Greene, D., Conway, M., Carthy, J., Cunningham, P.: Down the (white) rabbit hole: the extreme right and online recommender systems. Social Sci. Comput. Rev. **33**(4), 459–478 (2015)
21. Poole, K.T., Rosenthal, H.: A spatial model for legislative roll call analysis. Am. J. Polit. Sci. 357–384 (1985)
22. Poole, K.T., Rosenthal, H.: Congress: a political-economic history of roll call voting. Oxford University Press on Demand (2000)
23. Ramaciotti Morales, P., Cointet, J.P.: Auditing the effect of social network recommendations on polarization in geometrical ideological spaces. In: 15th ACM Conference on Recommender Systems, RecSys' 21 (2021)
24. Ramaciotti Morales, P., Cointet, J.P., Benbouzid, B., Cardon, D., Froio, C., Metin, O.F., Ooghe, B., Plique, G.: Atlas multi-plateformes d'un mouvement social: Le cas des gilets jaunes. Statistique et Société (2021)
25. Ramaciotti Morales, P., Cointet, J.P., Laborde, J.: Your most telling friends: propagating latent ideological features on twitter using neighborhood coherence. In: 2020 IEEE/ACM International Conference on Advances in Social Networks Analysis and Mining (ASONAM), pp. 217–221. IEEE (2020)
26. Ramaciotti Morales, P., Cointet, J.P., Froio, C.: Posters and protesters. J. Comput. Social Sci. 1–29 (2022)
27. Ramaciotti Morales, P., Cointet, J.P., Muñoz Zolotoochin, G., Fernández Peralta, A., Iñiguez, G., Pournaki, A.: Inferring attitudinal spaces in social networks (2022)
28. Ramaciotti Morales, P., Cointet, J.P., Muñoz Zolotoochin, G.: Unfolding the dimensionality structure of social networks in ideological embeddings. In: Proceedings of the 2021 IEEE/ACM International Conference on Advances in Social Networks Analysis and Mining, pp. 333–338 (2021)
29. Ramaciotti Morales, P., Muñoz Zolotoochin, G.: Measuring the accuracy of social network ideological embeddings using language models. In: 2022 International Conference on Information Technology & Systems (ICITS22) (2022)
30. Ramaciotti Morales, P., Tabourier, L., Ung, S., Prieur, C.: Role of the website structure in the diversity of browsing behaviors. In: Proceedings of the 30th ACM Conference on Hypertext and Social Media, pp. 133–142 (2019)
31. Sokal, R.R.: A statistical method for evaluating systematic relationships. Univ. Kansas, Sci. Bull. **38**, 1409–1438 (1958)
32. Uscinski, J.E., Enders, A.M., Seelig, M.I., Klofstad, C.A., Funchion, J.R., Everett, C., Wuchty, S., Premaratne, K., Murthi, M.N.: American politics in two dimen-

sions: partisan and ideological identities versus anti-establishment orientations. Am. J. Polit. Sci. **65**(4), 877–895 (2021)
33. Xiao, Z., Song, W., Xu, H., Ren, Z., Sun, Y.: Timme: twitter ideology-detection via multi-task multi-relational embedding. In: Proceedings of the 26th ACM SIGKDD International Conference on Knowledge Discovery & Data Mining, pp. 2258–2268 (2020)

Classical and Quantum Random Walks to Identify Leaders in Criminal Networks

Annamaria Ficara[1(✉)], Giacomo Fiumara[1], Pasquale De Meo[2], and Salvatore Catanese[1]

[1] MIFT Department, University of Messina, 98166 Messina, Italy
aficara@unime.it, gfiumara@unime.it, scatanese@unime.it
[2] DICAM Department, University of Messina, 98168 Messina, Italy
pdemeo@unime.it

Abstract. Random walks simulate the randomness of objects, and are key instruments in various fields such as computer science, biology and physics. The counter part of classical random walks in quantum mechanics are the quantum walks. Quantum walk algorithms provide an exponential speedup over classical algorithms. Classical and quantum random walks can be applied in social network analysis, and can be used to define specific centrality metrics in terms of node occupation on single-layer and multilayer networks. In this paper, we applied these new centrality measures to three real criminal networks derived from an anti-mafia operation named Montagna and a multilayer network derived from them. Our aim is to (i) identify leaders in our criminal networks, (ii) study the dependence between these centralities and the degree, (iii) compare the results obtained for the real multilayer criminal network with those of a synthetic multilayer network which replicates its structure.

Keywords: Classical random walks · Quantum random walks · Centrality · Criminal networks

1 Introduction

The term "random walk" was firstly introduced in 1905 by Pearson [20]. Random walks can be useful to solve practical problems. In fact, the randomness of objects can be analyzed and simulated through them. Also the correlation among objects can be computed using random walks. They are fast becoming key instruments in various fields such as physics, economics, computer science, biology, or chemistry [25].

A simple random walk model in the mathematical space is a random process on a regular lattice where at each step one point jumps to another position according to a certain probability distribution [25]. Random walks can be applied to a network (or graph) defined as a set of nodes with edges between them. Graphs can be organized into multiple layers, which represent different types of nodes or edges. In this kind of application, the transition probability

H. Cherifi et al. (Eds.): COMPLEX NETWORKS 2016 2022, SCI 1077, pp. 190–201, 2023.
https://doi.org/10.1007/978-3-031-21127-0_16

between nodes is greater when the strength of the association between nodes is stronger. After an appropriate number of steps, a random path able to describe the structure of a network can be obtained. In computer science, the PageRank is the most known algorithm based on random walks [19]. This algorithm randomly walks among web pages to compute their importance.

Aharonov, Davidovich and Zagury were the first to propose quantum walks in 1993 [1]. These walks can be considered as the counter part of classical random walks in quantum mechanics. The main difference between quantum walks and classical random walks is that the former do not converge to some limiting distributions [25]. Quantum interference can cause significantly faster or slower spreading of quantum walks compared to classical random walks. Algorithms based on quantum walks have lower time complexity with respect to the ones based on classical random walks, and they provide an exponential speedup over any classical algorithm [11].

Random walks can find an application in computer vision, semi-supervised learning, network embedding, and social network analysis [17,22]. Some researchers also studied the application of random walks on knowledge discovery, graphs, science of science, and text analysis [2,12,22]. Quantum walks can be used instead to element distinctness, search problems, and decision trees [3,23]. Moreover, classical algorithms are often accelerated by using quantum walks.

Wald and Böttcher [24] at first introduced a framework for classical, quantum, and hybrid random walks with stochastic resetting on networks. Then, Böttcher and Porter [6] used continuous-time classical and quantum random walks to formulate occupation, PageRank, betweenness, and closeness centralities for multilayer networks.

In this paper, we decided to apply classical and quantum occupation centralities on very specific kinds of real networks, i.e. Mafia networks. These networks are different from traditional social networks because first of all they consist entirely of suspected criminals, and then they are considered as organizations of criminal groups and activities. For this reason, research on criminal networks is focused on their structure, efficiency, resilience and capacity to achieve criminal objectives. To the best of our knowledge, classical and quantum walks have never been used to identify leaders in criminal networks. Criminal leaders are strategic brokers within criminal networks [8]. Moreover, to reinforce our finding, we want to compare the results of classical and quantum walks on Mafia networks to the ones obtained on Barabási-Albert (BA) [5] models, which according to our previous work [9], are the one to better replicate criminal networks.

Mafia networks are in fact a good example of real-world networks with respect to the geometry of connections [16]. Indeed, these connections are the building blocks of the entire Mafia, and more generally of organized crime. The Sicilian Mafia is one of the major native mafia-like organizations. It is composed by loose confederations of about one hundred families (or cosche). Since at least the 19th century, the social and economic life especially in Southern Italy is affected by these families [18]. Social network analysis and graph theory can be applied to Mafia networks to describe their structure and functioning, to identify leaders [15], to evaluate police interventions aimed at disrupting them [10], or to construct crime prevention systems [7].

2 Classical and Quantum Random Walk Centrality Measures

2.1 Classical Random Walks

Random walks on graphs are sequences of nodes which start from a source node by selecting an edge, crossing the edge to reach a new node, and repeating the process. An equivalent concept can be found in the Markov chain which is a stochastic process that assumes values in a discrete set, where the next state of the chain only depends on the current state and not on the past states [21]. The Markov chain can be viewed as a graph where nodes are represented by the states and edges by the possible next states, which are randomly determined. The set of states is discrete, but their evolution in terms of time can be discrete or continuous.

In a discrete-time Markov chain, each step is associated with a probability distribution, i.e. a set of probabilities that a walker is on a node or in a state. This probability distribution can be described with a vector, after choosing an order for the states. Given a graph $G(V, E)$ with set of nodes $V = \{v_1, \ldots, v_n\}$ ($|V| = n$) and set of edges E, the probability distribution is described by the vector $[p_1(t) \cdots p_n(t)]$, where $p_1(t)$ is the probability that a walker is on node v_1 at time t and so on. If the process begins with the walker on the first node, $p_1(0) = 1$ and $p_i(0) = 0$ for $i = 2, \ldots, n$.

The future position of the walker cannot be precisely identified, but it is possible to determine a probability distribution knowing the transition matrix M. By knowing the probability distribution at the time t, the distribution at time $t+1$ can be obtained by the following formula:

$$p_i(t+1) = \sum_{j=1}^{n} M_{ij} p_j(t) \,. \tag{1}$$

In vector form, we have $\vec{p}(t+1) = Mp(t)$.

The entry M_{ij} of the transition matrix is the probability of the walker, who is on node v_j, to go to node v_i. The transition matrix is related to the adjacency matrix A of the graph, and it can be defined as:

$$M_{ij} = \frac{A_{ij}}{k_j} \,, \tag{2}$$

where k_j is the degree of node v_j, $A_{ij} = 1$ when two nodes v_i and v_j are connected or $A_{ij} = 0$ if they are not. If there is no edge from v_j to v_i, the walker goes to one of the adjacent nodes and transition probability is the same for all of them.

When time is a continuous variable, the problem of a walker who can go from node v_j to a neighbour v_i at any time can be seen as a liquid seeping from v_j to v_i. At the beginning, it is likely that the walker is in the node v_j. As time goes by, the probability that the walker is in v_j decreases, but the probability he is in one of its neighbors increases. The transition between neighbors occurs with a probability γ per unit time, which is a transition rate presumably constant for all times and for all nodes. An infinitesimal time interval ϵ is generally used to set up and solve the differential equation of the problem with continuous variables. Then, the probability that a walker goes from node v_j to v_i is given by $\gamma\epsilon$. If node v_j has a degree k_j, it means it has k_j neighbors. Therefore, the probability that the the walker is in one of the neighbors of v_j after time ϵ is $k_j\gamma\epsilon$, while the probability of finding him in v_j is $1 - k_j\gamma\epsilon$. At this point, the entry $M_{ij}(t)$ of the transition matrix at time t is defined as the probability that the walker on node v_j goes to node v_i in the time interval t:

$$M_{ij}(\epsilon) = \begin{cases} 1 - k_j\gamma\epsilon + O(\epsilon^2), & \text{if } i = j; \\ \gamma\epsilon + O(\epsilon^2), & \text{if } i \neq j. \end{cases} \tag{3}$$

An auxiliary matrix called generating matrix H, is also defined as:

$$H_{ij}(\epsilon) = \begin{cases} k_j\gamma, & \text{if } i = j; \\ -\gamma, & \text{if } i \neq j \text{ and adjacent}; \\ 0, & \text{if } i \neq j \text{ and non-adjacent}. \end{cases} \tag{4}$$

The next state of a Markov chain only depends on the current configuration of the chain. The transition matrix can be multiplied at different times as follows:

$$M_{ij}(t+\epsilon) = \sum_k M_{ik}(t)M_{kj}(\epsilon)\,. \tag{5}$$

The index k runs over all the neighbours of node v_j. In fact, if the walker is in node v_j, the probability that he goes to v_k in the time interval is $M_{kj}(\epsilon)$, independently of the value of k. If there is no edge between v_j and v_k for a specific k, $M_{kj}(\epsilon) = 0$.

By isolating the term $k = j$, using the Eqs. 3 and 4, and then by moving the first term on the right-hand side to the left-hand side and dividing it by ϵ, the following differential equation is obtained:

$$\frac{dM_{ij}(t)}{dt} = -\sum_k H_{kj}M_{ik}(t). \tag{6}$$

The solution of Eq. 6 with initial condition $M_{ij}(0) = \delta_{ij}$ is:

$$M(t) = e^{-Ht}\,. \tag{7}$$

After the definition of the transition matrix, the probability distribution at time t can be easily obtained. If the initial distribution is $\vec{p}(0)$, $\vec{p}(t) = M(t)\vec{p}(0)$.

2.2 Quantum Random Walks

The electron spin is described by quantum mechanics as a unit vector in the Hilbert space $\mathbb{C}^2$ [21]. The spin up and the spin down are respectively described by the vectors $|0\rangle$ and $|1\rangle$:

$$|0\rangle = \begin{bmatrix} 1 \\ 0 \end{bmatrix}, \quad |1\rangle = \begin{bmatrix} 0 \\ 1 \end{bmatrix}. \tag{8}$$

The notions of spin up and spin down refer to $\mathbb{R}^3$. Quantum mechanics describes in fact the behavior of the electron before entering the magnetic field. In this case and if the electron is somehow isolated from the macroscopic environment, its spin state is described by a linear combination of vectors $|0\rangle$ and $|1\rangle$. Therefore, $|\psi\rangle = a_0|0\rangle + a_1|1\rangle$, where the coefficients a_0 and a_1 are complex numbers which satisfy the constraint $|a_0|^2 + |a_1|^2 = 1$.

The time evolution of an isolated quantum system is described by a unitary transformation. If the state of the quantum system at time t_1 is described by vector $|\psi_1\rangle$, the system state $|\psi_2\rangle$ at time t_2 is obtained from $|\psi_1\rangle$ by a unitary transformation U, which depends only on t_1 and t_2. Therefore, we have $|\psi_2\rangle = U|\psi_1\rangle$.

The evolution postulate is to be written in the form of a differential equation which provides a method to obtain operator U once given the physical context. It is called Schrödinger equation and it is as follows:

$$\frac{d|\psi\rangle}{dt} = -iH|\psi\rangle. \tag{9}$$

In the passage from the continuous-time classical walk model to the continuous-time quantum walk model, the transition matrix is simply converted to an equivalent unitary operator, and the vector which describes the probability distribution to a state vector. Matrix H, given by Eq. 4 is Hermitian, and therefore matrix M given by Eq. 7 is not unitary. M can be easily made unitary within the context of Hilbert spaces by multipling h with the imaginary unit, that is by replacing H with iH. Then, the evolution operator of the continuous-time quantum walk can be defined as follows:

$$U(t) = e^{-iHt}. \tag{10}$$

If the initial condition is $|\psi(0)\rangle$, the quantum state at time t is:

$$|\psi(t)\rangle = U(t)|\psi(0)\rangle, \tag{11}$$

and the probability distribution is:

$$p_k = |\langle k|\psi(t)\rangle|^2, \tag{12}$$

where k runs over all nodes of the graph (or states of the Markov chain) and $|k\rangle$ is the state of the computational basis which correspond to the node v_k. The computational basis of space $\mathbb{C}^2$ is the set $\{|0\rangle, |1\rangle\}$.

2.3 Random Walk Occupation Centralities

Recently continuous-time classical and quantum random walk centrality measures have been defined for single-layer and multilayer networks [6,24].

Starting from Eq. 7, the explicit Euler integration scheme can be used to simulate a classical random walker:

$$p_{t+1} = 1 - H_c \Delta t \, p_t \,, \tag{13}$$

where H_c is the classical Hamiltonian defined as $H_c = LD^{-1}$, where $L = D - A$ is the Laplacian matrix and D is the degree matrix with $D_{ii} = k_i$ if $i = j$ and $D_{ij} = 0$ if $i \neq j$.

The classical random walks approaches for connected networks and sufficiently long times a stationary probability distribution p^* such that:

$$p_i^* = \sum_{j=1}^{n} \frac{A_{ij}}{k_j} p_j^* = \frac{k_i}{\sum_{i=1}^{n} k_i} \,. \tag{14}$$

The stochastic vector p_i^* indicates the stationary classic random walk occupation probability OP_c for a node v_i, and it can be used to define a random walk centrality measure for graphs.

The Schrödinger equation (see Eq. 9) can be instead solved using the Crank-Nicholson integration scheme [4,24]. Expanding the evolution operator in Eq. 10 into a Taylor series, we have:

$$|\psi_{t+1}\rangle = -2i\Delta t H_q |\psi_t\rangle + |\psi_{t-1}\rangle \,, \tag{15}$$

where H_q is the Hermitian quantum Hamiltonian defined as $H_q = \mathcal{L}$, where $\mathcal{L} = D^{-\frac{1}{2}} L D^{-\frac{1}{2}}$ is the normalized version of the Laplacian matrix $\mathcal{L}$.

For continuous-time quantum walks, unitary time evolution does not lead to a stationary state. Instead, the long-time behavior of a quantum walk is characterized by its long-time mean:

$$q_i^* = \lim_{T \to \infty} \frac{1}{T} \int_0^T \langle i || \psi(t) \rangle \langle \psi(t) || i \rangle dt. \tag{16}$$

where dt is an infinitesimal time step, $|\psi(t)\rangle\langle\psi(t)|$ is a density operator, and $|i\rangle \in \mathbb{C}^{\mathcal{N}}$ is an orthonormal basis vector.

Even in this case, q_i^* denotes the continuous-time quantum random walk occupation probability OP_q of a node v_i, and therefore it gives a type of random walk centrality.

3 Methods and Results

We used the classical and quantum occupation centralities described in Sect. 2.3 to study the importance of nodes in three real single-layer criminal networks

(Meetings, Phone Calls and Crimes) related to an anti-mafia operation called Montagna [7,10,13], from which we derived a fourth network in the form of a flattened multilayer network. The Montagna operation focused on the Mistretta and Batanesi clans who monopolized the sector of public contracts in the Tyrrhenian strip and in the nebroidal district of the province of Messina, by working with some entrepreneurs associated to the Sicilian Mafia. Meetings contains 256 physical meetings between 101 suspected criminals observed during the police physical surveillance. Phone Calls contains 124 phone calls between 100 suspects wiretapped during the police audio surveillance. Crimes contains 74 connections between 25 suspected criminals who committed crimes together. The three networks share 20 nodes. We also created an undirected and weighted multilayer network with 226 actors, 454 intralayer edges and 3 layers corresponding to the three single-layer criminal networks Meetings, Phone Calls and Crimes. In this work, a single-layer network (flattened) is created by merging the three layers of the multilayer network. In the flattened network, there is one node for each actor of the starting multilayer network and an edge between two nodes if the corresponding actors are linked in one layer. Meeting, wiretap and co-offending networks provide fundamentally different pictures of the criminal organization, and therefore are all particularly useful in finding leaders. Also the flattened multilayer network can be useful to have a more complete picture of the organization. Figure 1 shows Meetings, Phone Calls, Crimes and the flattened multilayer network.

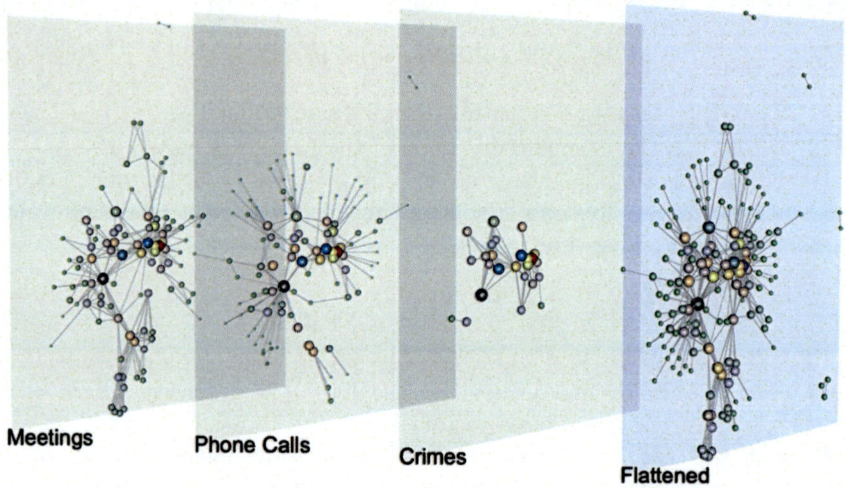

Fig. 1. Meetings, Phone Calls and Crimes and the flattened multilayer network from the Montagna anti-mafia operation.

Our analysis was conducted using the mathematical framework[1] of continuous-time classical and quantum centrality measures created by Böttcher and Porter [6].

The top 20 nodes ranked by their classical and quantum occupation probabilities are shown in Fig. 2. In the analysis of central nodes in a Mafia network, we have to consider the hierarchical structure of a Mafia family in which individuals can be categorized as boss, underboss, consigliere (i.e., an advisor to the boss), messaggero (i.e., a messenger among families), caporegime (i.e., who heads a local crew of soldiers), soldier (*i.e.*, average type criminal) and associate (e.g., entrepreneurs, pharmacists, police officers, politicians etc.). The results for the Meetings and Phone Calls networks confirm the one obtained through the application of the degree as we did in [15], revealing that the most important actors are nodes 18 and 47. These nodes are respectively Caporegime of the Mistretta family and deputy Caporegime of the Batanesi family. The application of OP_c and OP_q on the Crimes network reveals instead different central nodes who are not key members of the Mafia hierarchy but are associates such as entrepreneurs (i.e., nodes 54 and 64), construction workers (i.e., node 72) or soldiers (i.e., nodes 48 and 49). Differently from the application of degree on the multilayer network, in this case, we obtain different central nodes such as 103, 136 and 111. Nodes 103 and 111 were intercepted because they were in phone contact respectively with a Caporegime of the Mistretta family (i.e., node 18) and a non-existent entrepreneur, created for the purpose of fraud (i.e., node 109). Node 136 is instead a cohabitant of a Caporegime of the Batanesi family (i.e., node 27). The role of these figures in the mafia business in unclear but their centrality can direct investigations and lead to deeper investigation of them. Classical and quantum occupation probabilities identify in most cases the same leaders in criminal networks. Therefore, when a micro approach is adopted in SNA application to organized crime focusing, as in our case, on one organization or network with a relatively low number of nodes, there is no clear advantage to using quantum random walks instead of classical ones. However, quantum walks can be useful to speed up the investigations, and to find leaders when a macro approach is adopted to study crime through SNA. In this case, in fact, larger networks focusing on specific national or regional criminal markets or offender categories are analyzed.

In Fig. 3, we show the probabilities that a node with a certain degree is occupied by a classical and quantum random walker. As explained in [6], we can observe a linear dependence between OP_c and OP_q which is due to the fact that quantum walks satisfy Eq. 14 because they do not approach a stationary state. Their long-time behavior is instead characterized by the long-time mean as shown in Eq. 16. Quantum walks are not proportional to the degree which is not able to capture node occupation properties of this kind of walks. These properties, in fact, also depend on other structural features of the underlying networks. The minimum and maximum degrees of the Meetings, Phone Calls, Crimes networks are respectively 1 and 24, 1 and 25, 1 and 13, as shown in the

[1] Available at: https://gitlab.com/ComputationalScience/ctqw_centrality.

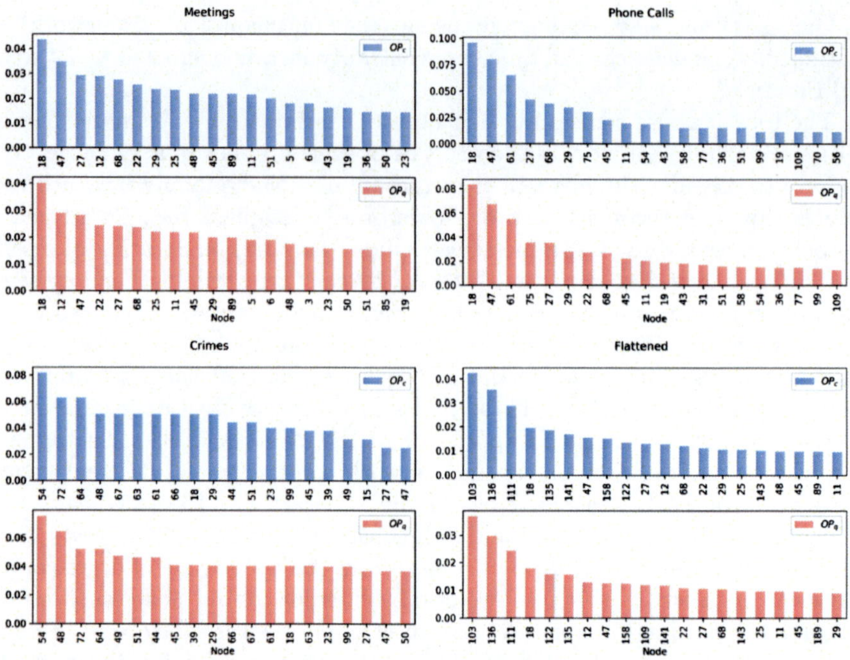

Fig. 2. 20 top ranked nodes computed with classical OP_c and quantum OP_q random walk occupation probabilities in the single layer networks meetings, phone calls and crimes, and the flattened Montagna multilayer network.

plot of the degree distribution in [14]. The differences between these classical and quantum centralities seem to be greater in the flattened multilayer network.

In [6], the authors asserted that the differences between the classical and quantum occupation centralities were (i) larger in BA single-layer networks [24] or in synthetic multilayer networks which include BA layers, and (ii) smaller for empirical networks. The second results is not valuable for real criminal networks, and this can be explained just through the first finding. In our previous work [9], we used some popular network models such as Erdös-Rényi, Watts-Strogatz and different configurations of BA to replicate the topology of a criminal network. Our experiments identified the BA model as the one which better represents a criminal network. To prove our thesis, we decided to build two flattened multilayer networks with three BA layers. Each layer is created with the same number of nodes of the Meetings, Phone Calls, Crimes networks (i.e., 101, 100, and 25, respectively) and a number of edges as close as possible to that of these networks (i.e., 256, 124, and 74, respectively). BA models are built based on a specific parameter m, which indicates the number of edges that are preferentially attached to existing nodes with high degree. For the first layer, we tried two different configurations of the BA graphs with $m = 2$ (obtaining 198 edges) and $m = 3$ (obtaining 294 edges). For the second layer, we used $m = 1$ (obtaining

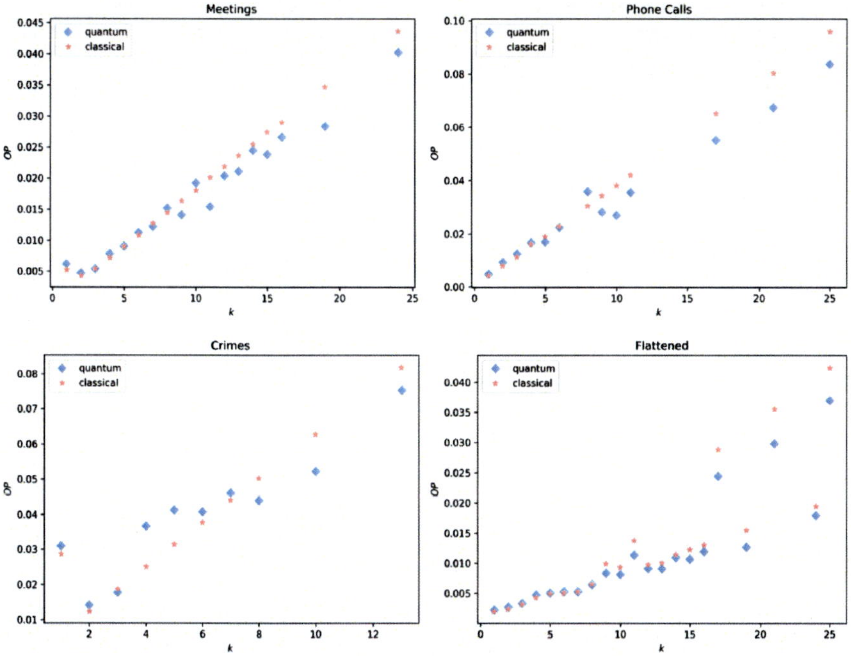

Fig. 3. Classical and quantum random walk occupation probabilities for the single layer networks meetings, phone calls and crimes, and the flattened Montagna multilayer network that a node of degree k is occupied by a classical and quantum random walker, respectively.

99 edges) and $m = 2$ (obtaining 196 edges). For the third layer, we choose $m = 3$ (obtaining 66 edges) and $m = 4$ (obtaining 84 edges). Figure 4 shows classical and quantum occupation centralities versus the node degrees on the described synthetic multilayer networks.

4 Conclusions and Discussion

In this paper, we used classical and quantum continuous-time random walk centrality measures to identify leaders in (i) three single-layer criminal networks related to an anti-Mafia operation called Montagna, and (ii) a flattened multilayer network created from them.

In the formulation of these measures, different Hamiltonians describe the evolution of classical and quantum walks. Other classical and quantum continuous-time random walk centralities, such as random walk betweenness or closeness, can be derived by modifying these Hamiltonians, and tracking different properties of absorbing random walks. In our future works, we want to apply these other measures on our criminal networks to identify leaders, and to define a new disruption framework based on random walks, comparing the targeted attacks

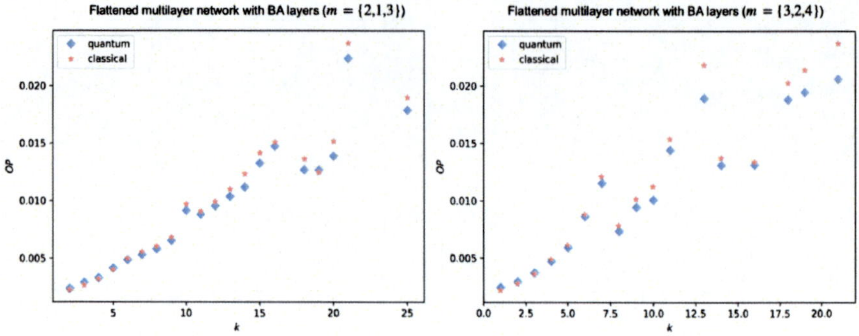

Fig. 4. Classical and quantum random walk occupation probabilities for a flattened multilayer network composed by three Barabási-Albert layers with 101 nodes and 198 edges, 100 nodes and 99 edges, 25 nodes and 66 edges (left column); 101 nodes and 294 edges, 100 nodes and 196 edges, 25 nodes and 84 edges (right column).

based on the classical centrality measures with those based on their random walk versions.

Moreover, we are planning to extend the mathematical framework by Böttcher and Porter [6] for whole multilayer networks considering the layered structure, and not just the flattening function. In fact, we can obtain a complete picture of the activities and the structure of criminal organizations only considering the various layers not as single-layer networks but as a whole multilayer network [14]. Then, we want to explore the possibility of developing a disruption framework for multilayer networks.

References

1. Aharonov, Y., Davidovich, L., Zagury, N.: Quantum random walks. Phys. Rev. A **48**(2), 1687 (1993)
2. Amancio, D.R.: Comparing the topological properties of real and artificially generated scientific manuscripts. Scientometrics **105**(3), 1763–1779 (2015)
3. Ambainis, A.: Quantum walk algorithm for element distinctness. SIAM J. Comput. **37**(1), 210–239 (2007)
4. Askar, A., Cakmak, A.S.: Explicit integration method for the time-dependent Schrodinger equation for collision problems. J. Chem. Phys. **68**(6), 2794–2798 (1978)
5. Barabási, A.L., Albert, R.: Emergence of scaling in random networks. Science **286**(5439), 509–512 (1999)
6. Böttcher, L., Porter, M.A.: Classical and quantum random-walk centrality measures in multilayer networks. SIAM J. Appl. Math. **81**(6), 2704–2724 (2021)
7. Calderoni, F., Catanese, S., De Meo, P., Ficara, A., Fiumara, G.: Robust link prediction in criminal networks: a case study of the Sicilian Mafia. Expert Syst. Appl. **161**, 113666 (2020)
8. Calderoni, F., Superchi, E.: The nature of organized crime leadership: criminal leaders in meeting and wiretap networks. Crime Law Social Change **72**(4), 419–444 (2019)

9. Cavallaro, L., Ficara, A., Curreri, F., Fiumara, G., De Meo, P., Bagdasar, O., Liotta, A.: Graph comparison and artificial models for simulating real criminal networks. In: Benito, R.M., Cherifi, C., Cherifi, H., Moro, E., Rocha, L.M., Sales-Pardo, M. (eds.) Complex Networks and Their Applications IX, pp. 286–297. Springer, Cham (2021)
10. Cavallaro, L., Ficara, A., De Meo, P., Fiumara, G., Catanese, S., Bagdasar, O., Song, W., Liotta, A.: Disrupting resilient criminal networks through data analysis: the case of Sicilian Mafia. PLOS ONE **15**(8), 1–22 (2020)
11. Childs, A.M., Cleve, R., Deotto, E., Farhi, E., Gutmann, S., Spielman, D.A.: Exponential algorithmic speedup by a quantum walk. In: Proceedings of the Thirty-fifth Annual ACM Symposium on Theory of Computing, pp. 59–68 (2003)
12. de Arruda, H.F., Silva, F.N., da F. Costa, L., Amancio, D.R.: Knowledge acquisition: a complex networks approach. Inf. Sci. **421**, 154–166 (2017)
13. Ficara, A., Cavallaro, L., Curreri, F., Fiumara, G., De Meo, P., Bagdasar, O., Song, W., Liotta, A.: Criminal networks analysis in missing data scenarios through graph distances. PLOS ONE **16**(8), 1–18 (2021)
14. Ficara, A., Fiumara, G., Catanese, S., De Meo, P., Liu, X.: The whole is greater than the sum of the parts: a multilayer approach on criminal networks. Future Internet **14**(5), 123 (2022)
15. Ficara, A., Fiumara, G., De Meo, P., Catanese, S.: Multilayer network analysis: the identification of key actors in a Sicilian Mafia operation. In: Perakovic, D., Knapcikova, L. (eds.) Future Access Enablers for Ubiquitous and Intelligent Infrastructures, pp. 120–134. Springer, Cham (2021)
16. Ficara, A., Saitta, R., Fiumara, G., De Meo, P., Liotta, A.: Game of Thieves and WERW-Kpath: two novel measures of node and edge centrality for Mafia networks. In: Teixeira, A.S., Pacheco, D., Oliveira, M., Barbosa, H., Gonçalves, B., Menezes, R. (eds.) Complex Networks XII, pp. 12–23. Springer, Cham (2021)
17. Li, X., Han, Z., Wang, L., Lu, H.: Visual tracking via random walks on graph model. IEEE Trans. Cybern. **46**(9), 2144–2155 (2016)
18. Paoli, L.: Mafia Brotherhoods: Organized Crime, Italian style. OUP (2008)
19. Paparo, G.D., Müller, M., Comellas, F., Martin-Delgado, M.A.: Quantum google in a complex network. Sci. Rep. **3**, 2773 (2013)
20. Pearson, K.: The problem of the random walk. Nature **72**, 294 (1905)
21. Portugal, R.: Quantum Walks and Search Algorithms. Quantum Science and Technology, Springer, New York (2013)
22. Sarkar, P., Moore, A.W.: Random walks in social networks and their applications: a survey. In: Aggarwal, C. (ed.) Social Network Data Analytics, pp. 43–77. Springer, Boston, MA (2011)
23. Shenvi, N., Kempe, J., Whaley, K.B.: Quantum random-walk search algorithm. Phys. Rev. A **67**(5), 052307 (2003)
24. Wald, S., Böttcher, L.: From classical to quantum walks with stochastic resetting on networks. Phys. Rev. E **103**(1), 012122 (2021)
25. Xia, F., Liu, J., Nie, H., Fu, Y., Wan, L., Kong, X.: Random walks: a review of algorithms and applications. IEEE Trans. Emerg. Topics Comput. Intel. **4**(2), 95–107 (2020)

Random Walk for Generalization in Goal-Directed Human Navigation on Wikipedia

Dániel Ficzere(✉), Gergely Hollósi, Attila Frankó, and András Gulyás

Budapest University of Technology and Economics, 1111 Műegyetem rkp. 3, Budapest, Hungary
ficzere.daniel@vik.bme.hu

Abstract. Models of human navigation have been investigated in many ways on complex networks. These findings suggest that the characteristics of human navigation change during the navigation from the start to the destination. However, it is not fully clear to what extent the navigation is defined by the human navigator or the graph and the environment. Our work examines the early phase of human navigation, where we investigate the impact of the graph structure on human navigation with a random walk model based on PageRank. Our results suggest that a very high portion of human navigation in the early generalization phase can be modeled with random navigation.

Keywords: PageRank · Wikipedia · Human navigation · Random walk

1 Introduction

Navigation in complex networks is an important topic that has been investigated by researchers in a number of areas of computer science and beyond. Directed scale-free graphs have a couple of nodes with high incoming degrees (called hubs), which are seem to be responsible for navigating through complex graphs. It is observed e.g. that human navigation in information networks is a two-phase process combines of the phenomena they call the exploitation of the known and the exploration of the unknown [4]. Their results suggest that humans either follow specific links on purpose (exploitation) whenever they are confident enough that those links bring them closer to their particular target, or they select links almost arbitrarily at random (exploration), whenever they do not possess enough knowledge to relate the candidate links to their target. In the exploration part, the user often visits high incoming degree nodes in the graph. Similar phase definition are also discussed by [5] and [9]. [9] states that most subjects navigate through high-degree hubs in the early phase, while their search is guided by content features thereafter, while [5] mentions that their findings confirm that there is a zoom-out and a homing-in phase, where users are guided by generality at first and textual similarity to the target later [1].

H. Cherifi et al. (Eds.): COMPLEX NETWORKS 2016 2022, SCI 1077, pp. 202–213, 2023.
https://doi.org/10.1007/978-3-031-21127-0_17

The main question is whether the early part—especially when the user is lost—can be described by random navigation, i.e. choosing the next edge by a uniform distribution? Human navigation can be random in certain circumstances, e.g. when lost in the forest, humans choose mainly random directions until finding some path or trail. If it is so, then the early part is not only about human behavior but is the *property* of the graph itself. A couple of works apply the incoming degree as the degree of importance of a node in finding the navigation path through the graph [2,5], i.e. it might be correlated with the relative frequency of the node in navigational paths. To model random navigation, and acquire the relative frequency of nodes in random navigation, a first-order Markov chain is applied, supposing a uniformly chosen next link from each node. However, while calculating the stationary distribution of a graph with a couple of million nodes is computationally hard; but the PageRank algorithm helps us to estimate it efficiently.

To investigate random navigation on a complex graph, the whole Wikipedia is used along with Wikigame[1] goal-directed navigation game. Wikipedia is the largest encyclopedia created ever. Besides the topic description, Wikipedia pages contain several hyperlinks to other topics, making Wikipedia an excellent information network for evaluating human navigation behavior. Furhtermore, Wikipedia is a scale-free network, so the node degree distribution follows a power law. In Wikigame, players randomly get a start and a destination article, and the goal is to solve the task of navigation from the start to the destination via as few Wikipedia articles as possible. Players have no knowledge of the global network structure besides semantical knowledge. Thus, they must rely solely on the local information—the outgoing links connecting the current article to its neighbors—and on their expectations about which articles are likely to be interlinked. Thus, Wikigame provides ground-truth human navigational patterns to be compared to random navigation on Wikipedia.

As the main contribution of our paper, we define a model based on PageRank to model the initial phase of human navigation on a scale-free network such as Wikipedia. To validate our model, we create the graph representation of Wikipedia and use the navigation of Wikigame users as a ground truth. We compared the results of our model to the indegree properties of the network as the indegree is also a good measure of generalization in a network. The paper is structured as follows: Sect. 2 introduces the Markov chains for random navigation, the role of node degree for stationary distributions and the applicability of PageRank algorithm for such a problem. The main properties of the used Wikipedia and Wikigame datasets are presented in Sect. 3. We introduce the key metrics and the model validation process for the generalization phase in human-goal directed navigation in Sect. 4. Section 5 concludes our paper and identify further research topics.

[1] https://www.thewikigame.com.

2 Methods

In order to describe random navigation, first-order discrete-time Markov chains with finite state space were used, where the transition probability distribution can be represented by the **P** transition matrix, with the (i, j)th element defined by Eq. 1.

$$p_{ij} = Pr(X_{n+1} = j | X_n = i) \tag{1}$$

In a navigation graph, p_{ij} is the probability of a node transition from node i to node j. A stationary distribution π is a (row) vector, whose entries are non-negative and sum to 1 and it satisfies Eq. 2 for a given **P** transition matrix.

$$\pi \mathbf{P} = \pi \tag{2}$$

π is a normalized ($\sum_i \pi_i = 1$) multiple of a left eigenvector of the transition matrix P with an eigenvalue of 1. Besides that, the used Markov chain is time-homogeneous also, so the **P** can be calculated also by Eq. 3.

$$\lim_{k \to \infty} \mathbf{P}^k = \mathbf{1}\pi \tag{3}$$

In case of a navigation graph, the π distribution gives the relative frequencies of the node visits after an infinite number of steps from a random starting article.

Suppose some graph with N nodes, with adjacency matrix **A** and let $\mathbf{D} = \text{diag}(d_1, \ldots, d_N)$, where $d_i = \sum_j a_{ij}$ is the outgoing degree of the node i. In case of a *random walk*, the transition probabilities can be calculated as

$$\mathbf{P} = \mathbf{D}^{-1}\mathbf{A}. \tag{4}$$

For undirected graphs (i.e. $\mathbf{A} = \mathbf{A}'$), the stationary distribution is proportional to the degrees of the nodes, meaning, that $\pi_i \propto d_i$ [6]. However, for directed graphs (i.e. $\mathbf{A} \neq \mathbf{A}'$), no such simple, closed form solution can be found.

Supposing a large, asymmetric **P** transition matrix, it is computationally challenging to calculate the stationary distribution of a Markov chain, e.g. in case of Wikipedia, **P** is a matrix with 20 million rows and columns, for which the eigen decomposition is an unviable problem. The PageRank algorithm can help to solve such a problem and offers a good approximation for the stationary state of the Markov chain.

The PageRank algorithm was initially implemented in Google's search engine [8]. In PageRank, node's importance can be interpreted as the more a node is pointed by important nodes, the more it is important. PageRank is equivalent to the stationary distribution of a random surfer following a memory-less Markov process. The PageRank algorithm defines the transition matrix from the **P** matrix as

$$\mathbf{R} = (1 - \epsilon)\mathbf{P} + \frac{\epsilon}{N}\mathbf{1}_N, \tag{5}$$

where ϵ is the so-called damping factor, and $\mathbf{1}_N$ is the matrix from ones with size $N \times N$. In the case of many real graph, the transition matrix is not ensured to be ergodic and the ϵ damping factor helps to make the Markov chain irreducible and aperiodic. So during the random walk from node j, with probability $(1-\epsilon)$ the user choose the next article uniformly with probability $1/d_j$ or with probability ϵ the user teleport uniformly towards a arbitrary node of the network. These teleportations ensure that the user cannot be stuck in the network and that the steady state probability distribution is unique.

The stationary distribution is then calculated iteratively, starting from an $\pi^{(0)}$ initial distribution as

$$\pi^{(t+1)} = \pi^{(t)}\mathbf{R}, \tag{6}$$

where the implementation of the PageRank algorithm can exploit the sparsity of matrix $\mathbf{P}$. The convergence to the stationary distribution is governed by the second eigenvalue of the $\mathbf{R}$ matrix, which is less than or equal to the $(1-\epsilon)$ factor [3]. Thus, the error is decreasing with each step depending on the dumping factor as

$$\text{Err}(n+1) \leq (1-\epsilon)\text{Err}(n). \tag{7}$$

By applying the recursive formula and a decent dumping factor, quick convergence can be reached to the stationary distribution. E.g., according to (7) the error decreases after 30 iterations with $\epsilon = 0.15$ by $(1-0.15)^{30}$, which is less then 1%.

3 Evaluation

To investigate the random navigation over complex graphs we used the publicly available Wikipedia [7] graph and the well-known Wikigame dataset for accessing the ground-truth human navigation patterns. The latest version of the whole English Wikipedia[2] have been processed and used, based on its XML description, the graph representation of the Wikipedia have been created. Special cases—such as articles starting with ,,Category:"—have been eliminated from the dataset. Also, there are redirect links in the XML Wikipedia representation, where the next hyperlink of an article redirects to another article. However, the redirection process is invisible for the user, therefore during navigation these redirection steps are also irrelevant. The final size of the graph consisted of more than 21 million nodes and 350 million edges.

For the goal-directed navigation samples we used the private dataset of Wikigame. The main purpose of our work is to analyze and characterize the average human navigation behavior, therefore we collected 150 000 navigation paths from the Wikigame dataset. Each navigation are produced by different users to eliminate user biases and characteristics. Also, there are multiple game types in Wikigame, so we sampled only from the basic game type, so-called "speed-race", where the main goal is to reach the destination article from a random start article via minimal number of articles. Besides that, only finished

[2] English dump, 2022.01.01.

games have been collected. A game is finished only when the destination article is the same as the last step of the user's path in Wikigame. These are quite important steps as the results could be very distorted if e.g. the "5-click-to-Jesus" game type is included to the analysis.

The normalized histogram of the game lengths (user navigation path lengths) is presented in Fig. 1. The figure depicts that more 99% of the game lengths are between 3 and 18 number of steps. In some cases it is important to consider the game lengths, because some characteristics of the navigation change observing shorter or longer games.

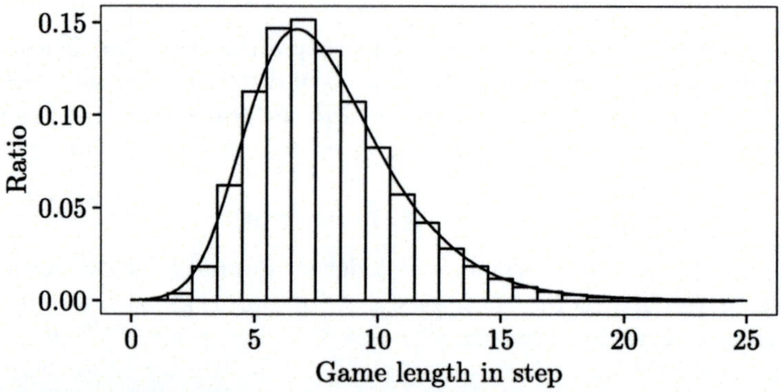

Fig. 1. Histogram of the game length in Wikigame and the calculated density function with 0.74 bandwidth value. It shows that a significant portion of game lengths are between 3 and 18 steps (more than 99%).

We created a C++ application to process and calculate the PageRank values efficiently, while RocksDB was applied to store the dataset and for optimized lookups.

4 Discussion

At first glance, the zoom-out and homing-in phases can be identified on the indegree distribution of the nodes used in a step in Wikigame (Fig. 2). In the zoom-out phase (the generalization phase), the indegrees are increasing, then in the semantical phase (where the user tries to navigate towards the destination node), the indegrees of the nodes are decreasing. However, while the indegree of a node has a great importance in a directed complex graph, it is not clear, how to interpret the indegree regarding graph navigation. In contrast, the stationary distribution of a Markov chain has a well interpretable and probability-based meaning. In this section, we use both the indegree and the PageRank value (i.e. stationary probability) of the nodes to investigate the behavior of human navigation. Our analysis focused on the first generalization phase, more specifically, our hypothesis is that the graph structure has quite significant role in this generalization process, and even the random navigation would offer a great model to this phase, so a user could reach to a "general" node using even random navigation.

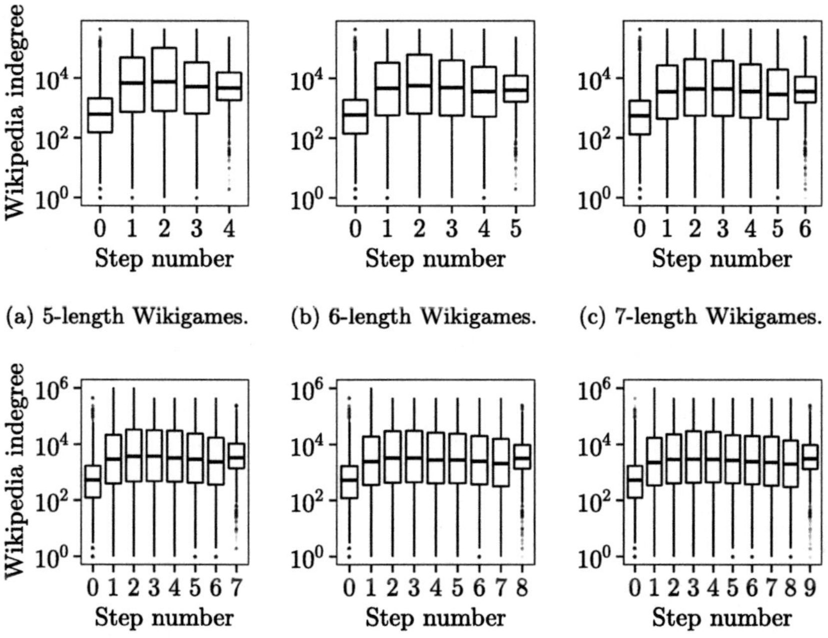

(a) 5-length Wikigames. (b) 6-length Wikigames. (c) 7-length Wikigames.

(d) 8-length Wikigames. (e) 9-length Wikigames. (f) 10-length Wikigames.

Fig. 2. Wikipedia articles' indegree distribution per step for different Wikigame path lengths. The zoom-out and homing-in phases can be identified as in the zoom-out (generalization) phase the indegrees are increasing, then in the homing-in phase, the indegrees of the nodes are decreasing.

4.1 Metrics

To compare ground-truth values and the result of the evaluation (the indegree and the PageRank), three different metrics were used.

The first metric is the *page intersection ratio* (PIR), which is the number of common pages in the top-N pages sorted separately by the PageRank, the indegree and the relative-frequency of the ground-truth usage values. To establish the relevance of this metric, Fig. 3a shows that high portion of all click by Wikigame users comes from a very small portion of Wikipedia articles. Therefore the top articles based on Wikigame usage are quite characteristic properties of the navigation. Based on that we used the top-N graph nodes in our metrics during the further analysis of user generalization problem.

The second metric is the *histogram intersection ratio* (HIR) similarly for the top-N PageRank and indegree articles compared to the ground-truth usage values. As the top-N used nodes of the Wikigame and the top-N articles of indegree are not the same set, the analyzed histograms are based on the union of this two set of article names. The evaluation process was similar for the PageRank articles also. So, after the creation of the normalized histograms we analyzed the „similarity" of the PageRank and indegree histograms compared to the ground-truth usage histogram with the HIR.

The last used metric is the *Jensen-Shannon divergence*, which is a method of measuring the similarity between two probability distributions based on the *Kullback-Leibler divergence*. For discrete probability distributions P and Q defined on the same probability space, χ, the Jensen-Shannon divergence is a symmetric metric and it always has a finite value as defined by Eq. 7.

$$D_{JS}(P \parallel Q) = \frac{1}{2}D_{KL}(P \parallel M) + \frac{1}{2}D_{KL}(Q \parallel M), \tag{8}$$

where M is defined by $M = \frac{1}{2}(P+Q)$, and $D_{KL}(P \parallel M)$ is the Kullback-Leibler divergence of the distributions P and M.

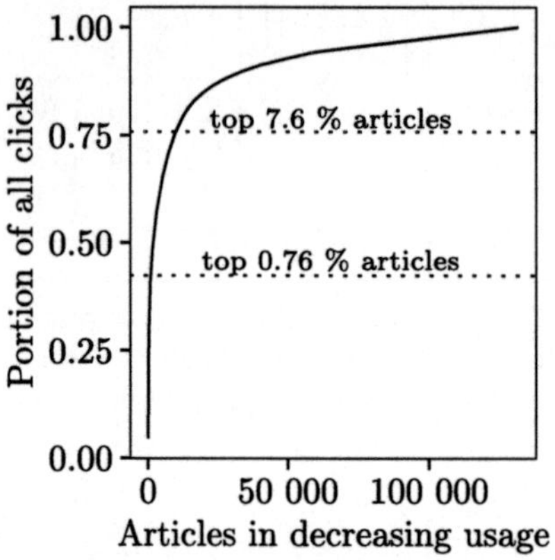

(a) Complementary cumulative usage of Wikipedia articles by Wikigame users. It shows that 42.4 % of all clicks comes from only 0.76 % (1000 articles) of all used articles and 75.8% comes from 7.6% (10000 articles).

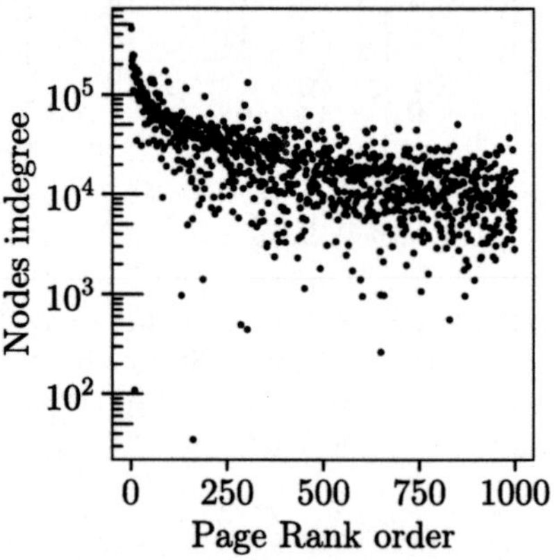

(b) Wikipedia indegrees of the top 1000 Wikipedia articles according to PankRank order.

Fig. 3. General Wikipedia statistics

4.2 Model Validation and Comparison

Using the metrics presented above, the two measures, i.e. the indegree and the PageRank of the nodes are compared to the ground-truth Wikigame navigation patterns. As it can be seen in Fig. 3b, the indegree and the pagerank are somewhat correlated, however, there are fluctuations and subtle differences, so the expected results are going to be different for the indegree and the PageRank.

Figure 4 shows the PIR metric at different N values in the top-N pages. As it can be seen, sorting the articles by the PageRank gives more overlap in the top-N used articles than the indegree at almost any N value. The figure shows that for example around 40% of the top 1000 used Wikigame articles are the same as the top 1000 highest indegree nodes and around 50% of the top 1000 Pagerank articles. In the context of how many articles are in the Wikipedia evaluation (more than 21 million), the fact is quite impressive. Also, PageRank seems to describe the behavior more precisely, meaning that random navigation results in a very similar behavior to user navigation.

However, the most used nodes are likely to be hubs in the network, so it's worth to investigate the metrics at different Wikigame step values, as our hypothesis is that users tend to navigate to hubs in the first phase of the game, even unconsciously. As we are trying to characterize the generalization phase not the whole game, we examine the PIR per step for three different number of top articles (50, 200, 1000) in Fig. 5. Overall the results are similar to Fig. 4, as the PageRank model has higher intersection ratios with the Wikigame, but another important fact is that the PIR ratio is the highest in the first few steps, and decreasing quasi-monotonously.

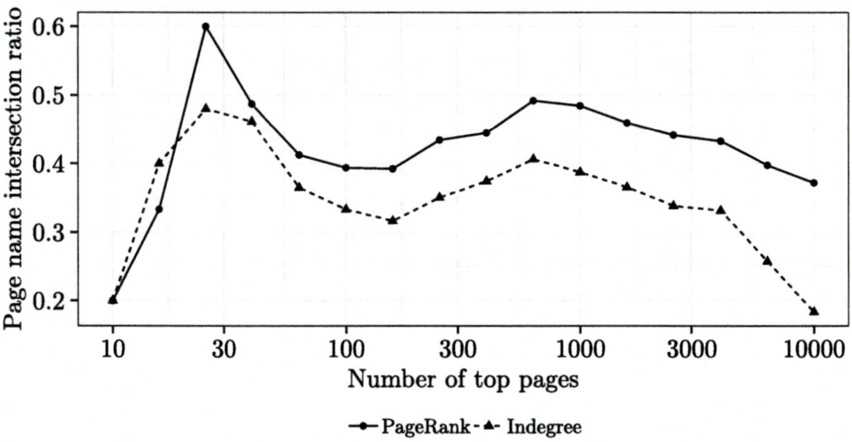

Fig. 4. The intersection ratio of the top PageRank and top indegree articles compared to Wikigame usage for different number of top articles. It shows that the articles by the PageRank gives more overlap in the top-N used articles than the indegree at almost any N value.

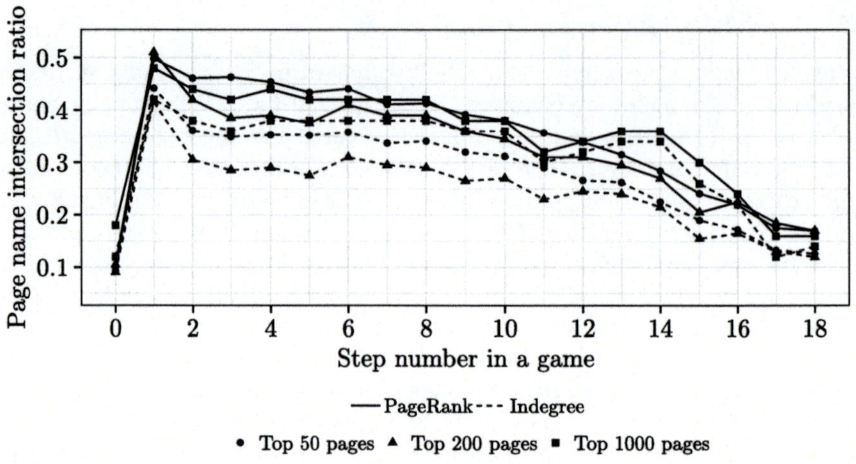

Fig. 5. The intersection ratio of the top PageRank and top indegree articles compared to Wikigame usage for different number of top articles (50, 200, 1000) per step. It highlights that the PageRank model has higher intersection ratios with the Wikigame than the Indegree model.

The HIR metric is presented in Fig. 6. The results show that the Pagerank model has a higher histogram intersection ratio for every examined cases, as well. The ratio is significantly better for the first 2–3 steps of the game, which indicates different behavior from the users in the generalization phase of the game.

Figure 7 confirms the results above that the probability distribution based on the Pagerank model has better fit thanks to the lower Jensen-Shannon-divergence compared the indegree model. This means that the PageRank model describes better the navigational process than the indegree properties of the articles. The main trends are similar to the previous cases.

Looking at the results, it is clearly visible, that the game has basically two phases, the generalization phase and the home-in phase. The generalization phase is generally a short part of the game, consisting of 1 or 2 steps. Here, the user finds hubs very easily and quick. However, results show, that by applying random navigation on the Wikipedia graph, hubs are reached with high probability, so the generalization phase is really similar to a random navigation (yet not the same).

Also, the results show that the well interpretable random navigation based on Markov chains, and computed by the PageRank algorithm describes the generalization phase more precisely compared to the indegree distribution of the nodes. PageRank has also the advantage of taking the structure of the graph into account compared to the bare indegree number of the nodes. However, random navigation is applicable only to the generalization phase, when users reach a hub with a semantically connected link to the target node, it fails to behave randomly.

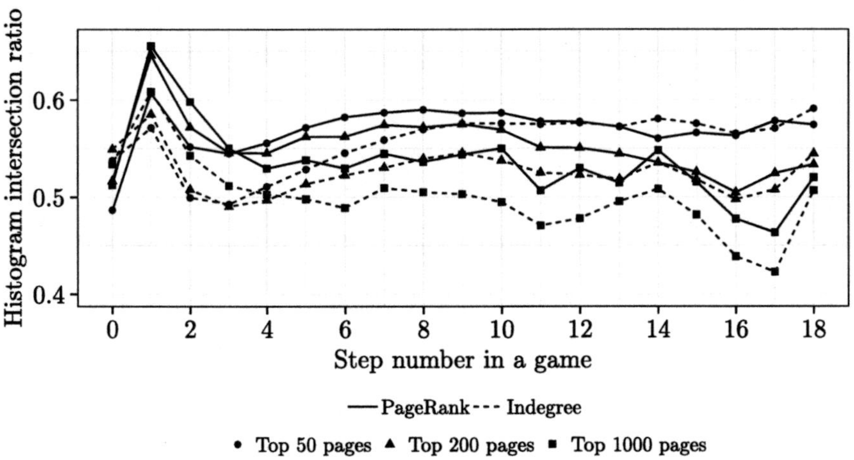

Fig. 6. The histogram intersection ratio of the top PageRank and top indegree articles compared to Wikigame usage for different number of top articles (50, 200, 1000) per step. First the union of the top 50/200/1000 articles were determined both for the PageRank–Wikigame usage and indegree–Wikigame usage, then the pdfs of the union set of articles have been calculated and compared as histogram intersection ratios. (Higher ratio means better fit.)

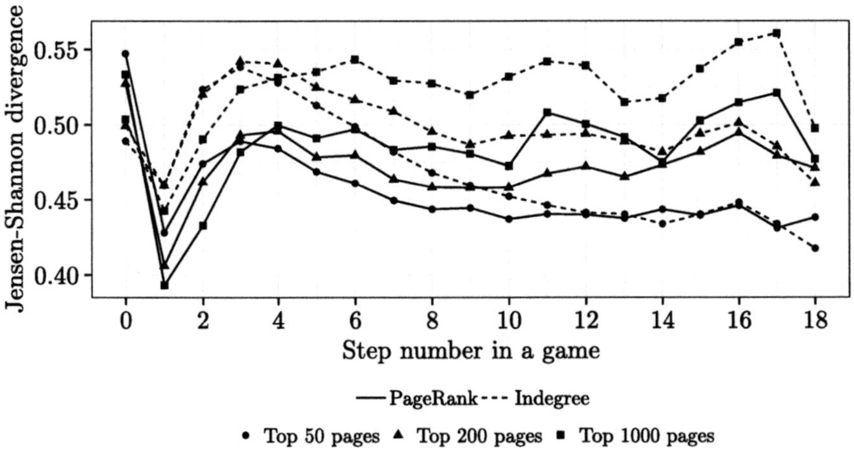

Fig. 7. The Jensen-Shannon divergence of the top PageRank and top indegree articles compared to Wikigame usage for different number of top articles (50, 200, 1000) per step. First the union of the top 50/200/1000 articles were determined both for the PageRank–Wikigame usage and indegree–Wikigame usage, then the pdfs of the union set of articles have been calculated and compared as Jensen-Shannon divergence. (Lower value means better fit.)

5 Conclusion

The paper investigated the human navigation behavior in complex graphs using the publicly available Wikipedia graph. The ground-truth human navigation patterns are samples from the Wikigame application. The main question was, that how different is the human navigation pattern in the generalization phase from random navigation? Or differently, is the human navigation determined in generalization phase by the human behavior, or by the graph structure?

In our work, the previously stated conjecture that the human navigation has at least two phases, was confirmed using first-order Markov-chains. Also, it was shown that the generalization phase is rather short and humans find hubs quickly. However, this quickness is not only a human behavior, the structure of the graph has a great impact on the generalization phase. We defined a model based on PageRank which has better properties according to the examined metrics than indegree which is also quite a good measure of generality. Our work do not state that users only navigate randomly to generalize, but a very high portion of their navigational behavior can be model with random navigation in complex graphs. In our opinion, random navigation by humans is not really surprising, moreover, it is a general behavior. When humans are lost in a navigation process (like in a forest or on an unknown street without a map), they try different directions to find some familiar place (handhold) from where they can continue the navigation purposeful. Obviously, semantical characteristics have a role in the navigation process even in the first phase, the creation of a comprehensive model where these characteristics are also considered is our current research topic. Moreover, we plan to investigate and identify user-specific characteristics as this paper only covers average human navigational behavior.

References

1. Berahmand, K., Nasiri, E., Forouzandeh, S., Li, Y.: A preference random walk algorithm for link prediction through mutual influence nodes in complex networks. J. King Saud Univ. Comput. Inf. Sci. (2021)
2. Gabrilovich, E., Markovitch, Shaul: Wikipedia-based semantic interpretation for natural language processing. J. Artif. Int. Res. **34**(1), 443–498 (2009)
3. Haveliwala, T., Kamvar, S.: The second eigenvalue of the google matrix. Technical Report 2003–20, Stanford InfoLab (2003)
4. Helic, D., Strohmaier, M., Granitzer, M., Scherer, R.: Models of human navigation in information networks based on decentralized search. In: Proceedings of the 24th ACM Conference on Hypertext and Social Media, HT '13, pp. 89–98, New York, NY, USA. Association for Computing Machinery (2013)
5. Lamprecht, Daniel, Lerman, Kristina, Helic, Denis, Strohmaier, Markus: How the structure of Wikipedia articles influences user navigation. New Rev. Hypermedia Multimed. **23**(1), 29–50 (2017)
6. Lovász, László.: Random walks on graphs. Combinatorics, Paul Erdos is eighty **2**(1–46), 4 (1993)
7. Meta-Wiki. Data dump torrents—Meta

8. Page, L., Brin, S., Motwani, R., Winograd, T.L: The pagerank citation ranking: Bringing order to the web. Technical Report 1999-66, Stanford InfoLab, November (1999). Previous number = SIDL-WP-1999-0120
9. West, R., Leskovec, J.: Human wayfinding in information networks. In: Proceedings of the 21st International Conference on World Wide Web, WWW '12, pp. 619–628, New York, NY, USA. Association for Computing Machinery (2012)

Sometimes Less Is More: When Aggregating Networks Masks Effects

Jennifer M. Larson[1(✉)] and Pedro L. Rodríguez[2]

[1] Vanderbilt University, Nashville, TN 37240, USA
jennifer.larson@vanderbilt.edu
[2] Meta, Menlo Park, CA 94025, USA

Abstract. A large body of research aims to detect the spread of something through a social network. This research often entails measuring multiple kinds of relationships among a group of people and then aggregating them into a single social network to use for analysis. The aggregation is typically done by taking a union of the various tie types. Although this has intuitive appeal, we show that in many realistic cases, this approach adds sufficient error to mask true network effects. We show that this can be the case, and demonstrate that the problem depends on: (1) whether the effect diffuses generically or in a tie-specific way, and (2) the extent of overlap between the measured network ties. Aggregating ties when diffusion is tie-specific and overlap is low will negatively bias and potentially mask network effects that are in fact present.

Keywords: Diffusion · Social networks · Tie aggregation · Multiplexity · Bias

1 Introduction

A large body of research aims to detect the spread of something through a social network. Be it voting behavior [18], news [12], new technology [8], disease [5], or patronage benefits [7], social networks are thought to matter because their ties—the relationships connecting pairs of people—serve as conduits that can spread ideas, information, goods, germs, and social judgment.

When researchers conduct an empirical test of this kind of spread, they collect a measure of the relevant social network [13]. Because "social relationship" is nebulous, the standard approach is to measure a few concrete types of social interactions. For instance, [8] use surveys to inquire about respondents' friends, family, potential money lenders, and potential problem solvers. Jennifer et al. [12] ask respondents about seven types of interactions, including visits to others' homesteads and shared meals. Banerjee [4] as about twelve, including borrowing rice and kerosene.

H. Cherifi et al. (Eds.): COMPLEX NETWORKS 2016 2022, SCI 1077, pp. 214–223, 2023.
https://doi.org/10.1007/978-3-031-21127-0_18

With multiple measures of social ties in hand, researchers typically take the union of these ties to construct one social network to be used for their analyses [3,4,8,11,12]. A tie in this aggregate network indicates that one or both of the tied individuals (depending on the coding rule the authors employ) reported having at least one of the measured relationships with the other. The logic that drives researchers to aggregate all measured ties is straightforward: each additional type of interaction contains weakly more information about the social relationship between two people, so the more interactions included, the better the measure of the social network.

However, we argue that this logic does not necessarily hold for all instances of something spreading through a network ("diffusion"). Some things may indeed diffuse generically, in which case one type of tie is as good as another for their spread. On the other hand, things may diffuse specifically along only certain kinds of ties. Take, for instance, the diffusion of useful, factual information, such as that a university is likely to declare a snow day tomorrow. We may expect this information to spread equally well along a tie that indicates shared membership in the rugby club and along a tie that indicates a twelve-year friendship. Now consider the diffusion of more sensitive information, such as that a student is considering reporting a professor for misconduct. In the latter case, students may opt to use their more trusted ties exclusively; one type of social tie would not be as good as another.

In short, ties that indicate a different kind of interaction between two people may also *work* differently to spread something, depending on the context. When diffusion is tie-specific, aggregating ties effectively adds measurement error to the network which, in an analogue to the regression context, can attenuate and mask true effects.

We begin by showing that different measures of social interactions among the same group of people can in fact pick up quite distinct views of the social network, using data from [12]. Then we demonstrate the problem of aggregating ties using a set of Erdős-Rényi random networks that vary in key ways. We show that when diffusion is specific, aggregating networks substantially attenuates the estimate of true effects. We further show that whether aggregation problems will arise depends on two factors: the amount of new information that that a candidate new tie type would add—a property we call "network overlap"—and the relative sizes of the networks being aggregated.[1]

Our results highlight the importance of careful theory and knowledge of context to underpin a network study. Determining whether the diffusion at hand is likely to be specific or generic relies heavily on theory and a deep qualitative understanding of local context. Without a guiding understanding of how something should, in principle, pass from person to person, researchers risk failing to detect it empirically.[2]

---

[1] This is distinct from the notion of overlap that refers to the extent to which two nodes share the same neighbors (see [15,17]).

[2] For a case in point, see [14], which shows that evidence of behavior spreading through the network is masked when the seven types of ties are aggregated. When disaggre-

2 Empirical Studies of Social Network Diffusion

Scholars are interested in "network effects" when they believe ties in a network can transmit something. In principle, theory should dictate precisely which ties are of interest. When theory is strong and the tie of theoretical interest has one clear operationalization, researchers are able to measure only a single type of tie. For instance, to study peer influence in adopting insurance among rice farmers, [6] measure one link type: household heads' close friends with whom they most frequently discuss rice production or other financial issues. To examine the advantage conferred to politicians by familial ties, [7] gather data exclusively on marriage connections between families in the Philippines.

Measuring a single tie type is the exception rather than the rule. When theory is not strong enough to identify one precise tie type that should be interest, or when a single type of tie in concept does not have a single obvious operationalization, researchers collect data on multiple types of ties.

For instance, in their study of farmers' decisions about sunflower crops in Mozambique, [3] aim to detect a relationship between one's peers' decisions and one's own by recording respondents' family members, friends, and neighbors. To study whether peers affect one's choice to give deworming medication to children in Kenya, [11] ask respondents to name the five friends and the five relatives they speak to most frequently, other social contacts whose children attend the local schools, and people with whom they speak about child health matters.

To learn about how social information spreads through Indonesian hamlets, [1] record both blood relatives and shared membership in social organizations. Jennifer et al. [12] measure the channels through which news may spread from person to person by asking Ugandan villagers about people with whom they spend time, share meals, exchange household visits, discuss religion, discuss politics, share secrets, and speak on the phone. Banerjee [4] measure twelve types of ties that may be responsible for spreading the word about a microfinance program: people who visit the respondent's home, people the respondent visits, kin, non-relatives with whom the respondent socializes, sources of borrowed money, sources of borrowed material goods, potential recipients of lent money, potential recipients of lent material goods, sources of medical advice, sources of general advice, receivers of advice, and prayer partners. Ferrali et al. [8] operationalize the ties that may spread information about new technology with friendship, family, potential money lenders, and potential solvers of problems about public services.

In all of the above cases of measuring multiple types of ties, the researchers take the union of the different types of ties to construct a measure of the "social network."[3] Doing so is the standard approach. At first blush, aggregating ties in this way seems justified since the resulting network contains the data's maximum information about social ties.

gated, the authors find evidence consistent with specific diffusion along the most intimate types of ties.

[3] In rare instances, researchers instead or also look at the different networks separately; see [2, 14].

Moreover, a large literature has made clear that social relationships are complex. A relationship between two people can manifest in many ways, featuring a variety of kind of interactions. This "multiplexity" is a key feature of human networks (see [9,16]). In this view, social ties are multi-faceted and so a person who is my friend may also be my coworker as well as the person with whom I would discuss politics. We might even say that this relationship is especially rich because it features these three dimensions.

The difficulty arises when we flip this logic and look for evidence of a social tie by measuring these dimensions separately. Of course if everyone in a group of interest who are friends are also coworkers and are also political discusison partners, then there is no issue. Measuring one is the same as measuring all and taking their union. However, as we show below, when we pick a single dimension of a relationship—just political discussion partners, say—the network formed by those ties alone can look quite different from a network formed by a different dimension—working together, for instance. Given that different dimensions can add different information, it is worth considering whether aggregating across the different dimensions is always best.

To preview our argument, consider the setting of [12] in which villagers can spread news to one another, including news that the researchers inserted into the network about an upcoming local event that would give out soap. Suppose that the true way that villagers pass on this kind of news is by telling anyone with whom they have any kind of social relationship. In such a case, measuring a variety of different social relationships and aggregating them would indeed maximize the relevant information.

Now suppose instead that villagers find events hosted by outsiders to be a political matter, worthy of discussion with one's politically-minded social ties and no others. Or perhaps villagers are concerned about their safety at an unusual event and will only discuss it with their elders. Different still, imagine that villagers fear that their participation, if discovered, would offend the local political elites, and so only spread word to their most trusted contacts. In each of these three scenarios, the true spread of information about the soap event only occurs along a specific kind of social tie—political discussion partners, elders, and the most trusted, respectively.

When this is the case, aggregating different kinds of ties can undermine the detection of network effects. In fact, by including the irrelevant information, the data may no longer reveal the true spread through the network.

3 Different Tie Types Generate Different Networks

To see that different measures of social ties can contain quite different information, consider the social network from one of the Ugandan villages in [12]. Seven different types of relationships are measured among the villagers by survey. Figure 1 highlights four of them. Each villager is asked to name other villagers with whom they regularly spend time (top left), discuss politics (top right), share secrets with (bottom left), and speak to on the phone (bottom right).

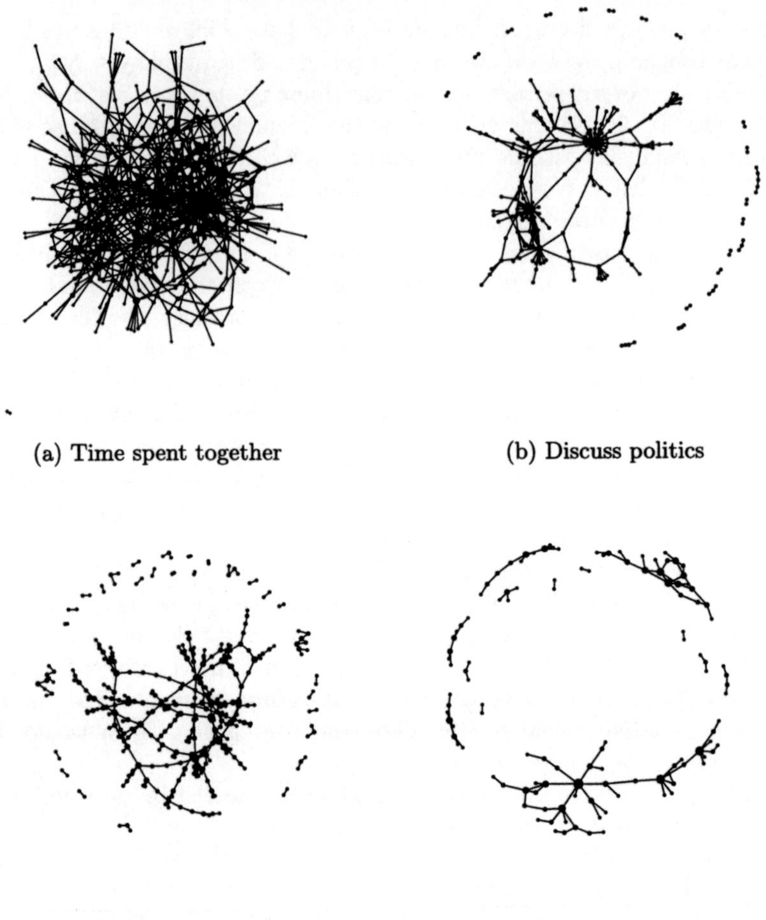

(a) Time spent together (b) Discuss politics

(c) Share secrets (d) Speak over the phone

Fig. 1. Different structural characteristics are apparent when social ties are measured with different kinds of interactions among the same people. Data from [12].

Although these questions are asked of the same villagers, each of the four dimensions of their social relationships looks very different when visualized as its own network. Taking their union to construct a single social network results in each dimension adding a different amount and type of information. If all of these relationships function the same for something spreading from villager to villager, then aggregation needs no further thought; they all provide a view of the same thing. But if diffusion in fact only occurs along ties based on a particular relationship (or even a strict subset of the relationships), then aggregation can mask our ability to detect it.

4 Theory

Suppose we are interested in evaluating whether peer effects play a role in diffusing a given piece of information in a rural village in Uganda. We hypothesize that once supplied to a random selection of "seed" villagers, the information will spread through each of these seed's contacts to the contacts of their contacts and so on. Contacts may vary in the type of tie that binds them. For example, contacts with whom an ego shares meals may differ from contacts with whom she works. In order to capture this heterogeneity suppose we measure two types of ties, ties of type A and ties of type B, the sets of which we denote S_A and S_B respectively. Some time after seeding the information we survey (non-seeded) villagers and inquire whether they have knowledge of the seeded information.

There is evidence of peer-effects if the likelihood of having knowledge of the information is positively correlated with the proportion of an ego's neighbors that were seeded. To evaluate this relationship we first need to determine which ties are relevant to quantifying the proportion of an ego's neighbors that were seeded. Ties of type A, of type B or both?

Suppose our seeded information did diffuse through the village's underlying social networks but only through ties of type A. In this case only the presence of a tie of type A between any two villagers in our sample is informative of peer effects. Taking the union of S_A and S_B is equivalent to introducing *noise* into our covariate measure. Given this equivalence we can appeal to the extensive literature on covariate measurement error to enumerate the consequences of aggregating networks in the presence of a tie-specific diffusion process. These are:

1. *Attenuation bias*: coefficient estimates are biased toward zero.
2. *Downward biased test statistics*: resulting in a higher probability of falsely failing to reject the null (type-II error).

The severity of these effects will depend on the true magnitude of the coefficient and the *noise-to-signal ratio.*[4] Unlike traditional covariate measurement error however—over which the researcher often has no control and little information—error resulting from mistakenly aggregating networks can be both avoided and quantified. Denote η as the noise-to-signal ratio and $n(\cdot)$ as the cardinality of a set. Returning to our hypothetical example we can show that:

$$\eta_{AB} = \frac{n(S_B - S_A)}{n(S_A)} \tag{1}$$

That is, the noise-to-signal ratio is equivalent to the number of ties in S_B not in S_A as a proportion of the number of ties in S_A. Notice, generally $\eta_{AB} \neq \eta_{BA}$, hence the subscript. It is useful to decompose η_{AB} into two components which we label size- and overlap-ratio. We define these as follows:

[4] For the derivation of this result for OLS see [10] and for logistic regression see [19].

Definition 1. Given two types of ties A and B, with respective sets S_A and S_B, define the ***size-ratio*** of S_A to S_B, as the ratio of the number of ties in S_B and number of ties in S_A. Formally:

$$s_{AB} = \frac{n(S_B)}{n(S_A)} \tag{2}$$

Definition 2. Given two types of ties A and B, with respective sets S_A and S_B, define the ***overlap-ratio*** of S_A to S_B as the proportion of ties in S_A also found in S_B. Formally:

$$o_{AB} = \frac{n(S_A \cap S_B)}{n(S_A)} \tag{3}$$

As with η_{AB}, neither ratio is symmetric with respect to its arguments hence the subscripts. It can be shown that η_{AB} is a function of these two ratios. Specifically,

$$\eta_{AB} = \frac{n(S_B - S_A)}{n(S_A)} = \frac{n(S_B) - n(S_A \cap S_B)}{n(S_A)} = \underbrace{\frac{n(S_B)}{n(S_A)}}_{\text{size-ratio}} - \overbrace{\frac{n(S_A \cap S_B)}{n(S_A)}}^{\text{overlap-ratio}} = s_{AB} - o_{AB} \tag{4}$$

All else equal, the larger the size-ratio, the more noise we are adding by taking the union. However, some of the ties in S_B may also be found in S_A. These ties do not add any noise.

5 Simulated Networks

In this section we illustrate the network aggregation problem using simulated random networks. Specifically we generate hypothetical networks with two types of ties, type A and type B, with varying size- and overlap-ratios. For a given population of nodes of size N,

1. Generate a list L of all potential ties between pairs of nodes in N.
2. Randomly select a subset S from L.
3. From S, randomly select a subset S_A, to make up the set of links of type A.
4. Remaining links in S along with o percent of links in S_A make up S_B.

We follow these steps to generate random networks for each combination of size-ratio s and overlap-ratio o such that $s \in \{1, 2, 3, 4\}$ and $o \in \{0, 0.25, 0.5, 1\}$.

For each of our generated networks we then select a random subset of nodes to serve as seeds, 5% in each. We then simulate a one-period diffusion process in which we assume information can only diffuse through ties of type A (in other words, we assume that diffusion is tie-specific). Knowledge of the information by i, Y_i, is a Bernoulli random variable taking value 1 with probability:

$$Pr(y_i = 1) = logit^{-1}(\beta_0 + X_i^A \beta_1)$$

where X_i^A is the proportion of i's neighbors that are seeds according to ties of type A. We set $\beta_0 = -1$ and $\beta_0 = 0.2$ and simulate 500 one-period diffusions for each size/overlap-ratio combination.[5] After each simulation we estimate the "noisy model":

$$Pr(y_i = 1) = logit^{-1}(\beta_0 + X_i^{agg}\beta_1)$$

where X_i^{agg} is the proportion of neighbors that are seeds according to the union of S_A and S_B. In practice, adding noise to the covariate measure reduces its variance, thereby reducing its explanatory power. Table 1 provides evidence for this intuition. Each value corresponds to the ratio of the average standard deviation of X_i^{agg} and X_i^A specific to a size- and overlap-ratio combination. As the size-ratio (overlap-ratio) of S_A to S_B increases (decreases) the smaller the variance of X_i^{agg} relative to the variance of X_i^A.

Table 1. Ratio of the mean standard deviations of X_i^{agg} and X_i^A.

Overlap-ratio						
Size ratio		0	0.25	0.5	0.75	1
	1	0.63	0.73	0.82	0.91	1
	2	0.50	0.55	0.60	0.64	0.68
	3	0.43	0.46	0.49	0.52	0.54
	4	0.38	0.40	0.42	0.44	0.46

Now define the *bias ratio* as the ratio of the estimated $\hat{\beta}_1$ and the true β_1. An unbiased estimator applied to the correct model yields on average a bias-ratio of 1. Figure 2 plots the distribution of the bias ratio for varying levels of overlap-ratio, holding the size-ratio equal to 1 and 4 in the left and right sub-figure respectively. Both figures show how the estimated effect is increasingly attenuated as the overlap-ratio between the two networks decreases. Moreover, even in the presence of perfect overlap—formally $S_A \subseteq S_B$—, we still observe attenuation as the number of uninformative ties increases in proportion to informative ties (see Fig. 2b).

As noted above, adding noise to the covariate measure also results in downward biased test statistics thereby increasing the probability of making type-II inferential errors. For each set of simulations -500 for each size-/overlap-ratio combination- we computed the proportion of regressions for which β_1 is statistically significant at 0.05. Table 2 summarizes these results. It is clear that as the size-ratio increases and overlap-ratio decreases the proportion of regressions for which we correctly reject the null decreases.

[5] The magnitude of β affects the severity of the attenuation bias, not whether it is present or not.

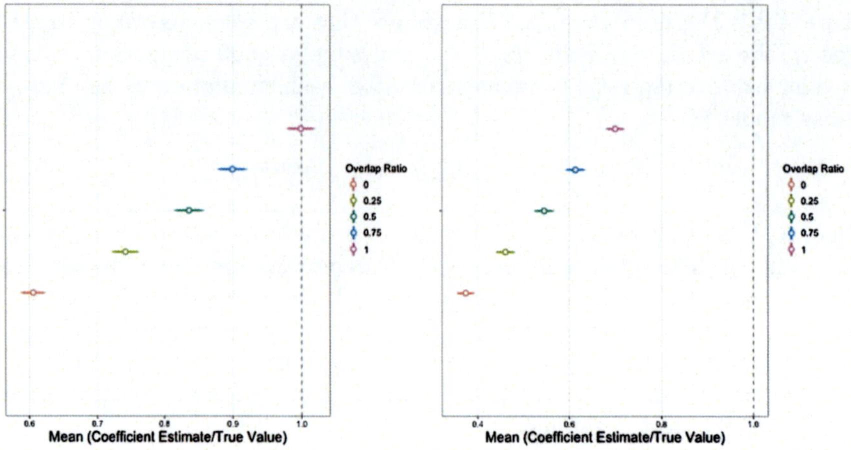

(a) Size-ratio = 1 (b) Size-ratio = 4

Fig. 2. Mean bias ratio as a function of size- and overlap-ratio.

Table 2. Proportion of simulations with significant β_1 (95% level)

Overlap-ratio						
Size ratio		0	0.25	0.5	0.75	1
	1	0.82	0.92	0.95	0.99	0.99
	2	0.58	0.73	0.86	0.92	0.96
	3	0.41	0.63	0.70	0.84	0.90
	4	0.35	0.53	0.67	0.74	0.86

6 Conclusion

Our article reexamines the standard approach of social network measurement, which defaults to taking the union of multiple types of social ties. This approach results in one aggregate "social network" that summarizes all information on social ties contained in the data. We point out that for many realistic cases, this logic is too simple, and can mask network effects that are in fact present.

We show that the decision of whether and how to aggregate ties should be informed by two factors: (1) whether the effect in question diffuses generically or in a tie-specific way, and (2) the extent of overlap between the measured networks. Aggregating ties when diffusion is tie-specific and overlap is low will negatively bias and potentially mask network effects that are in fact present.

Of course, knowing how the diffusion process should work is a substantial requirement. Our results highlight just how crucial strong theory and qualitative knowledge are for detecting meaningful network effects.

References

1. Alatas, V., Banerjee, A., Chandrasekhar, A.G., Hanna, R., Olken, B.A.: Network structure and the aggregation of information: theory and evidence from Indonesia. Am. Econ. Rev. **106**(7), 1663–1704 (2016)
2. Baldassarri, Delia: Cooperative networks: altruism, group solidarity, reciprocity, and sanctioning in Ugandan producer organizations. Am. J. Sociol. **121**(2), 355–395 (2015)
3. Bandiera, O., Rasul, I.: Social networks and technology adoption in northern Mozambique. Econ. J. **116**(514), 869–902 (2006)
4. Banerjee, A., Chandrasekhar, A.G., Duflo, E., Jackson, M.O.: The diffusion of microfinance. Science **341**(6144), 1236498 (2013)
5. Bearman, P.S., Moody, J., Stovel, K.: Chains of affection: the structure of adolescent romantic and sexual networks. Am. J. Sociol. **110**(1), 44–91 (2004)
6. Cai, J., De Janvry, A., Sadoulet, E.: Social networks and the decision to insure. Am. Econ. J. Appl. Econ. **7**(2), 81–108 (2015)
7. Cruz, C., Labonne, J., Querubin, P.: Politician family networks and electoral outcomes: evidence from the Philippines. Am. Econ. Rev. **107**(10), 3006–37 (2017)
8. Ferrali, R., Grossman, G., Platas, M., Rodden, J.: Peer effects and externalities in technology adoption: evidence from community reporting in Uganda. SSRN (2018)
9. Gondal, N.: Multiplexity as a lens to investigate the cultural meanings of interpersonal ties. Social Netw. **68**, 209–217 (2022)
10. Greene, W.H.: Econometric Analysis. Pearson Education India (2003)
11. Kremer, M., Miguel, E.: The illusion of sustainability. Quarter. J. Econ. **122**(3), 1007–1065 (2007)
12. Larson, J.M., Lewis, J.I:. Ethnic networks. Am. J. Polit. Sci. **61**(2), 350–364 (2017)
13. Larson, J.M., Lewis, J.I.: Measuring networks in the field. Polit. Sci. Res. Methods **8**(1), 123–135 (2020)
14. Larson, J.M., Lewis, J.I., Rodríguez, P.: From chatter to action: how social networks inform and motivate in rural Uganda. Brit. J. Polit. Sci. (2021). https://doi.org/10.1017/S0007123421000454
15. Mattie, H., Onnela, J.-P.: Edge overlap in weighted and directed social networks. Netw. Sci. **9**(2), 179–193 (2021)
16. Mesch, G., Talmud, I.: The quality of online and offline relationships: the role of multiplexity and duration of social relationships. Inf. Soc. **22**(3), 137–148 (2006)
17. Peng, J., Agarwal, A., Hosanagar, K., Iyengar, R.: Network overlap and content sharing on social media platforms. J. Market. Res. **55**(4), 571–585 (2018)
18. Sinclair, B., McConnell, M., Michelson, M.R.: Local canvassing: the efficacy of grassroots voter mobilization. Polit. Commun. **30**(1), 42–57 (2013)
19. Stefanski, L.A., Carroll, R.J.: Covariate measurement error in logistic regression. Ann. Stat. 1335–1351 (1985)

An Adaptive Network Model Simulating the Effects of Different Culture Types and Leader Qualities on Mistake Handling and Organisational Learning

Natalie Samhan[1], Jan Treur[1](✉), Wioleta Kucharska[2], and Anna Wiewiora[3]

[1] Social AI Group, Vrije Universiteit Amsterdam, Amsterdam, The Netherlands
j.treur@vu.nl
[2] Faculty of Management and Economics, Gdańsk University of Technology, Gdansk, Poland
wioleta.kucharska@pg.edu.pl
[3] QUT Business School, Queensland University of Technology, Brisbane, Australia
a.wiewiora@qut.edu.au

Abstract. This paper investigates computationally the following research hypotheses: (1) Higher flexibility and discretion in organizational culture results in better mistake management and thus better organizational learning, (2) Effective organizational learning requires a transformational leader to have both high social and formal status and consistency, and (3) Company culture and leader's behavior must align for the best learning effects. Computational simulations of the introduced adaptive network were analyzed in different contexts varying in organization culture and leader characteristics. Statistical analysis results proved to be significant and supported the research hypotheses. Ultimately, this paper provides insight into how organizations that foster a mistake-tolerant attitude in alignment with the leader, can result in significantly better organizational learning on a team and individual level.

Keywords: Mistake management · Hebbian learning · Organizational learning · Vicarious learning · Organizational culture · Transformational leadership

1 Introduction

Committing mistakes in a professional field is usually perceived as a complete aberration. It is often associated with adverse feelings of guilt, embarrassment, and even lower self-esteem. However, mistakes are an inevitable aspect of an organization whether it acknowledges it or not. Although, the organization's conception and reaction to making mistakes can affect its long-term learning and performance [6]. Tacit beliefs of individuals regarding committing mistakes are often a by-product of the environment, including the organization's culture and leader's attitudes [13]. It has been empirically established that mistake tolerance aid organizational learning and as a result its performance [23]. A mistake non-tolerant environment makes it difficult to report mistakes and that often

H. Cherifi et al. (Eds.): COMPLEX NETWORKS 2016 2022, SCI 1077, pp. 224–238, 2023.
https://doi.org/10.1007/978-3-031-21127-0_19

perpetuates the mistake and harms an organization's long-term capacity to learn and improve. Whereas a mistake-tolerant environment is warier of long-term repercussions and less concerned about the immediate costs of the mistake. This attitude toward managing mistakes not only develops a healthier psychological response to mistakes but also promotes deliberate experimentation and critical analysis, as opposed to hiding mistakes and trying to move past them swiftly [6].

An organization's culture, in addition to a leader's behavior regarding mistake tolerance, is crucial to organizational learning [13, 14]. The idea, however, is not to cultivate an environment that encourages making mistakes, but one that encourages 'fail intelligently' to maximize the learning value of mistakes and promote intelligent risk-taking [23]. In essence, this paper aims to carry out computational simulations to explore the effect on organizational learning twofold: in relation to the organizations' culture and leader's behavior, and in relation to mistake management which is influenced by the former.

2 Background Literature

This section provides background literature revolving around the relevant concepts and acts as a basis for the experimental simulations in the paper.

The company culture is a shared mindset or as Hofstede et al. [9] says a shared "software of the mind", that strongly determines the perception of company values, employees' attitudes, beliefs, social norms, and behaviors. Schein [17] distinguished three fundamental levels of organizational culture consisting of (1) observable artefacts such as physical layout, the dress code, the manner in which people address each other; (2) values such as organizational norms and believes manifesting in people's behaviors; and (3) basic underlying assumptions, least apparent and based on an organization's historical events that determine perceptions, feelings and behaviour. Organizational culture is typically evaluated based on organizational values that organizational members have in common.

Cameron and Quinn [2] identified the key competing company values framework determining company culture (See Fig. 1). According to this framework, organizational culture can be derived from a grid of four quadrants that vary in two dimensions. The first-dimension ranges from flexibility and discretion on one end, to stability and control on the other end. This dimension focuses on the distinction of flexibility which values individuality versus stability which values top-down control. The second-dimension ranges from external focus and differentiation on one end, to internal focus and integration on the other end. This dimension addresses whether an organization values progress within the organization (internal focus), versus progress in juxtaposition to other organizations or competitors (external focus). This is also sometimes referred to as the external and internal positioning of an organization respectively. The four types of cultures that arise from these dimensions are clan, adhocracy, hierarchy, and market.

Leaders have a powerful influence over the organization's environment and indirectly on individuals' behavior. Given certain leadership qualities, leaders can foster an environment that facilitates multilevel learning and institutionalization of desired behaviors. Leaders who empower their employees manifest better results. Certain leadership practices, such as delegation, coaching, and recognition, are complementary Leadership empowerment behaviors (LEB) and fulfill Lawler's [15] four basic conditions of empowerment [7]. All three of these practices can encourage employees to more improvement and feedback which can manifest better results.

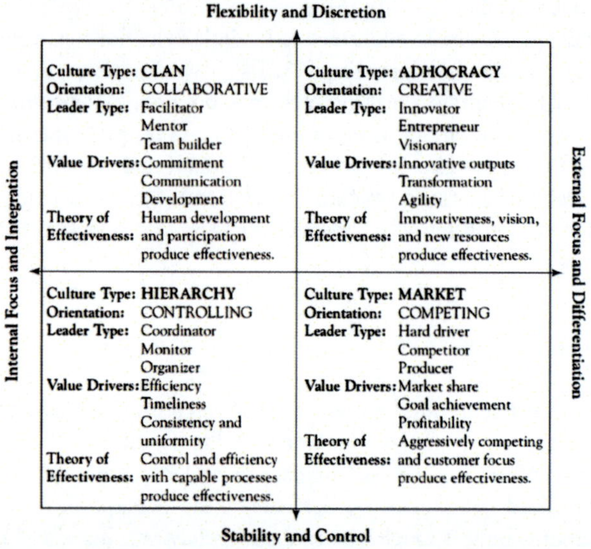

Fig. 1. The competing values framework

Moreover, delegating tasks to employees can improve task reproduction and also collaboration—team learning. Similarly, coaching behavior guides the employees more and has the same effect as a delegation. Finally, leader recognition empowers employees and improves reinforcement [7]. Furthermore, transformational leaders can contribute significantly to better organizational learning [13]. Next to professional skills, such attributes as social status and consistency in the behavior of the leader support their transformational power. A leader's status comprises both social status and formal organizational status. Having both types of status high can enhance the view and attention to the leader and improve vicarious learning. High social status is also important to captivate learners to capture attention to the desired learned behavior. Further, it is important that the leader acts consistently in order to increase the chance of employees memorizing behavior patterns via mirroring [15]. The assumptions and choices made for the computational model were mainly based on the already mentioned literature and [16, 18].

Organizational learning is a dynamic multi-level cyclical process that has learning flows directed from individual or team level to organizational level and vice versa. There are two learning flows in an organization: feed forward and feedback directions [8, 24]. Feedback learning is a common method where the learner receives information from the organization for improving ensuing behaviors in relation to the present performance. Contrastingly, feed-forward learning focuses on the organization receiving input from the learner or employee to improve future-oriented solutions or options. The multilevel learning theory introduced by Crossan et al. [8] is an interdependent theory suggesting learning flows are not necessarily linearly, from micro-level individuals to team level–known as meso-level—to systematic organizational levels–known as macro-level [24]. The learning from micro and meso levels to macro levels is categorized as feed-forward learning, whereas feedback learning occurs from macro to meso and micro levels, when individuals and teams learn from the organization.

There are numerous of learning theories, but one of the most renowned theories is the social learning theory. The theory is based on vicarious learning—also known as observational learning—where the individual learns simply by observing and imitating a model person's behaviors; this process is often referred to as mirroring the model. Vicarious learning has three distinct stages [1]. The first stage is the observational phase where the learner requires attention to observe the model in order to memorize the pattern of behavior—a process referred to as retention. The second stage is the imitation phase where the learner mirrors the model and reproduces the desired behavior, which can be strengthened by reinforcement. Reinforcement is a process that rewards proper reproduction and enhances learning. The final stage is the participation phase where the behavior is fully learned. These intricate learning stages can be modeled in an organizational context and adapted according to the existing culture and the behavior of the leader—model.

The neurophysiological basis of vicarious learning can be elucidated by understanding mirror neurons and learning processes in the brain. Mirror neuron activations in the brain extend to vicarious activations in the somatosensory cortex which is in the same region mirror neurons are located [10]. The presence of such activations during vicarious learning could be partially explained by the Hebbian learning theory.

Hebbian learning comes from Donald Hebb's theory which provides a neurophysiological account of learning and plasticity. The core concept of Hebbian learning emphasizes how one neuron that is continuously and persistently involved in firing a signal to another neuron elicits metabolic changes in one or both neurons. Hebbian learning mainly takes into account the causality of the neuron firing, indicating temporal precedence, but also contingency as the activity of the presynaptic neuron can be indicative of that of the postsynaptic activity [11]. The essence of temporal precedence in Hebb's learning theory can be translated to temporal-causal networks in an organizational context.

Based on the existing insights from the organizational culture, leadership and organizational learning literature, we argue that when organizational culture is in alignment with the leader it then manifests presumptions on mistakes, making individuals either try to cover up their mistakes or exploit the learning value and alter the shared mental model of the organization to learn from it. For example, when cultural values are focused on employee involvement, experimentation, learning from mistakes and favouring change and at the same time when the leader aligns with these values by modelling relevant culturally aligned behaviours, their subordinates are more likely to reveal their shortcomings and learn from them. Conversely, when the culture and the leadership style promote values and behaviours that discourage talking about and sharing mistake, employees are more likely to hide their shortcomings, resulting in not learning.

This paper aims to utilize adaptive network modeling to computationally demonstrate how mistake-tolerant organizations learn better when using mistake management as a source of learning. Ultimately, the research paper explores the effect of culture and leader characteristics on mistake management and individual and team learning. The paper aims to utilize- in silico- experiments and statistically compare the results to investigate the following proposed research hypotheses:

Hypothesis 1: Higher flexibility and discretion in organizational culture results in better mistake management and thus better organizational learning.
Hypothesis 2: Effective organizational learning requires a transformational leader to have both high social and formal status and consistency.
Hypothesis 3: Company culture and leader's behavior must align for the best learning effects.

3 Methodology: Self-Modeling Networks

This section delves into more detail about the network modeling approach along with the computational mechanisms behind it, in addition to the motives for the final adaptive network model used to run the simulations.

Network-oriented modeling can be used to better understand a myriad of social, cognitive, and biological phenomena. By graphically conceptualizing a complex process, the modeling approach acts as a basis of the conceptualization by representing the dynamic processes without separating or isolating different states. Temporal-causal networks specifically, incorporate the timing of dynamic or cyclical processes based on a continuous-time temporal dimension to time causal effects [19–21]. This network-oriented modeling approach has shown to be an appropriate method for modeling a complex phenomenon such as handling mental models in mental processes [22] and organizational learning and the roles of different contextual factors; e.g., [3–5].

Organizational learning [8, 12, 24] is a non-linear process where causal relations underlying knowledge can change over time. This has been modeled in [3–5] by adaptive reified network models, also referred to as self-modeling networks [21]. To elaborate, a complex process can be represented conceptually as a base-level network, but to add adaptivity of some of its network characteristics to the model, a self-model representing these network characteristics can adapt the network structure of the base-level network during the process. This next-higher level in a network for a base-level network is called a first-order self-model level. Along the same line, a second-order self-model level can be added to make the first-order self-modeling level model adaptive as well; this enables control of adaptation. Different context factors can influence higher-level nodes, such that a high extent of context-sensitivity of base processes and also of adaptation processes can be modeled.

In the current paper, a second-order adaptive self-modeling network model is used to conduct different simulation experiments (Sect. 4). The aim is to demonstrate the effect of the independent variables: organizational culture and leader's consistency and status, on the dependent variables: mistake management and organizational learning. A temporal-causal network takes into account the following characteristics:

- **Connectivity characteristics**: connections from a state X to a state Y have a connection weight value, represented by $\omega_{X,Y}$ which indicates the strength of the connection.
- **Timing characteristics**: for all states Y, a speed factor represented by η_Y indicates how fast the state is changing upon a causal impact.
- **Aggregation characteristics**: for all Y states, a combination function represented by $\mathbf{c}_Y(..)$ is referred to; it determines the way in which the causal impacts from other states combine for state Y; see Table 1 for them.

A dedicated software environment based on these network characteristics is utilized to generate the simulation output; see Chap. 9 of [20]. An overview of the role matrices description can be found in Sect. 2 of the Appendix (see the Linked Data at https://www.researchgate.net/publication/361481262. Based on the background literature provided, a generalized conceptual model (see Fig. 2) was first constructed to form a basis for the final self-modeling network representation used. This section discusses key decisions in the formation of the self-modeling network model. The concepts discussed in the background literature are translated into a computational network model using the software indicated above.

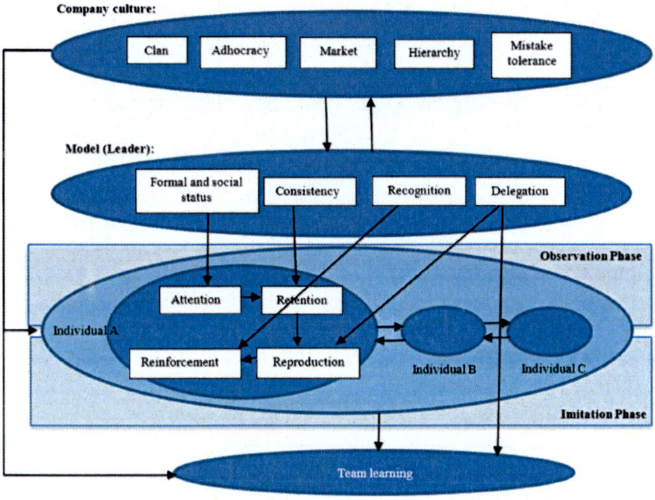

Fig. 2. Abstract conceptualisation for the adaptive network model

The basis of this model lies in the variation of culture types of organizations. This notion is captured by modeling context factors that affect the flexibility and focus in ways that align with the different culture types. The focus of the company is discussed in terms of positioning, where internally focused organizations have low positioning and externally focused organizations have high positioning. Each one of the four culture context factors has varying connection weights to the flexibility and positioning, allowing to model different types of cultures without manually altering the values of flexibility and positioning initially.

Table 1. Overview of combination functions selected to model the simulations

Name	Notation	Formula	Parameters
Identity	id(V_1, ..., V_k)	V_1	None
Advanced logistic sum	alogistic$_{\sigma,\tau}$(V_1, ..., V_k)	$\left[\frac{1}{1+e^{-\sigma(V_1+...+V_k-\tau)}} - \frac{1}{1+e^{\sigma\tau}}\right](1+e^{-\sigma\tau})$	p(1) = steepness σ p(2) = threshold τ
Step-once	steponce$_{\alpha,\beta}$ (V_1, ..., V_k)	Time t 1 if $\alpha \leq t \leq \beta$ else 0	p(1) = start time α p(2) = end time β
Max Hebbian composition	maxhebb$_{\mu}$ (V_1, ..., V_k)	$\max(\text{hebb}_{\mu}(V_1, V_2, V_3), V_4, \ldots, V_k)$ with $\text{hebb}_{\mu}(V_1, V_2, V_3) = V_1 V_2(1 - V_3) + \mu$ V_3	p(1) = persistence factor μ

Further, leader traits were considered in relation to the existing culture so that scenarios would not increase in complexity of the possible variations; as discussed in Sect. 2, the main inspiration for modelling choices and assumptions made has been based on [7, 15, 16, 18]. Table 2 below details how the context factors, when high, influence the culture and therefore leadership empowerment characteristics and openness in relation to the existing culture.

Table 2. The effect of context factors on culture and dependent leader qualities

	Flexibility	Positioning	Recognition	Delegation	Openness
C1	High	Low	Mid	Mid	High
C2	Low	High	Low	Low	Mid
C3	High	High	Low	Low	High
C4	Low	Low	Low	Low	Mid

Other leader qualities such as consistency and status were modeled independently of the culture. Consistency was a simple abstraction modeled through one state with the initial value varying in different simulations. Contrastingly, status varies across two dimensions where both social and formal organizational status are considered important in leader behavior. In this model both these concepts were considered as one state, that is stat_L. In scenarios where high status is modeled, the social and formal organizational status can be determined by the according culture type.

The final model constructed comprises 81 different states, an overview of all states with their according level can be seen in the table in Sect. 5 of the Appendix (Linked

Data at https://www.researchgate.net/publication/361481262). The second-order adaptive network model is depicted in Fig. 3 below with the key components highlighted in Table 3.

Table 3. Legend for important elements in the adaptive network model

Component	Description
Blue nodes	The blue base-level states represent reproduction
Yellow nodes	The yellow weight states represent attention to world observations
Light purple nodes	The green nodes represent the company culture
Dark purple nodes	The dark purple weight states represent attention to leader interactions
Pink nodes	The pink weight states represent retention and learning
Orange nodes	The orange persistence states represent reinforcement
Green nodes	The green nodes represent feed forward and feedback learning
Blue arrows	The blue arrows are upward connections from a lower-level plane
Red arrows	The red arrows are downward adaptive connections from a higher-level plane
Black arrows	The black arrows are connections from and to states on the same plane level

The base model represents two individual mental models along with a leader and a shared mental model, the individuals interact with the leader periodically and learn a set of tasks (a, b, c). The individual and team learning is influenced by the organizational culture and mistake management. The model was also constructed in a way in which context factors facilitate organizational culture change over time, specifically transitions from one culture to another. An overview of the key stages of the designed network are summarized below:

Individual mental models: Oval encapsulations in the base plane represent mental models of certain tasks. Individuals A and B have additional inputs from world observation states that can alter their mental model and reproduction of the tasks. These states are a result of observations and mirroring the leader's mental model combined with Hebbian learning which involves the **W**-states in the middle of the middle plane.

Shared mental model: Represented by the parallelogram encapsulation and is formed by taking input indirectly from individual learning of A and B in by connections between **W**-states in the first-order self-model level (middle) plane: feed forward learning. It occurs by transfer of knowledge from individual learning states (weight states) to shared learning weight states which directly affect the base-level shared mental model. In turn, individuals can learn from a shared mental model by connections between **W**-states in the opposite direction in the middle plane: feedback learning.

Mistake Handling: The dotted oval represents impartial knowledge for task c. This has a negative weight on mistake recognition (mistM1) so that if both b and c task states

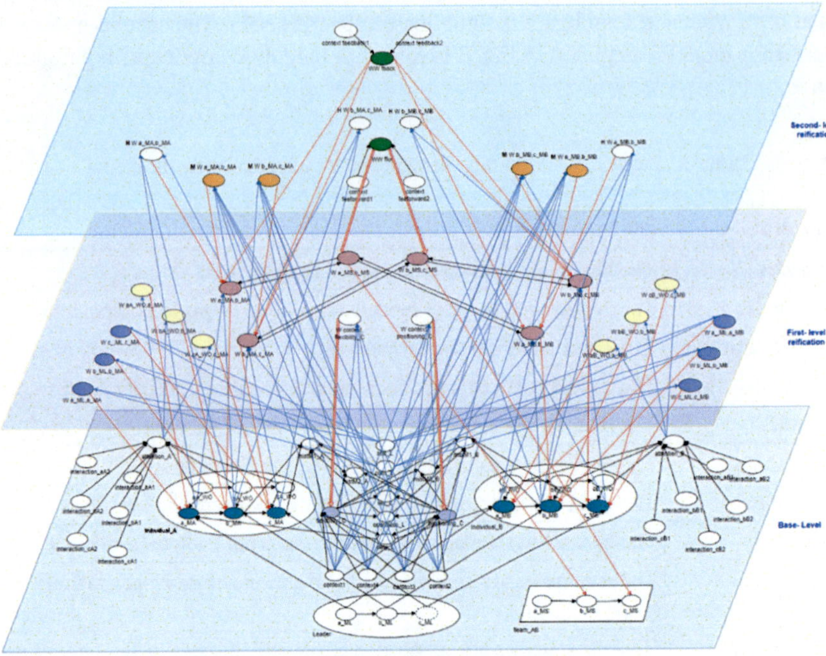

Fig. 3. The second-order adaptive network model

are high mistake noticing will be low. Mistake management (mistM2) depends on the mistake noticing state and organizational culture.

Leader interactions: Two interaction rounds of leader and individuals regarding each task takes place. All interaction states influence the attention of individuals and thus attention to observations states in the first-order self-model level.

Leader Qualities: All leader qualities (status, consistency, openness, recognition, and delegation) are represented individually. Recognition state in base-level influences persistence of Hebbian learning (in second-order level) as it reproduces the same effect of strengthening learning.

Attention: World observations along with interactions affect the base-level attention state. The attention base-level state indirectly influences individual's mental model by affecting attention first-order level weight states. First-order level weight states comprise attention to leader interactions and world observations. Leader attention weight states are influenced by status and consistency of the leader.

Organizational culture: Adaptiveness of culture is modelled by weight factors in first-order plane which are influenced by the context factors. The thick red lines represent four downwards connections, representing influences from each context factor.

Individual learning: Hebbian learning in first-order self-model plane takes place by weight states that influence the connections between mental model components in the base-level. The learning quality is also influenced by second-order states.

Learning quality: Persistence in the second-order self-model plane influence learning forgetfulness and learning rate; lower persistence values means more forgetfulness.

Learning flows: Feed forward and feedback learning control in the second-order self-model plane influences shared and individual weight states respectively in first-level self-model plane. They are affected by context states in second-order self-model plane which control timing of different learning flows. Thick pink lines from the **WW** feedforward state represent two downwards connections from individual learning weight states.

Through variation of settings of the model described above, it was possible to generate 24 different scenarios that model the 4 types of culture with constant factors as a baseline, and for each type of culture, additional simulations will vary in consistency and status of the leader. An overview of all simulated scenarios is highlighted in Table 4.

Table 4. Overview of scenarios modeled with varying culture type and leader qualities

#	Culture modeled	Context factors	Consistency	Status
1	Clan	High C1	High	High
2	Market	High C2	High	High
3	Adhocracy	High C3	High	High
4	Hierarchy	High C4	High	High
5	Clan	High C1	Low	Low
6	Market	High C2	Low	Low
7	Adhocracy	High C3	Low	Low
8	Hierarchy	High C4	Low	Low
9	Clan	High C1	High	Low
10	Market	High C2	High	Low
11	Adhocracy	High C3	High	Low
12	Hierarchy	High C4	High	Low
13	Clan	High C1	Low	High
14	Market	High C2	Low	High
15	Adhocracy	High C3	Low	High
16	Hierarchy	High C4	Low	High
17	Hierarchy; Adhocracy	Transition of high C4 to C3	High	High
18	Hierarchy; Clan	Transition of high C4 to C1	High	High
19	Market; Clan	Transition of high C2 to C1	High	High
20	Market; Adhocracy	Transition of high C2 to C3	High	High
21	Adhocracy; Hierarchy	Transition of high C3 to C4	High	High
22	Adhocracy; Market	Transition of high C3 to C2	High	High
23	Clan; Hierarchy	Transition of high C1 to C4	High	High
24	Clan; Market	Transition of high C1 to C2	High	High

4 Simulation Results

This section displays graphs for the simulation results of some of the scenarios. The graphs display a gradual increasing effect of learning and mistake management increasing as the flexibility of the organization increases and its positioning decreases. Prior to displaying the effect of culture on mistake management and learning. Temporal context will be given in terms of the learning flows and interactions. As indicated in Fig. 1 of Sect. 3 of the Appendix (see https://www.researchgate.net/publication/361481262), there are two rounds of the individual and leader interactions followed by feed forward learning with feedback learning after.

As seen in Fig. 4, interaction time extends from time point 50 to 150, which is followed by feed forward and feedback learning. The team learning peaks are aligned with the time of feed forward learning. That is because feed forward learning involves learning flows from micro-level (individual) to meso-level (team). Whereas individual learning does not only depend on the timing of feedback learning flows and is consistently higher because individuals can also learn from their environment, i.e. via world observations. Figure 4 shows the effect of flexible culture types on mistake acceptance and organizational learning.

Figure 5 displays the effect of less flexible cultures on the dependent variables when both consistency and status are low. In comparison to Figs. 2 and 3 of Sect. 3 of the Appendix, the same culture types when consistency and status are high display a big impact on individual learning and mistake management. Team learning on the other hand displays a smaller deviation where consistency and status of the leader are high. The most noticeable difference in team learning values was depicted in the hierarchy culture type.

Further, Sect. 3 in the Appendix displays the effect of the culture type on the dependent variables when consistency is low and status is high. These results are identical when status is low and consistency is high. This supports the notion that both status and consistency play an equally important role in organizational learning and mistake tolerance. Figure 5 displays the simulation where hierarchical culture transitions to clan culture. The graph is in line with the expected results as the transition to a culture with more flexibility has a positive correlation on the dependent variables. The hierarchy culture exists from the beginning till time point 450, whereas clan begins from that point on. Statistical analysis was carried out to compare the 24 scenarios according to the culture type and leader characteristics. The numerical simulation output was measured for the three dependent variables considered: individual learning, team learning, and mistake management. Since the scenarios only vary in culture and leader behavior, the simulations reproduce consistent mistake management, individual and team learning variances. Thus, the simulations were seen as one population that assumes homoscedasticity, that is equal variances among the dependent variables. Moreover, since each simulation is independent of the other simulation its compared to, an independent samples statistical test is used. A Wilcoxon-rank-sum nonparametric test, equivalent to Mann Whitney U-test, was utilized. The test considers the summed rank scores to derive a statistical score and a p-value which is considered with a 95% confidence level ($\alpha = 0.05$).

The results were processed using a Python-IDE and an open-source scientific library (SciPy) to carry out the rank-sum test. The graphical output was first stored as excel

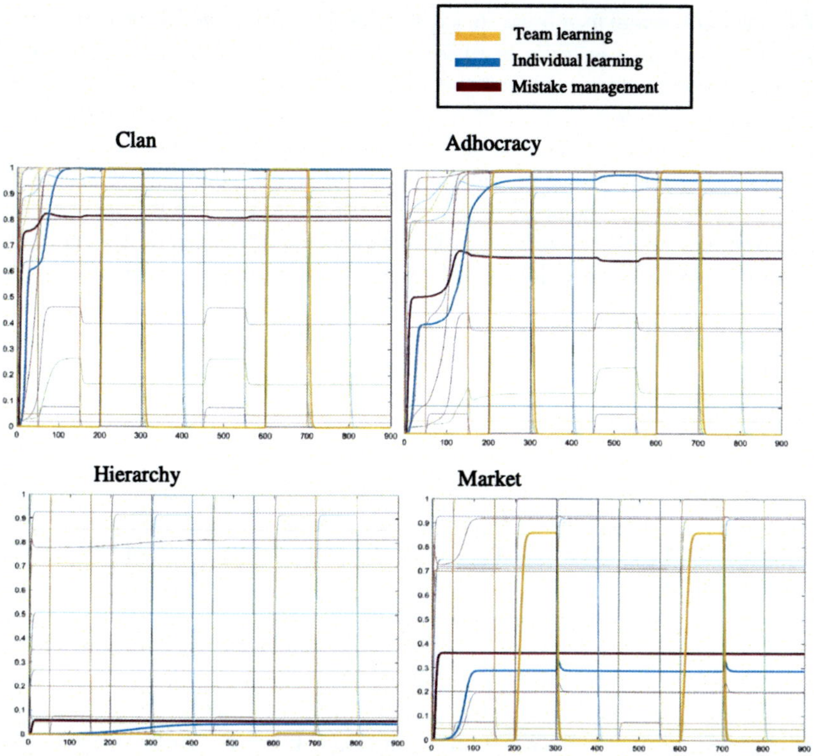

Fig. 4. Overview of effect of culture on dependent variables when consistency and status are high (upper graphs) and when consistency and status are low (lower graphs). In these graphs and the one in Fig. 5, the horizontal axis represents time (using an abstract time unit) and the vertical axis activation level (in the 0–1 interval)

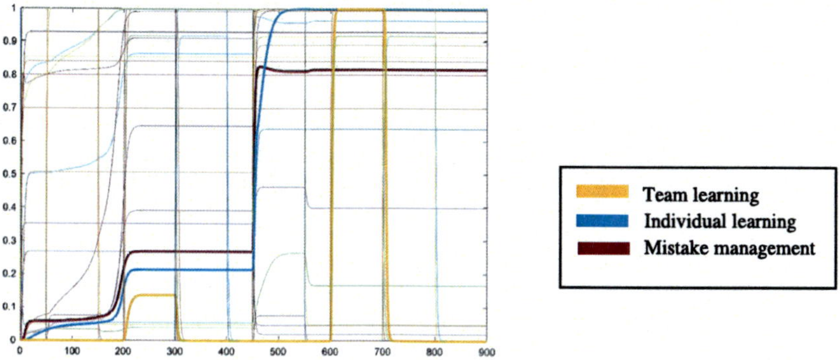

Fig. 5. Transition from hierarchy to clan culture when consistency and status are both high

files that show the state values at each time point ($\Delta t = 0.25$). The files were then converted to pandas data frame split into individual data frames for each dependent variable, which were then refined into a smaller time frame which excludes the first and last 50 time points in order to ensure significant data points. The scenarios were then compared with each other according to different cultures or the same culture and different leader behavior. A total of 54 comparisons were calculated measuring the statistical difference in individual learning, team learning, and mistake management. An overview of the statistical test scores can be summarized in the table of Sect. 4 in the Appendix. All statistical comparisons were significant as the p-value was much lower than the significance level (*p-value* ≤ 0.03, where $\alpha = 0.05$) except for the gray-colored cells which proved to be insignificant differences. Thus, we can reject the null hypotheses as the presence of these non-significant results are still inline and support the research hypotheseses.

5 Discussion

According to the simulation results and statistical analysis, it is clear that culture in accordance with certain transformational leadership qualities plays a role in mistake acceptance and organizational learning. The simulations provide a clear visual indicator of the effect on the dependent variables: individual learning, team learning, and mistake acceptance. But they also demonstrate significant statistical results from each other. The statistical analysis in Sect. 4 aimed to provide statistical significance in simulation comparisons to support the research hypotheses:

1. Higher flexibility and discretion in organizational culture results in better mistake management and thus better organizational learning.
2. Effective organizational learning requires a transformational leader to have both high social and formal status and consistency.
3. Company culture and leader's behavior must align for the best learning effects.

Twelve of the non-statistically significant results are in accordance with the second hypothesis. These non-significant values show that alternating low consistency or status when the other is high is not significantly different from each other; as one of them being low demonstrates the same effect as the other being low. Lack of high consistency or status hinder the imitation phase making the retention much lower in such scenarios. In that regard there is limited learning from the leader and only team learning and learning from world observations. This supports that for optimal results both consistency and status require to be high. Further, this aligns not only with the second hypothesis but also the third hypothesis as having a culture with high flexibility and discretion and high transformational leadership qualities, such as consistency and status, achieves the best results in terms of organizational learning.

Finally, all statistically significant results (non-gray cells) support the first hypothesis as higher flexibility always resulted in better individual learning and mistake management. Notably, team learning did not result in statistically significant differences. There are two instances, besides culture transitioning scenarios, in which team learning had insignificant difference (*p-value* ≥ 0.05). This can be due to the fact that team learning values are completely dependent on the incoming connections from individual learning values. Thus, when individual learning is significantly high, team learning values will consistently result in the same pattern displaying higher values than individual learning. This pattern however was not evident in all culture simulations. Hierarchy with low status or consistency, for instance, resulted in significantly low level of individual learning which was not enough to surpass the threshold to increase team learning level more.

References

1. Boone, T., Reilly, A.J., Sashkin, M.: Social Learning Theory. Prentice-Hall, Albert Bandura Englewood Cliffs, N.J. (1977)
2. Cameron, K., Quinn, R.: Diagnosing and Changing Organizational Culture: Based on the Competing Values Framework, 3rd edn. Jossey-Bass (2011)
3. Canbaloğlu, G., Treur, J., Roelofsma, P.H.M.P.: Computational modeling of organisational learning by self-modeling networks. Cogn. Syst. Res. **73**, 51–64 (2022a)
4. Canbaloğlu, G., Treur, J., Wiewiora, A.: Computational modeling of the role of leadership style for its context-sensitive control over multilevel organisational learning. In: Proceedings of the 7th International Congress on Information and Communication Technology, ICICT'22 (Lecture Notes in Networks and Systems). Springer Nature Switzerland AG (2022b) https://www.researchgate.net/publication/355186556
5. Canbaloğlu, G., Treur, J., Wiewiora, A. (eds).: Computational Modeling of Multilevel Organisational Learning and its Control Using Self-Modeling Network Models. Springer Nature Switzerland AG, forthcoming (2023). https://www.researchgate.net/publication/355582177
6. Cannon, M.D., Edmondson, A.C.: Failing to learn and learning to fail (Intelligently). Long Range Plan. **38**(3), 299–319 (2005). https://doi.org/10.1016/j.lrp.2005.04.005
7. Chénard Poirier, L.-A., Morin, A.J.S., Boudrias, J.-S.: On the merits of coherent leadership empowerment behaviors: a mixture regression approach. J. Vocat. Behav. **103**, 66–75 (2017)
8. Crossan, M.M., Lane, H.W., White, R.E.: An organizational learning framework: from intuition to institution. Acad. Manag. Rev. **24**, 522–537 (1999)
9. Hofstede, G., Hofstede, G.J., Minkov, M.: Cultures and Organizations. McGraw Hill (2010)
10. Jackson, P.L., Brunet, E., Meltzoff, A.N., Decety, J.: Empathy examined through the neural mechanisms involved in imagining how I feel versus how you feel pain. Neuropsychologia **44**(5), 759–761 (2006). https://doi.org/10.1016/j.neuropsychologia.2005.07.015
11. Keysers, C., Gazzola, V.: Hebbian learning and predictive mirror neurons for actions, sensations and emotions. Philos. Trans. R. Soc. B: Biol. Sci. **369**(1644), 20130175 (2014)
12. Kim, D.H.: A Framework and Methodology for Linking individual and organizational learning: applications in TQM and product development Massachusetts Institute of Technology Doctoral Dissertation (1993)
13. Kucharska, W.: Leadership, culture, intellectual capital, and knowledge processes for organizational innovativeness across industries: the case of Poland. J. Intellect. Cap. **22**(7), 121–141 (2021)
14. Kucharska, W.: Do mistakes acceptance foster innovation? Polish and US cross-country study of tacit knowledge sharing in IT. J. Knowl. Manag. **25**(11), 105–128 (2021)
15. Lawler, E.E.: The ultimate advantage: creating the high-involvement organization. Choice Rev. Online **30**(01), 30–038730–0387 (1992). https://doi.org/10.5860/choice.30-0387
16. Lawler, E., Boudreau, J., Mohrman, S.: Achieving Strategic Excellence: An Assessment of Human Resource Organizations. Standford University Press, Stanford, CA (2006)
17. Schein, E.H.: Organizational culture. Am. Psychol. **45**(2), 109–119 (1990)
18. Sims, H.P., Manz, C.C.: Social learning theory: The role of modeling in the exercise of leadership. J. Organ. Behav. Manag. **3**(1981), 55–63 (1981)

19. Treur, J.: Modeling higher-order adaptivity of a network by multilevel network reification. Netw. Sci. **8**, 110–144 (2020a)
20. Treur, J.: Network-Oriented Modeling for Adaptive Networks: Designing Higher-Order Adaptive Biological, Mental and Social Network Models. Springer Nature (2020b)
21. Treur, J.: Modeling Multi-Order Adaptive Processes by Self-Modeling Networks (Keynote Speech). In: Tallón-Ballesteros, A.J., Chen, C.-H. (eds.) Proceeding of the 2nd International Conference on Machine Learning and Intelligent Systems, MLIS'20. Frontiers in Artificial Intelligence and Applications, vol. 332, pp. 206–217. IOS Press (2020c)
22. Treur, J., Van Ments, L. (eds.): Mental Models and their Dynamics, Adaptation, and Control: a Self-Modeling Network Modeling Approach. Springer Nature, Cham (2022)
23. Weinzimmer, L.G., Esken, C.A.: Learning from mistakes: how mistake tolerance positively affects organizational learning and performance. J. Appl. Behav. Sci. **53**(3), 322–348 (2017). https://doi.org/10.1177/0021886316688658
24. Wiewiora, A., Smidt, M., Chang, A.: The how of multilevel learning dynamics: a systematic literature review exploring how mechanisms bridge learning between individuals, teams/projects and the organization. Eur. Manag. Rev. **16**(1), 93–115 (2019)

Biological Networks

Modeling of Hardy-Weinberg Equilibrium Using Dynamic Random Networks in an ABM Framework

Riccardo Tarantino[1], Greta Panunzi[2], and Valentino Romano[3(✉)]

[1] Department of Humanities, University of Palermo, Palermo, Italy
[2] Department of Statistical Sciences, University of Rome "La Sapienza", Rome, Italy
[3] Department of Biological, Chemical, and Pharmaceutical Sciences and Technologies, University of Palermo, Palermo, Italy
valentino.romano@unipa.it

Abstract. Hardy-Weinberg equilibrium is the fundamental principle of population genetics. In this article, we present a new NetLogo model called "Hardy-Weinberg Basic model v 2.0", characterized by a strict adherence to the original assumptions made by Hardy and Weinberg in 1908. A particularly significant feature of this model is that the algorithm does not make use of the binomial expansion formula. Instead, we show that using a procedure based on dynamic random networks, diploid equilibrium can be achieved spontaneously by a population of agents reproducing sexually in a Mendelian fashion. The model can be used to conduct simulations with a wide range of initial population sizes and genotype distributions for a single biallelic autosomal locus. Moreover, we also show that without any mathematical formalism the algorithm is also able to confirm the prediction of Kimura's diffusion equations on the time required to fix a new neutral allele in a population, due to genetic drift alone.

Keywords: Hardy-Weinberg Equilibrium · Wright-Fisher model · Dynamic random networks · Agent-based model · NetLogo

1 Introduction

1.1 Historical Background and Conceptual Framework

Hardy-Weinberg Equilibrium is the first cornerstone of population genetics. The name of this principle is due to its two theorists, the English mathematician Godfrey Hardy and the German physician Wilhelm Weinberg, who independently modelled it and linked, using simple mathematical formalism, allele and genotype frequencies in an ideal population with certain characteristics [1, 2]. More specifically, Hardy's inspiration for showing the existence of the Equilibrium was a contemporary debate over a medical example (i.e., human brachydactyly). He considered a character linked to a single biallelic autosomal locus, fairly large population numbers, random mating, equal fertility for all individuals and even distribution of sexes among the three possible genotypes. Then, he derived

H. Cherifi et al. (Eds.): COMPLEX NETWORKS 2016 2022, SCI 1077, pp. 241–250, 2023.
https://doi.org/10.1007/978-3-031-21127-0_20

the expected genotype ratios at each generation based on allele frequencies. Wilhelm Weinberg published the binomial square principle the same year as Hardy but, in distinction to Hardy, who did no further work in population genetics, Weinberg went on to many other discoveries. Like Mendel's, his work went unrecognized for many years [3]. From the time of these achievements onwards, the importance of applying mathematics to genetic theory has been widely acknowledged, and it is still an active area of research [4, 5].

After these independent formulations, between 1928 and 1931 some other founders of populations genetics, namely Ronald Fisher, John Haldane and Sewall Wright, began to study the specific departures from the equilibrium, like those resulting by sampling distortions caused by limited population numbers [6]. The latter distortion is also known as random *genetic drift*, which was first investigated by the so-called Wright-Fisher model [7, 8]. In its most basic form, the Wright-Fisher model considers random drift as the only disruptive force of the HWE, due to the stochastic oscillations of allele frequencies resulting from random mating [9].

The simplest version of the Wright-Fisher model can easily simulate both HWE and drift, keeping in mind their sharing of almost all assumptions. Indeed, the model considers a randomly mating population consisting of a constant number of diploid hermaphroditic individuals which reproduce over discrete generations. New individuals are formed, at each generation, by random sampling with replacement of gametes produced by the parents, who die immediately after reproduction [10]. It is an example of a Markov process, where the future state ($t + 1$) is only determined by its present state t (i.e., it is a memory-less process). In a biallelic model, there are two absorbing states, corresponding to the irreversible fixation or loss of one of the two alleles (i.e., when the frequency of one allele is 1 and that of the other one is 0) [11].

However, calculating the exact solution of the Wright-Fisher model without any simplification has been proved to be hard in practice. For this reason, starting from the 50s, the father of the Neutral Theory of Evolution Motoo Kimura developed a solution consisting in approximating the discrete WF process with a continuous time, continuous space diffusion process [12], which later brought him and Ohta [13] to develop the method of diffusion equations to estimate the *conditional* mean time of fixation for a single mutant neutral allele in a homogeneous allele pool ($4N_e$, with N_e = effective population size). The estimation of the time until fixation (expressed in number of generations) of a new allele can be achieved in different ways, and several studies have tried to confirm or improve the original Kimura's results with alternative approaches, such as those that make use of the coalescent theory [14, 15].

1.2 Agent-Based Modeling

NetLogo is a widely used software based on Agent-based Modeling (ABM) principles [16] and, during the past twenty years, many models have already been developed for processes and phenomena studied in a wide spectrum of disciplines ranging from Psychology to Biology and Chemistry. Such models are archived in a large Models Library and/or published in specialized journals. Nevertheless, few NetLogo models have attempted to properly implement HWE, and none of these models has fully implemented the assumptions made in the original mathematical formulation proposed by Hardy and Weinberg.

2 Materials and Methods

2.1 Implementation of HWE NetLogo Model

We have created the "Hardy-Weinberg Basic Model v. 2.0" using version 6.2.0 of NetLogo, which was downloaded from the NetLogo Home page of Northwestern University (https://ccl.northwestern.edu/netlogo/). Details on the implementation and features of this model can be found on the info and code tabs of the model and in the Results section of this article. For further information on the use of "Hardy-Weinberg Basic Model v. 2.0", the interested readership can contact the authors. The model has been uploaded on the NetLogo User Community. It can also be downloaded from the following link: https://www.dropbox.com/s/8sxvovb99bb04yg/HARDY-_1.NLO?dl=0.

2.2 Statistical Tests and Computer Simulations

To verify that the population generated in NetLogo is in Hardy-Weinberg Equilibrium, we used the Chi square goodness-of-fit test with continuity correction [17], X_c^2, for c = 0.5 [18]. The test is applied at each generation to compare the genotype frequencies obtained computationally and those expected, calculated on the basis of allele frequencies. The Chi squares for the sample are then:

$$X_c^2 = \sum_{i \geq 1} \frac{\left(\left|n_{ij} - E_{ij}\right| - c\right)^2}{E_{ij}} \text{ for } c = 0.5$$

The degree of freedom is the number of classes minus 1 and then minus the number of parameters estimated from the data, so with two alleles we consider only one degree of freedom. Hence, if we use a significance level of 5%, the tabulated, Chi square is 3.84. We reject the null hypothesis if the computed X_c^2 is greater than 3.84.

To establish if the average value of generations that we need to fix a newly appearing allele "*a*" is not different from the theoretical value ($4N_e$), we performed six different simulations with different population sizes (50, 60, 70, 80, 90, 100, 200, and 500 individuals/agents). These values of population sizes have been chosen to highlight the effects of genetic drift, whose action is stronger with small populations. On the other hand, genetic drift is acting with any finite population size. We did not try population sizes higher than 500 because this would take many more generations to reach fixation as well as requiring a longer computational time. Each simulation keeps track of the frequency evolution of the new allele (q) whose starting value is 1/2N (N = census number). All simulations were performed using the *BehaviorSpace* tool of NetLogo. Simulations stopped when $p = 0$ or $q = 0$, where p is the frequency of the more diffused allele "*A*". The results are stored in an Excel spreadsheet, where the number of generations required to fix or extinguish the "*a*" allele is reported for each iteration/replica. The cases of our interest are only the fixation events. We then compare by the one sample t-test the average value of generations calculated from a minimum of 20,000 replicates needed to fix the originally rare allele "*a*", with the theoretical value ($4N_e$) of the Kimura's model based on diffusion equations [13]. One-sample t-test is derived under the assumption that the population of data is normally distributed. Fortunately, even when data are not normally distributed, this method works well when the sample size is large enough. As we can observe in the following table, for each simulation we have a large sample size ($n > 100$).

Population size	50	60	70	80	90	100	200	500
Sample size (*n* of fixations)	168	193	116	105	130	107	58	42

With the one-sample t-test we compare, in each simulation, the average number of generations necessary to fix the allele "*a*" with the theoretical number $4N_e$ obtained with Kimura's diffusion approximation. Hence, for each simulation we have the following hypothesis test:

$$\begin{cases} \mu = 4Ne \\ \mu \neq 4Ne \end{cases}$$

where:

- $4N_e = 400$ if population size $= 50$
- $4N_e = 480$ if population size $= 60$
- $4N_e = 560$ if population size $= 70$
- $4N_e = 640$ if population size $= 80$
- $4N_e = 720$ if population size $= 90$
- $4N_e = 800$ if population size $= 100$
- $4N_e = 1600$ if population size $= 200$
- $4N_e = 4000$ if population size $= 500$.

Looking at the p-value, we reject the null hypothesis if $p \leq 0.05$ [19].

For each simulation we calculated the test statistic, the p-value and the confidence intervals considering the Student's t-distribution [20]. All results are computed using R 4.0 software [21]. We calculate the one sample t-test using the R software's function "t.test" [21].

3 Results

3.1 The Hardy-Weinberg Basic Model

The algorithm of the NetLogo HW model is presented in the code tab of the model. The agents in our model are diploid organisms with separated sexes. Each subset of agents is characterised by one of the three possible genotypes for a single autosomal locus (DD, DR, and RR). Two variables are considered for each agent: "sex" and "partner". Before running the model, the user may change the initial number of agents for each genotype, while the number of individuals for each sex is always kept constant with 1:1 ratio throughout simulations. In the setup phase, agents occupy fixed positions in the NetLogo world and are arranged in a circle. The random mating among all agents is implemented by links connecting, at each generation, two agents of opposite sex. In other words, the mating procedure was implemented by a dynamic random network whose topology continuously and synchronously change over time. The networks have many connected components with each component made of two nodes (N) corresponding to a mating pair. The process continues until no more unpaired agents are present (i.e., when the total number of edges is equal to the number of biunivocal pairs, N/2). Each mating pair generates two offspring, one male and one female. Gametes are not modelled, since reproduction is implemented by the expected genotypes reflecting Mendelian segregation ratios (e.g., there is a 50% probability to produce a DR offspring from a DR x RR cross). At the end of each cycle of mating and reproduction, both parents instantly die, that is, all the parental nodes are deleted and replaced by offspring nodes, that form a new random network. In the code, loop formation and elimination of random networks is implemented by executing the procedure "create-pairs" as follows:

```
to create-pairs
  ;; "singles" do not have any partner. Each single female creates
a link with a random single male. At the end of this procedure, all
females and males are paired
  let singles turtles with [ partner = nobody ]
  ask turtles with [ sex = "female" ] [
   if any? singles [
   if ( partner = nobody ) and ( any? other turtles with [ sex = "male"
] with [ partner = nobody ] ) [
      set partner one-of other turtles with [ sex = "male" ] with [
partner = nobody ]
     create-link-with partner
     ask partner [
     set partner myself
     ]
   ]
   ]
   ]
   tick
end
```

The choice of producing two sibs of opposite sex for each mating guarantees that sex ratio (1:1) and population size remain constant through generations. Thus, the population of agents that can be recreated in our model resembles a HW population because it is diploid, panmictic, large enough to maintain the equilibrium over time, has a 1:1 sex ratio, and has not overlapping generations.

A screenshot of the interface is shown in Fig. 1. Four buttons are used: (i) "setup" to initialize a simulation, (ii) "go" to run the simulation in a continuous mode, (iii) "create-pairs" to run it in a discrete mode, and (iv) "reproduce" to make couples generate their offspring. Three sliders are used to regulate the initial number of individuals for each of the three genotypes. Several monitors allow to keep track of (i) population size, (ii) number of agents for each sex, (iii) number of generations, (iv) observed frequencies for each genotype, (v) expected frequencies, and (vi) Chi square values. Expected genotype frequencies are estimated by the classic HWE formula $p^2 + 2pq + q^2 = 1$, where p and q are the frequencies of the "D" and "R" allele, respectively. p^2, $2pq$ and q^2 are, in this order, the expected number of "DD" (homozygotes), "DR" (heterozygotes), and "RR" (homozygotes). Two plots are used to display the evolution of allele and genotype frequencies. A third plot keeps track of change of the Chi-square values at each generation. In the latter plot, a red line is used to indicate the critical Chi square value of 3.84 for $\alpha = 0.05$ and $df = 1$.

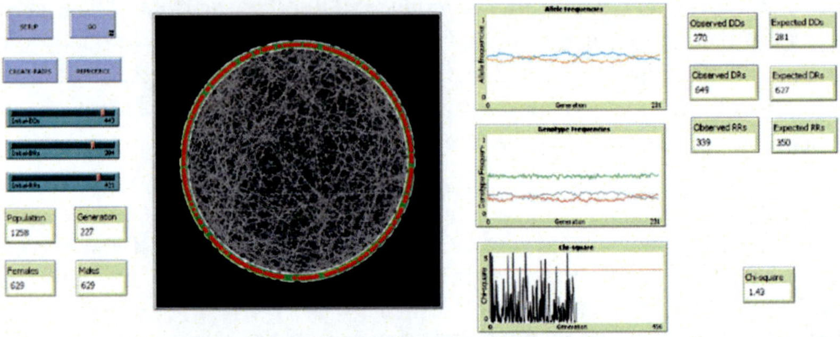

Fig. 1. Interface of the NetLogo "Hardy-Weinberg basic model".

3.2 Time to Fixation of a New Neutral Allele

After calculating the mean time to fix the "*R*" allele and considering only simulations where the rare allele was ultimately fixed, we compare each of these average values with the theoretical value of $4N_e$. If variance in offspring number is greater than random expectation, Ne is smaller than N (i.e., the census number). On the contrary, if there is less random variation, N_e is greater than N. Finally, if all individuals have equal reproductive success, as in our case, then $N_e = 2N$ [10]. Thus, N_e values used for calculations are 2N. By applying the one-sample t-test for each N_e and sample size (i.e., Number of fixation events), we obtain the results shown in Table 1.

As in each simulation the p-value is greater than 0.05, we do not reject the null hypothesis (H_0: $\mu = 4Ne$) (see Methods). We can come to the same conclusion by observing the confidence intervals shown in Table 1 and Fig. 2; the theoretical value of $4N_e$ belongs to the confidence intervals at the 95% level. Table 1 and Fig. 2 also show that, as the effective population size increases, the width of the confidence intervals increases as well. This result is consistent with our expectation because, as the effective sample size increases, the sample size (i.e., number of fixations) decreases. In other words, the frequency of fixation events of the newly arisen neutral allele shows an inverse relationship with the total number of individuals in the population [14, 22].

Table 1. Statistical analysis of simulated fixation time observed in NetLogo for a new neutral allele.

DR/(DR + DD)	$4N_e$	t-test (df)	Average number of generations	95% confidence interval
1/50	400	−0.204 (167)	396.101	358.288 – 433.914
1/60	480	0.476 (192)	489.010	451.644 – 526.377
1/70	560	−0.198 (115)	554.129	495.439 – 612.820
1/80	640	0.398 (104)	655.086	579.914 – 730.258
1/90	720	1.554 (129)	773.069	705.488 – 840.651
1/100	800	−0.217 (106)	790.421	702.991 – 877.850
1/200	1600	0.805 (57)	1692.534	1462.277 – 1922.792
1/500	4000	0.617 (42)	4210.595	3521.06 – 4900.13

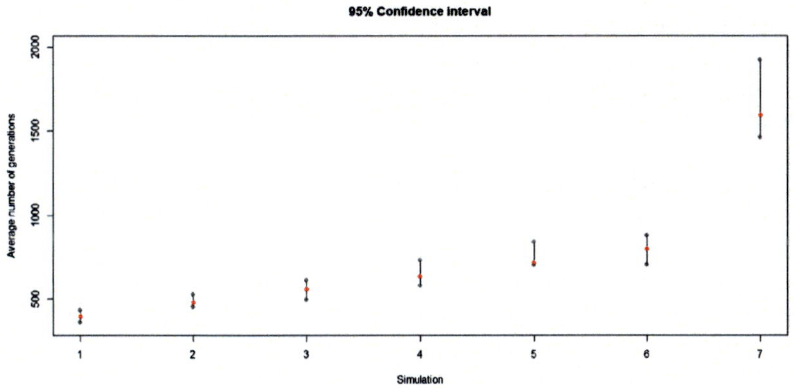

Fig. 2. The simulations are represented on the abscissa, while the number of generations are represented on the ordinate axis. Segments in (black) represent the confidence intervals at a 95% level. The red points represent the theoretical values $4N_e$, corresponding to N = 50, 60, 70, 80, 90, 100, 200. The point corresponding to N = 500 has not been included to improve C.I. visualization. For each simulation the theoretical values are included in the range, so we accept the null hypothesis in all cases.

Based on the above results, we accept the hypothesis that the outcomes of the simulations performed with our NetLogo model agree with the theoretical values estimated by Kimura and Ohta [13].

4 Discussion

In this study, our main aim was to create a NetLogo model of the Hardy-Weinberg Equilibrium using a complex system's perspective. Our approach was to develop an algorithm that does not make use of the binomial expansion formula of $(p + q)^2$. In this way, we show that—by using an ABM framework—diploid equilibrium can be achieved spontaneously as an emergent property of the complex system represented by the collective behaviour of agents reproducing in a Mendelian fashion. This feature distinguishes our model from other NetLogo models developed so far and also dealing with HWE (more about this below). Beside this peculiarity, the model is also strictly compliant to the implicit and explicit assumptions originally made in 1908 by the English mathematician Godfrey Harold Hardy and the German physician Wilhelm Weinberg. These assumptions can be summarized as follows:

- the organism is diploid; the considered gene is autosomal and bi-allelic (e.g., Brachydactyly);
- reproduction is sexual;
- generations are rigorously non-overlapping;
- sexes are evenly distributed;
- the allele and genotype frequencies are the same in males and females;
- mating is random (the population is panmictic);
- population size is large;
- migration is negligible, since gene flow by a population of different genetic structure would potentially change the allele frequencies in the original population;
- mutation can be ignored;
- natural selection does not affect the alleles under consideration.

Consistent with these premises, the simulation shows that equilibrium of genotype and allele frequencies is reached in one generation, and the population remains in equilibrium in successive generations, maintaining a constant sex ratio (1:1). Compliance to HWE can be verified by watching at the Chi square values displayed in the corresponding monitor and plot. Moreover, using the sliders, the number of individuals can be decreased up to a value where genetic drift is so strong to cause a rapid deviation from HWE, as can be verified in the Chi square plot when it repeatedly displays values above 3.84.

We shortly discuss here the recent "Hardy-Weinberg Equilibrium" model developed by Dabholkar and Wilensky [23], whose aims appear more similar to ours. These Authors aimed to model HWE in a population of rock pocket mice displaying alternative fur coat colors and state, in the Info tab, that their model performs according to the assumptions of HWE. However, by running their model, it can be easily verified that several agents survive for more than one tick/generation, that is, they have overlapping generations. This behavior makes that model non-compliant to the original HW model. In our model, we successfully captured the original level of abstraction of the HWE by using random dynamic networks. Therefore, a network approach can improve both adherence to the assumptions of population genetics' models and graphical representation of mating patterns in species with separate sexes, since all the current couples of nodes can be easily visualized by running the model in the discrete mode. The model does not rely on more specific network concepts, such as node activation or weight adaptation, since they are neither necessary nor relevant to analyze the collective behavior of our reproducing agents. However, it must be noted that the integration of ABM and network theory is presently a very active research area [24, 25].

Models dealing with finite populations must also deal with the effects of genetic drift. Thus, to explore further the performance of our model, we have used it to estimate the time of fixation for a new neutral allele because of genetic drift alone for different values of N_e. We have found that the results obtained with the simulations are consistent with those deriving from the approximated method of Kimura and Ohta [13], since the difference between the hypothesized value of $4N_e$ (with N_e = effective population size) and our calculated values is statistically not significant. Since the first formulation of this fundamental resolutive method of the Wright-Fisher model by Kimura and Otha [13], other independent approaches (e.g., coalescence theory [15]) have been used but, to our knowledge, ABM has never been used for this purpose. We are not reporting results for N values > 100 because, above these values, simulations with NetLogo gradually become computationally intensive.

The extensive educational use of NetLogo today gives us hope that our model will be a useful tool for students undertaking college-level courses in population genetics. Using this model, teachers in this discipline have the opportunity not only to conduct their lessons with a multidisciplinary approach integrating genetics, networks, statistics, and agent-based modeling, but they can also encourage their students' systems thinking in studying and learning population genetics.

We plan to develop new versions of the model allowing to test for the effects of other evolutionary factors apart from genetic drift (i.e., selection, migration, mutation) in finite populations.

Acknowledgements. This work was funded by the Italian Ministry of Research, Teaching and University (MIUR) within the activities of the Prin project # 2020TACEZR_004 entitled: "Crossing the Sea: ancient and modern human genomes to study the evolutionary dynamics of Sicily, Sardinia and Corsica".

References

1. Hardy, G.H.: Mendelian proportions in a mixed population. Science **28**, 49–50 (1908)
2. Weinberg, W.: Uber den Nachweis der Vererbungbeim Menschen. Jahresh. Ver. Vaterl. Naturkd. Wurttemb **64**, 369–382 (1908)
3. Crow, J.F.: Population genetics history: a personal view. Annu. Rev. Genet. **21**, 1–22 (1987)
4. Sun, L., Gan, J., Jiang, L., Wu, R.: Recursive test of Hardy-Weinberg equilibrium in tetraploids. Trends Genet. **37**, 504–513 (2021)
5. Jagathesan, T.: Mathematical analyses in genetics and evolution. JDMS **09**, 138–142 (2022)
6. Mayo, O.: A century of Hardy-Weinberg equilibrium. Twin Res. Hum. Genet. **11**(3), 249–256 (2008)
7. Fisher, R.A.: On the dominance ratio. Bull. Math. Biol. **52**, 297–318 (1990)
8. Wright, S.: Evolution in Mendelian populations. Genetics **16**, 97–159 (1931)
9. Tataru, P., Simonsen, M., Bataillon, T., Hobolth, A.: Statistical Inference in the Wright-Fisher model using allele frequency data. Syst. Biol. **66**(1), e30–e46 (2017)
10. Charlesworth, B.: Effective population size and patterns of molecular evolution and variation. Nat. Rev. Genet. **10**, 195–205 (2009)
11. Hartl, D.L.: A Primer of Population Genetics and Genomics, p. 150. Oxford University Press, Oxford (2020)

12. Messer, P.: Neutral models of genetic drift and mutation. In: Kliman, R.M. (ed.) Encyclopedia of Evolutionary Biology, pp. 119–123. Academic Press, Oxford (2016)
13. Kimura, M., Ohta, T.: The average number of generations until fixation of a mutant gene in a finite population. Genetics **61**(3), 763–771 (1969)
14. Otto, S.P., Whitlock, M.C.: Fixation probabilities and times. In: Encyclopedia of Life Sciences. Wiley, Hoboken (2005)
15. Greenbaum, G.: Revisiting the time until fixation of a neutral mutant in a finite population—a coalescent theory approach. J. Theor. Biol. **7**(380), 98–102 (2015)
16. Wilensky, U.: NetLogo. http://ccl.northwestern.edu/netlogo/. Center for Connected Learning and Computer-Based Modeling, Northwestern University, Evanston, IL (1999)
17. Emigh, T.: A comparison of tests for Hardy-Weinberg equilibrium. Biometrics **36**(4), 627–642 (1980)
18. Yates, F.: Contingency table involving small numbers and the x2 test. J. R. Stat. (Suppl.) **1**(2), 217–235 (1934)
19. Thisted, R.A.: What is a P-value. Departments of Statistics and Health Studies. The University of Chicago, Chicago (1998)
20. Bulpitt, C.: Confidence intervals. Lancet **329**(8531), 494–497 (1987)
21. R Core Team: R: A Language and Environment for Statistical Computing. R Foundation for Statistical Computing, Vienna, Austria (2018). http://www.R-project.org/
22. Kimura, M.: On the probability of fixation of mutant genes in a population. Genetics **47**(6), 713–719 (1962)
23. Dabholkar, S., Wilensky, U.: NetLogo Hardy-Weinberg equilibrium model. http://ccl.northwestern.edu/netlogo/models/HardyWeinbergEquilibrium. Center for Connected Learning and Computer-Based Modeling, Northwestern University, Evanston, IL (2020)
24. Bouadjio-Boulic, A., Amblard, F., Gaudou, B.: Dynamic agent-based network generation. In: 9th International Conference on Agents and Artificial Intelligence (ICAART 2017), Feb 2017. pp. 599–606. Porto, Portugal (2017)
25. Anderson, T.M., Dragicevic, S.: Network-agent based model for simulating the dynamic spatial network structure of complex ecological systems. Ecol. Model. **389**, 19–32 (2018)

COMBO: A Computational Framework to Analyze RNA-seq and Methylation Data Through Heterogeneous Multi-layer Networks

Ilaria Cosentini[1], Vincenza Barresi[2], Daniele Filippo Condorelli[2], Alfredo Ferro[3], Alfredo Pulvirenti[3], and Salvatore Alaimo[3](✉)

[1] Department of Physics and Astronomy, University of Catania, 95124 Catania, Italy
[2] Department of Biomedical and Biotechnological Sciences, Section of Medical Biochemistry, University of Catania, 95123 Catania, Italy
[3] Department of Clinical and Experimental Medicine, Bioinformatics Unit, University of Catania, 95123 Catania, Italy
salvatore.alaimo@unict.it

Abstract. Multi-layer Complex networks are commonly used for modeling and analysing biological entities. This paper presents a new computational framework called COMBO (Combining Multi Bio Omics) for generating and analyzing heterogeneous multi-layer networks. Our model uses gene expression and DNA-methylation data. The power of COMBO relies on its ability to join different omics to study the complex interplay between various components in the disease. We tested the reliability and versatility of COMBO on colon and lung adenocarcinoma cancer data obtained from the TCGA database.

Keywords: Multi-layer networks · Multi-omics analysis · Biological data integration · Next-Generation sequencing · Graph database

1 Introduction

Understanding tumor phenotype is a challenging task. Indeed, independently investigating each component is not enough to characterize its entire biology. In addition, studying the interactions and relations between the different omics (i.e., genomics, transcriptomics, and proteomics) in the cancer microenvironment is crucial to stratify case subtypes or distinguish cases from controls. For example, the study of single gene perturbations is insufficient to explain the cause of abnormal behavior in a tissue that triggers the disease. The usage of Complex Networks to describe natural phenomena such as biological systems has taken over [1]. This approach has been demonstrated to help reveal unknown mechanisms behind the different diseases. Indeed, the interdependence between multiple biological elements is responsible for disease onset and progression [2, 3]. Here, the network represents a practical approach [4, 5]. The analysis of omics functional interaction networks (e.g., genes, proteins, and ligands) is promising for unveiling their regulatory mechanisms. However, due to the vast number of elements and information provided by the biological components, using networks to highlight the main

H. Cherifi et al. (Eds.): COMPLEX NETWORKS 2016 2022, SCI 1077, pp. 251–264, 2023.
https://doi.org/10.1007/978-3-031-21127-0_21

aspects of a condition is not straightforward. Nevertheless, it represents a significant milestone for science [6]. However, in the biological field, simple networks (where there is at most one edge between each pair of nodes) are not suitable to represent the multiple types of relationships among various constituents [7–9]. On the other hand, multi-layer networks (every node is assumed to belong to every layer, each layer represents a specific kind of relation, along with intralayer edges type for each layer) fit nicely in real-world system interactions avoiding the loss of essential properties [8, 10, 11]. There are several examples of multi-layer network usage in Biology [12, 13]. As the integration of different sources of genetic data is still challenging, modeling a multilayer network in which each layer represents specific omics can improve cancer characterization [14–17]. However, state-of-art methods focus on visualization and analysis of multi-layer networks, leaving the pre-processing phase to the user [16, 18–21]. This study proposes a bioinformatics framework to infer important information about a specific disease by integrating the transcriptional and epigenetic (methylation) features. We propose a novel pipeline, COMBO (Combining Multi Bio Omics), for multi-layer network inference and analysis to identify relevant pathways in the studied case. Taking advantage of the Boolean implication method, both transcriptomic and epigenomic data were analyzed through StepMiner and BooleanNet systems [22, 23]. The goal was to identify the implication between the different transcripts and methylated CpGs. The obtained results were used to generate heterogeneous multi-layer graphs. Subsequently, Neo4J was exploited to query the multi-layer network with properly defined Cypher queries. The model has been tested on colon cancer samples (TCGA-COAD) having the amplification of chromosome 20, also called GAIN, compared to colon tumoral samples without that chromosomal alteration, identified as DISOMIC. Only the GAIN multi-layer network exhibited interesting patterns when searching for all the paths that link the expression layer with the methylation layer, including genes located on chromosome 20. In addition, the enriched analysis of the genes composing the path shows the alteration of pathways that play an essential role in cancer. To demonstrate the plasticity of the pipeline to any human disease, we tested it in TCGA-LUAD samples. The comparison focused on samples having upregulated IWS1 transcript (upIWS1) versus those with downregulated IWS1 (downIWS1). As shown for COAD, the paths that involve IWS1 and upregulated genes are available only for the multi-layer generated for the upIWS1 condition. Furthermore, a Reactome enrichment analysis shows the alteration of crucial genes in lung adenocarcinoma. Our bioinformatics approach is a flexible and easy-to-use method in preliminary cancer studies to identify significant disease signatures.

2 Materials and Methods

2.1 Pipeline Design

The pipeline comprises five phases (Fig. 1). First, it generates multi-layer networks for each condition under investigation. Next, networks are stored in Neo4j. Then, through custom Cypher queries, the multi-layer is interrogated to generate sub-networks that can be analyzed differently (i.e., enrichment pathway analysis).

STEP 1: Input Pre-preprocessing. The input files for the analysis consist of the expression and methylation matrices containing pre-processed counts and B-values,

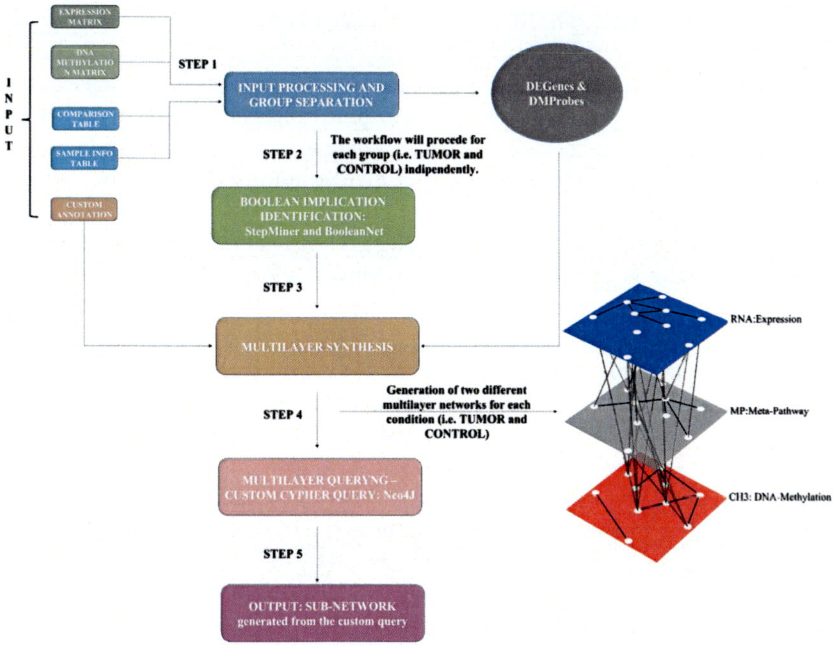

Fig. 1. The COMBO pipeline workflow.

respectively. The initial matrix contains all the samples representing the conditions to compare. Two tables are needed for the differential expression and methylation analysis. The first one contains the name of each sample and its group. The second table represents the comparisons between groups. The number of columns in the comparison table equals the number of multi-layer networks. The first step is processing the input matrices (expression and methylation) to identify differentially expressed genes and differentially methylated CpGs. The differentially expressed genes and the methylated probes are generated using the R package limma [24]. The user can choose the threshold for the identification of differentially expressed genes (default = 1.0, corresponding to logFC < -1 and logFC > 1) and differentially methylated probes (default = 0, equivalent to logFC < 0 and logFC > 0) along with the adjusted p-value (default < 0.05). In addition, the count and the B-value in the expression and methylation matrices are converted in $\log_2$FPKM and M-value, respectively. These will be provided as input for the next phase.

STEP 2: Boolean Implication. The second step is the generation of boolean implications with the StepMiner [22] and BooleanNet [23] algorithms. **StepMiner Algorithm** was initially developed to extract binary patterns in time-series microarray gene expression data. After sorting values in ascending order, StepMiner fits the data by a step function to discretize the jump from low to high values. In this step, it finds the best threshold t for each gene. Then, values above $t + 0.5$ are classified as "High," values below $t - 0.5$ are classified as "low," and values between $t - 0.5$ and $t + 0.5$ are classified as "intermediate." Two different discretized matrix was generated for expression and methylation data. $\log_2$FPKM for expression data and M-value for methylation data were used as input for StepMiner. The M-value is calculated as the $\log_2$ ratio of the intensities of the methylated probe versus the unmethylated probe, as reported in [25]. **The Boolean Values** obtained from Step Miner are converted in the gene-pairs relationship using BooleanNet. The Boolean implications between gene expression and methylation levels are extracted as "if-then" relationships. All the potential pairs of values of genes in each sample are plotted in a scatter plot which is divided into four quadrants "low-low", "low-high", "high-low", and "high-high." For each pair of genes, the algorithm checks which quadrant is significantly sparsely populated with points compared to the other quadrants. In the scatter plot, each point represents the expression value of the pair of genes for a specific sample. Based on the empty quadrant in the scatter plot, the algorithm detects six types of Boolean implications. The Boolean relationships can be symmetric or asymmetric. Symmetric relationships include the equivalent type, which corresponds to highly positively correlated genes, or the opposite type, which corresponds to the highly negatively correlated genes. The asymmetric ones correspond to every sparse quadrant and are Divided into low—low, high—low, low—high, and high—high. An implication between two genes is significant when one or more quadrants is sparsely populated according to a statistical test, and there are enough high and low values for each gene. Although both software packages are pretty old, recent papers confirm that StepMiner and BooleanNet are the state-of-art methods to infer Boolean implications between biological entities [26–29]. The user can select the delta threshold for StepMiner, the BooleanNet statistic parameter (s), and the *p*-value threshold. The statistical parameters used for Boolean analysis in COAD samples were $s = 4$ for methylation data and $s = 2$ for RNA-seq data. Whereas $s = 2$ was used for expression and methylation data in LUAD samples. In the COMBO pipeline the user can choose the *p*-value (default $= 0.01$) and the BooleanNet statistic (default $= 6$).

STEP 3: Multi-layer Network Synthesis. Networks (or graphs) are mathematical tools to represent relationships between entities or objects. For example, interactions, reactions, or structural relations are depicted as the links between the interacting elements (genes, proteins, metabolites) in a biological network. Multi-layer networks have different layers containing a set of homogeneous nodes equal or not between the layers [30]. The edges between nodes in the same layer are indicated as intra-layer edges, whereas the edges linking different layers are called inter-layer edges. When inter-layer edges connect nodes in two different layers representing the same entity, the network is classified as a multiplex network [31].

The third step is the multi-layer generation. We show an example in Fig. 2. In the example, we include three layers: expression, meta-pathway, and methylation. The expression level contains nodes and directed intra-layer edges selected from BooleanNet [23] analysis conducted with RNA-seq data. Similarly, the nodes and directed intra-layer edges that participate in the methylation level are selected from BooleanNet methylation output. The meta-pathway layer was added to connect the other two layers. The meta-pathway is a network obtained by merging all KEGG pathways through their common nodes, annotated with experimentally validated miRNA-target and Transcription Factor-miRNA interactions [32–34]. Consequently, the meta-pathway

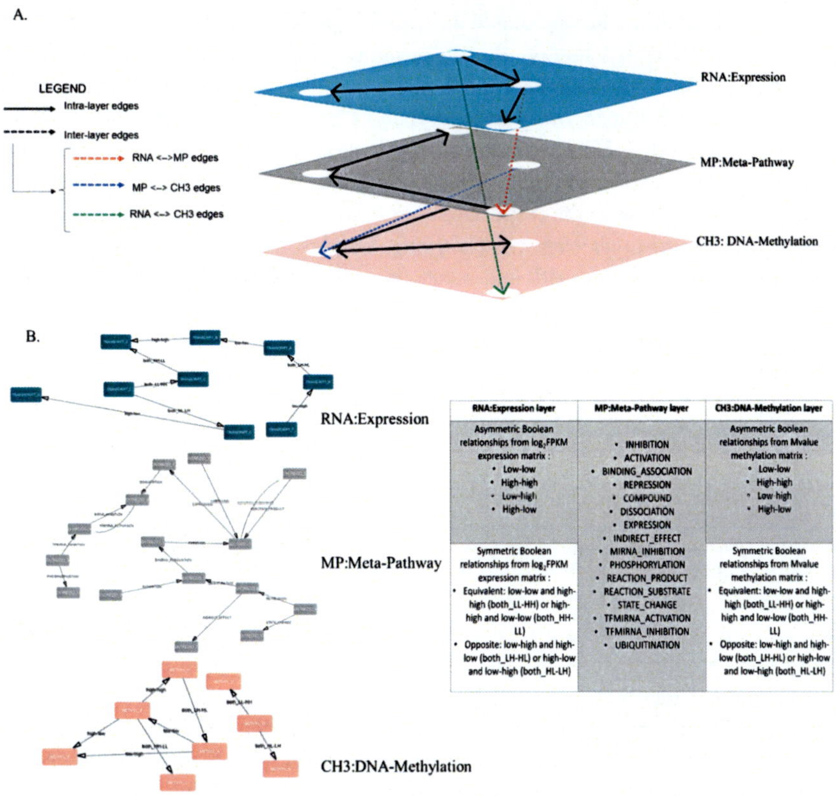

Fig. 2. (A) A heterogeneous multi-layer network example (generated with pymnet python library http://www.mkivela.com/pymnet/). The interlayer edges are depicted with different colors. (B) Example of the structure of the single layer. The table shows the edges attribute for each layer.

layer includes genes, miRNAs, and metabolites linked with several edges, which represent a validated biological interaction (i.e., activation, repression, binding association, compound, dissociation, expression, indirect effect, inhibition, miRNA inhibition, phosphorylation, reaction product, reaction substrate, state change, transcription factor—miRNA activation, ubiquitination). We followed the procedure described in Alaimo et al. [32, 33] for the KEGG meta-pathway and the one described in Alaimo et al. [34] for the Reactome-KEGG meta-pathway, which integrates REACTOME pathways to the meta-pathway environment, yielding a richer and more comprehensive model. A unique nodeID is generated for each node in the three layers. If a node is present in two or more layers, the ID will differ because each layer reports different aspects of the same gene. In addition, directed inter-layer edges are inserted between two nodes representing the same gene to connect the other layers. Each node has an annotation that the users will employ for querying the network. The default annotations are divided into four categories: (i) nodes details: nodeID, layer, entrezId, symbol, and chromosomal localization of the gene; (ii) expression of the genes generated in the first step: using the parameters defined by the user, the upregulated genes are indicated as "UP_REG," whereas the downregulated genes are annotated as "DOWN_REG." The remaining genes do not have any annotation; (iii) methylation information: two columns are included in this section, one for the hypomethylated probes and the other for the hypermethylated one. In addition, for each hyper- or hypo-methylated CpG site, we also report the location (Body, Exon, 3'UTR, 5'UTR, Promoter); (iv) custom annotation: the user can choose to add custom annotations for each node, loading a table for each annotation reporting the name of the genes with the respective annotation in the different conditions.

STEP 4–5: Multi-layer Network Querying. The last two steps involve using Neo4j to store the multi-layers network and query its content. Neo4j (https://neo4j.com/) is a native graph database that implements a true graph model down to the storage level. Once the multi-layer is stored on Neo4J, the user can query the network using Cypher language to find a correlation between the nodes in the transcription and methylation layers. Cypher is Neo4j's query language to retrieve data from the graph and was inspired by SQL. Based on specific node annotation, the user can choose custom Cypher queries.

2.2 Colon Cancer Case Study

The Gene expression and methylation data of tumor samples and normal mucosa were obtained from The Cancer Genome Atlas (TCGA; https://www.cancer.gov/tcga) database. The samples were grouped into three classes, the "GAIN-chr20" with the amplification of chromosome 20, the "DIS-chr20" without amplification in that chromosome, and the mucosae group ("MU"). The corresponding clinical annotations and chromosomal assets were obtained from the cBioPortal database. The samples having both the RNA-seq and methylation data were selected: DIS-chr20 ($n = 68$), GAIN-chr20 ($n = 89$), MU ($n = 19$). Samples presenting microsatellite instability were inserted in this analysis to have a significant comparison.

2.3 Lung Cancer Case Study

TCGA-LUAD project samples were downloaded from The Cancer Genome Atlas (TCGA; https://www.cancer.gov/tcga) databases. We divided the samples into "upIWS1" and "downIWS1," identifying the distribution of IWS1 transcript across the samples. Since the distribution is normal (Kolmogorov-Smirnov test, p-value $= 0.1309$), we selected the samples using the threshold calculated as follows: $averageFPKM \pm \sigma(standarddeviation)$ (mean $= 0.8$, $\sigma = 0.2$). Gene expression and methylation data of tumor samples and normal tissue were downloaded. The samples having both the RNA-seq and methylation data were selected: upIWS1 ($n = 95$), downIWS1 ($n = 47$), normal tissues ($n = 21$).

2.4 Data Analysis

The raw counts and B-values were downloaded and pre-processed using the TCGAbiolinks [35] package. The statistical parameters used for the Boolean analysis in COAD samples were $s = 4$ for methylation data and $s = 2$ for RNA-seq data. Whereas $s = 2$ was used for expression and methylation data in LUAD samples. We used the default parameters for the analysis. For each node, we added some custom attributes to query the network. For example, we added the results of the DEG analysis of GAIN-chr20vsDIS-chr20 and upIWS1vsDownIWS1 in the two case studies, respectively. The genes are indicated as "OVER_T" or "DOWN_T." In addition, the protein expression, mutation status, and the transcription factor annotation for each node were also inserted. The protein expression data were retrieved from The Cancer Protein Atlas—TCPA https://tcpaportal.org/tcpa/) for the selected TCGA-COAD samples. The expression was determined by calculating the unpaired two-sample t-test and the mean between the sample in the same group. The mutation data were collected in cBioPortal (https://www.cbioportal.org/) for each condition. The complete list of transcription factors was retrieved from the Human Transcription Factor database (http://humantfs.ccbr.utoronto.ca/) [36]. The insertion of custom information allowed us to query the multi-layer networks more comprehensively. The multi-layer networks were generated using Neo4J (https://neo4j.com/). The query results from the two conditions were compared to maintain only the exclusive nodes and edges. Then, the remaining nodes were subjected to Reactome enrichment analysis using the ReactomePA R package [37].

3 Results and Discussion

Our multi-layer approach is an easy-to-use bioinformatics pipeline to generate and analyze multi-layer networks combining multi-omics data. To show the capability of this method, we performed two case studies with colon and lung adenocarcinoma cancers data retrieved from TCGA. The user can use this method for a preliminary study to identify critical genes in a specific disease and evaluate the interplay between DNA methylation and transcription. Finally, the multi-layer approach is also helpful for the biomedical and research lab to filter the vast amount of RNA-seq and methylation data.

3.1 TCGA-COAD Results

The chromosome-20 amplification is a hallmark of colorectal cancer, along with the amplification of chromosome 13, chromosome 8, and chromosome 7 [38, 39]. In recent years, scientists have identified the importance of chr-20 amplification in colon cancer, mainly since it is the location of multiple oncogenes such as BCL2L1, AURKA, TPX2, and SRC [40–44]. However, despite numerous functional studies of various genes located in chromosome 20, the amplification of this chromosome in colon cancer remains under-investigated, among other critical molecular features [45]. By constructing a multi-layer network, we could identify remarkable pathways in COAD having the chr20 amplification.

Initially, we selected the COAD samples deposited on TCGA, distinguishing the samples having the amplification of chromosome 20 "GAIN-chr20" (Fig. 3A) and "DIS-chr20" (Fig. 3B), which do not show aberration in that chromosome. In the GAIN-chr20 group, chromosome 13 amplification appears in 72% of samples because of the correlation between the two chromosomal abnormalities [38, 39]. In addition, normal tissues were also analyzed as a control. To demonstrate how the usage of Boolean implication to generate the layer and, in general, how our approach is representative of disease

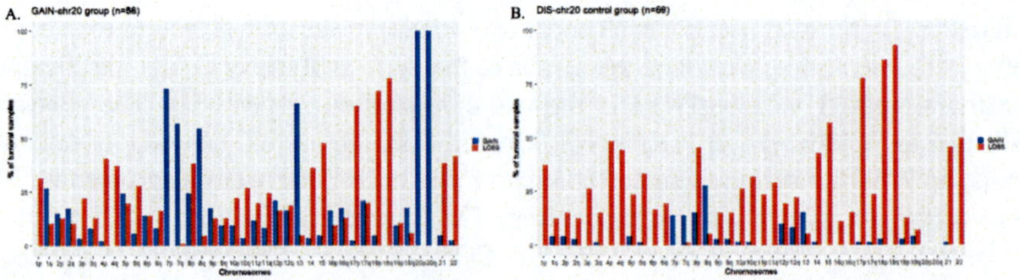

Fig. 3. (A) Chromosomal aberration distribution on TCGA GAIN-chr20 COAD samples (n = 89). 100% of samples show the amplification of chromosome 20. A small group also includes the amplification of chromosome 13. (B) Chromosomal aberration distribution on TCGA DIS-chr20 COAD samples (n = 69). The samples were selected without the amplification of chromosome 20.

profile, we compared the Over-UpT genes located on chromosome 20 listed in Condorelli et al. [38] with the nodes included in the RNAlayer of both multi-layers (GAIN-chr20 and DIS-chr20). A gene located on chromosome 20 is defined as Over-UpT if it is upregulated in cancer and further overexpressed by gene dosage ("positive caricature transcriptomic effect") [38, 39]. Firstly, all the nodes on chromosome 20 selected using the Boolean implication were compared with the Over-UpT genes reported in Condorelli et al.. 11 nodes are shared between the GAIN-chr20 RNAlayer and the Over-UpT list. These genes are absent in the DIS-chr20 RNAlayer. Then, more accurate research was conducted, selecting all the nodes with annotations "UP_REG" and "OVER_T." Although Over-UpT genes were detected using a different methodology (microarray in Condorelli et al. and RNA-seq in our approach) and a different logFC cutoff, 9 Over-UpT genes are in common (Fig. 4A). The list of shared genes, reported in Fig. 4B, shows bona fide cancer driver genes, already implicated in cancer pathogenesis [46–54]. The enrichment analysis of the common Over-UpT genes is reported in Fig. 4C. These genes are involved in cancer-critical pathways.

Using the proper Neo4j queries, it is possible to extrapolate detailed information about the phenotype studied.

Since the molecular signature that distinguishes the GAIN-chr20 from the DIS-chr20 samples is the amplification of chromosome 20, we searched in the two multi-layers for all the paths that involve nodes located on chromosome 20 (Fig. 5A). The DIS-chr20 network did not show any path. This result demonstrates how the multi-layer network method reproduces perfectly the condition studied. Figure 5 shows the sub-network structure resulting from the query (B) and the enriched pathways heat plot (C). For each pathway, the gene involved is indicated colored based on logFC generated from the comparison between GAIN-chr20 and DIS-chr20. One of the main pathways influenced is the "Signal Regulatory family interaction," in which the SRC transcript seems upregulated while the SIRP protein family results downregulated. The Src family kinase has been studied as a therapeutic target in colon cancer since it is usually overexpressed and associated with a worse prognosis [55, 56]. The multi-layer nodes annotation shows that the SRC protein is overexpressed in the analyzed data. In contrast, the downregulation of the SIRP family has been demonstrated fundamental for inhibiting cell apoptosis in prostate cancer [57].

A.

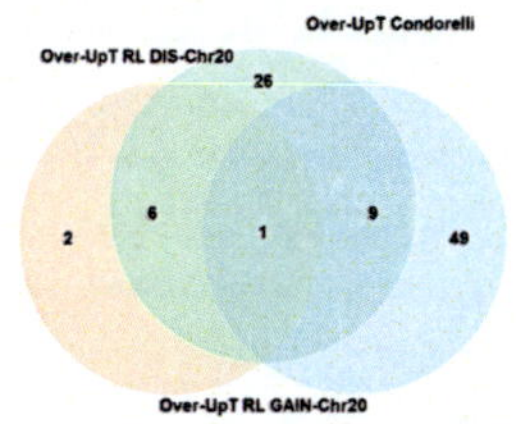

B.

Genes Over-UpT	logFC GAINvsMU	logFC GAINvsDIS
MYBL2	1,88	0,65
NELFCD	1,34	0,79
FERMT1	1,80	1,05
PLCB4	1,67	1,89
AHCY	1,48	0,78
CSEL1	1,60	0,64
DDX27	1,28	1,07
RPS21	1,30	0,96
CPNE1	1,56	1,20

C.

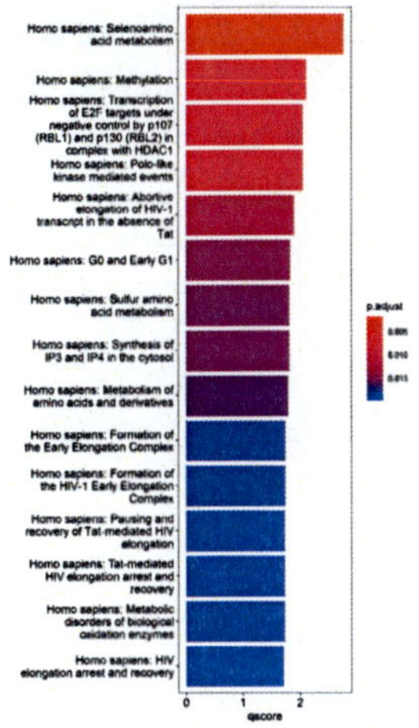

Fig. 4. (A)Venn diagram showing the common Over-UpT genes located on chromosome 20 between DIS-chr20 RNAlayer (RL) (orange circle),GAIN-chr20 RL (green circle) and Condorelli et al. list (purple circle). (B) List of Over-UpT common genes with the logFC comparing GAIN and Mucosae samples (logFC GAINvsMu) and GAIN and DIS samples (logFC GAINvsDIS). (C) Bar plot of enriched terms.

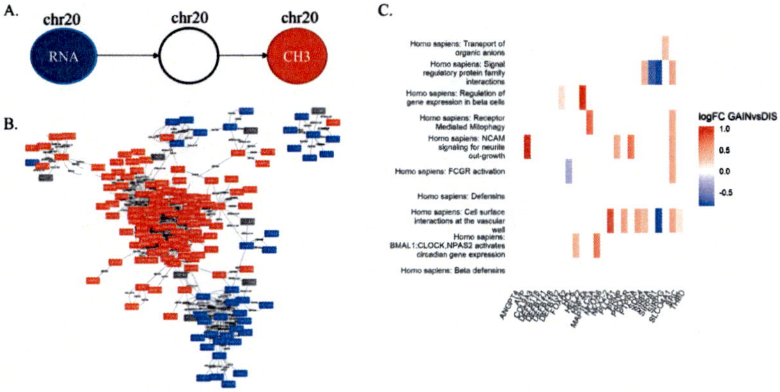

Fig. 5. (A) Cypher query pattern includes all the paths connecting nodes located on chromosome 20. (B) Schematic Cytoscape representation of paths selected by a custom query. The node colors depict the belonging layers: blue for the RNA layer, red for the CH_3 layer, and grey for the MP layer (C). The Heatmap reports the enrichment analysis starting from the nodes included in the sub-network. The logFC is generated from the difference between GAIN-chr20 and DIS-chr20.

3.2 TCGA-LUNG Results

IWS1 is one of the critical elements in lung adenocarcinoma since its phosphorylation influences the splicing and, consequently, the transcription of genes involved in cell proliferation [58]. Usually, the tumoral cells regulate the gene's transcription, promoting the alternative splicing to generate multiple transcripts and control cancer pathobiology

[59, 60]. Therefore, studying the multi-layer, we compared samples having upregulated IWS1 versus downregulated samples.

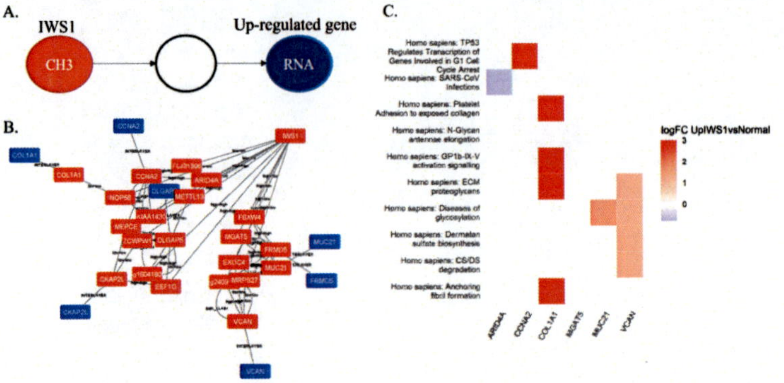

Fig. 6. (A) Custom query representation which retrieves the connection between the source node IWS1 in the CH_3 layer (red) and any target node annotated as "UP_REG" in the RNA layer (blue). Sub-network UpIWS1 obtained by Cypher query is depicted using Cytoscape in (B) No nodes lay in MP layer. (C) Heatmap reports the enrichment analysis starting from the nodes included in the sub-network. The logFC is generated from the difference between UpIWS1 and Normal tissue.

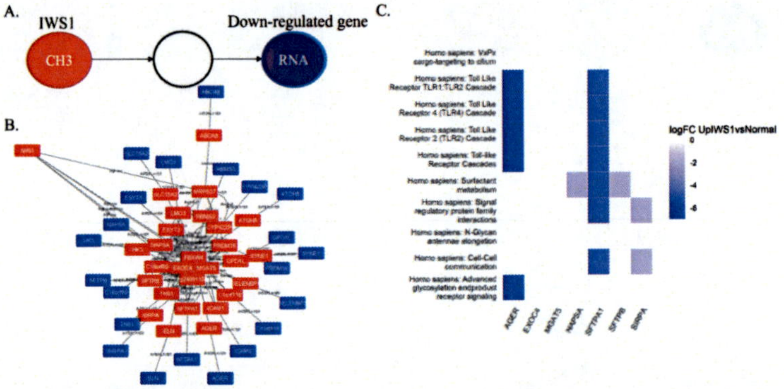

Fig. 7. (A) Designed Cypher query, which combines IWS1 node on CH_3 layer (red) with down-regulated RNA layer (blue) genes. (B) Schematic representation of the subnetwork resulting from the before cited query. (C) The enrichment analysis result for the selected nodes is reported as Heatmap. The logFC is generated from the difference between UpIWS1 and Normal tissue.

To identify the influence of IWS1 over other genes, the multi-layers were queried, looking for all the paths that connect IWS1 in the methylation layer with the upregulated genes (Fig. 6A) or the downregulated genes (Fig. 7A) in the expression layer. The query retrieved a result only for the upIWS1 multi-layer and not for the downIWS1 condition. The enrichment analysis (Fig. 6C) of the genes involved in the paths that connect IWS1 with upregulated genes shows an alteration of *CCNA2*, *COL1A1* and *MUC21* identified as key molecules in the lung adenocarcinoma progression [61–63]. On the other hand, evaluating the enrichment analysis (Fig. 7C) on the genes obtained by Cypher query looking for the connection between IWS1 and downregulated genes, *AGER* and *SFTPA1* are highlighted. Both genes are usually downregulated in lung cancer and are validated as risk factors in LUAD [64–66].

4 Conclusions

In conclusion, the COMBO multi-layer approach is a new pipeline to query epigenomic and transcriptomic data to identify relevant signatures in a specific disease. Furthermore, the method's versatility allows the user to analyze several conditions focusing on different aspects of the pathology.

References

1. Boccaletti, S., Latora, V., Moreno, Y., Chavez, M., Hwang, D.: Complex networks: structure and dynamics. Phys. Rep., 175–308 (2006)
2. De Domenico, M.: Multilayer network modeling of integrated biological systems: comment on "network science of biological systems at different scales: a review" by Gosak et al. Phys. Life Rev., 149–52 (2018)
3. Rai, A., Pradhan, P., Nagraj, J., Lohitesh, K., Chowdhury, R., Jalan, S.: Understanding cancer complexome using networks, spectral graph theory and multilayer framework. Sci. Rep. **7**, 41676 (2017)
4. Barabási, A.-L., Gulbahce, N., Loscalzo, J.: Network medicine: a network-based approach to human disease. Nature Rev. Genet., 56–68 (2011)
5. Goh, K.-I., Cusick, M.E., Valle, D., Childs, B., Vidal, M., Barabási, A.-L.: The human disease network [Internet]. In: Proceedings of the National Academy of Sciences, pp. 8685–90 (2007)
6. Lv, Y., Huang, S., Zhang, T., Gao, B.: Application of multilayer network models in bioinformatics. Front. Genet. **12**, 664860 (2021)
7. Zheng, W., Wang, D., Zou, X.: Control of multilayer biological networks and applied to target identification of complex diseases. BMC Bioinform. **20**, 271 (2019)
8. Boccaletti, S., Bianconi, G., Criado, R., del Genio, C.I., Gómez-Gardeñes, J., Romance, M., et al.: The structure and dynamics of multilayer networks [Internet]. Phys. Rep., 1–122 (2014)
9. Mangioni, G., Jurman, G., De Domenico, M.: Multilayer flows in molecular networks identify biological modules in the human proteome [Internet]. IEEE Trans. Netw. Sci. Eng., 411–20 (2020)
10. Kivela, M., Arenas, A., Barthelemy, M., Gleeson, J.P., Moreno, Y., Porter, M.A.: Multilayer networks [Internet]. J. Compl. Netw., 203–71 (2014)
11. Kenett, D.Y., Perc, M., Boccaletti, S.: Networks of Networks—An Introduction [Internet]. Chaos, Solitons and Fractals, pp. 1–6 (2015)
12. Hammoud, Z., Kramer, F.: Multilayer networks: aspects, implementations, and application in biomedicine. Big Data Anal. (2020)
13. McGee, F., Ghoniem, M., Melançon, G., Otjacques, B., Pinaud, B.: The state of the art in multilayer network visualization. Comp. Graph. Forum. 125–49 (2019)
14. Bersanelli, M., Mosca, E., Remondini, D., Giampieri, E., Sala, C., Castellani, G., et al.: Methods for the integration of multi-omics data: mathematical aspects. BMC Bioinform. (2016)

15. Tordini, F., Aldinucci, M., Milanesi, L., Liò, P., Merelli, I.: The Genome Conformation As an Integrator of Multi-Omic Data: The Example of Damage Spreading in Cancer. Frontiers in Genetics (2016)
16. Zitnik, M., Leskovec, J.: Predicting multicellular function through multi-layer tissue networks. Bioinformatics., i190–8 (2017)
17. Gligorijević, V., Pržulj, N.: Methods for biological data integration: perspectives and challenges. J. Roy. Soc. Interface., 20150571 (2015)
18. Domenico, M.D., De Domenico, M., Porter, M.A., Arenas, A.: MuxViz: a tool for multilayer analysis and visualization of networks. J. Compl. Netw., 159–76 (2015)
19. De Bacco, C., Power, E.A., Larremore, D.B., Moore, C.: Community detection, link prediction, and layer interdependence in multilayer networks. Phys. Rev. E. **95**, 042317 (2017)
20. Škrlj, B., Kralj, J., Lavrač, N.: Py3plex toolkit for visualization and analysis of multilayer networks. Appl. Netw. Sci. (2019)
21. Hammoud, Z., Kramer, F.M.: An R Package to Create, Modify and Visualize Multilayered Graph. Genes, p. 519 (2018)
22. Sahoo, D., Dill, D.L., Tibshirani, R., Plevritis, S.K.: Extracting binary signals from microarray time-course data. Nucl. Acids Res., 3705–12 (2007)
23. Sahoo, D., Dill, D.L., Gentles, A.J., Tibshirani, R., Plevritis, S.K.: Boolean implication networks derived from large scale, whole genome microarray datasets. Genome Biol. **9**, R157 (2008)
24. Ritchie, M.E., Phipson, B., Wu, D., Hu, Y., Law, C.W., Shi, W., et al.: limma powers differential expression analyses for RNA-sequencing and microarray studies. Nucl. Acids Res. **43**, e47 (2015)
25. Du, P., Zhang, X., Huang, C.-C., Jafari, N., Kibbe, W.A., Hou, L., et al.: Comparison of Beta-value and M-value methods for quantifying methylation levels by microarray analysis. BMC Bioinform. BioMed. Central **11**, 1–9 (2010)
26. Sgariglia, D., Conforte, A.J., Pedreira, C.E., de Carvalho, L.A.V., Carneiro, F.R.G., Carels, N., et al.: Data-Driven Modeling of Breast Cancer Tumors Using Boolean Networks. Frontiers in Big Data [Internet]. Frontiers Media SA (2021)
27. Xu, X., Zhu, L., Yang, Y., Pan, Y., Feng, Z., Li, Y., et al.: Low tumour PPM1H indicates poor prognosis in colorectal cancer via activation of cancer-associated fibroblasts. Br J Cancer. Nature Publ. Group **120**, 987–995 (2019)
28. Dabydeen, S.A., Desai, A., Sahoo, D.: Unbiased Boolean Analysis of Public Gene Expression Data for Cell Cycle Gene Identification. The American Society for Cell Biology, Mol Biol Cell (2019)
29. Sahoo, D., Wei, W., Auman, H., Hurtado-Coll, A., Carroll, P.R., Fazli, L., et al.: Boolean analysis identifies CD38 as a biomarker of aggressive localized prostate cancer. Oncotarget. Impact J. **9**, 6550–6561 (2018)
30. da Mata, A.S., da Mata, A.S.: Complex networks: a mini-review [Internet]. Brazilian J. Phys. 658–72 (2020)
31. Kinsley, A.C., Rossi, G., Silk, M.J., VanderWaal, K.: Multilayer and multiplex networks: an introduction to their use in veterinary epidemiology. Front. Vet. Sci. **7**, 596 (2020)
32. Alaimo, S., Giugno, R., Acunzo, M., Veneziano, D., Ferro, A., Pulvirenti, A.: Post-transcriptional knowledge in pathway analysis increases the accuracy of phenotypes classification. Oncotarget **7**, 54572–54582 (2016)
33. Alaimo, S., Marceca, G.P., Ferro, A., Pulvirenti, A.: Detecting disease specific pathway substructures through an integrated systems biology approach. Noncoding RNA. **3** (2017)
34. Alaimo, S., Rapicavoli, R.V., Marceca, G.P., La Ferlita, A., Serebrennikova, O.B., Tsichlis, P.N., et al.: PHENSIM: phenotype simulator. PLoS Comput. Biol. **17**, e1009069 (2021)

35. Silva, T.C., Colaprico, A., Olsen, C., D'Angelo, F., Bontempi, G., Ceccarelli, M., et al.: TCGA workflow: analyze cancer genomics and epigenomics data using bioconductor packages. F1000 Res., 1542 (2016)
36. Lambert, S.A., Jolma, A., Campitelli, L.F., Das, P.K., Yin, Y., Albu, M., et al.: The human transcription factors. Cell **175**, 598–599 (2018)
37. Yu, G., He, Q.-Y.: ReactomePA: an R/Bioconductor package for reactome pathway analysis and visualization. Mol. Biosyst. 477–9 (2016)
38. Condorelli, D.F., Spampinato, G., Valenti, G., Musso, N., Castorina, S., Barresi, V.: Positive Caricature Transcriptomic Effects Associated with Broad Genomic Aberrations in Colorectal Cancer. Scientific Reports (2018)
39. Condorelli, D.F., Privitera, A.P., Barresi, V.: Chromosomal density of cancer up-regulated genes, aberrant enhancer activity and cancer fitness genes are associated with transcriptional cis-effects of broad copy number GAINs in colorectal cancer. Int. J. Mol. Sci. **20** (2019)
40. Sillars-Hardebol, A.H., Carvalho, B., Beliën, J.A.M., de Wit, M., Delis-van Diemen, P.M., Tijssen, M., et al.: *BCL2L1*has a functional role in colorectal cancer and its protein expression is associated with chromosome 20q GAIN. J. Pathol. 442–50 (2012)
41. Carvalho, B., Postma, C., Mongera, S., Hopmans, E., Diskin, S., van de Wiel, M.A., et al.: Multiple putative oncogenes at the chromosome 20q amplicon contribute to colorectal adenoma to carcinoma progression. Gut **58**, 79–89 (2009)
42. Sillars-Hardebol, A.H., Carvalho, B., Tijssen, M., Beliën, J.A.M., de Wit, M., Delis-van Diemen, P.M., et al.: TPX2 and AURKA promote 20q amplicon-driven colorectal adenoma to carcinoma progression. Gut **61**, 1568–1575 (2012)
43. Ptashkin, R.N., Pagan, C., Yaeger, R., Middha, S., Shia, J., O'Rourke, K.P., et al.: Chromosome 20q amplification defines a subtype of microsatellite stable, left-sided colon cancers with wild-type RAS/RAF and better overall survival. Mol. Cancer Res. (2017)
44. Voutsadakis, I.A.: Chromosome 20q11.21 amplifications in colorectal cancer. Cancer Genom. Proteom. **18**, 487–96 (2021)
45. Bui, V.M.H., Mettling, C., Jou, J., Sun, H.S.: Genomic amplification of chromosome 20q13.33 is the early biomarker for the development of sporadic colorectal carcinoma. BMC Med. Genom. **13**, 149 (2020)
46. Liu, Q., Guo, L., Qi, H., Lou, M., Wang, R., Hai, B., et al.: A MYBL2 complex for RRM2 transactivation and the synthetic effect of MYBL2 knockdown with WEE1 inhibition aGAINst colorectal cancer. Cell Death Dis. Nature Publ. Group **12**, 1–11 (2021)
47. Song, S., Li, D., Yang, C., Yan, P., Bai, Y., Zhang, Y., et al.: Overexpression of NELFCD promotes colorectal cancer cells proliferation, migration, and invasion. Oncol. Targ. Ther. Dove Press **11**, 8741 (2018)
48. Li, L., Li, P., Zhang, W., Zhou, H., Guo, E., Hu, G., et al.: FERMT1 contributes to the migration and invasion of nasopharyngeal carcinoma through epithelial–mesenchymal transition and cell cycle arrest. Cancer Cell Int. BioMed. Central **22**, 1–14 (2022)
49. Yang, C., Li, D., Bai, Y., Song, S., Yan, P., Wu, R., et al.: DEAD-box helicase 27 plays a tumor-promoter role by regulating the stem cell-like activity of human colorectal cancer cells. Oncol. Targ. Ther. Dove Press **12**, 233 (2019)
50. Wu, S., Zhang, W., Shen, D., Lu, J., Zhao, L.: PLCB4 upregulation is associated with unfavorable prognosis in pediatric acute myeloid leukemia. Oncol. Lett. Spandidos Publ. **18**, 6057 (2019)
51. Belužić, L., Grbeša, I., Belužić, R., Park, J.H., Kong, H.K., Kopjar, N., et al.: Knock-down of AHCY and depletion of adenosine induces DNA damage and cell cycle arrest. Sci. Rep. Nature Publ. Group **8**, 1–16 (2018)
52. Pimiento, J.M., Neill, K.G., Henderson-Jackson, E., Eschrich, S.A., Chen, D.T., Husain, K., et al.: Knockdown of CSE1L gene in colorectal cancer reduces tumorigenesis in vitro. Am. J. Pathol. (2016)

53. El Khoury, W., Nasr, Z.: Deregulation of ribosomal proteins in human cancers. Biosci Rep. **41** (2021)
54. Wang, Y., Pan, S., He, X., Wang, Y., Huang, H., Chen, J., et al.: CPNE1 Enhances Colorectal Cancer Cell Growth, Glycolysis, and Drug Resistance Through Regulating the AKT-GLUT1/HK2 Pathway, Vol. 14. Onco Targets Ther. Dove Press, p. 699 (2021)
55. Chen, J., Elfiky, A., Han, M., Chen, C., Saif, M.W.: The role of Src in colon cancer and its therapeutic implications. Clin. Colorectal Cancer. **13**, 5–13 (2014)
56. Jin, W.: Regulation of Src Family Kinases during Colorectal Cancer Development and Its Clinical Implications. Cancers, pp. 12 (2020)
57. Yao, C., Li, G., Cai, M., Qian, Y., Wang, L., Xiao, L., et al.: Prostate cancer downregulated SIRP-α modulates apoptosis and proliferation through p38-MAPK/NF-κB/COX-2 signaling. Oncol. Lett. **13**, 4995–5001 (2017)
58. Sanidas, I., Polytarchou, C., Hatziapostolou, M., Ezell, S.A., Kottakis, F., Hu, L., et al.: Phosphoproteomics screen reveals akt isoform-specific signals linking RNA processing to lung cancer. Mol. Cell. **53**, 577–590 (2014)
59. Paronetto, M.P., Passacantilli, I., Sette, C.: Alternative splicing and cell survival: from tissue homeostasis to disease. Cell Death Differ. **23**, 1919–1929 (2016)
60. Coomer, A.O., Black, F., Greystoke, A., Munkley, J., Elliott, D.J.: Alternative splicing in lung cancer. Biochim. Biophys. Acta Gene Regul. Mech. **1862**, 194388 (2019)
61. Yoshimoto, T., Matsubara, D., Soda, M., Ueno, T., Amano, Y., Kihara, A., et al.: Mucin 21 is a key molecule involved in the incohesive growth pattern in lung adenocarcinoma. Cancer Sci. **110**, 3006–3011 (2019)
62. Hou, L., Lin, T., Wang, Y., Liu, B., Wang, M.: Collagen type 1 alpha 1 chain is a novel predictive biomarker of poor progression-free survival and chemoresistance in metastatic lung cancer. J. Cancer. **12**, 5723–5731 (2021)
63. Ruan, J.S., Zhou, H., Yang, L., Wang, L., Jiang, Z.S., Wang, S.M.: CCNA2 facilitates epithelial-to-mesenchymal transition via the integrin αvβ3 signaling in NSCLC. Int. J. Clin. Exp. Pathol. **10**, 8324–8333 (2017)
64. Serveaux-Dancer, M., Jabaudon, M., Creveaux, I., Belville, C., Blondonnet, R., Gross, C., et al.: Pathological implications of receptor for advanced glycation end-product gene polymorphism. Dis. Markers. **2019**, 2067353 (2019)
65. Zhang, W., Fan, J., Chen, Q., Lei, C., Qiao, B., Liu, Q.: SPP1 and AGER as potential prognostic biomarkers for lung adenocarcinoma. Oncol. Lett. **15**, 7028–7036 (2018)
66. Yuan, L., et al.: SFTPA1 is a potential prognostic biomarker correlated with immune cell infiltration and response to immunotherapy in lung adenocarcinoma. Cancer Immunol. Immunother. **71**(2), 399–415 (2021). https://doi.org/10.1007/s00262-021-02995-4

A Network-based Approach for Inferring Thresholds in Co-expression Networks

Nicolás López-Rozo(✉), Miguel Romero, Jorge Finke, and Camilo Rocha

Department of Electronics and Computer Science, Pontificia Universidad Javeriana, Cali, Colombia
nicolaslopez@javerianacali.edu.co

Abstract. Gene co-expression networks (GCNs) specify binary relationships between genes and are of biological interest because significant network relationships suggest that two co-expressed genes rise and fall together across different cellular conditions. GCNs are built by (i) calculating a co-expression measure between each pair of genes and (ii) selecting a significance threshold to remove spurious relationships among genes. This paper introduces a threshold criterion based on the underlying topology of the network. More specifically, the criterion considers both the rate at which isolated nodes are added to the network and the density of its components when the threshold varies. In addition to Pearson's correlation measure, the biweight midcorrelation, the distance correlation, and the maximal information coefficient are used to build different GCNs from the same data and showcase the advantages of the proposed approach. Finally, a case study presents a comparison of the predictive performance of the different networks when trying to predict gene functional annotations using hierarchical multi-label classification.

Keywords: Gene co-expression network · Hierarchical multi-label classification · Gene function prediction · Network density · Correlation metrics.

1 Introduction

A gene co-expression network (GCN) of an organism specifies binary relationships between genes that are likely controlled by the same transcriptional regulatory program, are functionally related, or are members of the same pathway or protein complex. They are of biological interest because significant co-expression relationships show a similar expression pattern across different experimental conditions, meaning that two co-expressed genes rise and fall together across different samples. GCNs are usually built in a two-step approach: first, by calculating a co-expression measure between each pair of genes, and second, by selecting a significance threshold to remove spurious relationships between genes. In practice, a correlation metric between any two genes is unlikely to be zero (even if their expression behavior is completely independent), thus resulting in a GCN

H. Cherifi et al. (Eds.): COMPLEX NETWORKS 2016 2022, SCI 1077, pp. 265–276, 2023.
https://doi.org/10.1007/978-3-031-21127-0_22

that is a complete weighted graph with some spurious edges. However, dealing with a dense network can render its computational analysis intractable. The challenge is thus to use a correlation metric that identifies significant relationships and makes sense from a biological perspective, while making the resulting GCN accessible and tractable for computational analysis.

This paper proposes a new hard-threshold criterion for discarding relationships from gene correlation networks. The main novelty of the criterion is that it considers the underlying topology of the network, more specifically, the rate at which isolated nodes are connected to the network and the density of the network components as the threshold varies. Unlike other authors that define several threshold ranges based on network density minima and the number of edges and nodes [1,19], the proposed approach selects between two specific values of the selected topological properties of the network.

The Pearson and Spearman correlation measures are commonly used for finding correlations among genes [18]. However, these correlation coefficients are designed to identify linear or monotonic relationships, respectively, and must therefore be interpreted carefully. Moreover, Spearman correlation has been found to achieve a lower accuracy than Pearson when analyzing continuous data [6]. In this paper, different correlation metrics are evaluated to showcase the proposed approach. In particular, the biweight midcorrelation (BICOR) [8], the distance correlation (DCORR), and the maximal information coefficient (MIC) are used to build and compare different GCNs from the same data.

A case study is presented, which analyses the performance of hierarchical multi-label classification (HMC) for predicting gene functions based on different generated GCNs. The analysis focuses on the biological processes of the Gene Ontology (GO) hierarchy [5]. Functional information is imported from the DAVID Bioinformatics Resources [7]. The experimental results suggest that, although performance of the functional prediction for all metrics is similar, an improvement greater than the variance can be observed for large hierarchies using BICOR to construct the network. Ultimately, these results highlight the importance of characterizing the relationships among genes in elucidating biological functionality of organisms and suggest how the proposed thresholding criterion can be extended to other biological contexts.

The remainder of the paper is organized as follows. Section 2 presents preliminary concepts. Section 3 presents the proposed threshold criterion and shows its impact on building a gene co-expression network from a dataset of rice. Section 4 presents a case study in gene function prediction based on a hierarchical multi-label classification model and compares the performance of the resulting co-expression networks. Section 5 summarizes related work and draws some conclusions.

2 Preliminaries

2.1 Correlation Metrics

Correlation metrics are used for measuring the degree of association (i.e., linear relationship) between two random variables. The most common correlation metric is perhaps the Pearson correlation coefficient (PCC). It is computed as the covariance ratio of two variables and the product of their standard deviations. Given two variables X and Y, the PCC is defined as

$$\mathrm{PCC}(X, Y) = \frac{\sum_{i=1}^{n}(x_i - \bar{x})(y_i - \bar{y})}{\sqrt{\sum_{i=1}^{n}(x_i - \bar{x})^2}\sqrt{\sum_{i=1}^{n}(y_i - \bar{y})^2}},$$

where n is the number of points for each variable, x_i refers to the i-th data point, and $\bar{x}$ is the sample mean of X. The definition is analogous for y_i and $\bar{y}$.

The biweight midcorrelation (BICOR), an alternative to PCC, is a measure of similarity between samples that relies on the median rather than the mean for sample centrality [8]. For two random variables X and Y, BICOR is defined as

$$\mathrm{BICOR}(X, Y) = \sum_{i=1}^{n} \tilde{x}_i \tilde{y}_i,$$

where $\tilde{x}_i$ and $\tilde{y}_i$ are normalized versions of the original values x_i and y_i, which in turn, are defined as

$$\tilde{x}_i = \frac{(x_i - \mathrm{med}(x))w_i^{(x)}}{\sqrt{\sum_{j=1}^{n}[(x_j - \mathrm{med}(x))w_j^{(x)}]^2}},$$

where $\mathrm{med}(x)$ corresponds to the median of x and $w_i^{(x)}$ is a weight assigned to each value x_i based on its deviation from the median. The definition of $\tilde{y}_i$ and $w_i^{(y)}$ are analogous (more details on computing these weights can be found in [15]).

Distance correlation (DCORR) is a correlation metric that provides a better performance than PCC regarding complex relationships and outlier effect. In contrast to PCC, DCORR can measure non-linear relationships and is less affected by outliers [6]. Furthermore, the DCORR between a pair of variables is equal to zero if and only if they are independent. The metric does not assume that data is normally distributed and results from the distance covariance between X and Y, DCov^2. In particular, the distance covariance is defined as

$$\mathrm{DCov}^2(X, Y) = \frac{1}{n^2}\sum_{i,j=1}^{n}[A_{ij}B_{ij}],$$

where A_{ij} and B_{ij} are defined as $A_{ij} = a_{ij} - \bar{a}_{i\cdot} - \bar{a}_{\cdot j} + \bar{a}_{\cdot\cdot}$ and $B_{ij} = b_{ij} - \bar{b}_{i\cdot} - \bar{b}_{\cdot j} + \bar{b}_{\cdot\cdot}$, given that $a_{ij} = \|x_i - x_j\|$ and $b_{ij} = \|y_i - y_j\|$ are Euclidian norms. The

DCORR between X and Y is defined as

$$\mathrm{DCORR}(X,Y) = \frac{\mathrm{DCov}^2(X,Y)}{\sqrt{\mathrm{DCov}^2(X,X)\mathrm{DCov}^2(Y,Y)}}.$$

Other type of metrics that can be used to assess correlation have their origins in information theory. They are based on the concept of mutual information of two variables to measure mutual dependence between them. The maximal information coefficient (MIC) quantifies the degree in which a discrete variable can give information about another variable. It is based on computing a grid and observing how well the grid encapsulates the relationship between the two variables [12]. MIC has been applied in the context of gene co-expression network analysis. It has been shown to outperform other models (such as WGCNA [8]) on finding non-linear relationships [11].

2.2 Gene Co-expression Networks

Gene co-expression networks are represented as undirected graphs where each node corresponds to a gene and a pair of nodes is connected with an edge if there exists a significant co-expression relationship between them.

Definition 1. Let V be a set of genes, E a set of edges that connect pairs of genes, and $w : E \to \mathbb{R}_{\geq 0}$ a weight function. A *(weighted) gene co-expression network* is a weighted graph $G = (V, E, w)$.

Gene co-expression networks are of biological interest since co-expressed genes usually belong to the same regulatory pathway or protein complex and can elucidate biological functionality for unlabeled genes [8]. High-throughput gene expression profiling technologies such as microarrays or RNA-Seq yield datasets that can be used for generating co-expression networks [13]. If the expression profiles for several genes under different experimental conditions are measured, then the network can be constructed by connecting with an edge those pairs of genes showing similarity in their expression patterns.

For the purpose of this paper, a hierarchical multi-label classification model is used to fairly evaluate the performance of the different correlation metrics and their resulting GCNs in predicting gene functional annotations.

2.3 Hierarchical Classification

Classification problems may be defined using binary, multi-class, or multi-label prediction tasks, where predictions consist of a single class, a single class from a set of mutually exclusive classes, and a subset of classes, respectively. Hierarchical multi-label classification (HMC) addresses the task of structured output prediction where the classes are organized into a hierarchy and an instance may belong to multiple classes. In many problems, such as gene function prediction, classes inherently satisfy these conditions [9]. The problem of predicting gene functions refers to the task of identifying associations between genes and functions based on biological information, such as gene co-expression networks. This problem is usually addressed using HMC.

Moreover, it is often the case in HMC datasets that individual classes have few positive instances. In gene function prediction, typically only a few genes are associated to specific functions. This implies that for most classes, the number of negative instances by far exceeds the number of positive instances. Hence, the focus is on recognizing the positive instances, i.e., on identifying the associations between genes and functions. For this reason, the area under the average precision-recall (PR) curve introduced by [16] is used for evaluation, denoted as AU($\overline{\text{PRC}}$). This metric transforms the multi-label problem into a binary one by computing the precision and recall for all functions A' together. This corresponds to micro-averaging the precision and recall. Multiple thresholds are used to create a PR curve, where each point represents the precision and recall for a give threshold that can be computed as:

$$\overline{\text{Prec}} = \frac{\sum_i TP_i}{\sum_i TP_i + \sum_i FP_i} \quad \text{and} \quad \overline{\text{Rec}} = \frac{\sum_i TP_i}{\sum_i TP_i + \sum_i FN_i}.$$

3 Co-expression Network Construction

This section introduces the method for constructing co-expression networks based on the expression profiles of genes.

According to [14], two possible approaches are feasible to filter spurious edges, namely, *hard-* and *soft-thresholding*. In the former, a threshold value that 'cuts' the network is defined to discard the edges whose weight falls below. In the latter, a function is designed to map each correlation value to an edge weight, suppressing the influence of weak edges and enhancing the influence of strong ones. Here, a hard-thresholding approach is defined based on the observed co-expression values, which considers the topology of the network.

3.1 Relationship between Threshold and Network Density

Defining the threshold to remove spurious relationships is not a trivial task [4]: setting the threshold too high results in many disconnected small network components; setting it too low introduces noise in the topology of the network. Furthermore, it is impossible to define a universal correlation coefficient threshold because underlying biochemical processes change from one organism to another, among different tissues of the same organism, or even among different experimental designs [18].

Some network properties are relevant for analyzing the network growth and should be introduced for further discussions. The average degree of a network represents the average number of connections of any node in a network. A normalized version of this metric represents a measure of network *density* (which ranges from 0 to 1). Note that a network with no edges has density of 0, while a fully connected clique has a density of 1. Since a network may be composed of several connected components, consider the measure of *average density* $\bar{\rho}$ as the average density of its connected components (an isolated node is a component with $\bar{\rho} = 1$; similarly, a network consisting of cliques also has $\bar{\rho} = 1$). The work in [1,17,19] provides an important insight on how the network behaves with respect to changes in the threshold. In the proposed method, the number of nodes with at least 1 edge, the number of edges, and the average density of the network are relevant properties.

Consider a threshold that is just a fraction higher than the highest co-expression value across the entire network. At this starting point, the network will consist only of isolated nodes and therefore $\bar{\rho} = 1$. As the threshold starts to decrease, some edges are sparsely added to the network and, up to a certain point, tree-like structures start to emerge. Trees are graphs with low density and therefore the average density of the network decreases. As the threshold continues to decrease, more connections are expected to appear between nodes already connected than between disconnected nodes. Further decrease in the threshold increases the average network density until all edges are covered and the network becomes a clique.

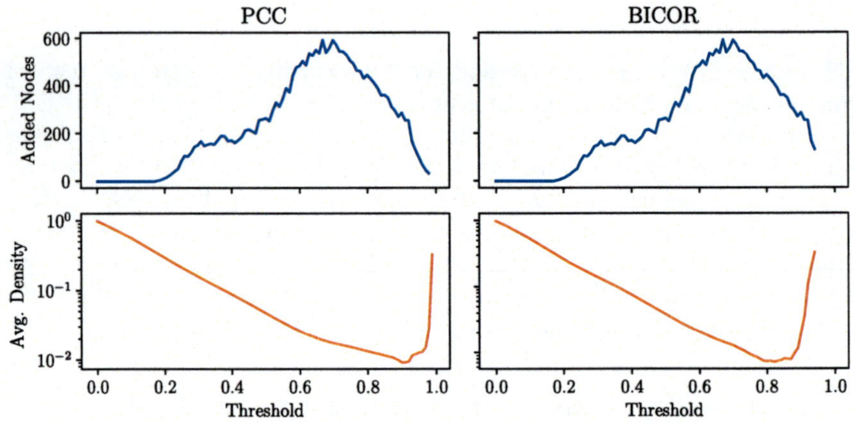

Fig. 1. Example of the behavior of the number of added nodes and average density for PCC and BICOR.

Figure 1 depicts the aforementioned behavior by setting different thresholds to build a co-expression network for rice. The co-expression values are computed using PCC and BICOR (only nodes with a least one edge are considered). In this case, the behavior does not depend on the correlation metric; nevertheless, the dominance of tree-like structures or the densification process depends completely on the underlying correlation values and the intrinsic biological processes involved.

3.2 Threshold Inference

Note that there exists a threshold for which the average network density is minimal (this is, mainly, because with a high threshold the average network density is maximal and equal to 1, and with a low one the resulting network is a clique). This observation is key because it means a transition from a behavior dominated by tree-like structures to one dominated by densification of the existing components. From a biological perspective, at this point, the most intrinsically relevant clusters should have been already established.

Additionally, an important aspect on the network growth is how fast isolated nodes become connected to other components, which contributes to the emergence of the tree-like structures. If the rate at which nodes are being added is growing, network growth is yet unstable and clusters are still forming. In contrast, if the rate of node aggregation is decreasing, then weaker connections are appearing among the existing clusters. In this regard, the turning point of node aggregation is also an important clue to select the threshold, since the most relevant clusters are already formed.

The process of inferring the threshold assumes as input an expression matrix $\mathbf{E}$ containing e expression values for n genes and a correlation metric m. The expression profile for each gene consists of e experimental levels of expression under certain conditions (i.e., $|\mathbf{E}| = n \times e$). The correlation metric m is used to compute the co-expression matrix $\mathbf{C}$ containing the pairwise correlation values between n genes (i.e., $|\mathbf{C}| = n \times n$). The proposed definition of the threshold $t_{\mathbf{C}}$ is given by

$$t_{\mathbf{C}} = \min\left(\operatorname{argmax}(|\Delta V_{I,\mathbf{C}}|), \operatorname{argmin}(\bar{\rho}_{\mathbf{C}})\right),$$

where $\Delta V_{I,\mathbf{C}}$ is the change in the number of isolated nodes as the threshold on the correlation values in $\mathbf{C}$ moves down and $\bar{\rho}_{\mathbf{C}}$ is the average network density.

3.3 Co-expression in Rice

This sections evaluates the selected metrics for computing the co-expression between genes and describes the resulting networks after applying the proposed approach. In each experiment, a discretization of the range of each metric in steps of 0.01 is applied, as in [1].

Data. Expression values for 23,374 genes of rice (*Oryza sativa Japonica*) across 2,678 accessions are retrieved from the NCBI Gene Expression Omnibus (GEO) Datasets [2]. Expression values range from 0.009 up to 280,584, with an average expression value of 1,252. Additionally, the original gene identifiers (AffyID) are converted to Entrez-IDs using the DAVID bioinformatics resources [7].

Analysis. Table 1 presents the main properties of the resulting networks after computing the correlation of rice data using each metric and applying the proposed method for finding the threshold. The column labeled *Nodes* refers to the number of nodes with at least 1 edge.

Table 1. Properties of the resulting networks after applying the proposed threshold method.

Network	Nodes	Edges	Components	Avg. degree	Density
PCC	11,241	1,195,905	473	212.78	0.0189
BICOR	7,344	382,202	586	104.09	0.0142
DCORR	13,069	2,184,554	12	334.31	0.0256
MIC	14,233	5,790,949	255	813.74	0.0572

It can be observed that the correlation metric applied to the expression profiles highly influences the resulting network topology, hereby represented with the number of non-isolated nodes, number of edges, number of connected components, average degree, and density. Also note that the amount of nodes and edges greatly changes from metric to metric. Although DCORR and MIC differ on the number of nodes by around one thousand, the number of edges and the network density of MIC increases by more than twice. In contrast, BICOR has around half of the nodes as DCORR, but less than a fifth of the edges. Finally, note that there exists a great variation in the number of connected components: DCORR is not as dense as MIC, but has less components, connected in a sparse manner.

4 Case Study: Gene Function Prediction

This section presents a comparison of the predictive performance of the generated networks on the gene function prediction problem using HMC. This case study is focused on the biological processes of the Gene Ontology (GO) hierarchy [5]. The functional information is imported from DAVID Bioinformatics Resources [7]. It comprises 3,531 biological processes and 6,367 hierarchical relations, which are part of the GO hierarchy, and a total of 289,407 associations between genes and functions.

To fairly compare the generated networks, it is necessary to use the same functions and genes for all co-expression networks. For that purpose, (i) the greatest connected component (GCC) of each network is computed, (ii) the common genes between the GCCs of all networks are selected, and (iii) a subset of GO sub-hierarchies is selected such that at least 10 genes are associated to each biological process. As a result, 10 sub-hierarchies of biological processes are used, represented by their root. Table 2 depicts the sub-hierarchies sorted by their number of biological processes.

The intersection of all GCCs is computed and extracted, so that the same set of genes is used to train the HMC model for each dataset. The HMC models are built based on random forests of decision trees, where all functions of the hierarchy are considered at once. Additionally, k-fold cross validation is used to avoid overfitting in training (the number of folds is $k = 5$). The parameter values used for random forest classifiers are: 200 estimators (*n_estimators*) and

Table 2. Sub-hierarchies of biological processes, represented by their root. Each sub-hierarchy is considered an independent dataset.

Root	Description	Functions
GO:0002376	Immune system process	9
GO:0044419	Biological process involved in interspecies interaction between organisms	13
GO:0032501	Multicellular organismal process	18
GO:0022414	Reproductive process	24
GO:0032502	Developmental process	53
GO:0051179	Localization	86
GO:0050896	Response to stimulus	102
GO:0065007	Biological regulation	184
GO:0008152	Metabolic process	399
GO:0009987	Cellular process	517

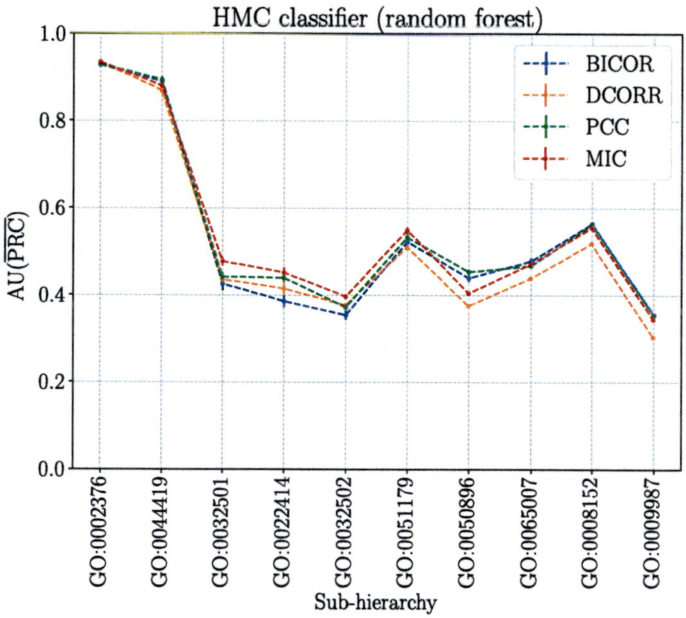

Fig. 2. Prediction performance for the gene function prediction problem on rice, measured with $AU(\overline{PRC})$, for the networks generated with all correlation metrics, namely, PCC, BICOR, DCORR and MIC.

minimum number of samples of 5 (*min_samples_split*). Note that one model is build for each network and dataset.

Figure 2 shows the prediction performance measured with $AU(\overline{PRC})$ for the networks generated with all correlation metrics, namely, PCC, BICOR, DCORR, and MIC. It illustrates the variance of the performance for 50 runs of the models. Although the performance for all metrics is similar, MIC outperforms the other metrics (considering the variance) in the smaller sub-hierarchies, with the exception of GO:0044419. In a similar way, BICOR outperforms the other metrics for the larger hierarchies, i.e, GO:0065007, GO:0008152 and GO:0009987. The remaining hierarchies have inconclusive results.

5 Related Work and Concluding Remarks

A significant effort has been put in comparing different correlation metrics on biological datasets, considering that PCC is used in the majority of co-expression network analysis [18]. BICOR itself appears as part of the WGCNA package of R for gene co-expression analysis and has proved to be robust to outliers. The work in [15] compared BICOR and some mutual information estimators like MIC for Gene Ontology (GO) enrichment analysis. The results suggest that MIC tends to overfit the data from brain cancer, blood lymphocyte, yeast, mouse adipose tissue, and mouse muscle. In fact, the Topological Overlap Matrix method, based on BICOR, dominates 7 of 8 datasets in terms of GO enrichment [10]. Similar to the results in [15], our own findings suggest that the benefits of applying BICOR instead of PCC or Spearman coefficient are marginal.

The recent work in [6] compared the efficacy of using DCORR, PCC, MIC, and Spearman as part of the WGCNA framework for gene co-expression network analysis. Four datasets of expression samples including macrophage, liver, cervical cancer, and pancreatic cancer were used to compared the performance of these metrics on GO enrichment of biological processes. The results show that DCORR is stable for highly correlated pairs of genes as the size of the datasets grows, while PCC only presents a comparable, yet decaying stability in the macrophage dataset. Furthermore, when analyzing the module preservation of GO clusters, DCORR has higher one-to-one correspondence.

Hard- and soft-thresholding procedures have been proposed in the literature [14,18]. A notorious soft-thresholding approach corresponds to WGCNA, which raises the computed PCC to a power $\beta \geq 1$ to ensure a scale-free behavior [8]. This procedure enhances high correlations at the expense of low correlations. Hard-thresholding approaches define a significance value and keep the edges with a correlation value above the threshold. The work in [3] proposed an approach that uses permutation-based significance tests: by independently permuting the components of a gene expression profile, they estimate a p-value for each observed correlation. The authors of [1,19] selected the threshold based on the behavior of the number of edges, number of nodes, and network density, resulting in a range in which the threshold can be selected. The work in [17] extended [1] by adding another consideration: once a range is defined upon the previous criteria, a clustering analysis is carried out to verify the cluster stability and choose the most appropriate end of the range to select as threshold. Unlike [17], the proposed method is based on the growing mechanisms and the topology of the underlying network. Understanding that the appearance of biologically significant modules is closely related to how tree-like structures appear and how the existing components densify, gives an insight on an emergent behavior among genes.

The results in this paper suggest that it is possible (and in some cases convenient) to use robust correlation metrics other than Pearson for obtaining gene co-expression networks. Also, a characterization of the growing mechanisms of co-expression networks in terms of network density and rate of aggregation of isolated nodes has been presented. Although the differences in the performance among some metrics are marginal, the use of BICOR for gene function prediction yields, in terms of three evaluation metrics, promising results for larger GO sub-hierarchies (consisting of more than 180 functions). Similarly, MIC can be used for enhancing gene function prediction performance when the size of the GO sub-hierarchy is small (consisting of less than 90 functions). The hard-thresholding identification was used to build co-expression networks and was used to showcase a solution to the gene function prediction problem for *Oryza sativa Japonica*, a variety of rice.

As future work, the characterization of other weighted networks needs to be developed and the elucidation of similar network growing mechanisms in other contexts, as well as the exploration of other correlation metrics (e.g., information-based) for generating co-expression networks, needs to be pursued. Furthermore, this study applies a fixed range discretization with steps of 0.01 to compute the evolution of network density and isolated node aggregation, but alternative discretization approaches need to be assessed.

Acknowledgments. This work was funded by the OMICAS program: Optimización Multiescala In-silico de Cultivos Agrícolas Sostenibles (Infraestructura y Validación en Arroz y Caña de Azúcar), anchored at the Pontificia Universidad Javeriana in Cali and funded within the Colombian Scientific Ecosystem by The World Bank, the Colombian Ministry of Science, Technology and Innovation, the Colombian Ministry of Education, and the Colombian Ministry of Industry and Tourism, and ICETEX, under GRANT ID: FP44842-217-2018.

References

1. Aoki, K., Ogata, Y., Shibata, D.: Approaches for extracting practical information from gene co-expression networks in plant biology. Plant Cell Physiol. **48**, 381–90 (2007)
2. Barrett, T., Wilhite, S.E., Ledoux, P., Evangelista, C., Kim, I.F., Tomashevsky, M., Marshall, K.A., Phillippy, K.H., Sherman, P.M., Holko, M., Yefanov, A., Lee, H., Zhang, N., Robertson, C.L., Serova, N., Davis, S., Soboleva, A.: NCBI GEO: archive for functional genomics data sets-update. Nucl. Acids Res. **41**(D1), D991–D995 (2012). https://doi.org/10.1093/nar/gks1193
3. Carter, S.L., Brechbühler, C.M., Griffin, M., Bond, A.T.: Gene co-expression network topology provides a framework for molecular characterization of cellular state. Bioinformatics **20**(14), 2242–2250 (2004). https://doi.org/10.1093/bioinformatics/bth234
4. Couto, C.M.V., Comin, C.H., Costa, L.D.F.: Effects of threshold on the topology of gene co-expression networks. Molecular Biosyst. **13**, 2024–2035 (2017)
5. Gene Ontology Consortium: The gene ontology resource: 20 years and still going strong. Nucl. Acids Res. **47**(D1), D330–D338 (2019)

6. Hou, J., Ye, X., Feng, W., Zhang, Q., Han, Y., Liu, Y., Li, Y., Wei, Y.: Distance correlation application to gene co-expression network analysis. BMC Bioinform. **23**, 81 (2022)
7. Huang, D.W., Sherman, B.T., Lempicki, R.A.: Systematic and integrative analysis of large gene lists using david bioinformatics resources. Nature Prot. **4**, 44–57 (2009)
8. Langfelder, P., Horvath, S.: WGCNA: an R package for weighted correlation network analysis. BMC Bioinform. **9**, 559 (2008)
9. Levatić, J., Kocev, D., Džeroski, S.: The importance of the label hierarchy in hierarchical multi-label classification. J. Intell. Inform. Syst. **45**(2), 247–271 (2015). http://link.springer.com/10.1007/s10844-014-0347-y
10. Margolin, A.A., Nemenman, I., Basso, K., Wiggins, C., Stolovitzky, G., Dalla Favera, R., Califano, A.: ARACNE: an algorithm for the reconstruction of gene regulatory networks in a mammalian cellular context. BMC Bioinform. **7**(Suppl 1), S7 (2006)
11. Rau, C., Wisniewski, N., Orozco, L., Bennett, B., Weiss, J., Lusis, A.: Maximal information component analysis: a novel non-linear network analysis method. Front. Genet. **4** (2013). https://www.frontiersin.org/article/10.3389/fgene.2013.00028
12. Reshef, D.N., Reshef, Y.A., Finucane, H.K., Grossman, S.R., McVean, G., Turnbaugh, P.J., Lander, E.S., Mitzenmacher, M., Sabeti, P.C.: Detecting novel associations in large data sets. Science (New York, N.Y.) **334**, 1518–1524 (2011)
13. Schaefer, R.J., Michno, J.M., Jeffers, J., Hoekenga, O., Dilkes, B., Baxter, I., Myers, C.L.: Integrating co-expression networks with gwas to prioritize causal genes in maize. The Plant Cell **30**, 2922–2942 (2018)
14. Silverman, E.K., Schmidt, H.H.H.W., Anastasiadou, E., Altucci, L., Angelini, M., Badimon, L., Balligand, J.L., Benincasa, G., Capasso, G., Conte, F., Di Costanzo, A., Farina, L., Fiscon, G., Gatto, L., Gentili, M., Loscalzo, J., Marchese, C., Napoli, C., Paci, P., Petti, M., Quackenbush, J., Tieri, P., Viggiano, D., Vilahur, G., Glass, K., Baumbach, J.: Molecular networks in network medicine: development and applications. Wiley Interdisciplinary Reviews. Syst. Biol. Med. **12**, e1489 (2020)
15. Song, L., Langfelder, P., Horvath, S.: Comparison of co-expression measures: mutual information, correlation, and model based indices. BMC Bioinform. **13**, 328 (2012)
16. Vens, C., Struyf, J., Schietgat, L., Džeroski, S., Blockeel, H.: Decision trees for hierarchical multi-label classification. Mach. Learn. **73**(2), 185–214 (2008)
17. van Verk, M.C., Bol, J.F., Linthorst, H.J.M.: Prospecting for genes involved in transcriptional regulation of plant defenses, a bioinformatics approach. BMC Plant Biol. **11**, 88 (2011)
18. Wang, Y.X.R., Li, L., Li, J.J., Huang, H.: Network modeling in biology: statistical methods for gene and brain networks. Stati. Sci. Rev. J. Inst. Math. Stat. **36**, 89–108 (2021)
19. Zhang, L., Yu, S., Zuo, K., Luo, L., Tang, K.: Identification of gene modules associated with drought response in rice by network-based analysis. PLOS ONE **7**, e33748 (2012)

Building Differential Co-expression Networks with Variable Selection and Regularization

Camila Riccio(✉), Jorge Finke, and Camilo Rocha

Pontificia Universidad Javeriana, Cali, Colombia
criccio35@gmail.com

Abstract. This work introduces a technique for the inference of differential co-expression networks. The approach takes as input a matrix of differential expression profiles, where each entry corresponds to the Log Fold Change of a gene expression between control and stress conditions for a specific sample. It outputs a matrix of coefficients, where each non-zero entry represents a pairwise connection between genes. The proposed approach builds on Lasso, and is applied to differential expression profiles of rice between control and salt-stress conditions. A total of 25 genes were identified to respond to salt stress and as differentially expressed. About half of these genes (11) were reported with a statistically significant number of different GO annotations relevant to salt stress response.

Keywords: Network inference · Lasso-based inference · Overlapping clustering · Salt-stress · Rice · *Oryza sativa*

1 Introduction

Gene regulatory networks specify how biological systems respond to perturbations through the rewiring of molecular interactions. Co-expression networks provide a framework to better understand molecular mechanisms and gene regulation. Thanks to the increasingly high availability of transcriptomic data, robust gene co-expression networks are becoming more widely available. A differential co-expression network represents a particular type of network, which is used to identify changes in response to external stimuli (e.g., changes in activity of gene expression regulators or signaling [3,7,28]). Differential co-expression network analysis is an approach for identifying modules of genes with meaningful variation between different experimental conditions (e.g., control and stress). Such an analysis uses as input gene expression data, representing gene expression between control and stress conditions, and output a set of genes that are likely to be involved in the biological response to the specific stress.

Technically, differential co-expression analysis builds a network from the relationships between genes differential expression profiles, that is, the log fold

H. Cherifi et al. (Eds.): COMPLEX NETWORKS 2016 2022, SCI 1077, pp. 277–288, 2023.
https://doi.org/10.1007/978-3-031-21127-0_23

change (LFC) between control and stress expression across multiple observations for each gene. LFC represents a logarithm of the ratio between the control expression and the expression under stress and involves two key steps. First, setting a correlation measure between the genes. Second, filtering the list of pairs using a threshold value for the correlation score [29]. The Pearson correlation coefficient is the most popular measure used for step 1. It assumes linear correlation, normally distributed values, and is sensitive to outliers [17]. A major limitation of using this approach, for building a differential co-expression network, is that it demands large enough sample sizes for statistically reliable results [6,20,22]. However, large samples sizes are often prohibitive due to time and computational constraints. For instance, a differential co-expression network with N genes across M control samples and M stressed samples is built by first building a matrix $X \in \mathbb{R}^{N \times M}$, where each entry $X(n, m)$ corresponds to the LFC value of gene n for sample m. Then, the Pearson's coefficient is computed for each pair of distinct genes n_1 and n_2 with the input vectors $X(n_1, \_)$ and $X(n_2, \_)$, each of length M. That is, assuming as input X, building the differential co-expression network takes $O\left(\binom{N}{2}M\right) = O(N^2M)$ time, where $O(M)$ is the usual time for computing the Pearson's coefficient on M samples.

This paper introduces a method based on the penalized least absolute shrinkage and selection operator (Lasso) [33] for building differential co-expression networks. It overcomes the above-mentioned computational limitation by computing a regression of the differential expression profile of one gene against all others, instead of using a correlation metric to identify significant edges between genes. The resulting coefficients represent the strength of the relationship between the corresponding pair of genes, where zero strength indicates no edge between them. Additionally, Lasso has the advantage of yielding accurate parameter estimates even with small sample sizes [13]. For N genes and M samples (both under control and stress), building the differential co-expression network with Lasso takes $O(NM^2)$ time. However, for the case where there are many more variables than observations (i.e., $N \gg M$), as the usual biological expression datasets, it takes $O(NM)$ time [11]. While the proposed approach is used in this work for building *differential* co-expression networks, it can also more generally be used for building co-expression networks.

Lasso simultaneously performs variable selection and regularization by forcing the least significant coefficients to be zero, which naturally favors the inference of a sparse network. It performs ℓ_1 regularization by forcing the sum of the absolute value of the regression coefficients to be less than a fixed value. Lasso iteratively searches for a degree of penalty λ that minimizes the mean square error of the regression. At the optimal value of λ, it performs variable selection, which results in a reduced number of non-zero coefficients. The variables with a zero value coefficient are excluded from impacting the regression, which prevents the model from over-fitting. Lasso properties are particularly useful in the construction of co-expression networks since these type of networks are expected to be sparse [36]. Moreover, it avoids the need to define a threshold for selecting meaningful edges from the pairwise relationships between genes.

Finally, since Lasso supports small sample sizes, it is well suited for most typical transcriptomic data sets.

The proposed approach is used in combination with a slightly modified version of the workflow proposed in [26] (but with a different module detection algorithm to identify genes that respond to salt stress, namely, the ANGEL algorithm [27]). RNA-seq data was accessed from the GEO database [8], accession number GSE98455. It represents 57,845 gene expression profiles of shoot tissues measured under control and stress conditions in 92 accessions of the Rice Diversity Panel 1 [12]. A total of 25 genes are identified to respond to salt stress and as differentially expressed genes (DEG). About half of these genes (11) are reported with a statistically significant number of different GO annotations relevant to salt stress response.

2 Methodology

This section presents a description of differential co-expression networks, introduces the Lasso-based approach to build differential co-expression networks, summarize the Angel algorithm to detect overlapping modules in the network, and explains an enrichment technique to evaluate the biological significance of the detected modules.

2.1 Differential Co-expression Network

A *network* is an undirected graph $G = (V, E)$ where $V = \{v_1, v_2, \ldots, v_N\}$ is a set of N *vertices* (or *nodes*) and $E = \{e_1, e_2, \ldots, e_Q\}$ is a set of Q *edges* (or *links*) between vertices. $G = (V, E)$ can be represented by an adjacency matrix $A \in \{0, 1\}^{N \times N}$ that is symmetric. A matrix entry in positions (v_i, v_j) and (v_j, v_i) is equal to 1 whenever there is an edge connecting vertices v_i and v_j, and equal to 0 otherwise. *Differential co-expression* is the altered co-expression patterns of genes between two particular conditions (e.g., control and stress). In a differential co-expression network, each vertex corresponds to a gene. A link indicates a common alteration in the expression pattern between two genes when changing from one condition to the other. Differential co-expression networks are of biological interest because adjacent nodes in the network represent genes that jointly respond to similar stress conditions.

2.2 Co-expression Network Construction with Lasso

Lasso regressions [33] can be seen an advantageous approach for the Constructing co-expression networks using Lasso [33] offers key advantages compared to other methods that rely on the Pearson correlation coefficient (or any other correlation metric). Several assumptions behind computing Pearson limit effectiveness [17] and statistical significance especially if sample sizes are small [6]. Such condition is common across numerous efforts to build co-expression networks [20,22]. The majority of typical transcriptome datasets tend to be small in terms of the number of samples. Differential co-expression networks are built using differential expression profiles instead of the expression profiles themselves as in co-expression networks. This section explains build differential co-expression networks using Lasso can lead to robust networks even in the presence of small sample sizes.

Furthermore, note that building a network $G = (V, E)$, that is, a representation of pairwise relationships E over a set of vertices V, is equivalent to inferring a neighborhood for each vertex (i.e., the set of vertices to which it is connected). Given M observations on N genes (vertices) represented in a data matrix $X \in \mathbb{R}^{N \times M}$, the set of neighbors of vertex $v_i \in V$, denoted,

$$V(v_i) := \{v_j : (v_i, v_j) \in E\}$$

is inferred by regressing x_i against all other variables

$$x_{\setminus i} := [x_1, \ldots, x_{i-1}, x_{i+1}, \ldots, x_N]^T \in \mathbb{R}^{N-1}.$$

The result is a matrix $B \in \mathbb{R}^{N \times N}$ whose diagonal is zero and the remaining $N-1$ entries of a row i correspond to the coefficients of the regression of x_i against $x_{\setminus i}$. Each entry $B(i, j)$ represents the strength of the relationship between vertices v_i and v_j, where zero strength indicates no connection.

For each variable x_i the regression problem has the form:

$$\underset{\beta_i}{\text{minimize}} \left\|X_i - X_{\setminus i}\beta_i\right\|_2^2 + \lambda \left\|\beta_i\right\|_1, \tag{1}$$

where X_i and $X_{\setminus i}$ represent the observations on x_i (i.e., the transpose of the first row of X) and the rest of the variables, respectively. The vector $\beta_i \in \mathbb{R}^{N-1}$ is a vector of coefficients for x_i. In Eq. 1, the first term can be interpreted as a local log-likelihood of β_i and the ℓ_1 penalty is added to enforce sparsity. The regularization parameter λ balancing the two terms. Lasso is repeated for all the variables leading to a set of $N \times N$ coefficients that are computed from $\beta_1, \ldots, \beta_N$. Note that there is no guarantee that $B(i, j) \neq 0$ implies $B(j, i) \neq 0$. Therefore, the information in $V(v_i)$ and $V(v_j)$ is combined to enforce symmetry: an edge (v_i, v_j) is meaningful, if $B(i, j)$ and $B(j, i)$ are both non-zero.

Note also that including the ℓ_1 penalty allows Lasso to identify the variables that are strongly associated with the response variable (i.e., variable selection). Since the value of the regularization parameter λ determines the degree of penalty and the accuracy of the model, cross-validation is used to select a regularization parameter that minimizes the mean-squared error. If the degree of penalty λ is equal to zero, the solution is the same as least-squares (LS) linear regression [5]. For larger values of λ, larger number of coefficients are shrunk towards zero. Compared to LS, Lasso offers the following advantage. Unlike LS, Lasso does not yield non-zero estimates, which would results in a fully connected network, and giving rise to the problem of setting a threshold above which and edge is considered significant. Lasso avoids this additional step as it simultaneously performs parameter estimation and variable selection by forcing the least significant coefficients to zero through the ℓ_1 penalty.

2.3 Overlapping Clustering with ANGEL

ANGEL [27] is a static node-centric algorithm for detecting potentially overlapping modules in networks. It takes as input a graph G, a merging threshold ϕ, and an empty set of communities C. The algorithm's main loop cycles over each node, extracts the corresponding ego-minus-ego network, and computes the local communities it contains using Label Propagation (LP) [23]. During LP, every node is initialized with a unique label. In following steps, each node adopts the current label of the majority of its neighbors. In case of bow-tie situations, the classic LP formulation randomly selects a single label for the contended node. Here however soft community memberships are allowed, that is, each node can belong to multiple communities for the case of a bow-tie configuration. Once the outer loop on the network nodes is completed, the algorithm compacts the community set to avoid the presence of fully contained communities.

Finally, note that compared to HLC [1], the computational cost of ANGEL is significant less and can be approximated by $O(|V|)$.

2.4 Functional Enrichment

Our analysis of differential co-expression networks relies on the detection of gene modules. Such modules are used to investigate relationships occurring between genes performing similar biological functions [34]. Functional enrichment of each module is a critical step to understand the underlying processes contributing to phenotype or stress responses. This section describes how to evaluates the quality of the modules using Gene Ontology (GO) [2,9] enrichment.

Given a gene module, an enrichment analysis finds which GO terms are over-represented or under-represented by using annotations for that gene set. Gene module enrichment analysis is performed using the Fisher's Exact Test [32] in combination with a robust False Discovery Rate (FDR) [19] correction for multiple testing. Fisher's exact test is a statistical significance test used in the analysis of contingency tables. The FDR control is a statistical method used in multiple hypothesis testing to correct for multiple comparisons. In a list of statistically significant findings, FDR is used to control the expected proportion of incorrectly rejected null hypotheses ("false discoveries"). Here, a Benjamini-Hochberg correction is used [4]. The result for each module is a list of statistically significant GO terms ranked by their adjusted p-values.

For each GO term in each module, a contingency table is built (see Table 1). The hypothesis statement is the following for each module:

H_0: The module is a random sample from network.
H_1: The module has more genes with the GO term than expected by chance.

Following the configuration in Table 1, the p-value is calculated:

$$p = \frac{\binom{a+b}{a}\binom{c+d}{c}}{\binom{N}{a+c}} = \frac{\binom{a+b}{b}\binom{c+d}{d}}{\binom{N}{b+d}} \tag{2}$$

Table 1. Contingency table configuration for each module.

	Genes in module	Genes out the module	Total
Annotated genes	a	b	$a+b$
Not annotated genes	c	d	$c+d$
Total	$a+c$	$b+d$	N

The p-value represents the probability (or chance) of seeing at least a genes out of the total $a+c$ genes in the module annotated with a particular GO term, given the proportion $(a+b)/N$ of genes in the whole genome that are annotated with that GO term. That is, the GO terms shared by the genes in each module are compared to the background distribution of the annotations. The closer the p-value is to zero, the more significant is the association of the particular GO term with the module of genes (i.e., the less likely that the observed annotation of the GO term to the module occurs by chance). In other words, if all of the genes in a module were associated with, say "DNA repair", this term would be optimally significant. However, since all genes in the genome (with GO annotations) are indirectly associated with the top level term "biological process", this would not be significant if all the genes in a module were associated with this high-level term.

If a module has at least one GO term with a significant p-value, the module is said to be enriched. This binary classification of a module between enriched and non-enriched allows us to evaluate, in a general way, the biological significance of the modules. The higher the proportion of enriched modules in a differential co-expression network, the better they capture the biological interactions of genes that jointly respond to a specific stress condition.

3 Case Study

This section presents a case study in the identification of genes that respond to saline stress in rice. The differential co-expression network is built using the approach presented in Section 2.2 and the framework in [26], with the ANGEL module detecting algorithm. The goal of this case study is to evaluate whether the proposed approach, in addition to being less computationally costly, finds a number of differentially expressed genes (DEG) and genes with GO annotations relevant to salt stress that is statistically significant.

3.1 Association Network Construction with Lasso

Consider the input data $X \in \mathbb{R}^{N \times M}$, which represents the matrix of differential expression profiles (corresponds to L_1 in [26]). The matrix X results from pre-processing RNA-seq data of rice both under control and salt stress conditions (GEO database [8] accession number GSE98455). Therefore, X contains the LFC of $N = 8,929$ genes under control and salt stress conditions in $M = 92$ samples.

The differential co-expression network, inferred using Lasso regressions, is composed of $|V| = 7,474$ vertices and $|E| = 67,061$ edges. All the genes in this network are part of the network previously constructed in [26] and preserves 21,123 out of 39,850,128 of the original connections. IN other words, the resulting network is a subnetwork of the one constructed in [26].

3.2 Identification of Co-expression Modules

In [26], the approach for module detection requires finding a threshold for Pearson correlations to define the adjacency matrix. Here the proposed Lasso-based approach bypasses this step since it is able to directly infer network connections without additional parameters.

The ANGEL algorithm distributes a total of 5,577 genes across 1,462 modules with at least 3 genes each. Using the module enrichment technique described in Section 2, a total of 65% of all modules are identified as enriched, meaning that they have some over-represented GO annotations. In other words, the modules identified by ANGEL are biologically relevant. Figure 1 compares the threshold Pearson network (thP) in [26] with Lasso-based network (nbL), both in terms of the proportion of enriched modules and gene overlaps. Regarding module enrichment, note that the proposed approach surpasses the approach of [26].

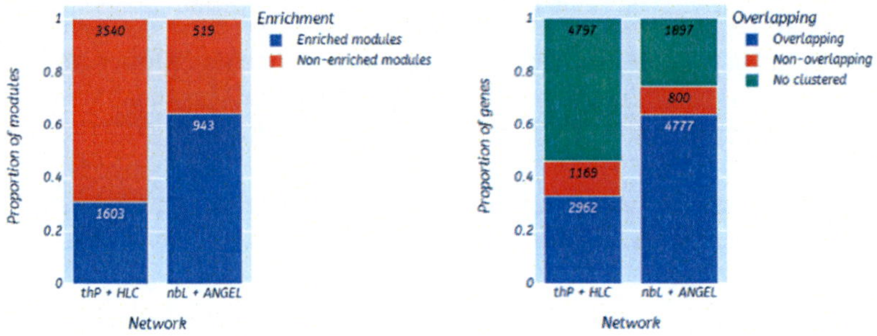

Fig. 1. Modules enrichment proportion and overlapping proportion of genes for thP and nbL networks.

Regarding the overlapping modules of genes for the nbL network, notice that the amount of transcription factors (TF) in the gene set belonging to multiple modules is statistically significant (p-value less than 0.05 for the Fisher's Exact Test). This supports the biological relevance of the overlapping modules. TFs regulate the expression of multiple genes and hence affect multiple pathways of varying functions [25]. Since TFs control different functions, they are expected to be found in overlapping modules. Another interesting finding is that, according to an enrichment analysis in ShynyGO [14], one of the pathways with the highest over-representation in the set of overlapping genes that corresponds to "response to stress" (GO:0006950).

3.3 Gene Selection

Based on the modules detected with ANGEL in the nbL network (following the workflow proposed in [26]), a total of 25 genes are identified as responsive to salt stress. All 25 genes are also identified as DEG. Genes LOC_Os07g39390, LOC_Os04g35010, and LOC_Os01g33450 are selected by both approaches, in the thP and nbL network.

From the 25 identified genes, after individual gene enrichment with the RGAP [18] and UniProt [10] databases, 11 genes report 17 different GO annotations relevant to salt stress response (which is statistically significant based on Fisher's exact test, p-value$<$ 0.05). Table 2 lists these genes and the corresponding GO annotations relevant to salt stress response. The remaining 14 selected genes are:

- LOC_Os01g25920,
- LOC_Os01g33450,
- LOC_Os01g35789,
- LOC_Os01g35930,
- LOC_Os02g05790,
- LOC_Os05g36994,
- LOC_Os06g16050,
- LOC_Os06g22394,
- LOC_Os06g38210,
- LOC_Os07g22494,
- LOC_Os07g39390,
- LOC_Os09g06634,
- LOC_Os10g04050,
- LOC_Os10g24094.

The apoplast (GO:0048046) is the first subcellular compartment confronted with stress conditions when plants are subjected to salt stress [31]. Stress is first sensed by the receptors in membranes (GO:0016020), which then generates secondary signal messengers like calcium, reactive oxygen species, kinases (GO:0004672, GO:0016301, GO:0016740), and phosphates followed by the activation of transcription factor genes (GO:0003700) that eventually coordinates the plant's adaptive biochemical and physiological responses [16] (GO:0006950, GO:0009628, GO:0006952). Protein kinases regulate the phosphorylation and dephosphorylation of other proteins, and play a crucial role in stress signal transduction. In addition, serine/threonine protein kinases (GO:0004674) have also been known to be involved in multi-stress tolerance in plants [16].

Salt-induced toxicity negatively affects CO_2 fixation and thylakoid reactions of photosynthesis, which take place in thylakoids (GO:0009579) and the stroma of the chloroplast, resulting in poor plant growth and reduction in yield [15]. An essential process for growth, development, and homeostasis of organisms is the dynamic balance between ubiquitination and deubiquitination (GO:0071108, GO:1990380, GO:0004843, GO:1990380) [30]. In particular, inhibition of shoot and root development (GO:2000280) is the primary response to salt stress [35]. Other, independent studies confirm that the enhanced catalytic and transferase activities (GO:0016740) in salt-stressed rice plants, as well as the transport (GO:0006810) of salt and all related ions through the plant, reinforce salt stress tolerance [24].

Table 2. Selected genes with associated GO terms relevant to salt stress response.

LOC ID	GO term	GO name
LOC_Os03g44880	GO:2000280	Regulation of root development
LOC_Os03g63870	GO:0006950	Response to stress
	GO:0009628	Response to abiotic stimulus
LOC_Os04g35010	GO:0003700	DNA-binding transcription factor activity
LOC_Os06g09688	GO:0016020	Membrane
	GO:0009579	Thylakoid
LOC_Os07g15440	GO:0071108	Protein K48-linked deubiquitination
	GO:1990380	Lys48-specific deubiquitinase activity
	GO:0004843	Cysteine-type deubiquitinase activity
	GO:1990380	Lys48-specific deubiquitinase activity
LOC_Os07g37385	GO:0006810	Transport
LOC_Os07g43570	GO:0004674	Protein serine/threonine kinase activity
	GO:0004672	Protein kinase activity
	GO:0016301	Kinase activity
	GO:0016740	Transferase activity
	GO:0016020	Membrane
LOC_Os10g40520	GO:0006810	Transport
LOC_Os12g01290	GO:0006952	Defense response
LOC_Os12g14440	GO:0048046	Apoplast
LOC_Os12g27220	GO:0016740	Transferase activity

4 Concluding Remarks

This work proposes a novel approach for constructing co-expression networks based on the penalized least absolute shrinkage and selection operator (Lasso) [33]. Edges between genes are stablished based on the Lasso regression coefficients between the differential expression profile of one gene against all others. The approach extends the workflow described in [26] for identifying genes related to salt stress in rice. In particular, it uses the ANGEL algorithm, a static node-centric algorithm for detecting modules with overlaps, which improves effectiveness and time complexity. Note that the proposed approach can be used for building differential and non-differential co-expression networks.

The Lasso-based approach is computationally appealing as each of the N Lasso problems can be solved independently. This makes the proposed approach a good candidate to exploit parallelization [21]. Moreover, this method avoids the additional problem that often arises in constructing co-expression networks based on correlations: the definition of a threshold to identify the strongest connections that define the edges of the network. Using Lasso is especially convenient in inferring differential co-expression networks because it yields accurate parameter estimates even with small sample sizes [13], a common condition studying expression data under control and stress conditions.

The modified workflow was applied to a case study of rice under salt stress. The resulting network, inferred with the Lasso approach contained 7,474 vertices and 67,061 edges. 65% of the identified modules were enriched. The amount of transcription factors in the set of genes belonging to multiple modules (overlapping genes) was statistically significant. Finally, a total of 25 genes were selected as genes that respond to salt stress in rice. All 25 genes were also identified as DEG. Of these identified genes, 11 reported a statistically significant number of different GO annotations relevant to the salt stress response.

As future work, the proposed Lasso-based approach can be applied to infer co-expression networks of other organisms and other types of stresses. Moreover, the inferred networks can be used in comparisons and downstream analysis of co-expression networks. Developing a parallel implementation of the current workflow is an important research direction to further reduce time complexity. Further exploration of co-expression networks should provide valuable insights into the gene interactions and their joint response to stresses.

Acknowledgments. This work was funded by the OMICAS program: Optimización Multiescala In-silico de Cultivos Agrícolas Sostenibles (Infraestructura y Validación en Arroz y Caña de Azúcar), anchored at the Pontificia Universidad Javeriana in Cali and funded within the Colombian Scientific Ecosystem by The World Bank, the Colombian Ministry of Science, Technology and Innovation, the Colombian Ministry of Education, and the Colombian Ministry of Industry and Tourism, and ICETEX, under GRANT ID: FP44842-217-2018.

References

1. Ahn, Y.Y., Bagrow, J.P., Lehmann, S.: Link communities reveal multiscale complexity in networks. Nature **466**(7307), 761–764 (2010)
2. Ashburner, M., Ball, C.A., Blake, J.A., Botstein, D., Butler, H., Cherry, J.M., Davis, A.P., Dolinski, K., Dwight, S.S., Eppig, J.T., et al.: Gene ontology: tool for the unification of biology. Nature Genet. **25**(1), 25–29 (2000)
3. Aydin, B., Arga, K.Y.: Co-expression network analysis elucidated a core module in association with prognosis of non-functioning non-invasive human pituitary adenoma. Front. Endocrinol. **10**, 361 (2019)
4. Benjamini, Y., Hochberg, Y.: Controlling the false discovery rate: a practical and powerful approach to multiple testing. J. Roy. Stat. Soc. Ser. B (Methodol.) **57**(1), 289–300 (1995)
5. Björck, Å.: Least squares methods. Handbook Num. Anal. **1**, 465–652 (1990)
6. Bujang, M.A., Baharum, N.: Sample size guideline for correlation analysis. World **3**(1), 37–46 (2016)
7. Chowdhury, H.A., Bhattacharyya, D.K., Kalita, J.K.: (Differential) Co-expression analysis of gene expression: a survey of best practices. IEEE/ACM Trans. Comput. Biol. Bioinform. **17**(4), 1154–1173 (2019)
8. Clough, E., Barrett, T.: The gene expression omnibus database. In: Statistical Genomics, pp. 93–110. Springer (2016)

9. The Gene Ontology Consortium: The gene ontology resource: enriching a gold mine. Nucl. Acids Res. **49**(D1), D325–D334 (2021)
10. Consortium, U.: Uniprot: a worldwide hub of protein knowledge. Nucleic Acids Res. **47**(D1), D506–D515 (2019)
11. Efron, B., Hastie, T., Johnstone, I., Tibshirani, R.: Least angle regression. Ann. Stat. **32**(2), 407–499 (2004)
12. Eizenga, G.C., Ali, M.L., Bryant, R.J., Yeater, K.M., McClung, A.M., McCouch, S.R.: Registration of the rice diversity panel 1 for genomewide association studies. J. Plant Regist. **8**(1), 109–116 (2014)
13. Finch, W.H., Finch, M.E.H.: Regularization methods for fitting linear models with small sample sizes: Fitting the Lasso estimator using R. Pract. Assess. Res. Eval. **21**(1), 7 (2016)
14. Ge, S.X., Jung, D., Yao, R.: Shinygo: a graphical gene-set enrichment tool for animals and plants. Bioinformatics **36**(8), 2628–2629 (2020)
15. Hameed, A., Ahmed, M.Z., Hussain, T., Aziz, I., Ahmad, N., Gul, B., Nielsen, B.L.: Effects of salinity stress on chloroplast structure and function. Cells **10**(8), 2023 (2021)
16. Hossain, M.R., Bassel, G.W., Pritchard, J., Sharma, G.P., Ford-Lloyd, B.V.: Trait specific expression profiling of salt stress responsive genes in diverse rice genotypes as determined by modified significance analysis of microarrays. Front. Plant Sci. **7**, 567 (2016)
17. Hou, J., Ye, X., Feng, W., Zhang, Q., Han, Y., Liu, Y., Li, Y., Wei, Y.: Distance correlation application to gene co-expression network analysis. BMC Bioinform. **23**(1), 1–24 (2022)
18. Kawahara, Y., de la Bastide, M., Hamilton, J.P., Kanamori, H., McCombie, W.R., Ouyang, S., Schwartz, D.C., Tanaka, T., Wu, J., Zhou, S., et al.: Improvement of the oryza sativa nipponbare reference genome using next generation sequence and optical map data. Rice **6**(1), 1–10 (2013)
19. Korthauer, K., Kimes, P.K., Duvallet, C., Reyes, A., Subramanian, A., Teng, M., Shukla, C., Alm, E.J., Hicks, S.C.: A practical guide to methods controlling false discoveries in computational biology. Genome Biol. **20**(1), 1–21 (2019)
20. Liesecke, F., De Craene, J.O., Besseau, S., Courdavault, V., Clastre, M., Vergès, V., Papon, N., Giglioli-Guivarc'h, N., Glévarec, G., Pichon, O., et al.: Improved gene co-expression network quality through expression dataset down-sampling and network aggregation. Sci. Rep. **9**(1), 1–16 (2019)
21. Osborne, M.R., Presnell, B., Turlach, B.A.: A new approach to variable selection in least squares problems. IMA J. Num. Anal. **20**(3), 389–403 (2000)
22. Ovens, K., Eames, B.F., McQuillan, I.: The impact of sample size and tissue type on the reproducibility of gene co-expression networks. In: Proceedings of the 11th ACM International Conference on Bioinformatics, Computational Biology and Health Informatics. pp. 1–10 (2020)
23. Raghavan, U.N., Albert, R., Kumara, S.: Near linear time algorithm to detect community structures in large-scale networks. Phys. Rev. E **76**(3), 036106 (2007)
24. Rahman, A., Nahar, K., Al Mahmud, J., Hasanuzzaman, M., Hossain, M.S., Fujita, M.: Salt stress tolerance in rice: emerging role of exogenous phytoprotectants. Adv. Int. Rice Res. **9**(3), 139–174 (2017)
25. Renkawitz, R.: Transcription Factors and Regulation of Gene Expression, pp. 1886–1890. Springer Berlin Heidelberg, Berlin, Heidelberg (2006)
26. Riccio-Rengifo, C., Finke, J., Rocha, C.: Identifying stress responsive genes using overlapping communities in co-expression networks. BMC Bioinform. **22**(1), 1–17 (2021)

27. Rossetti, G.: Angel: efficient, and effective, node-centric community discovery in static and dynamic networks. Appl. Netw. Sci. **5**(1), 1–23 (2020)
28. Savino, A., Provero, P., Poli, V.: Differential co-expression analyses allow the identification of critical signalling pathways altered during tumour transformation and progression. Int. J. Molecul. Sci. **21**(24), 9461 (2020)
29. Serin, E.A., Nijveen, H., Hilhorst, H.W., Ligterink, W.: Learning from co-expression networks: possibilities and challenges. Front. Plant Sci. **7**, 444 (2016)
30. Snyder, N.A., Silva, G.M.: Deubiquitinating enzymes (dubs): Regulation, homeostasis, and oxidative stress response. J. Biol. Chem. **297**(3) (2021)
31. Song, Y., Zhang, C., Ge, W., Zhang, Y., Burlingame, A.L., Guo, Y.: Identification of NACL stress-responsive apoplastic proteins in rice shoot stems by 2d-dige. J. Proteom. **74**(7), 1045–1067 (2011)
32. Sprent, P.: Fisher Exact Test (2011)
33. Tibshirani, R.: Regression shrinkage and selection via the Lasso. J. Roy. Stat. Soc. Ser. B (Methodol.) **58**(1), 267–288 (1996)
34. Van Dam, S., Vosa, U., van der Graaf, A., Franke, L., de Magalhaes, J.P.: Gene co-expression analysis for functional classification and gene-disease predictions. Brief. Bioinform. **19**(4), 575–592 (2018)
35. Wang, Y., Zhang, W., Li, K., Sun, F., Han, C., Wang, Y., Li, X.: Salt-induced plasticity of root hair development is caused by ion disequilibrium in arabidopsis thaliana. J. Plant Res. **121**(1), 87–96 (2008)
36. Yeung, M.S., Tegnér, J., Collins, J.J.: Reverse engineering gene networks using singular value decomposition and robust regression. Proc. Nat. Acad. Sci. **99**(9), 6163–6168 (2002)

Inferring Probabilistic Boolean Networks from Steady-State Gene Data Samples

Vytenis Šliogeris[1], Leandros Maglaras[2], and Sotiris Moschoyiannis[1(✉)]

[1] School of Computer Science and Engineering, University of Surrey, Guildford, UK
v.sliogeris@surrey.ac.uk, s.moschoyiannis@surrey.ac.uk

[2] School of Computer Science and Informatics, De Montfort University, Leicester, UK
leandros.maglaras@dmu.ac.uk

Abstract. Probabilistic Boolean Networks have been proposed for estimating the behaviour of dynamical systems as they combine rule-based modelling with uncertainty principles. Inferring PBNs directly from gene data is challenging however, especially when data is costly to collect and/or noisy, e.g., in the case of gene expression profile data. In this paper, we present a reproducible method for inferring PBNs directly from *real* gene expression data measurements taken when the system was at a steady state. The steady-state dynamics of PBNs is of special interest in the analysis of biological machinery. The proposed approach does not rely on reconstructing the state evolution of the network, which is computationally intractable for larger networks. We demonstrate the method on samples of real gene expression profiling data from a well-known study on metastatic melanoma. The pipeline is implemented using Python and we make it publicly available.

Keywords: Steady-state data samples · Network structure · Dynamics · Discretisation · Predictor sets · Perceptron · Complex networks

1 Introduction

Rapid progress in the development of next-generation sequencing technologies for genomics has provided valuable insights into complex biological systems [12]. Modelling single-cell or gene networks is becoming increasingly important. The question of modelling complex molecular regulatory networks is an important one for bioinformatics. The goal of systems biology is to intervene on the state of the cell, using the dynamics of the underlying regulatory network. A model that could accurately represent such dynamics could be used for analysis, including control [14,19,26,27,36], steady-state distribution [8,18,24,31], observability [28, 37,38]. Such analyses aid the development of genetic therapies [11].

Vytenis Sliogeris and Sotiris Moschoyiannis were funded by UKRI grant 77032. Thanks are due to Evangelos Chatzaroulas for his help with optimising the codebase.

H. Cherifi et al. (Eds.): COMPLEX NETWORKS 2016 2022, SCI 1077, pp. 289–300, 2023.
https://doi.org/10.1007/978-3-031-21127-0_24

Boolean Networks (BNs) were introduced for this purpose by Kauffman [15]. In brief, a BN comprises a set of Boolean variables, each variable representing the on/off state of a gene, while interactions between genes are expressed by Boolean functions. It was found that even randomly generated BNs exhibit behaviour reminiscent of gene regulatory networks, with naturally arising attractor states which represent cell types or the phenotype [6,35]. This explains the popularity of BNs for modelling gene interactions [2,10].

However, with few exceptions, gene expression data suggests a number of possible successor states to any given state in a BN, thereby refuting the determinism inherent in BNs. Thus, a *probabilistic* BN (PBN) was introduced by Shmulevich et al. [30] in which the definition of a BN was adapted such that for each gene, at each time point, a Boolean function (and predictor gene set) is chosen with some conditional probability [29].

Inferring the PBN representation of a gene regulatory network (GRN) is quite involved. First, the directed graph expressing interactions between genes needs to be constructed; then, the Boolean functions need to be determined; followed by determining the probabilities of selecting a Boolean function as well as the number of candidate functions on each gene. Existing work (cf Sect. 2) tends to focus on inference from *time-series* gene expression data as the temporal aspect reveals the transition structure of the corresponding PBN. However, as already pointed out in [4], there are concerns over the number of (typically expensive to obtain) observations needed in such gene microarray data. Approaches based on ODEs (e.g., [21]) require lots of observations to tune the large number of parameters of the model, while in practice only a handful are available. More such observations are available when the underlying gene network is at a *steady state* [31], e.g, see gene expression profiles of melanoma by Bittner et al. [5].

In this paper, we propose a systematic method for inferring PBNs directly from real gene expression data measurements, collected using microarray technology, when the system is at a *steady-state*. The steady-state (long-run) behaviour of a PBN is of interest to system biology as it allows to determine the long-term influence of a gene on another gene or determine the long-term joint probabilistic behaviour of a few selected genes [31].

The key contribution of our paper is a reproducible pipeline for going from gene (steady-state) data samples to the PBN representation of the long-run behaviour of the underlying genetic network. We use a predictor gene set rather than temporal data to infer the "transition structure". Unlike other proposals, our method does not require the construction of the *probability transition matrix*, whose size grows exponentially on the number of nodes, and hence becomes computationally intractable for larger networks [1].

The remainder of the paper is structured as follows. Section 2 outlines related work. Preliminary background knowledge is presented in Sect. 3. The main algorithm for our inference method is in Sect. 4. PBNs are produced in Sect. 6 using the process described in Sect. 5. Concluding remarks are in Sect. 7.

2 Related Work

There have been various methods for PBN inference, focusing on causality, using different types of gene data [13]. Previous work on PBN inference from time series gene data includes [32], SCODE [21] with ODEs, and most recently the Stochastic Conjunctive Normal Form (SCNF) -based method by Apostolopoulou et al. [3] which can address larger networks.

Previous work on inference from steady-state data samples is relatively limited and goes back to Shmulevich et al. [31]. A tool for computing the steady-state distribution (ssd) probabilities has been proposed in [23]. Melkman et al. [22] infer *threshold* PBNs, a particular version of PBNs where every input threshold function of a node must have the same number of parameters and also satisfy certain stringent conditions. Kobayashi et al. [18] construct PBNs from BNs by casting inference as an integer linear programming problem and construct a PBN that fits the given steady-state distribution.

Kim et al. [17] use steady-state gene data samples from the study on *metastatic melanoma* by Bittner et al. [5] (we use the same data here). They choose the genes for their PBN using a combination of Coefficient of Determination (COD) analysis and biological background knowledge (we do not assume any prior knowledge). For the functions, they ternarise their data, and construct Lookup Tables in place of the functions for each gene. They also analyse the PBNs produced by analysing the steady-state distribution (ssd) of the resulting network.

Shmulevich et al. [30], who introduced PBNs, describe a method for determining functions for nodes in a PBN. This requires finding sets of input genes which have high COD with the target gene, and using the predictive model used for the calculation of the COD as the function for the particular set of input genes. The probability for choosing the particular input gene set is proportional to the COD of the input gene set.

Discretisation of gene data is an important factor for inference. Chen et al. [7] describe a method for quantising gene data using the expressions of *housekeeping* genes within the dataset. Housekeeping genes are genes which keep a constant expression, as they perform important functions within the cell. Since they have a constant expression, they can be used to estimate the probability distribution function (PDF) of the gene expressions within a microarray. The constructed PDF can be used for using a hypothesis test to determine whether or not a gene is over- or under-expressed. However, this method hinges on knowledge of which of the genes are housekeeping genes and this typically is not readily available.

As discussed in the introductory section, we focus on constructing PBNs from real, microarray gene data samples, collected while the system is in a steady-state, instead of simulated, time-series data or starting from BNs. We present a reproducible method to perform such a task.

3 Preliminaries

3.1 Boolean Networks

A BN [15] is a directed graph, $G = \{V, E\}$, comprised of vertices V and edges E. The vertices $v \in V$ represent the Boolean variables, which in this case represent genes in a gene regulatory network. The directed edges $\{v_i, v_j\} = e_{i,j} \in E$ represent that one variable, v_i, influences another, v_j. Each vertex is associated with a Boolean function f_i given by $f_i : \{0,1\}^{n_{in}} \mapsto \{0,1\}$. The input for f_i is a Boolean vector of length n_{in}, which represents the states of all of the input vertices, and the output is a single Boolean value, which is then used as the next state of the variable v_i. For a vertex i, the *input* vertices are the vertices from which all incoming edges originate, given by $\{v_j | \exists \{v_j, v_i\}\} = e_{j,i} \in E$.

3.2 Probabilistic Boolean Networks

Probabilistic Boolean networks are an extension of Boolean networks. They are directed graphs G, as in Boolean networks, except each function f_i for each node i in the case of Boolean networks is replaced by a set of Boolean functions $F_i = \{f_i^1, f_i^2, \ldots, f_i^{l_i}\}$, and probabilities $c_i = \{c_i^1, c_i^2, \ldots, c_i^{l_i}\}$. Hence, the logical function f_i has l_i possibilities, each with a corresponding conditional probability of being selected at every time step.

More formally, during run time, a function f_i^j for the node v_i is chosen with probability c_i^j, $j \in [1, l_i]$. PBNs are an extension to BNs in the sense that if each node within a PBN has a single function, it becomes identical to the BN.

3.3 State Transition Graphs

For each PBN there exists a state transition graph (STG). An STG is a directed graph $G = \{V, E\}$, where the vertices $v_i \in V$ represent the possible states of the PBN, and the edges $\{v_i, v_j\} = e_{i,j} \in E$ represent the possibility of a transition from state v_i to v_j. Since the probability of getting to another state v_j only depends on the current state v_i, we can say that the STG is a Markov chain.

By saying that the PBN has a steady state distribution (ssd), we mean that the STG of the PBN has a steady state distribution. For an STG to have an SSD, it needs to be *ergodic* - that is, every state can be reached from every other state. To guarantee that the STG is ergodic, random perturbations with low probability are introduced to the PBN.

3.4 Microarray Gene Data Samples

The data used to infer a PBN in our work was taken from the study of metastatic melanoma found in Bittner et al. [5], which has been extensively studied in the literature [17,25,27,33]. The study extracts and analyses the gene expression profiles of 31 melanoma cells using microarray technology. To make sure that the gene expression levels used in inferring the corresponding PBN are those of genes when the network is at a steady state, the Kolmogorov-Smirnov (KS) statistic is applied, as discussed in more detail in Sect. 5.

To utilise a particular gene in DNA, see [7], assuming the cell is at a steady-state, the relevant segment of the molecule must first be transcribed, producing messenger RNA (mRNA) which is accessible to the rest of the proteins. The quantity of mRNA in a cell signifies the degree of protein production associated with a particular gene.

DNA microarrays measure the presence of mRNA within a cell. The microarrays consist of a surface with an array of robotically placed complementary DNA for the genes to be analysed. mRNA tightly bonds with complementary DNA, hence the microarray can be used to isolate different mRNA molecules. The process is known as *hybridisation.*

The quantity of mRNA within a cell is measured by tagging the mRNA with fluorescent molecules, hybridising them with a microarray, and exciting the fluorescent molecules. The emitted brightness is proportional to the amount of mRNA present.

Since the amount of mRNA differs depending on the gene, the data is normalised by dividing the values recorded by the values recorded from a reference probe. Since values recorded are non-negative, the ratio values are in the range of $[0, \infty)$. Furthermore, since we would expect the values of within the reference probe and the sample to not be different, the median for the ratio values is expected to be 1. These are the values provided by Bittner et al. [5] in the form of a matrix of size 8,150 (number of genes) by 31 (number of samples). A small sample of the raw data is shown later in Fig. 1(a).

For demonstrating our method of inferring a PBN, we work with the subset of melanoma genes analysed by Datta et al. [9], which are extensively studied in the literature [17,18,25,27,33], namely *WNT5A, pirin, S100P, RET1, MART1, HADHB* and *STC2*. This offers straightforward validation for our approach since it produces the same PBN.

It is worth noting that larger PBNs may be constructed following the pipeline described in this paper, and we have constructed the 28 node PBN given in [33] as well as a 70 node PBN, which includes the 28 nodes already studied in [33] padded with the 42 nodes with the highest weighting of importance, using discriminative weights [5], which determine how a gene changes during the experiment compared to the control cells.

3.5 Coefficient of Determination

Coefficients of Determination (CODs) were described by Kim et al. [16] as a method to determine which gene determines the state of which other gene. A COD of a target variable, Y, with regards to an input variable, X, is a measure on how well the target variable can be predicted using the input variable. A predictive model f is used to predict the value of the target variable with and without the input variable, and compute the errors $\bar{e}$ and e respectively. The relative change of error of the predictive model is the COD θ, given by Eq. 1:

$$\theta = \frac{\bar{e} - e}{\bar{e}} \tag{1}$$

There are no constraints on what can be used as a predictive model. We opted for a perceptron. This is because there exists a closed-form solution for linear regression of the perceptron, described by Kim et al. [16], which can be used instead of training. This aids in lowering the computation time.

The weights of a perceptron, A, can be computed using the closed form solution:

$$\begin{aligned} A &= R^{+} \cdot C \\ R &= X \cdot X^{T} \\ C &= X \cdot Y \end{aligned} \tag{2}$$

3.6 Discretisation

Since PBNs use discrete values, the gene data which consists of real values has to be discretised. Discretisation is a process where values get mapped from the real value domain to the integer domain. For the problem at hand, since genes can be in one of two states, the range of the function should be either 0 or 1. Hence the function should take the form of:

$$f : G \rightarrow G_d, x \geq 0, \forall x \in G, y \in \{0, 1\} \forall y \in G_d \tag{3}$$

Such a method is described in detail in [34]. It consists of deciding upon a threshold value t with which all real values are compared. Each value then gets mapped to 0 if it is below the threshold, and to 1 otherwise, as given by Eq. 4.

$$G_d[x, y] = \begin{cases} 0 \; G[x, y] < t \\ 1 \; G[x, y] \geq t \end{cases} \tag{4}$$

The threshold may be any metric. Common metrics are means or medians. The threshold may also be the boundary between the top $x\%$ of entries and the rest.

Shmulevich et al. [29] describe a process of using k-means clustering to cluster the data, and assigning values to the data points depending on the cluster they belong to. However, since half of the data points lie in the range $(0, 1)$, and the other half is in the range $(1, \infty)$, the lower cluster ends up larger, resulting in a larger threshold that produces more zeros. This can be remedied by performing k-means clustering on the logarithms of the data points. This makes the ranges of both halves the same, producing more representative clusters.

4 Inference of PBNs

In this section we describe the inference method and how it can be implemented. Our approach to inferring a PBN starts with the real gene expression data in the form of a matrix G as input (see Fig. 1(a)), and produces a PBN (see Fig.

1(b)). The input matrix is of size $m \times n$, where m is the number of genes and n is the number of samples.

The method we apply for inferring PBNs draws upon work done by Shmulevich et al. [30]. First, it requires the dataset to be discretised (recall Section 3.6). This process is performed in Algorithm 1.

```
Data: Input array G, discretisation axis, calculation method
Result: Discretised array G
for row g in G over axis do
    t = method(G);
    for element in g do
        if element ¡ t then
            element = 0;
        else
            element = 1;
        end
    end
    return G ;
end
```

Algorithm 1: Discretisation algorithm

Given a target gene, n_p sets of genes with the highest CODs are found. This is down following Algorithm 2.

```
Data: Input array G, number of input genes per tuple k, number of input
      tuples n
Result: List of predictor-COD-input tuples
tuples = [];
for targetIndex in 1... G.geneAxisSize do
    indexCombinations =
    generateAllCombinations(1... G.geneAxisSize \ targetIndex, k);
    // Would return triplets of all possible k-combinations of
       the gene indexes within the array.
    buffer.init(n);
    for inputCombo in indexCombinations do
        COD, weights = calcCOD(G[targetIndex],G[inputCombo]);
        if COD > min(buffer.COD) then
            buffer.add(COD, weights, inputCombo);
            buffer.removeSmallestCOD() ;
        end
    end
    tuples[targetIndex] ← buffer;
end
return buffer;
```

Algorithm 2: Tuple list generator

A buffer of size n_p is initialised, and each possible combination of input genes have their CODs calculated. If a combination of inputs has a COD higher than at least one saved in the buffer, the buffer entry with the lowest COD gets replaced by the new combination of inputs. This results in a buffer full of input combinations with the highest CODs. One such buffer is initialised per target gene, resulting in n_p input combinations per target gene.

During run-time, a set of input genes is chosen with probability proportional to the COD of the set, and the next state is governed by the state of those input genes in conjunction with the predictive model that was saved. For all intents and purposes, the list with input gene, perceptron weights and probabilities are enough to construct a PBN, as the input genes convey the connectivity, and the perceptron weights convey the logic for that set of input genes. The process is summarised in Algorithm 3.

Data: Input array G, discretisation axis, threshold calculation method, number of input genes per tuple k, number of input tuples n
Result: List of predictor-probability-input tuples
$\hat{G}$ = discretise(G, axis, method);
predictorList = genPredictorList($\hat{G}$, k, n);
CODsum = 0 ;
for *tuple in predictor list* **do**
| CODsum ← CODsum + tuple[COD] ;
end
for *tuple in predictor list* **do**
| tuple[probability] ← $\frac{tuple[COD]}{CODsum}$;
end
return predictor list;

Algorithm 3: General algorithm for generating predictor lists

Ratio Data for Group of 12 Unclustered Cutaneous Melanomas

UACC-1012	UACC-1529	UACC-647	WM1791-C	UACC-827	HA-A	UACC-930	UACC-903
0.94	1.68	0.99	0.97	3.06	1.83	2.14	0.54
0.58	0.38	1.1	1.03	2.29	1.71	0.72	1.04
0.63	0.2	1.09	1.06	1.01	0.7	0.86	0.87
0.57	0.28	0.83	0.81	0.92	0.97	0.75	1.14
0.73	0.77	1	0.82	0.75	0.57	0.79	0.6
0.66	0.62	0.89	0.58	0.69	1.21	1.14	0.81
0.61	0.51	0.8	0.51	0.56	0.45	0.65	0.78
0.89	0.83	0.96	1.02	1.06	0.19	1.23	0.79
0.88	0.6	0.87	0.8	1	0.52	1.82	1.09
0.58	0.82	1.13	1.1	1.44	0.73	1.89	1.34
2.61	1.85	1.24	6.11	1.4	0.9	2.54	1.24
1.03	0.91	1.07	1.16	0.98	1.07	0.95	0.93
1.19	1.13	1.06	1.17	0.84	0.85	1.01	0.9
1.31	1.58	1.13	1.64	1.41	1.48	1.91	1.39
1.02	0.99	0.78	0.91	0.67	0.78	1.3	0.97
0.52	0.52	0.59	1.63	0.27	0.71	1.5	1.12
0.85	0.82	1.17	2.68	0.59	1.52	1.07	1.1

(a) The input, gene expression values for 31 samples taken at a steady-state, given in the form of a matrix

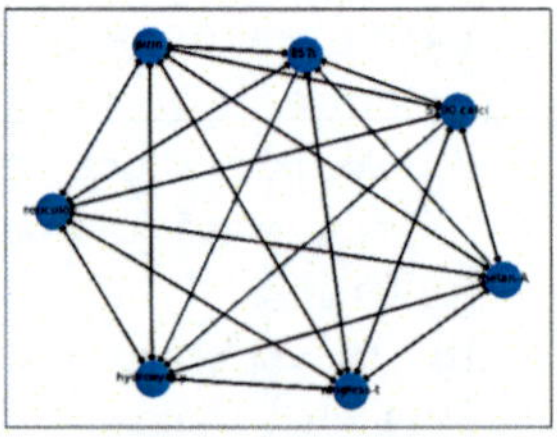

(b) The output is a PBN

Fig. 1. Input and output for the inference method

5 Analysis

The analysis of the generated PBNs in our approach are based on steady-state distribution, which is fairly standard, e.g., see [17]. The PBN is run for T steps in order to get it within a steady state. Then it is run for the next N steps, recording the state it is at. To confirm whether or not the PBN is in a steady state after T steps, the Kolmogorov-Smirnov (KS) statistic is calculated for the two halves of N.

The entries recorded in N are split in to two halves - one containing states $[0, \frac{N}{2}]$, the other containing $[\frac{N}{2} + 1, N]$. The entries are subsampled with the interval G. The histograms are converted to cumulative distribution functions (CDFs), and the maximum vertical distance between them is found, which is the KS statistic.

The significance test shows the probability of the two CDFs being drawn from the same distribution. If the PBN had not reached a steady state after T steps, the halves of N would be drawn from different distributions, which would be indicated by the KS test. The recorded states are a string of binary values. Therefore, for ease of analysis, they are used as gray-coded integers, and displayed on a histogram (cf. Fig. 2). This makes the horizontal distance on the histogram proportional to the Hamming distance between two network states.

6 Evaluation

We have implemented the pipeline using Python 3 and make it publicly available on https://github.com/UoS-PLCCN/pbn-inference.

We have constructed PBNs of size 7 from data produced by Bittner et al. [5] using different thresholds for the quantisation methods. The thresholds were (a) average of a gene expression; (b) median of a gene expression, and (c) k-means clustering of a gene expression. The data was quantised on a per-gene basis, with each gene having 10 triplets of input genes.

For the construction and validation of the histograms representing the steady-state distribution, we have chosen the parameters to be $T = 10^6$, $N = 4 \cdot 10^6$, $G = 10$ and $R = 100$. On a laptop with 32 GB of RAM and an Intel® Core^{TM} i7-7700HQ Processor, each histogram took around 9 hours to produce. The results are shown in Fig. 2.

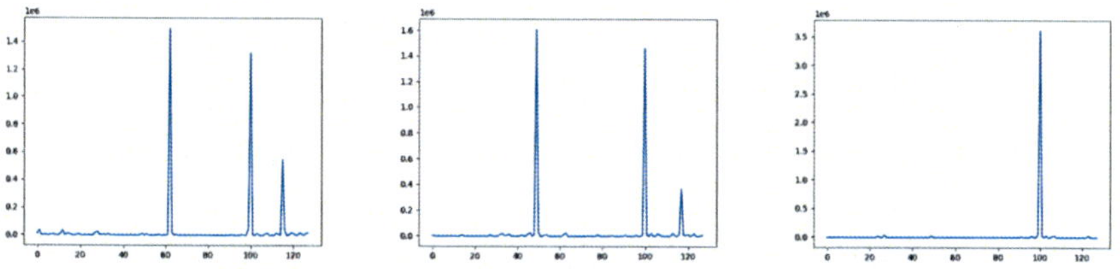

(a) Mean (b) Median (c) K-Means clustering

Fig. 2. SSDs of PBNs generated using different quantisation methods. States on the x-axis; SSD probability on the y-axis

It can be seen that the average and the median quantisation methods produce very similar histograms, with three peaks each, and the latter two peaks being in similar positions. The histogram generated using the PBN constructed from k-means clustering only has one prominent state, which can also be observed in the other two PBNs. It may be constructive to note that the few very prominent states in the histograms shown in Fig. 2 agrees with the assumption claimed by Kim et al. [17] that gene regulatory networks found in nature only occupy a small fraction of the possible state space.

For the purposes of direct comparison, we have trialled the proposed method in the DREAM (Dialogue on Reverse Engineering Assessment and Methods) challenge[1] which offers a benchmark for network inference (DREAM 3) [20] and scored 8th (out of 29).

7 Conclusion

In this work we described the inference a PBN directly from real gene data, collected using microrarray technology, which were taken when the system was at a steady-state. This kind of gene profiling is typically less costly to obtain than time series data, and includes more data points. Using the evaluation methods described in the literature, e.g., by Kim et al. [17], we have concluded that the pipeline works well for the examples provided. However, it is subject to fine-tuning the parameters. We have provided the method in a systematic pipeline which can be reproduced. We made it publicly available on github https://github.com/UoS-PLCCN/pbn-inference.

We note that the method scored 8th (out of 29) in the DREAM challenge and has been used to infer large PBNs (N = 200).

It is worth noting that the proposed method does not require a state transition probability matrix to be produced. It can be extracted from the PBN, however, the time required grows exponentially with the size of the PBN. This means that conventional mathematical methods in the literature that make use of the transition probability matrix may not always be applicable.

One concern is that the transitions get fitted to the quantised dataset. It is widely accepted that the states observed in the dataset are steady states of the cells. Since the transition rules get fitted to the steady states of the cells, the resulting PBN will be driven towards the steady states observed within the data. However, while it is certain that the method captures the long-run behaviour (steady-state) of the underlying gene regulatory network, there is little certainty that the PBN will behave with biological accuracy between the observed steady states. This concern could possibly be addressed by using time-series gene data to augment the method presented here, as this type of data captures the change of gene expression levels with respect to time. This promises to capture the behaviour at and between steady states, without reconstruction of the state evolution of the PBN, and is certainly worth exploring further in future work.

[1] https://dreamchallenges.org/project/dream-3-in-silico-network-challenge/.

References

1. Akutsu, T., et al.: Control of Boolean networks: hardness results and algorithms for tree structured networks. J. Theor. Biol. **244**(4), 670–679 (2007)
2. Albert, R., Othmer, H.G.: The topology of the regulatory interactions predicts the expression pattern of the segment polarity genes in Drosophila melanogaster. J. Theor. Biol. **223**(1), 1–18 (2003)
3. Apostolopoulou, I., Marculescu, D.: Tractable learning and inference for large-scale probabilistic Boolean networks. IEEE Trans. Neur. Netw. Learn. Syst. **30**(9) (2019)
4. Bar-Joseph, Z.: Analyzing time series gene expression data. Bioinformatics **20**(16), 2493–2503 (2004)
5. Bittner, M., et al.: Molecular classification of cutaneous malignant melanoma by gene expression profiling. Nature **406**, 536–40 (2000)
6. Chatzaroulas, E., Sliogeris, V., Victori, P., Buffa, F.M., Moschoyiannis, S., Bauer, R.: A structural characterisation of the mitogen-activated protein kinase network in cancer. Symmetry **14**(5) (2022)
7. Chen, Y.: Ratio-based decisions and the quantitative analysis of cDNA microarray images. J. Biomed. Opt. **2**(4), 364 (1997)
8. Ching, W.K., Zhang, M.K.N., Akutsu, T.: An approximation method for solving the steady-state probability distribution of probabilistic Boolean networks. Bioinformatics **23**(12), 1511–1518 (2007)
9. Datta, A., Choudhary, A., Bittner, M.L., Dougherty, E.: External control in markovian genetic regulatory networks. Mach. Learn. 4, 52, 3614 – 3619 (2003)
10. Davidich, M., Bornholdt, S.: The transition from differential equations to Boolean networks: a case study in simplifying a regulatory network model. J. Theor. Biol. **255**(3), 269–77 (2008)
11. Fumia, H.F., Martins, M.L.: Boolean network model for cancer pathways: Predicting carcinogenesis and targeted therapy outcomes. PLoS ONE **8**(7), e69008 (2013)
12. Gawad, C., Koh, W., Quake, S.: Single-cell genome sequencing: current state of the science. Nat. Rev. Genet. **17**, 175–188 (2016)
13. Glymour, C., Zhang, K., Spirtes, P.: Review of causal discovery methods based on graphical models. Front. Gene. **10**(524) (2019)
14. Karlsen, M.R., Moschoyiannis, S.: Evolution of control with learning classifier systems. Appl. Netw. Sci. **3**(1), 1–30 (2018)
15. Kauffman, S.A.: Metabolic stability and epigenesis in randomly constructed genetic nets. J. Theor. Biol. **22**(3), 437–467 (1969)
16. Kim, S., Dougherty, E., Bittner, M., Chen, Y., Sivakumar, K., Meltzer, P., Trent, J.: General nonlinear framework for the analysis of gene interaction via multivariate expression arrays. J. Biomed. Opt. **5**, 411–24 (2000)
17. Kim, S., Dougherty, E.R., Chen, Y., Bittner, M., Suh, E.: Can markov chain models mimic biological regulation? J. Biol. Syst. **10** (2003)
18. Kobayashi, K., Hiraishi, K.: Design of probabilistic Boolean networks based on network structure and steady-state probabilities. IEEE Trans. Neur. Netw. Learn. Syst. **28**(8), 1966–1971 (2017)
19. Liu, Y.Y., Slotine, J.J., Barabási, A.L.: Controllability of complex networks. Nature **473**(7346), 167 (2011)
20. Marbach, D., Prill, R.J., Schaffter, T., Mattiussi, C., Floreano, D., Stolovitzky, G.: Revealing strengths and weaknesses of methods for gene network inference. Proc. Nat. Acad. Sci. **107**(14), 6286–6291 (2010)

21. Matsumoto, H., et al.: SCODE: an efficient regulatory network inference algorithm from single-cell RNA-Seq during differentiation. Bioinformatics **33**(15), 2314–2321 (2017)
22. Melkman, A.A., Cheng, X., Ching, W.K., Akutsu, T.: Identifying a probabilistic Boolean threshold network from samples. IEEE Trans. Neur. Netw. Learn. Syst. **29**(4), 869–881 (2018)
23. Mizera, A., Pang, J., Yuan, Q.: Assa-pbn: An approximate steady-state analyser of probabilistic boolean networks. In: Automated Technology for Verification and Analysis. Springer International Publishing, Cham, pp. 214–220 (2015)
24. Moschoyiannis, S., Shields, M.: A set-theoretic framework for component composition. Fundamenta Informaticae **59**(4), 373–396 (2004)
25. Pal, R., Datta, A., Dougherty, E.R.: Optimal infinite-horizon control for probabilistic Boolean networks. IEEE Trans. Sign. Process. **54**(6), 2375–2387 (2006)
26. Papagiannis, G., Moschoyiannis, S.: Learning to control random Boolean networks: A deep reinforcement learning approach. In: Complex Networks 2019, Vol. 881. Springer, Cham, pp. 721–734 (2019)
27. Papagiannis, G., Moschoyiannis, S.: Deep reinforcement learning for control of probabilistic Boolean networks. In: Complex Networks 2020, Vol. 944. Springer, pp. 361–371 (2020)
28. Savvopoulos, S., Moschoyiannis, S.: Impact of removing nodes on the controllability of complex networks. In: Complex Networks (2017)
29. Shmulevich, I., Dougherty, E.R.: Probabilistic Boolean Networks: The Modeling and Control of Gene Regulatory Networks. SIAM (2010)
30. Shmulevich, I., Dougherty, E.R., Kim, S., Zhang, W.: Probabilistic Boolean networks: a rule-based uncertainty model for gene regulatory networks. Bioinformatics **18**(2), 261–74 (2002)
31. Shmulevich, I., et al.: Steady-state analysis of genetic regulatory networks modelled by probabilistic Boolean networks. Comp. Funct. Genom. **4**(6), 601–608 (2003)
32. Silescu, A., Honavar, V.: Temporal Boolean network models of genetic networks and their inference from gene expression time series. Compl. Syst. **13**(2001), 61–78 (2001)
33. Sirin, U., Polat, F., Alhajj, R.: Employing Batch Reinforcement Learning to Control Gene Regulation Without Explicitly Constructing Gene Regulatory Networks, pp. 2042–2048 (2013)
34. Velarde, C., Rubio-Escudero, C., Romero-Zaliz, R.: Boolean networks: a study on microarray data discretization. In: ESTYLF08, Cuencas Mineras (Mieres-Langreo), pp. 17–19 (2008)
35. Voukantsis, D., Kahn, K., Hadley, M., Wilson, R., Buffa, F.M.: Modeling genotypes in their microenvironment to predict single- and multi-cellular behavior. GigaScience **8**(3) (2019). https://doi.org/10.1093/gigascience/giz010
36. Wu, Y., Shen, T.: Policy iteration algorithm for optimal control of stochastic logical dynamical systems. IEEE Trans. Neur. Netw. Learn. Syst. **29**(5), 2031–2036 (2019)
37. Zhang, K., Johansson, K.H.: Efficient verification of observability and reconstructibility for large boolean control networks with special structures. IEEE Trans. Autom. Contr. **65**(12), 5144–5158 (2020)
38. Zhu, Q., Liu, Y., Lu, J., Cao, J.: Controllability and observability of Boolean control networks via sampled-data control. IEEE Trans. Control. Netw. Syst. **6**(4), 1291–1301 (2019)

Quantifying High-Order Interactions in Complex Physiological Networks: A Frequency-Specific Approach

Laura Sparacino[1(✉)], Yuri Antonacci[1], Daniele Marinazzo[2], Sebastiano Stramaglia[3,4], and Luca Faes[1]

[1] Department of Engineering, University of Palermo, 90128 Palermo, Italy
laura.sparacino@unipa.it
[2] Department of Data Analysis, Faculty of Psychological and Educational Sciences, Gent University, 9000 Gent, Belgium
[3] Dipartimento Interateneo di Fisica, Universitá degli Studi di Bari Aldo Moro, 70121 Bari, Italy
[4] INFN, Sezione di Bari, 70126 Bari, Italy

Abstract. Recent advances in information theory have provided several tools to characterize high-order interactions (HOIs) in complex systems. Among them, the so-called O-information is emerging as particularly useful in practical analysis thanks to its ability to capture the overall balance between redundant and synergistic HOIs. While the O-information is computed for random variables, its extension to random processes studied in the frequency domain is very important to widen the applicability of this tool to networks whose node exhibit rich oscillatory content, such as brain and physiological networks. This work presents the O-information rate (OIR), a measure based on the vector autoregressive and state space modelling of multivariate time series devised to assess the synergistic and redundant HOIs among groups of series in specific bands of biological interest. The new measure is illustrated in two paradigmatic examples of physiological networks characterized by coupled oscillations across a wide range of temporal scales, i.e. the network of cardiovascular and cerebrovascular interactions where redundant synchronized activity emerges around the frequencies of vasomotor and respiratory rhythms, and the network of scalp electroencephalographic signals where synergetic HOIs are detected among the alpha and beta waves recorded over the primary sensorimotor cortex.

Keywords: Brain connectivity · Cardiovascular oscillations · Information dynamics · Multivariate time series · Spectral analysis · Vector autoregressive models

1 Introduction

The increasing availability of recordings of biomedical signals is nowadays boosting the development of new methods for the data-driven modelling of complex biological systems. These methods typically describe the interactions between different systems (e.g.,

H. Cherifi et al. (Eds.): COMPLEX NETWORKS 2016 2022, SCI 1077, pp. 301–309, 2023.
https://doi.org/10.1007/978-3-031-21127-0_25

cardiovascular, cardiorespiratory, cerebrovascular or brain interactions) using pairwise measures of coupling or causality [19]. However, in spite of the widespread utilization of pairwise measures to describe interactions in a network, there is evidence that such measures cannot full capture the interplay among the multiple nodes of a complex system [5]. In fact, brain and physiological networks exhibit collective behaviors which are integrated at different hierarchical levels, thus displaying interactions that involve more than two network nodes. These so-called high-order interactions (HOIs) occur for instance when cardiovascular interactions are influenced by the effects of the respiratory activity [9], and identify brain sub-networks with distinct neurocognitive profiles [15]. HOIs have been analyzed in physiological and brain networks using information decomposition methods that evidence the synergistic or redundant nature of the interplay among multiplets of variability series [9]. One of such measures is the O-information [20], a metric capable of revealing synergy- and redundancy-dominated circuits in multivariate systems mapped by vector random variables. The concept underlying the definition of the O-information was applied to random processes quantifying, in a network of N processes, the synergistic or redundant causal contribution brought to the target by the remaining $N-1$ processes [22]. While the dynamic O-information [22] is an asymmetric (target-specific) measure of HOIs, a recent work defined a symmetric measure which generalizes the O-information to dynamic random processes [7]; this measure, denoted as O-information rate (OIR), was defined from the linear parametric representation of vector autoregressive (VAR) processes and was expanded in the frequency domain exploiting the framework of state-space (SS) models [3].

The possibility to assess HOIs in the frequency domain offered by the spectral expansion of the OIR [7] opens the way to the evaluation of redundant and/or synergistic interactions within specific frequency bands with physiological meaning. In this context, the present work applies the spectral OIR measures to networks of physiological time series and brain signals, with the aim of investigating the nature of multivariate interactions underlying the communication among different physiological systems, or among spatially distributed units of the same physiological systems. To this end, we focus first on multivariate physiological time series reflecting the dynamics of heart period, arterial pressure, breathing volume and cerebral blood flow to investigate the joint cardiovascular, cerebrovascular and respiratory regulation during postural stress [8], and then on multichannel EEG recordings measured in different scalp locations to assess brain connectivity during motor execution [1]. The two applications feature physiological time series rich of oscillatory content, which thus lend themselves to the spectral analysis of HOIs performed by the proposed tools.

2 O-Information Rate

The organizational structure of M stationary discrete-time stochastic processes $X^M = \{X_1,...,X_M\}$, which map the states visited by a network of dynamic systems, is assessed in the information-theoretic domain using the so-called O-information rate (OIR). The OIR quantifies the net synergistic or redundant information shared per unit of time by a subset of N processes taken from X^M. The OIR is null for pairs of processes ($\Omega_{X^2} = 0$), while for multiplets (i.e., groups of $N > 2$ processes) it is defined recursively as [7]:

$$\Omega_{X^N} = \Omega_{X^N_{-j}} + \Delta_{X^N_{-j};X_j}, \tag{1}$$

where X^N_{-j} is the subset without the process X_j, and where the OIR increment obtained adding X_j to X^N_{-j} is:

$$\Delta_{X^N_{-j};X_j} = (2-N) I_{X^N_{-j};X_j} + \sum_{\substack{m=1\\ m\neq j}}^{N} I_{X^N_{-mj};X_j}, \tag{2}$$

with $X^N_{-mj} = X^N \backslash \{X_m, X_j\}$. In (2), $I_{A;B}$ denotes the mutual information rate (MIR) between the vector stochastic processes A and B; the MIR quantifies the information shared by two processes per unit of time. The OIR can be either positive or negative for multiplets of order $N \geq 3$, with $\Omega_{X^N} > 0$ and $\Omega_{X^N} < 0$ denoting respectively redundant or synergistic interactions, i.e. interactions which can or cannot be retrieved from sub-groups of processes considered separately.

In this work, the OIR is computed in time and frequency domains following a linear parametric approach. The original vector X^M is represented by the vector autoregressive (VAR) model:

$$\mathbf{X}_n = \sum_{k=1}^{p} \mathbf{A}_k \mathbf{X}_{n-k} + \mathbf{U}_n, \tag{3}$$

where p is the model order, defining the maximum lag used to quantify interactions, $\mathbf{X}_n = [X_{1,n} \ldots X_{M,n}]^T$ is an M-dimensional vector collecting the present states of all processes, $\mathbf{A}_k$ is the $M \times M$ coefficient matrix describing the interaction from $\mathbf{X}_{n-k}$ to $\mathbf{X}_n$ at lag k, and $\mathbf{U}_n$ is a vector of zero-mean uncorrelated white noise processes with $M \times M$ positive definite covariance matrix Σ_U. While the VAR model (3) provides a global representation of the overall multivariate process, to describe the linear interactions relevant to the subset of processes $Z = \{Z_1, Z_2\}$, with $Z_1 = X_j$ and $Z_2 = X^N_{-mj}$ (m varying as in (2)), there is the need to define reduced VAR models involving only those processes. Hence, X^M is represented as a VAR process in the framework of state-space (SS) models, and the N submodels describing the joint dynamics of Z_1 and Z_2 are derived [3]. Their frequency domain representation allows to obtain the power spectral density (PSD) matrices describing the autonomous oscillatory content of Z_1 and Z_2 and their cross-spectra. A logarithmic spectral measure of the total coupling between Z_1 and Z_2 is then defined as [11]:

$$f_{Z_1;Z_2}(\omega) = \log \frac{|\mathbf{S}_{Z_1}(\omega)||\mathbf{S}_{Z_2}(\omega)|}{|\mathbf{S}_Z(\omega)|}, \tag{4}$$

where $\omega = 2\pi \frac{f}{f_s}$, with frequency $f \in [0, f_s/2]$ and f_s the sampling frequency, $\mathbf{S}_{Z_1}(\omega)$ and $\mathbf{S}_{Z_2}(\omega)$ are the autospectra of the two processes Z_1 and Z_2 while $\mathbf{S}_Z(\omega)$ is the PSD matrix of the joint process $\mathbf{Z} = [Z_1 Z_2]$, and $|\cdot|$ is the matrix determinant. Given that the integration over the whole frequency axis of the spectral measure (4) yields the corresponding time-domain measure $I_{Z_1;Z_2}$ in (2), the frequency-specific OIR increment can be then defined in analogy to (2) as [7]:

$$\delta_{X_{-j}^N;X_j}(\omega) = (2-N)f_{X_j;X_{-j}^N}(\omega) + \sum_{\substack{m=1 \\ m\neq j}}^{N} f_{X_{-mj}^N;X_j}(\omega), \tag{5}$$

and can be used to compute recursively a frequency-domain version of the OIR, in analogy to (1), i.e. [7]:

$$\nu_{X^N}(\omega) = \nu_{X_{-j}^N}(\omega) + \delta_{X_{-j}^N;X_j}(\omega). \tag{6}$$

It is easy to show that the spectral OIR increment in (5) and the spectral OIR in (6) satisfy the spectral integration property, i.e. the average over all frequencies of each of these spectral functions yields the corresponding information-theoretic function (respectively, (2) and (1)).

In this work, the VAR model fitting the M series was identified through the ordinary least squares method. Starting from the estimated VAR parameters (i.e., the VAR coefficients $\mathbf{A}_k$ and the innovation covariance Σ_U) and exploting the SS representation, the spectral functions in (5) and (6) were computed for all multiplets of orders 3 to N. Each spectral distribution obtained as in (6) was then integrated within specific frequency bands of the spectrum to retrieve information about high-order interactions within those bands. This procedure allowed us to evaluate redundant and/or synergistic interactions specific of band-limited stochastic oscillations with physiological meaning.

3 Application to Physiological Networks

3.1 Cardiovascular and Cerebrovascular Variability Series

We applied the framework to a database of physiological time series collected from 13 young healthy subjects during supine resting (REST) and passive standing in the 60° upright position reached after head-up tilt (TILT) [2]. For each subject and condition, $M = 5$ stationary sequences of 250 beat-to-beat values of heart period (tachogram, series T), systolic and diastolic arterial pressure (S, D), respiration (R), and mean cerebral blood flow velocity (F), were considered for the analysis. A VAR model was fitted on the time series measured for each subject and condition, with model order set according to the Akaike Criterion. Then, the spectral OIR was computed for each multiplet of order $N = 3, 4, 5$ and integrated within the low frequency (LF, 0.04–0.15 Hz) and high frequency (HF, 0.15–0.4 Hz) bands of the spectrum.

The results collected in Fig. 1 show that the OIR integrated in both LF and HF bands was positive in the large majority of subjects in both the analyzed conditions, and showed a tendency to increase with the number of series in the analyzed multiplets. This finding suggests that physiological networks probed by beat-to-beat variability series are dominated by redundancy. The result confirms similar findings observed in cardiovascular and cardiorespiratory networks [8, 18], and extends them to cerebrovascular and integrated physiological networks.

Multivariate interactions were found to be stronger for HF oscillations than in the LF band, suggesting a main role of respiration, whose oscillations are typically mostly confined within the HF band [9], in driving redundant interactions in the cardiovascular and cerebrovascular networks. This result is confirmed by the observation that in the LF band the highest redundancy was displayed by multiplets including the series T, S, D and F, while significantly lower values were found in multiplets including the series R, in both experimental conditions. For these multiplets, a tendency towards an increase in the redundancy (though not statistically significant) was observed moving from REST to TILT, suggesting a possible role of sympathetic activation in driving the redundancy of LF oscillations in cardiovascular and cerebral blood flow variables [18]. Conversely, results in the HF band showed some statistically significant differences between multiplets only during TILT; specifically, the multiplets containing the series R, S and F displayed the highest redundancy.

Overall, these results confirm for HOIs the redundant nature of cardiovascular and cerebrovascular interactions previously reported for triplets of physiological processes [7,9,18], and document the relevance of separating LF and HF contributions to elicit the role of respiration on cardiovascular and cerebrovascular interactions. Moreover, the tendency of the OIR to increase with tilt was not statistically significant, suggesting that these redundant effects are preserved during postural stress.

The understanding of the different and complex ways of dynamic integration of organ systems as a complex network remains one of the biggest problems in field of Network Physiology. Physiological systems exhibit complex dynamics, operate at different time scales and are regulated by multi-component mechanisms, which has been known to challenge the study of physiologic coupling and causality [4,13]. These aspects, together with the evidence that cardiovascular and cerebrovascular interactions occur through the coupling of rhythms in different frequency bands with different physiological meaning [19], make our spectral approach eligible to probe HOIs in these networks. [8]. Our results document that respiration acts as a major driver of multivariate redundant interactions in physiological networks, confirming that HOIs can have different nature for different rhythms because synergistic and redundant behaviors generally alternate in different bands of the frequency spectrum [1,8].

3.2 EEG Recordings

The analyzed dataset refers to EEG signals relevant to 20 healthy subjects randomly chosen from a database of 109 participants (https://physionet.org/content/eegmmidb), recorded for each subject from 64 electrodes referenced to both mastoids (international 10–20 system, $f_s = 160$ Hz) [12,21]. We analyzed a resting state condition in which the participants were relaxed (REST) and a condition in which they were asked to open and close the right fist cyclically (RIGHT). The raw signals were firstly detrended, then filtered (band-pass, 2–35 Hz; notch, 59–61 Hz) and finally epoched to extract 15 trials of 4 s each for each participant and condition. We performed the analysis on the signals recorded by the four electrodes depicted in Fig.2(a), i.e. $X_1 = C_3, X_2 = C_z, X_3 = C_4, X_4 = F_z$. For each subject and trial, a VAR model was identified from the four selected time series setting the model order according to the Bayesian Information criterion. Then, the estimated VAR parameters were used to compute the spectral OIR for each multiplet of order $N = 3,4$. Finally, values indicative of HOIs occurring for the α and β brain

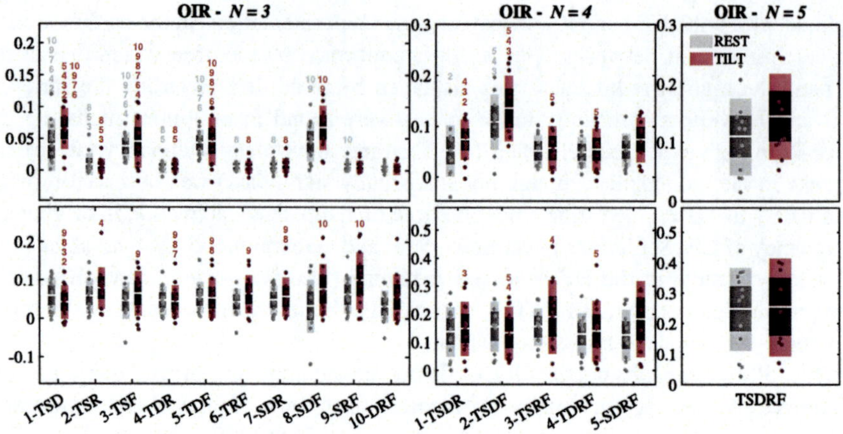

Fig. 1. Distribution across subjects and individual values of the OIR (top panels: spectral OIR integrated in the LF band, 0.04–0.15 Hz; bottom panels: spectral OIR integrated in the HF band, 0.15–0.4 Hz) computed at REST (gray) and during TILT (violet) for all possible multiplets of order 3,4,5 obtained grouping the time series of heart period (T), systolic pressure (S), diastolic pressure (D), respiration (R) and mean cerebral blood flow velocity (F). Numbers in the OIR(3) and OIR(4) panels indicate pairs of distributions for which the mean OIR differed significantly in a given condition (Student t-test for paired data, $p < 0.05$). No statistically significant differences between REST and TILT were detected.

rhythms were obtained by integrating the measures over the relevant frequency ranges (i.e., $\alpha = [8-12]$ Hz, $\beta = [16-26]$ Hz).

Figure 2(b,c,d,e,f) reports the grand-average over participants and trials of the frequency profiles obtained for each multiplet separately for the REST and RIGHT conditions in the frequency range 2–35 Hz. The trends reveal the prevalence of positive values of the OIR, denoting redundant interactions, for the triplets including the signal recorded at the electrode F_z (panels b,c,d), while the triplet $[C_3 - C_z - C_4]$ displays negative OIR values related to synergy within the α and β bands (panel e); synergistic interactions are detected, although weaker, also for the multiplet of order 4 including all the analyzed electrodes.

Fig. 2(g,h) depicts the distributions across participants and trials of the spectral OIR integrated over the α and β frequency bands, computed for each multiplet separately for the REST and RIGHT conditions. The execution of the motor task is generally associated with an increase of the OIR denoting higher redundancy for multiplets including the EEGs recorded at the central electrodes F_z and C_z, together with C_3, C_4, or both C_3 and C_4; the increase is observed particularly in the β band, and is statistically significant, according to a Wilcoxon paired test performed with 5% significance, for the multiplet of order 4 including the central electrodes F_z and C_z and both the lateral electrodes C_3 and C_4 (panel h). On the contrary, the OIR profiles exhibit a decrease in the α band when computed for the triplet $[C_3 - C_z - C_4]$ (panel g); such a decrease, though not statistically significant, denoties a tendency to higher synergy during task execution.

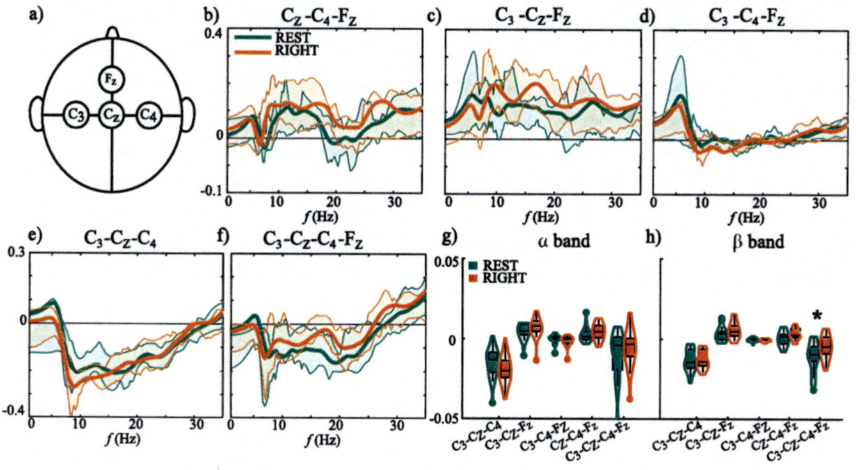

Fig. 2. (a) EEG electrodes montage highlighting the positions of the four selected electrodes. (b–f) Spectral profiles (mean (bold line) and 25th–75th percentiles (shades) over 20 subjects and 15 trials per subject) of the OIR computed for each multiplet during relaxation (REST, green lines) and while opening and closing the right fist (RIGHT, orange lines). (g–h) Violin plots of the distribution across participants and trials of the OIR values computed by integrating the spectral OIR within the α band (8–12 Hz (g)) and within the β band (16–26 Hz) during relaxation (REST) and opening/closure of the right fist (RIGHT). *, $p < 0.05$ Wilcoxon test.

The study of human brain activity during motor execution is very important in clinical contexts and in neuroscience. In fact, motor actions commonly derive from the involvement of several areas in the brain causing excitatory and inhibitory coupling among different regions in the two hemispheres [10]. To investigate the joint EEG activity of specific brain areas designated to the planning and execution of hand motor task [16], we selected four EEG signals recorded from channels located on the central line (F_z, C_Z) and on scalp areas contralateral and ipsilateral to the right-hand motor execution task (respectively, C_3 and C_4). In agreement with our previous work showing a widespread redundant behavior for this network of EEG interactions [1], we document the prevalence of redundancy for HOIs. However, we also show that combinations of signals measured from both central electrodes and electrodes located in ipsi- and contralateral locations give rise to synergistic HOIs that reflect the emergence of interaction mechanisms not retrievable from a pairwise analysis. From a physiological point of view, these functional mechanisms involve spatial locations (C_3 and C_4) and emerge in frequency bands (α and especially β) which are linked to the well-known phenomenon of event-related desynchronisation occurring during motor execution and imagery [17].

Nevertheless, we stress the preliminary nature of our results, which need confirmation on larger datasets [12], also after adopting methodological improvements (e.g., the consideration of the statistical significant of the detected OIR values and increments [22]) and addressing common issues of brain connectivity analyses (e.g., those related to the effects of volume conduction on pairwise and higher-order connectivity measures [14,23]).

4 Conclusions

Brain and physiological networks are not only composed by several functionally interconnected nodes, but are also endowed with oscillatory dynamics deployed across a wide range of temporal scales, which vary from the 20 s period of slow sympathetic and vasomotor rhythms up to a frequency of almost 100 Hz for the faster EEG waves. The spectral expansion of the O-information presented in this work allows to deal at the same time with the multivariate and multiscale nature of physiological dynamics, characterizing them by means of HOIs measures assessed within specific frequency bands.

Our results indicate that different networks whose collective dynamics are studied in different frequency bands display distinct peculiar patterns of HOIs. The physiological network probed by heart rate, arterial pressure and cerebral blood flow variability displays slower oscillations entrained by mechanical and autonomic influences, as well as faster rhythms driven by respiration [6], which give rise to mostly redundant interactions at different orders. On the other hand, a growing body of work is documenting how brain networks can display synergy as an emergent behavior, suggesting that synergistic interactions may serve to integrate and complement redundant sub-networks [15,24]. The results here obtained, albeit in a preliminary fashion, support this hypothesis, thus opening the way to the use of the frequency-specific O-information measures as a tool to uncover higher-order effects which otherwise, with current pairwise and time-domain approaches, would remain hidden.

References

1. Antonacci, Y., Minati, L., Nuzzi, D., Mijatovic, G., Pernice, R., Marinazzo, D., Stramaglia, S., Faes, L.: Measuring high-order interactions in rhythmic processes through multivariate spectral information decomposition. IEEE Access **9**, 149486–149505 (2021)
2. Bari, V., Marchi, A., De Maria, B., Rossato, G., Nollo, G., Faes, L., Porta, A.: Nonlinear effects of respiration on the crosstalk between cardiovascular and cerebrovascular control systems. Philos. Trans. Roy. Soc. A Math. Phys. Eng. Sci. **374**(2067), 20150179 (2016)
3. Barnett, L., Seth, A.K.: Granger causality for state-space models. Phys. Rev. E **91**(4), 040101 (2015)
4. Bashan, A., Bartsch, R.P., Kantelhardt, J., Havlin, S., Ivanov, P.C., et al.: Network physiology reveals relations between network topology and physiological function. Nature Commun. **3**(1), 1–9 (2012)
5. Battiston, F., Cencetti, G., Iacopini, I., Latora, V., Lucas, M., Patania, A., Young, J.-G., Petri, G.: Networks beyond pairwise interactions: structure and dynamics. Phys. Rep. **874**, 1–92 (2020)
6. Cohen, M.A., Taylor, J.A.: Short-term cardiovascular oscillations in man: measuring and modelling the physiologies. J. Physiol. **542**(3), 669–683 (2002)
7. Faes, L., Mijatovic, G., Antonacci, Y., Pernice, R., BarÁ, C., Sparacino, L., Sammartino, M., Porta, A., Marinazzo, D., Stramaglia, S.: A Framework for the Time-and Frequency-Domain Assessment of High-Order Interactions in Brain and Physiological Networks. arXiv preprint: arXiv:2202.04179 (2022)

8. Faes, L., Pernice, R., Mijatovic, G., Antonacci, Y., Krohova, J.C., Javorka, M., Porta, A.: Information decomposition in the frequency domain: a new framework to study cardiovascular and cardiorespiratory oscillations. Philos. Trans. Roy. Soc. A **379**(2212), 20200250 (2021)
9. Faes, L., Porta, A., Nollo, G., Javorka, M.: Information decomposition in multivariate systems: definitions, implementation and application to cardiovascular networks. Entropy **19**(1), 5 (2016)
10. Gerloff, C., Richard, J., Hadley, J., Schulman, A.E., Honda, M., Hallett, M.: Functional coupling and regional activation of human cortical motor areas during simple, internally paced and externally paced finger movements. Brain J. Neurol. **121**(8), 1513–1531 (1998)
11. Geweke, J.: Measurement of linear dependence and feedback between multiple time series. J. Am. Stat. Assoc. **77**(378), 304–313 (1982)
12. Goldberger, A.L., Amaral, L.A., Glass, L., Hausdorff, J.M., Ivanov, P.C., Mark, R.G., Mietus, J.E., Moody, G.B., Peng, C.-K., Stanley, H.E.: Physiobank, physiotoolkit, and physionet: components of a new research resource for complex physiologic signals. Circulation **101**(23), e215–e220 (2000)
13. Ivanov, P.C., Bartsch, R.P.: Network physiology: mapping interactions between networks of physiologic networks. In: Networks of Networks: The Last Frontier of Complexity. Springer, pp. 203–222 (2014)
14. Kotiuchyi, I., Pernice, R., Popov, A., Faes, L., Kharytonov, V.: A framework to assess the information dynamics of source eeg activity and its application to epileptic brain networks. Brain Sci. **10**(9), 657 (2020)
15. Luppi, A.I., Mediano, P.A., Rosas, F.E., Holland, N., Fryer, T.D., O'Brien, J.T., Rowe, J.B., Menon, D.K., Bor, D., Stamatakis, E.A.: A synergistic core for human brain evolution and cognition. Nature Neurosci. **25**(6), 771–782 (2022)
16. Passingham, R.: Supplementary motor cortex and self-initiated movement. Neur. Program., 13–24 (1989)
17. Pfurtscheller, G., Da Silva, F.L.: Event-related EEG/MEG synchronization and desynchronization: basic principles. Clin. Neurophysiol. **110**(11), 1842–1857 (1999)
18. Porta, A., Bari, V., De Maria, B., Takahashi, A.C., Guzzetti, S., Colombo, R., Catai, A.M., Raimondi, F., Faes, L.: Quantifying net synergy/redundancy of spontaneous variability regulation via predictability and transfer entropy decomposition frameworks. IEEE Trans. Biomed. Eng. **64**(11), 2628–2638 (2017)
19. Porta, A., Faes, L.: Wiener-granger causality in network physiology with applications to cardiovascular control and neuroscience. Proc. IEEE **104**(2), 282–309 (2015)
20. Rosas, F.E., Mediano, P.A., Gastpar, M., Jensen, H.J.: Quantifying high-order interdependencies via multivariate extensions of the mutual information. Phys. Rev. E **100**(3), 032305 (2019)
21. Schalk, G., McFarland, D.J., Hinterberger, T., Birbaumer, N., Wolpaw, J.R.: BCI 2000: a general-purpose brain-computer interface (BCI) system. IEEE Tran. Biomed. Eng. **51**(6), 1034–1043 (2004)
22. Stramaglia, S., Scagliarini, T., Daniels, B.C., Marinazzo, D.: Quantifying dynamical high-order interdependencies from the o-information: an application to neural spiking dynamics. Front. Physiol. **11**, 595736 (2021)
23. Van de Steen, F., Faes, L., Karahan, E., Songsiri, J., Valdes-Sosa, P.A., Marinazzo, D.: Critical comments on EEG sensor space dynamical connectivity analysis. Brain Topogr. **32**(4), 643–654 (2019)
24. Varley, T.F., Pope, M., Faskowitz, J., Sporns, O.: Multivariate information theory uncovers synergistic subsystems of the human cerebral cortex. arXiv preprint: arXiv:2206.06477 (2022)

A Novel Reverse Engineering Approach for Gene Regulatory Networks

Francesco Zito, Vincenzo Cutello, and Mario Pavone(✉)

Department of Mathematics and Computer Science, University of Catania, v.le Andrea Doria 6, 95125 Catania, Italy
francesco.zito@phd.unict.it, cutello@unict.it, mpavone@dmi.unict.it

Abstract. Capturing the rules that govern a particular system can be useful in any field where the causes of its effects are unknown. Indeed, discovering the causes that produced a particular effect is extremely useful in fields such as biology. In this paper, a reverse engineering method based on machine learning is proposed. This method was used to replicate real world behaviour and use this knowledge to generate the relative Gene Regulatory Network. The datasets from the DREAM4 Challenge were used to validate this method.

Keywords: Reverse engineering · Complex network · Machine learning · Gene regulatory network · Multivariate time-series forecasting

1 Introduction

Most real-world systems can be represented by a graph with a nontrivial topology, called a complex network, where the nodes represent real-world entities (or variables) and the arcs represent the interactions between the entities themselves [15]. The goal is to better understand the functioning of the real system by studying the interactions between the entities involved and the properties of the network [12]. To build the complex network associated with a real-world system we need to carefully reverse engineering it. In other words, we have to build the complex network representing the real system by measuring only the outputs of the system over a period of time [22].

Unlike forward problems, where the effects can be accurately predicted by the system model given one or more causes as inputs, inverse problems are the opposite [20]. Here, the mathematical formulation of the problem is not known and therefore it is not possible to assert anything about it [5]. The only known information come from the observations of the system. Explaining the actual causes of a particular phenomenon, however, can be an extremely difficult, if not impossible, because small differences in effects can lead to large differences in causes, or the same effect can result from more than one cause [21]. In reality, estimating the parameters of the model is a difficult and time-consuming process due also to the large dimensional space, although some efficient algorithms

H. Cherifi et al. (Eds.): COMPLEX NETWORKS 2016 2022, SCI 1077, pp. 310–321, 2023.
https://doi.org/10.1007/978-3-031-21127-0_26

were proposed [6,7]. Several approaches can be found in literature to solve such inverse problems. For example, in [9] different mathematical approaches to solve this problem are explained. However, as stated in [4], mathematical approaches may not be sufficient to find a solution in reasonable time with high accuracy. Therefore, in recent years, many studies have been focused on the application of deep learning techniques to inverse problems, e.g., in image processing [14], astronomy [8], physics [16], biology [1], civil engineering [3], and so on.

In this paper, we present a novel approach based on machine learning techniques that has two goals: create an artificial environment capable of replicating the behaviour of a real environment based solely on observations of the variables of interest, and generate a complex network that reveals the relationships among variables in the system.

One of the most interesting areas of computer science where the application of reverse engineering is on the rise, is in the field of genetics [17]. Micro-array technology allows researchers to examine the presence of multiple genes in a DNA sample and their levels of expression [11]. Using our method, we were able to firstly artificially reproduce this behaviour and secondly create a genetic regulatory network showing the iterations between genes.

In Sect. 2 we will discuss modelling and creating an artificial environment that can replicate the real environment. In Sect. 3, we will discuss a methodology that uses the artificial environment to facilitate the creation of a gene regulatory network. In Sect. 4, we present the results obtained.

2 Modeling

To solve the inverse problem the goal we need to find the best model m such that

$$d = G(m) \tag{1}$$

where $G(\cdot)$ is an operator describing the explicit relationship between the observed data d and the model parameters [2]. In this section, we present a novel approach based on machine learning techniques to determine the best model m that can replicate the behaviour of a real environment when only the observations are available.

2.1 Environment

An environment basically consists of several elements (such as entities or variables) that interact with each other according to well-defined rules established during the design process. The two main components of an environment are: the agents and the rules that govern their interactions. Each agent is responsible for predicting the future value of a variable based on the current state of the environment. The interaction rules between the agents, on the other hand, define the mechanism by which the state of the environment evolves. Figure 1 depicts the environment scheme under consideration.

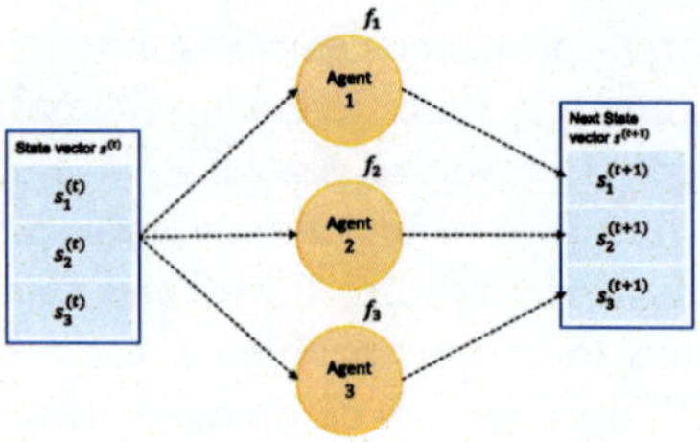

Fig. 1. Environment architecture

Given an environment with k variables, the state of the environment at time t, denoted by $s^{(t)}$, is defined as a k-vector in which the generic element $s_i^{(t)}$ is the value of the i-th variable at time t. The subsequent state of the environment $s^{(t+1)}$ is computed as follows:

$$s^{(t+1)} = f\left(s^{(t)}\right) \tag{2}$$

It can be seen from figure 1 that each agent in the environment is responsible for predicting the value of the variable to which it refers. Accordingly, there are as many agents as there are variables in the environment, so the equation 2 can also be written as follows.

$$s^{(t+1)} = f\left(s^{(t)}\right) = \left[f_1\left(s^{(t)}\right), f_2\left(s^{(t)}\right), \dots, f_k\left(s^{(t)}\right)\right] \tag{3}$$

where f_i denotes the agent function and represents the i-th component of the function f.

2.2 Agent

An agent can be formally described as a function $f_i : S^k \rightarrow S$ that predicts the i-th component of the state of the environment at time $t+1$, given its previous state. We use two different types of prediction scheme: simple agent prediction and delta agent prediction.

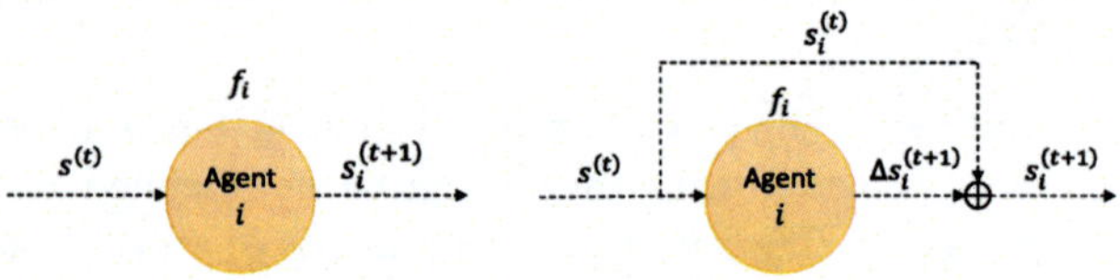

(a) Simple Agent prediction (b) Delta Agent prediction

Fig. 2. Agent prediction scheme

Simple Agent Prediction In the first prediction scheme (Fig. 2a), an agent directly estimates the future value of the variable according to Eq. 4.

$$s_i^{(t+1)} = f_i\left(s_1^{(t)}, s_2^{(t)}, \ldots, s_k^{(t)}\right) \tag{4}$$

Agent Delta Prediction On the other hand, Fig. 2b illustrates the second agent prediction scheme. In this case, an agent predicts by how much the current value should be increased or decreased. In other words, it predicts the offset between the value of the i-th component of the state at time $t+1$ and its value at time t. The output of the agent is calculated as follows.

$$s_i^{(t+1)} = s_i^{(t)} + \Delta s_i^{(t+1)} = s_i^{(t)} + f_i\left(s_1^{(t)}, s_2^{(t)}, \ldots, s_k^{(t)}\right) \tag{5}$$

Each agent in the environment can be considered as a black-box function with its own architecture and configuration. The basic model for each agent is shown in Fig. 3.

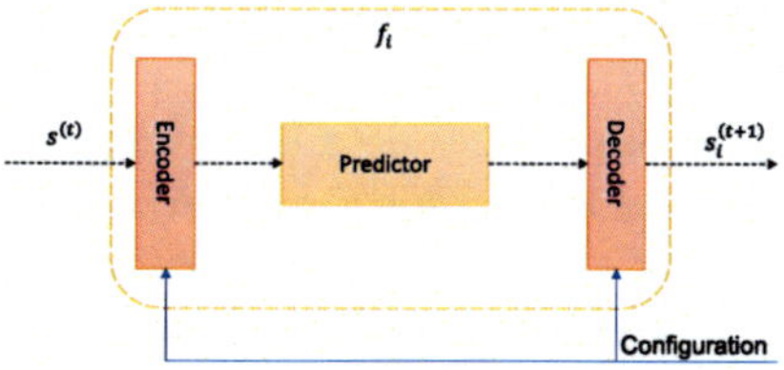

Fig. 3. The basic model of an agent.

The Predictor is the core of the agent and it is responsible for forecasting the value of a variable given a specific value as input. A predictor can be anything from a neural network to a regression model to a decision tree. Different types of predictors were tested to explore different possible configurations of the artificial environment. In Table 1 we report the available configurations for each predictor.

Table 1. Agent configurations for each type of predictor. 1) Predictor specifies the type of predictor used to predict the variable; 2) Machine learning tasks defines whether the predictor is a classifier and returns a class value or whether it is a regressor and predicts a real value; 3) Agent prediction scheme defines how the agent's output is composed; 4) Training parameters are used to train the predictor.

Predictor name	Machine learning task	Agent prediction scheme	Training parameters
Fully Connected Neural Network (FCNN)	Classification (L)/regression	Simple/delta prediction	Epochs, mini batch size, learning rate
Reccurent Neural Network (RNN-LSTM)	Classification (L)/regression	Simple/delta prediction	Epochs, mini batch size, learning rate
Convolutional Neural Network (CNN)	Regression	Simple/delta prediction	Epochs, mini batch size, learning rate
Simple Linear Regression (SLR)	Regression	Simple/delta prediction	

Since an environment may contain different types of agents, it is important to use additional layers to ensure agent interoperability. Therefore two additional blocks have to be added before and after the predictor: encoder and decoder. The encoder's role is to convert the agent's input into a format that the predictor can understand. The decoder's role, on the other hand, is to convert the predictor's output into a format that is compatible with the architecture of the environment ($s_i^{(t+1)}$).

3 Methodology

In a gene regulatory network, each gene can be activated or inhibited depending upon the expression level of another gene, called regulator or regulatory gene. To find the regulatory genes of each gene, we used a simple principle: if varying the expression level of gene G_i causes a significant change in the expression level of gene G_j, then G_i might be a good candidate as a regulator of gene G_j [10]. However, this approach can only be used if the mathematical formulation that determines the relationships between gene expression levels is known. Several methods for constructing a gene inference model have been proposed in the literature. For instance, continuous models such as the ordinary differential equations ODE, which are based on estimates of the inference level over time. Although the ODE approach provides detailed information about the dynamics of gene expression, it requires high quality data to build an accurate model [13]. In this section, we first look at how the model described above can be used to predict the expression level of genes, and then take advantage of the artificial environment to generate GRNs.

3.1 Gene Expression Level Prediction

We create an artificial environment by using: the observations (X) represented by a $k \times n$ matrix, where k is the number of variables and n is the number of observations over time; and the environment's configuration that consists of a collection of k tuples, each associated with an agent and containing the elements listed in Table 1, such as, the predictor, ML Task, prediction scheme, and training parameters. Algorithm 1 shows the procedure for creating an artificial environment and the evaluation of the environment thus created.

Since the model of the i-th agent depends on its associated configuration, each agent must be trained on its own training set consisting of pairs (X, Y_i), where Y_i is a n vector and contains the values of the i-th variable shifted forward by one time unit. The artificial environment E is composed by all the agents created according to the architecture described in Sect. 2.1.

Ideally, an optimal artificial environment without any perturbation and with identical initial conditions should be able to produce the same response as a real environment. Figure 4 depicts the artificial environment's response over a time interval equal to twice the observation time. As we can see, the response of the artificial environment before 1000 time units is similar to the observation in the real environment. On the other hand, after 1000 time units, the artificial

Algorithm 1: Environment-creation algorithm

```
A ← ∅ ;                                              /* collection of agents */
for i = 1, 2, ..., k do
    Y_i = createTargetData (X,Config_i) ;
    A_i = createAgent (X,Y_i,Config_i) ;
    A = A ∪ A_i
end
E = makeEnvironment (A) ;
X̂ = simulation(E,0,n) ;    /* the simulation starts at time 0 and ends
    after n time steps, the resulting matrix is a k × n matrix and
    contains all predicted values for each variable for each time point.
    */
Q_% = evaluation(X, X̂) ;                            /* using equation 6 */
```

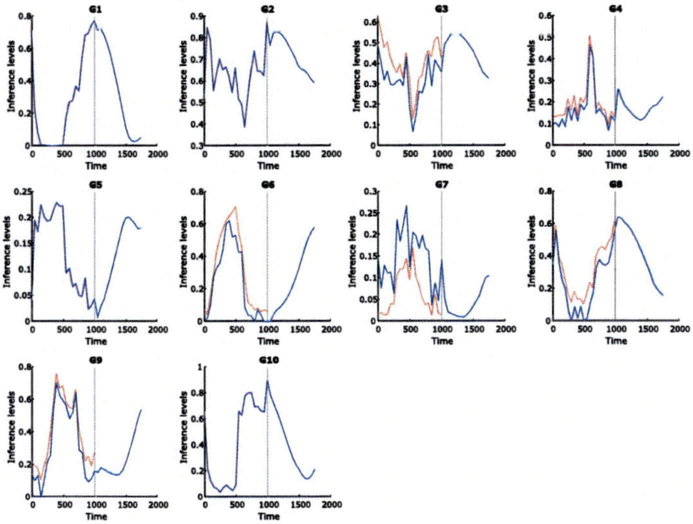

Fig. 4. Prediction of gene expression levels of ten genes over time. The red line represents the actual observations, while the blue line shows the predicted values over time.

environment uses its knowledge to forecast the inference level of the genes for the subsequent time points.

To validate the effectiveness of the prediction, the cosine similarity measure is used. Given two matrices, $X \in \mathbb{R}^{k \times n}$ and $\hat{X} \in \mathbb{R}^{k \times n}$, representing, respectively, the actual and predicted data and the generic element x_{ij}, corresponding to the value of the i-th variable at the j-th time, the similarity index $Q_{\%}$ denotes how close the prediction is to the target values in percentage terms and it is defined as follows:

$$Q_{\%} = \frac{1}{k}\left(\sum_{i=1}^{k}\left|\frac{\sum_{j=1}^{n} x_{ij}\hat{x}_{ij}}{\sqrt{\sum_{j=1}^{n} x_{ij}^2}\sqrt{\sum_{j=1}^{n} \hat{x}_{ij}^2}}\right|\right) * 100, \tag{6}$$

To choose the appropriate configuration of the agents so to maximize the similarity index, we used a local search algorithm [19,24] that continuously improves a possible solution until the best configuration for the environment is found.

3.2 Gene Regulatory Network

Algorithm 2 shows the pseudo-code of the procedure used to determine the regulatory matrix by the artificial environment. Suppose we want to determine whether the gene G_i is a regulatory gene of G_j. Using the artificial environment, the expression value of G_i can be manually set to V_i at time t (lines 5–6), and it will be possible to observe how the gene G_j responds to this change after time t+1 (lines 7–8). Clearly, we cannot know whether or not the measurements obtained in this way are correct. For that, we would need a real dataset and would have to validate the measurements in the field using micro-array technology.

We denote by "initial time" $t-1$ the time when the environment does not react to perturbations and evolves naturally, by "transition time" t we denote the time when the perturbation acts on the gene G_i and changes its expression level manually, and finally by "final time" $t+1$ we denote the time after the perturbation when the artificial environment reacts to it and the expression level of the genes is calculated taking this change into account. According to Eqs. 3, 7, 8, and 9 represent the state of the environment calculated at the initial time, at the transition time, and at the final time, respectively.

Algorithm 2: Building a GRN using the artificial environment

1 E_s = `findStability` (E) ; `/* Simulates the system until the variables no longer greatly fluctuate and are within a certain range. At the end of this procedure, the current time point of the environment` E_s `will be equal to the initial time. */`
2 **for** $i = 1, 2, \ldots, k$ **do**
3     $\hat{E}$ = `clone` (E_s) ;
4     $s^{(t-1)}$ = `getCurrentState` ($\hat{E}$) ;
5     $\hat{E}$ = `nextStep` ($\hat{E}$,i,V_i) ;
6     $s^{(t)}$ = `getCurrentState` ($\hat{E}$) ;
7     $\hat{E}$ = `nextStep` ($\hat{E}$) ;
8     $s^{(t+1)}$ = `getCurrentState` ($\hat{E}$) ;
9   **for** $j = 1, 2, \ldots, k$ **do**
10   calculate the regulatory value $r_{i,j}$ according to the equation 10
11   **end**
12 **end**

The initial time defined above is determined by the procedure in line 1 of the Algorithm 2. Basically, we used a linear regression model to determine when all variables in our artificial environment have reached stability. In other words, they have no fluctuations and move in a finite interval. When all variables exhibit this behaviour, the system is considered stable and the current time point corresponds to our initial time.

Although the environment state is calculated as usual using Eqs. 2, 8 used to determine the state of the environment at the transition time is slightly different. Indeed, in this case, the i-th component of the state must be set to a constant V_i.

$$s^{(t-1)} = \left[f_1\left(s^{(t-2)}\right), f_2\left(s^{(t-2)}\right), \ldots, f_i\left(s^{(t-2)}\right), \ldots, f_k\left(s^{(t-2)}\right)\right], \quad (7)$$

$$s^{(t)} = \left[f_1\left(s^{(t-1)}\right), f_2\left(s^{(t-1)}\right), \ldots, V_i, \ldots, f_k\left(s^{(t-1)}\right)\right], \quad (8)$$

$$s^{(t+1)} = \left[f_1\left(s^{(t)}\right), f_2\left(s^{(t)}\right), \ldots, f_i\left(s^{(t)}\right), \ldots, f_k\left(s^{(t)}\right)\right]. \quad (9)$$

By comparing the expression levels at the initial and at the final time, it is possible to determine which genes have been affected by the perturbation (Fig. 5).

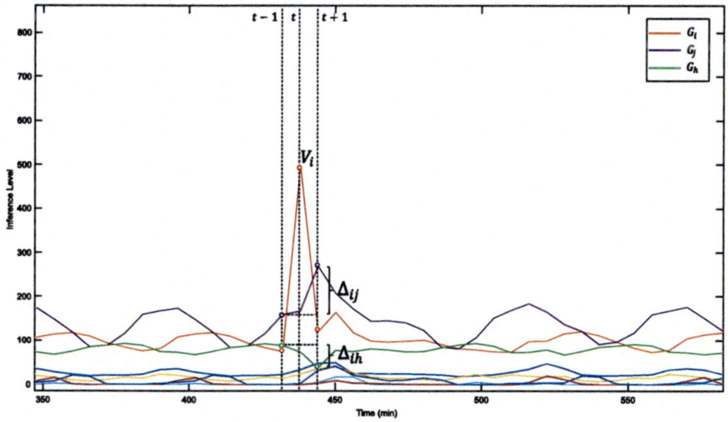

Fig. 5. Example of calculating the regulatory value of two genes G_j and G_h using G_i as their regulatory gene. $\Delta_{i,j}$ and $\Delta_{i,j}$ represent the variation between the expression level of genes at the final time and at the initial time. V_i is the value used to determine which genes are regulated by the i-th gene.

We define the *regulatory value* $r_{i,j}$ as the offset between the expression level of the gene G_j at the final time and at the initial time, taking into account a perturbation of the gene G_i at the transition time (line 10). The regulatory value

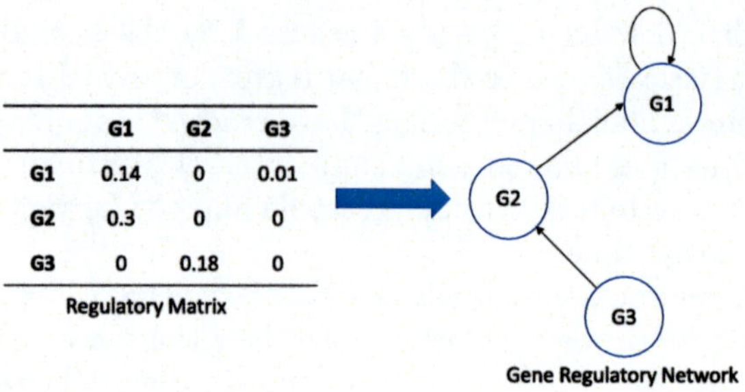

Fig. 6. Method for creating a gene regulatory network starting from the corresponding regulatory matrix with a threshold value of 0.1.

thus obtained is normalized using the maximum value M_j observed for the i-th gene during all the observation time.

$$r_{i,j} = \left|\frac{\Delta_{i,j}}{M_j}\right| = \left|\frac{s_j^{(t+1)} - s_j^{(t-1)}}{M_j}\right|. \tag{10}$$

The comparison between the expression levels of the G_j at the final and at the initial time could provide us with important information about which genes are correlated with each other. Accordingly, G_i is a gene regulator of G_j if and only if the regulatory value $r_{i,j}$ is much larger quantity of a threshold value (> 0). A *regulatory matrix* $R \in \mathbb{R}^{k\times k}$ contains all regulatory values discovered for each gene pair. A gene regulatory network can be extracted using a regulatory matrix. Given a threshold value, it is possible to determine which genes are essentially the regulators of each gene. Figure 6 shows, as an example, how a regulatory matrix with only three genes is transformed into a gene regulatory network using as threshold value of 0.1.

4 Results

In this section, we discuss the results obtained considering ten datasets from DREAM4 Challenge, available in [23]. Each dataset consists of five or ten experiments, each with 21 observations of ten or one hundred genes recorded every fifty minutes. Since each dataset contains multiple experiments, the first half of the experiments were considered as a training set, while the second half served as a validation and testing set.

The obtained results are shown in Table 2. To examine the effectiveness of our methodology, two different groups of metrics were used. The similarity index was used to measure the reliability of the artificial environment. Performance metrics (such as accuracy, precision, sensitivity and specificity) were used to compare the predicted gene regulatory network with the target network. In the field of gene regulation, the typical elements of the confusion matrix also have a biological significance, as reported in [23].

Table 2. The results table shows the metrics used to validate our methodology. $Q_{\%}$ is the similarity index, Thr. stands for threshold, Acc. for accuracy, Prec. for precision, Sens. for sensitivity, and Spec. for specificity.

Dataset name	$Q_{\%}$	**Thr.**	**TP**	**TN**	**FP**	**FN**	**Acc.**	**Prec.**	**Sens.**	**Spec.**
insilico_size10_1	0.97856	0.17	13	72	3	2	0.9444	0.8125	0.8667	0.9600
insilico_size10_2	0.9614	0.25	11	72	2	5	0.9222	0.8462	0.6875	0.9730
insilico_size10_3	0.9794	0.1	12	73	2	3	0.9444	0.8571	0.8000	0.9733
insilico_size10_4	0.9705	0.09	10	75	2	3	0.9444	0.8333	0.7692	0.9740
insilico_size10_5	0.7287	0.17	4	73	5	8	0.8556	0.4444	0.3333	0.9359
insilico_size100_1	0.8990	0.11	146	9621	103	30	0.9866	0.5863	0.8295	0.9894
insilico_size100_2	0.9361	0.08	157	9631	20	92	0.9887	0.8870	0.6305	0.9979
insilico_size100_3	0.9295	0.32	112	9675	30	83	0.9886	0.7887	0.5744	0.9969
insilico_size100_4	0.9239	0.03	118	9599	90	93	0.9815	0.5673	0.5592	0.9907
insilico_size100_5	0.9520	0.07	113	9677	30	80	0.9889	0.7902	0.5855	0.9969

- True Positive (TP) denotes the number of regulatory mechanisms correctly predicted by our approach.
- True Negative (TN) represents the number of arcs that are not present in both the predicted GRN and the target GRN.
- False Positive (FP) denotes the number of regulatory mechanisms predicted by our approach that are incorrect.
- False Negative (FN) denotes the number of regulatory mechanisms not detected by our approach.

As mentioned in the previous section, a threshold needs to be defined to generate a gene regulator from the regulatory matrix. In our experiments, we tested several threshold values ranging from 0 to 1 with a resolution of 0.01. However, due to space limitations, we have only listed the experiments with the highest accuracy and precision for each dataset in the results table.

As it can be observed, the higher the similarity index, the better the accuracy and precision of the generated gene regulatory network. Conversely, the precision decreases when the similarity index is lower, although the accuracy of the predicted gene regulatory network is still quite good. Another aspect that needs to be examined is the number of false negatives. When the number of genes increases, the number of regulatory mechanisms discovered is lower than expected. Conversely, when the number of genes is lower, the number of false negatives is acceptable and the overall sensitivity is higher than when the number of genes is higher. Similarly, false positives are the number of regulatory mechanisms added by our method but not included in the target regulatory network. The resulting gene regulatory network will therefore have more arcs than expected if false positives are high and that will affects the specificity.

5 Conclusions

In this paper, we presented a new method for genetic inference problem. Unlike other methods already present in literature such as Boolean Network [18], our technique allows us not only to perform experiments with our artificial environment, but also to determine the actual interaction between genes with a good accuracy and precision.

In fact, the ability to simulate how the expression levels of a collection of genes change over time is one of the most interesting features of our approach. However, as mentioned above, we are currently unable to demonstrate this part of the work, as all our experiments require to be validated in the laboratory.

Other interesting future work is to compare our method, which is based on an artificial environment, with other existing techniques that use instances with more than a thousand genes.

We are also aware that the threshold defined in Sect. 3.2 was not computed by any method and, therefore, this could be an obstacle. However, we are already working on a solution that consists of using a probabilistic algorithm to define the threshold directly.

Finally, we would like to thank the anonymous reviewers for their careful reading of our manuscript and their insightful comments and suggestions.

References

1. Alvarez, J.M., Brooks, M.D., Swift, J., Coruzzi, G.M.: Time-based systems biology approaches to capture and model dynamic gene regulatory networks. Ann. Rev. Plant Biol. **72**(1) (2021). https://par.nsf.gov/biblio/10231631
2. Aster, R.C., Borchers, B., Thurber, C.H.: Parameter Estimation and Inverse Problems. Elsevier (2018)
3. Avci, O., Abdeljaber, O., Kiranyaz, S., Hussein, M., Gabbouj, M., Inman, D.J.: A review of vibration-based damage detection in civil structures: from traditional methods to machine learning and deep learning applications. Mech. Syst. Sign. Process. **147**, 107077 (2021). https://www.sciencedirect.com/science/article/pii/S0888327020304635
4. Bai, Y., Chen, W., Chen, J., Guo, W.: Deep learning methods for solving linear inverse problems: Research directions and paradigms. Sign. Process. **177**, 107729 (2020). https://www.sciencedirect.com/science/article/pii/S0165168420302723
5. Chikofsky, E., Cross, J.: Reverse engineering and design recovery: a taxonomy. IEEE Softw. **7**(1), 13–17 (1990)
6. Cutello, V., Krasnogor, N., Nicosia, G., Pavone, M.: Immune algorithm versus differential evolution: A comparative case study using high dimensional function optimization. In: 8th International Conference on Adaptive and Natural Computing Algorithms (ICANNGA), Vol. LNCS 4431, pp. 93–101 (2007)
7. Cutello, V., Morelli, G., Nicosia, G., Pavone, M., Scollo, G.: On discrete models and immunological algorithms for protein structure prediction. Nat. Comput. **10**(1), 91–102 (2011). https://doi.org/10.1007/s11047-010-9196-y
8. Flamary, R.: Astronomical image reconstruction with convolutional neural networks. In: 2017 25th European Signal Processing Conference (EUSIPCO), pp. 2468–2472 (2017)

9. Groetsch, C.W., Groetsch, C.: Inverse Problems in the Mathematical Sciences, Vol. 52. Springer (1993)
10. Hecker, M., Lambeck, S., Toepfer, S., van Someren, E., Guthke, R.: Gene regulatory network inference: Data integration in dynamic models-a review. Biosystems **96**(1), 86–103 (2009). https://www.sciencedirect.com/science/article/pii/S0303264708002608
11. Huynh-Thu, V.A., Sanguinetti, G.: Gene regulatory network inference: an introductory survey. In: Gene Regulatory Networks, pp. 1–23. Springer (2019)
12. Kantarci, B., Labatut, V.: Classification of complex networks based on topological properties. In: 2013 International Conference on Cloud and Green Computing, pp. 297–304 (Sep 2013)
13. Karlebach, G., Shamir, R.: Modelling and analysis of gene regulatory networks. Nature Rev. Mol. Cell Biol. **9**(10), 770–780 (2008). https://doi.org/10.1038/nrm2503
14. Lucas, A., Iliadis, M., Molina, R., Katsaggelos, A.K.: Using deep neural networks for inverse problems in imaging: Beyond analytical methods. IEEE Sign. Process. Magaz. **35**(1), 20–36 (2018)
15. Mata, A.S.d.: Complex networks: a mini-review. Brazilian J. Phys. **50**(5), 658–672 (2020). https://doi.org/10.1007/s13538-020-00772-9
16. Pilozzi, L., Farrelly, F.A., Marcucci, G., Conti, C.: Machine learning inverse problem for topological photonics. Commun. Phys. 1(1), 57 (2018). https://doi.org/10.1038/s42005-018-0058-8
17. Rubiolo, M., Milone, D.H., Stegmayer, G.: Extreme learning machines for reverse engineering of gene regulatory networks from expression time series. Bioinformatics **34**(7), 1253–1260 (2017). https://doi.org/10.1093/bioinformatics/btx730
18. Shmulevich, I., Dougherty, E.R., Zhang, W.: Control of stationary behavior in probabilistic Boolean networks by means of structural intervention. J. Biol. Syst. **10**(04), 431–445 (2002). https://doi.org/10.1142/S0218339002000706
19. Talbi, E.G.: Metaheuristics: From Design to Implementation. Wiley Publishing (2009)
20. Vauhkonen, M., Tarvainen, T., Lähivaara, T.: Inverse Problems, pp. 207–227. Springer International Publishing, Cham (2016), https://doi.org/10.1007/978-3-319-27836-0_12
21. Yaman, F., Yakhno, V.G., Potthast, R.: A survey on inverse problems for applied sciences. Math. Prob. Eng., 976837 (2013). https://doi.org/10.1155/2013/976837
22. Yang, Y., Yang, H.: Complex network-based time series analysis. Phys. A Stat. Mech. Appl. **387**(5), 1381–1386 (2008). https://www.sciencedirect.com/science/article/pii/S0378437107011235
23. Zhao, M., He, W., Tang, J., Zou, Q., Guo, F.: A comprehensive overview and critical evaluation of gene regulatory network inference technologies. Brief. Bioinform. **22**(5) (2021). https://doi.org/10.1093/bib/bbab009
24. Zito, F., Cutello, V., Pavone, M.: Optimizing multi-variable time series forecasting using metaheuristics. In: 2022, 14th Metaheuristics International Conference (MIC), vol. LNCS (to appear), pp. 1–15 (2022)

Using the Duplication-Divergence Network Model to Predict Protein-Protein Interactions

Nicolás López-Rozo[(✉)], Jorge Finke, and Camilo Rocha

Department of Electronics and Computer Science, Pontificia Universidad Javeriana, Cali, Colombia
nicolaslopez@javerianacali.edu.co, jfinke@javerianacali.edu.co, camilo.rocha@javerianacali.edu.co

Abstract. Interactions between proteins are key to most biological processes, but thorough testing can be costly in terms of money and time. Computational approaches for predicting such interactions are an important alternative. This study presents a novel approach to this prediction using calibrated synthetic networks as input for training a decision tree ensemble model with relevant topological information. This trained model is later used for predicting interactions on the human interactome, as a case study. Results show that deterministic metrics perform better than their stochastic counterparts, although a random forest model shows a feature combination case with comparable precision results.

Keywords: Duplication-Divergence model · Protein Interaction Prediction · Edge Embeddings · Human Interactome.

1 Introduction

Interactions between proteins are key to most biological processes inside cells [12]. Protein-protein interactions (PPIs) underlie a variety of interdependent mechanisms, including signal transduction, homeostasis control, and stress responses. They play an important role in physiological and developmental processes such as protein phosphorylation, transcriptional co-factor recruitment, and transporter activation [32]. Based on the outcome of numerous Yeast-2-Hybrid testing (Y2H), networks of PPIs can be constructed [20,25]. However, identifying protein interactions through Y2H is costly both in terms of time and resources [11,15]. For example, constructing a PPI network for an organism with 2,000 proteins requires testing of about 2 million potential interactions. Moreover, it has been shown that a large amount of false negatives and positives (sometimes near 20%) are found in PPIs generated from several techniques such as Y2H, Tandem Affinity Purification Mass Spectrometry, and ChIP-Seq [14]. In that regard, computational tools for narrowing down the combinatorial search

H. Cherifi et al. (Eds.): COMPLEX NETWORKS 2016 2022, SCI 1077, pp. 322–334, 2023.
https://doi.org/10.1007/978-3-031-21127-0_27

space of such interactions offer a cost-effective alternative, especially for organisms that produce a large number of proteins. A further limitation is that many interactomes, such as the human one, lack many relations [27].

This paper explores the prediction of PPIs using a computational approach and information available from an existing PPI network (in-silico prediction). The proposed approach leverages the intrinsic topological information of the neighborhood of a protein using deterministic and non-deterministic method to train a classification model in charge of predicting the existence of interactions. Several decision-tree models are tested, in particular random forests [22]; they seem to perform better at PPI prediction. The trained model is used on a PPI constructed from experimental evidence; the precision of the model is validated against a more recent version of the PPI for the same organism.

This study applies the proposed approach to the human interactome. XGBoost and random forest classifiers are selected for the prediction task, and are trained using synthetic networks obtained from the duplication-divergence (DD) graph model [9]. The topological information from the synthetic networks is extracted with two different approaches: the first one computes higher degree neighborhood scores (closures), and the second one uses random walks and an unsupervised deep learning model (embeddings). The prediction PPI is the interactome from the Human Reference Protein Interactome Mapping Project [13] and the validation PPI is an assembly of 12 databases constructed by Gysi et al [5]. The results show that deterministic approaches achieve greater precision than stochastic ones, but also suggest that a greater spectrum of classification models and embedding techniques must be explored. Ultimately, these results highlight the need of broader exploration of in-silico approaches that, supported by biologically relevant information, elucidate the interactions of proteins in any given organism.

Outline. The remainder of the paper is organized as follows. Section 2 presents preliminary notions used throughout the paper. Section 3 presents the proposed approach, and introduces the steps required for the generation of the synthetic networks. In Section 4 a case study on the human interactome is developed. Section 5 summarizes related work and concludes the paper.

2 Preliminaries

This section presents preliminaries on closures on graphs, feature learning with node2vec, and ensemble learning.

2.1 Network Closures

Let $G = (V, E)$ be an undirected graph with $n = |V|$ proteins as *nodes* and $m = |E|$ pairwise interactions between nodes as *edges*. The graph G can be represented as a Boolean adjacency matrix A_G of size n^2: for $0 \leq i, j < n$, the entry $A_G(i, j)$ is 1 if there is a direct interaction between proteins i and j, and 0 otherwise. Since G is undirected, A_G is symmetric.

Computing the number of paths of length 2 and 3 in G can be done with $(A_G)^2$ and $(A_G)^3$, respectively. Another important metric in PPI analysis is the degree-normalized count of the paths of length 3, which is computed as in [10]:

$$L3(i,j) = \sum_{p,q \in V} \frac{A_G(i,p) \cdot A_G(p,q) \cdot A_G(q,j)}{\sqrt{k(p) \cdot k(q)}},$$

where $k(i)$ is the number of neighbors of v_i, known as its *degree*. Local-community information extracted from the induced graph of the paths of length 3 is also relevant to predict PPI networks [17]. This metric, known as $CH_2 - L_3$, maximizes internal links in local communities and minimizes external ones, according to a topology-driven mechanistic model. These four metrics will be used throughout this paper as *closures*.

2.2 Feature Learning of a Network and node2vec

The unsupervised algorithm node2vec for feature learning in networks is based on natural language processing [4]. It treats the problem of feature learning in a network as a maximum likelihood optimization problem. The function $f : V \to \mathbb{R}^d$ to be learned maps each node u to the d-tuple $f(u)$ of features of u. Therefore, f is represented as a matrix of size $|V| \times d$. The optimization problem becomes feasible when the technique assumes *conditional independence* and *spatial symmetry*. Conditional independence means that the likelihood of observing a node does not depend on any other node's observation. Spatial symmetry means that two neighbor nodes have a symmetric effect over each other in the d-dimensional feature space.

For each node $u \in V$, the expression $N(u)$ denotes the *network neighborhood* of u. This set is generated up to a fixed sampling strategy and a fixed size. The sampling strategy is a combination of depth-first (DFS) and breadth-first (BFS) sampling. Starting from vertex u, the former selects the nodes farthest in a path from u, while the latter selects the ones closest to u. Based on these two strategies, node2vec computes random walks for each node u in combination with user-provided weights for back-edges and forward edges to build $N(u)$. Finally, node2vec uses deep learning networks to learn the associated function $f : V \to \mathbb{R}^d$ of features of G from all the network neighborhood sets $N(u)$.

Since the goal of this work is link prediction, the function to be learned needs to have the edges E of G as its domain. In particular, the feature function has type $g : E \to \mathbb{R}^d$ and it is defined to obtain a d-tuple representation $g(u,v) = g(v,u)$ for each link $(u,v) \in E$ based on the d-tuples $f(u)$ and $f(v)$ computed by node2vec. Several binary operators can be considered for learning edge features with this approach. For instance, piece-wise average $\left(\frac{f_i(u)+f_i(v)}{2}\right)$, piece-wise multiplication $(f_i(u) * f_i(v))$, weighted-L1 $(|f_i(u) - f_i(v)|)$, and weighted-L2 $|f_i(u) - f_i(v)|^2$, where f_i denotes the i-th projection over $f(u)$. The piece-wise multiplication alternative, also known as Hadamard operator, is used in this study for obtaining the d-dimensional edge representation in g.

2.3 Random Forest, Gradient Boosted Trees, and XGBoost

Ensemble learning is a branch of supervised learning algorithms in which several base estimators are trained and their predictions combined in order to improve robustness over a single estimator. Two big groups are usually distinguished: *averaging methods* and *boosting methods*. Random forest is an example of averaging methods, while gradient boosted trees are representative for boosted methods.

In the random forest implementation of scikit-learn [19], each decision tree is built from a bootstrap sample from the training set (i.e. sampling with replacement). The node splitting process during the tree construction finds the best split either for a random subset of features or from all input features. On the other hand, the XGBoost library implements supervised learning models based on gradient boosted decision forests [2]. The main difference with random forest is that an added estimator tries to minimize the current emsemble error, instead of being an independent predictor.

A set $E' \supseteq E$ is fabricated for training purposes and a dataset T of size $|E'| \times d$: for each $e \in E'$, the corresponding row is $g(e)$, even if $e \notin E$. The algorithms in XGBoost are used to infer the Boolean function $l : T \rightarrow \mathbb{B}$ so that $l(g(e))$ is the supervisory signal for $e \in E'$. The labeling function l is then used to select candidate edges in $E' \setminus E$ to add to G.

The problem of learning the optimal structure of a decision tree is known to be NP-complete [7]. In practice, it is usual to define an objective function to be minimized for learning l, so that it has an internal structure approximating the optimal one. In this case, *training loss* and a *regularization term* are added for the objective function. Training loss is usually calculated as the mean squared error between prediction and observation. The regularization term is, intuitively, a measure penalizing the depth of the decision tree.

3 Training with the Duplication-Divergence Model

This section presents an overview of the duplication-divergence network model [8] and proposes an approach for using the model to predict links in protein-protein interaction networks.

3.1 The Model

From a biological perspective, it is widely accepted that a new protein appears as a copy of an existing one in the interactome [26]. This process is known as *duplication* and is the main driving mechanism of evolution in PPI networks. Random mutations also occur, leading to differences between the source and duplicated proteins; thus, a certain degree of *divergence* is expected.

The duplication-divergence (DD) model was first proposed by Kim et al. in [9] as a network model, but the variant used in this study can be attributed to Ispolatov et al [8]. The model takes as input a given (undirected) connected graph G_0 and a probability δ. For any given time $t \geq 0$, the graph G_t evolves to the graph G_{t+1} as follows:

Duplication. A node u from the graph G_t is chosen at random to create v, which is connected to all neighbors of u, so that $N(u) = N(v)$.

Divergence. Each of the edges from v to $N(u)$ is deleted with probability δ. If at least one edge remains, the replica is preserved; otherwise, the attempt is considered futile and the network does not change.

The criteria of failed attempts is biologically relevant, since it assures that the resulting graph is connected. Leaving nodes with no edges produces disconnected nodes whose survival during the evolution process is questionable. Critical biological pathways, such as degradation, are run by housekeeping proteins that are highly connected and in principle have links to all proteins in a species [8].

3.2 Parameter Estimation

It is key to define the removal probability δ to use the DD model. As addressed by several authors, the DD model preserves the most relevant topological properties of the PPI networks: average degree, power-law exponent, average clustering coefficient, average path length, and observed bipartite cliques [8,18,23]. Given that the number of nodes is already a direct parameter of the DD model, the number of edges is calibrated only with the removal probability δ.

Let $G_p = (V_p, E_p)$ be the PPI network to be predicted and $G_v = (V_v, E_v)$ the target PPI network. The idea is to generate a synthetic network $G_s = (V_s, E_s)$ by finding the parameter δ so that $G_s := \text{DD}(|V_p|, \delta)$ has a similar amount of edges (i.e., $|E_v| \approx |E_s|$). Due to the intrinsic stochastic nature of the model, the considered number of edges is the average of K models generated for the same parameters (i.e., $\overline{|E_s|}$).

Once parameter δ is estimated, a number of repetitions R is defined and then R synthetic networks are created, having in mind a maximum admissible deviation factor Δ_E from the expected number of edges $|E_v|$:

$$(1 - \Delta_E)|E_v| \leq \overline{|E_s|} \leq (1 + \Delta_E)|E_v|.$$

On average, the number of edges generated by the DD model has a monotonic behavior and is inversely proportional to the removal probability δ. However, variance among repetitions with the same parameters is considerable, so that it is impractical to find δ with a precision lower than 0.001.

3.3 The Approach

After using the DD model with the calibrated parameters to obtain the R networks G_i, $0 \leq i < R$, the following procedure is carried out for each of them. Keep in mind that this description is intended to be general. Details about the application of this approach on the human interactome are given in Section 4.

1. From the complete synthetic network G_i some edge features are extracted, deterministically (e.g., L_3) or stochastically (e.g., node2vec embeddings).

2. A binary classification model $M^{(i)}$ is trained with a subset of those features, using all existing edges as positive evidence (Class 1) and a random subset of non-existing edges as negative evidence (Class 0). To reduce bias, the number of edges of each class is the same (balanced dataset).
3. The same edge features computed for G_i are now computed for the PPI network to be predicted G_p.
4. The model $M^{(i)}$ is used on all the unlabeled edges of the PPI network G_p; that is, on $(V \times V) \setminus G_p(E)$, using the computed edge features.
5. The predicted edges are ranked in descending order by their probability of existence according to the model and the top B are selected.
6. The top B predictions are compared with the target PPI network G_v and the precision is then computed.

Figure 1 shows a visual representation of the approach. Note that for the real PPI network, node features are computed independent to the synthetic networks. The hyperdimensional representation of the edges on the real PPI network is used to feed the classification models trained on the synthetic networks.

4 Prediction of Interactions on the Human Interactome

This section presents the approach for predicting PPIs on the human interactome using synthetic networks. The goal of this case study is to exhibit the proposed approach, having in mind the predictive limitations due to the incompleteness of the current human interactome [5,27].

4.1 Data

Two versions of the human interactome are used: the interactome used for the prediction G_p corresponds to HI-union and is available at the website of the Human Reference Protein Interactome Mapping Project [13]. It consists of 9,094 proteins and 64,006 interactions. The second interactome G_v, which is used to validate the predicted interactions, corresponds to a dataset published by Gysi et al. consisting of 18,505 proteins and 327,924 interactions [5]. It is an assembly of 21 public databases compiling experimentally derived PPI data.

It is worth noting that each interactome has a different identifier type: HI-union (G_p) uses Ensembl GeneID and HI-2021 (G_v) uses EntrezID. Conversion of G_v from EntrezID to Ensembl GeneID was done using BioDBnet web services [16]. Not all proteins could be translated, so the resulting network after translation had 18,173 proteins and 321,360 interactions.

4.2 Data Pre-processing

Since prediction depends almost entirely on the PPI network being connected, the greatest connected component from G_p was used for this study. This leaves G_p with $|V_p| = 8,986$ proteins connected by $|E_p| = 63,203$ interactions. On the other hand, the greatest connected component in G_v has 321,360 edges.

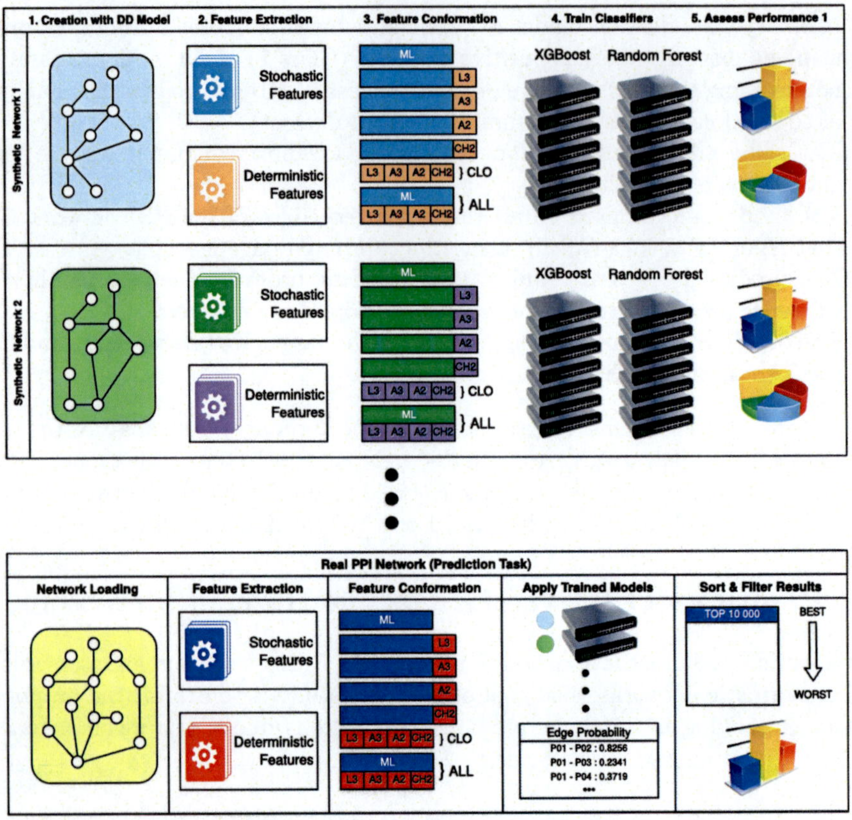

Fig. 1. Diagram of the proposed approach for PPI prediction using synthetic networks.

4.3 DD Model Calibration

The calibration process was carried out by a binary search of the parameter δ. The goal was a network with the expected amount of edges of the greatest connected component in G_v. The maximum admissible deviation factor was set to $\Delta_E = 0.1$ and the number of repetitions considered to average the number of edges was set to $K = 100$. The calibrated value is $\delta = 0.263$ and its computation took around 3 minutes.

4.4 Feature Conformations

The feature conformations used in this study are:

ML. After obtaining node2vec vector representations of the nodes, the Hadamard operator is used to obtain the edge embeddings. In total, 128 dimensions were retrieved with an unbiased search ($p = q = 1$ in node2vec) and default parameters. The implementation used in this study can be found in the karateclub library [21]. This model type is addressed as $M_{\mathrm{ML},\cdot}$.

L3. It uses the features from ML and includes the L_3 score. This model type is addressed as $M_{\mathrm{L3},\cdot}$.

A3. It includes the features from ML, as well as the raw count of paths of length 3 ($A_3 = (A_G)^3$). This model type is addressed as $M_{\mathrm{A3},\cdot}$.

A2. It includes ML features and the raw count of paths of length 2 ($A_2 = (A_G)^2$). This model type is addressed as $M_{\mathrm{A2},\cdot}$.

CH2. This conformation uses the features from ML and includes the score for the local communities of length 3 ($CH_2 - L_3$). This model type is addressed as $M_{\mathrm{CH2},\cdot}$.

CLO. It includes only the deterministic features: L_3, A_3, A_2 and $CH_2 - L_3$. This model type is addressed as $M_{\mathrm{CLO},\cdot}$.

ALL. This conformation uses all the features from ML and CLO. This model type is addressed as $M_{\mathrm{ALL},\cdot}$.

4.5 Classification Models

The classification models used in this study are XGBClassifier from the XGBoost library [2] and Random Forest Classifier from scikit-learn library [19], further addressed here as $M_{\cdot,\mathrm{XGB}}$ and $M_{\cdot,\mathrm{RF}}$, respectively.

4.6 Using the Proposed Approach

The prediction task consists of several steps, as mentioned in Section 3.3. A total of $R = 10$ DD model networks were generated using the calibrated parameters for a more robust assessment of PPI interactions. The creation of the synthetic networks took around 40 seconds. In total, 140 models $M^{(i)}{}_{F,C}$ were trained: $R = 10$ repetitions for each of the 7 feature conformations (F) and each of the classification models (C). After that, each model was used to predict the unlabeled edges of G_p and the top $B = 10,000$ unlabeled edges with the highest existence probability were selected. Finally, the evaluation metrics were computed. Each repetition of all models took around 9 hours of computing time.

The prediction results are presented in Table 1 as a 10-fold average precision for each of the 14 combinations of classification model and feature conformation. Precision values for the deterministic prediction are presented in the last column. Note that predictions overall present a low precision. The highest precision is achieved by the deterministic approach using CH_2-L_3. Out of the methods using machine learning predictors, the random forests model trained to consider only the information from the 4 closures ($M_{\mathrm{CLO,RF}}$) achieves the highest precision, which is comparable with the lowest precision from the deterministic subset of predictions. The set of features to give the lowest average prediction by using the XGBoost predictor is the combination of the 4 closure-based metrics. For the random forest classifier, combining all closure features and the embedding results in the lowest precision model.

Table 1. Precision results for top-10,000 edges for the combination of the 2 models and 7 feature conformations (average over 10 models trained on synthetic networks), as well as the deterministic prediction using each metric

Prediction:	Stochastic		Deterministic
Trained model:	XGBoost (%)	Random forest (%)	None (%)
L3	0.228	0.220	1.090
A3	0.230	0.230	0.450
A2	0.196	0.216	1.210
CH2	0.302	0.142	1.380
ML	0.223	0.214	–
ALL	0.227	0.106	–
CLO	0.115	0.481	–

5 Related Work and Concluding Remarks

Important work on modeling networks governed by duplication of nodes has been done under several approaches. Some authors find that the DD model is one of the models that best fits the distribution on some PPI datasets [24] and that it preserves the observed bipartite cliques [23]. Chung et al. analytically derive relationships associating the power-law distribution exponent β and the DD parameters using combinatorial probabilistic methods [3]. Their work involves both full and partial duplication of a node, and conclude that partial duplication (i.e., retaining only a fraction of the duplicated edges) generates networks consistent with biological networks. They also show that the power-law exponent for large graphs does not depend on the starting graph, but on the growth process itself.

Other important advances in the DD model are made by [8,9,18]. Kim et al. work on a DD model which –besides gene duplication– allows nodes to connect to the duplicated ones with a probability $\frac{r}{t}$ (process called *mutation*). A main finding is that in a mutation-dominated growth, disconnected components merge into larger ones (known as percolation), while in a duplication-dominated environment the process is not self-averaging and each outcome is itself singular [9]. Pastor-Satorras et al. find the characteristic degree distributions for such DD models, and compare the topological properties of the yeast PPI with a calibrated synthetic network in terms of average degree, power-law exponent, average clustering coefficient, and average path length [18]. Ispolatov et al. assess a model as presented in this paper, in which a duplicated node that disconnects from the network is discarded and the network does not change [8]. They analyze the properties of total duplication models, as well as highly divergent ones where the resulting networks are trees because only one edge is added to each duplicated node. They find a range of values for the DD parameter that generates a self-averaging network. To the best of the authors' knowledge, no use of synthetic networks with the DD network model has found a way for edge prediction in PPIs.

Prediction of PPIs is a highly active topic in bioinformatics. Several reviews make the effort of compiling different techniques and methods available [1,22,31]. Most techniques are based on one or several of the following approaches: to exploit the existing network connectivity [10,29], to leverage from the amino acid sequences [28], to use functional annotations [6,33], and to apply machine learning techniques to extract features or predict PPIs [30].

Two complementary approaches support this work. First, the work by Kovacs et al. elucidates the connectivity-based mechanics of PPIs by computing paths of specific distances, being paths of length 2 and 3 the most relevant [10]. Furthermore, they define the degree-normalized count of paths of length 3 (L_3), which performs consistently to predict novel PPIs on several organisms in a deterministic way. On the other hand, Xiao and Deng leverage the usage of deep learning models for the extraction of embeddings [29]. Based on that high-degree neighborhood information, they design a model for predicting PPIs using graph convolutional networks and consider CH_2-L_3 as an important deterministic approach to PPI prediction. These ideas are merged with the DD model for creating synthetic networks that mimic the structure of real PPI networks. The reason to use synthetic networks is that its connectivity is final. That is, an edge either exist or not, whereas real PPI networks mainly have certainty on the observed edges because the rest of the edges are considered unlabeled. On the other hand, decision tree classification methods usually have the best performance when predicting PPIs compared to other machine learning methods [22]. However, great care should be taken because they are prone to overfitting and sensitive to noise and correlated features given the nature of PPIs.

The results in this paper suggest that the deterministic approaches perform better for predicting PPIs. In general, metrics based on paths of length 3 (A3, L3, and CH2-L3) seem to perform better than the paths of length 2 (A2), which is consistent with [10]. Two of the most recent human PPI networks found in literature were used for the prediction (G_p) and performance evaluation (G_v), comprising 9.7% and 49.4% of the expected size of the human interactome ($\sim 650,000$ edges), respectively [27]. These datasets, in comparison with the HI-II-14 version used in other studies, may show a better picture on the real PPI prediction performance for any given model. Although G_v contains many interactions more than G_p, no total assessment can be done on the edges which are reported as non-existing, because some of those interactions might actually occur (but have not been documented). The main contribution is the proposed approach itself; to the best of the authors' knowledge, no PPI prediction task has attempted the use of synthetic networks. This seems a tractable approach from a resource- and computational-wise perspective.

As future work, the exploration of other tree-based classification models may show a more significant precision result. Namely, ensemble alternatives combining different types of models or models with different parameters may overcome the overfitting problems of random forest classifiers, as well as their sensitivity to noise. Furthermore, the exploration of datasets for other organisms might elucidate hidden biological mechanisms related to relevant closures or feature combinations.

Acknowledgments. This work was funded by the OMICAS program: Optimización Multiescala In-silico de Cultivos Agrícolas Sostenibles (Infraestructura y Validación en Arroz y Caña de Azúcar), anchored at the Pontificia Universidad Javeriana in Cali and funded within the Colombian Scientific Ecosystem by The World Bank, the Colombian Ministry of Science, Technology and Innovation, the Colombian Ministry of Education, and the Colombian Ministry of Industry and Tourism, and ICETEX, under GRANT ID: FP44842-217-2018.

References

1. Chang, J.W., Zhou, Y.Q., Ul Qamar, M.T., Chen, L.L., Ding, Y.D.: Prediction of protein-protein interactions by evidence combining methods. Int. J. Mol. Sci. **17** (2016)
2. Chen, T., Guestrin, C.: Xgboost: A scalable tree boosting system. In: Proceedings of the 22nd ACM SIGKDD International Conference on Knowledge Discovery and Data Mining, pp. 785-794. KDD '16, ACM, New York, NY, USA (2016)
3. Chung, F., Lu, L., Dewey, T.G., Galas, D.J.: Duplication models for biological networks. J. Comput. Biol. J. Comput. Mol. Cell Biol. **10**, 677–87 (2003)
4. Grover, A., Leskovec, J.: node2vec: Scalable feature learning for networks (2016). https://arxiv.org/abs/1607.00653
5. Gysi, D.M., Ítalo do Valle, Zitnik, M., Ameli, A., Gan, X., Varol, O., Ghiassian, S.D., Patten, J.J., Davey, R.A., Loscalzo, J., Barabási, A.L.: Network medicine framework for identifying drug-repurposing opportunities for COVID-19. Proc. Nat. Acad. Sci. **118**(19) (2021)
6. Halder, A.K., Bandyopadhyay, S.S., Chatterjee, P., Nasipuri, M., Plewczynski, D., Basu, S.: JUPPI: A multi-level feature based method for PPI prediction and a refined strategy for performance assessment. IEEE/ACM Trans. Comput. Biol. Bioinform. **19**, 531–542 (2022)
7. Hyafil, L., Rivest, R.L.: Constructing optimal binary decision trees is NP-complete. Inform. Process. Lett. **5**(1), 15–17 (1976)
8. Ispolatov, I., Krapivsky, P.L., Yuryev, A.: Duplication-divergence model of protein interaction network. Physical review. E-Stat. Nonlin. Soft Matt. Phys. **71**, 061911 (2005)
9. Kim, J., Krapivsky, P.L., Kahng, B., Redner, S.: Infinite-order percolation and giant fluctuations in a protein interaction network. Physical review. E-Stat. Nonlin. Soft Matt. Phys. **66**, 055101 (2002)
10. Kovács, I.A., Luck, K., Spirohn, K., Wang, Y., Pollis, C., Schlabach, S., Bian, W., Kim, D.K., Kishore, N., Hao, T., Calderwood, M.A., Vidal, M., Barabási, A.L.: Network-based prediction of protein interactions. Nat. Commun. **10**, 1240 (2019)
11. Laraia, L., McKenzie, G., Spring, D.R., Venkitaraman, A.R., Huggins, D.J.: Overcoming chemical, biological, and computational challenges in the development of inhibitors targeting protein-protein interactions. Chem. Biol. **22**, 689–703 (2015)
12. Lin, J.S., Lai, E.M.: Protein-protein interactions: Co-immunoprecipitation. In: Journet, L., Cascales, E. (eds.) Bacterial Protein Secretion Systems: Methods and Protocols, pp. 211–219. Springer, New York, New York, NY (2017)

13. Luck, K., Kim, D.K., Lambourne, L., Spirohn, K., Begg, B.E., Bian, W., Brignall, R., Cafarelli, T., Campos-Laborie, F.J., Charloteaux, B., Choi, D., Coté, A.G., Daley, M., Deimling, S., Desbuleux, A., Dricot, A., Gebbia, M., Hardy, M.F., Kishore, N., Knapp, J.J., Kovács, I.A., Lemmens, I., Mee, M.W., Mellor, J.C., Pollis, C., Pons, C., Richardson, A.D., Schlabach, S., Teeking, B., Yadav, A., Babor, M., Balcha, D., Basha, O., Bowman-Colin, C., Chin, S.F., Choi, S.G., Colabella, C., Coppin, G., D'Amata, C., De Ridder, D., De Rouck, S., Duran-Frigola, M., Ennajdaoui, H., Goebels, F., Goehring, L., Gopal, A., Haddad, G., Hatchi, E., Helmy, M., Jacob, Y., Kassa, Y., Landini, S., Li, R., van Lieshout, N., MacWilliams, A., Markey, D., Paulson, J.N., Rangarajan, S., Rasla, J., Rayhan, A., Rolland, T., San-Miguel, A., Shen, Y., Sheykhkarimli, D., Sheynkman, G.M., Simonovsky, E., Taşan, M., Tejeda, A., Tropepe, V., Twizere, J.C., Wang, Y., Weatheritt, R.J., Weile, J., Xia, Y., Yang, X., Yeger-Lotem, E., Zhong, Q., Aloy, P., Bader, G.D., De Las Rivas, J., Gaudet, S., Hao, T., Rak, J., Tavernier, J., Hill, D.E., Vidal, M., Roth, F.P., Calderwood, M.A.: A reference map of the human binary protein interactome. Nature **580**(7803), 402–408 (2020)
14. Ma, C.Y., Liao, C.S.: A review of protein-protein interaction network alignment: From pathway comparison to global alignment. Comput. Struct. Biotechnol. J. **18**, 2647–2656 (2020)
15. Macalino, S.J.Y., Basith, S., Clavio, N.A.B., Chang, H., Kang, S., Choi, S.: Evolution of In Silico Strategies for Protein-Protein Interaction Drug Discovery. Molecules. Basel, Switzerland, pp. 23 (2018)
16. Mudunuri, U., Che, A., Yi, M., Stephens, R.M.: biodbnet: The biological database network. Bioinformatics **25**, 555–6 (2009)
17. Muscoloni, A., Abdelhamid, I., Cannistraci, C.V.: Local-community network automata modelling based on length-three-paths for prediction of complex network structures in protein interactomes, food webs and more. BioRxiv (2018)
18. Pastor-Satorras, R., Smith, E., Solé, R.V.: Evolving protein interaction networks through gene duplication. J. Theor. Biol. **222**, 199–210 (2003)
19. Pedregosa, F., Varoquaux, G., Gramfort, A., Michel, V., Thirion, B., Grisel, O., Blondel, M., Prettenhofer, P., Weiss, R., Dubourg, V., Vanderplas, J., Passos, A., Cournapeau, D., Brucher, M., Perrot, M., Duchesnay, E.: Scikit-learn: machine learning in python. J. Mach. Learn. Res. **12**, 2825–2830 (2011)
20. Rajagopala, S.V.: Mapping the protein-protein interactome networks using yeast two-hybrid screens. Adv. Experiment. Med. Biol. **883**, 187–214 (2015)
21. Rozemberczki, B., Kiss, O., Sarkar, R.: Karate Club: An API oriented open-source python framework for unsupervised learning on graphs. In: Proceedings of the 29th ACM International Conference on Information and Knowledge Management (CIKM '20), pp. 3125-3132. ACM (2020)
22. Sarkar, D., Saha, S.: Machine-learning techniques for the prediction of protein-protein interactions. J. Biosci. **44** (2019)
23. Schweiger, R., Linial, M., Linial, N.: Generative probabilistic models for protein-protein interaction networks-the biclique perspective. Bioinformatics **27**, i142-8 (2011)
24. Shao, M., Yang, Y., Guan, J., Zhou, S.: Choosing appropriate models for protein-protein interaction networks: a comparison study. Brief. Bioinform. **15**, 823–38 (2014)
25. Shokri, L., Inukai, S., Hafner, A., Weinand, K., Hens, K., Vedenko, A., Gisselbrecht, S.S., Dainese, R., Bischof, J., Furger, E., Feuz, J.D., Basler, K., Deplancke, B., Bulyk, M.L.: A comprehensive drosophila melanogaster transcription factor interactome. Cell Rep. **27**, 955-970.e7 (2019)

26. Sreedharan, J.K., Turowski, K., Szpankowski, W.: Revisiting parameter estimation in biological networks: Influence of symmetries. IEEE/ACM Trans. Comput. Biol. Bioinform. **18**, 836–849 (2021)
27. Stumpf, M.P.H., Thorne, T., de Silva, E., Stewart, R., An, H.J., Lappe, M., Wiuf, C.: Estimating the size of the human interactome. Proc. Nat. Acad. Sci. U.S.A. **105**, 6959–64 (2008)
28. Sun, T., Zhou, B., Lai, L., Pei, J.: Sequence-based prediction of protein protein interaction using a deep-learning algorithm. BMC Bioinform. **18**, 277 (2017)
29. Xiao, Z., Deng, Y.: Graph embedding-based novel protein interaction prediction via higher-order graph convolutional network. PLOS ONE **15**, e0238915 (2020)
30. Yao, Y., Du, X., Diao, Y., Zhu, H.: An integration of deep learning with feature embedding for protein-protein interaction prediction. Peer J. **7**, e7126 (2019)
31. Zahiri, J., Emamjomeh, A., Bagheri, S., Ivazeh, A., Mahdevar, G., Sepasi Tehrani, H., Mirzaie, M., Fakheri, B.A., Mohammad-Noori, M.: Protein complex prediction: a survey. Genomics **112**, 174–183 (2020)
32. Zhang, Y., Gao, P., Yuan, J.: Plant protein-protein interaction network and interactome. Curr. Genom. **11**(1), 40–46 (2010)
33. Zhong, X., Rajapakse, J.C.: Graph embeddings on gene ontology annotations for protein-protein interaction prediction. BMC Bioinform. **21**, 560 (2020)

Machine Learning and Networks

SignedS2V: Structural Embedding Method for Signed Networks

Shu Liu(✉), Fujio Toriumi, Xin Zeng, Mao Nishiguchi, and Kenta Nakai

The University of Tokyo, 7-3-1 Hongo, Bunkyo-ku, Tokyo 113-8654, Japan
liu@torilab.net, tori@sys.t.u-tokyo.ac.jp, xzeng@g.ecc.u-tokyo.ac.jp, nishiguchi@torilab.net, knakai@ims.u-tokyo.ac.jp

Abstract. A signed network is widely observed and constructed from the real world and is superior for containing rich information about the signs of edges. Several embedding methods have been proposed for signed networks. Current methods mainly focus on proximity similarity and the fulfillment of social psychological theories. However, no signed network embedding method has focused on structural similarity. Therefore, in this research, we propose a novel notion of degree in signed networks and a distance function to measure the similarity between two complex degrees and a node-embedding method based on structural similarity. Experiments on five network topologies, an inverted karate club network, and three real networks demonstrate that our proposed method embeds nodes with similar structural features close together and shows the superiority of a link sign prediction task from embeddings compared with the state-of-the-art methods.

Keywords: Complex network · Signed network · Node embedding · Role-based embedding

1 Introduction

In real world networks, the relationship between two nodes might be heterogeneous. For instance, users can express their opinions on the content on news websites or video-sharing services by pressing the good or bad button or rate the products sold on EC sites from 1 to 5 stars based on their level of satisfaction [1]. In these cases, the constructed networks include multiple types of edges, binarily expressing both positive and negative edges. Such networks are classified as signed networks and contain richer information than standard (unsigned) networks.

Network embedding is a crucial field that embeds similar nodes close together in low-dimension vector space [2,3]. It is located in the center of data mining on networks, which allows vector-based machine learning methods to be applied to a network data structure. Several embedding methods have been proposed for signed networks [4–7]. The current methods generally focus on proximity similarity (neighbors connected by positive edges should be placed nearer each

H. Cherifi et al. (Eds.): COMPLEX NETWORKS 2016 2022, SCI 1077, pp. 337–349, 2023.
https://doi.org/10.1007/978-3-031-21127-0_28

other than those by negative edges) [7] and fulfilling social psychological theories (balance theory and status theory) [5,6].

On the other hand, the role-based embedding approach focuses on structural similarity [8] and embeds nodes together with similar connection patterns in vector space, even though they might be far away on a network [3]. However, as an important network embedding approach, no role-based embedding method has been proposed for signed networks to the best of our knowledge.

This research addresses this deficiency by first expressing a node's degree for signed networks and evaluating the structural similarity between nodes in signed networks. Next, we propose a role-based node-embedding method for signed networks. We conducted extensive experiments on five network topologies, an inverted karate club network, and three real networks. The results show that the proposed method embeds nodes with similar structural features in close proximity to each other, and that the structural embedding learned by the proposed method has the ability of the link sign prediction task.

The following are the main contributions of this paper. We propose a complex degree, which is a novel notion of degree in signed networks. We propose Exponential Biased Euclidean Distance (EBED) to measure the similarity between two complex degrees. Inspired by struc2vec [9], we propose SignedS2V, a role-based embedding method for signed networks that maps the node degrees to complex degrees and calculates the distance by EBED. We confirmed that the embeddings learned by SignedS2V are interpretable and can predict the signs of edges.

The rest of our paper consists of related works on node embedding and signed network embedding (Sect. 2), details about the proposed complex degree, Exponential Biased Euclidean Distance (EBED), and SignedS2V (Sect. 3), evaluation experiments on toy networks and real networks (Sect. 4), and conclusion (Sect. 5).

2 Related Work

2.1 Node Embedding

Embedding nodes into a low-dimension vector space is vital in network science, which transfers graph data to vector data and expands the possibilities for such downstream tasks as link prediction [10,11], node and link classification [12,13], anomaly detection [14], etc. Formally, given network $G = (V, E)$, node embedding's goal is to learn function $f : V \rightarrow \mathbb{R}^{|V| \times k}$ where $k \ll |V|$. The learned function clusters similar nodes closely together in embedded spaces. Node similarity can be divided into two major types: proximity similarity (community-based similarity) and structural similarity (role-based similarity) [15]. The former embeds nodes near neighbors or nodes in the same community. DeepWalk [16] is one example of a community-based embedding method that applies random walk to generate sequences of nodes and employs the Skip-Gram model to create node embeddings. node2vec [17] further extends DeepWalk by leveraging nodes' homophily and structural equivalence. LINE [18] uses first- and second-order neighbors to learn the representation vectors.

In contrast, the latter type is based on structural similarity, which embeds nodes close together if they share similar connection patterns (such as degrees and neighbors' degrees), ignoring the distance of the nodes in the network. struc2vec [9] calculates the similarities of nodes from the distribution of the neighbor's degree in multiple scales, applies biased random walk on multilayer graphs, and generates embeddings by the Skip-Gram approach. GraphWave [19] treats a node's local network neighborhood as a probability distribution and uses the empirical characteristic functions of the distribution to compose the structural embeddings of nodes.

Compared with community-based embedding, role-based embedding hasn't been well-researched yet. Our research focuses on the role-based embedding field because of its latent potential.

2.2 Signed Network Embedding

In many networks, the relationship between nodes is not homogeneous. If the type of edges in one network is binary, the edges are always represented as either positive or negative signs [1]. Such a system is called a signed network, which has superior expressiveness for richer information on edges in contrast to customary (unsigned) networks. The edges are sometimes labeled friend/trustworthy (positive) or foe/untrustworthy (negative).

Given signed network $G = (V, E^+, E^-)$, V denotes a set of nodes, and E^+ and E^- are sets of positive and negative edges. As in unsigned network embedding, the goal of signed network embedding is to learn function $f : V \to \mathbb{R}^{|V| \times k}$ to preserve the edges' sign information as well as the above-mentioned proximity and/or structural features.

Several embedding methods for signed networks have been proposed. SNE [20] adopted the log-bilinear model with random walk and used two signed-type vectors to capture the positive or negative relationship of each edge along the walking path. SiNE [7] is a deep learning framework that optimizes an objective function navigated by the assumption that neighbors connected by positive edges should be placed nearer than those by negative edges. SIDE [5] is a deep learning framework that formulates and optimizes likelihood over direct and indirect signed connections. BESIDE [6], another deep learning framework, takes "bridge" edges in a complementary manner and derives embedding from components by incorporating balance theory and status theory.

To the best of our knowledge, all of these schemes are community-based methods. Therefore, a role-based embedding method for signed networks is needed.

3 Proposed Method

3.1 Previous Method (struc2vec [9]) and Its Insufficiency for Signed Networks

struc2vec [9] is a proven, role-based embedding method that places structurally similar nodes together in an embedding space independent of their location in the network. Intuitively, struc2vec assumes that two nodes are structurally similar if they have a similar degree and a similar degree distribution in the neighbor's degree. The following four main processes are applied in struc2vec to create structural embeddings:

1. Calculate the structural similarity between each node pair in the network on multiple scales ranging from the degree of nodes to neighbors.
2. Construct a weighted multilayer graph where the weight in each layer represents the similarity in one scale.
3. Generate a context for each node by applying a biased random walk on a multilayer graph.
4. Learn the latent representations from the nodes' context by Skip-Gram [21].

In this research, we extend struc2vec to create the structural representations for signed networks. Since signed networks consist of two types of edges, positive and negative, the nodes in them have two degrees, positive degree d^+ and negative degree d^-, representing the numbers of neighbors connected through positive and negative edges. When considering the structural differences between two nodes, d^+ and d^- are independent, which means that the two types of degrees cannot be operated directly by four arithmetic operations, such as $d^+ + d^-$ or $d^+ - d^-$. Although $d^+ + d^-$ seems logical for indicating the number of interacting neighbors without considering the interaction type, operation $d^+ + d^-$ is equivalent to projecting a signed network to a normal network, which is not the purpose of this research.

Therefore, we must (1) deal with the relationship between d^+ and d^- and (2) accurately judge the distance between two nodes.

3.2 Degree in Complex Plane

To tackle the independence between d^+ and d^-, we apply complex numbers to propose a novel notion of degree d^C, called the d^C complex degree. For a node with d^+ (d^-) neighbors through positive (negative) edges,

$$d^C = d^+ + id^-, \tag{1}$$

where i is an imaginary unit. Each node in the signed networks can be replaced as a point on the complex plane. Since both d^+ and d^- are non-negative numbers, the nodes with both positive and negative neighbors are projected in the first quadrant, and those with only positive (or negative) neighbors are placed on the real (or imaginary) axis.

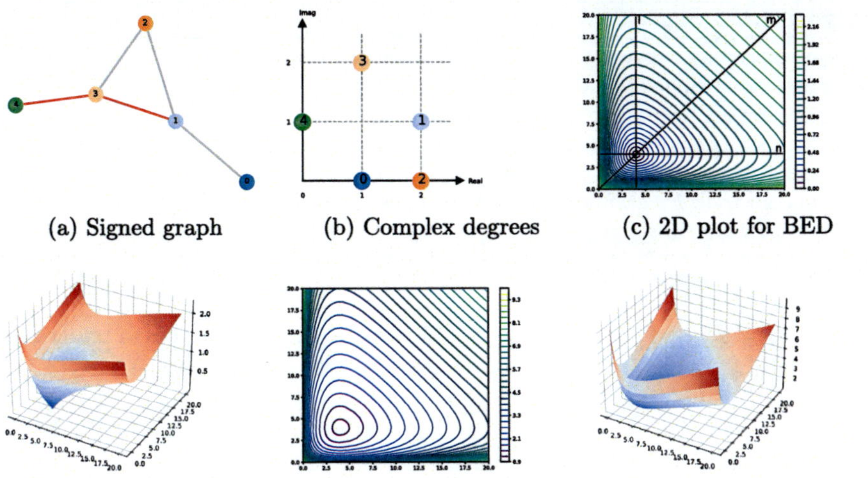

(a) Signed graph (b) Complex degrees (c) 2D plot for BED

(d) 3D plot for BED (e) 2D plot for EBED (f) 3D plot for EBED

Fig. 1. Example of signed network and its complex degree, 2D/3D plots for BED and EBED

Figures 1a and 1b show an example of a signed network and its nodes in a complex plane. The signed network in Fig. 1a contains five nodes and five edges (three positive gray edges and two negative red edges). These nodes can be placed in a complex plane in Fig. 1b according to the coordinates formed from their positive and negative degrees.

3.3 Exponential Biased Euclidean Distance (EBED)

After projecting the nodes on the complex plane, the next question is how to evaluate the structural similarity between two nodes. In this paper, we propose a novel distance, Biased Euclidean Distance (BED), to calculate the distance between two nodes on a complex plane. When comparing two pairs of nodes, complex degrees $(1 + i1)$ and $(2 + i2)$ should be more different than $(101 + i101)$ and $(102 + i102)$, although the Euclidean distances are both $\sqrt{2}$. Considering the power-law distribution nature of scale-free networks [22], we use logarithmic function $d' = \log d + 1$ to add a bias term to the complex degree: $d^{C\prime} = (\log d^{+} + 1) + i(\log d^{-} + 1)$. Then we use the Euclidean distance metric to calculate the distance between $d_A^{C\prime}$ and $d_B^{C\prime}$ for nodes A and B. BED is shown as follows:

$$BED(d_A^C, d_B^C) = \sqrt{\left(\log(d_A^+ + 1) - \log(d_B^+ + 1)\right)^2 + \left(\log(d_A^- + 1) - \log(d_B^- + 1)\right)^2}$$
$$= \sqrt{\left(\log \frac{d_A^+ + 1}{d_B^+ + 1}\right)^2 + \left(\log \frac{d_A^- + 1}{d_B^- + 1}\right)^2}, \quad (2)$$

Algorithm 1 SignedS2V

Require: signed network $G = (V, E^+, E^-)$, embedding dimension D
Ensure: node embedding $\theta_E = \mathbb{R}^{|V| \times D}$
1: get complex degrees $d_v^C = (d_v^+ + id_v^-)|v \in V$
2: calculate the multi-scales distances between node pairs by Formula 3 (or 4)
3: construct multi-layers graph M using multi-scales distances
4: random walk for each node to gain walk sequence set s
5: $\theta_E \leftarrow SkipGram(s,\ k)$

where $d_A^C = d_A^+ + id_A^-$ and $d_B^C = d_B^+ + id_B^-$ are the complex degrees of nodes A and B. $BED(d_A^C, d_B^C)$ is the Euclidean distance between $\log(d_A^+ + 1) + i\log(d_A^- + 1)$ and $\log(d_B^+ + 1) + i\log(d_B^- + 1)$. For an intuitive understanding of BED, we plot an example in Figs. 1c and 1d. The 2-dimensional contour map of the BED to the center node $(4+i4)$ is shown on the left side of Fig. 1c, and the x- and y-axes are the real and imaginary parts of d^C. Since the nodes alone horizontal line n (vertical line l) pass through the center node, the nodes on the left (bottom) side are denser than those on the right (top) side, indicating that BED is more sensitive to more minor degrees. Diagonal line m represents nodes with the same ratio between positive and negative degrees as a center node, and the contour lines are concave toward the center node. This indicates that among the nodes with a fixed number of edges, the node on line m gets the most miniature BED. The 3-dimensional contour plot is on the right side of Fig. 1d, where the z-axis shows the BED. There is a sharp point around the center node, which indicates that BED is extremely sensitive to a small difference around the center node. Since this phenomenon might complicate evaluation of the distance between two pairs of nodes, we further propose a derivation of BED, Exponential Biased Euclidean Distance (EBED):

$$EBED(d_A^C, d_B^C) = \exp\big(BED(d_A^C, d_B^C)\big) \tag{3}$$

Figures 1e and 1f show the 2- and 3-dimension plots. The distance increases exponentially as the node moves farther away from the center node in each direction.

We replace the distance function used in struc2vec with Formula 3 to extend struc2vec as a new method to gather structure-preserved embeddings for the nodes in the signed network.

Note that a logarithmic function term might not be necessary for simple networks whose degree distribution does not vary significantly or shows a transparent power-law distribution, such as in network topologies. For such networks, Exponential Euclidean Distance (EED) has potential:

$$EED(d_A^C, d_B^C) = \exp\left(\sqrt{(d_A^+ - d_B^+)^2 + (d_A^- - d_B^-)^2}\right). \tag{4}$$

3.4 Overall Algorithm of SignedS2V

The overall algorithm is shown in Algorithm 1. The input is embedding dimension D and signed undirected network $G = (V, E^+, E^-)$, where V is the node set and E^+ and E^- are the positive and negative edge sets. The output is $\theta_E = \mathbb{R}^{|V| \times D}$, where $|V|$ is the size of node set V. First, we create a complex degree of each node by Formula 1, $d_v^C = d_v^+ + id_v^-$, where $d_v^+ = |\{(u,v)|(u,v) \in E^+\}|$, $d_v^- = |\{(u,v)|(u,v) \in E^-\}|$. Then we calculate all the node pairs' distances $f_{(v,u)}^k, v \in V, u \in V$ from the k-hop neighbors' degree sequences by Formula 3 for scale-free networks (or Formula 4 for non-scale-free networks), where k ranges from 1 to k^* (the graph's diameter). Next we constructed k^*-layer graph M, where all nodes $v \in V$ exist in each layer, and the weight of node pair (v, u) in kth layer is derived from $f_{(v,u)}^k$. Next we generated context sequences of the nodes by applying biased random walk to graph M for each node. When the walker steps at node v on layer k, the probability of moving inside the current layer is q, and the probability of stepping from v to u follows the weight in M. $1 - q$ is the probability of changing the layer. Finally, we apply Skip-Gram to generate nodes' embedding $\theta_E = \mathbb{R}^{|V| \times D}$ from their context sequences.

4 Experiments

4.1 Network Topologies

Dataset As shown in Fig. 2, we created five simple signed networks from network topologies. Mesh (Fig. 2a) is a complete graph with ten nodes; star (Fig. 2d) is a star graph with ten peripheral nodes and one center node; tree (Fig. 2g) is a quadtree with a height of 2; ring (Fig. 2j) is a cycle graph with 20 nodes; path (Fig. 2m) is a graph with 20 linearly connected nodes. In each network, two edges with identical equivalence are assigned as negative edges (red edges), and the rest are positive (gray edges). Since all the nodes are colored to match their structure, nodes with identical color have equivalent structure.

Comparison methods In this research, we chose two embedding methods designed for signed networks (SIDE and BESIDE) and two role-based embedding methods for undirected unsigned networks (struc2vec and GraphWave). Note that in BESIDE, there is one component specified for predicting the sign of the edges in triangles; thus, the embedding learned by BESIDE directly contains the edges' sign information.

Experiment setting and hyper-parameters We embedded the networks into vector space with two dimensions and directly plotted the result on scatter plots. Each node is colored the same as the network figures , so the closer that nodes with the same color are placed, the better performance. For SignedS2V, we applied the EED in Formula 4 as the distance function for the simplicity of the degree distribution of those network topologies. We set the number of walks

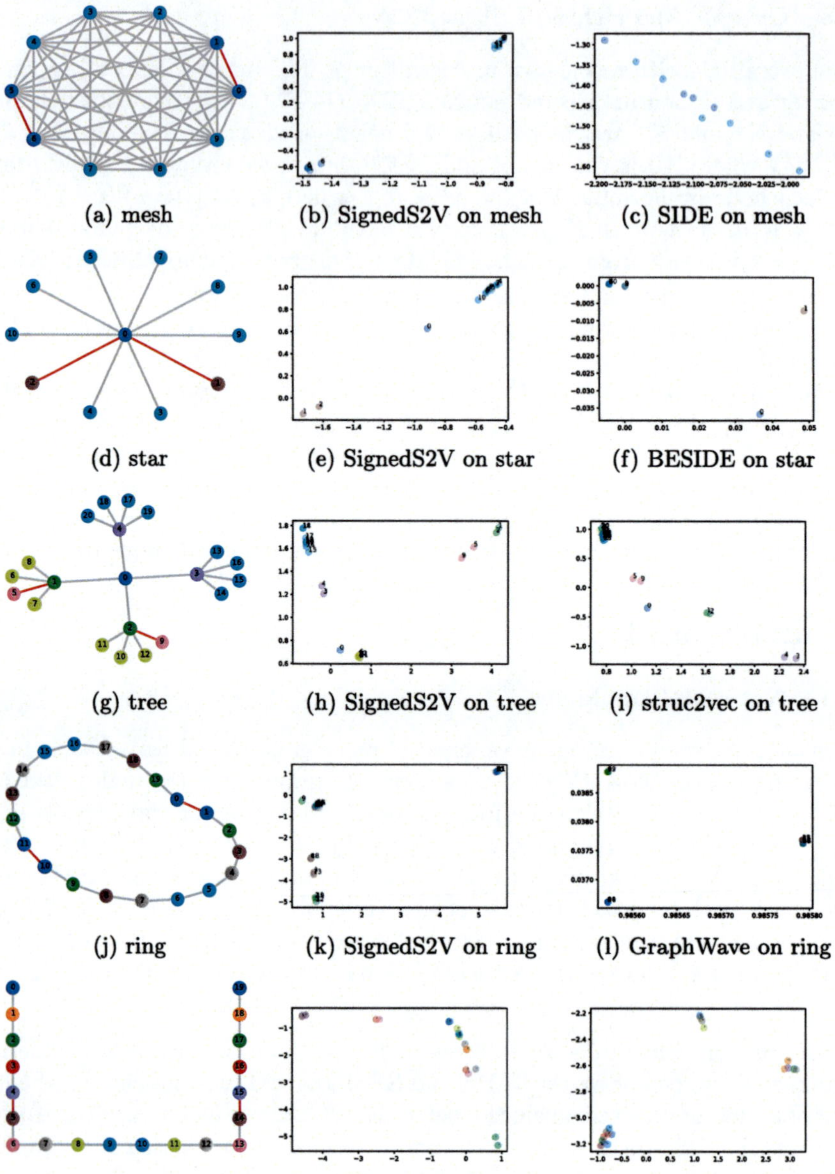

(a) mesh (b) SignedS2V on mesh (c) SIDE on mesh

(d) star (e) SignedS2V on star (f) BESIDE on star

(g) tree (h) SignedS2V on tree (i) struc2vec on tree

(j) ring (k) SignedS2V on ring (l) GraphWave on ring

(m) path (n) SignedS2V on path (o) SIDE on path

Fig. 2. Visualizations of embedding for network topologies

for each node to 20, 80 steps in each walk, and a window size of five for Skip-Gram. For SIDE, BESIDE, struc2vec, and GraphWave, we used the default hyper-parameter settings. Since the networks are all undirected, we used the undirected flag as a hyper-parameter for SIDE. As input networks for BESIDE, we created bidirectional networks where two bidirectional edges represent each edge in the original networks. To apply struc2vec and GraphWave for the signed networks, we split input signed network $G = (V, E^+, E^-)$ into two sub-networks, $G^+ = (V, E^+)$ and $G^- = (V, E^-)$, by the edges' signs, embedded G^+ and G^- as unsigned networks to θ_E^+ and θ_E^-, and finally created the embedding vectors for nodes by taking the element-wise mean from θ_E^+ and θ_E^- if one node exists in both embeddings.

Result The embedding plots are shown in Fig. 2. We show the SignedS2V result in the second column and randomly show the result for the comparison methods in the third column due to space limitations. SignedS2V captured the structural features from both the topologies' connection patterns and the edges' signs. In most cases, the nodes were placed together with equivalent structure and apart with different structural features. We even observed that slightly different nodes are placed nearer than those with a significant difference. For instance, nodes (1,2) and (5,9) in Fig. 2h were placed close to each other but far away from others because those four nodes have negative edges. On the other hand, the comparison methods failed to embed nodes at an interpretable level. SIDE embedded nodes with different colors together in Figs. 2c and 2o; BESIDE placed nodes 1 and 2 apart, but nodes 2 and 9 overlap in Fig. 2f. struc2vec failed to distinguish between nodes (7–12) and (13–20) in Fig. 2i since it is not applicable for signed networks. GraphWave had a good performance in Fig. 2l under this experiment's settings.

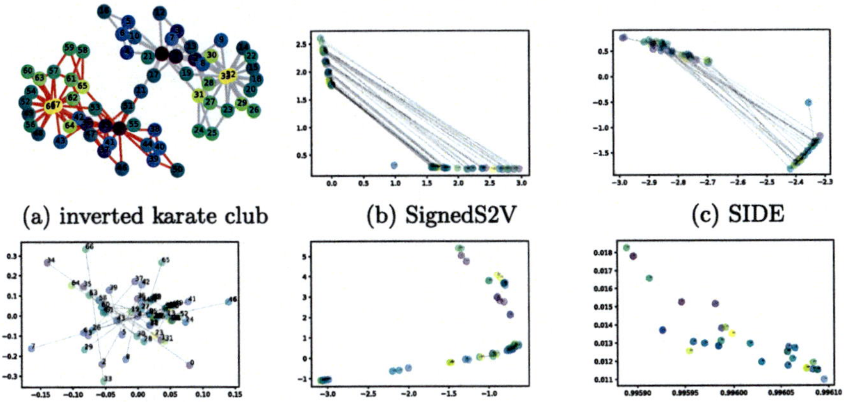

(a) inverted karate club (b) SignedS2V (c) SIDE

(d) BESIDE (e) struc2vec (f) GraphWave

Fig. 3. Visualizations of embedding for inverted karate club network

4.2 Inverted Karate Club Network

Dataset Zachary's Karate Club graph is an unsigned network with 34 nodes and 78 edges. To evaluate the embedding performance for both the positive and negative edges, we created an inverted karate club network from two Zachary's Karate Club graphs (Fig. 3a). One (node 0 to 33) has only 78 positive edges, and the other (nodes 34 to 67) has only 78 negative edges; the two graphs are connected through randomly selected node 11. We colored nodes the same color if two nodes are derived from the same original node in Zachary's Karate Club graph.

Experiment setting and hyper-parameters Since the inverted karate club network is a derivation of a social network that has a scale-free feature, we applied EBED in Formula 3 to calculate the distances between the nodes' degrees. The other settings were the same as in our previous experiment.

Result The embedding plots are shown in Fig. 3. A straight line links node pairs of the same color for a better visualization to check the relative location for the nodes with similar structures but completely different edge signs, such as nodes 0 and 34 and nodes 33 and 67.

SignedS2V embedded a clear result in Fig. 3b. The nodes formed three clusters: center node 11 in the middle, nodes with positive edges (nodes 0 to 33) at the right-bottom side, and nodes with negative edges (nodes 34 to 67) at the left-top side. Note that all 33 lines that linked the node pairs with identical structural features from two clusters were almost parallel. This result indicates that SignedS2V preserved both the structural features and the edge signs. On the other hand, the comparison methods failed to learn a sense-making embedding. In Fig. 3c, SIDE formed an embedding of two major clusters, and the lines are found both inside and outside the clusters; node 11 is not placed in a particular position. BESIDE failed to preserve either the structural features or the edges' signs in Fig. 3d. sturc2vec and GraphWave embedded nodes with similar structural features together, ignoring the edge signs.

4.3 Real Networks for Link Sign Prediction

Dataset We conducted a link sign prediction task with three real networks: slashdot, epinions, and wiki-RfA. Slashdot is a user relationship network on a news website and its community; its nodes denote users, and its edges are the relationship, either friend or foe. Epinions is an online social network on a consumer review site; its nodes are members, and its edges denote either a trustworthy or untrustworthy relationship. Wiki-RfA is a network that summarizes the voting result of the 'request for adminship (RfA)'; its nodes are Wikipedia editors, and its edges are the ballots that denote support, neutral, or opposing votes. The statistics are summarized in Table 1.

Table 1. Statistics of real networks

Dataset	# Node	# Edge	Positive edge (%)
slashdot	22,572	95,493	88.36
epinions	35,522	257,048	97.11
wiki-RfA	1,785	12,328	89.16

Experiment setting and hyper-parameters We embedded the real networks into a vector space with 64 dimensions and concatenated the two embedding vectors of the nodes on both ends as an edge-embedding vector with 128 dimensions. We applied EBED in Formula 3 for these real networks. The other hyper-parameters for each embedding method remained the same as in our previous experiment. We used the edge-embedding vectors as input to predict the edge signs. We sampled 80% of the edges as the training dataset to train a simple logistic regression model and a random forest classifier. The remaining 20% were used as the test dataset to evaluate the trained models' AUC scores. We set the max depth to 3, the number of trees to 100, and the Gini function as the quality measuring for the random forest classifier. The penalization norm of the logistic regression was l2; the other settings were default values.

Table 2. AUC results

Classifier	Dataset	SignedS2V	SIDE	BESIDE	struc2vec	GraphWave
Logistic regression	slashdot	**0.979**	0.612	*0.985*	0.730	0.652
	epinions	**0.973**	0.667	*0.976*	0.644	0.692
	wiki-RfA	**0.914**	0.585	*0.922*	0.743	0.663
Random forest	slashdot	**0.949**	0.640	*0.985*	0.769	0.673
	epinions	**0.906**	0.696	*0.975*	0.826	0.652
	wiki-RfA	**0.868**	0.609	*0.922*	0.765	0.703

Result Here we report the AUC scores as the results of the link sign prediction task in Table 2. As mentioned in Sect. 4.1, BESIDE is a deep learning based embedding method that consists of a specified component to predict the sign of the edges to generate embeddings. As such, the nodes' embeddings have already been learned to minimize the link sign prediction loss. Therefore, we used BESIDE as a benchmark against which to compare the performances of the other methods. We showed the BESIDE results in *italics* and marked the highest scores in **bold** in the rest of the methods.

SignedS2V achieved the best performance in all the datasets, surpassing the other methods by 0.1–0.3 in AUC scores and fairly close to BESIDE. SignedS2V has the capacity to generate edge signs that are related to embedding for the nodes, even though they are not directly forwarded to the embedding process, just through the positive and negative degrees. Therefore, as a role-based embedding method, SignedS2V captures the structural features as well as the features of the signs of the edges in the learned embeddings.

5 Conclusion

We proposed complex degree, which is a novel notion of degree in signed networks, and Exponential Biased Euclidean Distance (EBED) to measure the distance between two complex degrees. Following the struc2vec framework, we proposed SignedS2V, a role-based embedding method for signed networks by mapping the degrees of nodes to complex degrees and calculating the distance by EBED. We experimentally evaluated five network topologies, an inverted karate club network, and three real networks to demonstrate SignedS2V's outstanding performance on distinguished visualization and high accuracy on link sign prediction tasks.

Future tasks will improve the scalability of SignedS2V by simplifying the distance calculation process. Also, we might extend SignedS2V's capability to embed directed signed networks because of its considerable needs.

Acknowledgment. This work was supported by JST SPRING, Grant Number JPMJSP2108.

References

1. Leskovec, J., Huttenlocher, D., Kleinberg, J.: Signed networks in social media. In Proceedings of the SIGCHI conference on human factors in computing systems, pp. 1361–1370 (2010)
2. Zhou, J., Liu, L., Wei, W., Fan, J.: Network representation learning: from preprocessing, feature extraction to node embedding. ACM Comput. Surv. (CSUR) **55**(2), 1–35 (2022)
3. Ahmed, N., Rossi, R.A., Lee, J., Willke, T., Zhou, R., Kong, X., Eldardiry, H.: Role-based graph embeddings. IEEE Trans. Knowl. Data Eng. (2020)
4. Javari, A., Derr, T., Esmailian, P., Tang, J., Chen-Chuan, K., Rose, C.: Role-based signed network embedding. In: Proceedings of The Web Conference 2020, 2782–2788 (2020)
5. Kim, J., Park, H., Lee, J.-E., Kang, U.: Side: representation learning in signed directed networks. In: Proceedings of the 2018 World Wide Web Conference, pp. 509–518 (2018)
6. Chen, Y., Qian, T., Liu, H., Sun, K.: "Bridge" enhanced signed directed network embedding. In: Proceedings of the 27th ACM International Conference on Information and Knowledge Management, pp. 773–782 (2018)
7. Wang, S., Tang, J., Aggarwal, C., Chang, Y., Liu, H.: Signed network embedding in social media. In: Proceedings of the 2017 SIAM International Conference on Data Mining, pp. 327–335. SIAM (2017)
8. Rossi, R.A., Ahmed, N.K.: Role discovery in networks. IEEE Trans. Knowl. Data Eng. **27**(4), 1112–1131 (2014)

9. Ribeiro, L.F.R., Saverese, P.H.P., Figueiredo, D.R.: struc2vec: learning node representations from structural identity. In: Proceedings of the 23rd ACM SIGKDD International Conference on Knowledge Discovery and Data Mining, pp. 385–394 (2017)
10. Lü, L., Zhou, T.: Link prediction in complex networks: a survey. Phys. A Stat. Mech. Its Appl. **390**(6), 1150–1170 (2011)
11. Kumar, A., Sheshar Singh, S., Singh, K., Biswas, B. :Link prediction techniques, applications, and performance: a survey. Phys. A Stat. Mech. Its Appl. **553**, 124289 (2020)
12. Bhagat, S., Cormode, G., Muthukrishnan, S.: Node classification in social networks. In: Social Network Data Analytics, pp. 115–148. Springer (2011)
13. Tang, J., Aggarwal, C., Liu, H.: Node classification in signed social networks. In: Proceedings of the 2016 SIAM International Conference on Data Mining, pp. 54–62. SIAM (2016)
14. Akoglu, L., Tong, H., Koutra, D.: Graph based anomaly detection and description: a survey. Data Mining Knowl. Discov. **29**(3), 626–688 (2015)
15. Rossi, R.A., Jin, D., Kim, S., Ahmed, N.K., Koutra, D., Lee, J.B.: On proximity and structural role-based embeddings in networks: misconceptions, techniques, and applications. ACM Trans. Knowl. Discov. Data (TKDD) **14**(5), 1–37 (2020)
16. Perozzi, B., Al-Rfou, R., Skiena, S.: Deepwalk: online learning of social representations. In: Proceedings of the 20th ACM SIGKDD International Conference on Knowledge Discovery and Data Mining, pp. 701–710 (2014)
17. Grover, A., Leskovec, J.: node2vec: scalable feature learning for networks. In: Proceedings of the 22nd ACM SIGKDD International Conference on Knowledge Discovery and Data Mining, pp. 855–864 (2016)
18. Tang, J., Qu, M., Wang, M., Zhang, M., Yan, J., Mei, Q.: Line: large-scale information network embedding. In: Proceedings of the 24th International Conference on World Wide Web, pp. 1067–1077 (2015)
19. Donnat, C., Zitnik, M., Hallac, D., Leskovec, J.: Learning structural node embeddings via diffusion wavelets. In: Proceedings of the 24th ACM SIGKDD International Conference on Knowledge Discovery and Data Mining, pp. 1320–1329 (2018)
20. Yuan, S., Wu, X., Xiang, Y.: Sne: signed network embedding. In: Pacific-Asia Conference on Knowledge Discovery and Data Mining, pp. 183–195. Springer (2017)
21. Mikolov, T., Sutskever, I., Chen, K., Corrado, G., Dean, J.: Distributed representations of words and phrases and their compositionality (2013). arXiv preprint arXiv:1310.4546
22. Barabási, A.: Scale-free networks: a decade and beyond. Science **325**(5939), 412–413 (2009)

HM-LDM: A Hybrid-Membership Latent Distance Model

Nikolaos Nakis(✉), Abdulkadir Çelikkanat, and Morten Mørup

Section for Cognitive Systems, Technical University of Denmark, Kongens, Lyngby 2800, Denmark

nnak@dtu.dk, abce@dtu.dk, mmor@dtu.dk

Abstract. A central aim of modeling complex networks is to accurately embed networks in order to detect structures and predict link and node properties. The Latent Space Model (LSM) has become a prominent framework for embedding networks and includes the Latent Distance Model (LDM) and Eigenmodel (LEM) as the most widely used LSM specifications. For latent community detection, the embedding space in LDMs has been endowed with a clustering model whereas LEMs have been constrained to part-based non-negative matrix factorization (NMF) inspired representations promoting community discovery. We presently reconcile LSMs with latent community detection by constraining the LDM representation to the D-simplex forming the Hybrid-Membership Latent Distance Model (HM-LDM). We show that for sufficiently large simplex volumes this can be achieved without loss of expressive power whereas by extending the model to squared Euclidean distances, we recover the LEM formulation with constraints promoting part-based representations akin to NMF. Importantly, by systematically reducing the volume of the simplex, the model becomes unique and ultimately leads to hard assignments of nodes to simplex corners. We demonstrate experimentally how the proposed HM-LDM admits accurate node representations in regimes ensuring identifiability and valid community extraction. Importantly, HM-LDM naturally reconciles soft and hard community detection with network embeddings exploring a simple continuous optimization procedure on a volume constrained simplex that admits the systematic investigation of trade-offs between hard and mixed membership community detection.

Keywords: Latent space modeling · Community detection · Non-negative matrix factorization · Graph representation learning.

1 Introduction

Networks naturally arise in the vast majority of scientific domains from physics to biology in order to model interactions among diverse entities with numerous instances such as collaboration, protein-protein, and brain connectivity networks [23]. Hence, graph analysis tools have become crucial to extract and analyze

H. Cherifi et al. (Eds.): COMPLEX NETWORKS 2016 2022, SCI 1077, pp. 350–363, 2023.
https://doi.org/10.1007/978-3-031-21127-0_29

the underlying meaningful information from networks. In this direction, Graph Representation Learning (GRL) [36] approaches have become a dominant way to carry out various downstream tasks such as node classification, link prediction, and community detection thanks to their superior performance compared to the classical techniques. GRL models mainly aim to map similar nodes in the network to close latent positions in a low dimension space by automatically learning corresponding node features [7].

The initial GRL works aimed to learn representations or features by simulating random walks over networks, taking inspiration from the Natural Language Processing field [4,6,25,27,31]. They mainly extract embeddings by optimizing the co-occurrence probability of node pairs within a certain distance through random walks. In recent years, we have witnessed a tremendous increase in the number of Graph Neural Networks (GNN) [7] methods with their usage in supervised tasks. They primarily rely on iterative message-passing operations of node attributes and hidden features around the surroundings of nodes for a given task. The matrix decomposition-based models [26,27] are also a notable class of the GRL methods. They learn node embeddings by decomposing a designed target matrix based on first and higher-order proximity. However, few GRL methods rely on Non-negative Matrix Factorization (NMF), although it is a popular technique for unsupervised signal decomposition and approximation of multivariate non-negative data. NMF techniques have gathered lots of attention since they allow for structure retrieval through the latent factors of the imposed decomposition providing easy interpretable part-based representations[18].

Applications of NMF include network analysis allowing for efficient, unsupervised, and overlapping community detection, as well as GRL [2,20,33,34]. Within the NMF formulation, various works have sought to define mixed-membership frameworks for analysis and community detection purposes. A Mixed-Membership Stochastic Block Model (MM-SBM) [1] has been linked to the symmetric-NMF decomposition with uniqueness guarantees [20]. Standard least-squares NMF optimization was exchanged to a Poisson likelihood optimization for obtaining the propensity of nodes belonging to different communities [2]. In addition, a GRL approach for overlapping communities was presented in [33] where NMF was utilized to discover Poisson distributed mixed-memberships. These works, design mixed-memberships vectors for part-based representations [18] projected in an NMF constructed space where node similarity, as well as, position and metric properties, can be abstract.

The Latent Space Models (LSMs) are also one of the most powerful ways to learn latent representations [22]. These methods employ generalized linear models for constructing latent node embeddings which express important network characteristics. More specifically, the LDM [11] employs the Euclidean norm for positioning similar nodes closer in the latent space, which comes as a direct consequence of the triangular inequality, naturally representing transitivity (*"a friend of a friend is a friend"*) and homophily (*a tendency where similar nodes are more likely to connect to each other than dissimilar ones*) properties. The LDM can be generalized through the Eigenmodel [10] that can account for stochastic equivalence (*"groups of nodes defined by shared intra- and inter-group relationships"*) akin to the SBM [1] and the mixed membership SBM [1]. Furthermore, LDMs have been endowed with a clustering model imposing a Gaussian Mixture Model as prior forming the latent position clustering model [8,29].

In this study, we propose a novel unsupervised representation learning method over graphs, namely, the Hybrid-Membership Latent Distance Model (HM-LDM), by bringing together the strengths of LDM and NMF. Specifically, the HM-LDM offers a reconciliation between part-based representations of networks and low-dimensional latent spaces satisfying similarity properties such as homophily and transitivity. The choice of these similarity properties is of high significance and one of the key characteristics behind GRL since they allow for easily interpretable discovery of network structure. Additionally, our proposed method permits us to capture the latent community structure of the networks using a simple continuous optimization procedure over the log-likelihood of the network. Notably, unlike most existing approaches imposing hard community memberships constraints, the assignment of community memberships in our proposed hybrid model can be controlled and altered through the simplex volume formed by the latent node representations. We extensively evaluate the performance of the proposed method in the ability to perform link prediction, as well as, community discovery over various networks of different types. We demonstrate that our model outperforms recent methods.
Source code: *Hybrid-Membership Latent Distance Model.*

2 Problem Statement and Proposed Method

Let $\mathcal{G} = (\mathcal{V}, \mathcal{E})$ be an undirected graph where $\mathcal{V}$ shows the vertex set and $\mathcal{E} \subseteq \mathcal{V} \times \mathcal{V}$ the edge set. We use $\mathbf{Y}_{N \times N} = (y_{i,j}) \in \{0, 1\}^{N \times N}$ to denote the adjacency matrix of the graph where $y_{i,j} = 1$ if the pair $(i, j) \in \mathcal{E}$ otherwise it is equal to 0 for all $1 \leq i < j \leq N := |\mathcal{V}|$. Our main goal is to learn a representation, $\mathbf{w}_i \in \mathbb{R}^D$, for each node $i \in \mathcal{V}$ in a lower dimensional space ($D \ll N$) such that similar nodes in the network should have close embeddings. More specifically, we concentrate on mapping the nodes into the unit D-simplex set, $\Delta^D \subset \mathbb{R}_+^{D+1}$. Therefore, the extracted node embeddings can convey information about latent community memberships. Many GRL approaches also do not provide identifiable or unique solution guarantees, so their interpretation highly depends on the initialization of the hyper-parameters. In this study, we will also address the identifiability problem and seek identifiable solutions which can only be achieved up to a permutation invariance, as reported in Def. 1.

Definition 1 (*Identifiabilty*). An embedding matrix $\mathbf{W}$ whose rows indicating the corresponding node representations is called an *identifiable solution up to a permutation* if it holds $\widetilde{\mathbf{W}} = \mathbf{W}\mathbf{P}$ for a permutation $\mathbf{P}$ and a solution $\widetilde{\mathbf{W}} \neq \mathbf{W}$.

We define a Poisson distribution over the adjacency matrix $\mathbf{Y}$ of the network $\mathcal{G} = (\mathcal{V}, \mathcal{E})$ to be conditionally independent given the unobserved latent positions, and write the log-likelihood function as follows:

$$\log P(\mathbf{Y}|\boldsymbol{\Lambda}) = \sum_{\substack{i<j \\ y_{ij}=1}} \log(\lambda_{ij}) - \sum_{i<j} \Big(\lambda_{ij} + \log(y_{ij}!)\Big). \quad (1)$$

where $\boldsymbol{\Lambda} = (\lambda_{ij})$ is the Poisson rate matrix which has absorbed the dependency over the model parameters. We here adopted a Poisson regression model similar to the work in [9]. In this study, we make use of a Poisson likelihood for modelling binary networks, as validated in [33].

We propose the Hybrid-Membership Latent Distance Model (HM-LDM) with a log-rate based on the ℓ^2-norm as:

$$\log \lambda_{ij} = \Big(\gamma_i + \gamma_j - \delta^p \cdot ||\mathbf{w}_i - \mathbf{w}_j||_2^p\Big), \quad (2)$$

where $\mathbf{w_i} \in [0,1]^{D+1}$ and $\sum_{d=1}^{D+1} w_{id} = 1$, $\delta \in \mathbb{R}_+$ and $\gamma_i \in \mathbb{R}$ denotes the node-specific random-effects [9,16] describing essentially the tendency of nodes to sending and receiving connections, accounting for degree heterogeneity. In addition, the norm degree $p \in \{1,2\}$ controls the power of the ℓ^2-norm and combined with the latent embeddings sum-to-one condition constrains the latent space to the $D-$simplex with size equal to δ. A remarkable property of Eq. (2), for $p = 2$, is that it resembles a positive Eigenmodel with random effects: $\tilde{\gamma}_i + \tilde{\gamma}_j + (\tilde{\mathbf{w}}_i \boldsymbol{\Lambda} \tilde{\mathbf{w}}_j^\top)$ where $\boldsymbol{\Lambda}$ is a diagonal matrix having non-negative elements, i.e. $\tilde{\gamma}_i = \gamma_i - \delta^2 \cdot ||\mathbf{w}_i||_2^2$, $\tilde{\gamma}_j = \gamma_j - \delta^2 \cdot ||\mathbf{w}_j||_2^2$ and $\tilde{\mathbf{w}}_i \boldsymbol{\Lambda} \tilde{\mathbf{w}}_j^\top = 2\delta^2 \cdot \mathbf{w}_i \mathbf{w}_j^\top$ thus the squared Euclidean distance reconciles the conventional LDM and non-negativity constrained Eigenmodel. The squared Euclidean distance is not fully a metric but it still expresses homophily, leading to an interpretable latent space. Even though the triangle inequality is not exactly satisfied, it preserves the ordering of pairwise Euclidean distances, and it is highly preferred in applications since it is a strictly convex smooth function. By the well-known cosine formula, we have

$$||\mathbf{w}_i - \mathbf{w}_j||_2^2 = ||\mathbf{w}_i - \mathbf{w}_k||_2^2 + ||\mathbf{w}_k - \mathbf{w}_j||_2^2 + 2||\mathbf{w}_i - \mathbf{w}_k||_2 ||\mathbf{w}_k - \mathbf{w}_j)||_2 \cos(\theta),$$

where $\theta \in (-\pi/2, \pi/2)$ is the angle between the vectors $\mathbf{w}_i - \mathbf{w}_k$ and $\mathbf{w}_k - \mathbf{w}_j$. Notice that the third term also converges to 0 for similar nodes since we will have close representations. For the case where $\theta \in [-\pi, -\pi/2] \cup [\pi/2, \pi]$, it holds the triangle inequality: $||\mathbf{w}_i - \mathbf{w}_j||_2^2 \leq ||\mathbf{w}_i - \mathbf{w}_k||_2^2 + ||\mathbf{w}_k - \mathbf{w}_j||_2^2$.

The embedding vectors, $\{\mathbf{w}_i\}_{i=1}^N$ in Eq. (2), are constrained to non-negative values and to sum to one. Thereby, they reside on a simplex showing the participation of node $i \in \mathcal{V}$ over $D+1$ latent communities. Any LDM can be translated to the non-negative orthant without any loss in performance or in expressive capability. Non-negative embeddings do not affect the distance metric, as it is invariant to translation, as shown by Fig. 1 (a). In addition, the D-dimensional non-negative orthant can be reconstructed by a large enough D-simplex. Based on these arguments, it is trivial to show that for large values of the δ parameter in Eq. (2), despite the sum-to-one constraint on the embeddings $\mathbf{W}$, we obtain an unconstrained LDM, as distances are unbounded when $\delta \rightarrow +\infty$. In this

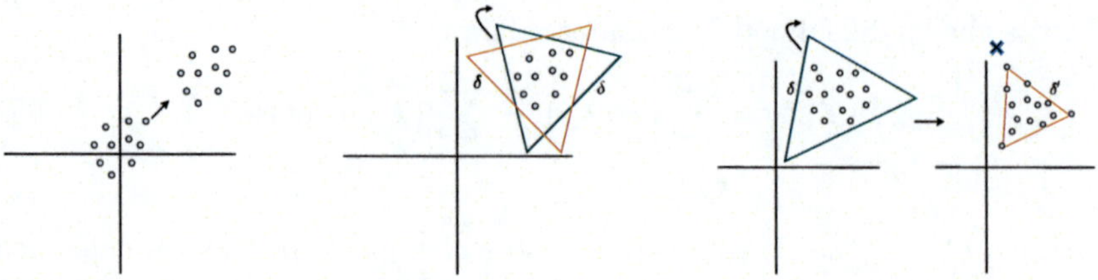

(a) Translation invariances.

(b) Rotation invariances.

(c) Decreased simplex volume ensuring identifiability.

Fig. 1. A 2-dimensional latent space with the 2-simplex given as the green and yellow triangles, the blue points denote embedding positions of the LDM and δ is the simplex size.

case, the memberships defined by $\mathbf{W}$ are not uniquely identifiable due to the distance invariance of rotation, as seen in Fig. 1 (b).

However, by shrinking the volume of the simplex (equivalent to decreasing δ), eventually the D-dimensional space of LDM will no longer be enclosed inside the D-simplex, forcing nodes to start populating the corners of this smaller simplex. We call a node *champion* if its latent representation is a standard binary unit vector.

Definition 2 (*Community champion*). A node for a latent community is called *champion* if it belongs to the community (simplex corner) while forming a binary unit vector.

The champion nodes are of great significance for identifiability because if every corner of the simplex is populated by at least one node (champion), then the solution of the model is identifiable (up to a permutation matrix) (Def. 1) as any random rotation does not leave the solution invariant anymore, as shown by Fig. 1 (c). We observe then, that the scalar, δ, controls the type of memberships of the model and its expressive capabilities. Large enough values lead to the basic LDM but inherits its rotational invariance. Small values of δ lead to identifiable solutions and ultimately hard cluster assignments. Thereby, for very small values of δ, nodes are solely assigned to the simplex corners. Lastly, we can also find regimes of values for δ that offer identifiable solutions but also performance similar to LDM, defining a silver lining.

A different take on the identifiability of the model for $p = 2$, can also be given under the Non-negative Matrix Factorization (NMF) theory. This is easily shown by a re-parameterization of Eq. (2) by $\tilde{\gamma}_i + \tilde{\gamma}_j + 2\delta^2 \cdot (\mathbf{w}_i \mathbf{w}_j^\top)$ as described in Eq. (2). In this formulation, the product $\mathbf{W}\mathbf{W}^\top$ defines a symmetric NMF problem which is an identifiable and unique factorization (up to permutation invariance) when $\mathbf{W}$ is full-rank and at least one node resides solely in each simplex corner, ensuring separability [12,20]. Under this NMF formulation, the product $\mathbf{w}_i \mathbf{w}_j^\top \in [0, 1]$ achieves its upper bound only if both nodes i and j reside in the same corner of the simplex. The parameter, δ, acts as a simple multiplicative factor in the first

Table 1. Network statistics; $|\mathcal{V}|$: # Nodes, $|\mathcal{E}|$: # Edges, $|\mathcal{K}|$: # Communities.

	AstroPh [19]	*GrQc* [19]	*Facebook* [19]	*HepTh* [19]	*Hamilton* [21]	*Amherst* [21]	*Rochester* [21]	*Mich* [21]
$\|\mathcal{V}\|$	17,903	5,242	4,039	8,638	2,118	2,021	4,145	2,933
$\|\mathcal{E}\|$	197,031	14,496	88,234	24,827	87,486	87,496	145,305	54,903
$\|\mathcal{K}\|$	–	–	–	–	15	15	19	13

term of the objective function of HM-LDM, given in Eq. (1), while in the second term acts as a power of the exponential function. For small values of δ, the model is biased towards hard latent community assignments of nodes since similar nodes achieve high rates only when they belong to the same latent community (simplex corner). On the other hand, nodes heading towards the simplex corners for large values of δ lead to an exponential change in the second term of the log-likelihood function given in Eq. (1). Thus, a possible hard allocation of dissimilar nodes to the same community penalizes the likelihood severely. For this reason, high order of δ benefits mixed-membership allocations.

3 Experimental evaluation

We proceed by evaluating the efficiency and performance of the proposed method. In our set-up, we make use of networks with unknown community structures, as well as, with ground-truth communities. We employ the former networks to validate the ability of our framework to discover identifiable latent structures and predict missing links. The latter networks are used to verify that the HM-LDM discovers communities successfully. We consider multiple social and scientific collaboration networks as shown by Table 1. We treat all networks as unweighted and undirected.

For the training of HM-LDM we optimize the log-likelihood function of Eq. (1) via the Adam optimizer [15] with learning rate $lr \in [0.01, 0.1]$. The node-specific random effects vector $\boldsymbol{\gamma} \in \mathbb{R}^N$ is randomly initialized and then tuned alone by optimizing a Poisson log-likelihood with a rate as $\log \lambda_{ij} = \gamma_i + \gamma_j$. Next, the latent embeddings matrix $\mathbf{W}$ is initialized based on the eigenvalues obtained by the spectral decomposition of the normalized Laplacian matrix of the network [13,24]. In all experiments, we compare against unsupervised methods, and we do not include GNNs since they perform poorly in unsupervised tasks due to the over-smoothing effect [35].

Link prediction: For the link prediction experiments, we follow the well-established strategy [6,25] and remove 50% of the network edges while keeping the residual network connected. The removed edges combined with a sample of the same number of node pairs (which are not the edges of the original network) construct the negative instances for the testing set. We utilize the residual network to learn the node embeddings.

We consider four networks with unknown community structures and asses performance across different dimensions. In Table 2, we compare the results of our method with other prominent GRL and NMF approaches in terms of the

Table 2. Area Under Curve (AUC-ROC) scores for varying representation sizes.

	AstroPh			*GrQc*			*Facebook*			*HepTh*		
Dimension (D)	8	16	32	8	16	32	8	16	32	8	16	32
DeepWalk [25]	0.945	0.950	0.952	0.919	0.916	0.929	0.986	0.986	0.984	0.874	0.867	0.873
Node2Vec [6]	0.950	<u>0.962</u>	<u>0.957</u>	0.897	0.913	0.930	<u>0.988</u>	<u>0.988</u>	<u>0.987</u>	0.881	0.882	0.881
LINE [31]	0.909	0.938	0.947	0.920	0.925	0.919	0.981	0.987	0.983	0.873	0.886	0.882
NetMF [27]	0.813	0.823	0.839	0.860	0.866	0.877	0.935	0.963	0.971	0.792	0.806	0.821
NetSMF [26]	0.891	0.901	0.919	0.837	0.858	0.886	0.975	0.981	0.985	0.809	0.822	0.836
LouvainNE [3]	0.813	0.811	0.819	0.868	0.875	0.873	0.958	0.961	0.963	0.874	0.867	0.873
ProNE [37]	0.907	0.929	0.947	0.885	0.911	0.921	0.971	0.982	0.987	0.827	0.846	0.859
NNSED [30]	0.861	0.882	0.891	0.792	0.808	0.828	0.908	0.927	0.935	0.756	0.779	0.796
MNMF [32]	0.893	0.925	0.0943	0.911	0.928	0.937	0.965	0.978	0.982	0.857	0.880	0.891
BigClam [34]	0.500	0.723	0.810	0.752	0.769	0.780	0.744	0.722	0.647	0.776	0.700	0.748
SymmNMF [17]	0.767	0.779	0.800	0.729	0.772	0.835	0.933	0.942	0.951	0.696	0.727	0.766
HM-LDM ($p = 1$)	<u>.956</u>	0.952	0.952	**0.944**	**0.948**	**0.951**	0.982	0.979	0.974	**0.916**	**0.921**	**0.924**
HM-LDM ($p = 2$)	**0.972**	**0.973**	**0.963**	<u>0.940</u>	<u>0.942</u>	<u>0.946</u>	**0.992**	**0.993**	**0.993**	<u>0.908</u>	<u>0.910</u>	<u>0.911</u>

Area Under Curve-Receiver Operating Characteristic (AUC-ROC). All baselines have been tuned and feature vectors for dyads are constructed based on binary operators (average, Hadamard, weighted-L1, weighted-L2) [6]. For these constructed feature vectors we further train a logistic regression model with L_2 regularization to make predictions. In particular, for the baselines we choose the hyperparameter settings for each model, as well as, the binary operator for which the logistic regression predictions return the maximum AUC-ROC score.

In contrast, for our models we adopt an unbiased evaluation, and we choose the first of the considered δ values which keeps the solution identifiable (at least one champion per community), as δ decreases. We note though, the existence of identifiable regimes with higher predictive power. Furthermore, predictions and AUC-ROC scores for HM-LDM, can be obtained directly (without the use of a logistic regression model) and are based on the learned Poisson rates λ_{ij} of the test set pairs $\{i, j\}$. The true dimensions for HM-LDM are $D + 1$ but reported as D since they express the true number of model parameters, for a fair comparison with the baselines. For our method, we show the mean performance over five independent runs (error bars were found to be in the scale of 10^{-3} and thus not presented).

Comparing now the results with the non-NMF models, we observe that our HM-LDM (either $p = 1$ or $p = 2$) outperforms the baselines and in most cases significantly, returning favorable results. For the NMF models, we see mostly a big performance gap with the HM-LDM, showcasing the existence of regimes for δ where we can successfully achieve identifiable community memberships while also exhibiting the link prediction power of the LDM. (AUC Precision-Recall scores are similar to the AUC-ROC scores and thus not presented)

Performance and simplex sizes: In Fig. 2, we provide the link prediction performance as a function of δ^2 in terms of the AUC-ROC scores across various latent dimensions, networks and for both $p = 1$ and $p = 2$. We here observe that small δ values provide the minimum scores. This phenomenon is anticipated due to the fact that homophily properties are not sufficiently met (except within clusters) due to the very small simplex volume that these low δ values define. Rethinking HM-LDM with $p = 2$ as a positive Eigenmodel, we can also notice how the positivity constraint on the Λ diagonal matrix does not allow for stochastic equivalence properties which would essentially boost performance even on low simplex volumes. As we increase the values of δ, we naturally reach the performance of an unconstrained LDM. Comparing now, the squared and simple ℓ^2-norm metric we observe that the former converges to performance saturation more rapidly.

Type and quality of latent memberships: In order to understand how the size of the simplex affects the membership types of HM-LDM, we provide in Fig. 3 the total network percentage of community champions as a function of δ^2 across various latent dimensions. As expected, for very small values of δ almost 100% of nodes are assigned solely to a unique simplex corner, yielding hard cluster assignments. As we increase δ, we observe that more and more nodes are assigned with mixed-memberships; on the other hand, the number of champions goes to zero across all dimensions for large values of δ. Contrasting again, the different powers p of the HM-LDM formulation, we notice that the decrease in community champions is steeper for $p = 2$. This also explains why the squared ℓ^2 choice leads to faster convergence in the AUC-ROC, as the model converges faster to the classic LDM. Overall, it is evident that the $p = 2$ HM-LDM needs smaller simplex volumes to be identifiable. We continue with assessing unique latent structures of HM-LDM. For that purpose, in Fig. 4 we provide the reorganized adjacency matrices with respect to the community allocations of HM-LDM (for mixed-memberships we assign a node based on the maximum membership). We witness how HM-LDM successfully discovers latent communities, facilitating part-based network representations while choosing appropriate δ regimes ensure identifiability.

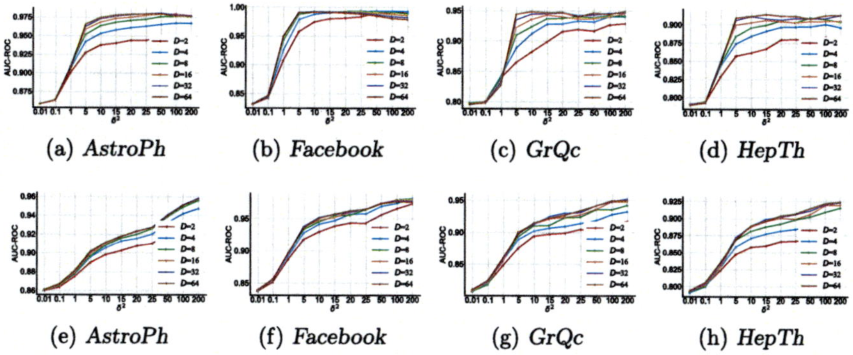

Fig. 2. AUC-ROC scores as a function of δ^2 across dimensions for HM-LDM. Top row: $p = 2$. Bottom row $p = 1$.

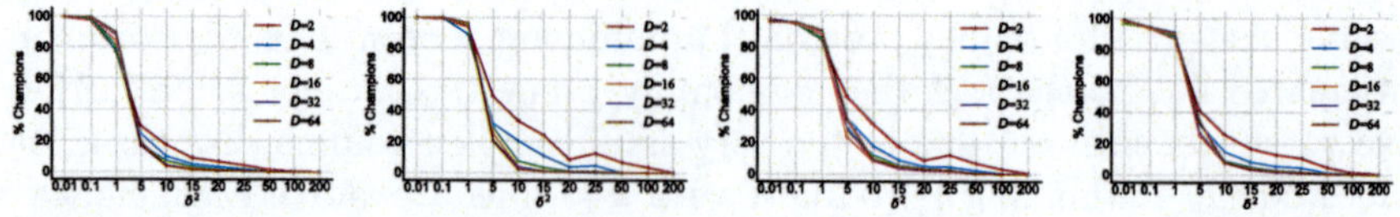

(a) *AstroPh* (b) *Facebook* (c) *GrQc* (d) *HepTh*

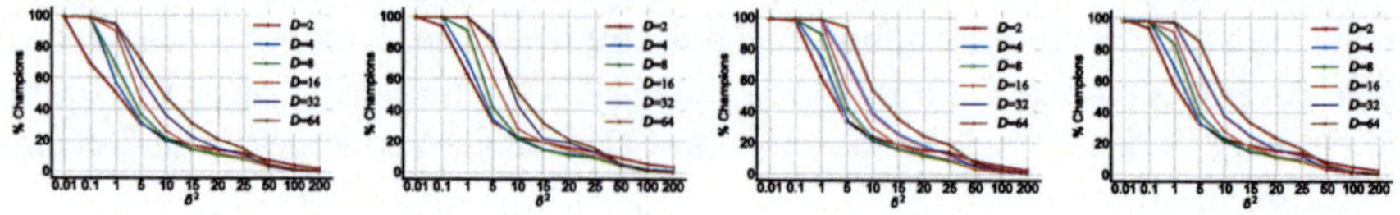

(e) *AstroPh* (f) *Facebook* (g) *GrQc* (h) *HepTh*

Fig. 3. Total community champions (%) in terms of δ^2 across dimensions for HM-LDM. Top row: $p = 2$. Bottom row $p = 1$.

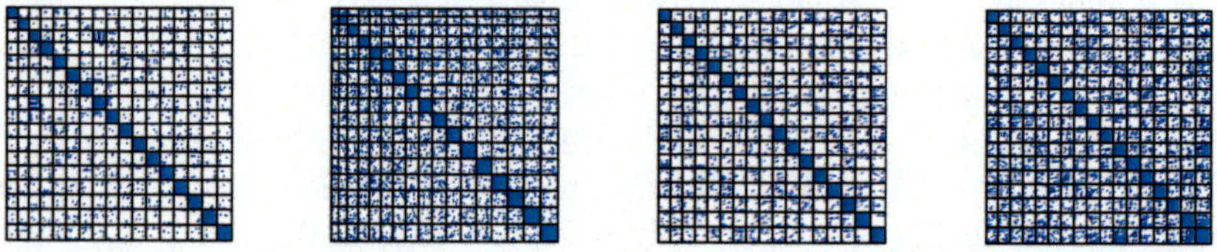

(a) *GrQc* ($p = 2$) (b) *HepTh* ($p = 2$) (c) *GrQc* ($p = 1$) (d) *HepTh* ($p = 1$)

Fig. 4. Ordered adjacency matrices based on the memberships of a $D = 16$ dimensional HM-LDM with δ values ensuring identifiability.

Experiments using real ground-truth communities: In order to assess the ability of HM-LDM to discover informative communities, we make use of four networks providing ground-truth community labels. For the NMF-based methods, including ours, we test the ability of the algorithms to detect valid structures by comparing the inferred memberships with the ground-truth community labels while we set the latent dimensions to be equal to the total number of communities. For the GRL approaches which do not define memberships, we extract latent embeddings and use *k-means* (average over 20 runs for robustness) to obtain memberships. We report the Normalized Mutual Information (NMI) score, as well as, the Adjusted Rand Index (ARI), both measures have been validated for community quality assessment in [5]. Again, all the baselines have been tuned individually for each network in terms of their hyperparameters. In contrast, for our HM-LDM, we do not perform any tuning and we just set $\delta = 1$ for all networks since this choice provides in general informative and mostly hard cluster assignments. For our method and the classic LDMs, we report scores averaged over five independent runs in each of which we run the algorithm five times extracting the model with the lowest training loss to remove the effect of local-minimas. We summarize our findings in Table 3, where we witness mostly favorable or on-par performance of HM-LDM with all of the competitive base-

Table 3. Normalized Mutual Information (NMI) and Adjusted Rand Index (ARI) scores for networks with ground-truth communities.

	Amherst		*Rochester*		*Mich*		*Hamilton*	
Metric	NMI	ARI	NMI	ARI	NMI	ARI	NMI	ARI
DeepWalk [25]	0.498	0.347	0.348	0.205	0.207	0.157	0.447	0.303
Node2Vec [6]	0.535	0.375	0.364	0.223	0.217	0.161	0.481	0.348
LINE [31]	0.549	0.452	0.365	0.217	**0.249**	0.192	0.499	0.411
NetMF [27]	0.491	0.330	0.377	0.243	0.237	0.136	0.456	0.297
NetSMF [26]	<u>0.562</u>	0.408	<u>0.381</u>	0.228	<u>0.242</u>	0.169	0.494	0.391
LouvainNE [3]	<u>0.562</u>	0.395	0.347	0.204	0.175	0.114	0.475	0.334
ProNE [37]	0.536	0.443	0.356	0.312	0.229	<u>0.200</u>	0.478	0.396
NNSED [30]	0.295	0.243	0.168	0.116	0.064	0.035	0.335	0.285
MNMF [32]	0.542	0.362	0.324	0.171	0.188	0.102	0.466	0.287
BigClam [34]	0.091	0.066	0.028	0.022	0.024	0.015	0.053	0.041
SymmNMF [17]	**0.596**	0.397	0.308	0.175	0.207	0.088	0.437	0.341
HM-LDM ($p = 1$)	<u>0.562</u>	<u>0.502</u>	**0.400**	**0.392**	0.228	**0.205**	**0.527**	<u>0.485</u>
HM-LDM ($p = 2$)	0.539	**0.506**	<u>0.384</u>	<u>0.373</u>	0.217	0.183	<u>0.507</u>	**0.504**

lines for the NMI metric. For the ARI metric, we observe that our framework outperforms significantly the baselines in all of the considered networks.

Comparison with the LDM: We further investigate the performance of HM-LDM against the LDM, including random effects for a fair comparison and for both normal and squared ℓ^2-norm LDM-Re and LDM-Re-$(\ell^2)^2$, respectively. Towards that aim, in Tables 4 and 5 we provide the performance scores for the link prediction and clustering tasks of each model. We here witness that constraining the latent space in identifiable simplex volumes leads to a minor decrease in the predictive power, in terms of the AUC-ROC. For the community detection task, we see favorable NMI scores while the HM-LDM leads to considerably higher ARI scores. Comparing the classical LDM with HM-LDM for $\delta^2 = 10^3$ provides on-par link-prediction performance but the clustering scores drop significantly. This is expected as for large simplex volumes the HM-LDM approximates almost exactly the LDM with the cost of identifiability.

Extension to bipartite networks: Finally, we showcase the extension of our HM-LDM framework to the analysis of bipartite networks. This is straightforward by introducing a different set of latent variables for the two disjoint sets of nodes, as defined by the bipartite structure. In particular, HM-LDM for $p = 2$, simply extends the symmetric NMF formulation, obtained for the undirected networks, to the non-symmetric NMF specification. In Fig. 5, we provide the re-ordered adjacency matrix with respect to the community allocations defined by the learned embeddings of HM-LDM for a *Drug-Gene* [19] network ($|\mathcal{V}| = 7,341|$, $|\mathcal{E}| = 15,138$) where we observe a clear block structure. Importantly, the HM-LDM offers identifiable joint embedding representations, mixed memberships, and community discovery for bipartite networks, tasks considered to be non-trivial and arduous.

Table 4. HM-LDM and LDM-RE comparison for the link prediction task.

	AstroPh			*GrQc*			*Facebook*			*HepTh*		
Dimension (D)	8	16	32	8	16	32	8	16	32	8	16	32
LDM-RE	0.973	0.974	0.979	0.949	0.952	0.954	0.993	0.994	0.992	0.920	0.923	0.923
HM-LDM ($p = 1, \delta^2$ = identifiable)	0.956	0.952	0.952	0.944	0.948	0.951	0.982	0.979	0.974	0.916	0.921	0.924
HM-LDM ($p = 1, \delta^2 = 10^3$)	0.967	0.967	0.965	0.956	0.955	0.951	0.985	0.986	0.987	0.932	0.931	0.926
LDM-RE-$(\ell^2)^2$	0.979	0.978	0.976	0.944	0.944	0.945	0.990	0.990	0.991	0.913	0.912	0.909
HM-LDM ($p = 2, \delta^2$ = identifiable)	0.972	0.973	0.963	0.940	0.942	0.946	0.992	0.993	0.993	0.908	0.910	0.911
HM-LDM ($p = 2, \delta^2 = 10^3$)	0.984	0.983	0.980	0.948	0.946	0.946	0.991	0.991	0.992	0.920	0.918	0.913

Table 5. HM-LDM and LDM-RE comparison for the clustering task.

	Amherst		*Rochester*		*Mich*		*Hamilton*	
Metric	NMI	ARI	NMI	ARI	NMI	ARI	NMI	ARI
LDM-RE	0.548	0.366	0.391	0.212	0.230	0.132	0.491	0.320
HM-LDM ($p = 1, \delta^2$ = identifiable)	0.562	0.502	0.400	0.392	0.228	0.205	0.527	0.485
HM-LDM ($p = 1, \delta^2 = 10^3$)	0.439	0.386	0.308	0.303	0.176	0.133	0.405	0.377
LDM-RE-$(\ell^2)^2$	.546	0.370	0.393	0.211	0.231	0.137	0.497	0.327
HM-LDM ($p = 2, \delta^2$ = identifiable)	0.539	0.506	0.384	0.373	0.217	0.183	0.507	0.504
HM-LDM ($p = 2, \delta^2 = 10^3$)	0.240	0.133	0.206	0.119	0.116	0.056	0.232	0.209

Complexity analysis: The HM-LDM framework requires the computation of the node pairwise distance matrix and consequently scales prohibitively as $\mathcal{O}(N^2)$ in time and space. Fortunately, there are various ways of scaling HM-LDM for the analysis of large-scale networks. One way is through unbiased estimators of the log-likelihood given by Eq. (1). This is possible through random sampling a set of network nodes S (per iteration) and taking a gradient step based on the log-likelihood of the block defined by the sampled node-set, returning an $\mathcal{O}(S^2)$ space and time complexity. Another option is through the case-control approach [28] scaling on the number of network edges as $\mathcal{O}(E)$. Lastly, the Hierarchical Block Distance Model (HBDM) [22] is an attractive option where gradient steps over the model parameters are based on a hierarchical approximation of the likelihood of the whole network. The HBDM model scales linearithmicly as $\mathcal{O}(N \log N)$ both in space and time while also offering hierarchical characterizations of structures at multiple scales.

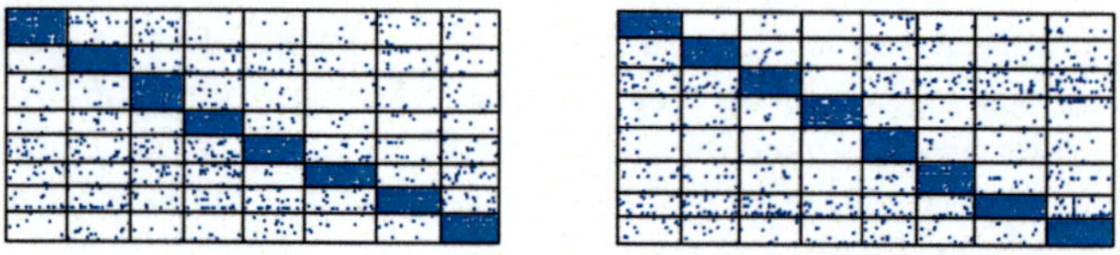

(a) $p = 1$, $\delta = 1$ (b) $p = 2$, $\delta = 1$

Fig. 5. *Drug-Gene* ordered adjacency matrices based on HM-LDM with $D = 8$.

4 Conclusion and Future Work

In this paper, we have proposed the HM-LDM that reconciles network embedding and latent community detection. The approach utilizes both the normal and squared Euclidean distance model where the latter integrated the non-negativity constrained Eigenmodel with the Latent Distance Model. We demonstrated that the model could be constrained to the simplex without losing expressive power. The reduced simplex provides unique representations, ultimately resulting in hard clustering of nodes to communities when the simplex is sufficiently shrunk. Notably, the proposed HM-LDM combines network homophily and transitivity properties with latent community detection enabling explicit control of soft and hard assignment through the volume of the induced simplex. We observed favorable link prediction performance in regimes in which the HM-LDM provides unique representations while enabling the ordering of the adjacency matrix in terms of prominent latent communities. Finally, we showed the ability of the model to extract correct community structures across multiple networks and showcased how the analysis extends to bipartite networks. Future work should compare the performance of HM-LDM against classical non-embedding methods such as the Degree Corrected Stochastic Block Model (DC-SBM)[14] or the Mixed Membership Stochastic Block Model (MM-SBM) [1]. Such a comparison is of particular interest since DC-SBM accounts for degree heterogeneity while MM-SBM for soft assignments, two important properties of HM-LDM.

Acknowledgements. We would like to thank the reviewers for the constructive feedback and their insightful comments. We would also like to thank Sune Lehmann, Louis Boucherie, Lasse Mohr Mikkelsen, and Giorgio Giannone for the useful and fruitful discussions. We gratefully acknowledge the Independent Research Fund Denmark for supporting this work [grant number: 0136-00315B].

References

1. Airoldi, E.M., Blei, D.M., Fienberg, S.E., Xing, E.P.: Mixed membership stochastic blockmodels. J. Mach. Learn. Res. **9**(65), 1981–2014 (2008)
2. Ball, B., Karrer, B., Newman, M.E.J.: An efficient and principled method for detecting communities in networks. CoRR abs/1104.3590 (2011)
3. Bhowmick, A.K., Meneni, K., Danisch, M., Guillaume, J.L., Mitra, B.: LouvainNE: Hierarchical louvain method for high quality and scalable network embedding. In: WSDM, pp. 43–51 (2020)
4. Çelikkanat, A., Malliaros, F.D.: Exponential family graph embeddings. In: AAAI, pp. 3357–3364 (2020)
5. Chakraborty, T., Dalmia, A., Mukherjee, A., Ganguly, N.: Metrics for community analysis: a survey (2016)
6. Grover, A., Leskovec, J.: Node2Vec: scalable feature learning for networks. In: KDD, pp. 855–864 (2016)
7. Hamilton, W.L., Ying, R., Leskovec, J.: Representation learning on graphs: methods and applications. IEEE Data Eng. Bull. **40**(3), 52–74 (2017)

8. Handcock, M.S., Raftery, A.E., Tantrum, J.M.: Model-based clustering for social networks. J. R. Stat. Soc. Ser. A Stat. Soc. **170**(2), 301–354 (2007)
9. Hoff, P.D.: Bilinear mixed-effects models for dyadic data. JASA **100**(469), 286–295 (2005)
10. Hoff, P.D.: Modeling homophily and stochastic equivalence in symmetric relational data (2007)
11. Hoff, P.D., Raftery, A.E., Handcock, M.S.: Latent space approaches to social network analysis. JASA **97**(460), 1090–1098 (2002)
12. Huang, K., Sidiropoulos, N.D., Swami, A.: Non-negative matrix factorization revisited: uniqueness and algorithm for symmetric decomposition. IEEE Trans. Sign. Proc. **62**(1), 211–224 (2014)
13. Jianbo Shi, Malik, J.: Normalized cuts and image segmentation. IEEE Trans. Patt. Anal. Mach. Intell. **22**(8), 888–905 (2000)
14. Karrer, B., Newman, M.E.: Stochastic blockmodels and community structure in networks. Phys. Rev. E **83**(1), 016107 (2011)
15. Kingma, D.P., Ba, J.: Adam: A method for stochastic optimization (2017)
16. Krivitsky, P.N., Handcock, M.S., Raftery, A.E., Hoff, P.D.: Representing degree distributions, clustering, and homophily in social networks with latent cluster random effects models. Soc. Netw. **31**(3), 204–213 (2009)
17. Kuang, D., Ding, C., Park, H.: Symmetric nonnegative matrix factorization for graph clustering. In: SDM (2012)
18. Lee, D.D., Seung, H.S.: Learning the parts of objects by nonnegative matrix factorization. Nature **401**, 788–791 (1999)
19. Leskovec, J., Krevl, A.: SNAP Datasets: Stanford large network dataset collection (2014)
20. Mao, X., Sarkar, P., Chakrabarti, D.: On mixed memberships and symmetric nonnegative matrix factorizations. In: ICML, vol. 70 (2017)
21. Mucha, P., Porter, M.: Social structure of facebook networks. Phys. A Stat. Mech. Appl. **391**, 4165-4180 (2012)
22. Nakis, N., Çelikkanat, A., Jørgensen, S.L., Mørup, M.: A hierarchical block distance model for ultra low-dimensional graph representations (2022)
23. Newman, M.E.J.: The structure and function of complex networks. SIAM Rev. **45**(2), 167–256 (2003)
24. Ng, A.Y., Jordan, M.I., Weiss, Y.: On spectral clustering: analysis and an algorithm. In: Proceedings of the 14th International Conference on Neural Information Processing Systems: Natural and Synthetic, pp. 849-856. NIPS'01, MIT Press, Cambridge, MA, USA (2001)
25. Perozzi, B., Al-Rfou, R., Skiena, S.: Deepwalk: Online learning of social representations. In: KDD, pp. 701-710 (2014)
26. Qiu, J., Dong, Y., Ma, H., Li, J., Wang, C., Wang, K., Tang, J.: NetSMF: Large-scale network embedding as sparse matrix factorization. In: WWW (2019)
27. Qiu, J., Dong, Y., Ma, H., Li, J., Wang, K., Tang, J.: Network embedding as matrix factorization: unifying DeepWalk, LINE, PTE, and Node2Vec. In: WSDM, pp. 459–467 (2018)
28. Raftery, A.E., Niu, X., Hoff, P.D., Yeung, K.Y.: Fast inference for the latent space network model using a case-control approximate likelihood. J. Comput. Graph. Stat. **21**(4), 901–919 (2012)
29. Ryan, C., Wyse, J., Friel, N.: Bayesian model selection for the latent position cluster model for social networks. Netw. Sci. **5**(1), 70–91 (2017)
30. Sun, B.J., Shen, H., Gao, J., Ouyang, W., Cheng, X.: A non-negative symmetric encoder-decoder approach for community detection. In: CIKM (2017)

31. Tang, J., Qu, M., Wang, M., Zhang, M., Yan, J., Mei, Q.: LINE: Large-scale information network embedding. In: WWW, pp. 1067–1077 (2015)
32. Wang, X., Cui, P., Wang, J., Pei, J., Zhu, W., Yang, S.: Community preserving network embedding. In: AAAI (2017)
33. Wind, D.K., Mørup, M.: Link prediction in weighted networks. In: 2012 IEEE International Workshop MLSP, pp. 1–6 (2012)
34. Yang, J., Leskovec, J.: Overlapping community detection at scale: a nonnegative matrix factorization approach. In: WSDM (2013)
35. Yang, L., Gu, J., Wang, C., Cao, X., Zhai, L., Jin, D., Guo, Y.: Toward unsupervised graph neural network: interactive clustering and embedding via optimal transport. In: ICDM (2020)
36. Zhang, D., Yin, J., Zhu, X., Zhang, C.: Network representation learning: a survey. IEEE Trans. Big Data **6**(1) (2020)
37. Zhang, J., Dong, Y., Wang, Y., Tang, J., Ding, M.: Prone: fast and scalable network representation learning. In: IJCAI (2019)

The Structure of Interdisciplinary Science: Uncovering and Explaining Roles in Citation Graphs

Eoghan Cunningham[1,2](✉) and Derek Greene[1,2]

[1] School of Computer Science, University College Dublin, Dublin, Ireland
eoghan.cunningham@ucdconnect.ie
[2] Insight Centre for Data Analytics, University College Dublin, Dublin, Ireland

Abstract. Role discovery is the task of dividing the set of nodes on a graph into classes of structurally similar roles. Modern strategies for role discovery typically rely on graph embedding techniques, which are capable of recognising complex local structures. However, when working with large, real-world networks, it is difficult to interpret or validate a set of roles identified according to these methods. In this work, motivated by advancements in the field of explainable artificial intelligence (XAI), we propose a new framework for interpreting role assignments on large graphs using small subgraph structures known as graphlets. We demonstrate our methods on a large, multidisciplinary citation network, where we successfully identify a number of important citation patterns which reflect interdisciplinary research.

Keywords: Role discovery · Node embedding · Citation network · Explainable artificial intelligence

1 Introduction

In light of the perceived importance of interdisciplinary research, many studies have been conducted that quantify article interdisciplinarity in an effort to identify relevant research trends and to explore their impact. The most widely-accepted methods for measuring interdisciplinarity assess *knowledge integration* or *knowledge diffusion* using citation information, thus measuring interdisciplinarity as some function of the balance, diversity, and dissimilarity of the disciplines identified in an article's cited papers [17,19] or citing papers [16,28]. To implement these metrics, each paper in a research corpus must be assigned to an explicit research topic or subject category, for which sources are numerous, inconsistent, and sometimes unavailable. Subject categories are most commonly assigned to papers according to the journals in which they are published. However, these assignments rarely agree with underlying citation community structure [17]. There is evidence that interdisciplinary research can be identified in a corpus according solely to the citation structure. Specifically, it has been shown

H. Cherifi et al. (Eds.): COMPLEX NETWORKS 2016 2022, SCI 1077, pp. 364–376, 2023.
https://doi.org/10.1007/978-3-031-21127-0_30

that frameworks that encode the *structural role* of articles in a citation graph can predict interdisciplinary interactions more accurately than those that encode only the proximity between papers [6]. In light of this, we explore the potential for modern graph learning methods to identify the citation structures associated with interdisciplinary research.

Numerous approaches have been proposed in the literature for the task of structural role discovery on graphs, where nodes on a graph are divided into classes of *structurally equivalent* nodes [21]. Early approaches in this area relied on graph sub-structures known as *graphlets* or *motifs* [12]. In large graphs, where we wish to identify higher-order structural features, counting large graphlets is very expensive. The majority of recent approaches employ a representation learning technique known as node embedding, which substitutes graphlets with alternative structural features like degree distributions [20] and diffusion wavelet patterns [8]. These methods are designed to learn higher-order structural relationships than those that can be discovered by small graphlets. However, in many cases, these alternatives approaches come at the cost of interpretability. When applied to graphs that are too large to be visualised reasonably, it is difficult to understand the substantive meaning of a given set of structural roles.

While embedding methods for role analysis have previously been shown to be capable of grouping nodes into known roles or structures (such as those in synthetic graphs or transport networks [8,20]), it remains unclear as to how these roles should be interpreted or validated when applied to real-world graphs with unknown roles. Moreover, different role discovery methods learn different sets of structural roles, depending on the application many or none of these clusterings may be valid. As such, it is critical that we can compare the roles discovered by different methods.

The core contribution of this work is a new framework for explaining a set of discovered roles using graphlet *orbits*, as described in detail in Sect. 3. Later in Sect. 4 we apply this framework to a large, multidisciplinary citation network to extract and interpret sets of structural roles. In the case of each paper, we compute Rao-Stirling diversity scores to indicate an articles interdisciplinarity. Crucially, in addition to interpreting a set of candidate roles, we explore the distributions of IDR scores assigned to papers in different clustering to assess if any of the candidate clusterings have grouped papers according to their interdisciplinarity.

2 Related Work

2.1 Measuring Interdisciplinarity

Interdisciplinary research is most commonly defined as research activity which integrates expertise, data or methodologies from two or more distinct disciplines. Accordingly, many studies assign an interdisciplinary research (IDR) score to a research article, which is calculated as a function of the balance, diversity and dissimilarity of the disciplines identified in the articles reference list [17,19]. Alternatively, some studies compute a similar score on the disciplines identified in an article's *citing* literature, instead measuring IDR according to an articles impact/influence across multiple disciplines [16,28]. A popular function for measuring IDR is the Rao-Stirling Diversity index [27]

$$D = \sum_{i,j(i \neq j)} p_i p_j d_{ij} \tag{1}$$

where IDR is measured as a pairwise summation over all pairs of disciplines identified in a set of articles cited by (or citing) some focal paper. Here p_i and p_j represent the proportion of references to disciplines i and j respectively, while d_{ij} refers to some precomputed distance between the disciplines. This metric, and its many variants, are reliant on explicit topic or discipline categories for research papers, such as those provided by Web of Science, Scopus, or Microsoft Academic. Such explicit categorisations for research papers, especially those assigned at a journal level, are problematic [1,13]. Moreover, the inconsistencies evident across many of these subject taxonomies [25] may confirm that no singular, correct categorisation exists. However, recent graph learning methods may be capable of identifying the citation structures associated with interdisciplinary research, without any knowledge of explicit discipline categorisation.

2.2 Local Structure and Role Embeddings

Research developed in social science studied local graph structures using small graph patterns such as triads, cycles and stars [15]. More recent research in computational science further developed these methods and proposed the term 'motif' – a subgraph pattern (or *graphlet*) which is significantly over-represented in a graph [12]. Motif and graphlet counts represent powerful features for expressing graph structure, and have been employed in graph learning tasks such as node classification and anomaly detection [7]. When using graphlet counts to describe nodes, it is more information to consider the position in the graphlet at which the node occurs. Equivalent positions within graphlets are called *automorphism orbits* or simply orbits. Figure 1 illustrates a subset of graphlets with 2, 3, 4 and 5 nodes, and includes each of the distinct orbits on these graphlets, as they were enumerated by [18].

Role discovery is the task of grouping nodes which share similar structural patterns in a graph into distinct classes [21]. Many modern approaches to role discovery rely on graph embedding, where nodes are transformed into low-dimensional vector representations [3,8,10,20]. Specifically, graph embedding methods for the purpose of role discovery (often called "role embedding") learn dense vector representations for nodes such that nodes that are structurally similar will be embedded nearby in the embedding space (i.e., will have similar representations). A clustering of the role embedding space thus represents a set of discovered roles. However, if this network cannot be visualised, it is difficult to interpret the roles. Moreover, with numerous approaches to role embedding (e.g. [3,8,10,20]), and many possible clusterings of each embedding space, we require some approach to explain a set of discovered roles so that they can be compared and validated.

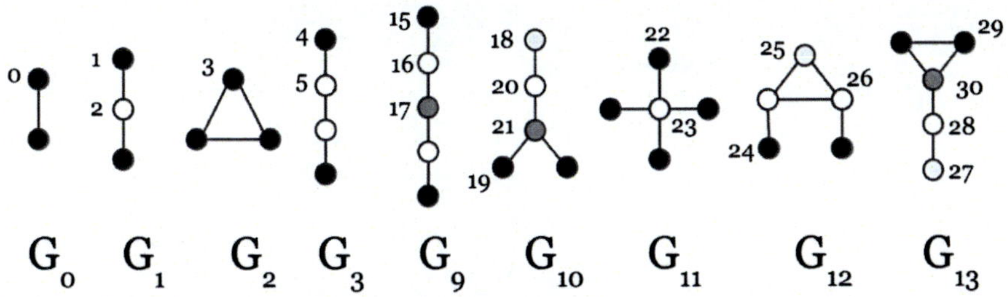

Fig. 1. Examples of graphlets with 2, 3, 4, and 5 nodes. Graphlets and orbits are enumerated according to [18]

2.3 Explanation Via Surrogate Models

In recent years, much research has been conducted in the field of *explainable artificial intelligence* (XAI) towards a goal of understanding and explaining so-called '*black box*' models [2]. One popular approach is to use a global surrogate model: "an interpretable model that is trained to approximate the predictions of a black box model" [14]. As such, the decisions made by a previously uninterpretable system can be explained through interpretations coming from the surrogate model.

Some classification models (such a logistic regression models or decision tree-based classifiers) are interpretable by definition, as any input feature's effect on the models classification can be measured (for example, using regression coefficients). However, many model-agnostic methods of interpretation have also been developed. Generally, these methods propose different means of perturbing the input data, and exploring its effect on the models output. For example, *Partial Dependence Plots* (PDPs) [9] offer a graphical representation of a models prediction as a function of its arguments or inputs. Broadly speaking, a PDP shows how a feature influences the prediction of a model, on average, for every value that it can take. The *Accumulated Local Effects* plot (ALE) [4] provides a similar illustration of a features effect, except it accounts for possible correlations between features, and as a result is more robust than the PDP in some applications. In 2021, global surrogate models were employed to provide feature-based explanations for communities detected on graphs [24]. Specifically, this work assessed the importance of different graph features using *permutation importance* [5], where the values for a feature are permuted throughout the dataset and the effect on model performance is measured to indicate its importance.

3 Methods

In this section we outline a general framework for uncovering and evaluating structural roles on a graph. The complete process is illustrated in Fig. 2.

We begin with a graph $G = (V, E)$, from which we wish to identify a set of discrete structural roles. We employ a role embedding algorithm to map each

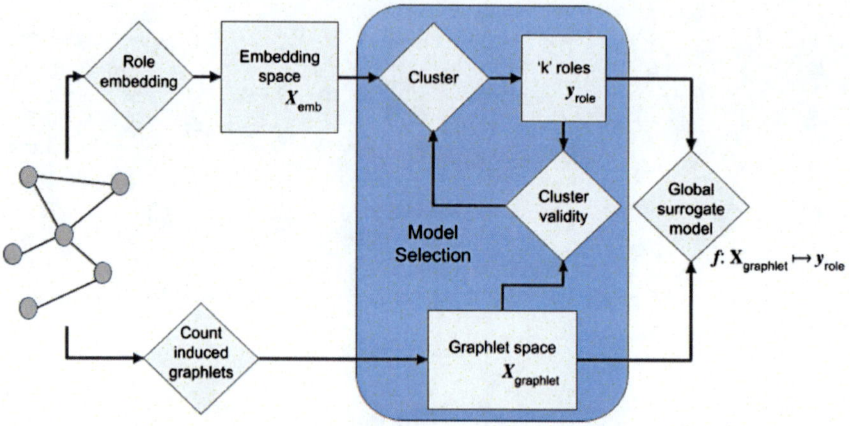

Fig. 2. Overview of the workflow for the proposed framework.

node $v \in V$ to an embedding space $X_{emb} = \mathbb{R}^{128}$, thus a clustering of X_{emb} is considered a set of discovered roles y_{role}. Additionally, we represent the same nodes in the graphlet space $X_{graphlet} = \mathbb{Z}^{73}$ derived from G, where each node v is represented by a *bag-of-orbits* vector $x^v = \{x_0^v, x_1^v, ..., x_{72}^v\}$, with x_i^v denoting the number of times node v is involved in induced graphlet orbit i. We use a vocabulary of 73 orbits, which we count on graphlets of size 2–5 using the ORCA method [11]. We refer to graphlets and orbits according to the enumeration by [18]. The graphlet space is first used to validate any set of roles we identify in the graph. By clustering the embedding space, we group nodes into k discrete roles, which we can evaluate using cluster validity metrics calculated on the graphlet space. Employing various role embedding algorithms and clustering methods, we identify a set of candidate clusterings (or roles) according to the separation they achieve in the graphlet space.

Graphlet-orbit counts are a powerful language for describing the structural features of nodes on a graph [7,18]. As such, a clustering (according to node embeddings) which is valid in the graphlet-orbit space should offer a reasonable set of structural roles. Additionally, we employ graphlet orbits to interpret a set of discovered roles. For a candidate clustering y_{role}, we fit a surrogate model $f : X_{graphlet} \mapsto y_{role}$. By modelling the role assignment in the graphlet space, we can explore the feature importance and effect of the different graphlet orbits in role assignments, according to many *model-agnostic* explanation techniques from the field of XAI [14]. In Sect. 4, we demonstrate graphlet-based explanation using permutation importance [5] and accumulated local effects plots [4]. As we will show, highlighting important or discriminatory orbits can offer a visual means of understanding the structure of a role in the graph.

4 Application

In this section we apply the methods described in Sect. 3 to extract and interpret sets of structural roles in a citation network. In the case of each paper, we compute Rao-Stirling diversity scores to indicate an articles interdisciplinarity and thus explore the distributions of IDR scores assigned to papers in different roles. We identify a set of roles which has grouped papers according to their interdisciplinarity. Finally, we use graphlets to interpret the structure of these more interdisciplinary roles, and highlight certain citation structures which are specific to interdisciplinary research.

4.1 Data

In order to discover the citation structures of interdisciplinary research, we require a large, dense citation network that contains research from a diverse set of disciplines. In addition, we require that each paper can be assigned to a subject category or discipline, according to an established taxonomy. We construct a novel citation network using Microsoft Academic Graph [26] citation data from a seed set of journal papers. This sets consist of samples of articles from Scopus indexed journals, stratified according to their All Science Journal Categories (ASJC). The graph contains samples of 1,500 articles published between 2017 and 2018 in Scopus indexed journals with the ASJCs *'Computer Science'*, *'Mathematics'*, *'Medicine'*, *'Chemistry'*, *'Social Sciences'*, *'Neuroscience'*, *'Engineering'*, and *'Biochemistry, Genetics and Molecular Biology'*. We maximise the completeness of the graph by including all available referenced articles that are published in Scopus indexed journals. In this manner, we produce a dense, multidisciplinary citation network, such that each article can be categorised according to the ASJC of the journal in which it was published. Later, these discipline categories can be used to compute article interdisciplinarity according to Rao-Stirling diversity of disciplines identified in both an articles citing and cited papers. In total, the citation graph contains 41,895 papers (nodes) and 129,159 citations (undirected edges).

4.2 Role Discovery

For each article in the citation graph, we learn role embeddings according to 4 approaches: (i) *Role2Vec* [3], (ii) *Struc2Vec* [20], (iii) *RolX* [10], (iv) *GraphWave* [8], and we cluster each embedding space using k-means clustering for values of k between 2 and 19. Articles clustered according to their role embeddings represent a set of structural roles.

4.3 Role Interpretation

Validation. Figure 3 shows the cluster validity of the roles discovered according to the 4 role embedding methods. Each set of roles represents a k-means clustering of the embedding space, which is then transformed to the graphlet space, where we measure the validity of the clustering using silhouette score [23]. We plot the silhouette score for each embedding for values of k in the range $[2, 19]$. Silhouette scores can take a value in the range [−1,1]. A high score represents dense, separate clusters, a score of 0 indicates an overlapping clustering, while a negative score indicates an incorrect clustering. According to these scores, we choose 3 candidate roles to demonstrate interpretation and explanation: (i) Struc2Vec ($k = 6$) which is an outlier in the Struc2vec roles and achieves an overlapping clustering, (ii) RolX ($k = 3$) which has the highest silhouette score for all approaches with more than 2 clusters, and (iii) GraphWave ($k = 3$) which achieves a positive score.

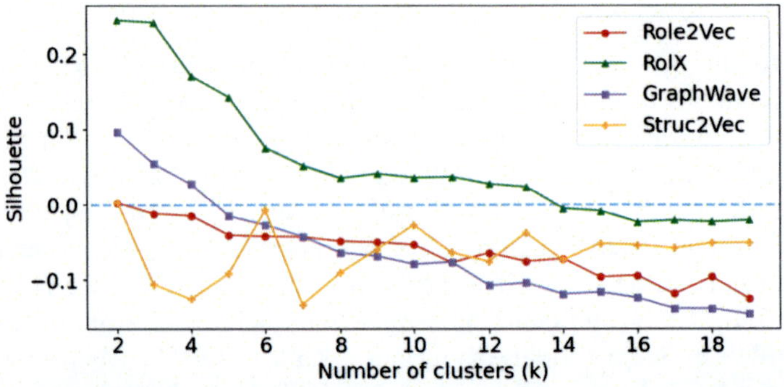

Fig. 3. Silhouette score for k-means clusterings of the different embedding spaces. Silhouettes are calculated according the clusterings if we describe each node by their bag-of-orbits vector, rather than their embedding.

Surrogate models. We fit a random forest classifier to model role assignments according to the bag-of-orbits vectors for each node. That is, we learn a function $f : X_{graphlet} \mapsto y_{role}$ for each of the candidate roles y_{role}. We choose a random forest classifier as we anticipate that structural roles may be non-linear in the graphlet space and may rely on combinations and interactions of features to model higher-order structures. As the orbit counts follow a power-law distribution, we log-transform all features in the graphlet space. Table 1 reports the 5 most informative features for each model according to permutation importance. In the case of the Struc2Vec ($k = 6$) roles, the overlap between clusters in the graphlet space is evident. The only informative features (with non-zero permutation importance) are small, local orbits – the approach is blind to deeper, more complex structures. The RolX ($k = 3$) approach, which best separates the nodes in the graphlet space, has grouped nodes according to larger, simple structures. The most informative orbits: 15, 4, and 1, each refer to chains of varying length. Finally, GraphWave ($k = 3$) appears to have grouped the nodes according to more complex, higher-order structures. Many of the features important to role classification in the GraphWave case (27, 24, and 18) contain combinations of chains, stars and cliques. Depending on the domain or application in which we employ role discovery, any one of these sets of roles may be the most valid or useful. However, without modelling the role assignments in the graphlet space, we are unable to understand which structures are being grouped in the discovered roles. We will use the GraphWave roles to demonstrate further explanation and explore the structure of interdisciplinary papers.

Table 1. Top 5 most important features for each surrogate model. Orbits are ranked by permutation importance, which is included in parentheses.

	Struc2Vec ($k = 6$)	RolX ($k = 3$)	GraphWave ($k = 3$)
1	0 (0.111 ± 0.0005)	15 (0.113 ± 0.0016)	27 (0.022 ± 0.0006)
2	2 (0.010 ± 0.0001)	4 (0.007 ± 0.0002)	15 (0.014 ± 0.0002)
3	3 (0.006 ± 0.0001)	1 (0.006 ± 0.0003)	17 (0.013 ± 0.0006)
4	5 (0.002 ± 0.0001)	27 (0.005 ± 0.0002)	24 (0.013 ± 0.0004)
5	16 (0.000 ± 0.0000)	19 (0.004 ± 0.0003)	18 (0.013 ± 0.0004)

Roles and their structure. The GraphWave method clusters 35,136 papers into role 0, 16,453 papers into role 1, and 306 papers into role 2. Figure 4 contains Accumulated Local Effects (ALE) [4] plots for 3 features/orbits and their effect on classification to each of the 3 GraphWave roles. We illustrate the ALE of orbits 27 (the end of a chain adjacent to a clique) and 17 (the middle of a long chain) as two of the most important structures (according to permutation importance). We also include the ALE of orbit 0 (node degree) as a valuable reference; it is useful to confirm that the roles are indeed separating nodes according to more complex features and not simply the number of edges. The ALE plot for orbit 27 shows that for low-to-mid values of that orbit count, a node will be classified as role 0. However, if a node's count for orbit 27 exceeds a threshold, it will be classified as role 1. We suppose 2 scenarios when a focal node's count for orbit 27 (a chain adjacent to a clique) will become large: (i) the node is adjacent to a large community – each triangle in which the node at position 30 participates will increase the count, (ii) the node exists at the center of a barbell graphlet, i.e. on a longer chain between two or more communities communities. We illustrate these scenarios in Fig. 5. There should exist some threshold value of 27 beyond which a node must exist on the chain between two communities. For example, if a focal node has a count for orbit 27 that is greater than the count of triangles (orbit 3) for the node at position 30, then the focal node must be adjacent to a second community (scenario (iii)). This threshold will be represented by the greatest value of orbit 3 in the graph. We include this value for reference in Fig. 4. Beyond this threshold, a node is more likely to be classified in role 1. Accordingly, we conclude that a node/paper that is on the end of a chain adjacent to a community will be assigned to role 0, while a node that exists on a bridge between two communities will be assigned to role 1.

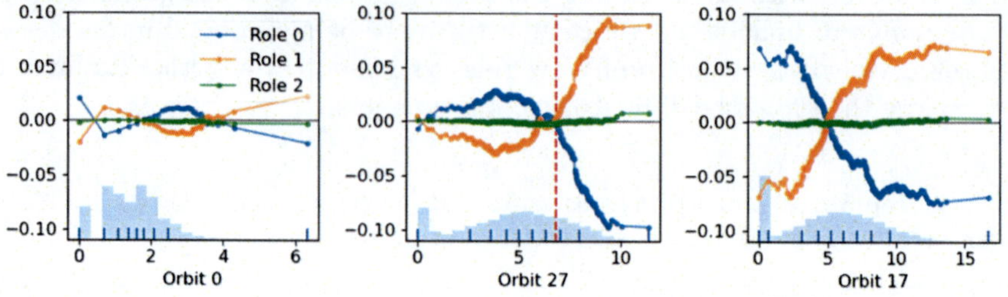

Fig. 4. Accumulated Local Effect [4] plots for a surrogate model which classifies nodes to GraphWave roles according to graphlet orbit counts. The figure shows the effect of 3 features on role classification: orbits 27 and 17; the most important features as measured by permutation importance, and orbit 0 (node degree); which we include for reference. In the case of orbit 27, we highlight the maximum value of orbit 3 that was observed in the graph.

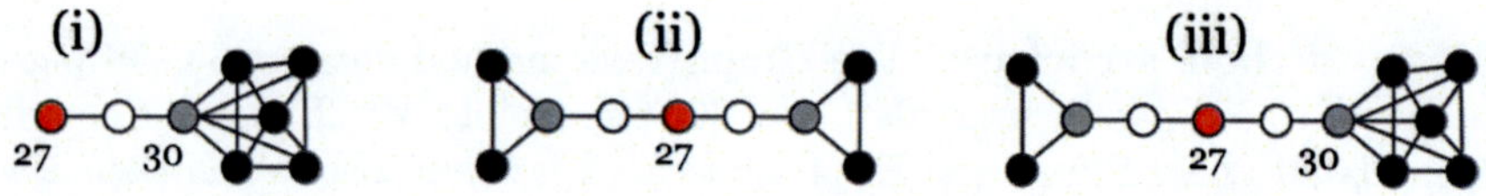

Fig. 5. Higher order structures containing orbit 27. Here (i) and (ii) represent likely structures for nodes with high counts for orbit 27. When the count for orbit 27 exceeds a threshold, we infer structure (iii).

In order to identify the structures specific to the smallest role (role 3), we can fit another surrogate model on only the nodes in cluster 1 and 2. ALE plots for this model are included in Fig. 6. In this case we find that orbit 27 does not meaningfully distinguish between the two roles. Instead, orbit 18 (the end of a chain adjacent to a star) is the most informative feature, and, for very high values of this orbit count, nodes will be assigned to role 2. Such nodes likely represent the centre of a bridge between large communities that are less densely connected (i.e., containing many open triads). We conclude this to be an important structure for role 3.

4.4 Interdisciplinary Roles

Figure 7 plots the IDR scores for the papers in each of the GraphWave roles. IDR is calculated according to the Rao-Stirling diversity of the ASJC categories identified in an article's citing papers. As Rao-Stirling IDR scores may be biased

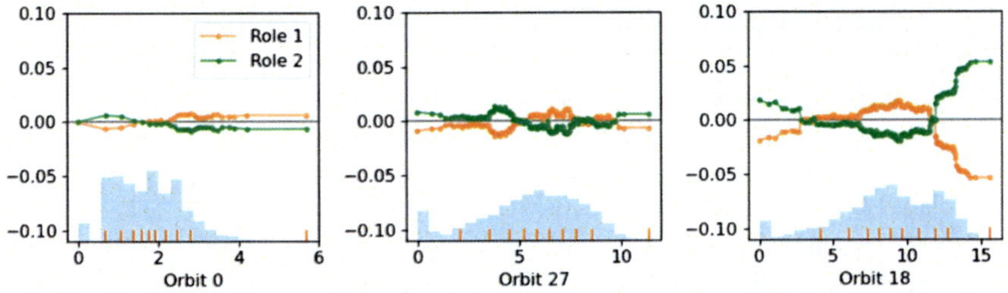

Fig. 6. Accumulated Local Effect [4] plots for a surrogate model which classifies nodes to GraphWave roles 1 or 2 according to graphlet orbit counts. The figure plots the effect of 3 features: orbits 0; node degree and 27; previously the most important feature for the global model, and orbit 18; the most important feature in the surrogate model.

according to the number of articles in the summation, we bin nodes by degree, and plot IDR distributions for each role, within each bin. Specifically, we log-transform the node degrees and group nodes into 10 bins of equal width, within which we plot IDR distributions for each role if the bin contains more than 50 papers from each role. According to these plots, we note that the structural roles identified by GraphWave have grouped papers into clusters with different IDR distributions. Even when we account for node degree (a potential bias of the Rao-Stirling IDR score), papers assigned to structural roles 1 and 2 have consistently greater IDR distributions than those assigned to role 0. We recall some of the important structures that were identified for roles 1 and 2: (1) a bridge between densely connected communities, and (2) a bridge between large, sparsely connected communities. We conclude these to be citation important structures associated with interdisciplinary research.

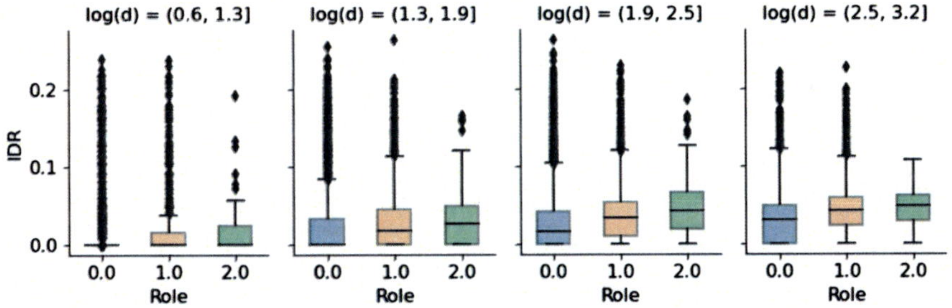

Fig. 7. Interdisciplinarity scores (IDR) computed for papers in each of the 3 GraphWave clusters. IDR is computed as the Rao-Stirling [27] diversity of the research disciplines identified in an articles citing papers. We bin papers/nodes according to the log of their degree, and compare distributions within each bin.

5 Discussion and Conclusions

Many modern methods for role discovery in graphs rely on node embeddings [22]. We have applied a number of these methods to learn structural roles on a research citation network. While these methods have previously been shown to be capable of grouping nodes into known roles (e.g., in synthetic graphs or transport networks [8,20]), it remains unclear how roles could be understood or validated when applied to graphs with unknown roles. In this work we outlined a framework for interpreting a set of discovered roles using graphlets and orbits. We leveraged methods from the field of explainable AI to explore the subgraph structures that are specific to discovered roles. We demonstrated our approach on a citation network, where we identified important structures specific to interdisciplinary research. It is evident in our analysis that different role discovery methods learn different sets of structural roles. In different applications many or all of these clusterings may be valid, but it is critical that we can compare the roles discovered by different methods. While our framework is general, and applicable to explanation and validation in all role discovery tasks, we highlight the utility of structural role embeddings in mapping interdisciplinary research.

For the task of identifying and mapping IDR interactions, structural paper embeddings could be augmented by considering additional, non-structural information, such as article or abstract text. This could provide a richer paper representation, without imposing a predefined or static disciplinary classification on the graph. There is also scope for improving upon our proposed framework. Many *model-agnostic* approaches have been developed for explaining surrogate models [14], which could be applied to interpret the role assignments in the graphlet space. For example, second-order effects of pairs of features can be calculated in a similar manner to the ALE analysis we have included [4]. Combinations of graphlets could be highly effective in modelling higher-order, more complex graph structures. One possible limitation of our current framework is the number of correlated features in the graphlet space. In future applications it may be possible to reduce the set of graphlet orbits to a minimal set of uncorrelated features via traditional feature selection techniques.

References

1. Abramo, G., D'Angelo, C.A., Zhang, L.: A comparison of two approaches for measuring interdisciplinary research output: the disciplinary diversity of authors versus the disciplinary diversity of the reference list. J. Inform. **12**(4):1182–1193 (2018)
2. Adadi, A., Berrada, M.: Peeking inside the black-box: a survey on explainable artificial intelligence (Xai). IEEE Access **6**, 52138–52160 (2018)
3. Ahmed, N.K., Rossi, R.A., Lee, J.B., Willke, T.L., Zhou, R., Kong, X., Eldardiry, H.: role2vec: Role-based network embeddings. In: Proceedings of the DLG KDD, pp. 1–7 (2019)
4. Apley, D.W., Zhu, J.: Visualizing the effects of predictor variables in black box supervised learning models. J. Roy. Stat. Soc. Ser. B (Stat. Methodol.) **82**(4), 1059–1086 (2020)

5. Breiman, L.: Random forests. Mach. Learn. **45**(1), 5–32 (2001)
6. Cunningham, E., Greene, D.: Assessing network representations for identifying interdisciplinarity (2022). arXiv preprint arXiv:2203.12455
7. Cunningham, P., Harrigan, M., Wu, G., O'Callaghan, D.: Characterizing ego-networks using motifs. Netw. Sci. **1**(2), 170–190 (2013)
8. Donnat, C., Zitnik, M., Hallac, D., Leskovec, J.: Learning structural node embeddings via diffusion wavelets. In: Proceedings of the 24th ACM SIGKDD International Conference on Knowledge Discovery and Data Mining, pp. 1320–1329 (2018)
9. Friedman, J.H.: Greedy function approximation: a gradient boosting machine. Ann. Stat. 1189–1232 (2001)
10. Henderson, K., Gallagher, B., Eliassi-Rad, T., Tong, H., Basu, S., Akoglu, L., Koutra, D., Faloutsos, C., Li, L.: Rolx: structural role extraction and mining in large graphs. In: Proceedings of the 18th ACM SIGKDD International Conference on Knowledge Discovery and Data Mining, pp. 1231–1239 (2012)
11. Hočevar, T., Demšar, J.: Combinatorial algorithm for counting small induced graphs and orbits. PloS ONE **12**(2), e0171428 (2017)
12. Milo, R., Shen-Orr, S., Itzkovitz, S., Kashtan, N., Chklovskii, D., Alon, U.: Network motifs: simple building blocks of complex networks. Science **298**(5594), 824–827 (2002)
13. Milojević, S.: Practical method to reclassify web of science articles into unique subject categories and broad disciplines. Quant. Sci. Stud. **1**(1), 183–206 (2020)
14. Molnar, C.: Interpretable Machine Learning. Lulu.com (2020)
15. Moreno, J.L.: Who shall survive?: A new approach to the problem of human interrelations (1934)
16. Porter, A., Chubin, D.: An indicator of cross-disciplinary research. Scientometrics **8**(3–4), 161–176 (1985)
17. Porter, A., Rafols, I.: Is science becoming more interdisciplinary? measuring and mapping six research fields over time. Scientometrics **81**(3), 719–745 (2009)
18. Pržulj, N.: Biological network comparison using graphlet degree distribution. Bioinformatics **23**(2), e177–e183 (2007)
19. Rafols, I., Meyer, M.: Diversity and network coherence as indicators of interdisciplinarity: case studies in bionanoscience. Scientometrics **82**(2), 263–287 (2010)
20. Ribeiro, L.F., Saverese, P.H., Figueiredo, D.R.: struc2vec: Learning node representations from structural identity. In: Proceedings of the 23rd ACM SIGKDD International Conference on Knowledge Discovery and Data Mining, pp. 385–394 (2017)
21. Rossi, R.A., Ahmed, N.K.: Role discovery in networks. IEEE Trans. Knowl. Data Eng. **27**(4), 1112–1131 (2014)
22. Rossi, R.A., Jin, D., Kim, S., Ahmed, N.K., Koutra, D., Lee, J.B.: On proximity and structural role-based embeddings in networks: misconceptions, techniques, and applications. ACM Trans. Knowl. Discov. Data (TKDD) **14**(5), 1–37 (2020)
23. Rousseeuw, P.J.: Silhouettes: a graphical aid to the interpretation and validation of cluster analysis. J. Comput. Appl. Math. **20**, 53–65 (1987)
24. Sadler, S., Greene, D., Archambault, D.: Selecting informative features for post-hoc community explanation. In: International Conference on Complex Networks and Their Applications, pp. 297–308. Springer (2021)
25. Shen, Z., Chen, F., Yang, L., Wu, J.: Node2vec representation for clustering journals and as a possible measure of diversity. J. Data Inform. Sci. **4**(2), 79 (2019)
26. Sinha, A., Shen, Z., Song, Y., Ma, H., Eide, D., Hsu, B.-J., Wang, K.: An overview of microsoft academic service (mas) and applications. In. Proceedings of the 24th International Conference on World Wide Web, pp. 243–246 (2015)

27. Stirling, A.: A general framework for analysing diversity in science, technology and society. J. Roy. Soc. Interf. **4**(15), 707–719 (2007)
28. Van Noorden, R., et al.: Interdisciplinary research by the numbers. Nature **525**(7569), 306–307 (2015)

Inferring Parsimonious Coupling Statistics in Nonlinear Dynamics with Variational Gaussian Processes

Ameer Ghouse(✉) and Gaetano Valenza

Department of Information Engineering and Bioengineering and Robotics Research Center E. Piaggio, School of Engineering, University of Pisa, Pisa, Italy
a.ghouse@studenti.unipi.it

Abstract. Falsification is the basis for testing existing hypotheses, and a great danger is posed when results incorrectly reject our prior notions (false positives). Though nonparametric and nonlinear exploratory methods of uncovering coupling provide a flexible framework to study network configurations and discover causal graphs, multiple comparisons analyses make false positives more likely, exacerbating the need for their control. We aim to robustify the Gaussian Processes Convergent Cross-Mapping (GP-CCM) method through Variational Bayesian Gaussian Process modeling (VGP-CCM). We alleviate computational costs of integrating with conditional hyperparameter distributions through mean field approximations. This approximation model, in conjunction with permutation sampling of the null distribution, permits significance statistics that are more robust than permutation sampling with point hyperparameters. Simulated unidirectional Lorenz-Rossler systems as well as mechanistic models of neurovascular systems are used to evaluate the method. The results demonstrate that the proposed method yields improved specificity, showing promise to combat false positives.

Keywords: Dynamical systems · Causal analysis · Robust methods

1 Introduction

Coupling measures derived from data allow discovery of system characteristics and information flow when the experiment does not permit intervention in the physical process. Traditionally, pairwise coupling metric such as mutual information or correlation are used for assessing coupling between systems [1], such as for assessing functional connectivity in the brain [2]. Dynamical system methods have been popular for uncovering coupling, such as Granger causality and its nonparametric extension, transfer entropy [3–5]. However, in their standard form, they assess linear relationships through multivariate autoregression models [4], though extensions have been made for assessing nonlinear coupling through kernel methods [6]. Nonetheless, such methods have been shown to report spurious detections of causality, i.e. false positives [7,8]. Straining out the possibilities

H. Cherifi et al. (Eds.): COMPLEX NETWORKS 2016 2022, SCI 1077, pp. 377–389, 2023.
https://doi.org/10.1007/978-3-031-21127-0_31

of false positive is imperative in hypothesis driven analysis to avoid improper conclusions based on a priori expectations [9]—in causal analysis, the prior notion is non causal relationship between two variables.

Convergent cross-mapping was introduced to account for cases when synergistic nonlinear coupling may emerge from complex dynamical systems [10]. This phenomenon is possible to be exploited as, holistically, the history of state space is assessed (nonparametrically) rather than being purely an analysis of the lag values' predictive power as with Granger causality. Formally, convergent cross-mapping (CCM) suggests that, in deterministic systems, there should be high correlation in predicting a cross-mapped system by taking a state from a state space reconstruction (as in [11]) of a putative variable to the state space reconstruction of a caused variable [10]. Subsequent ameliorations introduced Gaussian processes with the notion that the reconstructed state space has uncertainty, thus a prior Gaussian process mean and covariance functions were applied, either as in [12] to optimize the free parameters for state space construction, in [13] to analyze residuals in cross-mapping, or to assess a probability ratio as in [14]. We shall refer to the latter method in our paper.

Gaussian process models are characterized by mean and covariance (often referred to as the kernel) functions [15]. Though they provide nonparametric inference models for a posteriori analysis of random processes conditioned on observed data, they are sensitive to the hyperparameters of the kernel function that describes the covariance between observations. Point estimates of hyperparameters are optimized by maximum marginal likelihood of the observations. However, these point estimates may provide degenerate distributions when generating null distributions derived from permutation sampling for hypothesis testing as empirically seen in [14] and other contexts of Bayesian modeling [16]. We thus propose placing prior distributions on the hyperparameters and subsequently obtain an approximate posterior distribution conditioned on observations (who a priori are zero mean Gaussian processes) using variational Bayesian methods, hereafter referred to as Variational Gaussian Process Convergent Cross-Mapping (VGP-CCM). This method avoids expensive Markov Chain Monte Carlo integration methods for conditional distributions [17]. With the approximate posterior we integrate out hyperparameters effects from the coupling statistics and derive nondegenrate null distributions for hypothesis testing. We test this amelioration on unidirectional coupled Lorenz-Rossler systems and neurovascular systems.

2 Materials and Methods

2.1 Cross Mapping

A central tenet of nonlinear analysis in state space using a single time series, which we refer to in lower case letters such as x, hinges on state space reconstruction, where a time series of a variable in a system is viewed as a projection of system's topology to a single dimension (focusing on reals, $\mathbb{R}^{N\times 1}$ where N is number of time points), and where using delay-coordinate embedding maps

($\phi(x_i) = \{x_i, x_{i-\tau}, x_{i-2\tau}, ..., x_{i-m\tau}\}$ for i in $1..N - \tau m$, where m is an "embedding dimension" and τ delay time) of said time series can reconstruct a state space with isomorphisms to the generating system's structure in terms of differential geometry [11,18]. Cross-mapping exploits this theoretical feature of reconstructed state spaces and the nature of time series being low dimension projections of systems, such that a state space reconstructed by a putative variable ($\phi(x)$) should be independent of a caused variable ($\phi(y)$, implying its reconstructed state space contains no dynamical information of the caused variable leading to the independence of predictions of a caused variable by a putative variable's state. On the other hand, a prediction of a putative variable from a caused variable's state space should not be independent of the putative variable as its reconstructed geometry needs to have topology consistent with that of the putative variable to predict itself. Convergent cross-mapping in its original form approaches this problem as a regression task, particularly as performed by simplex regression [10,19].

2.2 Gaussian Process Convergent Cross Mapping

Introduced in [14], GP-CCM utilizes probability ratio to determine the most likely coupling direction based on spatial analysis in state space through incorporating a delay-coordinate map into the kernel function that defines the a priori covariance function of the Gaussian process, i.e. $Cov(x_i, x_j) = K(\phi(x_i), \phi(x_j))$. In other words, rather than assess predictive powers through regression analysis, GP-CCM proposed a probabilistic approach that relates the cross-mapping power as a multivariate probability distribution. Let X and Y be stochastic processes that generate the observed time series of x and y respectively. Probability functions for X and Y are derived by posterior analysis from conditioning a joint Gaussian Processes by the observed time series y and x respectively. Consider $\tilde{X}$ and $\tilde{Y}$ the null distributions for uncoupled processes (e.g. through permutation sampling). Then, the null GP-CCM probability ratio test would be:

$$\tilde{\kappa}_k(X,Y) = \frac{P(\tilde{Y}|X;\theta_x)}{P(\tilde{X}|Y;\theta_y)} \tag{1}$$

θ_x or θ_y are considered nonrandom point values, $\tilde{\kappa}$ is a function of the null distribution, and k is the permutation sample iteration. For sake of conciseness, we shall abuse the notation, e.g. $P(X|Y)$, for a conditioned distribution, e.g. $P(X|Y = y)$, for the rest of this paper.

Information Theoretic Results In [14], the maximum a posteriori (MAP) of the distributions $P(\tilde{Y}|X;\theta_x)$ and $P(\tilde{X}|Y;\theta_y)$ were used to provide scalar results. Given the form of a normal distribution with dimensionality d, mean μ and covariance Σ:

$$f_x(X) = ((2\pi)^d|\Sigma|)^{-\frac{1}{2}}\exp((X-\mu)\Sigma^{-1}(X-\mu)^\top) \tag{2}$$

The MAP would be where $X = \mu$, making the exponential an identity matrix, thus:

$$MAP(\mathcal{N}(\mu, \Sigma)) = ((2\pi)^d|\Sigma|)^{-\frac{1}{2}} \tag{3}$$

Which we can see, from taking the negative logarithm, is equivalent to the differential entropy of a Gaussian distribution with an offset of $\frac{d}{2}$:

$$\begin{aligned} H(\mathcal{N}(\mu, \Sigma)) &= \frac{d}{2} + \frac{d}{2}log(2\pi) + \frac{1}{2}log(|\Sigma|) \\ &= -log(MAP(\mathcal{N}(\mu, \Sigma))) + \frac{d}{2} \end{aligned} \tag{4}$$

Let $K_k(\tilde{X}, \tilde{Y}) = log(\frac{MAP(P(\tilde{Y}|X;\theta_x)}{MAP(P(\tilde{X})|Y;\theta_y)})$, we then obtain:

$$K_k(\tilde{X}, \tilde{Y}) = -H(\tilde{Y}|X;\theta_x) + H(\tilde{X}|Y;\theta_y) = log\left(\frac{|\Sigma_{\tilde{X}|Y}|}{|\Sigma_{\tilde{Y}|X}|}\right) \tag{5}$$

An interpretation of our probability ratio test in information theory can be the difference of the amount of information X provides given we were informed of Y a priori vs how much information Y provides given we were informed of X a priori. Caused variables should provide minimal information to the causing variable's state space, thus $H(X|Y) < H(Y|X)$ if X drives Y.

Standardizing Results and Null Distributions N samples of $K_k(X, Y)$ can be generated from N permutations. The values $K_k(X, Y)$ are normalized by the number of samples in the time series and then passed to a hyperbolic tangent function in order to standardize the values between (-1,1). From the samples of $K_k(X, Y)$, an empirical cumulative distribution is derived and a 1 sided test is performed to see whether the probability of the causal statistic is less than an α in the null distribution. α is often chosen to be 0.05. Concretely, to test whether X GP-CCM causes Y, the following is calculated:

$$p = \frac{\sum_{j=1}^{N} U(K(X, Y) - K_j(\tilde{X}, \tilde{Y}))}{N} \tag{6}$$

where U is the Heaviside function which is only 1 when the measured statistic for the actual observations $K(X, Y)$ is greater than the permutation samples $K(\tilde{X},\tilde{Y})$, otherwise 0.

2.3 Variational Gaussian Process Convergent Cross Mapping

Variational Approximation of the Hyperparameter Distributions The proposed method, Variational Gaussian Process Convergent Cross-Mapping (VGP-CCM), proposes to treat hyperparameters as random effects to be integrated out of the model, via Monte Carlo integration, exploiting a proposed a posteriori distribution of hyperparameters $Q(\theta_x|X)$ that approximate the true a posteriori distribution $P(\theta_x|X)$ in order to exploit independence structures that make integration computational feasible rather than resorting to more expensive methods such as Gibbs conditional sampling of the true posteriori [20].

To elaborate, hyperparameters are considered point estimates with no uncertainty in eq. 1. However, studies have shown that point estimates may not be stable in a variational sense and can correspond to models that overfit [21]. Calculus of variations provides a method to discover models that exhibit minimal "action" from perturbations in their function. We utilize this framework to discover functions of our hyperparameters from which we can draw samples to integrate out the effects of the hyperparameters in the model subsequently. We modify eq. 1 accordingly:

$$\begin{aligned}\tilde{\kappa} &= \frac{\int P(\tilde{Y}|\theta_x)P(\theta_x|X)d\theta_x}{\int P(\tilde{X}|\theta_y)P(\theta_y|Y)d\theta_y} \\ &= \frac{P(\tilde{Y}|X)}{P(\tilde{X}|Y)}\end{aligned} \tag{7}$$

The hyperparameter posterior distribution tend to be intractable to compute as they depend on the marginal distributions of $P(X)$ or $P(Y)$ (model evidence) respectively:

$$P(\theta_x|X) = \frac{P(X|\theta_x)P(\theta_x)}{\int P(X|\theta_x)p(\theta_x)d\theta_x} \tag{8}$$

We, instead, consider a simpler model $Q(\theta_x)$ that approximates the posterior $P(\theta_x|X)$ based on mean field approximations, eliminating the dependency of conditioning samples on X while minimizing the statistical distance in the Kullback-Leibler sense [22] from the true posterior distribution which is conditioned on X. In other words, we want to substitute a simple distribution $Q(\theta_x)$ with independence structures for $P(\theta_x|X)$ in eq. (7) to permit computationally efficient Monte Carlo integration:

$$\tilde{\kappa} \approx \frac{\int P(\tilde{Y}|\theta_x)Q(\theta_x)d\theta_x}{\int P(\tilde{X}|\theta_y)Q(\theta_y)d\theta_y} \tag{9}$$

The best approximation Q can be derived by maximizing the lower bound for the evidence (ELBO) provided the approximate posterior via the KL divergence.

$$\begin{aligned}KL(Q||P) &= E_{\theta_x\sim Q}(log(\frac{Q(\theta_x)}{P(\theta_x|X)})) = E_{\theta_x\sim Q}(log(\frac{Q(\theta_x)P(X)}{P(X|\theta_x)P(\theta_x)}) \\ &= log(P(X)) + KL(Q(\theta_q)||P(\theta_q)) - E_{\theta_x\sim Q}(P(X|\theta_x))\end{aligned} \tag{10}$$

As the KL divergence is strictly greater than 0, its minimization thereby is also a maximization of the ELBO. This can be seen by multiplying both sides by -1 and taking the evidence ($log(P(X))$) and moving it to the left hand side:

$$\begin{aligned} ELBO &= E_{\theta_x \sim Q}(log(P(X|\theta_x))) - KL(Q(\theta_x)||P(\theta_x)) \\ log(P(X)) - KL(Q||P) &= ELBO \\ log(P(X)) &\geq ELBO \end{aligned} \tag{11}$$

Thus our objective to obtain the optimal hyperparameter distributions $Q(\theta_x)$ is to maximize the above ELBO. The next question becomes how to choose the form of our distribution $Q(\theta_x)$.

Approximate Posterior Form The form of the posterior distribution is the art of the practitioner based on their domain knowledge. For the sake of this paper, we choose its form to follow a mean-field approximation that factorizes as $Q(\theta) = \prod_{k=1}^{K} Q(\theta_k)$ for K hyperparameters. θ comprises the hyperparameters of an exponential squared distance kernel with automatic relevance detection σ and l [23], and the inducing pseudopoints for a sparse kernel z [24]. In order to maintain strictly positive parameters for the kernel parameters as well as having a distribution from which we can reparameterize to draw samples to automatically compute gradients [25] and in which we can derive easily the KL-divergence, we choose log normal distributions as their priors. The distribution for the inducing points, instead follows a normal distribution. For the KL divergences in the ELBO, analytic expressions can be derived. It happens that log normal $log\mathcal{N}(\mu, \sigma)$ and normal distribution $\mathcal{N}(\mu, \sigma)$ [25] share the same expression for KL divergences.

Note that if the prior P(X) were chosen independent point distributions, i.e. $P(X) = \delta(X - \mu_p)$, the optimal $Q(X)$ distribution would have a collapsed mode on μ_p, arriving at the original formulation of GP-CCM in eq. (1) in the case that the prior μ_p were chosen as the maximum marginal likelihood solution to $P(X; \theta_x)$.

A priori the hyperparameters are ansatz from the data. We standardize the data to be zero mean unit variance, thus the covariance amplitude σ has parameters $\mu_\sigma = \exp(1)$ and $\sigma_\sigma = 1$. Length factor l of the kernel has $\mu_l = \exp(1)$, $\sigma_l = 1$. The inducing points are a priori mean equal to a random selection of a subset of the observations with $\sigma_z = 1$. Calculus of variations could be used to derive a coordinate ascent method of obtaining the optimal distribution Q [26]. We, instead, choose to maximize the ELBO using stochastic gradient ascent as in [25] and automatic differentiation in PyTorch [27] to avoid manually deriving the update equations. The expectation $E_{\theta_x \sim Q}(P(X|\theta_x))$ is approximated by 10 draws of θ_x from $Q(\theta_x)$ [25]. Furthermore, gradient clipping was used to minimize the effect of the stochastic gradient's variance [28]. The final result from VGP-CCM comes from applying eq. (5) to the distributions in eq. (7). Code is publicly available in Python at the author's Github profile.

2.4 Synthetic Data

Unidirectionally Coupled Lorenz and Rossler We simulate Rossler (Y) and Lorenz (X) systems with nonlinear unidirectionally coupling defined by either ϵ_x or ϵ_y and diffusion dynamics defined by a 6d Wiener process W. The joint system has the following form:

$$\begin{aligned} dX_0 &= (\sigma(X_1 - X_0) + \epsilon_y X_0(Y_0 - 1))dt)dt + \sigma_L(t)dW_{X1} \\ dX_1 &= (X_0(\rho - X_2) - X_1)dt + \sigma_L(t)dW_{X2} \\ dX_2 &= (X_0X_1 - \beta X_2)dt + \sigma_L(t)dW_{X3} \\ dY_0 &= (-\omega_2 Y_1 - Y_2 + \epsilon_x Y_0(X_0 - 1))dt + \sigma_R(t)dW_{Y1} \\ dY_1 &= (\omega_2 Y_0 + aY_1)dt + \sigma_R(t)dW_{Y2} \\ dY_2 &= (b + Y_2(Y1 - c))dt + \sigma_R(t)dW_{Y3} \end{aligned} \quad (12)$$

The coupling values were chosen as $\epsilon = \{\epsilon_x, \epsilon_y\} = \{(0,0), (2,0), (4,0), (0,0.2), (0,0.5)\}$. Thirty realizations of the dynamical system was drawn, where subsequent permutation analysis was performed to obtain a p-value. The values of the parameters of the system are used in order to induce deterministic chaos as seen in Table. 1.

Table 1. Parameters of the Lorenz-Rossler systems.

Lorenz	σ_L	dt	σ	β	ρ	Rossler	σ_R	dt	ω_1	ω_2	a	b	c
	10^{-5}	0.1	10	$\frac{8}{3}$	28		0.1	0.1	1.015	0.918	0.15	0.2	10

Unidirectionally Coupled Neurovascular Responses The previous system, as mentioned earlier, exhibit highly oscillatory behaviors specified by its chaotic nature where nearby states may exhibit drastically divergent trajectories. We also simulate signals which may not be as drastically oscillatory, though may be emergent of nonlinear phenomenon as seen in neurovascular signals as observed in neuroscientific studies that exploit hemodynamics like functional magnetic resonance imaging (fMRI) [29]. From [29], stimuli generating neurovascular signal (x) is approximated by a bilinear state equation

$$\dot{x} = Ax + \sum_{j=1}^{m} u_j B^j x + Cu \quad (13)$$

where the A matrix describes the autonomous undriven dynamics of the system (here treated as a diagonal matrix with values -1), u are stimuli of a condition (where the element u_j is 0 when jth event is not occurring, and 1 when the jth event is occurring), where B^j describes the interconnectivity between voxels of interest (i.e. $\begin{bmatrix} \rho_{11} & \rho_{12} \\ 0 & \rho_{22} \end{bmatrix}$ for voxel 1 causing voxel 2, and each ρ_{ij} is an i.i.d. random variable $\rho_{ij} \sim \mathcal{N}(0, 1)$). The B matrices do not change in time, rather they are realized for each simulated time course. C describes which voxels the stimulus condition interacts with (in this paper, simulated as a diagonal matrix). In our simulations, we have two conditions u_1, u_2 and u_3. u_1 and u_2 are events that occur 10 time each over a period of 1000 s; each event lasts 6 s on intervals of 1 min. u_1 invokes voxel 1 to voxel 2 interactions via a random variable, $\rho_{12}^1 \sim \mathcal{N}(0, 1)$ while u_2 does not invoke interaction between the voxels, i.e. $\rho_{12}^2 = 0$. u_3 is a binomial point process with $p = 0.3$ that describes random neural activity that elicit no functional interconnectivity between voxels 1 and 2, i.e. $\rho_{12}^3 = 0$. This serves as background noise where the activities are occurring on top of.

After the neural state equations, regional blood flow f, blood volume v, and deoxyhemoglobin concentration q is simulated using the proposed parameters and mechanistic equations in [29]. The nonlinear response of deoxyhemoglobin, q, is the neurovascular variable observed in fMRI. Its emergence from nonlinear equations provides a complex, nonlinearly saturating signal whose causality may be affected by parameters of regional properties that make traditional causal methods hard to apply and begs dynamical systems characterizations [30], and whose characterizations is vital in connectivity analysis in psychophysiological data. We simulate q with observational white noise added to corrupt the SNR to 5dB. Figure 1 demonstrates an exemplary simulation of these interactions.

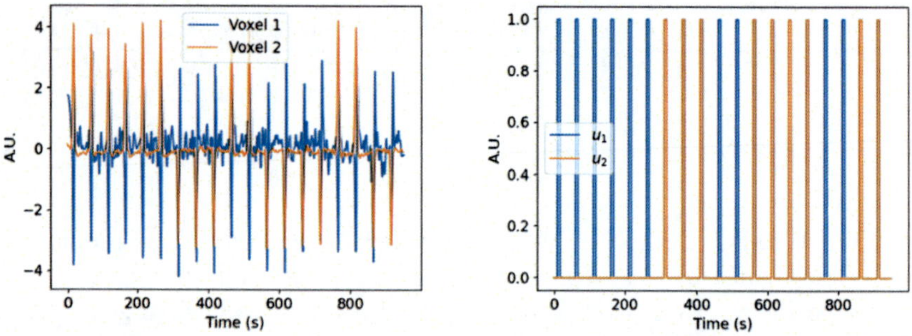

(a) Deoxyhemoglobin time courses

(b) Event time courses

Fig. 1. Exemplary Voxel time course without any SNR observation noise perturbations. **a** Demonstrates the deoxyhemoglobin signals that are triggered by the event stimuli shown in **b**.

2.5 Statistical Analysis

For VGP-CCM, we generate the null distribution of the coupling statistic using Eq. (2), substituting $P(\theta_x|X)$ and $P(\theta_y|Y)$ with our approximate distributions $Q(\theta_x)$ and $Q(\theta_y)$ respectively. $\tilde{Y}$ and $\tilde{X}$ are sampled from 30 permutations of the original time series X and Y respectively. GP-CCM, on the other hand, has its null distribution generated from eq. (1) with its hyperparameters obtained as the maximum a posteriori point estimate from Q obtained in Eq. (5). An $\alpha = 0.05$ is used as the threshold for significance in all tests when applied to eq. (6). Given the Rossler and Lorenz equations each have 3 variables, we perform 9 comparisons ($X_i \rightarrow Y_1, X_i \rightarrow Y_2, X_i \rightarrow Y_3$ for i in 1...3). Thus, we have 270 tests for each coupling value. We then calculate the specificity, a measure of proneness to type I errors, as $\frac{\text{N correctly accepted } H_0}{\text{N correctly accepted } H_0 + N \text{ incorrectly rejected } H_0}$ [31]. This same procedure is applied to the neurovascular system, except 100 time series are realized, thus 100 tests for coupling results are performed.

Table 2. Table displaying the number of rejected H_o for each coupling values and each directionality test for the chaotic systems (Rossler to Lorenz equations (R→L), Lorenz to Rossler equations (L→ R)) or the neurovascular system (Voxel 1 to Voxel 2 ($V_1 \rightarrow V_2$) or Voxel 2 to Voxel 1 ($V_2 \rightarrow V_1$). First 2 columns are results obtained from results either using point estimates of the hyperparameters (GP-CCM), while the last two columns are results using approximate posterior distributions over hyperparameters (VGP-CCM).

	GP-CCM		VGP-CCM	
Chaotic system	L→R	R→L	L→R	R→L
$\epsilon = (0.00, 0.00)$	172	94	0	3
$\epsilon = (0.00, 0.20)$	203	62	0	62
$\epsilon = (0.00, 0.50)$	22	248	1	158
$\epsilon = (2.00, 0.00)$	270	0	207	0
$\epsilon = (4.00, 0.00)$	270	0	210	0
Neurovascular system	$V_1 \rightarrow V_2$	$V_2 \rightarrow V_1$	$V_1 \rightarrow V_2$	$V_2 \rightarrow V_1$
$B = \begin{bmatrix} \rho_{11} & \rho_{12} \\ 0 & \rho_{22} \end{bmatrix}$	75	25	60	6

3 Results

Table 2 displays the set of significance rates for various couplings values for the unidirectionally coupled Lorenz-Rossler systems and for the neurovascular system. For zero coupling, VGP-CCM has only rejects the null hypothesis in 3 out of 270 realizations, whereas GP-CCM rejects the null hypothesis all but 4 of its test. Furthermore, VGP-CCM is more conservative than GP-CCM in dictating whether to make a claim of significant coupling at low coupling values, thus VGP-CCM very rarely made a false directionality claim. For the case of detecting Rossler driving Lorenz coupling, VGP-CCM reports a specificity of 99.6% while GP-CCM reports a specificity of 58.3%. In the direction from Lorenz driving Rossler, GP-CCM and VGP-CCM both report a specificity of 100%. For correctly rejecting the hypothesis in either direction, VGP-CCM has a specificity of 99.8% compared to GP-CCM's 79.1%. Similarly for the neurovascular systems, the proposed variational VGP-CCM provides more specificity at 94% compared to GP-CCM's 75%.

Figure 2 displays cumulative distributions for simulating a Lorenz system of equations driving Rossler equations for coupling values $\epsilon = \{(0,0)\}$ between variables X0 and Y0. The cumulative distribution reveals a robust null distribution when using the approximate posterior compared to using fixed values for the hyperparameters and performing permutations where results from only permuting samples results in a degenerate distribution with a collapsed mode as seen with zero coupling between the systems.

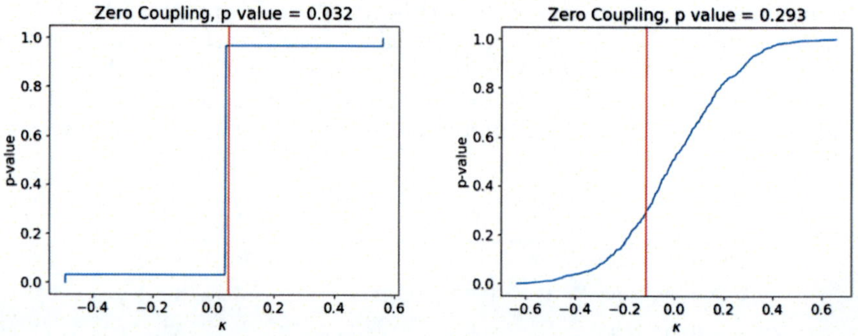

Fig. 2. Example null distributions and p values for zero coupling seen from the Lorenz system to the Rossler system using the original GP-CCM method (left) and the proposed variational GP-CCM method (right). The value for nonpermuted maximum a posteriori time series results is shown in red. It can be seen that VGP-CCM properly provides non significant results while GP-CCM does not.

4 Discussion

We propose the VGP-CCM framework, which extends the GP-CCM to use a variational approximation to the posterior distribution of hyperparameters. The presented study aimed to use a variational approximation to the posterior distribution of hyperparameters to develop a more parsimonious distribution for testing a null uncoupled hypothesis for GP-CCM. Tests were performed on simulated unidirectionally coupled Lorenz-Rossler systems to control the values of coupling between the variables of the system of equations to determine whether GP-CCM and VGP-CCM suffered from false positives in the significance tests and at what coupling could either make significance statements. Neurovascular systems were also simulated to illustrate VGP-CCM's efficacy at uncovering coupling in more slowly oscillating nonlinear signals compared to the chaotic maps studied in the Lorenz-Rossler system.

From the results over 30 realizations, as seen in Table 2, VGP-CCM has much improved results compared to GP-CCM, with nearly 20% greater specificity than GP-CCM. Furthermore, from the simulations of the neurovascular system, we demonstrated that the proposed variational extension continues to provide further robustness even for slow oscillating signals as observed in deoxy-hemoglobin signals, again with a similar substantial increase in specificity. This results on the neurovascular system was also impressive as it inferred the proper directionality even when there was intermittent forcing as described in Sect. 2.4. As mentioned in the introduction, indeed the degenerate nature of the null distribution, as seen in Fig 2 with GP-CCM allowed too much liberty to make claims of significance even in the wrong direction. VGP-CCM proves promising in the case that even in low coupling it will not make a claim to the wrong direction, giving researchers confidence in the results that if the raw statistic is insignificant but with high value, it may be a matter of needing to collect more realizations, as seen in the low couplings values in Table 2.

The prior distribution used in this study were indeed uninformative naively, though they allowed ease of computations, reparameterizations of means and variations with unit Gaussian random sample draws for automatic differentiation, and provided domain constraints. We suggest in studies where domain knowledge of how samples should a priori correlate is available, such as a kernel derived from mechanistic dynamic equations—e.g. time constants in linear ordinary differential equations as performed in dynamic causal modeling [32] or temperature in thermodynamic heat systems [33]–could be used. Otherwise, a separate dataset with labels can be used to infer optimal distributions through cross-validations. Even further directions may consider generative models for the posterior distribution conditioned on data, such as neural networks [17,23].

From this study, we can conclude that integrating out the hyperparameters from the GP-CCM probability ratio may result in a more parsimonious statistic giving robust null distributions for hypothesis testings. The computational aspects are lessened by avoiding Markov Chain Monte Carlo integration on the true conditional a posteriori distribution, instead opting for variational approximations of the hyperparameter distribution with independent mean field structures. The robustness was verified on simulated data from unidirectional Lorenz-Rossler systems, where VGP-CCM achieved much higher specificity than GP-CCM. Further studies will look into applying this methodology on real world data such as neurophysiological systems.

Acknowledgments. The research leading to these results has received partial funding from the European Commission — Horizon 2020 Program under grant agreement n 813234 of the project "RHUMBO", and by Italian Ministry of Education and Research (MIUR) in the framework of the CrossLab project (Departments of Excellence).

References

1. Stankovski, T., Pereira, T., McClintock, P.V.E., Stefanovska, A.: Coupling functions: universal insights into dynamical interaction mechanisms. Rev. Mod. Phys. **89**, 045001 (2017)
2. Friston, K.J.: Functional and effective connectivity: a review. Brain Connectivity **1**(1), 13–36 (2011)
3. Granger, C.: Investigating causal relations by econometric models and cross-spectral methods. Econometrica **37**(3), 424–38 (1969)
4. Schreiber, T.: Measuring information transfer. Phys. Rev. Lett. **85**, 461–464 (2000)
5. Barnett, L., Barrett, A.B., Seth, A.K.: Granger causality and transfer entropy are equivalent for gaussian variables. Phys. Rev. Lett. **103**, 238701 (2009)
6. Marinazzo, D., Pellicoro, M., Stramaglia, S.: Kernel method for nonlinear granger causality. Phys. Rev. Lett. **100**, 144103 (2008)
7. Smirnov, D.A., Bezruchko, B.P.: Spurious causalities due to low temporal resolution: towards detection of bidirectional coupling from time series. **100**, 10005 (Oct 2012)
8. Smirnov, D.A.: Spurious causalities with transfer entropy. Phys. Rev. E **87**, 042917 (2013)
9. Popper, K.R.: Science as falsification. Conjectures Refutations **1**(1963), 33–39 (1963)
10. Sugihara, G., May, R., Ye, H., Hsieh, C.-H., Deyle, E., Fogarty, M., Munch, S.: Detecting causality in complex ecosystems. Science **338**(6106), 496–500 (2012)
11. Takens, F.: Detecting strange attractors in turbulence. In: Rand, D., Young, L.-S. (eds.) Dynamical Systems and Turbulence, Warwick 1980, pp. 366–381. Springer, Berlin, Heidelberg (1981)
12. Feng, G., Yu, K., Wang, Y., Yuan, Y., Djurić, P.M.: Improving convergent cross mapping for causal discovery with gaussian processes. In: ICASSP 2020 — 2020 IEEE International Conference on Acoustics, Speech and Signal Processing (ICASSP), pp. 3692–3696 (2020)
13. Feng, G., Quirk, J.G., Djurić, P.M.: Detecting causality using deep gaussian processes. In: 2019 53rd Asilomar Conference on Signals, Systems, and Computers, pp. 472–476 (2019)
14. Ghouse, A., Faes, L., Valenza, G.: Inferring directionality of coupled dynamical systems using gaussian process priors: application on neurovascular systems. Phys. Rev. E **104**, 064208 (2021)
15. Rasmussen, C.E., Williams, C.K.I.: Gaussian Processes for Machine Learning (Adaptive Computation and Machine Learning). The MIT Press (2005)
16. Chung, Y., Gelman, A., Rabe-Hesketh, S., Liu, J., Dorie, V.: Weakly informative prior for point estimation of covariance matrices in hierarchical models. J. Educ. Behav. Stat. **40**(2), 136–157 (2015)
17. Blei, D.M., Kucukelbir, A., McAuliffe, J.D.: Variational inference: a review for statisticians. J. Am. Stat. Assoc. **112**(518), 859–877 (2017)
18. Sauer, T., Yorke, J.A., Casdagli, M.: Embedology. J. Stat. Physics. **65**, 579–616 (1991)
19. Sugihara, G., May, R.M.: Nonlinear forecasting as a way of distinguishing chaos from measurement error in time series. Nature **344**(6268), 734–741 (1990)
20. Geman, S., Geman, D.: Stochastic relaxation, Gibbs distributions, and the Bayesian restoration of images. IEEE Trans. Pattern Anal. Mach. Intell. **6**, 721–741 (1984)

21. Beal, M.J.: Variational Algorithms for Approximate Bayesian Inference. Ph.D. thesis, Gatsby Computational Neuroscience Unit. University College London (2003)
22. Kullback, S., Leibler, R.A.: On information and sufficiency. Ann. Math. Stati. **22**(1), 79–86 (1951)
23. MacKay, D.J.C.: Bayesian Methods for Backpropagation Networks, pp. 211–254. Springer, New York, NY (1996)
24. Snelson, E., Ghahramani, Z.: Sparse gaussian processes using pseudo-inputs. In: Proceedings of the 18th International Conference on Neural Information Processing Systems, NIPS'05, (Cambridge, MA, USA), pp. 1257-1264. MIT Press (2005)
25. Kingma, D.P., Welling, M.: Auto-encoding variational Bayes. In: 2nd International Conference on Learning Representations, ICLR 2014, Banff, AB, Canada, April 14–16, 2014, Conference Track Proceedings (2014)
26. Lee, S.Y.: Gibbs sampler and coordinate ascent variational inference: a set-theoretical review. Commun. Stat. — Theory Methods. 1–21 (2021)
27. Paszke, A., Gross, S., Chintala, S., Chanan, G., Yang, E., DeVito, Z., Lin, Z., Desmaison, A., Antiga, L., Lerer, A.: Automatic differentiation in Pytorch (2017)
28. Kim, Y., Wiseman, S., Miller, A.C., Sontag, D.A., Rush, A.M.: Semi-amortized variational autoencoders. In: Dy, J.G., Krause, A. (eds.) ICML. Proceedings of Machine Learning Research, vol. 80, pp. 2683–2692. PMLR (2018)
29. Stephan, K.E., Weiskopf, N., Drysdale, P.M., Robinson, P.A., Friston, K.J.: Comparing hemodynamic models with DCM. Neuroimage **38**(3), 387–401 (2007)
30. Friston, K.: Dynamic causal modeling and granger causality comments on: the identification of interacting networks in the brain using fMRI: model selection, causality and deconvolution. Neuroimage **58**(2–2), 303 (2011)
31. Gaddis, G.M., Gaddis, M.L.: Introduction to biostatistics: part 3, sensitivity, specificity, predictive value, and hypothesis testing. Ann. Emerg. Med. **19**(5), 591–597 (1990)
32. Friston, K., Harrison, L., Penny, W.: Dynamic causal modelling. NeuroImage **19**(4), 1273–1302 (2003)
33. Berline, N., Getzler, E., Michèle, V.: Heat Kernels and Dirac Operators. Springer (1996)
34. Papoulis, A., Pillai, S.U.: Probability, Random Variables, and Stochastic Processes, 4th edn. McGraw Hill, Boston (2002)

Detection of Sparsity in Multidimensional Data Using Network Degree Distribution and Improved Supervised Learning with Correction of Data Weighting

Shinya Ueno[1,2](✉) and Osamu Sakai[1]

[1] Department of Electronic Systems Engineering, The University of Shiga Prefecture, Hassaka-cho 2500, Hikone, Shiga 522-8533, Japan
s.ueno@sakigakes.co.jp

[2] Checkers Co., Ltd., 50, Nishishichijoonmaedacho, Shimogyo, Kyoto 600-8897, Japan

Abstract. Multidimensional data are representatives in a wide range of applications, from those in the latest state-of-the-art science and technology to specific social issues. And they have been subject to analysis using methods such as regression analysis and machine learning. However, they are rarely obtained as complete data and contain more or less biases and deficiencies. In this study, we form a network from a multidimensional dataset and use its degree distribution to detect data sparsity. Although model analysis based on the degree distribution has been conducted for many years, sparsity detection has not been a target of the degree distribution analysis. Furthermore, we attempt to increase the accuracy and precision of supervised learning by applying regressive weighting according to node grouping in the degree distribution spectrum. By making use of this algorithm, we can expand the range of utilization of incomplete data together with other promising progresses in complex networks.

Keywords: Network analysis · Multidimensional data · Sparsity · Supervised learning

1 Introduction

Multidimensional data are essential elements for various types of analysis and its analysis method has been developed for more than a quarter of a century. They are representative datasets in various fields and applications, such as medical applications [1–4] like clinical trials for new drug development and determination of physical condition based on skin color, social applications like road and railroad maintenance [5,6], industrial applications like product inspection [7–10], and chemical applications like synthesis of material compounds [11]. We have also been performed on the evaluation of plasmas with multiple dimensions where multidimensional data is inevitable for the evaluation of plasmas for industrial applications [12–14].

H. Cherifi et al. (Eds.): COMPLEX NETWORKS 2016 2022, SCI 1077, pp. 390–401, 2023.
https://doi.org/10.1007/978-3-031-21127-0_32

They have been subjected to analysis using various methods such as regression analysis [15–17], data mining, and supervised learning [18–20], both linear and nonlinear. In supervised learning, multidimensional data are often used as teacher data, and in our previous work, we have also used a multidimensional dataset as Training Data for calibrating optical sensors using neural networks [21].

However, all data that are targeted cannot always be available as a complete dataset in which their distributions in the multiple dimensions are quite uniform. For example, anomalous values may be observed, and it is quite possible that the amount of data may be biased or missing. We will discuss this in more detail below, but we have confirmed that bias or lack of data exists even in the Training Data we have used in our past studies [22].

Analysis on a dataset with missing data may reduce the accuracy and precision of the analysis results; for example, if the missing data contain information that is crucial for the analysis, it may result in a linear approximate model, whereas the original data distribution might be actually nonlinear. Therefore, how to treat incomplete data with missing data as if it were complete data is an important item in multidimensional data analysis.

Seeking more complete data can be expensive, and this hinders the use of multidimensional statistical data in various fields, not just for machine learning. In other words, when the statistical data have some degree of incompleteness, the data-analysis procedure with capable technology to perform accurate analysis based on such incomplete data with sparsity will expand the possibilities of using data analysis. Here, we note that on literature [23] defines "sparse" as areas where edges are not connected, and another report [25] defines it as areas where data do not exist. We treat sparsity as the absence of data in a region where it should be present.

In the previous studies, complete-case analysis or listwise deletion was used to address missing or biased data in multidimensional data. When the number of variables is large and the proportion of missing data is high, methods called available-case analysis and pairwise deletion have been used. These methods are used in various situations because they are intuitive and easy to implement. For example, methods such as assigning data to missing parts to prepare pseudo-complete data [26,27] and approximating missing parts by weighting formulas [28–30] have been studied. On the other hand, since machine learning has become widely used, supervised learning is often performed with missing and biased data [31].

In contrast, in our previous study [22], we applied the model of locally linear embedding (LLE) [23] and attempted to ensure the reliability of the target region by emphasizing the weight of the data surrounding the missing-data or sparse area. Here, the missing data were checked not automatically but manually, and the missing data were evaluated by simple doubling and/or tripling of the number of missing data. Although this study achieved certain results, such intuitive methods were only valid for this model data, and the method was not universally applicable to other datasets. In the field of machine learning for complex networks, rebalancing has been done by undersampling, which reduces the amount of data, and oversampling, which increases the amount of data [32]. In contrast, the method we are going to describe here is a method that attempts to supplement data regions that either do not exist or are extremely scarce by increasing the surrounding data.

Therefore, in this study, we propose a model based on a complex network that takes into account mutual positions of a large number of data points, and based on methods such as degree-distribution derivation and clustering, we select the amount of data that should be handled for rebalancing. Here, we propose a new algorithm that applies regressive weighting in the model. In contrast to LLE [23], which is based on distributions of local data points, our research is based on topology of the derived network and its statistical property. Over the years, many studies have been conducted on models based on complex networks, such as a random network in which each node is connected randomly and its degree distribution follows a Poisson distribution, and scale-free networks [24] in which the degree distribution follows a power-law distribution. That is, there have been many studies on degree distributions, but this study is unique since decomposition of the degree distribution is used for data-deficiency correction to improve accuracy of analysis, in particular, for supervised learning which requires training datasets preferably without sparsity.

2 Calculation Methods

2.1 Two Dimensional Color Coordinate and Target Task in Our Supervised Learning

Datasets in the multidimensional space are found in various scientific and technological areas, and in this study, color data on xy coordinates, which have been used in previous studies [21,22], were used in a model for examining the algorithm we propose here. The color data are expressed in three parameter red (R), green (G), blue (B), which are converted to points on the xy color coordinates by the following formulas defined in CIE1931 [33]: $X = 2.7689\ R + 1.7517\ G + 1.1302\ B$, $Y = 1\ R + 4.5907\ G + 0.0601\ B$, $Z = 0\ R + 0.0565\ G + 5.5943\ B$, $x = X\ /\ (X + Y + Z)$, $y = Y\ /\ (X + Y + Z)$.

When we consider data points on the xy color coordinates, color data are reduced from the three-dimensional attributes to the two-dimensional variables. Through this conversion, all R, G, and B color data are mapped on the xy color coordinates, and this conversion is not completely linear to R, G and B values; the data that were spaced at every equal step along the R, G, and B axes are not similarly spaced on the xy color coordinates. Furthermore, if we detect color values using any optical sensors, they include error rates due to poor matching coefficients to R, G, and B wavelength spectra. Thus, we always adapt a nonlinear calibration procedure to obtain sufficient accuracy for such multidimensional data. In our previous study [21], we successfully performed one way of suitable calibration, but we have not been sure about its validity with universality. In this study, we aim to accomplish a suitable and universal calibration method that can be performed automatically according to a simple algorithm, specifically for tuning weight coefficients for sparse training datasets in the multidimensional space.

2.2 Method for Network Diagram Formation and Analysis Algorithm

We show the flow chart about the data processing proposed in this study in Fig. 1. For the dataset described below, we calculated Euclidean distance between all the nodes with the specific xy color coordinate values (x, y) to find out connections or edges between nodes. A threshold for a connection was set for this distance value ranging from 0.01 to 0.05, and the adjoining points with a distance that fell below the threshold were connected to form an edge, and the resulting edge list was used to configure a network using Cytoscape [34]. The network created by this step is then evaluated for deficiencies and bias using the following two point of views. First, from the generated network diagram or visualized network topology, we detect differences from the case with the complete dataset without sparsity, based on an intuitive evaluation of its shape. Second, the degree k of each node is detected, the node count is summed up to create a spectrum for every 10°C, and the differences in the distributions among the datasets clarify sparsity in the datasets, being useful to determine if there is any bias or missing data. Finally, to make use of such unveiled features, for the training dataset with sparsity, supervised learning including regressive weighting is performed based on this degree-distribution analysis, and the effects of the weighting are evaluated using mean absolute errors (MAEs) in the test dataset.

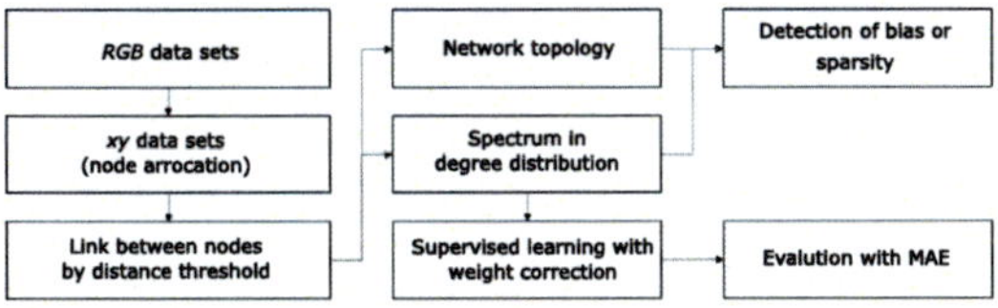

Fig. 1. Flow chart of data processing in this study.

2.3 Datasets on the Color Coordinate

Three datasets were prepared for the model used in this study; one of them is our target set that is the training dataset for calibration based on supervised learning, and the other two sets are for comparative data processing to clarify how our model with algorithm works.

In the first dataset, points are randomly placed on the xy color coordinates within the restricted area that can exhibit color data values. On the xy color coordinates, depending on formulae of R, G, and B variables, the color data points should be located inside the triangle whose vertices correspond to R, G, and B as shown in Fig. 2a. "Random Data" refers to this dataset in this study, and the total count of data points is 1892.

In the second dataset, the data values covering all combinations of R, G, and B variables with every equal value step were transformed to the xy color coordinates, and here this dataset was defined as "RGB All Data" and the total count of data points is 1728. The profile of data points representing these data in the color coordinates is shown in Fig. 2b. As explained in Sect. 2.1, when the dataset of R, G, and B with every equal value step is mapped to the data points on the xy color coordinates, the corresponding x and y coordinates of these points are in inequal spatial steps. While the density of data points in the center of the triangle and in the light blue direction is sufficient, certain gaps exist in the purple and yellow directions. In addition to these gaps, we find isolated data points around the red area.

The third dataset is our target in this study; this is referred to as "Training Data," and the total count of data points is 1631. We used the entire set of color data from the color catalog given on more than 1600 sample sheets (PANTONE FORMULA GUIDE, Pantone LLC, X-rite Inc.), and converted all data into the xy color coordinates, which we used in our previous study [21]. The distribution on the xy color coordinates shown in Fig. 2c reveals sparse areas or regions of missing data in the circled areas.

Thus, the obtained dataset is not necessarily comprehensive, and in many cases, the dataset contains some elements of bias or missing data. Therefore, we aim to identify such biases and deficiencies using network diagrams, reinforcing these incomplete elements by adjusting weight coefficients of the existing data, and to augment and calibrate the data using supervised learning.

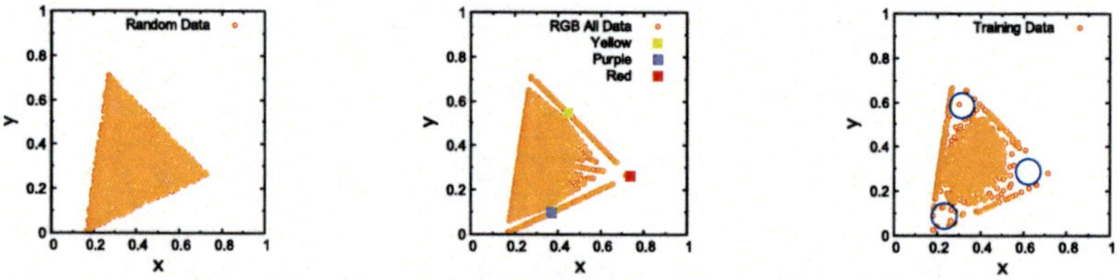

(a) (b) (c)

Fig. 2. Distributions of data points on the xy color coordinates for **a** "Random data", **b** "RGB all data", and **c** "Training data". The circles in blue line indicate sparse areas.

2.4 Data Calibration Methods and Weight Tuning

In Training Data, the values given in the color catalog on sample sheets are supervisory output signals whereas the counter values are obtained in the measurements of the sample sheets by the color sensor ("color checker" CC-01, Checkers

Co.) as measured values. Supervised learning with a neural network that is a simple one-hidden-layer perceptron was performed with R, G, and B in the measured values as inputs and x and y of the catalog values as outputs. Supervised learning was performed using the R package [35]. In our previous study, weight tuning among the given data points was performed simply by duplicating the target data; details are described in Ref. [21] and here we briefly review in the following. We categorized edge regions as areas with relatively small amounts of data, located on the edge of the given data space, and enhanced accuracy of their calibration by this simple weighting; we call this method "Edge-region". In this study, we propose a method to classify data points with lower densities based on the degree distribution of a complex network derived from distributed points in the multi-dimensional space, and to assign weights to these data points. We define this method as "byDegree". The difference detected over the catalog data, the measured data, and the calibrated data was quantified by MAE, and the MAE levels are representatives for the evaluation of this network-based weighting.

3 Calculation Results

3.1 Network Diagrams and Degree Distributions

For each dataset, from data points scattered on the xy color coordinates in Fig. 2, we perform calculations and derive distributions measured by varying the distance threshold between 0.01 and 0.05 and a network diagram when the threshold is set to 0.05, as shown in Figs. 3, 4 and 5. In the case of Random Data, the network diagram is triangular as shown in Fig. 3a, just like the distribution of spatial data points on the xy color coordinates. In addition, when the threshold is changed linearly from 0.01 to 0.05, as shown in Fig. 3b, the degree distribution is getting close to that of the Poisson's distribution as the threshold increases. Although the shape is slightly distorted due to the triangular arrangement of the points, this is similar to the cases with $p = 0.003 - 0.038$ in a random graph, where, in general random graphs, the probability of connecting each node is defined by p. In our case of this graph in Fig. 3, since the distance between each node is random, it can be said to be a derivative of a random graph.

On the other hand, in the network diagram of RGB All Data shown in Fig. 4a, while maintaining the triangle outlook to some extent, we find a group of data that almost deviated from the triangle pattern and other nodes completely deviated from it. This automatically indicates presences of gaps or isolated data groups that were visible on the xy color coordinates in Fig. 2b. In the degree distribution shown in Fig. 4b, we find a clear difference from that in the case of Random Data; these differences are more outstanding than those observed in the network diagram. In this dataset, the degree distribution includes multi spectra with several maxima and minima, and their widths are large with long tails. This difference from the degree distribution of Random Data indicates that the distribution of color information in RGB All Data on the xy color coordinates is a model with completely different characteristics from that of Random Data,

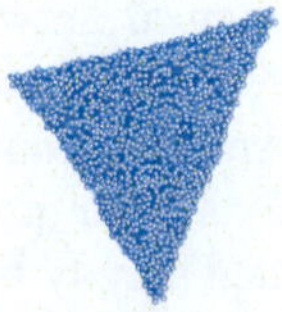

(a) Geographical network

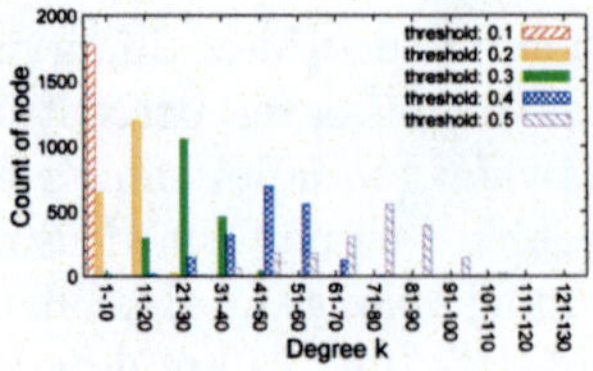

(b) Degree distribution

Fig. 3. Geographical network and degree distribution for random data. The distance threshold for networking is set to 0.5 on **a**, and from 0.1 to 0.5 on **b**.

arising from regularity in equal R, G and B steps in their components. It also has a different shape from the power-law distribution, leading to the fact that it is not a general scale-free network.

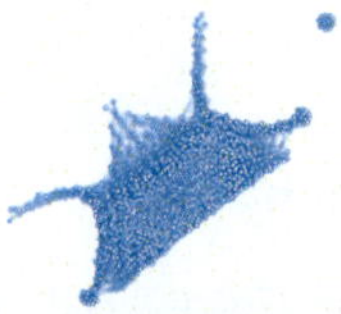

(a) Geographical network

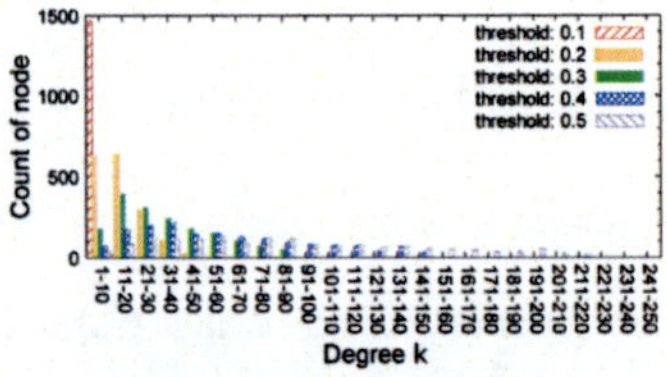

(b) Degree distribution

Fig. 4. Geographical network and degree distribution for RGB all data. The distance threshold for networking is set to 0.5 on **a**, and from 0.1 to 0.5 on **b**.

Finally, we consider the network diagram of Training Data, as shown in Fig. 5a. Although a rough footprint of the triangle remains, its shape is significantly broken, suggesting the existence of some kinds of data bias or deficiency, even when compared to the RGB All data. In other words, it is possible to diagnose and visualize data deficiencies and biases to some extent automatically and intuitively by comparing topology of network diagrams. Figure 5b shows the degree distribution of the Training Data. Figure 6a shows comparison of the degree distribution of the RGB All Data and the Training Data. We note that the count of data points in Training Data is somewhat higher approximately by 220, whereas the total k of RGB All Data is smaller. Despite of this discrepancy, from this comparison, it is outstanding that there is a large bias and sparse areas in Training Data. Furthermore, for RGB All Data, the degree distribution reaches at maximum around 51 and 60, from which the data points are decreasing (with some variation) as the k is raised. However, in Training Data, the number of data points that once decreased increases after 331, suggesting that the data can be divided into several clusters or spectra in the degree distribution. From this outlook investigation, we tentatively classified the data point into three spectra: Spectrum I (from 0 to 130), Spectrum II (from 131 to 330), and Spectrum III (from 331 to 440). Over these spectra, we search for suitable calibration methods by weighting (or replicating) Spectrum I, which is a lower degree spectrum and is estimated to correspond areas of lower data density or sparse regions.

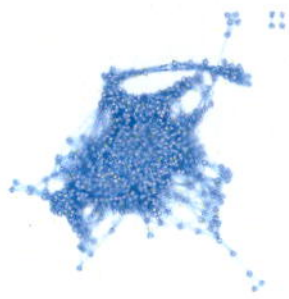

(a) Geographical network

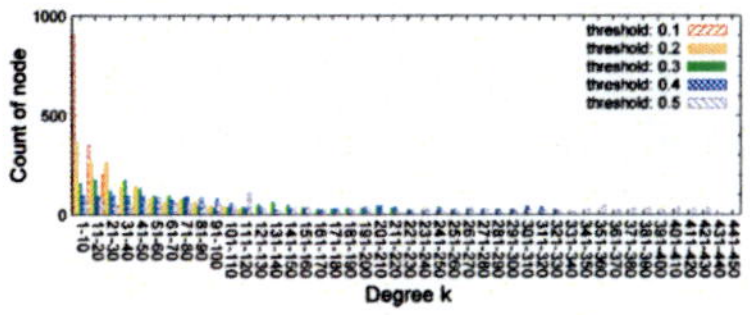

(b) Degree distribution

Fig. 5. Geographical network and degree distribution for training data. The distance threshold for networking is set to 0.5 on **a**, and from 0.1 to 0.5 on **b**.

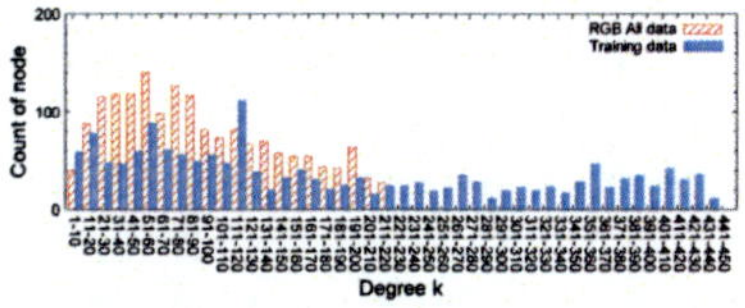

(a) Comparison of degree distribution

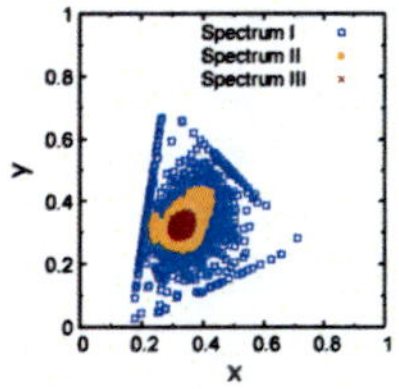

(b) Distribution on xy color coordinates.

Fig. 6. Classification of data using degree-distribution spectra. **a** Comparison of degree distributions of RGB all data (red diagonal lined) and training data (blue hatched) when the threshold is 0.5. **b** Distribution of datasets on the xy color coordinates. Blue squares indicate Spectrum I ($k = 1 - 130$), orange circles indicate Spectrum II ($k = 131 - 330$), and brown crosses indicate Spectrum III ($k = 331 - 440$).

3.2 Validity for Supervised Learning: Calibration Using Neural Networks

After applying our algorithm for supervised learning [21], the estimated effects on MAE before and after calibration are shown in Fig. 7b, where the varying spatial scattering of data points on the xy color coordinates is shown in Fig. 7a. In Fig. 7b, the method proposed in this paper reduces MAE as much as or more than "Edge Region," indicating that this method proposed in this study is more effective on Training Data. The improvement can also be intuitively recognized from the visualized data-point spatial distributions based on this methods on the xy color coordinates, with smaller spatial gaps between the catalog and the calibrated data [21].

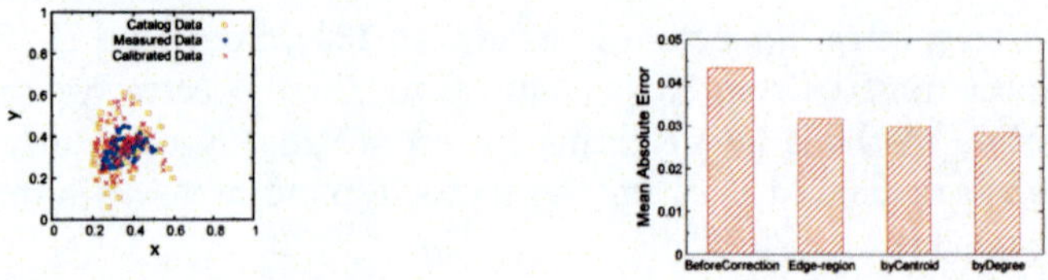

(a) Distribution on the xy color coordinates

(b) Comparison of MAE

Fig. 7. Validation of our method for supervised learning using multidimensional data. **a** Profiles of data points on two-dimensional space where orange squares indicate training data, blue circles indicate test values, and peach crosses indicate calibrated values. **b** Comparison of MAE of calibrated values. "BeforeCorrection" indicates the MAE before correction of data weighting, and "Edge-region," "byCentroid," and "byDegree" indicate the cases using edge region, R package (described in discussion), and degree distribution, respectively.

4 Discussion

As a comparative experiment using a more complicated method, we performed clustering of data points using the R package [35], and weights for supervised learning were set based on this clustering. Among the pre-installed methods, *Centroid* was used to optimize the clustering, and the dendrogram was used as the basis for dividing the clusters into several clumps. We assumed the cluster with the least number of data to be around the sparse area, and we attempted to calibrate them by assigning weights to them. This method is called "byCentroid" and the calibration results are shown in Fig. 7a. In this "byCentroid" experiment, we used the intuitive and easy-to-understand centroid method for creating clusters, but other clustering methods might be better, depending on the model, although it is sufficient for our aim here. Performance in comparing the learning results for each method is shown in Fig. 7b. All of the methods exhibit improvements over the results before weighting, and the best one for MAE is obtained with the learning based on "byDegree", proposed here.

Although sufficient results were obtained even with a low magnification factor in this model, it is necessary to select appropriate conditions based on the obtained results, such as a steeper gradient or a more detailed segmentation method for a more complicated dataset. In this case, the spectra were manually sorted by surveying quantitative data balancing, but it is easily possible to perform it automatically by deriving an approximate curve of the degree distribution and applying spectrum deconvolution, considering the maximum and minimum points discriminated.

Finally, we show results on entropy analysis, which represents macroscopic features of datasets visualized and configured in complex networks. When we assume hypothetical information flow from a given node s in a complex network which is the ensemble of all-node group S is induced into linked nodes t belonging to the group T, like Figs. 3a, 4a and 5a, accumulated effects of its flow probability $p(t|s)$ lead to the conditional entropy $H(T|S)$ [36], given as:

$$H(T|S) = \sum_{s \in S} p(s) H(T|S = s) = -\sum_{s \in S} p(s) \sum_{t \in T} p(t|s) \log_2 p(t|s), \quad (1)$$

where $p(s)$ is the existence probability at the node s. Assuming equal $p(s)$ over S and equal $p(t|s)$ at the node s over T, the calculated results of $H(T|S)$ are: 6.10 for Random Data, 6.17 for RGB All Data, and 6.88 for Training Data. These results suggest that, in comparison with cases of balanced node density, the network of Training Data takes larger conditional entropy, which indicates wider choices for linkages between nodes in hypothetical information flow of Training Data. This seems to be contradictory to occurrences of sparsity, in which such flow mobilities through linkages are somewhat limited. However, in our case of Training Data, it is not always valid since, around sparse areas, some of nodes are rather condensed, as indicated in Fig. 6. Furthermore, this high entropy also points out existences of areas with high density of data points, which is located in the central aggregation in Fig. 2c. Thus, entropy estimation can support another insight of the algorithm proposed here and reinforce understandings in terms of complex networks.

5 Conclusion

In this study, we successfully detected sparsity in given data points using network diagrams constructed from their multidimensional locations and their degree distributions. We also succeeded in improving the accuracy and precision of supervised learning by applying regressive weights to the spectrum group in the degree distributions. In a comparative experiment based on a typical clustering method, our result for supervised learning shows a better accuracy, which indicates that the simpler method proposed in this study gave better results. By utilizing this complex network technique, it is possible to obtain more accurate and precise analysis results in supervised-leaning model even if the acquired multidimensional data contain biases and deficiencies, which will lead to a wider range of utilization of multidimensional data.

Acknowledgements. The authors thank the members of in Checkers Co., Ltd., in particular Dr. K. Taguchi, for his useful comments. This work is partially supported by the Regional ICT Research Center of Human, Industry and Future at The University of Shiga Prefecture, by the Cabinet Office, Government of Japan, and by a Grant-in-Aid for Scientific Research from the Japanese Ministry of Education, Culture, Sports, Science and Technology (MEXT/JSPS KAKENHI) with Grant No. 22K18704.

References

1. Fitzmaurice, F.M., Laird, N.M., Ware, J.H.: Applied Longitudinal Analysis. Wiley, New York (2011)
2. Blomeke, R.C., Elliott, J.S., Senjaya, B., Hales, G.T.: A comparison of fingerprint image quality and matching performance between healthcare and general populations. In: Proceedings of 2009 IEEE 3rd International Conference on BTAS, vol. 9, pp. 1-4, IEEE, Washington DC (2009)
3. Morris, D., Coyle, S., Wu, Y., Lau, T.K., Wallace, G., Diamond, D.: Bio-sensing textile based patch with integrated optical detection system for sweat monitoring. Sens. Actuators B Chem. **139**, 231–236 (2009)
4. Jiang, Z., Hu, M., Gao, Z., Fan, L., Dai, R., Pan, Y., Tang, W., Zhai, G., Lu, Y.: Detection of respiratory infections using RGB-infrared sensors on portable device. IEEE Sens. J. **20**, 13674–13681 (2020)
5. Lee, S.J., Kim, H.M., Kim, S.I., Lee, H.M.: Evaluation of structural integrity of rail-way bridge using acceleration data and semi-supervised learning approach. Eng. Struct. **239**, 1–16 (2021)
6. Shim, S., Kim, J., Lee, S.W., Cho, G.C.: Road damage detection using super-resolution and semi-supervised learning with generative adversarial network. Autom. Constr. **135**, 1–16 (2022)
7. Chandy, R.P., Scully, P.J., Thomas, D.: A novel technique for online measurement of scaling using a multimode optical fibre sensor for industrial applications. Sens. Actuators B Chem. **71**, 19–23 (2000)
8. Zhou, Z.-K., Wang, U.-K., Gong, H.-G., Shi, Y., Wang, Z., Zhang, B.: A fully-integrated optoelectronic detector with high gain bandwidth product. IEEE Access **7**, 53032–53039 (2019)
9. Wotrnba, H.: Sensor sorting technology-is the minerals industry missing a chance? In: Proceedings XIII IMPC Istanbul 2006, pp. 21-29. IMPC, Istanbul (2006)
10. Leelasattarathkul, T., Liawruangrath, S., Rayanakorn, M., Liawruangrath, B., Oungpipat, W., Youngvises, N.: Greener analytical method for the determination of copper(II) in wastewater by micro flow system with optical sensor. Talanta **72**, 126–131 (2007)
11. Ramprasad, R., Batra, R., Pilania, G., Mannodi-Kanakkithodi, A., Kim, C.: Machine learning in materials informatics: recent applications and prospects. Comput. Mater. **3**(54), 1–13 (2017)
12. Sakai, O., Morita, T., Ueda, Y., Sano, N., Tachibana, K.: Chemical filters by non-thermal atmospheric pressure plasmas for reactive fields. Thin Solid Films **519**, 6999–7004 (2011)
13. Urabe, K., Hiraoka, Y., Sakai, O.: Hydrazine generation for the reduction process using small-scale plasmas in an argon/ammonia mixed gas flow. Plasma Sources Sci. Technol. **22**, 032003-1-4 (2013)
14. Urabe, K., Sakai, O.: Multiheterodyne interference spectroscopy using a probing optical frequency comb and a reference single-frequency laser. Phys. Rev. A **88**, 023856-1-5 (2013)
15. Girolami, M., Mischak, H., Krebs, R.: Analysis of complex, multidimensional datasets. Drug Discovery Today: Technol. **3**(1), 13–19 (2006)
16. Song, X., Wu, M., Jermaine, C., Ranka, S.: Statistical change detection for multi-dimensional data. In: KDD'07, SIGKDD, pp. 667-676. California (2007)
17. Dempster, A.P.: An overview of multivariate data analysis. J. Multivar. Anal. **1**, 316–346 (1970)

18. Zaidan, M.A., Motalagh, N.H., Fung, P.L., Lu, D., Timonen, H., Kuula, J., Niemi, J.V., Tarkoma, S., Petaja, T., Kulmala, M., Hussein, T.: Intelligent calibration and virtual sensing for integrated low-cost air quality sensors. IEEE Sens. J. **20**, 13638–13652 (2020)
19. Goodacre, R., Neal, M.J., Kell, D.B.: Quantitative analysis of multivariate data using artificial neural networks: a tutorial review and applications to the deconvolution of pyrolysis mass spectra. Zentralbl Bakteriol **284**, 516–539 (1996)
20. Fang, J., Yang, F., Tong, R., Yu, Q., Dai, X.: Fault diagnosis of electric transformers based on infrared image processing and semi-supervised learning. Glob. Energy Interconnection **4**, 596–607 (2021)
21. Ueno, S., Sakai, O.: Data driven calibration of color-sensitive optical sensor by supervised learning for botanical application. IEEE Sens. J. **22**, 11915–11927 (2022). https://doi.org/10.1109/JSEN.2022.3171221
22. Ueno, S., Sakai, O.: Low-cost color-sensitive optical sensor calibrated by sparse training data. In: Proceedings of the 2021 IEEE 10th GCCE, pp. 402-403. IEEE Consumer Technology Society, Kyoto (2021)
23. Roweis, S.T., Saul, L.K.: Nonlinear dimensionality reduction by locally linear embedding. Science **290**, 2323–2326 (2000)
24. Albert, R., Barabasi, A.-L.: Statistical mechanics of complex networks. Rev. Mod. Phys. **74**, 47–97 (2002)
25. Hasan, M.A., Chaoji, V., Salem, S., Zaki, M.: Link prediction using supervised learning. In: SDM06: Workshop on Link Analysis. Counter-Terrorism and Security, pp. 798–805. SIAM, Maryland (2005)
26. Rubin, D.B.: Multiple Imputation for Nonresponse in Surveys. Wiley, New York, NY, USA (1987)
27. Tanner, M.A., Wong, W.H.: The calculation of posterior distributions by data augmentation. J. Am. Stat. Assoc. **82**, 528–550 (1987)
28. Horvitz, D.G., Thompson, D.J.: A generalization of sampling without replacement from a finite universe. J. Am. Stat. Assoc. **47**, 663–685 (1952)
29. Rosenbaum, P.R., Rubin, D.B.: The central role of the propensity score in observational studies for causal effects. Biometrika **70**, 41–55 (1983)
30. Scharfstein, D.O., Rotnitzky, A., Robins, J.M.: Adjusting for nonignorable drop-out using semiparametric nonresponse models. J. Am. Stat. Assoc. **94**, 1096–1146 (1999)
31. Ma, M., Korniss, G., Szymanski, B.K.: Learning parameters for balanced index influence maximization. In: Processing 9th International Conference on Complex Networks and Their Applications, pp. 167–177. Springer, Madrid (2020)
32. Xue, J.-H., Hall, P.: Why does rebalancing class-unbalanced data improve AUC for linear discriminant analysis? IEEE Trans. Pattern Anal. Mach. Intell. **37**(5), 1109–1112 (2015)
33. Itten, J.: The Elements of Color. Van Nostrand Reinhold, New York, USA (1970)
34. Cytoscape open API. https://cytoscape.org/
35. The R Project for Statistical Computing. https://www.r-project.org/
36. Cover, T.M., Thomas, J.A.: Elements of Information Theory, 2nd edn. Wiley, Hoboken (2006)

Network Structure Versus Chemical Information in Drug-Drug Interaction Prediction

George Kefalas[1(✉)] and Dimitrios Vogiatzis[2]

[1] National Technical University of Athens, Athens, Greece
georgioskefalas@mail.ntua.gr
[2] American College of Greece—Deree, National Centre for Scientific Research, "Demokritos", Athens, Greece
dimitrv@acg.edu

Abstract. We apply two embedding mechanisms, node2vec & mol2vec, on the problem of predicting Drug-Drug Interactions (DDIs). These mechanisms, respectively, convert drugs into vectors using the chemical information of the underlying chemical compound and the network information from the graph of drug interactions. Our goal is to compare Single Link Prediction models that are based on each embedding method by exploring the topological features of the drug interactions graph that make each approach more efficient in making correct predictions. We base our experiments on the DrugBank data set and use various computational chemistry tools such RDKit and PubChem, along with NetworkX, in order to create the chemical and structural embeddings for each drug.

Keywords: Drug-drug interaction prediction · DrugBank · Network embeddings · Chemical embeddings · Graph topological features

1 Introduction

The efficacy of a drug can be improved when coadministered with other drugs [4]. Also, it is very common for patients that suffer from comorbidities to follow a multiple drug scheme. Yet, there can be adverse effects in those combinations, which occasionally might be toxic. It is an especially serious problem as the number of drugs that people get for a disease continues to increase [10,19].

While many drug interactions have been discovered, there are potentially new ones that could be predicted with computational methods, before their laboratory confirmation [3]. Thus predicting Drug-Drug Interactions (DDIs) is important for the well-being of patients. The problem of drug interaction prediction with computational methods is usually reduced to the problem of link prediction in a network of interactions. Additional features, such as structural, physicochemical and biochemical characteristics of chemical compounds could

H. Cherifi et al. (Eds.): COMPLEX NETWORKS 2016 2022, SCI 1077, pp. 402–414, 2023.
https://doi.org/10.1007/978-3-031-21127-0_33

potentially increase the accuracy of predictions; the usefulness of such properties is already proven in the field of drug discovery [30].

Our work is focused on single link prediction between drugs. We study the embedding of a drug's chemical formula towards vector representations that encapsulate the properties of chemical compounds. We also study the embedding of the network information that we extract when we consider each drug to be a node in a graph of drug interactions; interactions are denoted as edges in this type of graph. We utilize these embedding methods in creating machine learning models that attempt to predict interactions between drugs. Our interest is not just to isolate those methods and separately evaluate their efficacy in creating link prediction systems; we delve into the network of drug interactions and compare the two embedding mechanisms in order to discover which are the network properties that make each approach more efficient than the other when it comes to predicting DDIs. For example, we could assume that a drug for which there is plenty of network information (e.g. a high degree node in our graph) is a good candidate for network embedding methods that n exploit this kind of information. On the other hand, a drug that may seem isolated in the network of drug interactions (perhaps a newly discovered compound with only a few known interactions), may be more suitable for chemical embedding methods. We study various graph properties, such as the degree, core difference and betweenness centrality of nodes and edges in order to discover parts and characteristics of the interactions graph that will help us compare in detail the chemical and the structural methods of embedding drug information.

Contribution In this work we make the following contributions:

- Comparison of two drug interaction prediction methods. One based on network information and another on chemical information. The node2vec and mol2vec were used for embedding network and chemical information respectively.
- In-depth study of topological features that influence the accuracy of the two link prediction models. In particular, when is chemical information more useful than network information when it comes to drug interaction discovery.
- Experimental evaluation of the link prediction models on a graph created from DrugBank's [25] drug interactions.

The rest of the paper is structured as follows: In Sect. 2 we refer to methods of link prediction. Section 3 describes the embedding processes, the prediction, and the topological features that were used. A presentation of the dataset we extracted from DrugBank is provided in Sect. 4. The experimental results are discussed in Sect. 5 and conclusions and future work suggestions are drawn in Sect. 6.

2 Literature Review

Link prediction in graphs is a well researched area [26], with many applications in social networks analysis [22,27] and drug interaction prediction [21], to name just two prominent ones. Link prediction can be based on graph features, upon which matrix and tensor factorization can be applied [18,22]. Such methods have also been applied to graphs that evolve in time [5]. Node and edge embedding methods are widely used to create features for classification [12]. Lately graph neural networks have become popular in link prediction [17,28].

Link prediction has also been used for drug interactions, with methods varying from predicting the presence or absence of interactions, to methods that predict the type of an interaction in a multi-relational network [1,11]. Moreover, many methods have been proposed that employ additional features apart from graph based features, for instance the usage of chemical information has been proposed in [14]. Also a hybrid method that combines multiple types of information, including network information has been proposed in [28]. For multi-relational link prediction there are approaches based on graph neural networks [31]. A rather recent approach focused explicitly on chemical information to predict drug-drug interaction types [20]. Essentially, the authors performed multi-relation link prediction, where the links are the adverse side effects of drugs. This study went further into predicting alternative drug interactions that are not toxic.

This paper focuses on the difference between chemical and network information in DDIs prediction, and in particular how the graph topological features could render one of these types of information more useful for discovering new interactions.

3 Methodology

For each drug, we created two embedding vectors: a network and a chemical based representation. Next, we used three neural network classifiers to predict DDIs: a classifier based on node embeddings, another classifier for chemical embeddings, and a hybrid classifier that is based on both network and chemical embeddings. We evaluated the performance of each classifier, and proceeded to compare their DDI prediction behavior against various topological features on the network of drug interactions.

Mol2vec encodes chemical compounds as vectors with an unsupervised machine learning approach on a corpus of compounds that consists of all available chemical matter [9]. The vector representations of molecular substructures are close for chemically related substructures. In the experiments a pre-trained mol2vec model was used to produce the chemical embeddings of the drugs.[1]

Node2vec is a machine learning algorithm for producing node embeddings, by mapping the nodes of a graph to a low-dimensional space of features that maximizes the likelihood of preserving their network neighborhoods [7]. After harvesting DrugBank's drug interaction graph and then extracting a sample subgraph (see Sect. 4), we trained a node2vec model on the sampled graph and used it to produce the structural embeddings of the drugs.

[1] https://github.com/samoturk/mol2vec/tree/master/examples/models.

Training Samples The experiments focus on predicting drug interactions, we consider two categories of samples: positive and negative. Positive samples refer to observed interactions between drugs, and negative samples refer to interactions that have not been observed so far, i.e. a closed world assumption. The input to a classifier is pair of drugs represented as a vector, both for the mol2vec and the node2vec embeddings. The output is the presence or absence of an interaction. Figure 1 displays the logic of the classifiers for drugs v_i and v_j.

Note that each DDI leads to two different vector concatenations depending on the order the concatenation is done; in the following experiments we included both options to generate positive samples. Acquisition of positive samples was straightforward. However, to create negative interactions it was assumed that unobserved interactions do not exist, and thus the graph was sampled for pairs of nodes that do not form edges.

In order to have a balanced data-set for the experiments, we produced a set of negative samples equal in size to the set of positive samples.

We evaluated three classifiers based on the same neural network architecture of a feed forward model with two hidden layers: A *mol2vec* classifier based on mol2vec embeddings, a *node2vec* classifier based on node2vec embeddings, a *hybrid* classifier based on both mol2vec and node2vec embeddings. The aim was to compare chemical and structural information when used for DDI prediction. For this purpose, along with commonly used metrics in classification, we also used network topological features such as node degree, k-core values and betweenness centrality, in order to identify which topology characteristics make each approach more efficient in the problem of link prediction.

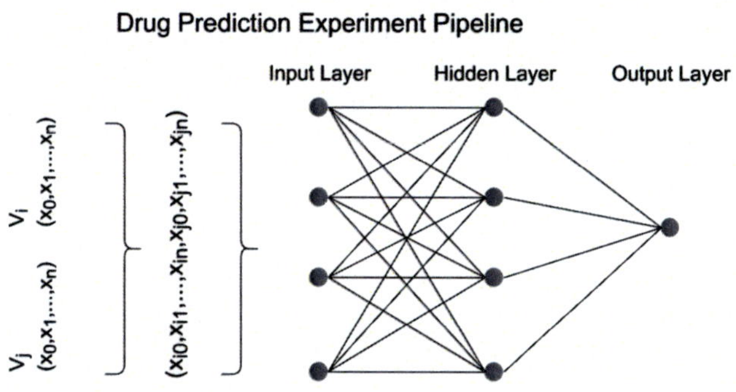

Fig. 1. DDIs, in the form of concatenated vectors of drug embeddings, are used as training samples for our classifiers.

4 Data Harvesting and Processing

Interactions Graph Data were harvested from DrugBank [25] (version 5.1.9), a drug interactions repository that is human curated. The data included 14,624 drug entries and 1,389,184 unique DDIs. Only the drugs that have at least one known interaction were kept, and we also excluded some drugs that have chemical compounds that are incompatible with the tools (i.e. PubChem) that we used . After the data cleaning, a set was obtained that comprised of 3,753 nodes and 1,207,953 edges, and it will be referred as the *full graph.*

When we refer to structural (or network) information regarding DDIs, it is critical to make a distinction between direct edges between drugs, which denote drug interactions, and information that derives from the rest of the graph properties that two interacting drugs (nodes) share. There is no point in creating node2vec embeddings over a graph that holds all the direct edges between interacting drugs, because that information—which is essentially the train and test set of the following experiments—will infiltrate in the structural embeddings. What we do instead, is take a small sample of the *full graph* that contains the same number of nodes and only 1% of its edges. We will refer to the result of this process as *sampled graph.* We train the node2vec model on the *sampled graph* and then use it to create the structural embeddings for our experiments.

Graph Sampling Sampling 1% of the edges of the *full graph*, serves two purposes. First, it showcases the ability of node2vec to capture graph properties of the original graph, from a subset of edges that is smaller by two orders of magnitude; this fact is presented in detail in Sect. 5. The second and most important purpose is to have node embeddings that capture information beyond the one-hop neighbours, thus minimizing the effect of direct links between the corresponding nodes.

Closeness centrality in Table 2 confirms that the sampled graph has been stripped off most of the direct edges, raising the average distance between nodes of the graph and, therefore, setting a level of difficulty for our model to identify interacting drugs through graph information that does not include the direct edges. Figure 3a provides a more detailed view on closeness centrality [6] for both graphs; in the case of more than one connected components we use the Wasserman and Faust formula [23].

Table 1. Basic properties for full graph and sampled graph.

Graph	Nodes	Edges	Avg. degree	Clustering Co.	Conn. components	Diameter
Full	3753	1,207,953	643.73	0.621	1	5
Sampled	3753	12,080	6.44	0.005	799	∞

Table 1 depicts some basic properties for the full and the sampled graph, and Table 2 contains the average centrality measures. Also, DrugBank labels drugs with the categories shown in Table 3; note that some drugs belong to more than one category. Interestingly, drugs that belong to different categories also seem to differentiate on a graph/topological perspective (avg. node degrees and std. of node degrees).

Table 2. Average centrality measures for full graph and sampled graph.

Graph	Degree C.	Eigenvector C.	Closeness C.	Betweenness C.
Full	0.1715	0.012	0.5030	0.00027
Sampled	0.0017	0.011	0.1556	0.00049

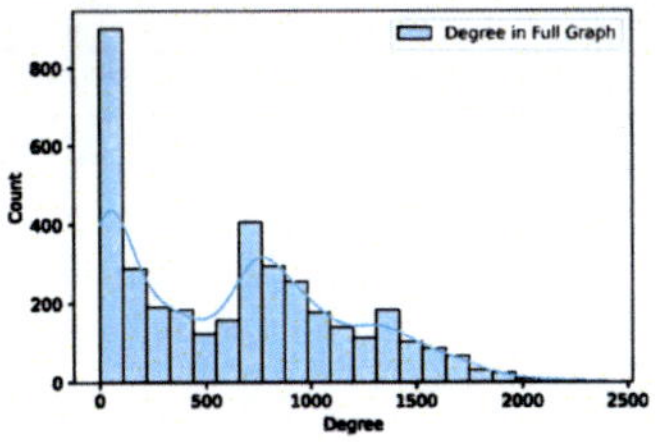

Fig. 2. Histogram and KDE for node degrees in *full graph*.

The average node degree of the *full graph* reflects a high density of drug-interactions, thus an average drugs interacts with 600 other drugs. Also, Fig. 2 reveals that there is a considerable number of drugs that have only few known interactions, as well as many drugs with thousands of identified interactions.

Although reducing closeness centrality in the *sampled graph* is one of our goals in order to exclude most of the direct edges' influence from our node2vec embeddings, we also need to keep enough edges of the *full graph* to maintain its core characteristics and, therefore, produce useful vectors that will efficiently train the node2vec classifier. Thus we considered the eigenvector centrality of the *full graph* and the *sampled graph*. Edges originating from high-scoring nodes contribute more to the score of a node than connections from low-scoring nodes. A high eigenvector score means that a node is connected to many nodes who themselves have high scores [2], denoting the transitive influence of nodes. Figure 3b shows that sampling the *full graph* did not result in a big difference at the *sampled graph's* eigenvector centrality distribution. We can say that nodes in the *full*

Table 3. *Full graph's* drug categories.

Category	Frequency	Degree (Mean, std.)
Approved	2179	787.54 ± 540.90
Investigational	1585	635.09 ± 541.51
Experimental	853	480.82 ± 447.84
Vet Approved	309	731.11 ± 578.32
Withdrawn	190	871.08 ± 492.24
Illicit	123	870.91 ± 503.60
Nutraceutical	63	282.95 ± 337.05

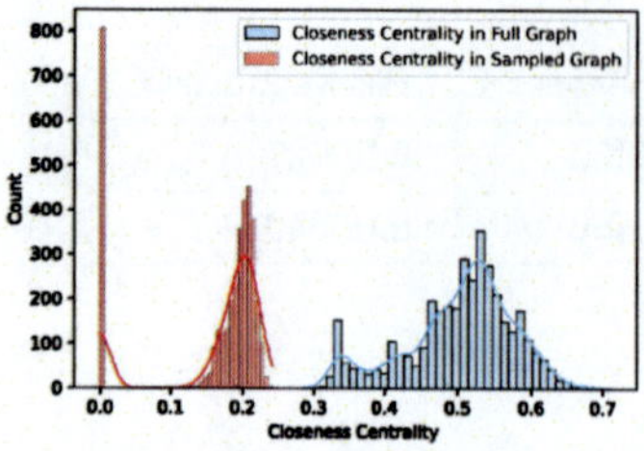

(a) Histogram and KDE of Closeness Centrality in *Full graph* and *Sampled graph*.

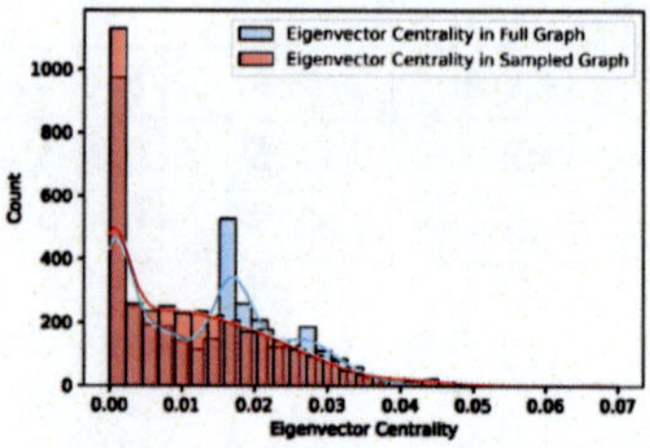

(b) Histogram and KDE of Eigenvector Centrality in *Full Graph* and *Sampled Graph*.

Fig. 3. Closeness centrality and eigenvector centrality.

graph that not only have many connections to other nodes, but are also connected with other nodes of importance—in the sense of eigenvector centrality—continue to hold this characteristic in the *sampled graph*.

Sampling a greater percentage than 1% of edges of the *full graph* to create the *sampled graph* leads to similar closeness centrality distributions between the two graphs, and sampling less than that causes great differences in their eigenvector centrality distributions. This is the reason behind the choice to sample 1% of the *full graph's* edges before training the node2vec model and producing the network embedding for each drug.

Data pre-processing Starting from DrugBank's drug IDs, we used various tools in order to obtain the chemical and the structural embeddings (see Fig. 4). With the NetworkX library [8] a graph data structure was created out of DrugBank data, and it was also used to create the *sampled graph*. We trained node2vec on the *sampled graph*, setting the algorithm to embed nodes to vectors of 128 dimensions. Once node2vec [7] was trained, we applied it on each drug (node) to produce the corresponding node2vec (structural) embedding.

To compute the mol2vec (chemical) embeddings, first we used PubChem's services and acquired the isomeric SMILES (Simplified Molecular Input Line Entry System) notation for each drug [13]. SMILES is a chemical notation system based on molecular graph theory [24]. Then, we used RDKit [16] to convert each SMILES to a MOL data structure [29], a widely-used chemical structure file format in which adjacent lists and adjacent matrices are used to describe a chemical compound's structure. Finally, mol2vec [9] was applied on the MOL structures, using a pre-trained model to embed drugs to vectors of 300 dimensions.

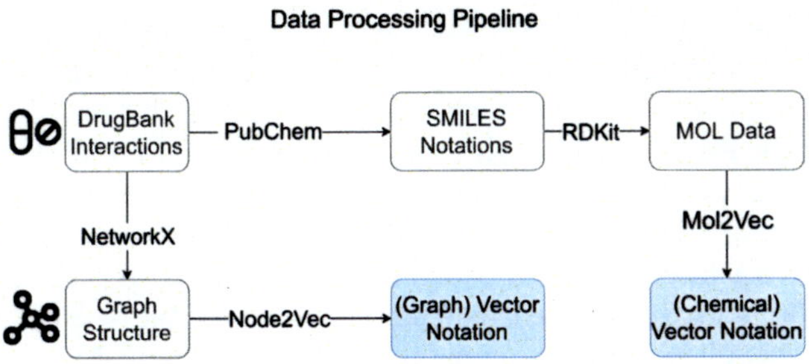

Fig. 4. Data processing pipeline: steps between DrugBank data and final embeddings for each drug.

5 Experimental Results

Classifier Evaluation Figure 5 displays the classification report of the three models on the test set; negative interactions are denoted by class 0 and positive interactions are denoted by class 1. We used a 65/5/30 split on all samples (positive and negative) of the *full graph* for training, validation and testing—we applied a random split, with the exception that all of the edges in the *sampled graph* (which are also present in the *full graph*) ended up on the training set. Also, both variations of an interaction sample (regarding the order of the concatenation of the corresponding drug embeddings) are always included in the same set (i.e., train, validation or test). Figure 8 compares the average value of useful graph properties of the test samples, and Fig. 8 provides a comparison between the Kullback-Leibler divergence of the distributions of test sample properties for the correct and false predictions of mol2vec and node2vec classifiers. *Average mean degree* refers to the average of all means of the node degrees of the test samples; similarly *average min degree* denotes the average of all minimum degrees between all test interactions, and *average max degree* denotes the average maximum degree. Also, *betweenness* centrality in this section is calculated only for positive samples, and refers to the corresponding edge betweenness centrality (and not the node betweenness centrality that is reported in Sect. 4).

The higher performance of the hybrid classifier compared to the other models in the classification report suggests that there is knowledge on DDIs that is unique for both the chemical and the structural embedding methods. Combining the embeddings to train the hybrid model—trading this abundance of information with higher vector dimensions that are known to hinder the learning capabilities of neural networks—leads to a better predictor. The only exception here is the higher precision on positive interactions for the node2vec classifier; hinting that the node2vec classifier shows a greater ability to identify negative interactions properly and maintain a lower number of false positives choices through the evaluation phase. Figure 6a confirms our assumption that drugs with few known interactions (possibly newly discovered compounds) make better candidates for chemical based predictors when recall is more important than accuracy; identifying more true interactions by trading some false positives may be a good trade off for a chemical researcher. The figure even sets a threshold value at a *min degree* of 350 where node2vec begins to perform better than mol2vec in terms of recall. Figures 6b and 7b show that for low core difference values[2] of sample interactions the node2vec classifier performs as well as the hybrid classifier in terms of recall and accuracy. The fact that node2vec classifier reaches hybrid model's efficiency means that chemical embeddings, when it comes for test interactions with low core difference, show no unique knowledge to add to structure embeddings' learning capabilities. Also, all models seem to perform better for lower values of core difference.

Class	Classifier	Precision	Recall	F1 Score
Negative Samples (Class 0)	mol2vec	0.91	0.89	0.90
	node2vec	0.90	0.94	0.92
	Hybrid	**0.96**	**0.94**	**0.95**
Positive Samples (Class 1)	mol2vec	0.88	0.91	0.89
	node2vec	**0.96**	0.93	0.94
	Hybrid	0.93	**0.95**	**0.94**

(a) Classification Report.

Category	mol2vec	node2vec	Hybrid
Experimental	87	94	96
Approved	85	93	94
Investigational	87	94	95
Vet Approved	88	94	97
Withdrawn	87	96	97
Illicit	87	95	96
Nutraceutical	91	90	97

(b) Accuracy percentage comparison for each drug category.

Fig. 5. Classification report and accuracy comparison per drug category.

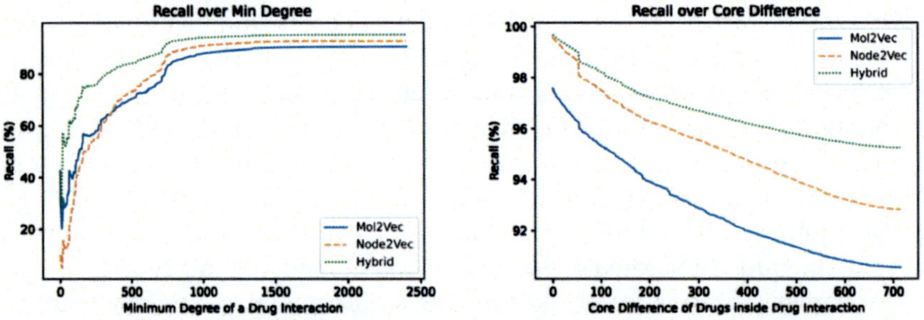

(a) Recall over Min Degree. (b) Recall over Core Difference.

Fig. 6. Recall plots for mol2vec, node2vec and hybrid classifiers.

[2] We consider the core difference of a sample interaction as the absolute difference of the k-core values of two nodes that are connected by the corresponding graph edge.

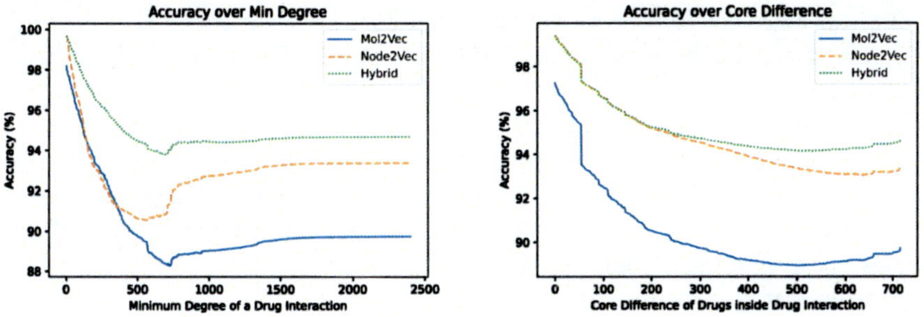

(a) Accuracy over Min Degree. (b) Accuracy over Core Difference.

Fig. 7. Accuracy plots for mol2vec, node2vec and hybrid classifiers.

	Metric	mol2vec	node2vec	Hybrid
False Prediction	Average Min Degree	518.06	470.07	517.01
	Average Max Degree	1137.02	1143.09	1163.44
	Average Mean Degree	827.54	806.58	840.22
	Average Core Difference	237.17	266.10	242.34
	Average Betweenness Centrality	4.57e-06	6.99e-06	6.76e-06
Correct Prediction	Average Min Degree	519.66	593.45	512.09
	Average Max Degree	1083.25	1143.96	1078.67
	Average Mean Degree	801.45	868.713	795.38
	Average Core Difference	228.50	196.01	231.80
	Average Betweenness Centrality	1.39e-06	1.29e-06	1.44e-06

(a) Comparison of average graph properties of interactions grouped by each model's prediction.

Classifier	Correct	False
Average Degree	0.018	0.047
Min Degree	0.019	**0.199**
Max Degree	0.010	**0.110**
Core Difference	0.013	**0.103**
Betweenness C.	**0.109**	0.071

(b) KL Divergence between mol2vec and node2vec sample distributions (values over 0.1 in bold).

Fig. 8. Result metrics by classifier choice (Correct or false).

6 Conclusions and Future Work

Regarding the prediction of presence or absence of edges/interactions we have observed the following results. Recall for the min-degree criterion is better for mol2vec up to a certain threshold, after that node2vec takes over. Also regarding the core-differences criterion, node2vec is always better than mol2vec. Moreover, the distribution of the betweeness centralities of the pairs of nodes for which the presence or absence of edges were correctly predicted, were very different for the mol2vec and the node2vec model.

We have also shown that low core differences between a pair of nodes make structural information based models a better candidate to predict interactions achieving the same accuracy and recall even with the hybrid approach. But for higher core differences, chemical information can boost the information that is provided by structural information and thus enhance the performance of interaction prediction.

By and large the hybrid model that combines structural and chemical information of drugs leads to more efficient predictors of DDIs than using either of the models.

After comparing structural and chemical approaches for predicting simple interactions between drugs, a question arises about what conclusions we would reach if we expanded our research on the broader field of predicting multi-labeled drug interactions. We can also expand the graph to include other types of nodes besides drugs (e.g. proteins).

Further study on techniques for negative samples generation could improve the quality of the training set and, possibly, allow for better predictors. When it comes to multi-relational link prediction, [15] provides a notable analysis on the importance of negative sampling, as well as useful methods for negative sample generation, such as the "corruption of positive samples" or "nearest neighbor sampling".

Acknowledgements. The authors would like to thank the Software and Knowledge Engineering Lab (SKEL) of the Institute of Informatics and Telecommunications, NCSR "Demokritos"; and the American College of Greece, Deree for their support.

References

1. Bagherian, M., Sabeti, E., Wang, K., Sartor, M.A., Nikolovska-Coleska, Z., Najarian, K.: Machine learning approaches and databases for prediction of drug-target interaction: a survey paper. Briefings Bioinformatics **22**(1), 247–269 (2021)
2. Bonacich, P.: Some unique properties of eigenvector centrality. Soc. Networks **29**(4), 555–564 (2007)
3. Cheng, F., Kovács, I.A., Barabási, A.L.: Network-based prediction of drug combinations. Nature Commun. **10**(1), 1–11 (2019)
4. Crystal, A.S., Shaw, A.T., Sequist, L.V., Friboulet, L., Niederst, M.J., Lockerman, E.L., Frias, R.L., Gainor, J.F., Amzallag, A., Greninger, P., et al.: Patient-derived models of acquired resistance can identify effective drug combinations for cancer. Science **346**(6216), 1480–1486 (2014)
5. Dunlavy, D.M., Kolda, T.G., Acar, E.: Temporal link prediction using matrix and tensor factorizations. ACM Trans. Knowledge Discovery From Data (TKDD) **5**(2), 1–27 (2011)
6. Freeman, L.C.: Centrality in social networks conceptual clarification. Soc. Networks **1**(3), 215–239 (1978)
7. Grover, A., Leskovec, J.: node2vec: scalable feature learning for networks. In: Proceedings of the 22nd ACM SIGKDD International Conference on Knowledge Discovery and Data Mining. pp. 855–864 (2016)
8. Hagberg, A., Swart, P., Chult, S.D.: Exploring Network Structure, Dynamics, and Function Using Networkx (Jan 2008). https://www.osti.gov/biblio/960616
9. Jaeger, S., Fulle, S., Turk, S.: Mol2vec: unsupervised machine learning approach with chemical intuition. J. Chem. Inf. Modeling 0(ja), null (0). https://doi.org/10.1021/acs.jcim.7b00616
10. Kantor, E.D., Rehm, C.D., Haas, J.S., Chan, A.T., Giovannucci, E.L.: Trends in prescription drug use among adults in the united states from 1999–2012. JAMA **314**(17), 1818–1830 (2015)

11. Kastrin, A., Ferk, P., Leskošek, B.: Predicting potential drug-drug interactions on topological and semantic similarity features using statistical learning. PloS one **13**(5), e0196865 (2018)
12. Kazemi, S.M., Poole, D.: Simple embedding for link prediction in knowledge graphs. Advan. Neural Inf. Proc. Syst. 31 (2018)
13. Kim, S., Chen, J., Cheng, T., Gindulyte, A., He, J., He, S., Li, Q., Shoemaker, B.A., Thiessen, P.A., Yu, B., et al.: Pubchem 2019 update: improved access to chemical data. Nucleic Acids Res. **47**(D1), D1102–D1109 (2019)
14. Koge, D., Ono, N., Huang, M., Altaf-Ul-Amin, M., Kanaya, S.: Embedding of molecular structure using molecular hypergraph variational autoencoder with metric learning. Molecular Informatics **40**(2), 2000203 (2021)
15. Kotnis, B., Nastase, V.: Analysis of the impact of negative sampling on link prediction in knowledge graphs. arXiv preprint arXiv:1708.06816 (2017)
16. Landrum, G., et al.: Rdkit: A Software Suite for Cheminformatics, Computational Chemistry, and Predictive Modeling (2013)
17. Long, Y., Wu, M., Liu, Y., Fang, Y., Kwoh, C.K., Chen, J., Luo, J., Li, X.: Pretraining graph neural networks for link prediction in biomedical networks. Bioinformatics **38**(8), 2254–2262 (2022)
18. Menon, A.K., Elkan, C.: Link prediction via matrix factorization. In: Joint European Conference on Machine Learning and Knowledge Discovery in Databases, pp. 437–452. Springer (2011)
19. Qato, D.M., Wilder, J., Schumm, L.P., Gillet, V., Alexander, G.C.: Changes in prescription and over-the-counter medication and dietary supplement use among older adults in the United States, 2005 versus 2011. JAMA Internal Med. **176**(4), 473–482 (2016)
20. Ryu, J.Y., Kim, H.U., Lee, S.Y.: Deep learning improves prediction of drug-drug and drug-food interactions. Proc. Nat. Acad. Sci. **115**(18), E4304–E4311 (2018)
21. Segura-Bedmar, I., Martínez, P., Herrero-Zazo, M.: Lessons learnt from the DDIExtraction-2013 shared task. J. Biomed. Inf. **51**, 152–164 (2014)
22. Wang, P., Xu, B., Wu, Y., Zhou, X.: Link prediction in social networks: the state-of-the-art. Sci. China Inf. Sci. **58**(1), 1–38 (2015)
23. Wasserman, S., Faust, K., et al.: Social Network Analysis: Methods and Applications (1994)
24. Weininger, D.: Smiles, a chemical language and information system. 1. introduction to methodology and encoding rules. J. Chem. Inf. Comput. Sci. **28**(1), 31–36 (1988)
25. Wishart, D.S., Feunang, Y.D., Guo, A.C., Lo, E.J., Marcu, A., Grant, J.R., Sajed, T., Johnson, D., Li, C., Sayeeda, Z., Assempour, N., Iynkkaran, I., Liu, Y., Maciejewski, A., Gale, N., Wilson, A., Chin, L., Cummings, R., Le, D., Pon, A., Knox, C., Wilson, M.: DrugBank 5.0: a major update to the DrugBank database for 2018. Nucleic Acids Res. **46**(D1), D1074–D1082 (Nov 2017). https://doi.org/10.1093/nar/gkx1037
26. Yang, Y., Lichtenwalter, R.N., Chawla, N.V.: Evaluating link prediction methods. Knowl. Inf. Syst. **45**(3), 751–782 (2015)
27. Yuliansyah, H., Othman, Z.A., Bakar, A.A.: Taxonomy of link prediction for social network analysis: a review. IEEE Access **8**, 183470–183487 (2020)
28. Zhang, M., Chen, Y.: Link prediction based on graph neural networks. In: Advances in Neural Information Processing Systems 31 (2018)
29. Zhang, W., Hou, T., Qiao, X., Xu, X.: Some basic data structures and algorithms for chemical generic programming. J. Chem. Inf. Comput. Sci. **44**(5), 1571–1575 (2004)

30. Zhang, W., Chen, Y., Liu, F., Luo, F., Tian, G., Li, X.: Predicting potential drug-drug interactions by integrating chemical, biological, phenotypic and network data. BMC Bioinf. **18**(1), 1–12 (2017)
31. Zitnik, M., Agrawal, M., Leskovec, J.: Modeling polypharmacy side effects with graph convolutional networks. Bioinformatics **34**(13), i457–i466 (2018)

Geometric Deep Learning Graph Pruning to Speed-Up the Run-Time of Maximum Clique Enumerarion Algorithms

A. Arciprete, V. Carchiolo[✉], D. Chiavetta, M. Grassia, M. Malgeri, and G. Mangioni

Dipartimento di Ingegneria Elettrica, Elettronica e Informatica, Università degli Studi di Catania, Sicily, Italy
vincenza.carchiolo@unict.it, marco.grassia@unict.it, michele.malgeri@unict.it, giuseppe.mangioni@unict.it

Abstract. In this paper we propose a method to reduce the running time to solve the Maximum Clique Enumeration (MCE) problem. Specifically, given a network we employ geometric deep learning in order to find a simpler network on which running the algorithm to derive the MCE. Our approach is based on finding a strategy to remove from the network nodes that are not functional to the solution. In doing so, the resulting network will have a reduced size and, as a result, search times of the MCE is reduced. We show that our approach is able to obtain a solver speed-up up to 42 times, while keeping all the maximum cliques.

Keywords: Cliques · Geometric deep learning · MCE

1 Introduction

Networks (or graphs) are widely used in many contexts and disciplines [3] as they help to understand real–world problems. In fact, one can "encode" a system in a network, study the phenomena happening on that system in the new representation, which is easier to work on, and then, if needed, "decode" the solution to the problem. For example, networks are used to encode social systems, like online communities and citation networks, technical systems and economic systems. The solution to many real-world problems requires the application of some algorithm to the network that models the phenomenon. Unfortunately, many algorithms are computationally difficult (i.e., NP-Hard) and do not scale with the input size. Furthermore, due do the increasingly large amount of data that is available nowadays, the size of these networks is growing fast. As a result, solving such problems is computationally challenging or even intractable in some cases. Among the problems on networks that are NP-Hard, in this paper we focus on the Maximum Clique Enumeration (MCE) problem, e.g., the enumeration version of the Maximum Clique problem. It consists in finding and listing all the maximum cliques in a network. A clique is a fully connected sub-graph of a finite and simple

H. Cherifi et al. (Eds.): COMPLEX NETWORKS 2016 2022, SCI 1077, pp. 415–425, 2023.
https://doi.org/10.1007/978-3-031-21127-0_34

graph. The problem of MCE is one of the most well-known and widely studied combinatorial problems [20]. Other than the pure theoretical aspects, finding the maximum cliques has several practical applications. For example, in [17] authors use the MCE to understand the principles of cell organizations, in [13] authors derive structural information and correlations between different metabolite levels by enumerating all the maximum cliques. Other application are presented in [7], where Fukagawa et al. proposed a method for transforming the Tree Edit Distance (TED) between unordered trees problem into a maximum vertex weight clique problem and Mori et al. into a maximum clique problem [18].

As mentioned above, the MCE problem is particularly challenging, especially for huge networks. In the literature there are several heuristic solvers (see [11,26]) that often provide suboptimal solutions in reasonable time. Alternative approaches to cope with NP–Hard problems involve machine learning [5,8]. For instance, in [9,14,15] the authors proposed a technique aiming at reducing the MCE search space by pruning the input networks, i.e., by removing those nodes that probably doesn't belong to any maximum cliques. More precisely, each node v is removed if a classic machine learning algorithm predicts the probability it belongs to some maximum clique is lower than some threshold. Prediction are performed by using both network measures and statistical properties as features for each node in order to capture structural properties of the network.

In this paper we build upon the previous approaches and propose a pre-processing step for the MCE problem inspired from those proposed in [9,14,15]. More specifically, we employ Geometric Deep Learning [25] to implement a network pruning strategy aiming at reducing the running time required to enumerate the maximum cliques. Geometric Deep Learning is a new field of Machine Learning that deals with the realization of models able to learn from non-Euclidean data, such as graphs. Exploiting the capabilities of Geometric Deep Learning models, we show that our method is able to drastically prune real-world networks by removing up to ∼88% of nodes and ∼77% of edges while preserving all the maximum cliques. Such pruning ratios translate into speed-ups up to 40 times for a variation of the Bron-Kerbosch algorithm [1] implemented in the *igraph* [4] network analysis software.

The paper is organized as in the following. Section 2 introduces the MCE problem characteristics and briefly presents a related work on the algorithm of the literature. Our solution is presented in Sect. 3, whereas in Sect. 4 we report the experimental setup and results. Finally, Sect. 5 discusses the results and future works.

2 Maximum Clique Enumeration Problem

In an undirected graph $G(V, E)$, where V is the set of the nodes (vertices) and E is the set of the edges, a clique is defined as the subset $S \subseteq V$ of nodes such that every couple of nodes in S are adjacent. In other words, a clique is a sub-graph where each node is connected by an edge to all the other nodes of the same sub-graph. A *maximum* clique is a clique with the largest possible number of nodes, whereas a *maximal* clique is a clique that cannot be enlarged. The Maximum Clique Enumeration problem is a combinatorial problem that consists in finding all the maximum cliques in a given graph, and was proven to be NP-hard (i.e., computationally difficult). The algorithms for finding and enumerating all the maximum cliques can be divided into the following three families:

1. **Exact solvers**: they are able to find an optimal solution with exponential complexity, and usually take a branch and bound approach. This approach consists in taking the entire research space and dividing it, iteratively, into a number of subspaces called branches. Iteration after iteration, a pruning operation is carried out by means of which branches that are not useful for the purpose of the final solution are pruned and therefore are no longer taken into account in the branching phase. The operation by which you decide whether a branch should be removed or not is called bounding.
2. **Heuristic solvers**: these algorithms are based on probabilistic models, which try to explore in the most efficient way the least possible number of subsets of vertices of the graph, getting as close as possible to the solution, without ever having the certainty of its accuracy. Unlike exact solvers, heuristic solvers require polynomial time to find a (often suboptimal) solution. This means that a heuristic solver cannot always find all the cliques of maximum size of a graph just as it is not even said that the size of every single clique found by the algorithm is the real one.
3. **Domain specific solver**: these algorithms are meant to solve the problem on graphs from specific domains, as they take advantage of the specific domain properties.

In the literature there are numerous algorithms that fall into the three families. For instance, the Bron-Kerbosch algorithm [12], used to determine all cliques of the maximum size of an undirected graph, is characterized by a computational complexity of $O(3^{(|V|/3)})$, where $|V|$ is the number of nodes of the graph. Fast Max-Clique Finder algorithm [21] is a heuristic solver whose computational complexity is $O(|V|^2)$, where it is the maximum degree within the graph. Cliquer is another very popular implementation of the algorithm defined by Östergård in 2002 [19], and adopts the branch and bound strategy. Since the performance of this algorithm depends on node sorting, a heuristic approach based on vertex coloring is used. EmMCE [2] is an algorithm that uses external memory to store those networks that cannot be allocated within RAM memory due to excessive size. To improve running time, this algorithm also runs in parallel. For further discussion about the MCE algorithms, we refer the Reader to [16].

3 Our Approach

In this work, we propose a pre-processing strategy based on the works from Lauri et al. [9,15] and on Geometric Deep Learning. The approach consists in augmenting the state-of-the-art solvers by pruning the nodes of the input networks, thus reducing the search space. More specifically, the key idea is to remove each node such that the probability (predicted by a Machine Learning model) it does not belong to some maximum clique is lower than a threshold, called *confidence threshold*. The confidence threshold can be used to trade-off between pruning rates and accuracy, and makes the approach more flexible, since some applications may need accuracy over performance gains or vice-versa. Of course, while the computational complexity of the solvers is not affected, the pre-processed input graphs are smaller, thus the computational time and memory usage are improved.

The main differences with the previous approaches are that we employ Graph Neural Network (GNN) layers, whereas the others use classic Machine Learning algorithms (like Multi-Layer Perceptrons and random forests), and that we only perform a single pruning stage. In particular, we use a Geometric Deep Learning model with up to three Graph Attention Network (GAT) [10,25] layers stacked together with a Multi-Layer Perceptron that, given the node embedding, performs regression on the nodes and returns a value between 0 and 1. We choose the Graph Attention Network layers since they are particularly suitable for our task: they do not perform neighborhood sampling, and they leverage the attention mechanism, proposed in [24] for Natural Language Processing tasks, to assign a relative importance score (i.e., the *attention coefficient*) to each neighboring node. Moreover, GAT layers also use self-attention (through self-loops) and allow multiple attention-heads to improve the prediction performance. As in [9], after the pre-processing stage, we feed the pruned network to an exact clique solver (specifically, to the one implemented in the Python *igraph* library) to retrieve all the maximum cliques.

We train our model in a supervised manner on relatively small real-world networks, where the maximum cliques can be found in reasonable time, and generalize to larger networks. The training target for each node is either 1 if the node belongs to some maximum clique, or 0 otherwise. Regarding the input node features we feed to our models, they are a subset of the ones used in [9] since the topology of the network is already exploited by the GNN layers. In particular, for each node we compute the normalized node degree, the Local Clustering Coefficient (LCC), and the Chi-squared χ^2 of normalized degree and of the Local Clustering Coefficient w.r.t. the neighbors'.

The computational complexity of our pre-processing step is linear with the number of edges. In fact, our approach consists in two steps: we first compute the node features, and then feed the network and the node features to the model. The most expensive node feature to compute is the Local Clustering Coefficient, which is:

$$O(\text{LCC}) = O(|V| \langle k^2 \rangle) \tag{1}$$

where $\langle k^2 \rangle$ is the second moment of the degree distribution [22]. On the other hand, the GAT layers have the following computational complexity:

$$O(\text{GAT}) = h \cdot (|V| + |E|) \tag{2}$$

where h is the number of attention heads used, and $|V|$ and $|E|$ are the number of nodes and edges respectively.

4 Experiments

In this Section we describe the experimental settings and the results obtained. We note that all the networks used in our experiments were downloaded from NetworkRepository.com [23], and that the models are implemented using the PyTorch Geometric (PyG) library v2.0.1 [6].

4.1 Traning and Validation

We train our models in a supervised manner for 250 epochs using an Adam optimizer. Being the label of the nodes binary (1 if they belong to some maximum clique or 0 otherwise), we choose the Binary Cross Entropy loss function. We build a balanced train set from the nodes of the 64 training networks reported in Table 1. Specifically, for the positive examples, we extract all the nodes that belong to some maximum clique, and, for the negative ones, sample the same number of nodes that do not belong any maximum clique. After each training epoch, we validate the model on the 9 networks reported in Table 2, and, at the end of the training, select the model that provides the best balanced accuracy. Regarding the confidence threshold T, we choose $T = 0.6$ in order to keep all the maximum cliques in the validation set.

4.2 Performance Measures

We define the following measures to evaluate the performance of our approach:

$$\text{similarity} = \begin{cases} 0 & \text{if } |\text{MC}|_p < |\text{MC}| \\ \#\text{MC}_p/\#\text{MC} \cdot 100 & \text{otherwise} \end{cases}$$

$$\text{speedup} = \text{time}(G)/\text{time}(G_p)$$

$$\text{node\_pruning\_rate} = (|V| - |V_p|)/|V| \cdot 100$$

$$\text{edge\_pruning\_rate} = (|E| - |E_p|)/|E| \cdot 100$$

where $|\text{MC}|_p, |\text{MC}|, \#\text{MC}_p, \#\text{MC}$ are the size and number of maximum cliques respectively in the pruned $(G_p(V_p, E_p))$ and original graph $G(V, E)$.

In particular, the similarity measures the percentage of cliques of maximum size held after the pruning stage, the speedup measures the improvement in computational time (i.e., the time spent by the solver on the original graph over the time spent on the pruned graph), and node_pruning_rate and edge_pruning_rate are the percentage of removed nodes and edges (as a consequence of removing the nodes).

Table 1. Train networks. Columns $|V|$ and $|E|$ are the number of nodes and edges respectively, while #MC and |MC| are the number and the size of the maximum cliques.

#	Network	$\|V\|$	$\|E\|$	#MC	|MC|	#	Network	$\|V\|$	$\|E\|$	#MC	|MC|
1	bio-CE-GT	924	3K	2	8	33	rt_uae	5K	6K	4	4
2	bio-CE-HT	2K	2K	1	4	34	rt_tlot	4K	4K	1	4
3	bio-CE-LC	1K	1K	2	7	35	rt_oman	5K	6K	6	4
4	bio-DM-HT	62K	4K	2	4	36	rt_lebanon	4K	4K	1	4
5	bio-DM-LC	658	1K	1	5	37	rt_damascus	3K	4K	11	4
6	bio-HS-HT	2K	13K	10	38	38	rt_bahrain	5K	8K	7	5
7	bio-SC-LC	2K	20K	68	29	39	rt_assad	2K	3K	1	5
8	bio-celegans	453	2K	6	9	40	rt_alwefaq	4K	7K	26	8
9	bio-celegans-dir	453	2K	6	9	41	ia-infect-dublin	410	3K	1	16
10	bio-diseasome	516	1K	1	11	42	ia-email-univ	1K	5K	1	12
11	bio-dmela	7K	25K	1	7	43	ia-fb-messages	1K	6K	8	5
12	bio-grid-fission-yeast	2K	25K	1	12	44	ia-reality	7K	8K	5	5
13	bio-grid-mouse	1K	3K	1	7	45	ia-email-EU	32K	54K	24	12
14	bio-grid-plant	1K	6K	9	9	46	email-dnc-corecipient	906	16K	1	75
15	bio-grid-worm	3K	13K	27	7	47	email-univ	1K	5K	1	12
16	bio-yeast	1K	1K	1	6	48	econ-mahindas	1K	8K	106	4
17	bio-yeast-protein-inter	2K	4K	1	6	49	econ-poli-large	16K	17K	34	4
18	soc-dolphins	62	159	3	5	50	econ-poli	4K	4K	7	4
19	soc-firm-hi-tech	36	147	1	6	51	ca-Erdos992	6K	8K	1	8
20	soc-hamsterster	2K	16K	1	25	52	ca-GrQc	5K	14K	1	44
21	soc-karate	34	78	2	5	53	aves-weaver-social	445	1K	1	12
22	soc-wiki-Vote	889	2	K1	7	54	aves-thornbill-farine	62	1K	7	27
23	soc-tribes	16	58	3	5	55	aves-wildbird-network-3	126	2K	2	25
24	soc-pages-food	620	2K	3	10	56	mammalia-voles-rob-trapping	1K	5K	2	8
25	soc-pages-tvshow	4K	17K	5	57	57	mammalia-voles-bhp-trapping	2K	5K	3	8
26	socfb-Reed98	962	18K	6	16	58	mhd3200b	3K	8K	199	6
27	web-EPA	4K	8K	47	4	59	sstmodel	3K	10K	2K	4
28	web-edu	3K	6K	1	30	60	tols4000	4K	6K	72	19
29	web-polblogs	643	2K	22	9	61	as20000102	6K	13K	17	10
30	web-webbase-2001	16K	25K	1	33	62	gene	1K	2K	1	13
31	tech-pgp	11K	24K	12	25	63	DD242	1K	3K	22	5
32	tech-routers-rf	2K	7K	5	16	64	cora	3K	5K	9	5

Table 2. Validation networks. Columns $|V|$ and $|E|$ are the number of nodes and edges respectively, while #MC and |MC| are the number and the size of the maximum cliques.

Network	$\|V\|$	$\|E\|$	#MC	|MC|
ca-AstroPh	19K	198K	1	57
ca-HepPh	12K	118K	1	239
rec-amazon	92K	126K	294	5
socfb-nips-ego	3K	3K	4	4
soc-pages-media	28k	206K	3	31
soc-pages-sport	14K	87K	15	29
tech-internet-as	40K	85K	15	29
tech-p2p-gnutella	63K	148K	16	4
web-frwikinews-user-edits	25K	194K	5	7

Table 3. Test networks. Columns $|V|$ and $|E|$ are the number of nodes and edges respectively, while #MC and |MC| are the number and the size of the maximum cliques.

Network	$\|V\|$	$\|E\|$	#MC	|MC|
ca-citesee	227K	814K	1	87
ca-dblp-2010	226K	716K	1	75
ca-dblp-2012	317K	1000K	1	114
rt-retweet-crawl	1000K	2000K	26	13
srt-higgs	425K	734	7	12
soc-delicious	536K	1000K	9	21
soc-sign-Slashdot081106	77K	517K	52	26
soc-themarker	69K	1000K	4	22
web-baidu-baike-related	416K	3000K	2	95
socfb-OR	63K	817K	2	30
socfb-wosn-friends	64K	1000K	2	30
tech-RL-caida	191K	608K	31	17

4.3 Results

After training our model and fixing the confidence threshold $T = 0.6$, we test our approach on the 12 networks in Table 3. The results, reported in Table 4, show that our model is able to prune up to 88.85% of nodes and up to 80.75% edges, which translate into a solver speed-up of up 42 times, while keeping all the maximum cliques (i.e. 100% in column "%" in Table 3). We also note that the size such networks is at least one order of magnitude larger than the ones in the train and validation set. Again, we stress that the confidence threshold can be used as a trade-off between desired speedup and *similarity*, and if the loss of maximum cliques is acceptable in turn of faster computational times, it can be further reduced. As an example, in Fig. 1 we show the speedup and similarity values obtained when using increasing confidence thresholds (with step 0.05), only for a subset of the test networks for sake of space. As the Reader may notice, higher confidence thresholds lead to larger pruning rates and speed-up (up to ~500 times), but may also lead to the loss of some (or all) the maximum cliques.

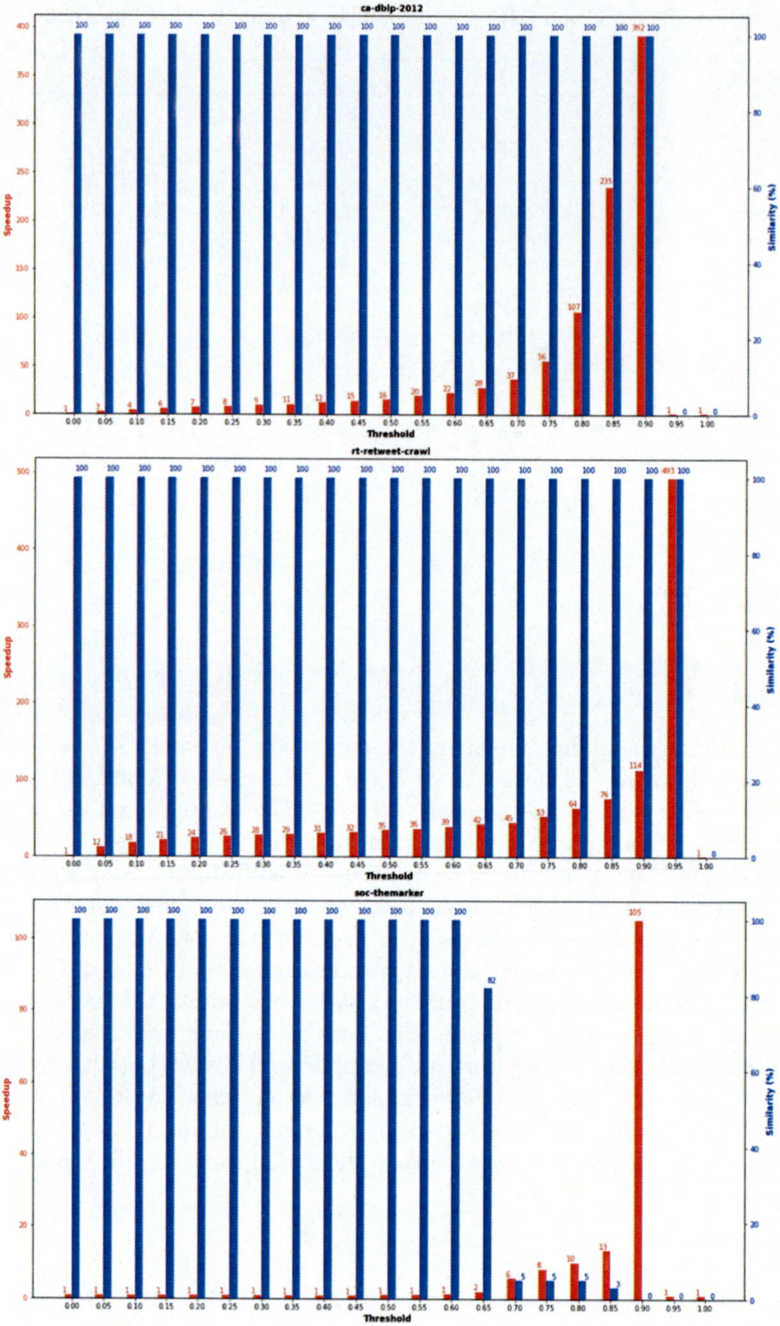

Fig. 1. Comparison of similarity (blue) and speedup (red) for different threshold on a subset of the test networks.

Table 4. Node pruning results.

Graph	Pruning rate		Solver time			Cliques		
	Node (%)	Edge (%)	Before (s)	After (s)	s	%	#MC	|MC|
ca-citeseer	88.85	74.54	6.05	0.19	31.84	100.0	1	87
ca-dblp-2010	88.37	77.38	5.96	0.21	28.38	100.0	1	75
ca-dblp-2012	84.69	74.72	11.76	0.49	24.0	100.0	1	114
rt-retweet-crawl	87.03	76.44	143.03	3.36	42.57	100.0	26	13
rt-higgs	84.04	70.40	18.58	0.61	30.46	100.0	7	12
soc-delicious	81.73	62.78	33.57	2.05	16.38	100.0	9	21
soc-sign-Slashdot081106	84.61	68.75	4.81	2.43	1.98	100.0	52	26
soc-themarker	73.48	55.29	811.55	600.98	1.35	100.0	40	22
web-baidu-baike-related	88.53	71.39	48.69	13.69	3.56	100.0	2	95
socfb-OR	78.0	53.63	9.91	7.12	1.39	100.0	2	30
socfb-wosn-friends	71.39	45.35	10.08	7.71	1.31	100.0	2	30
tech-RL-caida	87.74	80.75	5.60	0.49	11.43	100.0	31	17

The node pruning rate is the ratio of nodes predicted to not be part of a maximum clique, while the edge pruning rate is the ratio of edges removed as a direct consequence of the removal of nodes. The solver time columns are the time taken by the exact clique solver before and after the pruning (in seconds), while the s column is the speedup. The % column is the similarity measure defined in Sect. 4.2, whereas $\#MC$ and $|MC|$ are the number of maximum cliques found after the pruning phase.

5 Conclusions and Future Work

In this work we present a geometric deep learning based network node pruning pre–processing step aiming at reducing the computation time of Maximum Clique Enumeration solvers. We show that our approach is able to obtain speed-up up to 42 times, while keeping all the maximum cliques. Moreover, we show that even better performance can be obtained if we choose a confidence threshold which depends specifically on the network under consideration. This aspect deserve further investigations that will be object of future works.

Acknowledgment. This work has been partially supported by the project of University of Catania PIACERI, *PIAno di inCEntivi per la Ricerca di Ateneo.*

References

1. Bron, C., Kerbosch, J.: Algorithm 457: finding all cliques of an undirected graph. Commun. ACM **16**(9), 575–577 (1973). https://doi.org/10.1145/362342.362367
2. Cheng, J., Ke, Y., Fu, A.W.C., Yu, J.X., Zhu, L.: Finding maximal cliques in massive networks. ACM Trans. Database Syst. (TODS) **36**(4), 1–34 (2011)
3. Costa, L.D.F., Oliveira, O.N., Travieso, G., Rodrigues, F.A., Villas Boas, P.R., Antiqueira, L., Viana, M.P., Correa Rocha, L.E.: Analyzing and modeling real-world phenomena with complex networks: a survey of applications. Adv. Phys. **60**(3), 329–412 (2011)
4. Csardi, G., Nepusz, T.: The igraph software package for complex network research. InterJ. Complex Syst. 1695 (2006). https://igraph.org

5. Fan, C., Zeng, L., Sun, Y., Liu, Y.Y.: Finding key players in complex networks through deep reinforcement learning. Nature Mach. Intel. **2**(6), 317–324 (2020). https://doi.org/10.1038/s42256-020-0177-2
6. Fey, M., Lenssen, J.E.: Fast graph representation learning with PyTorch Geometric. In: ICLR Workshop on Representation Learning on Graphs and Manifolds (2019)
7. Fukagawa, D., Tamura, T., Takasu, A., Tomita, E., Akutsu, T.: A clique-based method for the edit distance between unordered trees and its application to analysis of glycan structures. BMC Bioinform. **12**(1), 1–9 (2011)
8. Grassia, M., De Domenico, M., Mangioni, G.: Machine learning dismantling and early-warning signals of disintegration in complex systems. Nature Commun. **12**(1), 5190 (2021). https://doi.org/10.1038/s41467-021-25485-8
9. Grassia, M., Lauri, J., Dutta, S., Ajwani, D.: Learning multi-stage sparsification for maximum clique enumeration (2019)
10. Grassia, M., Mangioni, G.: wsGAT: weighted and signed graph attention networks for link prediction. In: International Conference on Complex Networks and Their Applications, pp. 369–375. Springer (2021). https://doi.org/10.1007/978-3-030-93409-5_31
11. Jin, Y., Xiong, B., He, K., Zhou, Y., Zhou, Y.: On fast enumeration of maximal cliques in large graphs. Expert Syst. Appl. **187**, 115915 (2022)
12. Johnston, H.: Cliques of a graph-variations on the Bron-kerbosch algorithm. Int. J. Comput. Inf. Sci. **5**(3), 209–238 (1976)
13. Kose, F., Weckwerth, W., Linke, T., Fiehn, O.: Visualizing plant metabolomic correlation networks using clique-metabolite matrices. Bioinformatics **17**(12), 1198–1208 (2001)
14. Lauri, J., Dutta, S.: Fine-grained search space classification for hard enumeration variants of subset problems. Proc. AAAI Conf. Artif. Intel. **33**(01), 2314–2321 (2019). https://ojs.aaai.org/index.php/AAAI/article/view/4070
15. Lauri, J., Dutta, S., Grassia, M., Ajwani, D.: Learning fine-grained search space pruning and heuristics for combinatorial optimization (2020)
16. Li, X., Zhou, R., Chen, L., Zhang, Y., Liu, C., He, Q., Yang, Y.: Finding a summary for all maximal cliques. In: 2021 IEEE 37th International Conference on Data Engineering (ICDE), pp. 1344–1355. IEEE (2021)
17. Liu, G., Wong, L., Chua, H.N.: Complex discovery from weighted PPI networks. Bioinformatics **25**(15), 1891–1897 (2009). https://doi.org/10.1093/bioinformatics/btp311
18. Mori, T., Tamura, T., Fukagawa, D., Takasu, A., Tomita, E., Akutsu, T.: A clique-based method using dynamic programming for computing edit distance between unordered trees. J. Comput. Biol. **19**(10), 1089–1104 (2012)
19. Östergård, P.R.: A fast algorithm for the maximum clique problem. Discrete Appl. Math. **120**(1–3), 197–207 (2002)
20. Pardalos, P.M., Xue, J.: The maximum clique problem. J. Global Opt. **4**(3), 301–328 (1994)
21. Pattabiraman, B., Patwary, M.M.A., Gebremedhin, A.H., Liao, W.K., Choudhary, A.: Fast algorithms for the maximum clique problem on massive graphs with applications to overlapping community detection. Internet Math. **11**(4–5), 421–448 (2015)
22. Peixoto, T.P.: The graph-tool python library. figshare (2014), http://figshare.com/articles/graph_tool/1164194
23. Rossi, R., Ahmed, N.: The network data repository with interactive graph analytics and visualization. In: Twenty-Ninth AAAI Conference on Artificial Intelligence (2015)

24. Vaswani, A., Shazeer, N., Parmar, N., Uszkoreit, J., Jones, L., Gomez, A.N., Kaiser, L.u., Polosukhin, I.: Attention is all you need. In: Guyon, I., Luxburg, U.V., Bengio, S., Wallach, H., Fergus, R., Vishwanathan, S., Garnett, R. (eds.), Advances in Neural Information Processing Systems, vol. 30. Curran Associates, Inc. (2017). https://proceedings.neurips.cc/paper/2017/file/3f5ee243547dee91fbd053c1c4a845aa-Paper.pdf
25. Velickovic, P., Cucurull, G., Casanova, A., Romero, A., Lio, P., Bengio, Y.: Graph attention networks. STAT **1050**, 20 (2017)
26. Zhang, C., Zhang, Y., Zhang, W., Qin, L., Yang, J.: Efficient maximal spatial clique enumeration. In: 2019 IEEE 35th International Conference on Data Engineering (ICDE), pp. 878–889 (2019)

Graph Mining and Machine Learning for Shader Codes Analysis to Accelerate GPU Tuning

Lin Zhao[1(✉)], Arijit Khan[2], Robby Luo[3], and Chai Kiat Yeo[1]

[1] Nanyang Technological University, Singapore, Singapore
lin018@e.ntu.edu.sg asckyeo@ntu.edu.sg
[2] Aalborg University, Aalborg, Denmark
arijitk@cs.aau.dk
[3] Advanced Micro Devices Co., Ltd., Beijing, China
robby.luo@amd.com

Abstract. The graphics processing unit (GPU) has become one of the most important computing technologies. Disassembly shader codes, which are machine-level codes, are important for GPU designers (e.g., AMD, Intel, NVIDIA) to tune the hardware, including customization of clock speeds and voltages. Due to many use-cases of modern GPUs, engineers generally find it difficult to manually inspect a large number of shader codes emerging from these applications. To this end, we develop a framework that converts shader codes into graphs, and employs sophisticated graph mining and machine learning techniques over a number of applications to simplify shader graphs analysis in an effective and explainable manner, aiming at accelerating the whole debugging process and improving the overall hardware performance. We study shader codes' evolution via temporal graph analysis and structure mining with frequent subgraphs. Using them as the underlying tools, we conduct a frame's scene detection and representative frames selection. We group the scenes (applications) to identify the representative scenes, and predict a new application's inefficient shaders. We empirically demonstrate the effectiveness of our solution and discuss future directions.

Keywords: Graph analysis · Machine learning on graphs · GPUs · Shader code

1 Introduction

The graphics processing unit (GPU) was formerly a specialized microprocessor for offloading graphically intensive tasks, such as online video games that adopt real-time 3D graphics, special effects including virtual reality. Recently, GPUs are involved in many general-purpose, programmable, and multi-core applications [1,2]. One of the most exciting applications of GPUs consists of deep learning and machine learning. GPUs can process tons of data and train deep

H. Cherifi et al. (Eds.): COMPLEX NETWORKS 2016 2022, SCI 1077, pp. 426–439, 2023.
https://doi.org/10.1007/978-3-031-21127-0_35

neural networks for image and video analytics, speech recognition, natural language processing, and self-driving cars. In high-performance computing (HPC), scientific workflows, and Internet-of-Things (IoT) domains, modern GPUs are widely adopted to speed-up a variety of computing applications, e.g., gene mapping, virus tracing. The computer-aided design (CAD) software utilizes GPUs to visualize 3D objects in real time.

1.1 Challenges in Conventional GPU Tuning

From GPU designers' (e.g., AMD, Intel, NVIDIA) perspective, designing high-performance, low-power, and small-area chips is crucial for their products to seize market shares. Based on feedback from our industry collaborators, they experience three major challenges. **First**, since GPU designers are not the same as application designers (e.g., game developers), it is difficult for them to obtain the application source code and adjust their products accordingly. **Second**, as the field of GPU applications expands, a vast number of different programs are executed on the GPUs. However, engineers often tune GPUs (e.g., customization of clock speeds and voltages, VRAM tuning, memory timing, fan tuning) based on a few benchmark applications for hand-crafted key portions according to past experiences. This manual tweaking may result in missing critical information and sub-optimal performance. GPU tuning also requires an advanced level of technical knowledge. Incorrect tuning can make the system unstable, e.g., resulting in crashes, hangs, and/or graphical corruption [3]. **Third**, since new games and applications are emerging continuously, it is tedious to design a new tuning method for each new application.

1.2 Our Solution and Motivation

To solve the aforementioned challenges, we build a unified graph model that aims to accelerate the GPU performance tuning leveraging graph mining and machine learning tools. Our model analyzes the characteristics of the application's code executed on GPUs to find opportunities for speeding up the tuning process. In this work, we shall consider gaming programs and their disassembly shader codes, since graphic functions are still the primary and most complicated applications on GPUs. Our work can be extended to other applications on GPUs – when these applications are mapped onto the GPU, they are translated to machine-level disassembly codes by drivers [5].

The application's code is referred to *shader* (shader code), a piece of program to instruct the GPU how to render each pixel. Since shaders support complex, low-level, generic programming, and due to their proximity to the hardware, they are an important tool for GPU designers [7–10]. When shader codes are not available from application developers, we generate *disassembly shaders* by running those graphics applications on GPUs using a publicly available graphics debugger, such as RenderDoc [11]. Graph analytics has been adopted in code analysis in recent years [12–15]. In shader code analysis, it can simplify the codes to structural level. In Sect. 3.2, we demonstrate that such graph structure is a good proxy for the shader code . Thus, various graph mining and machine learning algorithms on graphs are beneficial to the analysis of shader codes. Our framework can process machine-level codes to assist in GPU's performance improvement by reducing the numbers of frames from one application and the total number of applications that need to be investigated. We also identify frequently occurring low-performance code structures that can lead the way for efficient debugging.

1.3 Our Contributions and Roadmap

Our contributions can be summarized as follows.

- We are the first to introduce a graph mining framework for analyzing the shader structures within video games (Sect. 3). Section 3.1 shows the process of preparing shader graph datasets from disassembly shader codes. In Sect. 3.2, we compare the code content similarity with the graph structural similarity and establish that graph representations can well encode the structural characteristics of disassembly shader codes. The statistics of the graph datasets and their temporal evolutions within a game are discussed in Sect. 3.3.
- In Sect. 4, we demonstrate the functionality which is to predict an unknown frame's scene via the Graph Convolutional Neural Network [17] with explanations.
- Section 5 demonstrates that when there is a new application, first we find similar existing applications via frequent subgraphs-based clustering methods. Since the GPU has already been tuned for these similar existing applications, one can quickly find the optimal tuning for the new application. We next introduce a method to predict shader's efficiency using frequent subgraphs, and identify a new application's low-efficiency shaders, thus locating its frequent weak points.

With results from our analysis, domain experts can accelerate the optimization process by only tuning hardware parameters using representative frames and representative applications. In addition, hardware engineers can understand common patterns (i.e., frequent subgraphs) in shader codes and hence, improve the hardware architecture in the next generation of products by adding specific hardware structures or re-designing the driver to process frequent sub-patterns more efficiently. Our code is open-sourced [18] and its demonstration [45] video is available at the YouTube — https://youtu.be/0WxvrJ6KuqY.

2 Related Work

With the advances in GPU architectures, researchers compared different generations of GPU designs [28]. Moya et al. [29] discussed the performance improvement due to various shader processing configurations. Many toolkits are developed by GPU designers and third-parties. AMD offers the

Radeon™Developer Tool Suite, including Radeon™GPU Profiler, Memory Visualizer, GPU Analyzer, and Developer Panel [31], using which game developers obtain in-depth access to the GPU and they can profile the computing usage, analyze pipeline bottlenecks, and other inefficiencies while developing the games. NVIDIA also provides a few performance analysis tools for CUDA C/C++ optimization [30]. **Intel VTune™** is capable of monitoring CPUs, GPUs, and FPGAs to locate the most time-consuming part of the developer's code. Additionally, the third-party tools, e.g., the feed-forward measurement performance tool model in [32] and **HPCToolKit** [33] build the call path profiles for GPUs to let developers know their code performance translated from a high-level language model. The aforementioned tools optimize GPUs from a single application's point of view. They focus on a specific application's optimization and require the resource supports such as the application's source code. In contrast, our model is developed for GPU designers, who would like to optimize their GPUs for multiple applications, including the emerging ones. To the best of our knowledge, there is no publicly available work that uses graph mining and machine learning for shader code analysis. Adopting graph analysis for GPU performance improvement process from the code's perspective is a novel paradigm which we investigated.

3 Datasets and Characteristics

A *scene* in the video refers to a series of motions that happen in a single location and continuous time. The majority of online games have multiple scenes. Within a scene, there are multiple frames. Within a frame, there are a number of *draw calls* which instruct the graphics APIs to draw object. There are different types of shaders for each draw call. Primarily used shaders in the games that we consider are vertex shader (**VS**), pixel shader (**PS**), geometry shader (**GS**), compute shader (**CS**), hull shader (**HS**), local shader (**LS**), and export shader (**ES**). Each of them is one piece of shader code. We collected about 4 million disassembly shaders, corresponding to 64 scenes of 29 games, from our industry partners in GPU manufacturing. Our dataset includes both benchmarks (**3D Mark** and **VRMARK series**) and online games.

3.1 Graph Data Extraction

We describe the data preparation in two steps.
Step 1: from video to shader code. There are a few open source tools that convert the video from a game into disassembly shaders such as AMD Radeon GPU Profiler [19] and RenderDoc [11].
Step 2: from shader code to graph format. The long-time running instructions are represented as nodes along with node-labels based on the GPU functional blocks involved. For the rest of instructions such as arithmetic calculations, the number of such successive instructions between two nodes are counted

Table 1. An example of disassembly shader and the graph created.

Disassembly shader instruction set
s_inst_prefetch 0x0003
s_load_dwordx8 s[16:23], s[2:3], null
v_mad_u32_u24 v2, s12, 16, v0
v_mad_u32_u24 v1, s13, 16, v1
s_waitcnt lgkmcnt(0)
image_load v[3:5], [v2,v1], s[16:23]
s_waitcnt vmcnt(0)
v_mov_b32 v6, v3
image_store v[3:6], [v2,v1], s[4:11]
s_endpgm

nodes and node-labels	
0,	*OTHERS*
1,	*SMEM_INSTR*
2,	*IMAGE_LOAD_INSTR*
3,	*IMAGE_STORE_INSTR*
4,	*OTHERS*

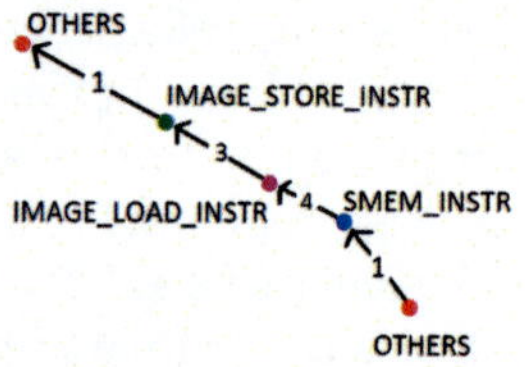

edges and edge-distances	
(0,1),	1
(1,2),	4
(2,3),	3
(3,4),	1

and stored as the edge-distance. As Table 1 shows, the instructions are categorized into a few groups (shown with different colors). In the node list, the first line s_inst_prefetch_0x003 is categorized in the group of *OTHERS* and is labeled as node 0 (red color). The second line is categorized as *SMEM_INSTR* which uses scalar memory and it becomes node 1 (blue color). The consecutive short-running instructions are grouped as edges. For this reason, a few instructions are skipped until *IMAGE_LOAD_INSTR* (line 6, magenta color, this becomes node 2). Notice that this graph is a chain graph. Complex structures, e.g., branches, cycles, and complex patterns also exist in our graphs. In Sect. 3.3, we show the number of cycles over different frames.

3.2 Effectiveness of Graph Structure

A piece of shader code can be considered as an algorithm that specifies where to take input data and how they will perform computation and data storage. Looking purely at the text contents of shader codes, we use the widely-used **TF-IDF Vectorizer** [20] method to represent shader codes as vectors. We then compute the cosine similarity (a value in [0,1]) between a pair of vectors. Finally, we report the similarity between two consecutive frames which is the average similarity value over all pairs (one from the first frame and another from the second frame) from all distinct shader codes (considering a specific type of shader, e.g., compute shader) in these two frames. Figure 1 demonstrates text contents-based similarity values for consecutive frames from two games: 3D Mark Time Spy (**3DM TS**) and Grand Theft Auto 5 (**GTA5**), considering specific types of shaders, e.g., **GS**, **CS**, and **HS**. Figure 1a–b show that frame structures change slightly considering all 600 frames for **3DM TS** and both **GS** and **CS** shader types. Figure 1c–d show text contents-based similarity for **GTA5** using **CS** and **HS** as examples. From this five-scene dataset, a clear change at the boundary between scenes can be observed for **CS**. Within each scene, the similarity value may vary because of addition or deletion of a few pieces of shader codes.

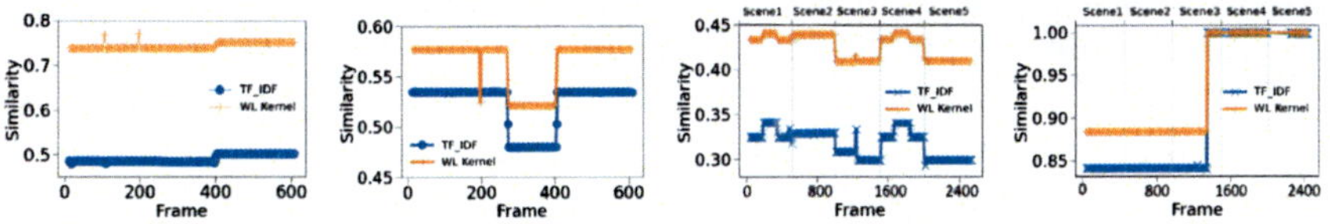

(a) **GS** shader, **3DM** (b) **CS** shader, **3DM** (c) **CS** shader, **GTA5** (d) **HS** shader, **GTA5**

Fig. 1. Content (**TF-IDF**) versus graph similarity (**WL kernel**).

Next, we convert shader codes into graphs (Sect. 3.1) and compute similarity between pairs of graphs via the *Weisfeiler-Lehman Graph Kernel* (**WL kernel**) [21]. The **WL kernel** maps an original graph into a sequence of graphs (kernel features), whose node attributes capture both structural and label information. We apply the **WL kernel** function from the **GraKel** library [22] to extract kernel features and compute the similarity between a pair of graphs. Analogously, we also report in Fig. 1 the similarity between two consecutive frames by computing the average **WL kernel** similarity value over all pairs (one from the first frame and another from the second frame) from all distinct shader code graphs (considering a specific type of shader, e.g., compute shader) in these two frames. Furthermore, the pairwise graph similarity values are normalized between 0 and 1, by dividing it with $\sqrt{\Pi_i \mathbb{K}(G_i, Gi)}$, where $\mathbb{K}(G_i, Gi)$ is the **WL kernel**-based self-similarity of graph G_i to itself, and we consider all G_i's from these two frames. We observe that the **WL kernel**-based similarity plots generally have the same trends as those based on **TF-IDF** text contents similarity. This indicates that the graph representation is a good proxy to reflect the contents in shader codes. Besides, graphs can represent both syntactic and semantic structures of code, compared to mainly syntactic notions in program text [16]. Therefore, we shall use graph structures of disassembly shader codes.

3.3 Graph Data Characteristics and Key Frames Selection

Each game consists of a large number of frames and draws for different rendering functions. We study the characteristics and evolution of shader code graphs across frames for each game and identify the *representative frames*. We show our results using 3D Mark Fire Strike (**3DM FS**, a benchmark game) and **GTA5** (an online game). We study graph changes across frames/scenes and show the variation of the number of nodes, edges, cycles, as well as the degree distribution. These provide an overview of the evolution of shader code graphs.

Figure 2a–b depict the number of cycles existing in each frame for the two games. The x-axis shows successive frames and the y-axis indicates the total number of cyclic structures in that frame. **3DM FS** has more loop structures in Scene 2. In **GTA5**, majority of the frames in Scene 3 have more cycles, compared to that in frames from other scenes. It indicates that there are more branches and loops implemented in this part. Fig. 2c–d show the degree distribution of nodes from every scene. The x-axis presents various node degrees for each scene. The y-axis is the percentage of nodes having the specific degree, with respect to the total number of nodes in that scene. For each scene, node degrees vary from 1 to 5. The majority of nodes have two edges, with one in-edge and another out-edge, implying the chain-like structure. The in-degree and out-degree values vary from 0 to 5. Most nodes have balanced in- and out-degrees.

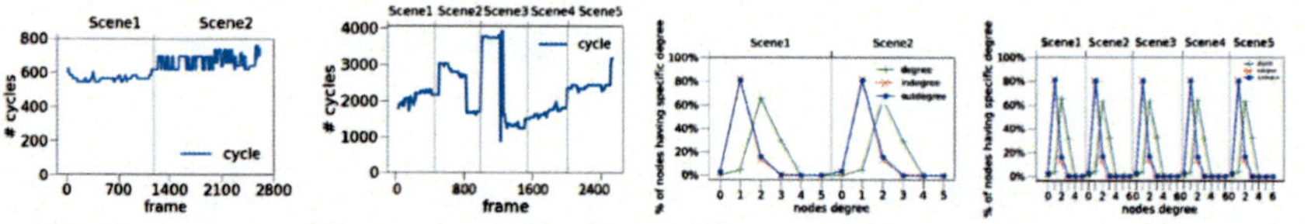

(a) **3DM FS**, # cycles over frames (b) **GTA5**, # cycles over frames (c) **3DM FS**, degree, in- and out-degree distributions (d) **GTA5**, degree, in- and out-degree distributions

Fig. 2. Evolution of structural properties over frames.

Structural changes across scenes. In Fig. 1, **WL kernel** similarity results reveal that the shader code graph structures remain almost same with a small variation within one scene, but often have noticeable changes between two scenes. We can conclude that within the same scene, the algorithms adopted are roughly the same. However, across different scenes, the algorithms used can be quite different. Sudden changes in frame structures can be detected based on the changes in **WL kernel** similarity across successive frames. Within a specific scene, we select the frame as the *representative* that has the highest **WL kernel** similarity with all other frames in that scene. In this way, by studying this frame, the GPU designers are able to cover most of the algorithmic patterns in that scene. If the algorithm structures do not vary too much within a scene, it is possible to reduce the number of times each algorithm is executed by storing the data into cache or local memory, and all these data can be retrieved in a very short time, compared to applying the algorithm again to render the image.

4 Predicting A Frame's Scene

In Sect. 3.3, we illustrate the similarities and difference of frame structures within scenes. Thus, it is possible to predict which scene a frame belongs to given an unknown frame, in a supervised manner. In particular, we employ a Graph Convolutional Neural Network (GCN) [17,23] for classification of frames and thereby predict an unknown frame's scene. Our GCN architecture includes two graph convolutional layers with the first layer having 128 units and the second layer having 64 units. Both layers use the ReLU activation function. The next layer is a global mean pooling layer, followed by a softmax classifier.

All distinct shader code graphs in each frame are merged to form one single, disjoint, large graph. The resulting graph is labeled based on the frame's scene number. In **3DM FS**, there are two classes corresponding to two scenes . **GTA5** contains five classes, denoting five scenes. We apply 4-fold cross validation. The learning rate is set to 0.0005 and the dropout rate is 0.5. The number of epochs is 50 for **3DM FS** and 900 for **GTA5**. For **3DM FS**, training and test accuracy quickly reach 100%, indicating that the frames are easily separable across different scenes. For **GTA5**, after 900 epochs, the training and test accuracy reach 90.64% and 88.46%, respectively, with average ROC_AUC score 97.04%. This shows that the **GTA5** dataset, with five scenes, is relatively harder to classify.

(a) Class=Scene 1 (b) Class=Scene 2 (c) Class=Scene 3 (d) Class=Scene 4 (e) Class=Scene 5

Fig. 3. **GTA5**: Scene prediction interpretability and important nodes.

Explainability study. We use the *Gradient-Weighted Class Activation Mapping* (**Grad-CAM**) interpretability method [24] to derive node importance in frame classification. In Fig. 3, we show the graphs corresponding to five test frames, one from each scene of **GTA5**. Clearly, as the frames are different, their graph structures are different too. The nodes highlighted in the red color indicate those nodes having higher positive contributions to the activation of the specific class label output (i.e., scene number). Finally, we identify the top-20 nodes from each of these five frames (each belonging to a different scene) having the highest positive contributions. We count the node-labels present in those top-20 nodes from each test frame and report these counts corresponding to each scene in Fig. 4a. The x-axis indicates the node-labels (i.e., GPU functional blocks involved in the shader code instruction) from these important nodes. Same color bars represent that they are the same functional block. The figure shows that the most important GPU functional blocks considered to distinguish each scene by our GCN-based classification method are quite different. These results demonstrate the effectiveness and interpretability of our frame classification approach.

5 Tuning for A New Application

We first provide an overview of our method for analyzing existing and new game datasets. With the existing game datasets, we obtain representative frames

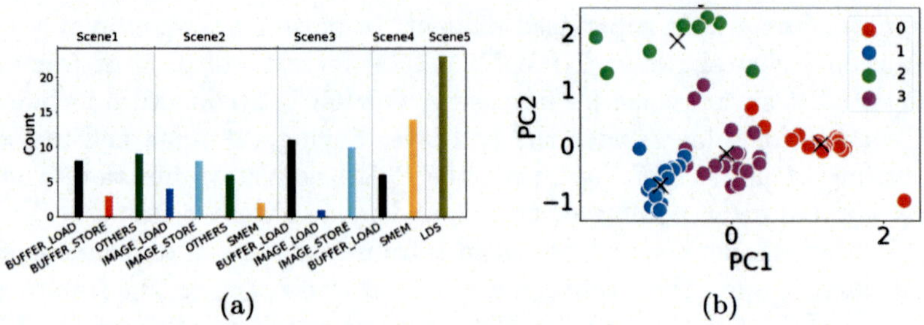

Fig. 4. **a** # GPU functional blocks in the top-20 nodes from each scene, having the highest positive contributions in scene prediction. **b** Scene clustering results with K-Means (shown with two principal components).

from various scenes by studying graph similarity over frames (Sect. 3.3). Next, frequent subgraphs are extracted from representative frames (Sect. 5.1). The game datasets and scenes are clustered into different groups based on similarity of their frequent subgraphs (Sect. 5.2). Thus, only representative games from each group need to be investigated to cover the majority of graph patterns across existing games. This minimizes the tuning effort for engineers. Furthermore, frequent patterns from representative games are annotated as high-efficiency or low-efficiency by domain experts. The engineers can upgrade and tune the GPU architecture and hardware parameters by knowing the common patterns. When a new game is given, its representative frames and frequent subgraphs are extracted. Next, we verify if the new game is similar to any of the groups of existing games. If it belongs to one of them, the tuning result from similar existing games can be re-used. If not, we predict its low-efficiency shaders in a supervised manner.

5.1 Frequent Subgraphs Mining

We mine frequent subgraphs with the gSpan algorithm [25] from each of 64 scenes spanning across 29 different games. The *minimum support* (minsup) value is set to 25% of the number of graphs in each representative frame. A frequent subgraph pattern is a subgraph that appears in at least minsup graphs in a graph database. For even higher support values, the frequent patterns get smaller and less in number. For lower support values, gSpan does not terminate in one day. Among output frequent subgraphs, we consider those with the number of nodes at least 5. In this way, we eliminate smaller subgraphs which appear frequently; however, they are less useful in hardware tuning.

5.2 Scenes Clustering

We identify representative scenes and games for GPU performance tuning, thus improving the tuning efficiency by eliminating duplicated works on games having similar characteristics. We consider all maximal frequent subgraphs as features for each scene to cluster them. Thus, we get 64 data points (one for every scene), each with dimensionality 825 (representing all maximal frequent subgraphs). For a specific scene S_i and dimension d_j, the value $S_i[d_j] \in (0, 1)$ is computed as the maximum pairwise **WL kernel** similarity between each frequent subgraph of S_i and the maximal frequent subgraph corresponding to d_j. Next, we apply K-Means [20], Gaussian Mixture [26], and Birch algorithms [27] for clustering these data points. Before clustering, Principal Component Analysis (PCA) is adopted to reduce the number of features so to avoid the curse of dimensionality. The optimal number of clusters is determined based on the Elbow method [43]. The Elbow method selects K where the within-cluster variance no longer decreases sharply (e.g., 4 for K-Means in our case).

Experimental results. We observe that K-Means and Gaussian Mixture produce clustering with 93.75% overlap, while BIRCH have 95.31% overlap with K-Means clustering. Scenes with different grouping results are near the boundary of two groups based on the three clustering methods. We visualize the K-Means clustering in Figure 4(b). Each group is labeled with a different color. The cross sign refers to the centroid of that group. Scenes in the same group have similar frequent subgraph structures, which means that the frequent algorithmic patterns used in these scenes are also similar. Therefore, selecting one scene, which is the nearest to the centroid of each group, for performance analysis and tuning will be both effective and efficient.

5.3 A New Game's Shader Efficiency Prediction

The frequently occurring sub-patterns from shader codes are valuable also for the code efficiency analysis and improvement. The overall GPU efficiency can be improved if the frequent patterns can be efficiently processed. The engineers can upgrade and tune the GPU architecture and hardware parameters by knowing the common patterns.

Due to their larger sizes and relatively complex shapes, by looking at a complete piece of shader code, it is often difficult even for domain experts to predict its efficiency or to identify the frequently occurring weaknesses. However, we observe that our industry collaborators can quickly recognize if a frequent shader subgraph is inefficient or not, due to its small size. To this end, we aim at predicting the efficiency of frequent shader subgraphs in a supervised manner. The mined frequent subgraphs are annotated as high (Label 1) or low efficiency/utilization (Label 0) by our industry collaborators based on GPU functional blocks and latency information. Moreover, we associate the counts of various GPU functional blocks in frequent subgraphs as their features. In this way, we get our data points (one for every frequent subgraph) having dimensionality 43 (representing all GPU functional blocks involved). The dataset is unbalanced with 88.02% in Label 1 and 11.98% having Label 0.

Experimental results. We employ the Random forest classifier, with 4-fold cross validation, to predict the efficiency of frequent shader subgraphs. We find the average accuracy to be 0.9865, AUC-PRC 0.9943, and AUC-ROC 0.9574. These results demonstrate that we can predict the efficiency of frequent shader subgraphs in a supervised manner. Moreover, given a new game, one can predict its performance by mining its own frequent subgraphs and using our supervised frequent subgraphs classification model.

Case study. Figure 5 presents two examples of low-efficiency shader patterns which are correctly predicted by our method. For the subgraph in Fig. 5a, there are many branches (indicated by LABEL) and long-time wait commands (indicated by WAITCT); it is likely that during jump or wait time, GPU resources are under-utilized. For the pattern in Fig. 5b, the structure, though simple, consists of many LOAD and STORE actions, which take longer time.

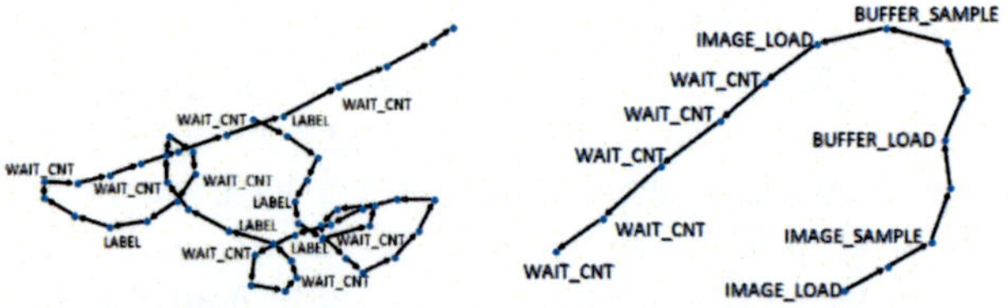

(a) **Subgraph 1** (b) **Subgraph 2**

Fig. 5. Two examples of low-efficiency shader patterns.

6 Conclusions

We proposed a unified graph model to denote shader codes as graphs, employed graph mining and machine learning to analyze disassembly shader code structures. We investigated a few research questions: Our single game analyses reveal that the algorithms used in each frame within a scene are similar, thus we can predict the scene for an unknown frame, and identify the representative frame from each scene. Our inter-game analyses demonstrate that we can group scenes and games based on their frequent sub-structures in the shader codes. As a result, only one or a few sample scenes are sufficient to cover the typical patterns in that group; hence, engineers can focus on the specific scenes and games for performance tuning. Finally, our novel shader efficiency prediction method based on frequent subgraphs accelerates the tuning process by identifying frequent algorithmic structures, including frequently occurring weaknesses. In future, we shall consider other portions of the shader code, such as input data, in addition to algorithmic structures, as well as disassembly machine codes from different applications, for a more comprehensive analysis. Besides, the frequent subgraph mining algorithm can be improved, e.g., SPMiner [42], to consider various attributes after additional information is added to the graphs, thereby extracting more sophisticated features.

Acknowledgement. This research is supported by Novo Nordisk Foundation grant NNF22OC0072415 and Singapore MOE tier-2 grant 2019-T2-2-042. Lin Zhao acknowledges support from Nanyang Technologies University, Advanced Micro Devices Co., Ltd.

References

1. Huang, S., Xiao, S., Feng, W.-c.: On the energy efficiency of graphics processing units for scientific computing. In: IPDPS (2009)
2. Owens, J.D., Houston, M., Luebke, D., Green, S., Stone, J.E., Phillips, J.C.: GPU computing. Proc. IEEE **96**(5), 879–899 (2008)
3. AMD.: How to Tune GPU Performance Using Radeon Software (n.d.). https://www.amd.com/en/support/kb/faq/dh2-020
4. Foster, I., Kesselman, C., Nick, J., Tuecke, S.: The Physiology of the Grid: an Open Grid Services Architecture for Distributed Systems Integration. Technical Report, Global Grid Forum (2002)
5. Zandbergen, P.: Machine Code and High-level Languages: Using Interpreters and Compilers (2022). https://study.com/academy/lesson/machine-code-andhigh-level-languages-using-interpreters-and-compilers.html/
6. Pipelines and Shaders with Direct3D 12 (2021). https://docs.microsoft.com/enus/windows/win32/direct3d12/pipelines-and-shaders-with-directx-12
7. de La Lama, C.S., Jääskeläinen, P., Kultala, H., Takala, J.: Programmable and scalable architecture for graphics processing units. Trans. High Perform. Embed. Archit. Compil. **5**(2019), 21–38 (2019)
8. Fam, S.T., Sowerby, A.: Shader Program Profiler (2015). https://patents.google.com/patent/US9799087B2
9. Ren, X., Lis, M.: CHOPIN: scalable graphics rendering in multi-GPU systems via parallel image composition. In: IEEE International Symposium on High-Performance Computer Architecture, HPCA, pp. 709–722. IEEE (2021)
10. Zaidi, S.I., Fam, S.T., Lotfi, P., Indukuru, V.R., Pan, J., Sowerby, A.M., Duprat, J.-L.: Shader Profiler (2018)
11. Karlsson, B.: RenderDoc (2021). https://renderdoc.org/docs/index.html
12. Gascon, H., Yamaguchi, F., Arp, D., Rieck, K.: Structural detection of android malware using embedded call graphs. In AISec (2013)
13. Alanazi, R., Gharibi, G., Lee, Y.: Facilitating program comprehension with call graph multilevel hierarchical abstractions. J. Syst. Softw. **176**(2021), 110945 (2021)
14. Tekin, U., Buzluca, F.: A graph mining approach for detecting identical design structures in object-oriented design models. Sci. Comput. Program. **95**(2014), 406–425 (2014)
15. Greenspector (2021). https://greenspector.com/en/home/
16. Allamanis, M., Brockschmidt, M., Khademi, M.: Learning to represent programs with graphs. In: ICLR (2018)
17. Kipf, T.N., Welling, M.: Semi-supervised classification with graph convolutional networks. In: ICLR (2017)
18. Zhao, L., Khan, A., Luo, R.: Our Code and Datasets (2022). https://github.com/forest2022/graphForest

19. Radeon™ GPU Analyzer (2021). https://gpuopen.com/rga/
20. Leskovec, J., Rajaraman, A., Ullman, J.D.: Mining of Massive Datasets, 2nd edn. Cambridge University Press (2014)
21. Shervashidze, N., Schweitzer, P., Leeuwen, E.J.V., Mehlhorn, K., Borgwardt, K.M.: Weisfeiler-Lehman graph kernels. J. Mach. Learn. Res. **12**(2011), 2539–2561 (2011)
22. Siglidis, G., Nikolentzos, G., Limnios, S., Giatsidis, C., Skianis, K., Vazirgiannis, M.: GraKeL: a graph kernel library in Python. J. Mach. Learn. Res. **21**(54), 1–5 (2020)
23. Hamilton, W.L., Ying, Z., Leskovec, J.: Inductive representation learning on large graphs. In: NeurIPS (2017)
24. Selvaraju, R.R., Cogswell, M., Das, A., Vedantam, R., Parikh, D., Batra, D.: Grad-CAM: visual explanations from deep networks via gradient-based localization. In: ICCV (2017)
25. Yan, X., Han, J.: gSpan: graph-based substructure pattern mining. In: ICDM (2002)
26. Lin, X., Yang, X., Li, Y.: A deep clustering algorithm based on gaussian mixture model. J. Phys.: Conf. Ser. **1302**, 032012 (August 2019)
27. Zhang, T., Ramakrishnan, R., Livny, M.: BIRCH: a new data clustering algorithm and its applications. Data Min. Knowl. Discov. **1**, 141–182 (June 1997)
28. Schaa, D., Kaeli, D.R.: Exploring the multiple-GPU design space. In: IPDPS (2009)
29. Moya, V., Gonzalez, C., Roca, J., Fernandez, A., Espasa, R.: Shader performance analysis on a modern GPU architecture. In: MICRO (2005)
30. NVIDIA Performance Analysis Tools (2021). https://developer.nvidia.com/performance-analysis-tools
31. Radeon™ Developer Tool Suite (2021). https://gpuopen.com/
32. Welton, B., Miller, B.P.: Diogenes: Looking for an Honest CPU/GPU Performance Measurement Tool (n.d.)
33. Zhou, K., Adhianto, L., Anderson, J., Cherian, A., Grubisic, D., Krentel, M., Liu, Y., Meng, X., Mellor-Crummey, J.: Measurement and analysis of GPU accelerated applications with HPCToolkit. Parallel Comput. **108**(2021), 102837 (2021)
34. Baddouh, C.A., Khairy, M., Green, R.N., Payer, M., Rogers, T.G.: Principal kernel analysis: a tractable methodology to simulate scaled GPU workloads. In: MICRO (2021)
35. Huang, J.-C., Nai, L., Kim, H., Lee, H.-H.S.: TBPoint: reducing simulation time for large-scale GPGPU kernels. In: IPDPS (2014)
36. Kambadur, M., Hong, S., Cabral, J., Patil, H., Luk, C., Sajid, S., Kim, M.A.: Fast computational GPU design with GT-pin. In: IISWC (2015)
37. Yu, Z., Eeckhout, L., Goswami, N., Li, T., John, L.K., Jin, H., Xu, C.-Z., Wu, J.: GPGPU-MiniBench: accelerating GPGPU micro-architecture simulation. IEEE Trans. Computers **64**(11), 3153–3166 (2015)
38. Sadyrin, D., Dergachev, A., Loginov, I., Korenkov, I.N., Ilina, A.: Application of graph databases for static code analysis of web-applications. In: MICSECS (2019)
39. Liu, J.: Enabling Static Program Analysis Using a Graph Database. Ph.D. Dissertation. Wright State University (2020)
40. Nair, A., Roy, A., Meinke, K.: FuncGNN: a graph neural network approach to program similarity. In: ESEM (2020)
41. Phan, A.V., Nguyen, M.L., Bui, L.T.: Convolutional neural networks over control flow graphs for software defect prediction. In: ICTAI (2017)
42. Wang, A.Z., You, J., Leskovec, J.: Frequent subgraph mining by walking in order embedding space. In: GRL Workshop (2020)

43. Thorndike, R.L.: Who belongs in the family? Psychometrika **18**(1953), 267–276 (1953)
44. Unreal Engine (2021). https://www.unrealengine.com/en-US/
45. Zhao, L., Khan, A., Luo, R.: ShaderNet: graph-based shader code analysis to accelerate GPU's performance improvement (Demonstration). In: GRADES-NDA workshop (2022)

Networks in Finance and Economics

Pattern Analysis of Money Flows in the Bitcoin Blockchain

Natkamon Tovanich[1(✉)] and Rémy Cazabet[2]

[1] Blockchain & B2B Platforms Chair, École Polytechnique, Institut Polytechnique de Paris, 91120 Palaiseau, France
natkamon.tovanich@polytechnique.edu
[2] Univ de Lyon, CNRS, Université Lyon 1, LIRIS, UMR5205, 69622 Villeurbanne, France

Abstract. Bitcoin is the first and highest valued cryptocurrency that stores transactions in a publicly distributed ledger called the blockchain. Understanding the activity and behavior of Bitcoin actors is a crucial research topic as they are pseudonymous in the transaction network. In this article, we propose a method based on taint analysis to extract taint flows—dynamic networks representing the sequence of Bitcoins transferred from an initial source to other actors until dissolution. Then, we apply graph embedding methods to characterize taint flows. We evaluate our embedding method with taint flows from top mining pools and show that it can classify mining pools with high accuracy. We also found that taint flows from the same period show high similarity. Our work proves that tracing the money flows can be a promising approach to classifying source actors and characterizing different money flow patterns.

Keywords: Bitcoin · Money flow · Taint analysis · Graph embedding

1 Introduction

Bitcoin is the oldest and most used cryptocurrency, attracting broad interest from the general public and researchers. In contrast to traditional financial networks, transactions can be observed by anyone on the public blockchain, on which users exchange Bitcoins pseudonymously. This data allows researchers to study economic activities in fine detail. One of the objectives of those research is to understand how the Bitcoin socio-technical system works, particularly, (1) Who are the important actors of the Bitcoin economy? [19,20,22]; (2) How is the network of transactions organized? [19,27,34]; and (3) How to identify and track illegal activity? [4,7,35].

Tracing the flow of money—where the money goes, to whom, and when—is also an essential task in cryptocurrencies and critical for financial forensics to trace money from suspicious sources and characterize different users' behaviors. We investigate two main questions regarding the relationship between the money source and subsequent transactions: (1) Does money flow differently in

H. Cherifi et al. (Eds.): COMPLEX NETWORKS 2016 2022, SCI 1077, pp. 443–455, 2023.
https://doi.org/10.1007/978-3-031-21127-0_36

the Bitcoin network depending on its source?; and consequently, (2) Can we characterize a source actor given the observation of the flow of its coins in the network?

We propose an original way to synthesize the money flow from a given source into a concise dynamic network called a *taint network*. We subsequently apply whole graph embedding methods to automatically assign taint networks to their origin actor. Beyond the demonstration that each actor has a characteristic flow allowing us to recognize its position in the network [1,22,32], our method can also be helpful for actor tracking as well as actor deanonymization tasks. The embedding of money flows from different actors is a promising feature for downstream tasks in machine learning models to classify the role of actors [15,38] or predict illegal transaction activities in the Bitcoin blockchain [35].

2 Related Work

Due to the pseudonymity of Bitcoin actors, the main research challenges concentrate on (1) deanonymizing those actors and (2) characterizing their roles in the transaction network. Early works proposed clustering heuristics to deanonymize addresses likely to belong to the same actors [6,9,29,37].

Our work focuses on identifying and characterizing actors in the transaction network. Previous works applied graph analysis and machine learning to classify the role of actors or whether the transactions are illicit or not. Most works derived a set of descriptive features (e.g., the number and frequency of transactions, in- and out-degrees, and the number of different addresses used) [2,5,11,20,23] or the high order moments of transaction time [18]. These approaches rely on the actor's behavior, which easily manipulates it to hide its activities.

Other approaches thus rely on graph motifs, i.e., the set of subgraph patterns describing the neighbors of an actor [15,28,36,38]. Due to computational reasons, those works construct the static graph features only from direct neighbors (2-motif) or neighbors of their neighbors (3-motif). They do not use the identity of these neighbors but simply numeric descriptions (e.g., the total amount sent or received and transaction fees). Node2vec has been used to embed the actor position in the address network [23]. Nonetheless, a few works include the temporal aspect to impose a temporal locality constraint on the motifs [35,36].

Contrary to these approaches, our proposed method does not rely on the actor's activity or its direct neighbors but on the temporal network describing the whole flow of coins sent by an actor. We rely on the principle of tainted flows to trace the coins from source actors. Taint analysis has been used most notably in the context of tracking money from illegal sources [1,3,8,24,32]. However, those works mainly focus on the destination of tainted coins. Our work expands this approach to analyze the full money flow from multiple sources. We are not merely interested in the destination of tainted coins but in characterizing the temporal networks created by those flows.

3 *Taint Flow* Extraction

Our objective in this work is to design a new method to characterize a **bitcoin source** based on the flow of its coins in the transaction network. The underlying hypothesis is that the way a coin travels in the transaction network is characteristic of its source. However, actors reached by tainted coins after an indefinite time and coins diffused in a more considerable amount of bitcoins in a transaction cannot be characteristics of their origin. After presenting the tainting process on Bitcoin's blockchain, we thus introduce two principles to keep only relevant information in the flow: *dissolved* coins and *tag actors.*

3.1 Bitcoin Taint Flow

Bitcoin uses the unspent transaction output (UTXO) transaction model [25]. According to this model, transactions do not transfer money from one account to another. Instead, each output—each UTXO—of the transaction represents an amount of coin belonging to a known bitcoin address—a cryptographic public key. The rightful owner of the UTXO uses the corresponding private key to claim the money. It spends the UTXO(s) by signing them as input(s) in a new transaction and sends new UTXO output(s) to recipients' addresses.

A Bitcoin *transaction network* can be modeled as a chain of UTXOs. A transaction (tx) is represented as a node. A directed edge represents a transfer of UTXO(s) from one transaction to another. We will refer to the in-edge and out-edge of tx as input ($tx.in$) and output ($tx.out$), respectively. Each UTXO edge (e) is characterized by the amount of Bitcoin ($e.amount$) and the owner's address of that UTXO ($e.address$). It also contains references to the receiving ($e.receive$) and spending transaction nodes ($e.spend$).

We define the *taint flow* as a directed acyclic graph (G_{flow}) tracing the sequence of transactions from a **source of interest** until **dissolution**. The source of interest can be one or several actors and limited to a given time interval based on the focus of the study. To construct a money flow, we recursively taint all UTXO outputs ($tx.out$) until the coins are **dissolved**.

Definition 1. (*Dissolution*). We consider that a tainted coin is **dissolved** when its future positions in the transaction network will no longer be characteristic of its original position. More formally, a coin is dissolved when it is spent in a transaction with a purity value below a minimum threshold.

Purity measure (ρ) has been used to determine when the money is dissolved and stop following the transaction outputs [8]. Purity is the percentage of tainted money from the origin transaction set, defined as:

$$\rho(tx) = \frac{\sum_{e \in tx.in} \rho(e.receive) \cdot e.value}{\sum_{e \in tx.in} e.value} \tag{1}$$

The purity of a transaction without inputs is 1 by definition because it is the root transaction in the transaction flow. In this study, we set a purity threshold

$\rho_{min} = 0.001$, which means that a coin is considered dissolved when it is spent in a transaction together with 1,000 times the amount of un-tainted coins. Besides, we stop following the flow when the transaction is > 1 year apart from the source transactions ($time_{max}$). Algorithm 1 describes the process of retrieving transaction outputs and adding them to the money flow graph.

Input : τ_o is a payout transaction as a seeding node of the payout flow.
Input : ρ_{min} is a minimum purity threshold.
Input : $time_{max}$ is a maximum time threshold.
Output: *edges* is the edge list of the payout flow.

```
queue ← PriorityQueue([τo]);
edges ← List();
while queue is not empty do
    tx ← queue.pop();
    if ρ(tx) ≥ ρmin and tx.time ≤ timemax then
        for e in tx.out do
            e.amount_flow ← e.amount × ρ(tx);
            edges.append(e);
            queue.append(e.spend);
        end
    end
end
```

Algorithm 1: Reward payout flow extraction

Our algorithm applies *haircut tainting*, which assumes that the tainted money is divided equally to all output transactions in proportion to their amount [1,32].

3.2 Actors and Tag Actors

We defined actors as a set of addresses corresponding to a person, a group of persons, an organization, or any other entity owning a set of private keys to claim the ownership of UTXOs from public key addresses. A simple but effective heuristic assumes that the input addresses in a transaction should belong to the same owner [12,29]. We use the input address clustering heuristic implemented in the BlockSci library [16] that also filter CoinJoin transactions [10] to discover a set of addresses (*e.address*) belonging to the same actor (also called address cluster, *e.cluster*). When analyzing a tainted flow, the relevant information is the actors involved in this flow. Therefore, a flow is summarized as a set of transactions between actors.

A taint flow can be large and sparse, making it difficult to compare with other flows [1]. To improve on this limit, we propose to work on variants of flows in which we keep only important actors named *tag actors*.

Definition 2. (*Tag Actor*). To characterize a flow, we can describe it using a subset of all encountered actors, called **tag actors**. Tag actors are prominent ones that are likely to stay constant in time and to be reached by many flows from different sources.

We propose two ways of defining tag actors: (1) **frequent actors** consist in keeping a fraction of the most frequent actors; and (2) **known actors** are chosen based on external data. In this work, we use WalletExplorer dataset [14] that provides a collection of 375 known actors, in particular services or companies, linking their addresses with the name and type of the service (e.g., exchange platform, gambling service, and marketplace).

4 Taint Flow Embedding

Since taint flows are represented as graphs, we use whole graph embedding approaches to assess the similarity of the taint flows and characterize their patterns. The principle of those methods, such as Graph2vec [26] or Anonymous walk embedding [13] is to assign a low-dimensional vector representation to each graph such as two graphs considered similar according to a chosen network structure representation are close in the resulting embedding space.

Recently, the Geo2DR methodology [31] was introduced to allow one to design custom embedding methods and construct whole graph embedding. The methodology consists of two phases: (1) induction of descriptive substructure patterns and (2) learning of vector representations. In this case, we deal with taint flows that have directed acyclic graphs, temporal nature, and different types of node labels. Therefore, we define our custom process to produce graph walks in the first phase.

4.1 Induction of Descriptive Substructure Patterns

We use a random walk-based approach to extract substructure patterns from taint flows. We compare different variants of random walks and nodes labels vocabularies:

- **RW**—Unbiased Random Walks. We generate random walks starting only from the source node, without considering weights, following edge temporal directions. A walk ends when encountering a dissolved node.
- **SPW**—Shortest Path Walk. To generate an instance of the shortest path walk, we randomly choose a leaf node (dissolved) and walk through the shortest path between the source and a dissolved node.

We prune the walks based on the following sets of tag actors:

- **All clusters**: We use no pruning and keep all actors.
- **Frequent clusters**: We keep only clusters that appear in more than 50% of all taint flows as tag actors. The objective is to increase the fraction of shared vocabulary between flows to make learning more efficient.

- **Known actors**: We use as tag actors only those known from an external source, WalletExplorer [14]. In the **name** variant, the node label corresponds to its name. In the **type** variant, the label corresponds to its type, one of exchange, wallet, service, marketplace, mixer, lending, and gambling.

We replace all mining pool clusters and names with a "mining" label to prevent the model from training the embeddings from the source actors.

Temporal Pattern In this variant, we use the same methods, but we integrate the time aspect into the walks, assuming that the time to reach different parts of the network might be characteristic of the source. We use a tuple (original label, time) where time is defined as $\lfloor \log_2(\Delta t) \rfloor$, with Δt the time elapsed between the encoded transaction and the source transaction, in days. Examples of temporal pattern labels are thus: (ID: 63566, day: 7), (Name: Bitstamp.net, day: 7), or (Type: Exchange, day: 7). We use rounded log values to avoid sparse vocabulary.

4.2 Learning Vector Representations

We use Distributed Memory Model of Paragraph Vectors (PV-DM) to train the embedding of the flow. PV-DM is one of the two variations of neural network models presented in the Doc2Vec paper [17]. The model is trained to maximize the prediction accuracy of the center vocable, given the surrounding vocables. We chose the PV-DM model because it preserves the order sequence of the walk rather than predicting a bag of words in a sentence in the PV-DBOW model. In our experiment, we set a typical embedding size ($n = 128$) to compare the different labeling strategies.

5 Flow-Based Actor Identification

In this section, we demonstrate how the taint flow embeddings can be leveraged to identify Bitcoin actors from transaction sources. We extract taint flows from Bitcoin mining pools and experiment with two actor-disambiguation tasks: (1) identification of source actors using supervised learning and (2) automatic discovery of actors based on clustering. Finally, we evaluate the capacity of the embedding to differentiate between temporal origins.

5.1 Taint Flows of Bitcoin Mining Pools

We focus our experiment on the identification of mining pools. Mining pools are among the most important actors in the Bitcoin ecosystem. They correspond to companies that regroup the activity of various miners—from individuals to mining farms—under a single entity, with the prospect of sharing the mining rewards obtained to mitigate the effect of chance on their source of revenue [30, 33]. We chose mining pools because they are well-studied actors, persisting long enough in time, for which we can be confident in the data for validation.

We extract taint flows from the top-3 mining pools for each month between 2013 and 2016 to represent different sources of flows from large representative mining pools. We chose this specific period because the WalletExplorer dataset stopped updating the actors used as known actors in 2016 [14].

For each top-3 pool in a month, we taint all coins received from coinbase transactions—i.e., newly generated coins—on a random day of that month as a set of source transactions. We then construct taint flows and embed them with different methods according to the process defined in the previous sections.

As a result, our dataset consists of 144 taint flows from 11 mining pools. They are of various sizes and can be too large to use traditional embedding methods on standard computers. The average number of transactions is 611,327 (sd: 362,073, median: 569,576, max: 2,181,876) while the average number of clusters is 303,955 (sd: 178,145, median: 285,991, max: 1,131,919). There are 3,697 frequent clusters existing in more than 50% of all flows.

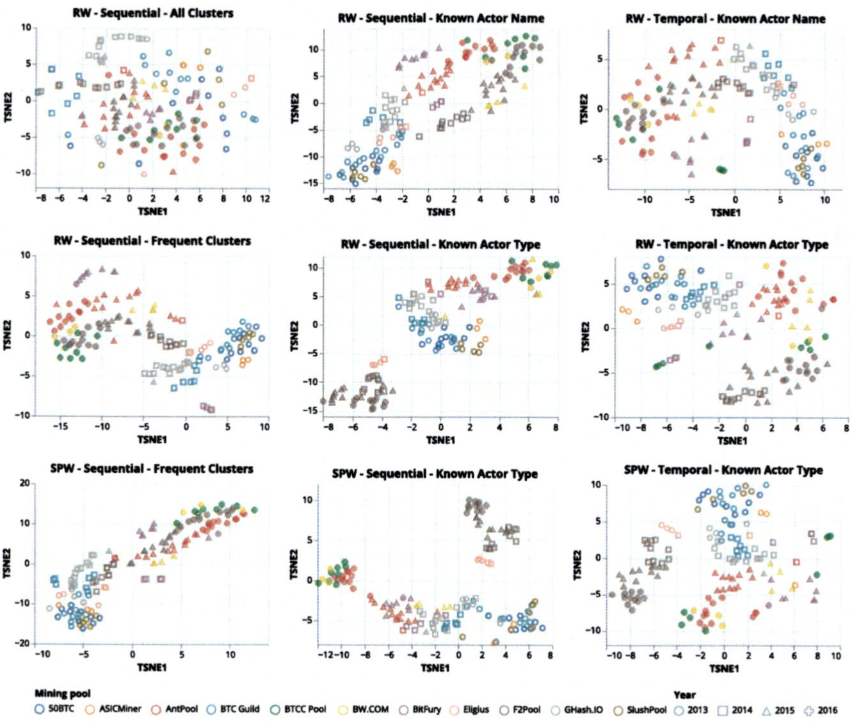

Fig. 1. T-SNE projections of selected taint flow embeddings.

The result of the representation learning, embedded in two dimensions using t-SNE [21], is visualized in Fig. 1. In most cases, the multiple taint flows of the same mining pool (nodes of the same color) seem close in the embedding space. The method thus captures at least some elements of the identity of the source

Table 1. Evaluation of taint flow embeddings of top-3 mining pools in 2013–2016.

Method	Accuracy		F1-score		NMI		ARI		AMI		Time Corr.	
Actor network features	0.250		0.152		0.120		0.096		0.017		0.118	
1. Graph2Vec												
Frequent clusters	0.146		0.086		0.195		0.173		0.117		0.263	
Known actor name	0.299		0.242		0.127		0.103		0.057		-0.085	
2. Sequential	RW	SPW	RW	SPW	RW	SPW	RW	SPW	RW	SPW	RW	SPW
All clusters	0.479	0.313	0.386	0.201	0.274	0.281	0.252	0.227	0.114	0.068	0.298	0.185
Frequent clusters	0.771	0.681	0.665	0.526	0.333	0.367	0.315	0.335	0.160	0.172	0.459	0.635
Known actor name	0.736	0.542	0.592	0.366	0.572	0.332	0.509	0.297	0.261	0.126	0.625	0.701
Known actor type	0.764	0.688	0.646	0.552	0.425	0.591	0.408	0.544	0.180	0.277	0.282	0.511
3. Temporal	RW	SPW	RW	SPW	RW	SPW	RW	SPW	RW	SPW	RW	SPW
All clusters	0.389	0.222	0.288	0.131	0.298	0.298	0.278	0.279	0.130	0.133	0.253	0.193
Frequent clusters	0.486	0.535	0.395	0.383	0.331	0.326	0.313	0.307	0.172	0.168	0.379	0.517
Known actor name	0.583	0.521	0.450	0.350	0.453	0.461	0.402	0.411	0.209	0.204	0.453	0.491
Known actor type	0.778	0.708	0.674	0.601	0.644	0.617	0.618	0.580	0.480	0.418	0.282	0.356

based on its taint flow. We also observe that the temporal aspect plays an important role and is well captured in the figures. For instance, in *RW—Temporal—Known Actor Name* and *RW—Sequential—Frequent Clusters* embedding, we see a shift—right to left—from circles to squares and triangles, and finally crosses, corresponding to the increasing years of the source.

5.2 Actor Identification Task

We train a k-Nearest Neighbor (k-NN) classification model with $k = 3$ to identify mining pools from the embedding space. k-NN is a simple model that can capture non-linear decision boundaries. We have to keep k small due to the relatively low number of observations. To evaluate the model performance, we use leave-one-out cross-validation (LOOCV), i.e., we consider that all sources are known but one and try to predict the identity of that unknown source from the others.

Baseline We define two baselines to compare the model performance with our embedding methods:

1. **Actor network features**: We extract a set of descriptive features from all cluster networks, including the number of nodes and edges, density, and degree assortativity. We also calculate nodes in- and out- degrees, clustering coefficient, and eigenvector centrality and report the minimum, 1st–3rd quantiles, maximum, mean, standard deviation, and mean absolute deviation for each feature.
2. **Graph2vec**: We train Graph2vec models [26] to embed frequent clusters and known actor name networks. We cannot train the model with the *All clusters* setting because the size of the graph can be enormous and make it impractical to compute the embedding.

Table 1 reports the Accuracy and F1 measure of the classification models. Our walk-based embeddings provide higher accuracy and F1-score than the baseline models, and pruning strategies improve performances as the *All clusters* approach consistently obtains the worst results. The best results are obtained using either *Temporal Known Actor Type* or *Sequential Frequent Clusters*. Overall, we found that *Random walks* provides better results than *Shortest Path Walks* in both sequential and temporal models.

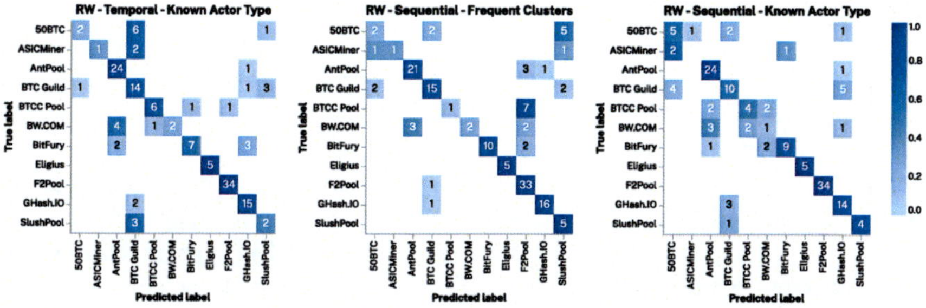

Fig. 2. Confusion matrix of the top-3 classification models.

Figure 2 presents the confusion matrices of the top-3 classification models: *Temporal Known Actor Type*, *Sequential Frequent Clusters*, and *Sequential Known Actor Type*. We observe that most of the actors for which we have many samples (e.g., AntPool, F2Pool, Ghash.IO) are well classified. Mining pools with few examples are more prone to errors, particularly ASICMiner, which has only three occurrences in our dataset and is mostly wrongly classified. This stresses the importance of having enough learning data in future works.

5.3 Actor Clustering Task

We assess if source actors can be discovered with the unsupervised approach by training k-means clustering on flow embeddings and checking the the cluster founds with the known sources. We train k-means models from 2 to 11 clusters and select the model with the highest *Silhouette* score. Table 1 reports the clustering evaluation with three standard scores: normalized mutual information (NMI), adjusted rand score (ARI), and adjusted mutual information (AMI).

We found that *Known Actor Type* clusters are the most efficient method for sequential and temporal embeddings. In contrast to the classification task, the *Shortest Path Walk* provides a better clustering for sequential embedding. Nonetheless, the three clustering metrics are relatively low. As source time and identity are highly correlated, we suspect the clustering might mix these two aspects and form the cluster containing pools in the same period.

5.4 Time Correlation

As we have observed in Fig. 1, the vectors of taint flow also embed a notion of time. This can be explained by Bitcoin's highly dynamic ecosystem in which actors appear, disappear, rise, and fall in popularity over time. Our approach depends on its source's identity and when this flow starts. Hence, it is possible that the embedding model group actors that frequently occur at the same period be closer in the embedding space.

To quantify how embeddings capture the evolution of taint flow patterns, we compute a *time correlation* score by computing the Spearman correlation coefficient between the distance in time expressed in months and the distance in the embedding for all pairs of taint flows. The time correlation in Table 1 shows a high correlation for *Sequential Known Actor Name* The correlation is lower in temporal embeddings compared with sequential ones.

Interestingly, the temporal aspect is poorly captured when using actor types instead of their identity. The correlation of *Known Actor Type* suggests that despite the rise and fall in popularity of specific actors, the type of actors reached by a flow from a particular source tends to stay constant, as shown in Fig. 1. This result highlights the importance of using time-independent vocabulary, in this case, the role of the actors, to train the unbias embedding model.

6 Discussion and Conclusion

In this work, we propose original methods to extract taint flows that take into account the temporal aspect of the Bitcoin transaction network and represent them using graph embedding techniques. We train classification and clustering models to evaluate how the tainted flows of coins can be used to identify the source actors in the Bitcoin transaction network.

Our experiments with mining pool taint flows show that although we could not reach a perfect precision in the actor identification task, a simple supervised approach yields a high accuracy. Unsupervised clustering is less convincing at this stage but could be improved by taking time into account and increasing the number of observations. The analysis of those results highlights what makes a taint flow characteristic of its source actor: (1) the starting time is significant as actors in the Bitcoin network emerge and disappear over time, and (2) the identity of encountered actors is not the only relevant element since we can also reach a good result using the characteristics of actors, especially their roles (actor type). This stresses the importance of using labeled data to improve model performance and raises another research direction to infer actor roles from on-chain data.

Our work demonstrates the relevance of using taint flows to characterize their source. However, we can achieve a better model performance with more taint flow data and more sophisticated classification and clustering models. Additional information such as the country of origin, network centrality, and more precise actor types, could also be a direction of improvement. Our method could be applied to other cryptocurrencies or other forms of diffusion, such as information in social media, by appropriately adapting the diffusion flow's construction. In future work, we intend to apply this approach to characterize money flows in other domains, particularly illegal and cybercrime activities, as well as propose a new technique to extract and explain meaningful patterns from those flows.

Acknowledgement. This project was partly founded by BITUNAM grant ANR-18-CE23-0004.

References

1. Ahmed, M., Shumailov, I., Anderson, R.: Tendrils of crime: visualizing the diffusion of stolen bitcoins. In: Graphical Models for Security, pp. 1–12. Springer (2019)
2. Akcora, C.G., Li, Y., Gel, Y.R., Kantarcioglu, M.: BitcoinHeist: topological data analysis for ransomware prediction on the bitcoin blockchain. In: The 29th International Joint Conference on Artificial Intelligence, pp. 4439–4445. IJCAI (2020)
3. Balthasar, T.D., Hernandez-Castro, J.: An analysis of bitcoin laundry services. In: Nordic Conference on Secure IT Systems, pp. 297–312. Springer (2017)
4. Bartoletti, M., Lande, S., Loddo, A., Pompianu, L., Serusi, S.: Cryptocurrency scams: analysis and perspectives. IEEE Access **9**, 148353–148373 (2021)
5. Bartoletti, M., Pes, B., Serusi, S.: Data mining for detecting bitcoin ponzi schemes. In: Crypto Valley Conference on Blockchain Technology, pp. 75–84. IEEE (2018)
6. Cazabet, R., Baccour, R., Latapy, M.: Tracking bitcoin users activity using community detection on a network of weak signals. In: International Conference on Complex Networks and Their Applications, pp. 166–177. Springer (2017)
7. Chainalysis Team: The 2022 Crypto Crime Report. Chainalysis Inc. (2022). https://go.chainalysis.com/2022-Crypto-Crime-Report.html
8. Di Battista, G., Di Donato, V., Patrignani, M., Pizzonia, M., Roselli, V., Tamassia, R.: Bitconeview: visualization of flows in the bitcoin transaction graph. In: IEEE Symposium on Visualization for Cyber Security, pp. 1–8. IEEE (2015)
9. Ermilov, D., Panov, M., Yanovich, Y.: Automatic bitcoin address clustering. In: IEEE International Conference on Machine Learning and Applications, pp. 461–466. IEEE (2017)
10. Goldfeder, S., Kalodner, H., Reisman, D., Narayanan, A.: When the cookie meets the blockchain: privacy risks of web payments via cryptocurrencies (2017). arXiv:1708.04748
11. Harlev, M.A., Sun Yin, H., Langenheldt, K.C., Mukkamala, R., Vatrapu, R.: Breaking bad: de-anonymising entity types on the bitcoin blockchain using supervised machine learning. In: The 51st Hawaii International Conference on System Sciences. ScholarSpace/AIS Electronic Library (2018)
12. Harrigan, M., Fretter, C.: The unreasonable effectiveness of address clustering. In: International IEEE Conferences on UIC/ATC/ScalCom/CBDCom/IoP/SmartWorld, pp. 368–373. IEEE (2016)
13. Ivanov, S., Burnaev, E.: Anonymous walk embeddings. In: International Conference on Machine Learning, pp. 2186–2195. PMLR (2018)
14. Janda, A.: Walletexplorer.com, https://www.walletexplorer.com/info
15. Jourdan, M., Blandin, S., Wynter, L., Deshpande, P.: Characterizing entities in the bitcoin blockchain. In: IEEE International Conference on Data Mining Workshops, pp. 55–62. IEEE (2018)

16. Kalodner, H., Möser, M., Lee, K., Goldfeder, S., Plattner, M., Chator, A., Narayanan, A.: BlockSci: design and applications of a blockchain analysis platform. In: 29th USENIX Security Symposium, pp. 2721–2738. USENIX Association (2020)
17. Le, Q., Mikolov, T.: Distributed representations of sentences and documents. In: International Conference on Machine Learning, pp. 1188–1196. JMLR (2014)
18. Lin, Y.J., Wu, P.W., Hsu, C.H., Tu, I.P., Liao, S.W.: An evaluation of bitcoin address classification based on transaction history summarization. In: IEEE International Conference on Blockchain and Cryptocurrency, pp. 302–310. IEEE (2019)
19. Lischke, M., Fabian, B.: Analyzing the bitcoin network: the first four years. Future Internet **8**(1), 7 (2016)
20. Liu, X.F., Ren, H.H., Liu, S.H., Jiang, X.J.: Characterizing key agents in the cryptocurrency economy through blockchain transaction analysis. EPJ Data Sci. **10**(1), 21 (2021)
21. Van der Maaten, L., Hinton, G.: Visualizing data using t-SNE. J. Mach. Learn. Res. **9**(11) (2008)
22. Meiklejohn, S., Pomarole, M., Jordan, G., Levchenko, K., McCoy, D., Voelker, G.M., Savage, S.: A fistful of bitcoins: characterizing payments among men with no names. Commun. ACM **59**(4), 86–93 (2016)
23. Michalski, R., Dziubałtowska, D., Macek, P.: Revealing the character of nodes in a blockchain with supervised learning. IEEE Access **8**, 109639–109647 (2020)
24. Möser, M., Böhme, R., Breuker, D.: An inquiry into money laundering tools in the bitcoin ecosystem. In: APWG eCrime Researchers Summit, pp. 1–14. IEEE (2013)
25. Nakamoto, S.: Bitcoin: a peer-to-peer electronic cash system. Tech. rep. (2008). http://bitcoin.org/bitcoin.pdf
26. Narayanan, A., Chandramohan, M., Venkatesan, R., Chen, L., Liu, Y., Jaiswal, S.: graph2vec: learning distributed representations of graphs (2017). arXiv:1707.05005
27. Nerurkar, P., Patel, D., Busnel, Y., Ludinard, R., Kumari, S., Khan, M.K.: Dissecting bitcoin blockchain: empirical analysis of bitcoin network (2009–2020). J. Netw. Comput. Appl. **177**, 102940 (2021)
28. Ranshous, S., Joslyn, C.A., Kreyling, S., Nowak, K., Samatova, N.F., West, C.L., Winters, S.: Exchange pattern mining in the bitcoin transaction directed hypergraph. In: International Conference on Financial Cryptography and Data Security, pp. 248–263. Springer (2017)
29. Reid, F., Harrigan, M.: An analysis of anonymity in the bitcoin system. In: Security and Privacy in Social Networks, pp. 197–223. Springer (2013)
30. Romiti, M., Judmayer, A., Zamyatin, A., Haslhofer, B.: A deep dive into bitcoin mining pools: an empirical analysis of mining shares. In: Workshop on the Economics of Information Security (2019)
31. Scherer, P., Lio, P.: Learning distributed representations of graphs with Geo2DR. In: ICML Workshop on Graph Representation Learning and Beyond (2020)
32. Tironsakkul, T., Maarek, M., Eross, A., Just, M.: Probing the mystery of cryptocurrency theft: an investigation into methods for cryptocurrency tainting analysis. In: Cryptocurrency Research Conference (2019)
33. Tovanich, N., Soulié, N., Heulot, N., Isenberg, P.: The evolution of mining pools and miners' behaviors in the bitcoin blockchain. IEEE Trans. Netw. Serv. Manage. (2022)
34. Vallarano, N., Tessone, C.J., Squartini, T.: Bitcoin transaction networks: an overview of recent results. Front. Phys. 286 (2020)

35. Weber, M., Domeniconi, G., Chen, J., Weidele, D.K.I., Bellei, C., Robinson, T., Leiserson, C.E.: Anti-money laundering in bitcoin: experimenting with graph convolutional networks for financial forensics. In: KDD Workshop on Anomaly Detection in Finance (2019)
36. Wu, J., Liu, J., Chen, W., Huang, H., Zheng, Z., Zhang, Y.: Detecting mixing services via mining bitcoin transaction network with hybrid motifs. IEEE Trans. Syst. Man Cybern. Syst. (2021)
37. Zhang, Y., Wang, J., Luo, J.: Heuristic-based address clustering in bitcoin. IEEE Access **8**, 210582–210591 (2020)
38. Zola, F., Eguimendia, M., Bruse, J.L., Urrutia, R.O.: Cascading machine learning to attack bitcoin anonymity. In: IEEE International Conference on Blockchain, pp. 10–17. IEEE (2019)

On the Empirical Association Between Trade Network Complexity and Global Gross Domestic Product

Mayank Kejriwal(✉) and Yuesheng Luo

University of Southern California, Los Angeles, CA 90292, USA
kejriwal@isi.edu luoyuesh@isi.edu
https://usc-isi-i2.github.io/kejriwal/

Abstract. In recent decades, trade between nations has constituted an important component of global Gross Domestic Product (GDP), with official estimates showing that it likely accounted for a quarter of total global production. While evidence of association already exists in macro-economic data between trade volume and GDP growth, there is considerably less work on whether, at the level of individual granular sectors (such as *vehicles* or *minerals*), associations exist between the complexity of trading networks and global GDP. In this paper, we explore this question by using publicly available data from the Atlas of Economic Complexity project to rigorously construct global trade networks between nations across multiple sectors, and studying the correlation between network-theoretic measures computed on these networks (such as average clustering coefficient and density) and global GDP. We find that there is indeed significant association between trade networks' complexity and global GDP across almost every sector, and that network metrics also correlate with business cycle phenomena such as the Great Recession of 2007–2008. Our results show that trade volume alone cannot explain global GDP growth, and that network science may prove to be a valuable empirical avenue for studying complexity in macro-economic phenomena such as trade.

Keywords: Computational economics · Network science · Global trade · Global economic growth · Product trade networks

1 Background and Motivation

In recent years, globalization has been a much discussed topic in both economic and political circles [12,20]. On the political front, globalization has sharply divided many observers, often along the lines of political affiliation, with some claiming that it has led to millions of people in vulnerable countries escaping poverty [15], while others have pointed out the damage to jobs and real wages (especially in sectors like manufacturing) in developed countries [5]. At the heart of this issue lies trade. Over the last decade, estimates show that trade accounted

H. Cherifi et al. (Eds.): COMPLEX NETWORKS 2016 2022, SCI 1077, pp. 456–466, 2023.
https://doi.org/10.1007/978-3-031-21127-0_37

for approximately a quarter of total global production [19]. Trade transactions include both goods and services, the latter including tourism and financial services, which do account in trade statistics and represent a flow in foreign capital. However, trade in goods is predominant from a global perspective, and production chains for these goods have become complex over time [18]. This complexity can also lead to fragility and supply chain disruption, as the pandemic recently showed [1].

Understanding the dynamics of trade, and its effect on global Gross Domestic Product (GDP), is a complex issue that is of great interest in macroeconomics [7], but which has not fully been addressed from a *network science* perspective. In part, the issue arises because constructing a rich network over time, and across multiple important sectors (such as agriculture and machinery), requires data that takes time to gather and process. Recently, such data was released as part of the ongoing research being done by the Atlas of Economic Complexity project at Harvard University [10]. Although the primary goal of this project is data visualization that allows research into the exploration of global trade flows across markets, the project has released fine-grained data that also allows us to construct sector-specific trade networks between countries over the last 25 years, leading up to a year before the pre-pandemic period. This period not only covers the financial crisis of 2007–2008 [9], but also China's entry into the World Trade Organization, and its emergence as a global economic force [21].

In this paper, we consider a simply stated, but empirically complex, research agenda: can network science, as applied to global trade networks between countries, be used to study the association between global GDP growth and increased trade complexity? Note that there is already evidence of such association in macro-economic data between trade *volume* and GDP growth [19]. However, complexity is not a phenomenon that can be studied through correlations between two simple time-series of trade volume and GDP. Instead, we need to first construct sector-specific trade networks over time (e.g., per annum) and empirically assess if changes in selected network science metrics, such as average degree and density, correlate with equivalent changes in global GDP growth. Theoretically, we note that changes in the global trade network (at least, in an unweighted representation) are not necessary for achieving higher volume or higher GDP. For example, one can imagine an extreme situation where the trade network that existed in 1995 continues to exactly be replicated by the trade network in 2015, and global GDP growth is explained purely through volume growth. In other words, if there is no *structural* change in the network at all, average degree and density (among a host of other network science metrics) would not change, and volume alone could explain trade's component in global GDP growth.

However, our results show that there has been considerable structural change in trade networks, with rapid changes occurring in the period of high growth leading up to the 2007–2008 financial crisis, and flattening soon after. There are interesting correlations between sectors: some sectors tend to be recession-proof while others are heavily influenced. In an absolute sense, the changes are small; however, viewed at the appropriate scale, are marked and significant.

In the economics literature, there has been considerable recent interest in applying network science to the study of complex phenomena [13,14]. Many such studies have involved the stock market, including modeling the stock market as a 'temporal' network; see, for example, representative work in [24,25]. Others have used network science to understand risk in the interbank system [22], among other complex systems research with applications in economics [8]. There has also been considerable literature on studies of economic complexity [2,11,17]. Although preliminary, the study presented in this paper adds to this knowledge by considering the correlation between trade complexity (as viewed and measured from a network science lens) and global GDP. To the best of our knowledge, such a study (even in a preliminary form) has been lacking in the literature thus far. Other complex phenomena have had a long history of being investigated through network science; numerous examples are presented in numerous books and reviews on the subject [3,4,6,23].

The rest of this paper is structured as follows. First, we detail materials and methods, including the manner in which the sector-specific and per-year trade networks were constructed and interpreted. Next, we report the findings of the study. Finally, we conclude the paper with a brief summary and some guidance on future research avenues.

2 Materials and Methods

2.1 Data

The data were obtained from the *Atlas*[1] project by Harvard's Growth Lab. Atlas includes trade statistics for 250 countries and regions, divided into 20 goods- categories and 5 service-categories. In total, 6,000 products are covered globally across the 25 categories. The raw trade data is collected from United Nations Statistical Division (COMTRADE) for goods, and the International Monetary Fund (IMF) for services. In this paper, we focus only on goods, which is described by two trade classification systems - Harmonized System (HS) 1992 and Standard International Trade Classification (SITC) revision 2.

HS data provides detailed classification of goods and their trade volumes between countries between 1995 to 2018. It covers about 5000 goods across 10 sectors (Table 1). HS categories are broken down into 1-, 2-, 4-, and 6-digit detail levels. Each level provides data for goods at a finer granularity. For example, the 1-digit level, which is considered for the studies in this paper, describes the 10 sectors in Table 1. SITC data provides a longer time-series from 1962 to 2018 but has far lower goods coverage (only about 15% of goods covered by HS), since many goods were not invented yet or had significant trade volumes before the 1990 s s (an example being the personal computer or laptop). SITC categories are also broken down into 1-, 2-, and 4- digit detail levels.

[1] https://atlas.cid.harvard.edu/data-downloads.

Table 1. The ten sectors covered in the HS dataset at the 1-digit level. Networks are constructed at this level of granularity, with one network typically constructed per sector as discussed in Sect. 2.2. The 'Other' sector is excluded from study due to its exceptional status.

Sector	# 4-Digit products	Examples
Textiles	181	Cotton sewing thread, globes
Agriculture	290	Crustaceans, peanuts
Stone	67	Gold, glass fibers
Minerals	67	Quicklime, nickel ore
Metals	157	Copper mattes, handsaws
Chemicals	219	Phosphides, sulfides of nonmetals
Vehicles	38	Railway service vehicles, wheelchairs
Machinery	174	Vacuum cleaners, thermionic
Electronics	48	Liquid crystal devices, games
Other	2	Trade data discrepancies, commodities not specified according to kind

For each of the two datasets, key variables include the country, trading partner, product and year. There is also detailed information on many other aspects of the trading relation, including export and import values (for a given good being traded between the two trading partners). While some of this information is used in our experiments, the primary information that we rely on for network construction (as subsequently described) is the set of goods being traded across countries over the time period 1995–2018 that is covered by the HS data.

2.2 Network Construction and Measures

As a first step toward network construction, we isolated the Country-Partner-Product-Year variables at the 4-digit product level (HS system). In total, the combined file has more than 113 million rows. We filtered the countries to only keep the ones that have data through all time periods, with 213 out of 250 locations being retained. For a specific product, or a union of products[2], we build a undirected graph for each year using the networkX[3] package, by first constructing a node for each country involved in the trading of the selected products (all 4-digit level products falling within a given 1-digit level sector), and linking two countries with an edge if a trading relation (i.e. import or export of the sector-specific goods being studied) exists between those countries. One graph, called a *trade network*, is constructed per sector. Six graph attributes are calculated and compared over time: number of nodes, average node degree, average node clustering coefficient, density, total number of triangles, and diameter.

[2] Since we construct graphs at the sector level (1-digit HS level, a union of products is involved since the data is available at the 4-digit product level, and multiple 4-digit level products comprise a 1-digit level HS 'sector' (similar to a taxonomy).

[3] https://networkx.org/.

3 Results

Figure 1 illustrates the average degree over time of the trade networks constructed for each of the five sectors, namely *Textiles, Agriculture, Stone, Minerals*, and *Metals.* A similar trend is observed for the other sectors (not shown herein). We find that the average degree of the network rapidly increases (by almost 50%) over a 15 year period (1995–2010) for all sectors and that the trends are largely consistent. Following the 2008 financial crisis, there is a plateauing effect, but it is unclear if this is due to long-term shifts due to the crisis itself or because of diminishing returns.

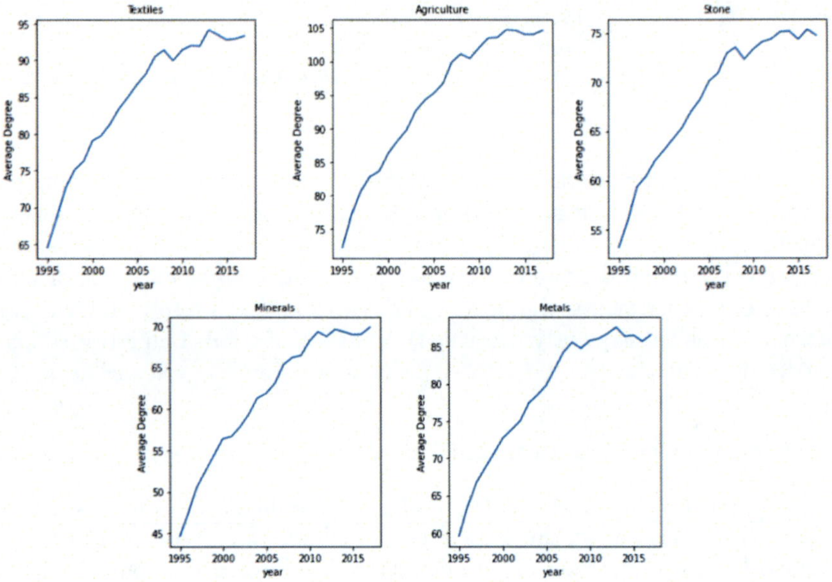

Fig. 1. Average degree over time of the trade networks constructed for the five sectors [Textiles, Agriculture, Stone, Minerals, Metals]. Distributions for the other four sectors were found to be qualitatively similar.

Figures 2, 3 and 4 plot some other common network metrics over time, such as average clustering coefficient, average density and total number of triangles, respectively. For the latter two, we see a near identical trend; however, it is important to note that the y-axis of the plots do not start at 0 for any of the plots. In comparing the numbers in 1995 and 2010, we find that the total number of triangles approximately doubles during that time, while the density increases by the same extent as the average degree. The metrics all suggest that significant increase in global trade was witnessed during that time. Importantly, this cannot be due only to the influence of a single country (such as China) or even a small set of countries, since we are not accounting for volume of trade in these networks. Since nodes in these networks are countries, and edges are only created if a minimum level of trade took place in a given sector between a pair of countries (as discussed earlier in *Materials and Methods*), each trade relation has a minimal degree of strength for it to be included as an edge in the network.

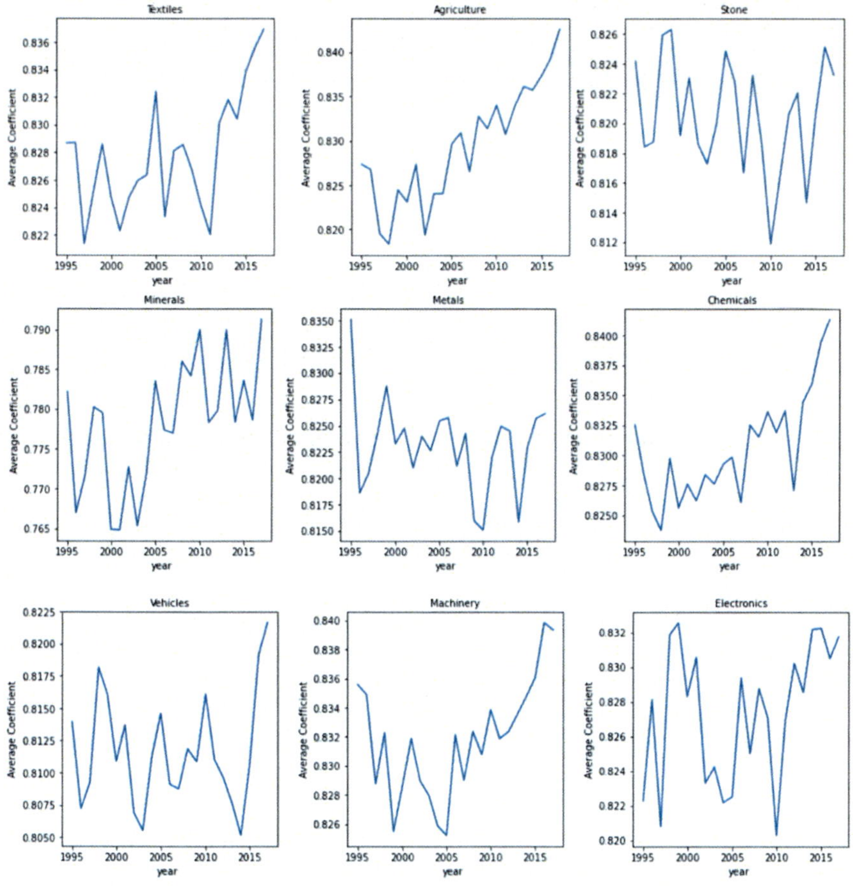

Fig. 2. Average clustering coefficients of nodes in the trade networks constructed for all nine sectors.

In contrast, the average clustering coefficient in Fig. 2 exhibits a more cyclical trend but only when viewed at a sufficiently fine-grained level (the y-axis has a range of only about 0.04 units at the maximum). Otherwise, the curves would look mostly flat, and tend to range between 75–85% depending on the sector. We do note that in the period leading up to 2010 (i.e. in the aftermath of the 2008 recession, which is when the ramifications on trade would tend to be felt more) there is usually a significant decline, especially in the electronics, textiles, metals and vehicles sectors. In contrast, sectors such as agriculture and machinery, which

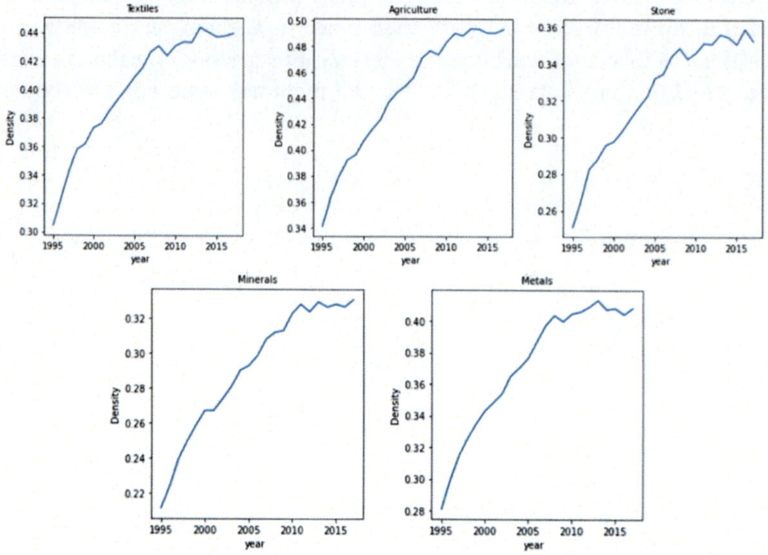

Fig. 3. Average density of the trade networks constructed for the five sectors [Textiles, Agriculture, Stone, Minerals, Metals]. Distributions for the other four sectors were found to be qualitatively similar.

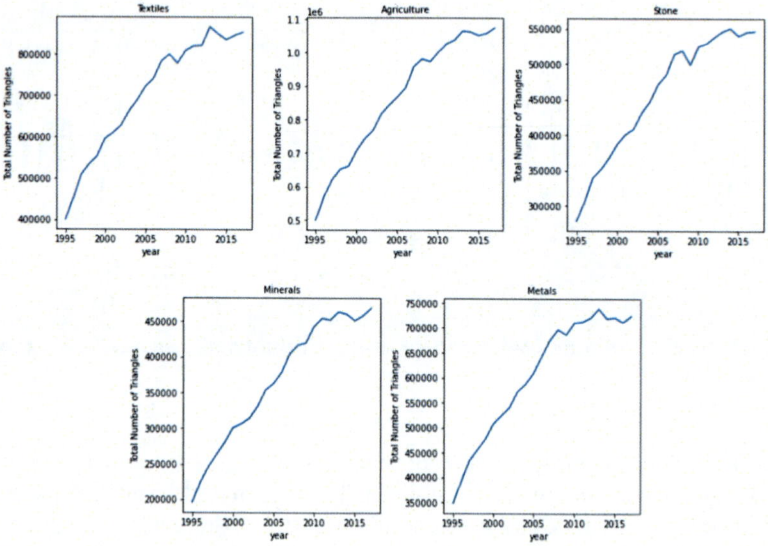

Fig. 4. Total number of triangles in the trade networks constructed for the five sectors [Textiles, Agriculture, Stone, Minerals, Metals]. Distributions for the other four sectors were found to be qualitatively similar.

tend to be capital resources that are needed in the manufacturing of a wide range of products (or in the case of agriculture, for servicing a basic need, such as food), the decline is less significant and continues to show an upward trend soon after. For electronics, we also see a similar precipitous decline in the immediate aftermath of the dot-com bubble in 2000. Hence, to a certain degree, cyclical economic fluctuations can be witnessed in network metrics, albeit with some lag (in other words, it would not be prudent to use such measures as predictive tools, only as analytical and quantitative measures of how trade is systemically affected by business cycles).

Although the plots are useful visual tools, we seek to better understand the correlation between these measures and the global GDP in Table 2. The table shows that there is indeed a high correlation (well above 0.9) for the series representing average degree, average density and total number of triangles, and far less of a correlation for the average clustering coefficient. Correlations for electronics, vehicles, chemicals, textiles and agriculture are most positive, and explain the critical role that trade and globalization plays in ensuring that these products are available at low prices for industries and markets worldwide. The role of agriculture in trade has not always been appreciated in recent decades; however, the ongoing war in Ukraine is already leading to commentary that food and agriculture (especially wheat) will be heavily affected as a consequence [16].

Table 2. Pearson's correlations of four metrics against the global GDP series for all ten sectors over the time period 1995–2017.

Sector	Average degree	Average clustering coefficient	Average density	Total # Trianges
Textiles	0.918	0.607	0.917	0.945
Agriculture	0.938	0.888	0.937	0.96
Stone	0.925	−0.147	0.925	0.946
Minerals	0.94	0.595	0.941	0.962
Metals	0.914	−0.208	0.914	0.936
Chemicals	0.918	0.607	0.917	0.945
Vehicles	0.938	0.888	0.937	0.96
Machinery	0.925	−0.147	0.925	0.946
Electronics	0.94	0.595	0.941	0.962
Other	0.914	−0.208	0.914	0.936

Finally, Fig. 5a and b show the average and percentage change, respectively, in trade volume (sum of exports and imports), with average taken over countries over the time period 1995–2018. For Fig. 5a, change in trade volume for a certain country is calculated from the difference in sum of export and import between a current year and its previous year. For Fig. 5b, percentage change is calculated by dividing change in trade volume with trade volume of the previous year[4].

[4] Theoretically, it is possible for the trade volume of the previous year to be 0 but this never occurred in practice in the dataset.

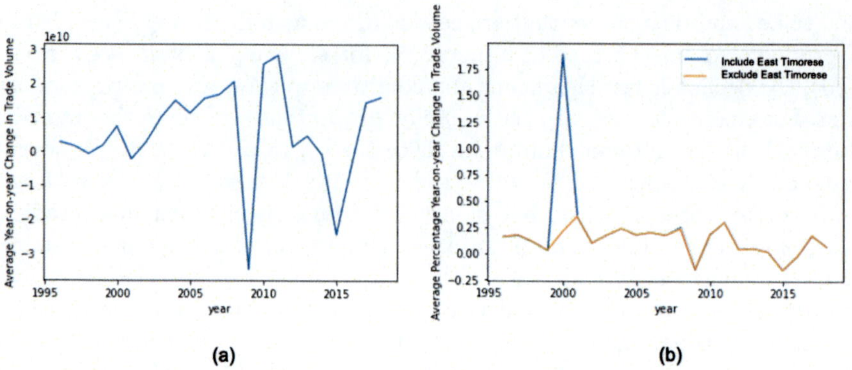

Fig. 5. a Average change in trade volume (sum of exports and imports), with average taken over countries over the time period 1995–2018. **b** Average change in trade volume (sum of exports and imports), with average taken over countries, compared to the previous year over the time period 1995–2018. Note that, the 1999 East Timorese crisis caused a spike in percentage change in trade volume in 2000.

The (percentage) change in trade volume was then averaged over all countries. In analyzing the results, we find, concurrent with the previous figures, that trade volume indeed declines precipitously leading up to, and shortly in the aftermath of the financial crisis, but rebounds briefly (perhaps due to long-duration contracts that needed to be fulfilled), before continuing its decline all the way till 2015. There is some growth thereafter, most likely attributable to overall economic growth but a plateauing effect can already be observed shortly thereafter. The percentage change in Fig. 5b shows that trade has remained largely stable year-over-year but there has been marked fluctuation since the financial crisis of 2008[5]. In the wake of populist movements since 2015, the prospects of global trade witnessing a revival seem uncertain at best.

Although we do not tabulate the results herein, we also computed the diameter of these international trade networks for each of the nine sectors. Diameters tended to be between 3–4 with the exception of minerals (maximum diameter of 4 was achieved in 2012), and in many cases, a diameter of 2 was already achieved by 1995. This illustrates that, while a trading relation does not exist between every pair of countries in any given sector, an intermediated relation does exist. This is a direct reflection of the degree of globalization that existed in international trade relations in the early part of this century.

[5] Note that we include two plots in Fig. 5b, since we found an outlier (East Timorese). The outlier, which occurs around 1999–2000 can be explained by the fact that, following the UN-sponsored act of self determination, East Timorese gained independence from Indonesia in 1999, and consequently became the first new sovereign state of the 21st century.

4 Conclusion

Much has been written in recent years about the benefits and costs of globalization and international trade. It was already well known that global trade and global GDP were highly associated. This paper looked to quantify globalization through a different lens: namely, through a network-theoretic study of trade networks between countries, constructed in a rigorous fashion. Rather than consider a single trade network, we relied on per-sector analyses using data from the Atlas project. We found that there is a heavy correlation between global GDP and multiple network science metrics such as average degree and number of triangles. The average clustering coefficient is less correlated, perhaps because trade networks were already well developed in the mid-1990s when the era of globalization was taking off in the aftermath of the dissolution of the Soviet Union and the liberalization of markets in Asia (most notably, China and India). In general, the average clustering coefficient remains high throughout, but some sectors show higher correlation than others. The data also confirms that globalization and trade relations have been in considerable flux since the 2008 financial crisis.

The data suggests many avenues for future research, not least of which is a study that takes the specific effects of COVID-19 into account. We suspect that there will be significant declines in all measures, and a complete disruption to the trade networks due to the supply chain, and other, bottlenecks caused by the pandemic. Indeed, a closer study of the volume of trade during, and after, this period may help us to quantify both the global and per-country impact of supply chain disruptions through a network science lens.

References

1. Akın Ateş, M., Suurmond, R., Luzzini, D., Krause, D.: Order from chaos: a meta-analysis of supply chain complexity and firm performance. J. Supply Chain Manage. **58**(1), 3–30 (2022)
2. Balland, P.A., Broekel, T., Diodato, D., Giuliani, E., Hausmann, R., O'Clery, N., Rigby, D.: The new paradigm of economic complexity. Res. Policy **51**(3), 104450 (2022)
3. Ben-Naim, E., Frauenfelder, H., Toroczkai, Z.: Complex networks, vol. 650. Springer Science & Business Media (2004)
4. Boguna, M., Krioukov, D., Claffy, K.C.: Navigability of complex networks. Nat. Phys. **5**(1), 74–80 (2009)
5. Butzbach, O., Fuller, D.B., Schnyder, G.: Manufacturing discontent: national institutions, multinational firm strategies, and anti-globalization backlash in advanced economies. Glob. Strategy J. **10**(1), 67–93 (2020)
6. Dorogovtsev, S.: Complex Networks. Oxford University Press, Oxford, UK (2010)
7. Golinelli, R., Parigi, G.: Tracking world trade and GDP in real time. Int. J. Forecast. **30**(4), 847–862 (2014)
8. Gomes, O., Gubareva, M.: Complex systems in economics and where to find them. J. Syst. Sci. Complex. **34**(1), 314–338 (2021)
9. Goodhart, C.A.: The background to the 2007 financial crisis. Int. Econ. Econ. Policy **4**(4), 331–346 (2008)

10. Hausmann, R., Hidalgo, C.A., Bustos, S., Coscia, M., Simoes, A.: The Atlas of Economic Complexity: Mapping Paths to Prosperity. Mit Press (2014)
11. Hidalgo, C.A.: Economic complexity theory and applications. Nat. Rev. Phys. **3**(2), 92–113 (2021)
12. Hoekman, B., Nelson, D.R.: Reflecting on populism and the economics of globalization. J. Int. Bus. Policy **1**(1), 34–43 (2018)
13. Jackson, M.O.: The Past and Future of Network Analysis in Economics. Available at SSRN 2520284 (2014)
14. Kenett, D.Y., Havlin, S.: Network science: a useful tool in economics and finance. Mind Soc. **14**(2), 155–167 (2015)
15. Khan, F.N., Majeed, M.T.: Globalization and poverty nexus: a panel data analysis. Forman J. Econ. Stud. 14 (2018)
16. Lang, T., McKee, M.: The Reinvasion of Ukraine Threatens Global Food Supplies (2022)
17. Metcalfe, J.S., Foster, J.: Evolution and Economic Complexity. Edward Elgar Publishing (2007)
18. Monostori, J.: Supply chains robustness: challenges and opportunities. Procedia CIRP **67**, 110–115 (2018)
19. Ortiz-Ospina, E., Beltekian, D., Roser, M.: Trade and Globalization. Our World in Data (2018)
20. Rodrik, D.: Why does globalization fuel populism? economics, culture, and the rise of right-wing populism. Annu. Rev. Econ. **13**, 133–170 (2021)
21. Rumbaugh, T., Blancher, N.: China: International Trade and WTO Accession. Available at SSRN 878859 (2004)
22. Simaan, M., Gupta, A., Kar, K.: Filtering for risk assessment of interbank network. Eur. J. Oper. Res. **280**(1), 279–294 (2020)
23. Strogatz, S.H.: Exploring complex networks. Nature **410**(6825), 268–276 (2001)
24. Zhao, L., Li, W., Cai, X.: Structure and dynamics of stock market in times of crisis. Phys. Lett. A **380**(5–6), 654–666 (2016)
25. Zhao, L., Wang, G.J., Wang, M., Bao, W., Li, W., Stanley, H.E.: Stock market as temporal network. Phys. A: Stat. Mech. Appl. **506**, 1104–1112 (2018)

Measuring the Stability of Technical Cooperation Network Based on the Nested Structure Theory

Wenhui Liu, Guoqiang Liang(✉), and Lizhi Xing

College of Economics and Management, Beijing University of Technology, Beijing, China
lianggq@bjut.edu.cn

Abstract. Nested structure is a structural feature that is conducive to system stability formed by co-evolution. In our opinion, it is just like what the biological species do in the mutualistic ecosystem that enterprises collaborate to apply for patents in the technical cooperation network, changing to form one dynamic equilibrium after another. In this paper, a nestedness-based analytical framework is built to reflect the topological stability of the technical cooperation network of Zhongguancun Science Park (Z-Park). We study why this technically mutualistic ecosystem can reach a stable equilibrium with time going by, as well as, we propose an index called Nestedness Disturbance Index (NDI) to study what the role park areas and technical fields play in the steady states.

Keywords: Technical cooperation · HETCN model · Nested structure · NODF · Nestedness disturbance index

1 Introduction

In the real world, the network is not in a static state but is constantly evolving. The same is true for the technical cooperation network. With the development of science and technology, the number of technical entities is increasing, and the technical cooperation network is in the process of continuous dynamic evolution. However, the present research on the evolution of the technical cooperation net-work mainly focuses on the change in the overall network structure characteristics, the change in the distribution of network entities, and the distribution of network time and space [1–3]. Few studies focus on the robustness of technical cooperation networks. So this paper aims to propose the robustness measurement of the technical cooperation network.

After learning the relevant researches, we found that "Nestedness" is a structural measure of the overall stability of the ecosystem that seeking evolutionary balance [4–6]. For technical cooperation networks which also seek evolutionary balance, nestedness can be used to measure the robustness of the network, too. While Z-Park as a significance Chinese innovation ecosystem, it's technical cooperation network can be a perfect example for this paper. Therefore, this paper takes Z-Park as an example, uses the nested structure to analyze the evolution trend of the robustness of the Z-Park technical cooperation network, and proposes the Nestedness Disturbance Index (NDI) to quantify the impact of different innovation entities on the robustness of the network.

H. Cherifi et al. (Eds.): COMPLEX NETWORKS 2016 2022, SCI 1077, pp. 467–477, 2023.
https://doi.org/10.1007/978-3-031-21127-0_38

2 Literature Review

2.1 Theoretical Development

In 1957, when Darlington studied the migration process of species, he found that when terrestrial species waded to the island, they would preferentially choose to inhabit the island closest to the land, and then some of them would reach the next nearest island through the shortest wading distance. As this process repeats, it naturally leads to a decline in the number of species geographically, reflecting the nesting characteristics 5. Based on previous research, Patterson et al. proposed the precise concept of nesting, that is, in a fully nested network, the neighborhood of a node with a lower node degree is a subset of the neighborhood of a node with a higher node degree 6. Dardanelli et al. found that the higher the degree of nesting, the stronger the resilience of the system after external disturbance, and the more stable the network structure [7]. The nested structure reflects the phenomenon that narrow-biological species (species with a lower degree of adaptation to the ecological environment) tend to interact with ubiquitous species (species with a higher degree of adaptation to the ecological environment) in the mutualistic network. The connection between the ubiquitous species constitutes the stable core of the entire ecological network, and the narrow-biological species surround this core so that the entire network structure is more stable [8].

Nested structures also exist in socioeconomic systems [9, 10], traditional and online social networks, and technological and information networks [11]. By reviewing the literature, it can be seen that the cooperation between various entities is an important prerequisite for the formation of the network nested structure.

2.2 Formation Mechanism

The reasoning behind the nesting structure was explained by the introduction of a self-organizing network model by Medan et al., who noted that nesting occurs when species switch their interactions in an attempt to increase the centrality of their neighbors [12]. Michael et al. proposed a network formation model in which nodes try to strategically maximize their centrality in the network in the presence of link decay. And both hierarchical organization and network nesting can be the result of a centrality-based network formation process [13]. Suweis et al. proposed a dynamic mechanism. He said that starting from a random species-species interaction matrix, species will reconnect their interactions to maximize their abundance. This dynamic will make the interaction network exhibit Nested structures [6].

After learning the researches, the following conclusions are drawn: (1) Most of the research on nested structure focuses on the formation of nested structure, which is more theoretical. (2) The related researches tend to study the formation process of nested structure in the network, but not enough attention is paid to the function of nested structure in the network.

3 Data and Modelling

3.1 Data Sources

This paper takes Z-Park as an example to build the technical cooperation network and selects the enterprises in the list of high-tech enterprises in 2020 that provided on the website of the Administrative Commission of Zhongguancun Science Park as the network nodes. In addition, since the patent is used as an index of R&D output, which better reflects the degree of mutual learning and technology integration between enterprises [14], this paper selected the paper as the index to reflect technical cooperation between enterprises. What's more, with the help of python, this paper capture the patent application information of the enterprises in the list in the global patent database of PatSnap, including patent publication number, application number, application date, title, application entity, park area, and technical field. Considering that there is a certain time lag between patent data disclosure and storage, the application time is set as 2008–2019. Besides, since the co-occurrence relationship of the patent application entities is used to map the cooperative relationship, this paper only retains the patent information of two or more application entities.

3.2 HETCN Model

Through investigation, it is found that some high-tech enterprises obtain the source of innovation through purchase of patent licenses, and even technology transfer, while some other enterprises achieve innovation independently with the original patent applicant being the current patentee. Combined with the analysis paradigm of the network motif, we classify the relationship between enterprises and patents into five situations. In Fig. 1, the yellow square node at the top is the patent and the blue circular node at the bottom is the relevant entities. The blue solid line between the two types of nodes represents the patent ownership, while the dotted line represents the patent propagation path. The black line indicates a strong patent correlation between enterprises while the dotted line indicates a weak correlation [15].

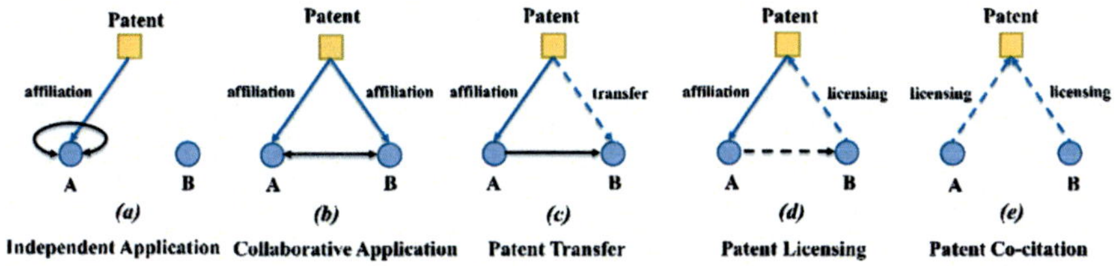

Fig. 1. Schematic diagram of technology diffusion process

After collecting and arranging the data, this paper selects the enterprises that meet the situation(b) as nodes to build a High-Tech Enterprise Technical Cooperation Network (HETCN). It can be represented by a graph $G = (V, E)$, where V is a node-set representing each enterprise, and E is an edge set, representing the relationship between technical cooperation among various enterprises. Since cooperation is coordinated with each other, the HETCN model constructed in this paper is an undirected network.

3.3 Nested Structure of HETCN

The growth mechanism of preferential links in complex networks makes each newly added node tend to be connected to the central node. Therefore, a small number of central nodes with larger node degrees form the core of the network, and other nodes with smaller node degrees are connected to the central node to form the periphery of the network, this "core-periphery" phenomenon is the nested structure discussed in this paper. Due to the heterogeneity of the network itself and the strengthening of the optimization mechanism, the nested structure is formed.

In technical cooperation networks, there are similar "core" and "periphery" entities. However, in such a network structure, although there are still absolute central nodes, the flow of resources leads to the formation of some "sub-center" nodes, which in turn form a balance with the center and conducive to the stability of the network structure. In the following, the node with larger node degrees will be called ubiquitous node while the node with smaller node degrees will be called narrow-biological node.

Figure 2 is a schematic diagram of the Z-Park technology cooperation network drawn by Gephi. As can be seen from the figure, in 2008, the technical cooperation network aggregated to a single ubiquitous node. By 2019, in addition to the largest ubiquitous node, multiple small aggregated ubiquitous nodes have gradually formed, and the nested structure of the technical cooperation network has gradually formed.

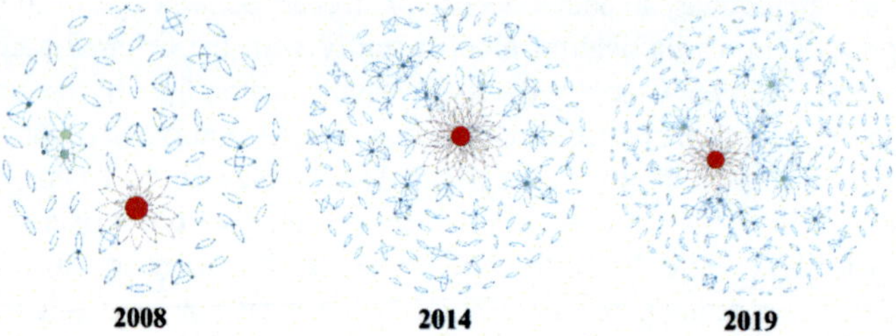

Fig. 2. The evolution trend of enterprise technical cooperation network

4 Methodology

4.1 Adjacency Matrix Rearrangement

This paper rearranges the adjacency matrix according to the size of the node degree to make it show the hidden nested features in the network, as shown in Fig. 3.

The blue dots in the figure represent the nodes in the network, while the blue squares indicate that there is an edge between the two nodes. What's more, the white indicates that there is no connection between the two nodes. In the rearranged adjacency matrix, the uppermost and leftmost nodes have the strongest ubiquity, and the lowermost and rightmost nodes have the strongest narrowness [16].

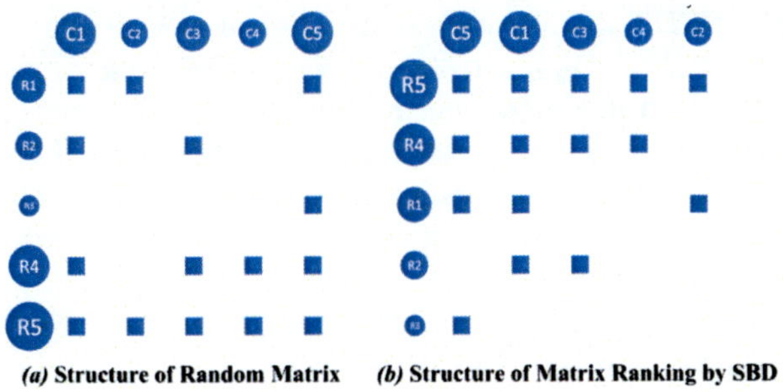

Fig. 3. Sorting of network adjacency matrix based on SBD algorithm

4.2 Measurement of the Nestedness

This paper use *NODF* index to calculate the nestedness of networks, which is based on two basic properties: Decreasing Fill (*DF*) and Paired Overlap (*PO*) [16].

Given that a matrix has *m* rows and *n* columns, and *MT* is the number of elements valued at 1 in any row or column. For any pair of rows $(i,j)(i<j)$, if $MT_i > MT_j$, then $DF_{ij} = 100$, otherwise $DF_{ij} = 0$; the same is true for columns.

For any pair of rows $(i,j)(i<j)$, $\left(PO_{ij}\right)$ refers to the percentage of 10s in a given row *j* that is located at identical column positions to the 10s observed in a row *i*; the same is true for columns. Therefore, for any up-to-down row pair, or any left-to-right column pair, the degree of paired nestedness $\left(N_{paired}\right)$ can be expressed by Eq. (1):

$$N_{paired} = \begin{cases} 0, & if\ DF_{paired} = 0 \\ PO, & if\ DF_{paired} = 100 \end{cases} \tag{1}$$

There are $m(m-1)/2$ row pairs in row *m*, and $n(n-1)/2$ column pairs in column *n*. Thus, the nestedness of the entire network can be calculated by "averaging all paired values of rows and columns"

$$NODF = \frac{\sum N_{paired}}{\left[\frac{m(m-1)}{2}\right] + \left[\frac{n(n-1)}{2}\right]} \times 100 \tag{2}$$

where the *NODF* value ranges from 0 to 100, with $NODF = 0$ indicating a non-nested network structure and $NODF = 100$ indicating a fully nested network structure.

4.3 Nestedness Disturbance Index

In this paper, by removing any entity in the sub-network and calculating the *NODF* of the network composed of the remaining entities, we can measure the contribution of the entity to the technical cooperation network nesting. On this basis, this paper proposes the Nestedness Disturbance Index (*NDI*) as an index to measure the role of different agents in maintaining network nesting, which is expressed by Eq. (3):

$$NDI = \frac{\left(NODF_{Original} - NODF_{Remove}\right)}{NODF_{Original}} \times 100 \tag{3}$$

Among them, $NODF_{Original}$ represents the original *NODF* value of the technical cooperation network, $NODF_{Remove}$ represents the *NODF* value of the network after removing an entity. A positive *NDI* value means that the *NODF* value of the technical cooperation network decreases after removing an innovation entity, that is, the entity has a positive impact on maintaining the nesting of the network structure where it is located, and vice versa. This index is universal for measuring the contribution of any node in the network to the nesting of the network.

5 Results and Discussion

5.1 Park Area Scale Empirical Analysis

5.1.1 Evolutionary Trend of Nestedness

According to the key field "affiliated park area" of the captured patent data, the HETCN is classified and aggregated, and finally formed Changping (CP) Park, Chaoyang (CY) Park, Dongcheng (DC) Park, Daxing-Yizhuang (DX-YZ) Park, Fangshan (FS) Park, Fengtai (FT) Park, Haidian (HD) Park, Huairou (HR) Park, Mentougou (MTG) Park, Miyun (MY) Park, Pinggu (PG) Park, Shijingshan (SJS) Park, Shunyi (SY) Park, Tongzhou (TZ) Park, Xicheng (XC) Park and Yanqing (YQ) Park such 16 nodes. Then this paper formed the technical cooperation networks from 2008 to 2019 and measured the *NODF* value of them, as shown in Fig. 4. Dividing the Z-Park into 16 parks may lead to area planning so that different parks may not integrate and complement each other. However, as can be seen from Fig. 4, from 2008 to 2019, the *NODF* value of HETCN at the park area scale fluctuated upward, that is, the degree of nesting among the parks gradually deepened.

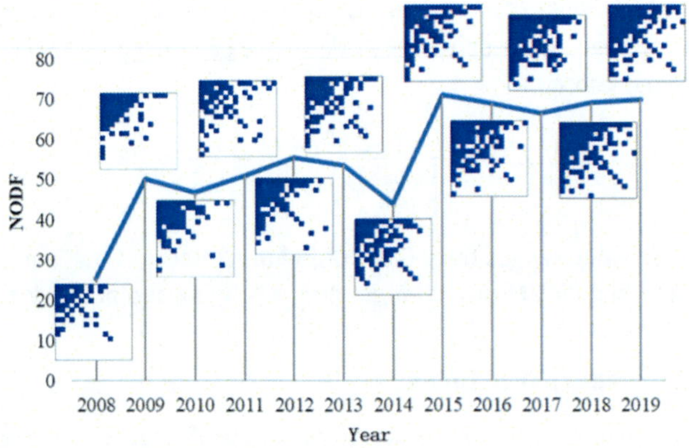

Fig. 4. Trends of the nested structure of HETCN at the park area scale from 2008 to 2019

The heat map in Fig. 4 can more clearly show the changes in the layout of technical cooperation in each park. It can be seen from the figure that the technical cooperation relationship between the parks gradually penetrates from the upper left corner of the adjacency matrix to the bottom, and the phenomenon of technology monopoly among a few parks gradually weakened by the diffusion of technology among the ubiquitous parks. While the interaction between the narrow-biological and ubiquitous parks is enhanced, the technology spreads from the network center to the outer layers, making the entire Z-Park more stable [17].

5.1.2 Simulation of the Robustness

This section calculates the *NDI* of each park in 2008, 2014, and 2019, as shown in Fig. 5. On the one hand, it analyzes the overall impact of each park on the robustness of the technical cooperation network from a global perspective. On the other hand, from the perspective of time-series changes, the role of each park in maintaining network stability is studied, then we can identify which park plays a key role in maintaining the stability of HETCN.

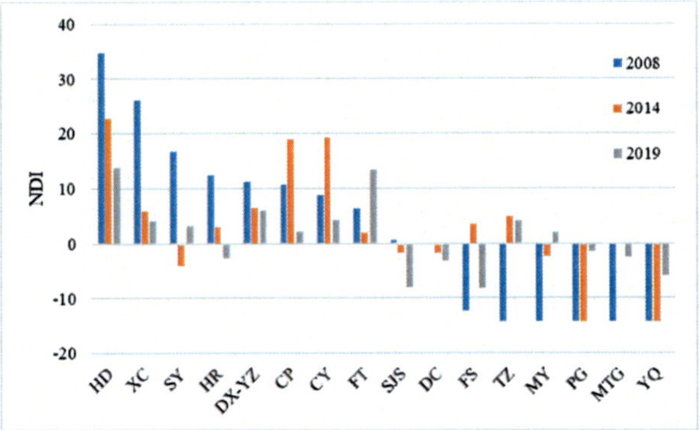

Fig. 5. *NDI* of each park in 2008, 2014 and 2019

On the whole, an important player in improving the nesting of the network is the ubiquitous park, which is also an important factor affecting the robustness of the technical cooperation network. In contrast, the perturbation of the narrow-biological park is weak and has the opposite effect in improving the nesting. Further observation of the time-series changes of the *NDI* index of the parks shows that the *NDI* of the ubiquitous parks is decreasing year by year, and the *NDI* of the narrow-biological parks is increasing year by year. That is to say, the influence of the ubiquitous parks on the robustness of HETCN is gradually weakening, while the importance of the narrow-biological parks is gradually increasing. This shows that the technical connections between the various parks are gradually intensive, and they are truly moving towards the balanced development of each park.

From an individual perspective, HD Park is the biggest factor affecting the robustness of the technical cooperation network at the park area scale. As the park with the most ubiquity, HD Park always ranks first in *NDI* value. Its removal will lead to a substantial reduction in the scale of the entire HETCN and rapid transmission of technical risks, which will seriously affect the technology-related enterprises in other parks. In turn, it will cause a strong disturbance to the robustness of HETCN.

In the figure, the NDI value of FT Park has risen significantly. From 2008 to 2019, its NDI ranking has risen to second place. In 2002, a large number of corporate headquarters have successively settled in, which has promoted the technical connection between FT Park and other parks, making FT Park play an increasingly critical role in maintaining the stability of HETCN.

5.2 Technical Field Scale Analysis

5.2.1 Evolutionary Trend of Nestedness

According to the key field "technical field" of the captured patent data, the HETCN is classified and aggregated, and finally formed Advanced Manufacturing Technology (AMT), Aerospace Technology (AT), Bioengineering and New Medicine (BNM), Electronics and Information Technology (EIT), Environmental Protection Technology (EPT), Modern Agricultural Technology (MAT), Maritime Engineering Technology (MET), Nuclear Application Technology (NAT), New Energy Technology (NET), New Materials and Applications (NMA) and Related to the top ten above (REL) such 11 nodes. Then this paper formed the technical cooperation networks from 2008 to 2019 and measured the *NODF* value of them, as shown in Fig. 6. It can be seen that from 2008 to 2019, the *NODF* value of HETCN at the technical field scale fluctuated continuously, but the overall value increased, and the degree of nesting in each technical field gradually deepened. This shows that high-tech enterprises in different fields have gradually formed a stable technical cooperation relationship.

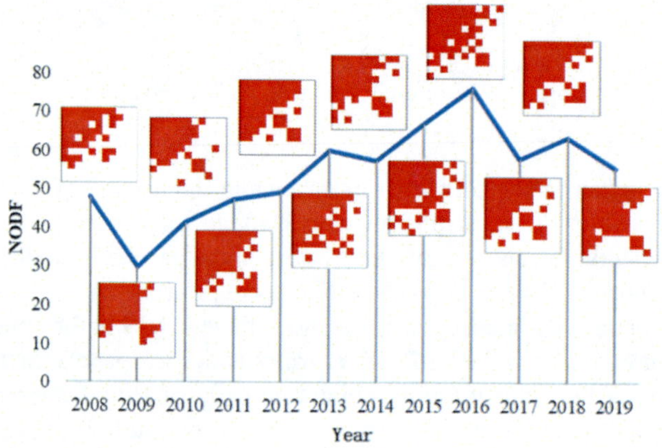

Fig. 6. Trends of the nested structure of HETCN at the technical field scale from 2008 to 2019

The heatmaps in Fig. 6 provide a clearer picture of the evolution of technical partnerships across technology domains. It can be seen from the figure that the technical cooperation relationship between enterprises gradually penetrates from the upper left corner of the adjacency matrix to the bottom, so we can learn that the scope of technology diffusion increases and the connection between different technical fields is gradually established. In addition, the integration of different technical fields shows certain modularity, which shows that a mature integration trend has been formed between some technical fields.

5.2.2 Simulation of the Robustness

This section calculates the *NDI* of each technical field in HETCN in 2008, 2014, and 2019. On the one hand, the overall impact of each technical field on the robustness of the technical cooperation network is analyzed from a global perspective. On the other hand, from the perspective of time-series changes, we studied the role of various technical fields in maintaining network stability so that we can identify the technical fields that play a key role in maintaining the stability of HETCN, as shown in Fig. 7.

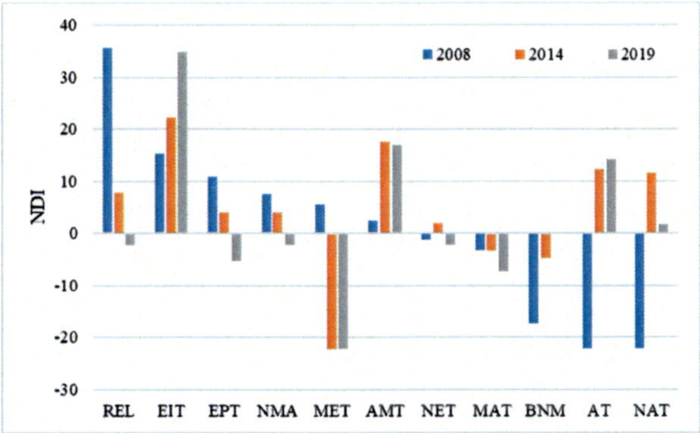

Fig. 7. *NDI* of each technology field in 2008, 2014 and 2019

On the whole, compared with the innovation network at the park area scale, the *NDI* value of each technology field in the technology field scale innovation network does not change significantly in time series. In 2019, the number of technical fields with a negative *NDI* increased, and the number of technical fields that supported the stability of HETCN decreased. Due to the strong technical versatility and wide range of technology diffusion, enterprises in the field of ubiquitous technology have technology cross-integration with most technical fields.

From an individual level, EIT is the technology field with the strongest ubiquity and the strongest disturbance to the robustness of HETCN. In the era of Industry 4.0, information technology drives the development of the manufacturing industry, and the manufacturing industry is deeply integrated into information technology and becomes a part of the information industry. EIT, as the most basic and necessary technical form in various fields, plays an important role in the robustness of the entire HETCN.

However, NAT, as a technical field with strong narrowness, has a strong disturbance to the robustness of HETCN, and this characteristic is not found in different parks. Since technological diversity is the premise of technological integration and diffusion, and also a necessary condition to achieve innovation, this paper makes certain conjectures. The reason why NAT has strong narrowness and strong nesting perturbation is that the strong professionalism makes it get rid of dependence and become an indispensable technical feature.

6 Conclusion

This paper takes Z-Park as an example to construct a network model of enterprise technology cooperation. Based on the two keywords of "affiliated park area" and "affiliated technical field" in the patent information, this paper uses the nesting structure as the evaluation index of the robustness of the technical cooperation network. On this basis, this paper proposed NDI to identify the key parks and technical fields with the strongest disturbance to the robustness of the technical cooperation network. The conclusions of the study are as followed.

With the increase in technical cooperation between ubiquitous and narrow-biological entities, technological resources spread from the core of the network to the periphery, and the technical relationship is more stable. Due to the scarcity of technical resources, there is competition for resources among innovation entities, and their ubiquity existence is one after another. However, the ubiquity of the main body has not changed much in a short period. In addition, due to the wide range of technical associations, the ubiquitous entities are easy to conduct risks and have a strong disturbing effect on the robustness of the technical cooperation network. While narrow-biological entities, due to the small scope of technological diffusion and the peripheral dependence of the network, has weak risk transmission ability, and will not disturb the robustness of the technical cooperation network.

Acknowledgements. This research was funded by the National Natural Science Foundation of China (Grant No. 71971006) and the 2021 High-level Technology Innovation Think Tank Youth Project (Project No. 2021ZZZLFZB1207016).

References

1. Meng, J., Qiu, C.B., Zuo, J.Y.: Research on the evolution of patent collaboration network of domestic institutions in China: based on the patent data of CNKI from 1999 to 2018. Inf. Stud: Theory Appl. **44**(12), 48–54 (2021)
2. Sun, Z.R., Fan, J., Sun, Y.: Influence of cooperation network evolution characteristics on innovation performance of scientific research institutions: taking Chinese academy of sciences as an example. Sci. Technol. Manage. Res. **41**(18), 131–139 (2021)
3. Jiao, M.Q., Du, D.B., Gui, Q.C., Hou, C.G.: The Spatio-temporal evolution and influencing factors of urban technical corporation networks: from the perspective of belt and road. Geogr. Res. **40**(04), 913–927 (2021)
4. Mariani, M.S., Ren, Z.M., Bascompte, J., Tessone, C.J.: Nestedness in complex networks: observation, emergence, and implications. Phys. Rep. **813**, 1–90 (2019)
5. Manville, R.H., Darlington, P.J.: Zoogeography: The Geographical Distribution of Animals. Wiley, Hoboken (1957)
6. Patterson, B.D., Atmar, W.: Nested subsets and the structure of insular mammalian faunas and archipelagos. Bot. J. Linn. Soc. **28**(1–2), 65–82 (1986)

7. Dardanelli, S., Bellis, M.L.: Nestedness structure of bird assemblages in a fragmented forest in Central Argentina: the role of selective extinction and colonization processes. Anim. Biodivers Conserv. **44**(01), 17–29 (2021)
8. Fontaine, C.: Ecology: abundant equals nested. Nature **500**(7463), 411–412 (2013)
9. Brintrup, A., Barros, J., Tiwari, A.: The nested structure of emergent supply networks. IEEE Syst. J. **12**(02), 1803–1812 (2018)
10. Ren, Z.M., Zeng, A., Zhang, Y.C.: Bridging nestedness and economic complexity in multilayer world trade networks. Palgrave Commun. **7**(01), 1–8 (2020)
11. Leskovec, J., Lang, K.J., Dasgupta, A., Mahoney, M.W.: Community structure in large networks: natural cluster sizes and the absence of large well-defined clusters. Internet Math. **6**(01), 29–123 (2008)
12. Medan, D., Perazzo, R., Devoto, M., Burgos, E.: Analysis and assembling of network structure in mutualistic systems. J. Theor. Biol. **246**(03), 510–521 (2007)
13. König, M.D., Tessone, C.J.: Network evolution based on centrality. Phys. Rev. E. **84**(05), 056108 (2011)
14. Sun, T.Y., Cheng, L.H.: Research on the structure characteristics and evolutionary pattern collaborative innovation network in China. Stud. Sci. Sci. **37**(08), 1498–1505 (2019)
15. Lin, L., Xu, J.Y., Xing, L.Z., Guan, J.: Analysis on the topological structure of inter-enterprise technology transfer network: taking Z-park as an example. Int. J. Mod. Phys. B. **34**(11), 28 (2020)
16. Ren, J.Q., Xing, L.Z., Han, Y., Dong, X.L.: Nestedness-based measurement of evolutionarily stable equilibrium of global production system. Entropy **23**(8), 1–28 (2021)
17. Zhuang, T.: Regional collaborative innovation relationship of industry-university-research institution and space-time evolution—based on evidence of patent cooperation. J. Tech. Econ. Manage. **10**, 123–128 (2020)

Dynamic Transition Graph for Estimating the Predictability of Financial and Economical Processes

Dmitriy Prusskiy, Anton Kovantsev(✉), and Petr Chunaev

ITMO University, 16 Birzhevaya Lane, 199034 Saint Petersburg, Russia
dimaprusskii@mail.ru, ankovantcev@itmo.ru, chunaev@itmo.ru

Abstract. The problem of time series predictability estimation often appears when one deals with the task of forecasting financial and economical processes, especially when the processes are not sustainable and presume critical transitions and/or behaviour changes in the generating complex dynamical system. In these cases it is important to notice the moment when transitions/changes start and to distinguish their direction as soon as possible in order to adjust the forecasting algorithm or, at least, properly evaluate the forecast accuracy. To deal with such effects, we propose a dynamic transition graph-based method for real-time incremental tracing of the changes in the predictability of time series describing financial and economical processes. Our method helps to filter some "noise" time series information and emphasize the significant aspects of the corresponding dynamical system behavior. Besides, we use several graph features such as centrality degree, number and size of loops, connectivity and entropy to evaluate the predictability. We also construct a graph neural network classifier and train it on specific artificial time series datasets to efficiently classify real-world time series by predictability in a real-time incremental tracing manner.

Keywords: Time series · Predictability · Network theory · Transition graphs

1 Introduction

Time series analysis and forecasting is an important task for many industries. Previously, this problem was solved mainly using statistical methods. Recently, machine learning methods have been increasingly used. Machine learning engineers spend a lot of time and computational resources to predict large numbers of time series. To optimize the costs of solving this problem, it makes sense to evaluate the predictability of a new time series before the moment of model building and forecasting.

Now there are methods for assessing the predictability of a time series using its internal features [1–3] and by constructing various graphs from the series and

H. Cherifi et al. (Eds.): COMPLEX NETWORKS 2016 2022, SCI 1077, pp. 478–489, 2023.
https://doi.org/10.1007/978-3-031-21127-0_39

evaluating the characteristics of these graphs [4–6]. However, the existing methods do not consider the volatile nature of time series and thus are not applicable when the processes are not sustainable and presume critical transitions and/or behaviour changes in the generating complex dynamical system. For example, the novel coronavirus pandemic has dramatically altered the nature of many time series associated with financial and economical processes. In these cases it is important to notice the moment when transitions/changes start and to distinguish their direction as soon as possible in order to adjust the forecasting algorithm or, at least, properly evaluate the forecast accuracy.

To deal with such effects, we propose in this paper a dynamic transition graph-based method for real-time incremental tracing of the changes in the predictability of time series describing financial and economical processes. Our method helps to filter some "noise" time series information and emphasize the significant aspects of the corresponding dynamical system behavior. Besides, we use several graph features such as centrality degree, number and size of loops, connectivity and entropy to evaluate the predictability. We also construct a graph neural network classifier and train it on specific artificial time series datasets to efficiently classify real-world time series by predictability in a real-time incremental tracing manner.

The paper is organized as follows. Section 2 is about the related works on the topic. Section 3 presents the outline of our research, Sects. 4 and 5 describe the process of graph construction from a time series and classification model which are mentioned in Sect. 3. Section 6 is about data we use in the study. In Sect. 7 we present the whole process of usage dynamic transition graph to classify the intervals of time series into predictability classes. Section 8 contains conclusions and our thoughts about future research.

The code, datasets, and experimental results are available on GitHub[1].

2 Related Works

When dealing with predictability concept, we should define what the predictability of time series is. In the study [1], the authors propose to divide the concept of predictability of a time series into two types: intrinsic predictability and realized predictability. In our work, by the predictability of a time series, we will mean the intrinsic predictability. It is defined as the maximum possible predictability that can be achieved regardless of the model. The problem with using internal predictability is that it cannot be measured, so only estimates can be used. On the other hand, the advantage of internal predictability is that it reflects the intrinsic nature of the time series and can be useful in any time series related problem.

There are many works devoted to assessing time series predictability before the moment of forecast. The main way to estimate time series predictability is to use their internal characteristics. The authors of the study [2] show that

[1] https://github.com/Anthony-Cov/Predictability-with-dynamic-graph.

the weighted entropy of permutations has a positive correlation with the forecast error. In another study [3], authors divide the time series into two clusters of different predictability using the correlation dimension, correlation entropy, Kolmogorov-Sinai entropy, Hurst exponent, noise and random walk measures invented by the authors. In one modern study [7] authors split transactional time series into predictability classes as we do in our work, but they use LSTM neural network.

In other works authors build graphs from time series and use graph characteristics, translating the problem of assessing predictability into the field of network theory. There are works that use various representations of time series in the form of graphs. The paper [4] uses visibility graphs to separate time series into randomness and deterministic. In another paper [5], authors build an ordinal partition transition graph and use the probability of self-transition loops to estimate the degree of randomness of the original time series.

Our way of time series transforming to the transition graph is based on the formulation of Takens' theorem which allows reconstructing the state space of a dynamical system by the time series produced by this series. This approach of state space clustering applied, for instance, in [6] for predictability analysis.

In our study we use graph neural network to classify graphs by predictability of its parent time series. A good example of using the graph neural network for graph classification problem is given by the authors of [8]. The dynamic approach which we wish to realize is reflected in [9] devoted to deep learning on spatio-temporal graphs. The experience of motion recognition by means of graph neural network, presented in [10] could also be useful for applying to time series analysis in our study as well as the theoretical analysis of social network graphs over time made by the authors of [11]. The last two papers do not concern the tasks of the time series, but nevertheless, we pay them our attention as they both are dealing with the issues of graphs and complicated dynamical systems.

3 Research Outline

We are going to find out, if it is possible and fruitful to use a time series graph transformation for predictability classification by means of graph neural network (GNN) in dynamic mode with incremental learning method. The GNN classifier trained in static on graphs made out of the set of artificial series of wide range predictability variations should be able to recognise the series or their parts which shows different forecasting quality for several forecasting models. If it is true then we could make a dynamic graph for a real-world time series and evaluate its predictability class on each step of the incremental process which could be useful for transitions or drift indication in real-world problems. Full experiment scheme is shown in Fig. 1

As a result we plan to show how the time series forecasting quality correlates with the predictability class estimated by the classifier on different pieces of the series which includes transition process or data drift.

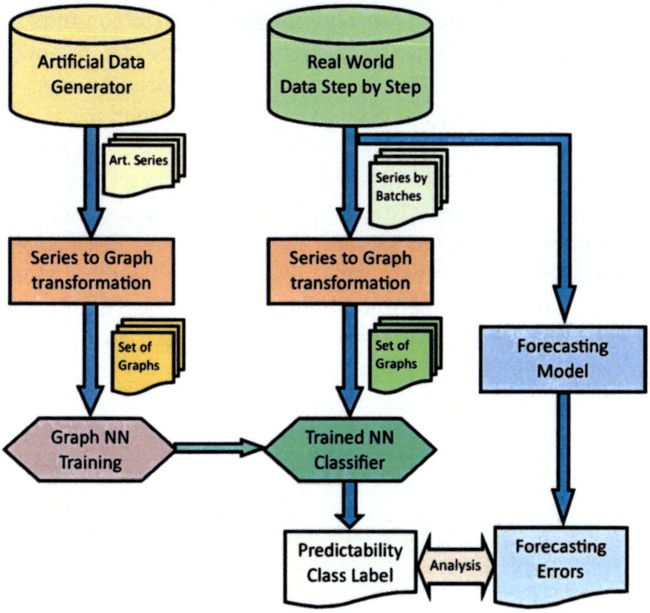

Fig. 1. Full scheme of experiment.

4 Dynamic Transition Graph Construction

To assess the predictability of time series, we use their representation in the form of a transition graph. As is known, graph models help to filter noise information and emphasize the significant aspects in the system behavior. It can also provide an explainable picture of the process and give an opportunity for making a forecast in the transition.

Our way to constructing transition graph is based on the Takens theorem and presented in [6]. We consider the time series as a result of some dynamical system evolution, so, due to Takens theorem we can reconstruct the trajectory of this evolution using the sequential fractions of the time series values. These fractions, which become the columns in our delay matrix, are the points of the system trajectory in its state space (1).

$$\{x_1, x_2, x_3, \ldots, x_N\} \rightarrow \mathrm{X}_{p\times(N-p+1)} = \begin{pmatrix} x_p & \cdots & x_N \\ \vdots & \ddots & \vdots \\ x_1 & \ldots & x_{N-p+1} \end{pmatrix}, \qquad (1)$$

where N is the time series length and $p = 2m+1$, where m is state space dimension size, which we estimate using Grassberger-Procaccia algorithm. The dimension of the matrix columns p is chosen according to the weak Whitney embedding theorem which states that any continuous function from an m-dimensional manifold to an p-dimensional manifold may be approximated by a smooth embedding provided $p > 2m$.

To build a transition graph, we divide delay matrix columns into clusters. We applied the hierarchical clustering algorithm with the "elbow" method to determine the number of clusters, which was automated to stop clustering at the point where the distances between the clusters occur to be less than 0.01 of the maximal distance between the state space points. It helps to get a reasonable number of clusters and hence the number of vertices in the graph for various time series. The time series flow, as well as the sequence of the matrix columns, determines the graph edges and their directions. Each edge of the graph is assigned its weight—the probability of transition between clusters. So, we get a fully connected directed and weighted graph. As the number of the edges strongly depends on the series properties, it can be dense or sparse. The graph construction process is shown in Fig. 2.

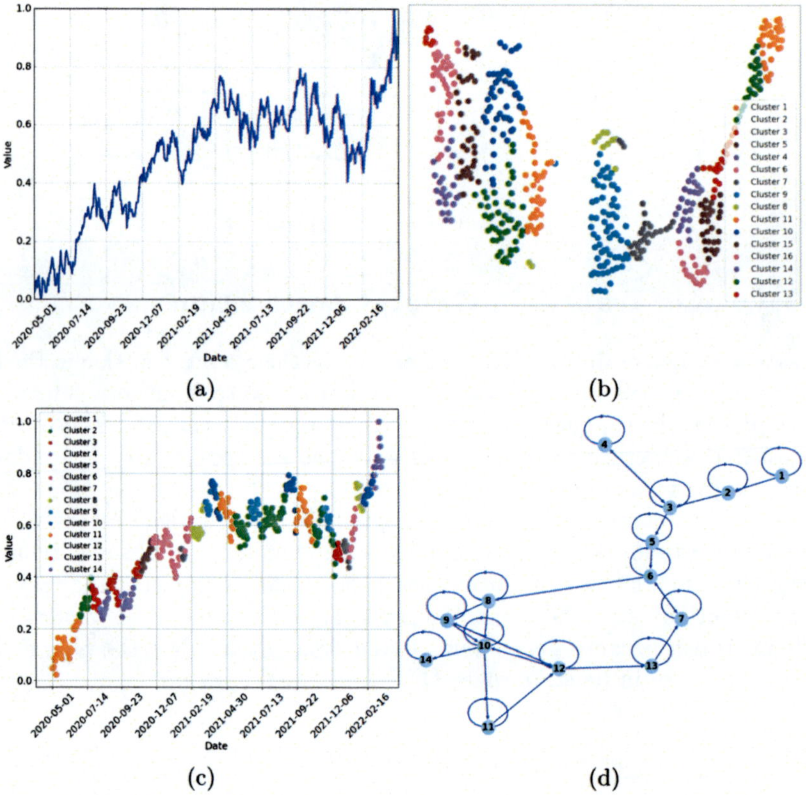

Fig. 2. Building a graph out of a time series by the state space reconstruction. *Initial series* **a**, *Phase trajectory in TSNE representation* **b**, *State space points matching the series observations* **c**, *Constructed graph* **d**.

To build a dynamic transition graph, we move along the time series with a rolling window and update the delay matrix. We extend the initial delay matrix

Algorithm 1 Dynamic transition graph construction algorithm

Require: $\{x_t\}_{t=1}^{N}$ — time series, N_{init} — steps for first graph, N_{rec} — steps for each graph iteration, p — embedding dimension, *dist* — minimal distance to add new cluster, *max_steps* — number of steps, after which the cluster is deleted
Construct cluster trajectory and initial graph from $\{x_t\}_{t=1}^{N_{init}}$
Calculate cluster centers
$step \leftarrow N_{init}$
while $step \leq N$ **do**
  Add new column $[x_{step-p+1}, \ldots, x_{step}]$ to delay matrix
  Calculate distances between new column and cluster centers
  if minimal distance is less than *dist* **then**
    Add new data point to the nearest cluster
  else
    Add new cluster and associate new data point with it
  end if
  if some of clusters have not been visited more than *max_steps* **then**
    Delete these clusters
  end if
  Update cluster trajectory
  if $step - N_{init}$ is divisible by N_{rec} **then** ▷ Every N_{rec} steps of the cycle
    Construct new graph from current cluster trajectory
  end if
  $step \leftarrow step + 1$
end while

with new columns, which are new points of the phase trajectory. We refer these points to the closest clusters or add a new cluster consisting of new points if the distance from the new points to any cluster is greater than D_{max}. In addition, at each step, we remove graph vertices that have not been visited for a long time. Therefore, while moving along the time series we obtain a sequence of transition graphs, which we call dynamic transition graph. Pseudo-code algorithm of dynamic transition graph construction is shown in Algorithm 1.

5 Predictability Classifier

In order to obtain a model for predictability classification in this study we use 1000 artificially generated time series of length 500 to train a model. All time series are labeled with three different predictability classes. All time series are composed from fully periodic series and random walk process generated from Bernoulli trails:

$$\begin{aligned} Y_i &= (1-\gamma_i)Y_p + \gamma_i Y_r, \\ \gamma_i &\in [0,1]; \quad i = 1, 2, \ldots, N, \end{aligned} \tag{2}$$

where Y_p and Y_r are fully periodic and random walk time series, respectively, and γ_i is random walk weight coefficient for series i. N is the number of series in the artificial data set. All time series are scaled, so maximum and minimum values are 1 and 0, respectively.

Time series of three classes differ from each other by random walk process coefficient in resulting time series. Examples of time series of different predictability classes are in Fig. 3. For the first attempt we divide the set of the artificial series into three groups according to their intrinsic (expected) predictability: the best, the middle and the worst. These are the three classes with the labels for training and evaluating our classifier. In fact, the series are labeled according to the number of series. In the range from 000 to 300—class '0' (determined series, the best predictability), from 301 to 700—class '1', from 701—class '2' (random walking, the worst predictability). We use 80% of these time series to train a model and 20%—to validate the model.

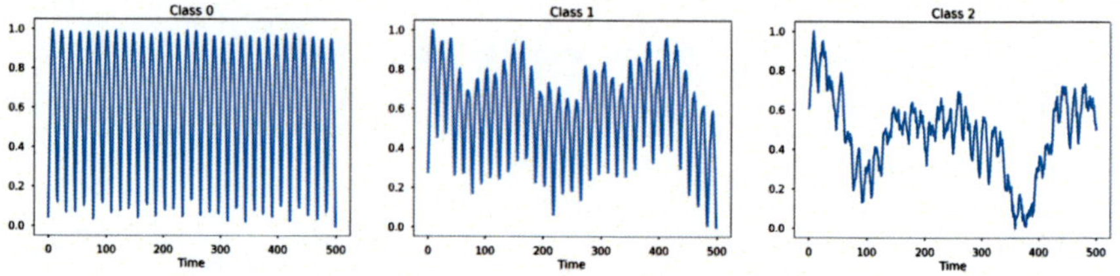

Fig. 3. Examples of artificial time series of different predictability classes.

We use these set of time series to fit a classifier which could help to distinguish the series of different predictability. The classifier is tuned for 3 classes of predictability. The graph for each of the series is constructed with Networkx and imported into Torch Geometric with the node features which are the number of state space trajectory points divided to all available points, the mean value of the series in the matching node and centrality degree of a node. These features occurred to provide the most assertive impact on the classifier quality.

The graph neural network (GNN) consists of three convolutional layers for node embeddings, a readout layer of global mean pooling and a final classifier with $\frac{1}{2}$ dropout and linear transform for the result.The GNN architecture and parameters were chosen in the experimental studies to achieve the sufficient quality of the classifier. Trained on $\frac{4}{5}$ of the set with Adam optimizer and cross entropy as a loss function through 100 epochs it shows 80–90% accuracy for the test set.

6 Real-world Data Description and Pre-processing

To evaluate the model on real data we use 87 economical and financial time series from website[2]. All time series are about economic or financial sector of USA with weekly frequencies. Time series have completely different nature. Among them there are periodic, monotonically increasing and series that look like random walk. Half of all time series reflect the crisis situation at the beginning of 2020.

[2] https://fred.stlouisfed.org/.

We use 620 last values of time series only. In order to remove outliers and "white" noise we use moving average smoothing with window width of 7 observations, and then rescale time series to (0,1) range. Examples of real time series used in experiment are shown on Fig. 4.

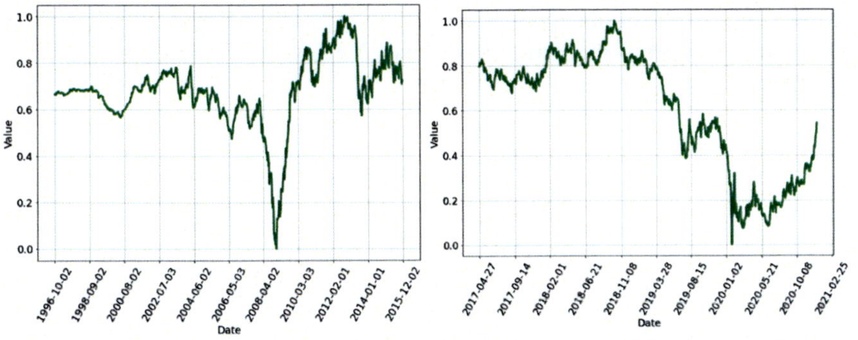

(a) (b)

Fig. 4. Examples of rescaled real economical time series used in the experiment. *Market Yield on U.S. Net unrealized gains on available-for-sale securities, large domestically chartered commercial banks* **a** ,*Treasury securities at 30-year constant maturity* **b**.

7 Experimental Results and Discussion

To assess the quality of the model on time series with varying predictability, artificial time series were generated, starting with a fully periodic time series, gradually turning into a random walk according to equation (2). The weight of periodic time series decreases over time as sigmoid function (3)

$$y_i = \alpha_i y_i^p + (1-\alpha_i) y_i^r, \\ \alpha_i = \frac{1}{1+\exp(t_i)}; \qquad t_i \in [-6,6]; \qquad i = 1,2,\ldots,L, \tag{3}$$

where y_i^p and y_i^r are elements of periodic and random walking components respectively, L is the length of the new artificial series. The range of t_i is chosen to make the weight α_1 close to 1 (0.998) and α_L to 0 (0.002) and at the same time to keep the variety of the values along the transition. The example of resulting time series is in Fig. 5. To determine the predictability of a time series, we formulate the task of classifying transition graphs built on time series. We consider three classes enough to test our hypothesis and train the classifier which is described in Sect. 5.

To use this model for time series with varying predictability, we build dynamic transition graph. Each time series is thus converted into a sequence of transition graphs. Each graph in this sequence is classified using a pre-trained model. As a result, we get a class label at each interval of the time series.

To evaluate the results, we analyze the relationship of the time series forecasting error at each step with the class label predicted by the model. For time series forecasting, we use the Multivariate Singular Spectrum Analysis (MSSA) model.

The result of the model evaluation is a set of predicted labels of predictability classes on a time series divided into intervals. Examples of such results are shown in Figs. 5 and 6.

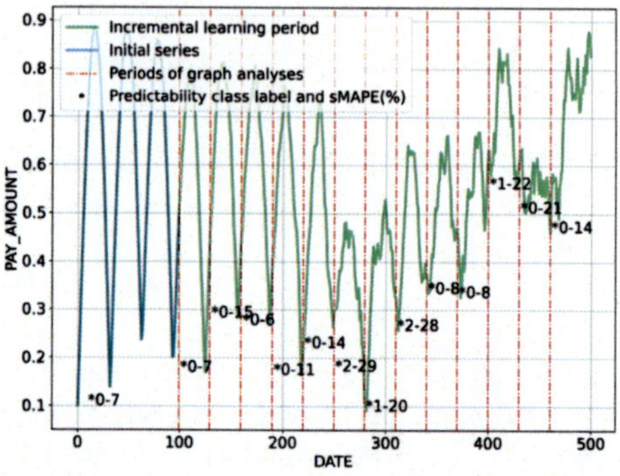

Fig. 5. The example of the series with varying predictability and the result of the model evaluation on it.

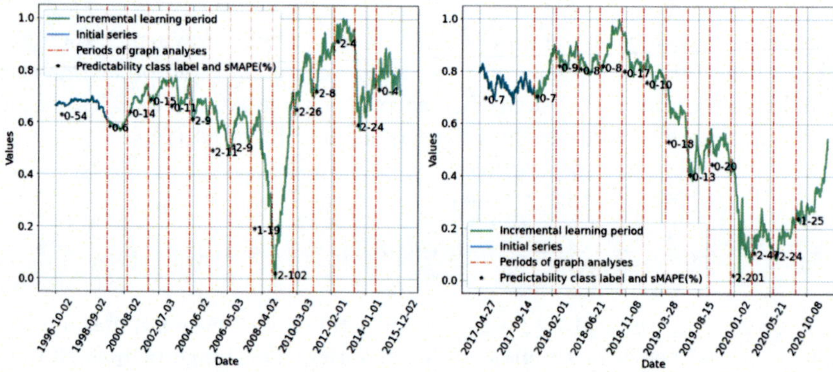

Fig. 6. The results of the model evaluation on real time series (Those shown in Fig. 4).

To check the quality of the model, we look at the relationship of predicted labels with forecast errors in the corresponding sections of the time series. We check the quality on real economic and financial time series from the website[3]. We

[3] https://fred.stlouisfed.org/.

use 30 time series of length 620. Each time series is divided into 21 intervals. The first interval of length 120 is to build the initial transition graph. The remaining 20 intervals of length 25. Dynamic transition graph is built with a step of 25. Trained on time series with static predictability, the model classifies each graph in a dynamic transition graph. As a result, all intervals are divided into three classes of predictability. The distributions of forecast errors in these classes of predictability are shown in Fig. 7.

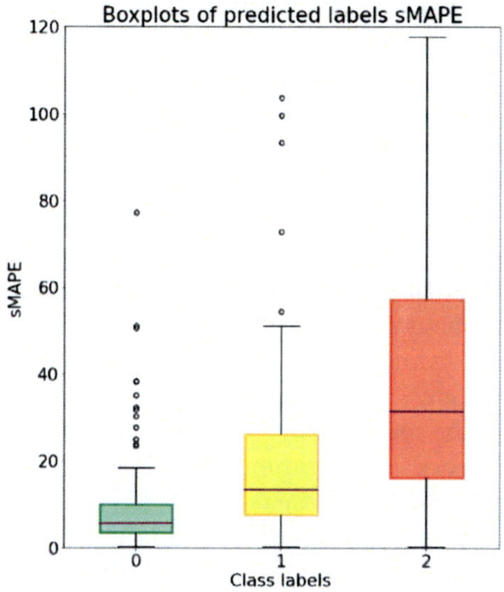

Fig. 7. Distribution of sMAPE by predictability classes for artificial series.

Table 1. Mann-Whitney U-test result for classification results.

Classes	U-score	p-value
0 and 1	6427	$7.47 \cdot 10^{-14}$
1 and 2	15500	$6.94 \cdot 10^{-17}$

The approach which we used to create our artificial data-set makes us expect that the symmetric mean absolute percentage error (sMAPE) of the forecasting refers to the intrinsic series predictability and hence to the labels which our classifier appoints to them. So, the mean sMAPE values of different classes should be statistically different and we can use the Mann-Whitney U test for the classifier quality evaluation. As a quantitative feature, we take the forecast error. The test results are presented in Table 1. Differences in the samples are statistically significant and Fig. 7 shows that the classifier separates the samples well in terms of predictability.

8 Conclusions and Future Work

As a result of our study, we have obtained a model that can track the predictability of time series real-time by converting the series into a graph and a graphical neural network classifier. The way we make graphs out of the time series gives the model of a dynamical system trajectory in its state space and helps to identify the patterns and tendencies in its movement. Besides, the graph models, as we know, help to filter some "noise" information and emphasize the significant aspects in the system behavior. It can also provide an explainable picture of the process and give some opportunity for making a forecast in the transition.

Our model, trained on the set of artificial time series, is able to deal with real-world series as well. We trained the classifier to recognize three classes of predictability: 0—the most predictable series, 1—medium, 2—the least predictable. The number of classes depends on real-world problems, which we deal with, and desired accuracy of the classifier. The more classes, the less accuracy due to greater similarity between the series in the neighboring classes.

In the process of real-time data appropriation the model rebuilds the graph and predicts the series behavior according to the predictability labels. This prediction could be used for forecasting model choice or its fine tuning, or for the forecast quality evaluation.

Thus, we combine the methods of time-series dynamic graph representation and graph neural network classification to estimate how the time series predictability changes in the dynamic mode.

For our future work we are intended to enhance the range of different kinds of "difficulties" in the artificial data set to see how the classifier performs. For instance, we can add the random walk series with normal or Cauchy distribution for the increments, or mix the data with the noise of various shapes and distributions. Experiments with the different numbers of predictability classes could also be interesting, as well as the using of some other graph parameters as the features for the classification model.

Acknowledgements. This research is financially supported by the Russian Science Foundation, Agreement 17-71-30029, with co-financing of Bank Saint Petersburg, Russia.

References

1. Pennekamp, F., Iles, A., Garland, J., Brennan, G., Brose, U., Gaedke, U., Jacob, U., Kratina, P., Matthews, B., Munch, S., Novak, M., Palamara, G., Rall, B., Rosenbaum, B., Tabi, A., Ward, C., Williams, R., Ye, H., Petchey, O.: The intrinsic predictability of ecological time series and its potential to guide forecasting. Ecol. Monogr. **89** (May 2019)
2. Garland, J., James, R., Bradley, E.: Model-free quantification of time-series predictability. Phys. Rev. E. **90**, 052910 (Nov 2014)
3. Kovantsev, A., Gladilin, P.: Analysis of multivariate time series predictability based on their features. Int. Conf. Data Min. Workshops (ICDMW) **2020**, 348–355 (2020)

4. Wang, M., Vilela, A., Du, R., Zhao, L., Dong, G., Tian, L., Stanley, H.: Exact results of the limited penetrable horizontal visibility graph associated to random time series and its application. Sci. Rep. **8**, 5130 (2018)
5. Borges, J., Ramos, H., Mini, R., Rosso, O., Frery, A., Loureiro, A.: Learning and distinguishing time series dynamics via ordinal patterns transition graphs. Appl. Math. Comput. **362**, 124554 (2019)
6. Kovantsev, A., Chunaev, P., Bochenina, K.: Evaluating time series predictability via transition graph analysis. Int. Conf. Data Min. Workshops (ICDMW) **2021**, 1039–1046 (2021)
7. Stavinova, E., Bochenina, K., Chunaev, P.: Forecasting of foreign trips by transactional data: a comparative study. Procedia Comput. Sci. 225–34 (2021)
8. Xu, K., Hu, W., Leskovec, J., Jegelka, S.: How powerful are graph neural networks?. In: International Conference On Learning Representations (2019). https://openreview.net/forum?id=ryGs6iA5Km
9. Jain, A., Zamir, A., Savarese, S., Saxena, A.: Structural-RNN: Deep Learning on Spatio-Temporal Graphs (Nov 2015)
10. Nicolicioiu, A., Duta, I., Leordeanu, M.: Recurrent space-time graph neural networks. In: Proceedings Of The 33rd International Conference On Neural Information Processing Systems (2019)
11. Leskovec, J., Kleinberg, J., Faloutsos, C.: Graphs over time: densification laws, shrinking diameters and possible explanations. In: Proceedings Of The Eleventh ACM SIGKDD International Conference On Knowledge Discovery In Data Mining (2005). https://doi.org/10.1145/1081870.1081893

A Network Analysis of World Trade Structural Changes (1996–2019)

Vu Phuong Hoang[1], Carlo Piccardi[2(✉)], and Lucia Tajoli[1]

[1] Department of Management, Economics, and Industrial Engineering, Politecnico di Milano, Milan, Italy
[2] Department of Electronics, Information, and Bioengineering, Politecnico di Milano, Milan, Italy
carlo.piccardi@polimi.it

Abstract. Global trade suffered a significant contraction in the value of trade flows as a result of 2008 financial crisis, and again in 2020 with the pandemic crisis. There are questions arising from these observations: What is the structure of international trade? How has it changed due to these shocks? Inspired by the advantages of the network method and using the most extensive coverage data in terms of time and geography, we explore the structure of international trade by presenting a comprehensive analysis of the World Trade Network (WTN) from several angles. Connectivity results suggest that countries' efforts towards multilateral trade relations have resulted in an increasingly dense network, highly reciprocal and clustered. However, the network has not been fully connected yet. While trade connections are distributed homogeneously among countries, trade value is concentrated in a small set of countries yielding a weighted core-periphery structure. We analyze the consequences of the past financial crisis of 2008 to infer the potential effects of the recent crisis. Although the shock did not affect the main connective overall trends, their tendencies were restrained after 2008. The financial downturn also marks a turning point in the clustering of the WTN from two main groups, led by the United States and Germany, to three, led by the United States, China, and Germany. Revisions in preferential trade partners are visible from 2008 onward. Our study provides an intuitive insight for policymakers in revising the medium term effects of a global shock.

Keywords: World trade · Network analysis · Communities

1 Introduction

International economics plays a vital role in the development at the level of nations, regions, and globally, and it has experienced crucial changes. For decades, world trade has increased dramatically and trade has been indicated as one of the engines of economic growth. For example, some developing economies in Asia have experienced a notable increase in imports and exports [16], together

H. Cherifi et al. (Eds.): COMPLEX NETWORKS 2016 2022, SCI 1077, pp. 490–501, 2023.
https://doi.org/10.1007/978-3-031-21127-0_40

with GDP growth rates much higher than the world average. Such a spectacular trend has been achieved not only intensively (i.e., through increases in trade flows between countries already trading in the past by an increase in the strength of countries' links) but also extensively (i.e., newly created trade links with an increase in the trade network density) [7].

After a decade of accelerated growth, since 2008 world trade was hit by two major shocks in a relatively short time span. The effects in terms of the overall value of trade flows and trend are visible in global statistics, but did such shocks changed the structural characteristics of globalization? This is the question we want to address in this paper by considering the world trade structure before and after the great financial crisis of 2008, that caused the largest downturn in trade flows since the end of World War II. We still have too little evidence to fully understand the consequences on trade of the much more recent pandemic shock, but we try to infer some possible effects from the previous experience.

The global financial crisis has had far-reaching repercussions on cross-border economic activity. Most literature agreed that the situation hurt economic growth in terms both of GDP and slowdown of the trade flow. For instance, after a sharp and sudden collapse in international trade in the last quarter of 2008, world trade flows declined by about 12% in 2009 [16]. This exceeded the estimated loss of 5.4% in world GDP [15]. The contraction in exports was especially acute for small open economies, several of which saw their trade volumes in the second half of 2008 fall by up to 30% year-on-year.

Since many previous contributions on the world trade system relied on data collected before 2010, the structural change of the World Trade Network (WTN) after 2008 was not thoroughly investigated. Also, analyses on more detailed levels (e.g. the mesoscale network structure) are limited. Given this context, studying the structure of WTN, its dynamics, and how it changed after the financial crisis is an exciting research question from our perspective. This strand of research not only contributes to our understanding of how the financial crisis affects the "shape" of international economic relations in both an intensive (trade value) and extensive manner (trade link), but also provides useful reference information for potential next crisis (for example the Covid-19 pandemic). Our paper applies network analysis to answer this question and contributes to the literature on two fronts: data use and methodological design. As for the former point, we consider an extended coverage of data (1996–2019) which allows us to document the evolution of WTN, including the pre- and post- financial crisis periods. Furthermore, unlike most existing literature [8,11,14], we do not filter out trade relations by their value nor weight them by the size of the exporter or importer, and we take into account all possible countries reported (206 countries). Considering a larger number of countries allows us to preserve the network's features and provides more accurate analyses of the international trade topological properties. As for the methods, we perform analysis at two levels of detail, to provide a throughout vision of WTN's structure and its dynamic: the global network, by visualizing the overall WTN and analyzing crucial topological indicators; and the mesoscale structure, by working on communities, Minimal Spanning Tree (MST), and core-periphery structures, to analyze the patterns of connectivity.

2 Methodology and Data

In the case of the WTN, the nodes are the countries and connections model the flows of goods from one country to another. Since the existence of exports from country A to country B does not imply exports from B to A, the WTN is portrayed as a directed graph. Also, even when exports are bidirectional, their values are in general different, thus a weighted directed graph is the most convenient representation. Since exports from a country to itself are not considered, the graph does not contain self-edges. If n is the number of countries, the structure of the WTN is described by the $n \times n$ adjacency matrix $A = [A_{ij}]$, with $A_{ij} = 1$ if there is an edge from i to j, and $A_{ij} = 0$ otherwise. Besides, the weighted adjacency matrix (or weight matrix) $W = [W_{ij}]$, with $W_{ij} > 0$ if and only if $A_{ij} = 1$, contains the monetary value of the export from i to j.

We construct the WTN using the BACI-CEPII data set built from data directly reported by each country to the United Nations Statistical Division (Comtrade). Two countries are considered to have a trade connection if there is a link between them in any of the about 5300 commodity sectors, and the total trade value is the aggregate of all sector values. The original dataset provides yearly data from 1996 to 2019. For our analysis, we convert them into biennial periods (i.e., 1996–1997, 1998–1999, ..., 2018–2019) by averaging two years accordingly. Using biennial periods halves the analytical burden while preserving the dynamics of world trade [4], and enables to maximize the number of links between countries. To avoid potential bias and make comparisons between periods straightforward, we keep the network size (i.e., the number of countries) constant by discarding those countries for which data is unavailable in any period of the dataset. After such a pre-processing, our sample contains 206 countries in 12 biennial periods.

3 Results

Figure 1 displays two biennial networks corresponding to the WTN in 1996–1997 and 2018–2019, the two extremes of the time interval under study. The figure reveals that a few countries concentrate most of the total trade value in all periods. Indeed, the same figure (bottom panel) shows that for some countries the relative strength (trade intensive margin) remained fairly constant over the entire period, while for others there was a remarkable change. Specifically, the United States (USA), Germany (DEU), and Japan (JPN) were the dominant trading nations in the first four biennial periods, whereas China (CHN) clearly emerges for the remaining periods: it has the eighth position in the total strength ranking in 1996–1997, but it becomes first since 2014–2015 by surpassing the United States.

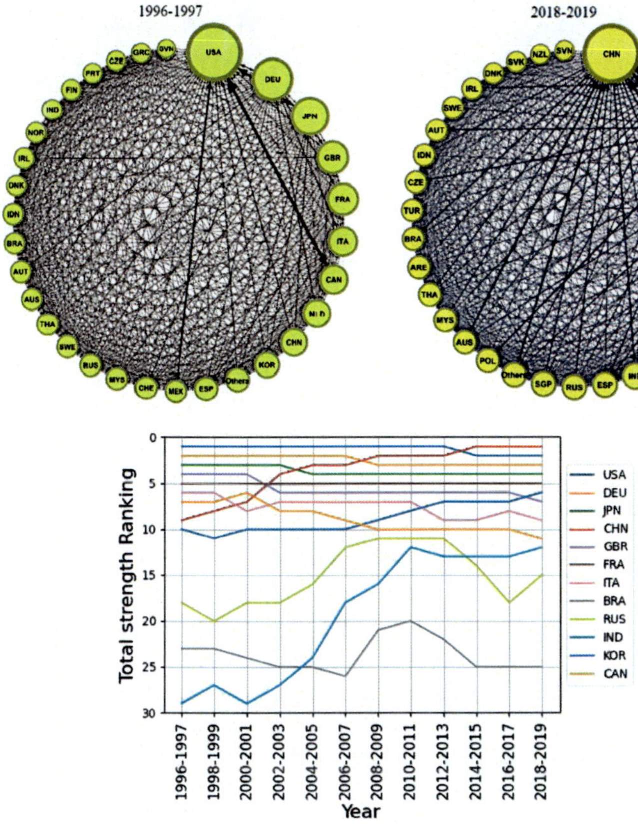

Fig. 1. Above: World Trade Network in 1996–1997 and 2018–2019. The diameter of each node corresponds to the magnitude of total strength (import + export) in each biennial period (only 15% of countries with the largest total strength are depicted for readability). Below: Ranking position for total strength (import + export) of a few selected countries.

Figure 2 displays the time evolution of a pool of network metrics for the WTN. Concerning how cohesively countries are connected, the figure shows that density is generally very large, with an average of 0.64 across the entire time span. Our findings agree with the literature about the overall increase (decrease) of density (mean geodesic distance) in 1996–2010, when density (trade extensive margin) increased consistently, but from 2010 to 2019 density changed slightly without a clear trend. Indeed, in the evolution of many topological indicators over time, a change is visible after 2008 (highlighted by a vertical line): for example, density stops increasing, as well as reciprocity, and centralization stops declining. This newly observed generalized change in the trend of topological indicators suggests a possible new evolution of the WTN after 2008.

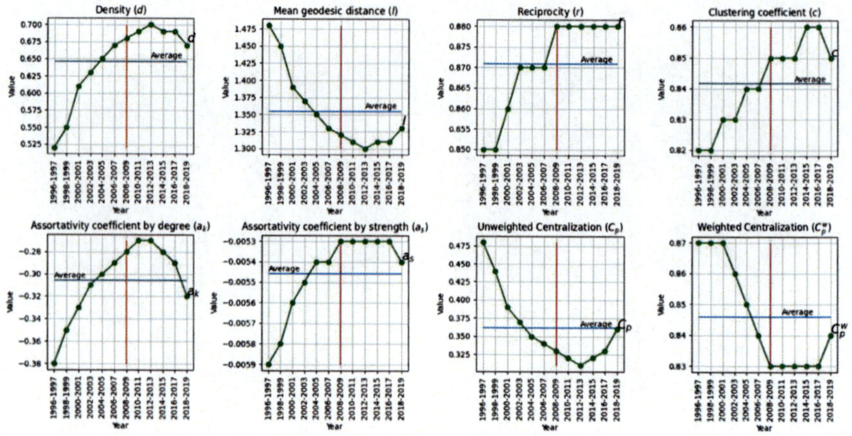

Fig. 2. Time evolution of a pool of topological metrics for the WTN. The vertical red bar marks the 2008 financial crisis.

Results in Fig. 2 also show that the relations in the WTN are reciprocal, with more than 8 out of 10 trade relations being bidirectional. From 1996 to 2019, reciprocity increased from 0.85 to 0.88, and this trend matches that of density. Likewise, the WTN is highly clustered with an average value of 0.84. This level of clustering suggests that it is very likely to find transitive relations (i.e., triads) among countries, and this likelihood has increased parallel to the increase in density; as new relations were built over time, new triads of trade partners were developed. This may be explained by immense world trade openness or new bilateral and multilateral trade agreements.

The evidence of negative assortative mixing by degree (i.e. disassortativity) reflects that countries with dissimilar numbers of connections may connect to each other. However, their correlations are relatively weak (−0.31 on average) and experienced an overall decrease in magnitude (from 0.38 to 0.32) toward uncorrelation: countries with fewer connections receiving more trade links may lead to a decreasing trend of correlation.

The assortativity mixing coefficient by strength is negative and close to zero. As agreed with existing contributions, there is no clear connective pattern driven by the intensity of countries' strength, which means that countries search for trading partners irrespective of their contribution to the total value of export. Again, it is arguable that the extensive trade margin prevails on the intensive one, as an increase in density drives this result: most countries maintaining a high number of trading partners should break any tendency to establish connections based on the strength of countries. Export diversification aims at increasing the number of trading partners to avoid concentrating trading relationships.

As depicted in Fig. 3 (top panels), the expectation that WTN may represent a real-world network with only a few heavily connected countries and many sparsely connected countries is unsatisfactory. On the contrary, the intensity—

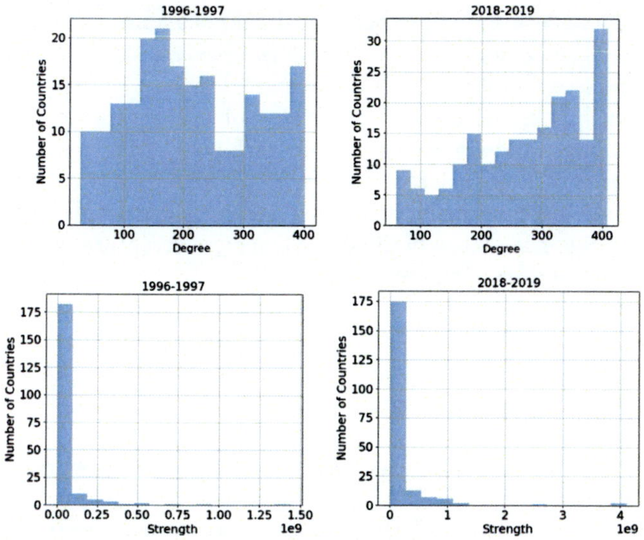

Fig. 3. Above: Total (import + export) degree distribution. Below: Total (import + export) strength distribution (1996–1997, 2018–2019).

rather than the number—of connections is dominated by a few nations. The degree distribution is not right-skewed but left-skewed, with most countries displaying many trading partners and a decline in the number of countries with very few connections. This suggests that the WTN does not fit the scale-free connective structure typical of many real-world networks. On the other hand, Fig. 3 (bottom panels) reveals that, as agreed with most contributions, the strength distribution is strongly right-skewed: most countries experience a low contribution in terms of total trade volumes, whereas a few countries contribute the most.

The scale-free connective structure of some real-world networks has been related to preferential attachment dynamics [1], in which nodes tend to connect to strongly connected nodes. Therefore, finding a dense and homogeneous distribution of links suggests that countries do not prefer to establish relations with a small set of well-connected countries but rather to maximize their trade partners. Likewise, the dense and homogeneous distribution of links contravenes a key driver of core-periphery network structures, namely the tendency of elements (i.e., countries) to restrict the set of potential trading partners due to decreasing returns to connectedness [9,10]. Furthermore, consistent with the reduction of trade costs and with the benefits inherent to international trade, it is arguable that establishing trade relations with an additional country does not necessarily require weakening or neglecting prior trade relations, thus maximizing the number of trading partners may be an optimal strategy. Consequently, from a network optimization viewpoint [3,10], our results suggest that the structure of the WTN is mostly driven by the benefits of establishing trading relations (e.g. fostering and diversifying exports, spurring economic growth), with those benefits not exhibiting a substantial marginal decrease as the number of trade partners increase amid falling trade costs and frictions.

Again from Fig. 2, the rather small value of the unweighted centralization index [6,13] shows no obvious sign of core-periphery structure, as far the pure connection pattern is considered. In a core-periphery network, nodes in the periphery should be minimally connected among them [5,9], and WTN high density is a signal that a core-periphery connective structure is rather unlikely. In contrast, the intensive trade relation confirms a high centralization if weights are accounted for, with a mean value of 0.84 across the time span. This is consistent with the right-skewed strength distribution, which shows that the WTN comprises a small group of countries with extensive inter- and intra-group trade connections, existing alongside small countries with low trade value connecting each other.

In summary, the WTN is increasingly dense but not fully connected, compact (low distance among countries), highly reciprocal and clustered, disassortative by degree, and very centralized in terms of intensity but not in topology. The centralization of the network, however, decreases over time due to the rising of emerging trade nations. The dynamics of the analyzed pool of WTN indicators are compatible with the globalization and integration of international trade, that is, the connective structure and dynamics of WTN may be explained by the potential benefits of increasing and diversifying exports outweighing the costs of establishing new trade relations. Unlike some strands of trading relations literature, increasing the number of linkages in international trade does not seem to entail an immediate increase in risk exposure or monitoring costs, or the depletion of finite resources. Therefore, high connectedness is a plausible and potentially optimal strategy. However, these trends have been restrained from 2008 onward: not too many new connections were established, and the network's centralization of trade value was even consolidated after 2008.

4 Cluster Analysis

In this section, we study the possible existence of communities in the WTN to understand the characteristics of economic integration. Moreover, we also investigate the WTN through hierarchical clustering.

One of the most widely used methods for community detection is modularity maximization. A popular approach is the Louvain method [2], which iteratively optimizes local communities with perturbations to the current partition, until modularity can no longer be improved. The result we obtain is depicted in Fig. 4 for three of the biennial periods analyzed.

In 1996–1997 the network is essentially formed by two communities, the largest one composed by Europe, Middle East and Central Asia, and the other one including North America, East Asia, and Asian Pacific countries. From 2002 to 2003 on, with the increasing role of China, the network shifts to a 3-community structure, with clusters essentially corresponding to Asia, Europe, and America.

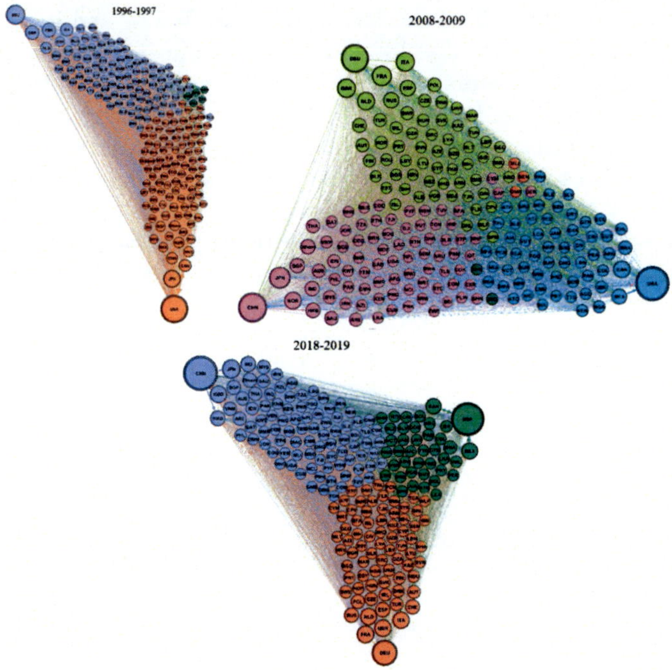

Fig. 4. Community structure of the WTN in 1996–1997, 2008–2009, 2018–2019.

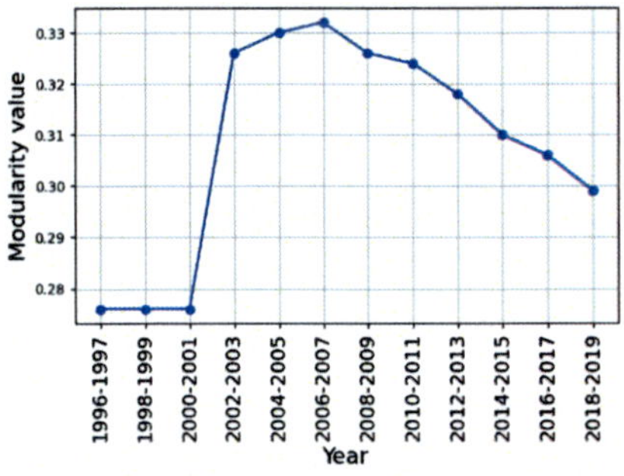

Fig. 5. Modularity values across the time interval under study.

However, although such communities are fully reasonable in geo-economic terms, the low modularity values (Fig. 5) reveal that the clusters are actually weak, i.e. communities are not isolated and have only a moderate prevalence of intra-community trade. Interestingly, the modularity value experienced a substantial increase until 2008–2009 with a sudden shift up in 2002, in coincidence with the admission of China in the World Trade Organization (WTO) in 2001. Such a sudden increase is synchronized with the passage from the 2- to the 3-community regime.

Further information about the structure of the WTN, and in particular its hierarchical organization, can be obtained by the Minimal Spanning Tree (MST), which highlights the skeleton, or backbone, of the network [17]. We extract the MST (actually a *maximal* tree, in our case) from the symmetrized weight matrix $W = \frac{1}{2}(W+W')$, a choice justified by the high reciprocity of the WTN [8,12,14]. By using Kruskal's algorithm, we first arrange the links in decreasing weight. Then we select the maximal one and insert it in the tree. Then we go on by repeatedly selecting the link with maximal weight, but we insert it in the tree only if it does not close loops, otherwise we discard it. At the end, the tree will contain exactly $n-1$ links.

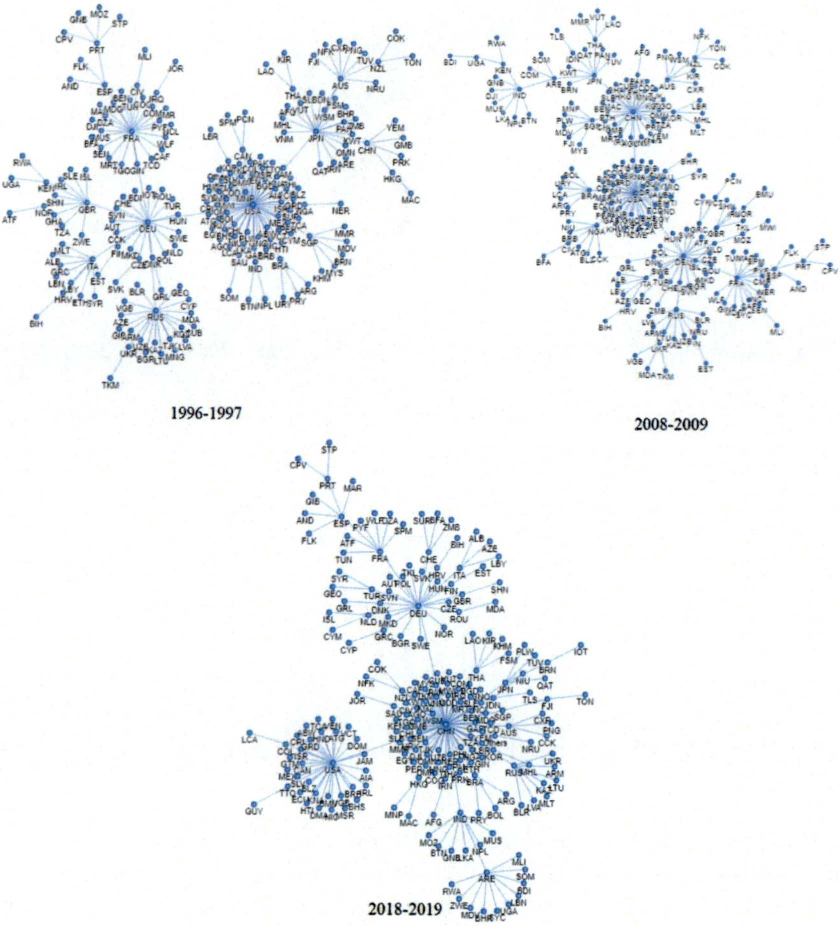

Fig. 6. World trade hierarchical clustering via MST.

Figure 6 displays the MST for three selected biennial periods (1996–1997, 2008–2009, 2018–2019). The most manifest feature of the first period (1996–

1997) is that the WTN exhibits two large clusters led by the United States and Germany, and five small clusters dominated by other traditional large economies (France, Great Britain, Italy, Japan, Russia). Geographically-based clustering is rather evident for this period. In 2008–2009, the hierarchical structure undergoes a noticeable shift, as China disrupts the 2-cluster structure and acquires a dominant position alongside the United States and Germany. These countries lead three easily recognizable geographical clusters, namely Southeast Asia and Pacific; America; and Europe and Central Asia. The revision of trade partners is easily recognized, while connections due to geographical similarity have been partially shaken. Remarkably, the group led by China has attracted several countries close to the United States before 2007, including some geographically closer to the United States than to China (e.g., Argentina, Brazil, and Chile). Likewise, some countries have fled the Europe and Central Asia group and clustered with China despite the larger distance (e.g., the subgroup led by Russia and some CIS countries, changing the dominant trade from Germany to China). Some features of this shift in the hierarchical structure thus contradict the assumption of a geographically based clustering. From 2008 to 2009 onward, the group dominated by China becomes the most prominent cluster, while the group influenced by United States gradually loses its leading position. Except for Germany, this most recent period also records the setback of clusters relevant to some other traditional large economies (Great Britain, France, Italy, Japan). On the other hand, another BRIC country, India, finds its role in the WTN structure as a visible subgroup connected to China.

Overall, despite the different methodologies, community analysis and hierarchical clustering by MST show similar results: across the years, the WTN undergoes a change in fragmentation from the two-way partition influenced by the US and Germany, to the three-way organization after the rise of China. A large trading partner revision is visible after 2008 for some regions, while, in contrast, the traditional large economies in Europe have remained strongly interconnected, despite experiencing a decline in the number of small countries depending on the trade with them.

5 Concluding Remarks

We analyze the structural change of WTN based on data available from 1996 to 2019 using network analysis. Unlike traditional methods used in trade analysis, our approach is able to capture the complexity arising from the numerous interaction among countries; hence it allows for a better description and analysis of world trade [8,14]. Using long coverage data in both time and geographical dimensions, we perform a comprehensive representation of the WTN's structure before and after the financial crisis from many different angles.

Our findings update and enhance the characterization of the connective structure of the WTN. From a network optimization viewpoint, we suggest that the connective structure of the WTN may be explained by the potential benefits of increasing and diversifying exports, outweighing the costs of establishing new trade relations. A novel result emerging from the analysis is that, compared with the dynamics before 2008, the crisis restrained or reversed to some extent the primary trends of WTN; the growth in the number of trade connections tended to slow down, the distance between countries was shortly widened, and the centralization of the network slightly strengthened after the crisis period. After the 2008 shock, it seems that the perceived level of risk in creating new links increases and, to some extent, it balances the push toward diversification, something not clearly emerging from previous analyses.

The WTN displays a weak community structure, with a tendency to become even weaker, suggesting that the push toward globalization prevails on regionalization of trade exchanges. Our result reveals that the WTN experienced a significant shift after 2008 when China disrupted the two-group hierarchical organization of world trade, led by the United States and Germany, and emerged as the most prominent cluster in the following period. Furthermore, China continues to be expected as the most attractive trade nation as it is receiving more connections. Preferential trading partner revision is visible and geographic alignment has become unsustainable in some regions (Asia-Pacific, South America). In contrast, the traditional large economies in Europe have remained strongly interconnected.

Regarding the policy implications, it is worth highlighting that our results provide new insights for analyzing and understanding the world trade at the global and regional level. For instance, results suggest that liberalization has produced an increasingly dense and homogeneous WTN, but they also indicate that most intense relations are still concentrated in a few countries. Many of the observed changes in the WTN structure are related to the rapidly growing role of some emerging countries, first of all China, more than to the effect of the financial crisis. Still, this shock left some marks in the WTN.

The impact of the Covid-19 pandemic on the architecture of the international trade network has not yet been investigated due to data availability issues at the time of writing this article, but the analysis of the propagation of this shock and of the resilience of the network could provide new information on how potential crises affect global production and supply chains, and some of the changes observed in the WTN after the 2008 shock could be consolidated. This is work to be done in future analysis.

References

1. Barabási, A.L., Albert, R.: Emergence of scaling in random networks. Science **286**(5439), 509–512 (1999)
2. Blondel, V.D., Guillaume, J.L., Lambiotte, R., Lefebvre, E.: Fast unfolding of communities in large networks. J. Stat. Mech Theory Exp. **2008**(10), P10008 (2008)
3. Cancho, R.F.I., Solé, R.V.: Optimization in Complex Networks, pp. 114–126. Springer Berlin Heidelberg (2003). https://doi.org/10.1007/978-3-540-44943-0_7
4. Cepeda-López, F., Gamboa-Estrada, F., León, C., Rincón-Castro, H.: The evolution of world trade from 1995 to 2014: a network approach. J. Int. Trade Econ. Dev. **28**(4), 452–485 (2019)

5. Craig, B., Von Peter, G.: Interbank tiering and money center banks. J. Finan. Intermediation **23**(3), 322–347 (2014)
6. Della Rossa, F., Dercole, F., Piccardi, C.: Profiling core-periphery network structure by random walkers. Sci. Rep. **3**, 1467 (2013)
7. Fagiolo, G.: The International Trade Network. The Oxford Handbook of Political Networks, vol. 1. Oxford University Press (2017)
8. Fagiolo, G., Reyes, J., Schiavo, S.: The evolution of the world trade web: a weighted-network analysis. J. Evol. Econ. **20**(4), 479–514 (2010)
9. Fricke, D., Lux, T.: Core-periphery structure in the overnight money market: evidence from the e-mid trading platform. Kiel Working Paper 1759 (2012). http://hdl.handle.net/10419/55868
10. Hojman, D.A., Szeidl, A.: Core and periphery in networks. J. Econ. Theory **139**(1), 295–309 (2008)
11. Kali, R., Reyes, J.: The architecture of globalization: a network approach to international economic integration. J. Int. Bus. Stud. **38**(4), 595–620 (2007)
12. Maeng, S.E., Choi, H.W., Lee, J.W.: Complex networks and minimal spanning trees in international trade network. Int. J. Mod. Phys. Conf. Ser. **16**, 51–60 (2012)
13. Piccardi, C., Tajoli, L.: Complexity, centralization, and fragility in economic networks. PLoS ONE **13**(11), 1–13 (2018)
14. Serrano, M.A., Boguna, M.: Topology of the world trade web. Phys. Rev. E **68**(1), 015101 (2003)
15. WTO: World Trade Report 2009—Trade Policy Commitments and Contingency Measures (2009). https://www.wto.org/english/res_e/booksp_e/anrep_e/world_trade_report09_e.pdf
16. WTO: World Trade Report 2013—Factors shaping the future of world trade (2013). https://www.wto.org/english/res_e/booksp_e/world_trade_report13_e.pdf
17. Wu, Z., Braunstein, L.A., Havlin, S., Stanley, H.E.: Transport in weighted networks: partition into superhighways and roads. Phys. Rev. Lett. **96**(14), 148702 (2006)

Green Sector Space: The Evolution and Capabilities Spillover of Economic Green Sectors in the United States

Hanin Alhaddad, Seyyedmilad Talebzadehhosseini, and Ivan Garibay(✉)

Complex Adaptive Systems Laboratory, Department of Industrial Engineering and Management Systems, University of Central Florida, Orlando, FL 32816, USA
ivan.garibay@ucf.edu
https://www.cs.ucf.edu/garibay/

Abstract. Countries' productive capabilities play a crucial role in effectively transitioning their economies towards becoming green. Current research does not address the productive capabilities in the green sectors. In particular, (a) the effect of green production capabilities on a country's green basket development and (b) whether the productive capabilities in its green sectors spillover to affect each other and its overall green growth. In this research, we use nonparametric statistics with network science techniques to analyze green sectors' evolution in the United States. The results of this research provide recommendations that could benefit the United States' green economic growth. In addition, it provides a methodology that can be used by countries' policymakers in building effective strategies that can accelerate their country's green economic growth.

Keywords: Green economy · Product space · Green growth · Path dependence

1 Introduction

Current production and consumption patterns create an ongoing demand for materials and products. They increase the dependence on natural resources and industrialization and intensify the stress placed on the environment [11]. Between the years 2010 to 2017, the material footprint increased by almost 15 billion metric tons, which accounts for a 17.4% increase [10]. Furthermore, the gas emissions in developing countries alone grew by 43% within the past two decades [10]. This growing impact on the environment flags an urgent need to create strategies and standardized policies to guide countries' shifts toward greener economies.

Green economy has diverse approaches and concepts tied to it [6]. This research defines green economy as an economy that focuses on producing and trading environmental products; which produce fewer emissions and need less energy in their production process. Although many countries are trying to shift their economies towards becoming green, they differ in their existing productive

H. Cherifi et al. (Eds.): COMPLEX NETWORKS 2016 2022, SCI 1077, pp. 502–513, 2023.
https://doi.org/10.1007/978-3-031-21127-0_41

capabilities (e.g., infrastructure, labor, knowledge, and capital) to implement such transformation [5,15]. While research argues that the productive capabilities of a country support or limit its transition toward a greener economy, identifying the actual effect of a country's productive capabilities is crucial to building a strategic transition plan for its green economy to thrive [7,15].

The United States (US), among the largest economies in the world, has carried significant growth in its green economy over the past years [3]. Current studies on the US economic growth show that the US expands its production basket without restricting its economy to the existing productive capabilities; non-path dependent economic growth [2]. However, studies on the US green economic growth show that the US expands its green production basket based on a mixed model of path and non-path-dependent as a result of initiating big jumps in its economy [15]. While Talebzadehhosseini et al. [15] analyzed the green economic growth in the US based on the evolution of its green production basket, their research did not address how the green sectors evolved and impacted the US green production basket.

This research aims to investigate the effect of productive capabilities in green sectors on the development of the US green production basket and to analyze the overlap in the productive capabilities between them. When applying non-parametric statistics and network science techniques, high heterogeneity in green sectors' productive capabilities and the capabilities spillover in between them is found. While the analysis in this research focuses only on the US, it can have multiple implications. Policymakers around the world can apply the same analysis to identify (1) green sectors' potential to grow in terms of their production basket, and (2) the most suitable strategies to achieve successful green economic growth based on the available capabilities in their country's green sectors, and (3) how available productive capabilities in one green sector spillover to other sectors. This research is structured first to provide a background on the topic in Sect. 2, then to present the data and analysis methodology in Sect. 3. Followed in Sect. 4 with the analysis results, and finally, in Sect. 5 conclusions with directions for future research are discussed.

2 Background

A country's capability is inferred in numerous studies based on its trade data and products that have a high Revealed Comparative Advantage (RCA). The RCA, introduced first by Balassa [1], provides a unified measure to compare countries' trading by determining whether a country exports a specific product competitively compared to the world. For a product, p, in a country, c, the RCA value is calculated using Eq. (1) by first finding the ratio of its export value from that country, E_{cp}, to the total export value of all products, P, in that country, $\sum_P E_{cP}$. Then by dividing the above-mentioned ratio by the ratio of the total export value of that product in the world, W, as shown in $\sum_W E_{Wp}$ to the export value of all products in the world, $\sum_{WP} E_{WP}$. If the calculated RCA value of a product, p, is equal to or larger than one, then that country, c , exports product, p , competitively.

$$RCA_{cp} = \frac{\frac{E_{cp}}{\sum_{P} E_{cP}}}{\frac{\sum_{W} E_{W_P}}{\sum_{WP} E_{WP}}} >= 1 \tag{1}$$

Based on the calculation of the RCA for a product, Hausmann et al. [4] introduced the concept of proximity. According to Eq. (2), the proximity between two products can be calculated as the conditional probability that a country exports product i competitively given it also exports product j competitively $P(x_{i,t}|x_{j,t})$ and vise versa. As shown in Eq. (3) for a country to be a competitive exporter of product i its RCA value should be equal to or larger than one. To only capture the products most related to each other Eq. (2) finds the minimum pairwise ρ of the pairs of the conditional probability referring to it as the distance between products or proximity as the authors introduce it. While the proximity between products could refer to a variety of factors in its essence it focuses on finding similarities in the technological requirements to produce a new product or improve the production of an existing product.

$$\rho_{i,j,t} = min\{P(x_{i,t}|x_{j,t}), P(x_{j,t}|x_{i,t})\} \tag{2}$$

$$X_{i,c,t} = \begin{Bmatrix} 1 \; ifRCA_{i,c,t} >= 1 \\ 0 \quad Otherwise \end{Bmatrix} \tag{3}$$

Hidalgo et al. [5] apply what is discussed earlier along with network analysis techniques to understand whether the type of products that a country exports affects its future economic performance in exporting other products (path-dependent economic growth). They construct a Product Space (PS) network from a country's export basket, using products as the nodes and the proximity between them as links. The authors infer countries' capabilities based on the density of their product space. They conclude that the denser the product space, the more productive capabilities that country shares in exporting the clusters of products, making that country's transitioning towards producing other products that are proximate within the Product Space more successful. Talebzadehhosseini et al. [15] apply a similar analysis to construct a product network from a list of environmental products that countries trade referring to it as the Green Product Space (GPS). Then they apply it to the US and China and test whether new products added to their green baskets followed a specific distribution (path-dependent) or were added randomly (non-path-dependent). Their results indicate mixed evidence of a path and non-path-dependent diversification process.

Mealy et al. [7] apply a similar analysis using the proximity calculation to investigate countries' capabilities and potential to transition toward greener economies. They present the Green Complexity Index (GCI) as a measure of countries' capabilities to export complex green products competitively, which they use to rank countries accordingly. Coniglio et al. [2] explore whether countries' export baskets developed following a path-dependent or path-defying diversification. They introduce a novel measure of the intensity of path dependence and find that the degree of path dependence between countries is highly heterogeneous. In addition, they use this measure to study the effect of path-dependent diversification on economic growth along with different factors. They find that countries with an abundance of natural resources will have a resource curse which they explain as the negative effect of resource abundance on diversifying their production basket. Nevertheless, they find that countries that strictly rely on capabilities following the PS framework have lower economic growth. While they find a negative relation between diversification following a path-dependent process and economic growth, especially in developing countries, they do not present it as an evaluation of policies that promote small jumps in the PS. With shifting toward a green economy being a persisting priority, more analysis is needed to understand how a country's capabilities, particularly in its green sectors, affect its structural transformation towards a greener economy.

3 Method

3.1 Analysis Data

In this analysis, worldwide export data dis-aggregated according to the Harmonized System Code at the 6-digit level (HS-6) is used [13]. The HS-6 is a unique identifier for each product based on an internationally standardized system code defined by the World Customs Organization. The trade data has been obtained for the years between 2003 and 2017 for all products[1]. The green products are mapped from the list of all products based on the (OECD) green products list that includes 246 green products from 11 different environmental sectors [12]. The green sectors are Air Pollution Control (APC), Cleaner or more Resource Efficient technologies and products (CRE), Environmentally Preferable Products based on end use or disposal characteristics (EPP), Heat and Energy Management (HEM), Environmental Monitoring (MON), Natural Resources Protection (NRP), Noise and Vibration Abatement (NVA), Renewable Energy Plant (REP), Management of Solid and Hazardous Waste and recycling systems (SWM), Clean up or Remediation of Soil and Water (SWR), Wastewater management and potable Water Treatment (WAT).

3.2 Finding Competitively Exported Products

To calculate the RCA value for all products in the world, Eq. (1) is used. Each product is then classified as either being competitively exported or not based on Eq. (3). In Fig. 1 a world choropleth map is presented to show the accumulated number of competitively exported green products between the years 2003 and 2017 by country. The US, with an accumulated total of 1985 green products, is among the countries with the highest number of competitively exported green products and is ranked third after Germany and Italy.

[1] Trade data is obtained from: https://oec.world/.

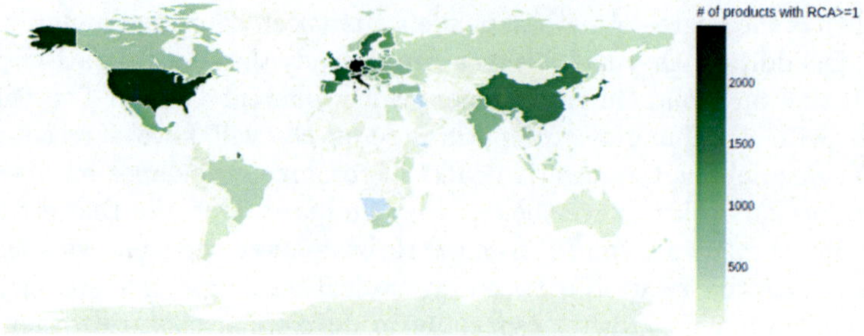

Fig. 1. World choropleth map colored by the accumulated number of competitively exported green products $RCA >= 1$ for the years between 2003–2017.

3.3 The Green Product Space

The product space is constructed using the same methodology used by Hidalgo et al. [5], where the RCA calculations from the previous step are used to construct a proximity matrix between all products. In Fig. 2 an example is shown to demonstrate the analysis that was applied. In Fig. 2(a) a matrix is constructed between all countries in the world and the products they export. If the value in the matrix is 1, the country is a competitive exporter of the product, and if it's 0 then it's not. Then the proximity matrix is constructed as shown in Fig. 2(b) between products that are exported competitively for all countries. The value in the matrix is the proximity between those products as explained earlier in Eq. (2). This is later used to obtain the product with the maximum relatedness (shaded in grey in the example) for each new product added to a country's export basket.

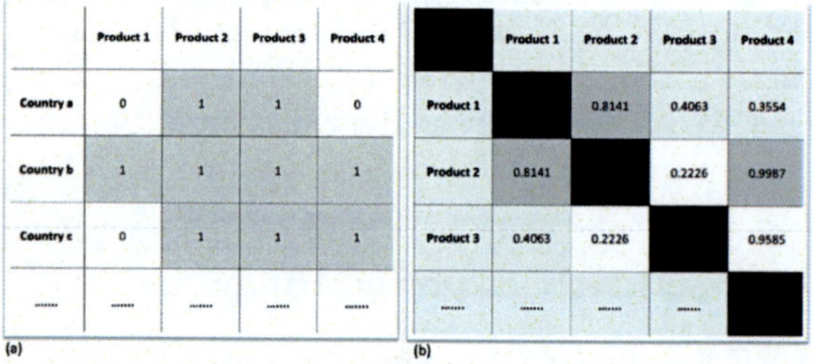

	Product 1	Product 2	Product 3	Product 4
Country a	0	1	1	0
Country b	1	1	1	1
Country c	0	1	1	1
....	....	....	....	....

(a)

	Product 1	Product 2	Product 3	Product 4
Product 1		0.8141	0.4063	0.3554
Product 2	0.8141		0.2226	0.9987
Product 3	0.4063	0.2226		0.9585
....	....	....	....	

(b)

Fig. 2. (a) An example of the matrix that connects countries to the products that each country exports. (b) An example of a proximity matrix where the products' relatedness is calculated for all countries.

4 Results

Observing the ratio between the average number of competitively exported green products yearly from the US to the total number of available green products. The ratio shows that the US is competitively exporting around 50% of all available green products by year. In Fig. 3 the ratio for each sector is found by dividing the mean number of competitively exported green products for each sector between the years 2003 and 2017 by the total number of available green products for that particular sector. The ratios in Fig. 3 can be used to infer the potential of green growth by sector that the United States can still accomplish in developing its green basket to include all possible green products. While the United States exports competitively nearly 80% of the possible green products from the MON and APC sectors, it has no products in the NRP sector which should be further investigated.

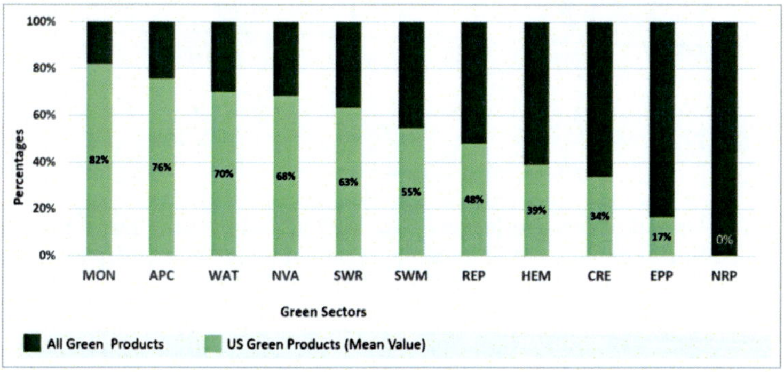

Fig. 3. The ratio between the average number of green products competitively exported by sector from the US by year to the total number of existing green products.

4.1 The United States Green Product Space

In Fig. 4(a) the US product space is presented[2]. Nodes that represent products are colored based on their category[3] while products that the US doesn't export are colored in light grey. The network links are filtered to only show proximity larger than 0.5. Shared productive capabilities between products are shown with clusters in the network. Products that are centered in PS share more capabilities compared to products that are located towards the edge of the PS [5]. In addition, it can be observed how specific categories overlap more than others. In Fig. 4(b)

[2] The US product space is visualized using NetworkX package in python: https://networkx.org.

[3] Product categories are obtained from Harvard University, T. G. L. Classifications Data version DRAFT VERSION. https://doi.org/10.7910/DVN/3BAL1O.

the US Green Product space is highlighted over the US product space to only show green products colored by their green sector. The green products from 11 green sectors are scattered across the product space sharing links with green and non-green products. This highlights the possible capabilities spillover between sectors in addition to non-green products.

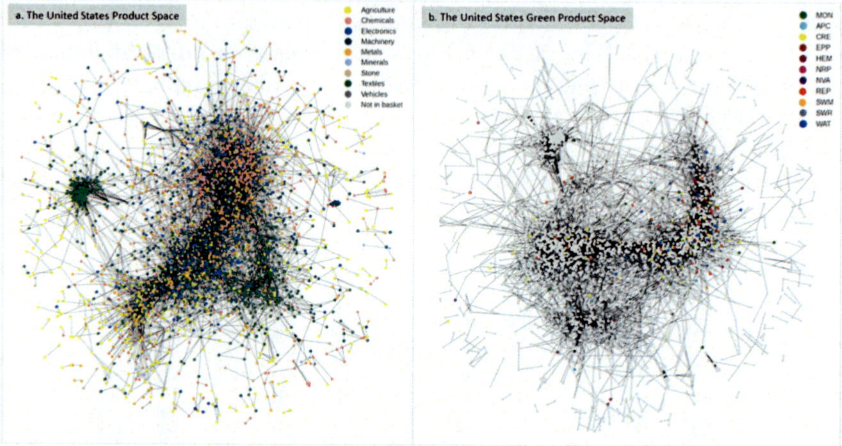

Fig. 4. a. The United States product space is highlighted over the world product space, b. The United States green product space is highlighted over the United States product space.

4.2 Calculating Products' Relatedness

To analyze how the US green sectors evolved, new green products added to the US export basket are first obtained following the methodology used by Mealy et al. and Talebzadehhosseini et al. [7,15]. Then based on the proximity matrix their relatedness to existing products that the US exports competitively are found. New products are defined as products with an RCA value in the initial year lower than a specified threshold given that at the end of the final year their RCA is larger or equal to 1. To find the best RCA threshold to use in the initial year, three different thresholds are tested and compared between countries with varying levels of green capabilities. Using the GCI ranking discussed earlier and introduced by Mealy et al [7], two countries other than the US are selected, one with an increasing GCI and another with a decreasing GCI over time e.g. Uganda and Algeria respectively. In Fig. 5 the number of new green products added on duration spans of 2 years between 2003 and 2017 and based on the three different thresholds is shown. The comparison shows that for a developed country like the US using a high RCA threshold gives more results which could be due to its economy being relatively competitive, it exports most of its products at a high revealed comparative advantage value. In Algeria, little to no difference is observed which can be due to it being a developing country that is conducting low development on its green basket. In Uganda, all thresholds indicate an increase in the number of new products added to their export basket competitively. This can be due to the increasing pattern of its GCI over the years as well as its having a diverse production basket in terms of RCA values. Based on the above using a higher RCA threshold (RCA = 0.75) value is better to capture new products becoming competitively exported to a developed country like the US. In Fig. 6 the count of new green products by category is shown. In the US a total of 38 new products were added to its green basket between 2003 and 2017 with the majority belonging to the CRE and REP green sectors. While the low number of new products added to the APC, WAT, and MON sectors can be explained by the existing high number of competitively exported products from it, the low number of new products added to the rest of the sectors needs further investigation, in particular, the absence of new products in the NVA and the EPP sectors.

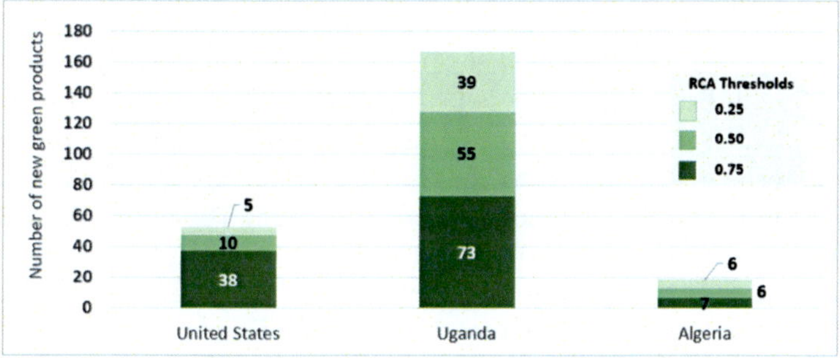

Fig. 5. Comparing the number of new green products added based on the RCA threshold at the initial year.

4.3 Green Sectors Evolution

To find whether the evolution of green sectors in the US followed a path-dependent process, the same methodology applied by Coniglio et al. [2] is used to test whether the distribution of relatedness followed a random (non-path dependent) distribution or not. We apply Shapiro-Wilk (SW) test as a nonparametric statistical test instead of the one-sided Kolmogorov-Smirnov (KS) test as it has a higher statistical power at the size of our data ($3 > n > 50$) [9]. Similar to the KS test the SW H_0 : is that the data is randomly distributed. When applying it to the distribution of relatedness for new products by sector we obtain the results shown in Table. 1. Given the p-value for each of the sectors, we cannot reject the null hypothesis at the significance level of 0.05 and conclude that there is no significant evidence that the data didn't follow a random distribution. We

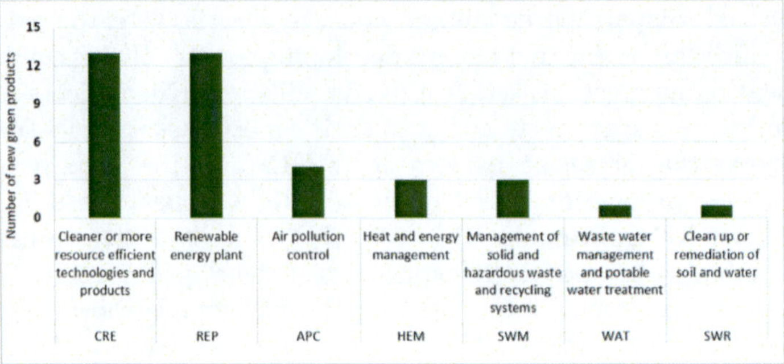

Fig. 6. The number of new green products in the US for years between 2003–2017 by green sector. New entries are defined as $RCA_{cp,t-2} < 0.75$ and $RCA_{CP,t} >= 1$.

further investigate this by measuring the intensity of path dependence by sector. The intensity of path dependence as it was first introduced by Coniglio et al. [2], is measured by finding the ratio between the number of new products classified as path dependent to the number of all products added to a country. To classify a product as being path dependent, Coniglio et al. [2] introduce two methods using randomly distributed counterfactual data. First, a product is considered path dependent if its maximum proximity value is higher than the mean value of the counterfactual distribution, or the mean plus one standard deviation for a more stringent threshold. Second, based on finding the percentile value of the maximum proximity in the counterfactual data. Applying the two methods the same results are obtained, with 71% of products being classified as path dependent when using the mean as a threshold, in addition to the majority of them falling on average in the 70th percentile. The intensity of path dependence by sector is presented in Fig. 7. High heterogeneity in the intensity of path-dependence between sectors is observed. While the (REP) sector is entirely developed based on a path-dependent process, the (SWR) and (WAT) sectors developed following a non-path-dependent process. The intensity of path dependence provides an insight into how each sector is evolving and how it can further grow. This can be used by policymakers to promote customized growth strategies that fit the individual sectors.

Table 1. Shaprio Wilk test on each of the green sectors.

Green sector	Sample size	Test statistic	p-value
APC	4	0.9656	0.8142
CRE	13	0.9704	0.8996
HEM	3	0.9722	0.6800
REP	13	0.9045	0.1542
SWM	3	0.9089	0.4144

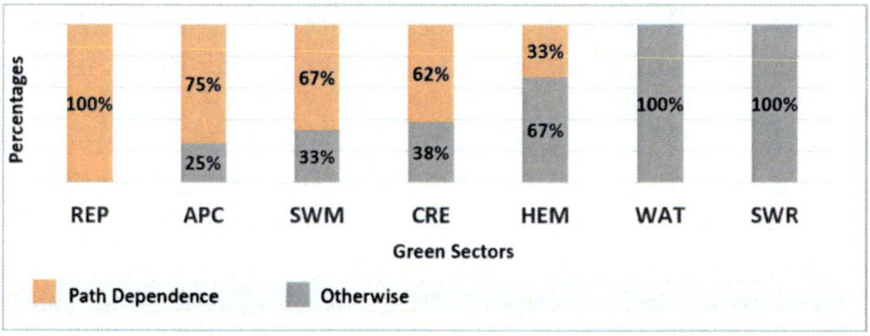

Fig. 7. Path dependence intensity by green sector of new products added to the US basket between 2003 and 2017.

4.4 The Green Sector Space

To construct the Green Sector Space, the sectors of products that share the maximum proximity to the list of new green products are obtained using the proximity matrix. In Fig. 8, the Sector Space is presented with a directed network connecting sectors of new products with sectors of products they share maximum proximity with[4] Network edges refer to new products added to a specific green sector (source node) that is connected to another sector based on it sharing the maximum proximity with a product from it (target node). The Network edges are sized according to the number of new products shared between the sectors' nodes. This provides a way to measure the level of the shared capabilities between the nodes' sectors. Nodes' sizes reflect the nodes' in-degree, the larger the node size the more shared capabilities new products added to the basket have with it. The nodes are colored by the betweenness centrality (BC) which indicates the central role a sector has in terms of sharing the most capabilities with other sectors, and its role in supporting a spillover of capabilities between sectors [8].

The non-green products node is the node with the largest size. In addition to having links with all nodes, it has the links with the maximum weights in the network directed to it from the CRE, SWM, SWR, and APC sectors respectively. This indicates the strong effect non-green products have on green sectors' development. In addition, given that the REP is among the sectors with the highest number of new products added, the highest BC in the network as well as the highest in-degree, we conclude that it is the sector with the highest dominance in the US green economy. Given that it entirely grew following a path-dependent process, in addition to the high intensity of path dependence of sectors that are connected to it like the APC, SWM, and CRE as seen in Fig. 7, improvements to the productive capabilities in this sector will not only grow it but will spill over to the sectors connected to it. Similarly, the SWM sector with its high intensity of path dependence and high potential to grow can benefit from enhancement to its capabilities, which can generate a spillover to the sectors connected to it like REP and the CRE.

[4] The Sector Space was visualized using Gephi software: https://gephi.org/.

Nevertheless, the HEM which is also connected to the REP, its low path dependence intensity, and its potential to grow as seen in Fig. 6, indicates that it will largely benefit from directly enhancing its productive capabilities, as it will not only grow the HEM sector but due to its relatively high centrality will likely create a spillover towards the APC and CRE sectors as well. Another sector with the same observation is the SWR having only a link with non-green products and products only related to the MON sector. Given that it is entirely growing following a non-path-dependent approach, focusing on enhancing the productive capabilities in this sector are needed for it to grow. While we use the term enhancing the productive capabilities in our recommendation interchangeably for sectors with low and high intensity of path dependence, it should be highlighted that sectors with products that have a low intensity of path dependence will likely benefit more from the enhancement of capabilities by initiating breakthroughs in its production process such as focusing on technological innovation activities in its production process [14].

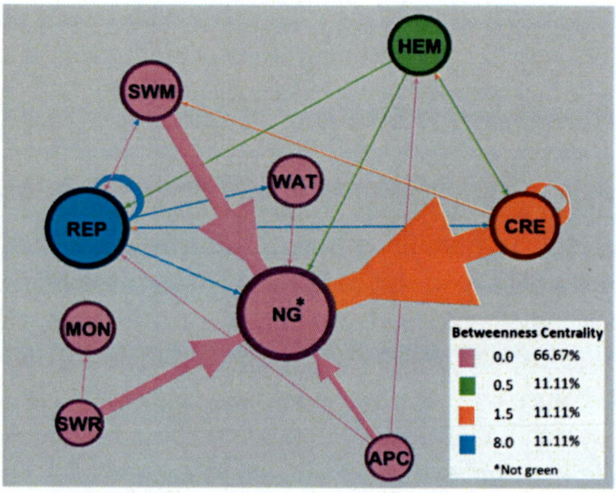

Fig. 8. Green sector space, sectors are connected based on the sector of new green products added to the US export basket (source node) and the sector of the product they share the maximum proximity with (target node).

5 Conclusions

Using nonparametric statistics and network science, we analyze green sectors' evolution in the United States and find high heterogeneity in the intensity of path dependence between its green sectors. We construct a network from green sectors based on the product space methodology and infer rich information on shared capabilities between sectors. In addition, we identify which sectors have the highest dominance in the network and provide recommendations that improvements to their productive capabilities will influence a spillover to other connected sectors. Furthermore, we highlight the dominant effect of non-green products on the sectors' network, which can be explored in future research. Finally, this study can be applied in any country to explore its green productive capabilities based on its sectors and help policymakers build effective recommendations to accelerate its green economic growth. Due to data limitations, only internationally exported data is used. However, upon data availability, the US internal market can be a direction for future research.

References

1. Balassa, B.: Trade liberalisation and "revealed" comparative advantage 1. The Manchester School **33**(2), 99–123 (1965)
2. Coniglio, N.D., Vurchio, D., Cantore, N., Clara, M.: On the evolution of comparative advantage: path-dependent versus path-defying changes. J. Int. Econ. **133**, 103522 (2021)
3. Georgeson, L., Maslin, M.: Estimating the scale of the us green economy within the global context. Palgrave Commun. **5**(1), 1–12 (2019)
4. Hausmann, R., Klinger, B.: The structure of the product space and the evolution of comparative advantage. CID Working Paper Series (2007)
5. Hidalgo, C.A., Klinger, B., Barabási, A.L., Hausmann, R.: The product space conditions the development of nations. Science **317**(5837), 482–487 (2007)
6. Loiseau, E., Saikku, L., Antikainen, R., Droste, N., Hansjürgens, B., Pitkänen, K., Leskinen, P., Kuikman, P., Thomsen, M.: Green economy and related concepts: an overview. J. Cleaner Prod. **139**, 361–371 (2016)
7. Mealy, P., Teytelboym, A.: Economic complexity and the green economy. Res. Policy 103948 (2020)
8. Newman, M.E.: A measure of betweenness centrality based on random walks. Soc. Netw. **27**(1), 39–54 (2005)
9. Razali, N.M., Wah, Y.B., et al.: Power comparisons of Shapiro-Wilk, Kolmogorov-Smirnov, Lilliefors and Anderson-Darling tests. J. Stat. Model. Anal. **2**(1), 21–33 (2011)
10. Sachs, J., Schmidt-Traub, G., Kroll, C., Lafortune, G., Fuller, G., Woelm, F., et al.: Sustainable Development Report 2020. Cambridge Books (2021)
11. Sarkodie, S.A., Owusu, P.A., Leirvik, T.: Global effect of urban sprawl, industrialization, trade and economic development on carbon dioxide emissions. Environ. Res. Lett. **15**(3), 034049 (2020)
12. Sauvage, J.: The Stringency of Environmental Regulations and Trade in Environmental Goods (2014)
13. Simoes, A.J.G., Hidalgo, C.A.: The economic complexity observatory: an analytical tool for understanding the dynamics of economic development. In: Workshops at the Twenty-Fifth AAAI Conference on Artificial Intelligence (2011)
14. Talebzadehhosseini, S., Garibay, I.: The interaction effects of technological innovation and path-dependent economic growth on countries overall green growth performance. J. Cleaner Prod. **333**, 130134 (2022)
15. Talebzadehhosseini, S., Scheinert, S.R., Garibay, I.: Global transitioning towards a green economy: analyzing the evolution of the green product space of the two largest world economies. In: International Conference on Complex Networks and Their Applications, pp. 633–644. Springer (2019)

Statistical Inference of Lead-Lag Between Asynchronous Time Series from P-Values of Transfer Entropy at Various Timescales

Christian Bongiorno(✉) and Damien Challet

Laboratoire de Mathématiques et Informatique pour la Complexité et les Systèmes, Université Paris-Saclay, CentraleSupélec, 91192 Gif-sur-Yvette, France
christian.bongiorno@centralesupelec.fr

Abstract. Symbolic transfer entropy is a powerful non-parametric tool to detect lead-lag between time series. Because a closed expression of the distribution of Transfer Entropy is not known for finite-size samples, statistical testing is often performed with bootstraps whose slowness prevents the inference of large lead-lag networks between long time series. On the other hand, the asymptotic distribution of Transfer Entropy between two time series is known. In this work, we derive the asymptotic distribution of the test for one time series having a larger Transfer Entropy than another one on a target time series. We then measure the convergence speed of both tests in the small sample size limits via benchmarks. We then introduce Transfer Entropy between time-shifted time series, which allows to measure the timescale at which information transfer is maximal and vanishes. We finally apply these methods to tick-by-tick price changes of several hundreds of stocks, yielding non-trivial statistically validated networks.

Keywords: Transfer entropy · Statistically validated networks · Order book

1 Introduction

Causality is the process from which an event influence the outcome of another process. In real-world data, detecting causal links may be hard. An intuitive way to infer causality consists of measuring lagged correlation between the states of two variables. Although this approach is in line with what is humanly perceived as a cause-effect relationship, these lagged correlations may, in many cases, come from the influence of hidden variables. Furthermore, observing significant lagged correlations is not a necessary condition for the inference of a cause-effect relationship because such a measure can be negligible in a highly non-linear and noisy system.

One of the early approaches to cause-effect detection is Granger causality [4]. The idea of Granger causality is to assume that B *causes* A is equivalent to B

H. Cherifi et al. (Eds.): COMPLEX NETWORKS 2016 2022, SCI 1077, pp. 514–525, 2023.
https://doi.org/10.1007/978-3-031-21127-0_42

forecasts A. Forecasting at that time was typically addressed with auto-regressive models. Therefore, Granger causality is able to detect causality only when a linear model is a good approximation to the dynamics of A and B. Needless to say, most real-world systems are not linear.

More generally, A can only be predicted from B if a significant amount of information is transferred from B to A. Transfer Entropy (TE) [10] measures the information gain from the past of B in predicting the future of A after discounting of self-information from the past of A. TE is more general than Granger causality. Both are equivalent for auto-regressive models with Gaussian variables [1]. While TE can be defined for continuous non-linear models, it becomes model-free when the states of A and B take or are approximated by discrete values. If the original models are not genuinely discrete, the longer the time series, the more refined the discrete description of the original time series can be. This is known as symbolic TE [12], which provides an unbiased estimator of the model-free TE. However, when only finite time series are available, the error of sample probabilities and thus of the sample TE increases with the chosen number of distinct discrete values. Such an irreducible error makes it difficult to distinguish a real signal from random noise fluctuations (for more spurious sources of TE, see [11]).

Testing the significance of TE can be done with data shuffling or resampling data, which is slow. Direct testing was proposed in Ref. [2], which introduced a null model for TE and shows that TE asymptotically converges to a chi-square distribution with 2 degrees of freedom for binary time series. A generalization of this test to more generic discrete time series for Markovian systems is found in Ref. [7]. Finally, TE between asynchronous time series was recently defined in Ref. [6].

Our main contributions are as follows: first, we derive an asymptotic statistical test to compare the amount of TE from B to A and from C to A, which allows us to significantly rank predictors pairs based on their respective information content. Secondly, we quantify the finite-size effects of the asymptotic tests by comparing their outcomes with those of bootstrap tests. Thirdly, we use time-shifting of the target time series in an asynchronous setting so as to infer significant time delays and the delay after which no statistically significant TE or TE difference takes place. Third, we demonstrate the effectiveness of our methods by inferring a statistically validated lead-lag network of tick-by-tick price changes from a large database of financial data and by characterising how TE changes as a function of time delays.

2 Significance Test for Transfer Entropy

Let us fix some notations. For any time t, information about the past of time series A is encoded by n_A different states. The simplest case is binary information (two symbols) about some previous time step $t' < t$; binary information about M different time steps amounts to $n_A = 2^M$ possible states. Note that the number of possible symbols per time step may change for each t' in a fixed way.

The number of possible states that describe the future of A is denoted by n_A^+. Similarly, the number of states that describe time series B at time t is n_B.

The transfer entropy from B to A measures the additional information provided by B with respect to the information about the past of A when one tries to predict A. It is defined by the difference between the conditional entropy $H(A^+|A)$ of the future of A, denoted by A^+, on its past, and the entropy $H(A^+|A,B)$ of A^+ conditional on the past of A and B, as follows

$$\mathrm{TE}^{B\to A} = H(A^+|A) - H(A^+|A,B). \tag{1}$$

References [2,7] show that the $\mathrm{TE}^{B\to A}$ converges to a chi-square distribution for binary states and Markovian systems with constant numbers of symbols per time step. As a warm-up for the next section, let us derive this result from the comparison between two decision tree models (see Fig. 1).

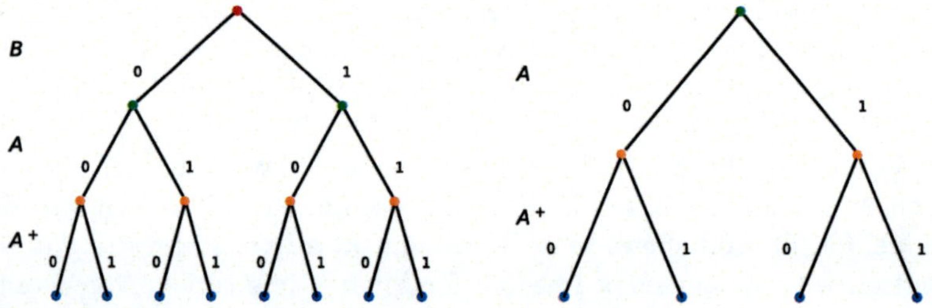

Fig. 1. Example of the maximum likelihood binary tree (on the left) that includes B, and the alternative model (on the right) that does not include B.

When testing for the presence of significant transfer entropy from B to A, the null hypothesis $\mathcal{H}_0$ and the alternative hypothesis $\mathcal{H}_1$ are respectively

$$\mathcal{H}_0 : \mathrm{TE}^{B\to A} = 0 \tag{2}$$
$$\mathcal{H}_1 : \mathrm{TE}^{B\to A} > 0. \tag{3}$$

The likelihood of the decision tree model that includes the series B equals

$$\log \mathcal{L}_{BAA^+} = T \sum_{b=1}^{n^B} \sum_{a=1}^{n^A} \sum_{a^+=1}^{n^{A^+}} \mathcal{P}(a^+, a, b) \log \mathcal{P}(a^+|a,b) = -T \cdot H(A^+|A,B), \tag{4}$$

where $\mathcal{P}(a^+|a,b)$ and $\mathcal{P}(a^+,a,b)$ are the sample probabilities.

The likelihood of the decision tree model that does not include the series B is given by

$$\log \mathcal{L}_{AA^+} = T \sum_{a=1}^{n^A} \sum_{a^+=1}^{n^{A^+}} \mathcal{P}(a^+, a) \log \mathcal{P}(a^+|a) = -T \cdot H(A^+|A). \tag{5}$$

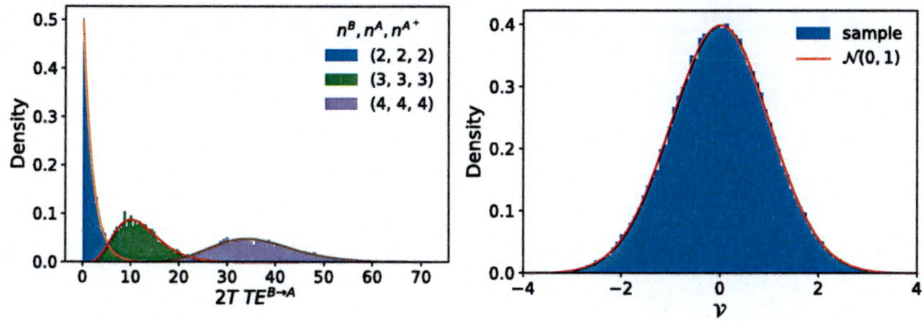

Fig. 2. The left panel shows the distribution of $2T \cdot \mathrm{TE}^{B\to A}$ obtained with reshuffling of the data-series B for different values of the number of symbols. The lines are the theoretical expectation from the $\mathcal{X}^2(\nu)$ distribution. The right panel Distribution of the Vuong metric obtained with independent 100,000 realizations of a time series of T=10,000 records for $S = 0.2$.

If the null hypothesis is true, the knowledge of B does not improves significantly predictability and thus the difference between both likelihoods will be due to over-fitting of the sampling error. In such a case, Wilks' theorem [15] states that

$$2(\log \mathcal{L}_{BAA^+} - \log \mathcal{L}_{AA^+}) = -2T\left(H(A^+|A,B) - H(A^+|A)\right)$$
$$= 2T \cdot \mathrm{TE}^{B\to A} \xrightarrow{d} \mathcal{X}^2(\nu) \qquad (6)$$

converges to a chi-square distribution in the limit of large sample size with ν degrees of freedom given by

$$\nu = n^A\left(n^{A^+} - 1\right)\left(n^B - 1\right). \qquad (7)$$

This number can be obtained by counting the difference in the branches of the two decision trees. Specifically the number of degrees of freedom of the model $\mathcal{P}(a^+|a,b)$ is $\nu^{AB} = n^A n^B (n^{A^+} - 1)$ where $\sum_{a^+=1}^{n^{A^+}} \mathcal{P}(a^+|a,b) = 1$, whereas from the model $\mathcal{P}(a^+|a)$ one finds $\nu^A = n^A(n^{A^+} - 1)$. Given that $\nu = \nu^{AB} - \nu^A$, (7) follows. If the number of symbols per time step is constant and if the maximum number of relevant time steps in the past is also constant, one recovers the result of Ref. [5]. Figure 2 (left) compares the distributions given by Eq. (6) and numerical simulations for various parameter values.

3 Comparing Two Transfer Entropy Measures

The same approach can be used to test if time series B is more informative than C to predict A. In other words, we aim to test if the real underlying TEs are such that

$$\mathcal{H}_0{:}\mathrm{TE}^{B\to A} = \mathrm{TE}^{C\to A} \qquad (8)$$
$$\mathcal{H}_1{:}\mathrm{TE}^{B\to A} > \mathrm{TE}^{C\to A}. \qquad (9)$$

Let us introduce the difference between the estimated TEs

$$Q = \mathrm{TE}^{B\to A} - \mathrm{TE}^{C\to A} = H(A^+|A,C) - H(A^+|A,B). \tag{10}$$

The likelihood of the model with B is

$$\log \mathcal{L}_{BAA^+} = T\sum_{x=1}^{n^B}\sum_{a=1}^{n^A}\sum_{a^+=1}^{n^{A^+}} \mathcal{P}(a^+,a,b)\log\mathcal{P}(a^+|a,b) = -T\cdot H(A^+|A,B) \tag{11}$$

while the likelihood of the model with C is

$$\log \mathcal{L}_{CAA^+} = T\sum_{c=1}^{n^C}\sum_{a=1}^{n^A}\sum_{a^+=1}^{n^{A^+}} \mathcal{P}(a^+,a,c)\log\mathcal{P}(a^+|a,c) = -T\cdot H(A^+|A,C). \tag{12}$$

The difference in likelihood is therefore

$$\Delta\mathcal{L} = \log\mathcal{L}_{BAA^+} - \log\mathcal{L}_{CAA^+} = T\cdot Q. \tag{13}$$

To obtain the distribution of the difference in likelihood of the maximum likelihood models, let us call ω the standard deviation of the point-wise likelihood difference

$$\omega^2 = \mathrm{Var}\left[\log\mathcal{P}(a^+|a,b) - \log\mathcal{P}(a^+|a,c)\right] \tag{14}$$

computed over the data-series of length T of elements (b, c, a, a^+).

Then for Vuong's closeness test [14] states that, if the null hypothesis is true,

$$\mathcal{V} := \frac{\Delta\mathcal{L} - \nu}{\sqrt{T}\omega} \xrightarrow{d} \mathcal{N}(0,1), \tag{15}$$

where $\nu = E(\Delta\mathcal{L})$, converges to a standardized normal distribution for $T \to \infty$, see Fig. 2 (right).

In fact, ν is the difference between the number of degrees of freedom of the two trees associated with $B \to A$ and $C \to A$. While the estimation of ν from the trees is hard for overlapping models, but given that the hypothesis of Vuong's test applies, it is possible to estimate it as follows: if the null hypothesis is true, $\mathrm{TE}^{B\to A} = \mathrm{TE}^{C\to A} = S > 0$ and $\Delta\mathcal{L}$ is the difference between two non-central chi distribution [2], divided by 2, with expected value

$$E[T\,\mathrm{TE}^{C\to A}] = \frac{1}{2}\left(\nu^C + S\right) = \frac{1}{2}\left[n^A\left(n^{A^+} - 1\right)\left(n^C - 1\right) + S\right] \tag{16}$$

$$E[T\,\mathrm{TE}^{B\to A}] = \frac{1}{2}\left(\nu^B + S\right) = \frac{1}{2}\left[n^A\left(n^{A^+} - 1\right)\left(n^B - 1\right) + S\right]. \tag{17}$$

The degrees of freedom ν^C and ν^B are given by (7). Therefore $E[\Delta\mathcal{L} - \nu] = 0$ if

$$\nu = (n^B - n^C)\frac{n^A}{2}(n^{A^+} - 1). \tag{18}$$

3.1 Bootstrap P-Value

Let us briefly describe two generic approaches to associate a p-value to TE from bootstrap resampling, which can of course be computationally intense. For example, significant p-values corrected by multiple comparison techniques must be orders of magnitude smaller than the intended significance threshold, which requires a very large number of bootstrap samples. For both approaches we must identify an event event matrix $E \in \mathbb{R}^{T \times 3}$ where $e_t = (a'_t, a_t, b_t)$ is a generic event. This matrix can be generalized to more than two variables.

Significant TE via Shuffling To estimate the p-value of sample $\text{TE}^{B \to A} = S$, it is enough to perform a permutation test of B, so as to break any link between the past of B and the future of A. This kind of approach requires many permutations of column B of matrix E. Let us denote by $E_{(r)}$ the data matrix E whose B column has been reshuffled, r being the index of a particular permutation of column B. This new matrix preserves both $H(A^+|A)$ and $H(B)$ but breaks any information flow between B and A. From $E_{(r)}$, one computes $\text{TE}_{(r)}^{B \to A} = S_{(r)}$. By performing q independent re-shuffling and TE estimations, the p-value can be defined as

$$p\text{-value} = \frac{\sum_{i=1}^{q} \delta(S > S_{(i)})}{q}, \tag{19}$$

where $\delta(x) = 1$ if the assertion is true and zero otherwise.

Comparing TEs via Bootstraps Assume that both $\text{TE}^{B \to A} = S^B$ and $\text{TE}^{C \to A} = S^B$ are significantly larger than zero. A different bootstrap procedure must be adopted. In this case, indeed, the bootstrap is used to associate a confidence interval to S^B and S^C. Let us start by defining the event matrix E with elements $e_t = (a'_t, a_t, b_t, c_t)$, then a bootstrap realization of $E_{(r)}$ is obtained by re-sampling with replacement the events such that the bootstrap event matrix $E_{(r)}$ has the same sample size of E, i.e., T. This bootstrap procedure does not break down the information flow from B and C to A. For each bootstrap realization we estimate $S_{(r)}^B$ and $S_{(r)}^C$. Finally the p-value of $\text{TE}^{B \to A} > \text{TE}^{C \to A} = S^C$ is defined as

$$p\text{-value} = \frac{\sum_{i=1}^{q} \delta(S_{(i)}^B > S_{(i)}^B)}{q} \tag{20}$$

where q is the number of independent bootstrap re-sampling and $\delta(x) = 1$ if the assertion is true, zero otherwise.

4 Finite Size Effects of the Statistical Tests

4.1 Testing for the Significant TE

In this section, we show that the analytical p-value converges to the bootstrap one for various numbers of the number of states, hence of the degrees of freedom.

In particular, we test first the convergence when the null hypothesis is true, then we test the power in detecting when the null hypothesis is false.

The true null hypothesis benchmark is obtained by sampling $\mathcal{P}(a', a)$ and $P(b)$ from two Dirichlet distributions with parameters set to α. Then a synthetic dataset of sample size T is obtained sampling a set of that from $\mathcal{P}(a', a, b) = \mathcal{P}(a', a)\mathcal{P}(b)$. In the upper panel left of Fig. 3 we show that both the analytical and bootstrap p-values converge to the same value irrespective of n^{A^+}, n^A, and n^B. However, we must be aware that this is not totally true. In fact, when n^{A^+}, n^A, n^B are large and when the sample size is small, the likelihood ratio test should not be applied due to slow convergence to the chi-squared distribution (see the upper right panel of Fig. 3). In particular, a small T leads to an underestimation of the analytical p-values. In the lower-left panel of Fig. 3, we show that small n^{A^+}, n^A, n^B are much more robust to finite sample size effects than the larger ones.

To test the false null hypothesis, the probabilities $\mathcal{P}(a', a, b)$ are sampled from a Dirichlet distribution with parameters set to α. In this set of simulation, the effect of α is explored in order to test the power of the test: small values of α produce very heterogeneous probabilities, while when $\alpha \gg 1$, all the probabilities become homogeneous, i.e., $\mathcal{P}(a', a, b) \to 1/(n^{A^+} n^A n^B)$. While the former is the ideal condition for every test, the latter is a difficult case. In the lower-right panel of Fig. 3, we show the fraction of times the analytical and bootstrap p-values reach values smaller than 0.01 over 1,000 independent simulations. Note that both the asymptotic analytical and the bootstraps results agree.

4.2 Comparing TEs

Benchmarking the true null hypothesis, in this case, is much harder. To do that, we should solve a non-linear minimization problem. Specifically, by using a Semidefinite Quadratic Linear Programming (SQLP), we build a set of probability $\mathcal{P}(a', a, b, c)$ such that $\left(\mathrm{TE}^{c\to a} - S^c)^2 + (\mathrm{TE}^{b\to a} - S^b\right)^2$ computed on the real benchmark probabilities is minimal. Since SQLP depends on the initial conditions, we randomize them by starting the optimization procedure from a set of probability obtained from a Dirichlet distribution with parameter $\alpha = 0.5$.

In the upper panel of Fig. 4 we show a scatter plot of the p-values obtained when the null hypothesis is true for a large sample size. Differently from Fig. 3, this time bootstrap and analytical p-values show a lower agreement, although on average unbiased, as depicted in the linear regression reported in the plot.

In the low sample size and large number of symbols regime (the upper right panel of Fig. 4), both p-values are biased but in a different way. The bootstrap p-values are too restrictive, while the analytical ones are too permissive. In fact, the expected p-value distribution when the null hypothesis is true should be a homogeneous distribution. In this case, since the Vuong test is two-sided, looking at the area of the CDF of p-values may be misleading. Therefore, to inspect the speed of convergence of the analytical p-values, we measure the average element-wise likelihood that the Vuong metric $\mathcal{V}$ is distributed as a $\mathcal{N}(0, 1)$ (see

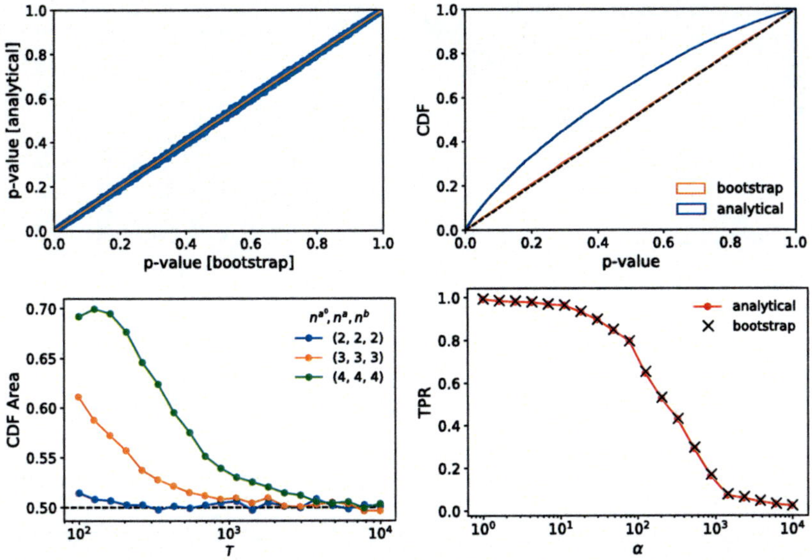

Fig. 3. Tests for $\mathrm{TE}^{B\to A} > 0$: *a*) comparison of bootstrap and analytical p-values for true null hypothesis with $T = 10,000$ and all possible combinations of $(n^{A^+} n^A n^B) \in [2, 4]$ repeated 100 times; *b*) cumulative distributions computed with 10,000 independent realizations of the bootstrap and analytical p-values for $(n^{A^+} n^A n^B) = (3, 3, 3)$ and $T = 100$, the unbiased model should lie on the bisector; *c*) Area under the CDF of analytic p-value for different sample size T, the area of the unbiased model is 0.5; *d*) comparison of the bootstrap an analytic true positive rate obtained with a p-value< 0.01 for false null hypothesis benchmark, the simulations refers to 1,000 simulation with $(n^{A^+} n^A n^B) = 2, 2, 2$ and $T = 10,000$, the parameter α controls the homogeneity of $\mathcal{P}(a', a, b)$. In all the plots the bootstrap p-value is computed with 10,000 bootstrap re-sampling.

the lower-left plot of Fig. 4). As expected, the bias is larger when the number of symbols is large and all the inspected cases converge to normality for $T > 1,000$.

Finally, in the lower right panel of Fig. 4, we compare the true positive rate of the bootstrap and analytical p-values. The analytical p-values, as expected, show a higher recall than the bootstrap ones. Although this advantage is larger in the small sample size regime, it is worth stressing that it might be compensated by a lower control of the first type errors.

5 Lagged Information Transfer Between Asynchronous Time Series: Limit Order Books

We apply the analytical asymptotic p-values described above to measure the information flow between changes in the order books of different stocks. Specifically, for each stock, we consider the time series of mid-prices, defined as the

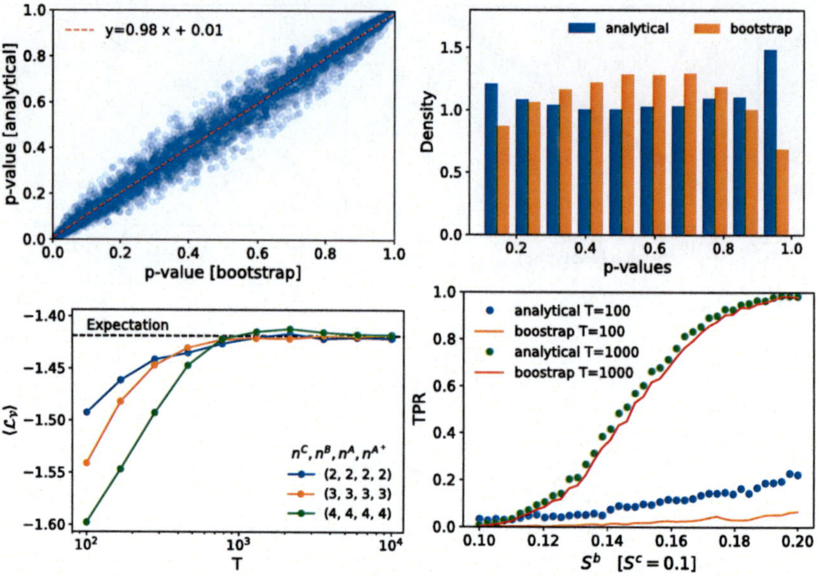

Fig. 4. Benchmarks performance of the Vuong test. Upper left plot: scatter plot of the bootstrap p-values and the analytical ones for all possible combinations of $n^{A^+}, n^A, n^B, n^C \in [2, 4]$ for $T = 10,000$ for $S^B = S^C = 0.1$, the dotted line is a linear regression. Upper right plot: distribution of 10,000 p-values obtained for $n^{A^+} = n^A = n^B = n^C = 4$ and $T = 100$ and $S^B = S^C = 0.1$, the expected distribution should be homogeneous. Lower left plot: average element-wise likelihood of the Vuong metric being distributed as a $\mathcal{N}(0, 1)$ and $S^B = S^C = 0.1$ while the dotted line is the likelihood corresponding to $\mathcal{N}(0, 1)$. Right lower plot: true positive rate computed over 1,000 simulations for $n^{A^+} = n^A = n^B = n^C = 4$ as a function of S^B for different values of T.

instantaneous average between the best buy and sell quotes, and encode their changes with their signs. The inter-event time between two consecutive price variations is not constant, and two time series are not synchronous. Ref. [6] generalized TE to asynchronous time series by matching a given event of time series A with the last known event of time series B. This type of matching has two consequences: i) one event of B may be matched to multiple events of A; ii) some event of B might be not matched to an event of A, especially if the typical event frequency is much larger for B than for A.

It is interesting to observe that the reaction time is not immediate in real applications. Therefore, we expect that an event on B might take some time to process and affect the future of A. To measure the amount of this transient, we apply a forward shift to the time series A of δt and apply the matching again.

In the left plot of Fig. 5, we show the TE from stock `NEM` to stock `ABX` computed on 2017-01-03, selecting only events between 09:35 and 16:25 (we removed 5 min at both ends of the trading day to reduce the effect of the auction phase).

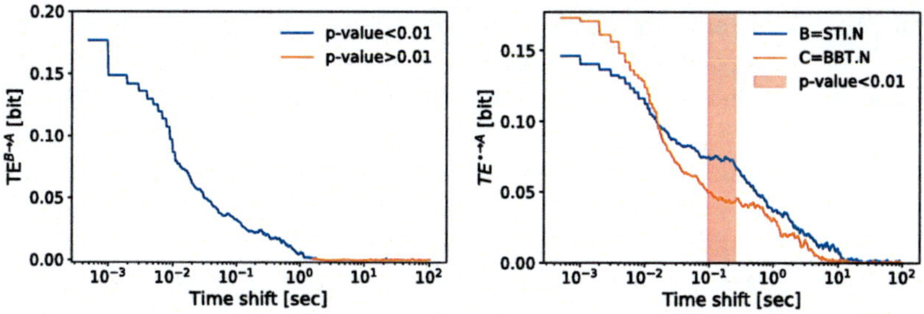

Fig. 5. Left plot: TE from `NEM` to `ABX` as a function of the forward shift of `ABX`; when the shift is larger than 1 second, TE is no more significant. The right plot: TE from `STI` and `BBT` to `RF` as a function of the time shift of `RF` computed over 3 d of data. The difference in TE is significant according a two-sided Vuong test only in the red band.

As expected, the TE drops down as the shift increases. By using the statistical test of Sect. 2 we can detect the maximum $\delta t \simeq 1s$ after which the binary price changes of `NEM` do not add any information to the predictability of those of `ABX`. Note that this delay does not allow one to make useful price predictions in a speculative setting, as prices move too little in 1 s to beat transaction costs. We emphasize here that the ability to perform this test on long time series without bootstrapping opens the way to large-scale studies of the timescales of TE and delays.

In the right plot of Fig. 5 we apply the TE comparison test with time shifts to similar data: we compare the TE from `STI` and `BBT` to `RF` as a function of the time shift δt. In this case, it seems that when $\delta t < 0.01$s, `BBT` transfers more information to `RF` than `BBT`, whereas on a larger time scales ($\delta t > 0.01$s) we observe the opposite. This, however, may simply be due to estimation noise. Indeed, the TE comparison test is only significant in the area highlighted in red at a 0.01 threshold, whereas the comparison for $\delta t < 0.01$ is not.

5.1 Network Inference

The analytic p-values of Sect. 2 can be used to extract the lead-lag network between different stocks. Specifically, we considered all the order books of NYSE and five German exchanges on a single day. Given that the information flow between the European and American exchange must be penalized due to physical distance, we expect to observe higher link homophily within a continent.

We restricted our analysis to a single day (2017-01-03) and selected only the price variations during the daytime when all the exchanges were simultaneously opened (14:30–16:30 UTC). In order to avoid small sample size biases of the analytic p-values, reported in Sect. 4.1, we removed all the stocks which have less than 1000 mid-price variations and we considered only binary variations, i.e., $P_{t+1} > P_t$ or $P_{t+1} < P_t$. After this data filtering, only $n = 1177$ stocks remain. Given that we are performing a multiple comparison test to control the number of false-positive selections, we adopted the Bonferroni correction [8] which prescribes reducing the null hypothesis threshold for each test to q/n_t where $1-q$ is the confidence level at which we can guarantee that no false link is included in the inferred network ($q = 0.01$) and $n_t = n(n+1)/2 = 693,253$ are the number of tests performed. We want to stress that for such a problem using a bootstrap p-value will be computationally infeasible due to the large number of tests performed and the low Bonferroni threshold.

On the resulting statistically validated network [13], shown in Fig. 6 left, only 0.3% of links were retained, and 767 nodes were not isolated. As we expected [3], we observed a high assortativity coefficient [9] by exchange, i.e., 0.56, which means that the number of links inside the same exchange is larger than the one expected from a configuration model.

In the right panel of Fig. 6, we display a coarse-grained network, where all the nodes that belong to the same exchanges are grouped into a single node with a size proportional to the log of the number of nodes, and all the links are grouped with a width proportional to the log the number of links. From this analysis, one observes that the largest interconnection is between NYSE and Frankfurt stock exchanges, which are the most important stock exchanges in our dataset.

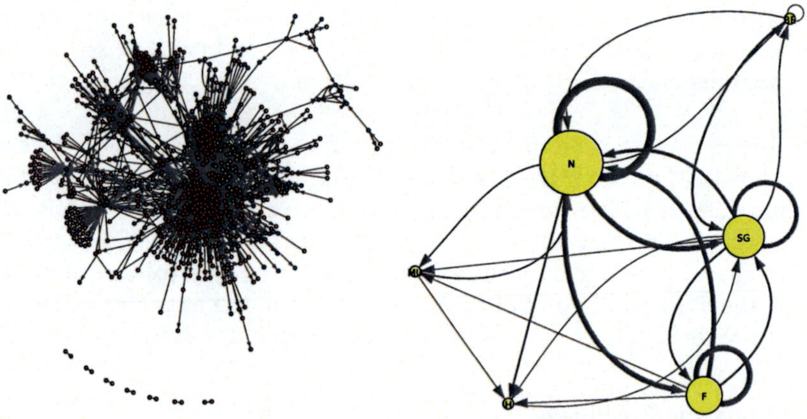

Fig. 6. Left plot: Bonferroni network obtained retaining only the significant TE among each pair of companies of NYSE and 4 German stock exchanges; the color code indicates the stock exchanges. Right plot: network representation of the information flow among the stock exchanges. The size of the nodes is proportional to the log of the exchange's size, and the links' width is proportional to the log of the number of links among the considered groups.

6 Conclusions

Having asymptotic tests for the presence of Transfer Entropy and a comparison of TE makes it possible to infer causality networks of a large number of nodes with long time series, especially since the minimum number of points required is relatively small. Using a time shift is especially relevant in systems in which latency is important, which deserves further study. Finally, being able to compare the TE from two sets of predictors of a given time series opens many possibilities, such as pair-wise ranking of feature importance.

References

1. Barnett, L., Barrett, A.B., Seth, A.K.: Granger causality and transfer entropy are equivalent for Gaussian variables. Phys. Rev. Lett. **103**(23), 238701 (2009)
2. Barnett, L., Bossomaier, T.: Transfer entropy as a log-likelihood ratio. Phys. Rev. Lett. **109**(13), 138105 (2012)
3. Bongiorno, C., London, A., Micciche, S., Mantegna, R.N.: Core of communities in bipartite networks. Phys. Rev. E **96**(2), 022321 (2017)
4. Granger, C.W.J.: Investigating causal relations by econometric models and cross-spectral methods. Econometrica 424–438 (1969)
5. Hadash, G., Kermany, E., Carmeli, B., Lavi, O., Kour, G., Jacovi, A.: Estimate and replace: a novel approach to integrating deep neural networks with existing applications. arXiv preprint: arXiv:1804.09028 (2018)
6. Harré, M.: Entropy and transfer entropy: the Dow Jones and the build up to the 1997 Asian crisis. In: Proceedings of the International Conference on Social Modeling and Simulation, Econophysics Colloquium 2014, pp. 15–25. Springer, Cham (2015)
7. Kontoyiannis, I., Skoularidou, M.: Estimating the directed information and testing for causality. IEEE Trans. Inf. Theory **62**(11), 6053–6067 (2016)
8. Miller, R.G.: Simultaneous Statistical Inference. Springer (1981)
9. Newman, M.: Networks: An Introduction, p. 225. Oxford University Press (2010)
10. Schreiber, T.: Measuring information transfer. Phys. Rev. Lett. **85**(2), 461 (2000)
11. Smirnov, D.A.: Spurious causalities with transfer entropy. Phys. Rev. E **87**(4), 042917 (2013)
12. Staniek, M., Lehnertz, K.: Symbolic transfer entropy. Phys. Rev. Lett. **100**(15), 158101 (2008)
13. Tumminello, M., Micciche, S., Lillo, F., Piilo, J., Mantegna, R.N.: Statistically validated networks in bipartite complex systems. PLoS ONE **6**(3), e17994 (2011)
14. Vuong, Q.H.: Likelihood ratio tests for model selection and non-nested hypotheses. Econ. J. Econ. Soc. 307–333 (1989)
15. Wilks, S.S.: The large-sample distribution of the likelihood ratio for testing composite hypotheses. Ann. Math. Stat. **9**(1), 60–62 (1938)

Networks and Mobility

Extracting Metro Passenger Flow Predictors from Network's Complex Characteristics

Athanasios Kopsidas[1](✉), Aristeides Douvaras[2], and Konstantinos Kepaptsoglou[1]

[1] School of Rural and Surveying Engineering, National Technical University of Athens, 9 Iroon Polytechniou Str, Zografou Campus, 157 73 Athens, Greece
akopsidas@mail.ntua.gr

[2] Deutsche Telekom Pan-Net, Fragkokklisias 9, 151 25 Marousi, Greece

Abstract. Complex network characteristics such as centralities have lately started to be associated with passenger flows at metro stations. Centralities can be leveraged in an effort to develop fast and cost-efficient passenger flow predictive models. However, the accuracy of such models is still under question and the most appropriate predictors are yet to be found. In this sense, this study attempts to investigate appropriate predictors, and develop a predictive model for daily passenger flows at metro stations, based exclusively on spatial attributes. Using the Athens metro network as a case study, a linear regression model is developed, with node degree, betweenness and closeness centralities of the physical network, node strength of the substitute network, and a dummy variable of station importance being as covariates. An econometric analysis validates that a linear model is suitable for associating centralities with passenger flows, while model's evaluation metrics indicate satisfying accuracy. In addition, a machine learning benchmark model is utilized to further investigate variable significance and validate the accuracy of the linear model. Last but not least, both models are utilized for predicting passenger flows at the new metro stations of the Athens metro network expansion. Findings suggest that node strength of the substitute network is a powerful predictor and the most significant covariate of both models; both models' accuracy and predictions converge to a great extent. The model developed is expected to facilitate medium-term disruption management through providing information about metro passenger flows at low cost and high speed.

Keywords: Passenger flow prediction · Complex network · Node centralities · Metro network · Linear regression model · Machine learning

1 Introduction

The importance of metro systems in metropolitan areas is undeniable, especially in an era of growing fuel costs and environmental awareness. As such, efficient planning of those systems is necessary to achieve high levels of performance and leverage all their capabilities. To do so, metro service planning must be based on realistic ridership (travel demand) estimates, which are naturally subject to several, often uncertain factors. On top of that, reliable insights about travel demand are essential for public transport disruption

H. Cherifi et al. (Eds.): COMPLEX NETWORKS 2016 2022, SCI 1077, pp. 529–540, 2023.
https://doi.org/10.1007/978-3-031-21127-0_43

management. When disruptions occur, mitigation plans should be prepared to alleviate negative impacts to public transport services. However, travel demand cannot always be accurately estimated using modern, smart-card based technologies, while short-term prediction models are data-intensive, and long-term predictions through Origin-Destination surveys are costly and time consuming.

Short-term metro passenger flow prediction has been in the spotlight for researchers, with efforts focusing mainly on statistical, machine learning and deep learning models [1]. This problem is usually approached by a spatial-temporal perspective, while the temporal component is treated as a time-series one [2]. Statistical time-series models have been based on autoregressive integrated moving average (ARIMA) and seasonal autoregressive integrated moving average (SARIMA) [3], while machine learning and deep learning-based efforts mainly include neural networks and support vector machine models [4–7]. Machine learning techniques have also been applied to long-term passenger flow prediction [8] and to predictions for passenger flows of new stations and lines [9, 10].

Complex network theory has already been utilized to represent public transport networks and investigate their topological properties [11]. Regarding disruption management in metro systems, complex networks have been exploited for robustness assessment [12, 13] and critical station identification [14]. However, only recently were passenger flows at metro stations associated with complex network characteristics. As suggested by Luo et al. [15], basic network centralities, like degree and betweenness, can indeed be related to passenger flows in urban rail transit. Driven by that finding, the complex characteristics of metro networks, such as node and edge centralities emerging through graph analysis can be considered as appropriate predictors of passenger flows. In this context, this study aims to develop an effective predictive model for passenger flows at metro stations using complex network characteristics as predictors. Apart from metro network centralities, possible metrics that could be appropriate predictors of passenger flows can be found among the centralities of the substitute network, the one representing the alternative routes and modes for moving from one metro station to another outside the metro system [14]. Such a model could be a crucial part of disruption management plans like bus bridging; thus, refers to medium-term disruptions, since during short-term disruptions "doing nothing" is the dominant strategy [16]. The remainder of this paper is structured as follows: in the next section, the paper's methodology is presented. The model developed using the Athens metro network as an application follows. Subsequently, findings are discussed and finally, the conclusions of the study are offered.

2 Methodology

As already mentioned, potential passenger flow predictors include not only the centrality measures of the physical (infrastructural) metro graph, but also on the respective measures of the graph incorporating information about the alternative modes (buses) and routes in a public transport system. The idea behind the latter is that there is a correlation between passenger flows at metro stations and the routes assigned to other modes serving those stations. A two-way causality vector is assumed; first, alternative bus services feeding a metro station means that more passengers can easily reach this station to possibly make a transfer. Second, more alternative services from a metro station can be an incentive for passengers to use the metro system as a first mode for reaching their destination. Either way, total passenger flows of the metro station tend to increase.

To materialize this idea and estimate the respective centrality measures, the methodology proposed in [14] is exploited. Two graphs G and G_s are constructed in order to represent an infrastructure and a substitute network of a metro system, respectively. The nodes of both networks account for the metro stations. The edges of the network of infrastructure account for the physical rail connection between two stations and they are unweighted. On the other hand, the edges of the substitute network account for alternative service (bus) routes being weighted by their performance in terms of quantity and travel time. As such, the substitute network represents the alternative bus routes serving the metro stations outside the metro system. An L-space and a P-space representations are used for the infrastructural and the substitute network, respectively.

To associate centralities with the passenger flows of the metro stations, a linear regression model is developed. The model estimation is based on the Ordinary Least Squares (OLS) method; to ensure that OLS is an appropriate estimator, its assumptions are tested through econometric analysis. The variables used in the model are appropriate centrality measures of aforementioned networks, as well as dummy variables incorporating the spatial dependencies of metro stations. The data about passenger flows at metro stations are daily, so that they can effectively support medium-term disruption management. The flows are calculated as the aggregate inflows and outflows of a station during the day.

2.1 Centrality Measures

Degree, betweenness and closeness centralities are considered as the most appropriate measures to capture station centrality and significance. Degree centrality of a node i measures the number of edges that i belongs to. In its normalized version, degree centrality of an unweighted network is given after divided by $(n - 1)(n - 2)/2$. As far as the graph of the infrastructural network is considered, a node's degree is a signal of whether a station is terminal, an intersection between multiple metro lines or an intermediate station of a single line. It is commonsense that stations of higher degree should serve more passengers than stations of lower degree.

Betweenness centrality shows how many of the total shortest paths of the network pass through a node i. Betweenness centrality can capture the importance of a station in terms of being a passthrough for many paths between other stations in the network. It is obvious that junction stations can carry large passenger volumes, only by their ability to connect different metro lines.

Furthermore, closeness centrality indicates the proximity of a node to all the other nodes of a network. Closeness centrality signifies the proximity of a station to the center of the network. As such, it is an appropriate measure of a metro station's passenger flows, since most of the real-world metro systems are radial, and ridership is higher in central business districts (CBDs) of metropolitan areas.

Last but not least, weighted degree centrality or strength of a node is a generalization of degree centrality in weighted networks. It is calculated as the sum of the weights of the edges that a node belongs to. When it comes to the substitute network, node strength is a measure of a metro station's connectivity to all the other stations, but outside the metro system. It is considered an important determinant of passenger flows, because of the double causality already explained. Centrality measures and the respective formulas are given in Table 1.

Table 1. Centrality measures used in model.

Centralities	Degree (C_D)	Betweenness (C_B)	Closeness (C_C)	Strength (S)
Formula	$\sum_{j}^{n} e_{ij}$	$\sum_{s \neq i \neq t \in N} \frac{\sigma_{st}^{i}}{\sigma_{st}}$	$\frac{n-1}{\sum_{i \neq j \in N} d_{ij}}$	$\sum_{j}^{n} w_{ij}$

where e_{ij} denotes the edge formed by the nodes I and j, n is the total number of nodes in the network, and w_{ij} denotes the weight of an edge e_{ij}.

2.2 Linear Regression Model and Evaluation Metrics

Let us assume that PF_i is an observation of passenger flows of station i. The centrality measures described (denoted as C and S), along with a dummy variable indicating the importance of the stations (denoted as I) are used as predictors of passenger flows. The linear model is developed as follows in Eq. (1):

$$PF_i = \beta_0 + \beta_1 C_D + \beta_2 C_B + \beta_3 C_C + \beta_4 S + \beta_5 I + \varepsilon_i, i = 1, \ldots, n \tag{1}$$

where ε_i is the error of the estimation I, β_0 is the intercept of the model, β_{1-5} are the coefficients of the independent variables, and $\varepsilon_i \sim N(0, \sigma^2)$. The linear model is based on major statistical assumptions [17]; that is, existence of a linear relationship between the dependent and independent variable(s), normality of the errors, homoscedasticity, lack of autocorrelation, and lack of perfect multicollinearity.

In order to evaluate the performance of the models developed, several statistical evaluation metrics are used. These metrics include Mean Absolute Error (MAE), Mean Absolute Percentage Error (MAPE), Root Mean Square Error (RMSE) and Normalized Root Mean Square Error (NRMSE). The models are developed in the training set and evaluated in the test set; these datasets consist of samples randomly extracted at the ratio of 75/25. The metrics are analytically given in Table 2.

Table 2. Evaluation metrics used for model assessment.

Metric	*MAE*	*MAPE*	*RMSE*	*NRMSE*
Formula	$\sum_{i=1}^{n} \frac{\lvert\hat{y}_i - y_i\rvert}{n}$	$\sum_{i=1}^{n} \frac{\frac{\lvert\hat{y}_i - y_i\rvert}{y_i}}{n}$	$\sqrt{\sum_{i=1}^{n} \frac{\lvert\hat{y}_i - y_i\rvert^2}{n}}$	$\frac{RMSE}{\bar{y}}$

3 Application

The methodology presented is applied to the Athens metro network in Greece. Using smart-card data for travel demand, passenger flows at the stations are firstly calculated, and the linear regression model is developed. Subsequently, the model is utilized to predict the passenger flows at the new stations of the Athens metro network expansion. The Athens metro network (Fig. 2) consists of 3 lines and 60 stations serving thousands of residents and tourists daily. The first part of the expansion (part A of Line 4) accounts for a new line of 15 new stations which is right now under construction. It is noted that the data about passenger flows and the performance of the alternative public transport routes correspond to a working-day, representative of the network's activity. Daily ridership in the Athens metro system accounts for 631,498 passengers, who create 2,955,446 inflows and equal outflows, that is, 5,910,892 total flows. From this information, it can also be derived that an average path includes approximately six stations. The five most crowded stations in terms of inflows, outflows and total flows are presented in Fig. 1.

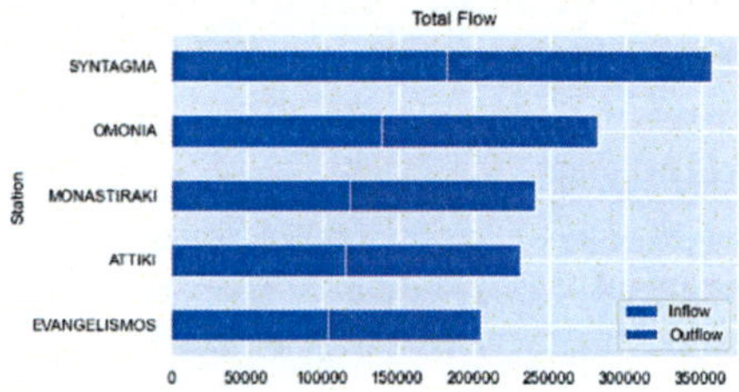

Fig. 1. Passenger flows of the top 5 stations in the Athens metro system.

Fig. 2. The Athens metro network (www.urbanrail.net/eu/gr/athens/athens.htm, retrieved on 14/7/2022).

Degree, betweenness and closeness centralities of the metro network, strength of the substitute network and station importance are used as predictors of passenger flows.

Three ways of importance are considered regarding a station's operation. In specific, a station is considered as important when: i) stands as a transfer station between metro and other rail networks (tram, suburban rail etc.), ii) is located in large commercial areas, and iii) is located near academic institutions. If one of these conditions is fulfilled, the station is considered to be "important". The variable takes the value of 1 in the presence of importance and the value of 0 in the absence of that. The model developed is presented in Table 3.

Table 3. Passenger flow linear regression model.

Variable	Attribute	Coefficient *(95% Confidence Interval)*	P-value	VIF
Constant		– 69,785 *(± 19,310)*	0.001	
Degree	Node degree centrality of the physical network	1,459,398 *(± 532,940)*	0.008	2.946
Betweenness	Node betweenness centrality of the physical network	157,523 *(± 44,061)*	0.001	3.189
Closeness	Node closeness centrality of the physical network	496,305 *(± 196,255)*	0.014	2.630
Sub_strength	Node strength of the substitute network	4.003 *(± 1.035)*	0.000	2.509
Important	1 if a station is important, 0 else	22,099 *(± 7,716)*	0.006	1.128
R^2	0.911	Durbin-Watson statistic	1.912	

According to Table 3, all covariates are significant at a level of statistical significance equal to 95%. On top of that, all independent variables except for closeness centrality are significant at a 99% level of statistical significance. At the same time, there is a positive correlation between all the explanatory variables and the dependent variable. That is, an increase of these variables leads to higher passenger flows at stations. For example, a one-unit increase of betweenness centrality is associated with an increase of approximately 157,523 passengers. Although these numbers may seem extreme (especially for the degree), it should be noted that centralities (except strength) are normalized meaning that the maximum increase of each of them cannot exceed 1. Thus, the real interpretation accounts for fractions of centrality deviations. Moreover, should a station be important, additional 22,099 passengers gather at it. Furthermore, a value of R^2 equal to 0.911 signals a high goodness of model's fit.

When it comes to linear model's assumptions, first, there is no evidence of strong multicollinearity since Variable Inflation Factors (VIF) are low (none of them exceeds 5), as well as Pearson correlation coefficients do not exceed 0.80. Pearson correlation matrix is depicted in Fig. 3. Moreover, Durbin-Watson statistic (1.912) indicates that there is no strong evidence of autocorrelation (a value near 2 indicates no autocorrelation). Furthermore, the Breusch-Pagan test indicates that there is no strong evidence of the existence of heteroscedasticity (p-value = 0.06 > 0.05). As for the distribution of the errors, the Shapiro-Wilk test (more appropriate for small samples) indicates that the errors are normally distributed (p-value = 0.948 > 0.05), thus validating the assumption of normality. Last but not least, linearity should be ensured by the fact that the residuals are distributed normally and they are simultaneously homoscedastic. All in all, all the assumptions of the linear model are validated, hence the model is appropriate for associating passenger flows with the explanatory variables.

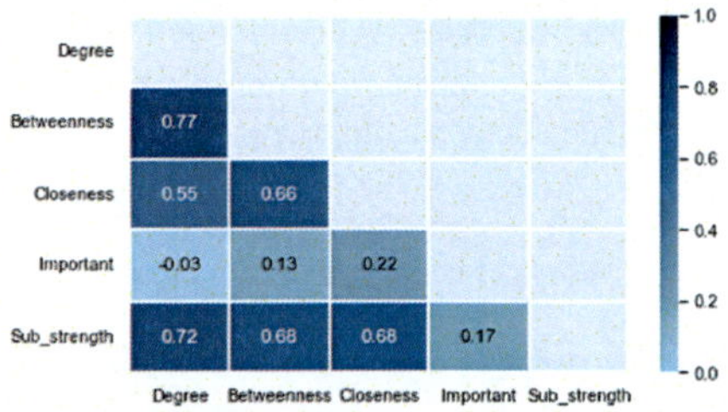

Fig. 3. Pearson correlation matrix for model covariates.

In addition, a machine learning benchmark model based on the Extreme Gradient Boosting algorithm (XGBoost) is developed using the same variables as predictors, in order to validate variable significance and linear model's accuracy. As XGBoost is one of the most appreciated machine learning models in terms of success rate [18], useful insight is expected to derive from a comparison to the linear model. The model's parameters (set as the most appropriate through a greed research fine-tuning process) are the following: "colsample_bytree" – 0.2, "eta" – 0.0005, "max_depth" – 1, "n_estimators" – 1100, "subsample" – 0.05. First, the significance of the independent variables in the model is evaluated by the feature importance F-scores, and the SHAP values. The former count how many times a variable is used for splitting a decision tree [19], while the latter indicate the impact of the features on individual predictions [20]. According to both feature importance and SHAP values, node strength of the substitute network is the most important feature regarding the XGBoost model, followed by closeness and betweenness centralities of the physical network. Feature importance and SHAP values are presented in Fig. 4.

When it comes to model evaluation, both regression and machine learning model accuracy is evaluated through the metrics described in the methodology, which are presented in Table 4. According to the evaluation metrics presented in Table 4, it can be inferred that linear model demonstrates rather satisfying accuracy and strong predictive power. Compared to the benchmark model, its accuracy is similar or even higher according to the RMSE and NRMSE.

Subsequently, both models are utilized to predict the daily passenger flows at the stations of the new line of the Athens metro network. It is noted that the real values are not available, since the stations are still under construction. The predictions of both models are presented in Table 5.

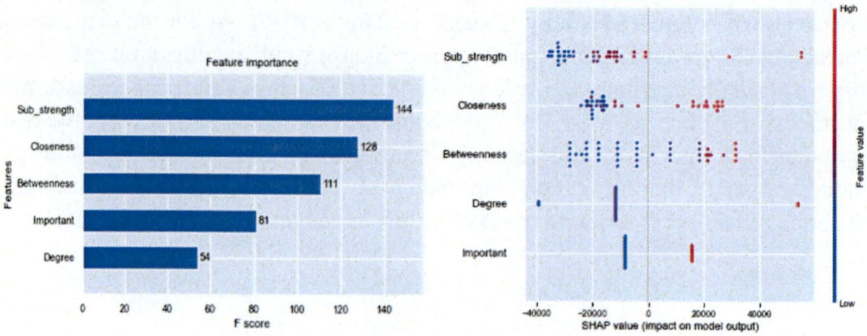

Fig. 4. XGBoost feature importance and SHAP values.

Table 4. Model evaluation metrics.

	MAE	MAPE (%)	RMSE	NRMSE (%)
Regression	14,318.81	28.18	18,449.49	18,73
XGBoost	16,030.42	17.00	22,318.83	21.08

Table 5. Predictions of the daily passenger flows at the new stations of the Athens metro network.

Station	Predicted flows		Station	Predicted flows	
	(Regression)	(XGBoost)		(Regression)	(XGBoost)
Alsos Veikou	21,207	16,779	Kolonaki	106,763	107,537
Galatsi	58,264	49,353	Evangelismos	265,438	297,971
Elikonos	62,033	60,722	Kessariani	98,446	103,011
Kipseli	76,548	52,766	Panepistimioupoli	94,118	90,084
Dikastiria	89,830	79,368	Ilisia	63,366	48,944
Alexandras	94,283	103,090	Zografou	84,482	75,496
Exarchia	110,697	123,362	Goudi	58,021	31,778
Academia	292,247	301,900			

It is noted that Evangelismos already exists as a station of another line in the Athens metro system. In addition, the Academia station will be constructed close to the existing station of Panepistimio, and an underground connection will be available for the passengers to make a transfer between lines. As such, when designing the graph of our methodology Academia and Panepistimio were considered as the same station/node. Thus, the passenger flows predicted consist of the flows from the existing network and the additional future flows from the new line. As a comparison between the real flows today and the predicted future flows, Evangelismos now serves 205,342 passengers daily and Academia/Panepistimio 166,080. It is obvious that the model developed predicts an increase in the flows, which is reasonable due to the fact that an additional line will feed these stations with passengers. However, the increase of the Academia station is rather high, mainly because of its location (close to the city center) and the proximity from the local universities.

4 Discussion

Although causality is not implied, there is strong evidence that complex network characteristics are correlated to passenger flows. However, the efforts based on developing predictive models for travel demand solely based on the complex network characteristics are still in their infancy. The model developed in this study suggests that characteristics of the substitute network can be useful for creating models of high accuracy. In particular, node strength of the substitute network is the most significant covariate in a linear model developed for this purpose. Its positive effect on passenger flows indicates the existence of a bidirectional causality between alternative routes and passenger flows. The assumptions that: i) the more alternative routes available to a metro station, the more passengers are fed to the metro system, and ii) the more alternative routes available from a metro station, the stronger the incentive to use this station as a transfer station, seem to be validated.

The evaluation metrics suggest that the model developed is of high predictive accuracy. In addition, the econometric analysis proves that a linear model is in fact appropriate for this purpose. When it comes to the comparison between the linear and the machine learning model, the latter seems to validate the performance of the former. Node strength of the substitute network is again the most important covariate. As such, there is even stronger evidence that node strength is in fact a valid predictor of passenger flows. Moreover, the evaluation metrics of the machine learning model are near the ones of the linear model. The convergence between the evaluation metrics of a machine learning model of high predictive power and the ones of the linear model, validates the appropriateness of the linear model.

Fast and accurate travel demand predictions are very useful at both strategic and tactical level regarding public transport planning. Complex network-based models can be extremely fast, but their reliability is still being tested. In this direction, the complex network-based model proposed in this study succeeds in reducing the gap between speed and reliability, by achieving high levels of predictive accuracy. The dependent variable of daily passenger flows indicates a medium-term horizon of the methodology proposed. That is, it is appropriate for daily predictions at a strategic level, design of mitigation strategies regarding planned disruptions, and contingency plans regarding unplanned disruptions of medium-size duration (one to three days). As for the disruption management occasions, passenger flow prediction can also support mitigation strategies such as bus bridging or frequency enhancing.

As an example of decision making at a strategic level, the model developed was applied to predict the daily passenger flows at the stations of a new metro line. The first section of the Athens new Line 4 was used as a case study. The predictions seem reasonable compared to the real data of other metro stations in the same system. In addition, the predictions of the XGBoost machine learning model, which was used as a benchmark, do not diverge to a great extent, but on some occasions are rather similar instead. The linear model seems to overestimate the flows at the peripheral stations and underestimate the flows at the central stations, compared to the machine learning predictions. This fact probably derives from the linear nature of the model; however, its suitability cannot be denied due to the aforementioned econometric analysis. When it comes to the stations shared by both the new and the existing lines, it is predicted that passenger flows will increase. Nevertheless, the magnitude of the increase is not similar for all the stations, since factors such as the proximity to the city center and to academic institutions add additional travel demand to a station. Unfortunately, the results cannot be validated by real data, as the stations are still under construction.

New lines are of particular interest as there is no intuitive knowledge of their operation, and thus many aspects lie in the realm of predictions. Reliable passenger flow predictions can facilitate the design and strategy of new lines in terms of both metro infrastructure characteristics (station capacity, line frequency etc.) and the interaction of the metro system with other public transport systems, such as buses, tram etc. When it comes to the latter, passenger flows at metro stations might affect the design of other public transport network in the framework of a multi-modal integrated public transport system. For instance, new routes should be assigned and existing route capacity should be enhanced in order to support the inflows and outflows of the new stations.

5 Conclusions

In this study, a linear regression model was developed to associate daily passenger flows at metro stations with appropriate network centralities. The most appropriate predictors used for this purpose are mainly characteristics of two complex networks: i) the physical metro network, and ii) the substitute network. The former captures the topology of the metro network, while the latter includes service attributes of the public transport system in which metro operates. In particular, node degree, betweenness and closeness centralities of the physical network, as well as node strength of the substitute network are used as predictors of passenger flows. Apart from the centralities, a dummy variable of station importance is also included in the model. After being applied to the Athens metro network, the results suggest that the linear model is suitable for prediction and its accuracy is rather satisfying. On top of that, through the comparison of the linear model to a machine learning benchmark model (XGBoost), its predictive power is validated.

Complex network centralities of a metro network such as degree, betweenness and closeness (both in terms of spatial and service attributes) have already been associated with station passenger flows. However, node strength of the substitute network had never been researched for this purpose. This study provided evidence that this complex network characteristic is a valid and powerful predictor. Complex network-based predictive models for passenger flows at metro stations are essential, because of their ability to provide fast and reliable estimations of the flows. This kind of predictions is crucial, both at strategic and tactical level. At a strategic level, accurate predictions of passenger flows facilitate the planning of new lines and stations, in terms of infrastructure design and routing or scheduling of the alternative public transport modes that support the metro system. At a tactical level, passenger flow predictions are essential for disruption management. The model developed in this study can successfully support medium-term disruption management including mitigation strategies such as bus bridging.

For future research, it is recommended that the model developed should be validated in additional public transport systems for testing its adaptability and universality. More complex networks than the Athens metro system would further highlight the strengths and weaknesses of the proposed methodology. Future efforts should especially be focused on node strength of the substitute network, which seems to be a very promising predictor. In addition, predictions of passenger flows at the stations of new metro lines can be utilized in order to estimate the criticality and robustness of those stations in cases of disruptions. Since passenger flows is a significant factor in such cases, accurate estimations can lead to an a priori efficient planning and management.

Acknowledgements. This work was supported by the Basic Research Program, PEVE 2021, National Technical University Athens.

References

1. Han, Y., Peng, T., Wang, C., Zhang, Z., Chen, G.: A hybrid GLM model for predicting citywide spatio-temporal metro passenger flow. ISPRS Int. J. Geo Inf. **10**(4), 222 (2021)
2. Ou, J., Sun, J., Zhu, Y., Jin, H., Liu, Y., Zhang, F., Huang, J., Wang, X.: STP-TrellisNets: Spatial-temporal parallel TrellisNets for metro station passenger flow prediction. In: Proceedings of the 29th ACM International Conference on Information & Knowledge Management, pp. 1185–1194 (2020)
3. Zheng, Z., Ling, X., Wang, P., Xiao, J., Zhang, F.: Hybrid model for predicting anomalous large passenger flow in urban metros. IET Intel. Transport Syst. **14**(14), 1987–1996 (2020)
4. Yang, X., Xue, Q., Yang, X., Yin, H., Qu, Y., Li, X., Wu, J.: A novel prediction model for the inbound passenger flow of urban rail transit. Inf. Sci. **566**, 347–363 (2021)
5. Sun, Y., Leng, B., Guan, W.: A novel wavelet-SVM short-time passenger flow prediction in Beijing subway system. Neurocomputing **166**, 109–121 (2015)
6. Li, H., Wang, Y., Xu, X., Qin, L., Zhang, H.: Short-term passenger flow prediction under passenger flow control using a dynamic radial basis function network. Appl. Soft Comput. **83**, 105620 (2019)
7. Guo, J., Xie, Z., Qin, Y., Jia, L., Wang, Y.: Short-term abnormal passenger flow prediction based on the fusion of SVR and LSTM. IEEE Access **7**, 42946–42955 (2019)
8. Toqué, F., Khouadjia, M., Come, E., Trepanier, M., Oukhellou, L.: Short & long term forecasting of multimodal transport passenger flows with machine learning methods. In 20th International Conference on Intelligent Transportation Systems (ITSC), pp. 560–566. IEEE (2017)
9. Gong, Y., Li, Z., Zhang, J., Liu, W., Yi, J.: Potential passenger flow prediction: a novel study for urban transportation development. In: Proceedings of the AAAI Conference on Artificial Intelligence, Vol. 34, No. 04, pp. 4020–4027 (2020)
10. Wang, K., Wang, P., Huang, Z., Ling, X., Zhang, F., Chen, A.: A Two-Step model for predicting travel demand in expanding subways. IEEE Trans. Intell. Transp. Syst. (2022)
11. Lin, J., Ban, Y.: Complex network topology of transportation systems. Transp. Rev. **33**(6), 658–685 (2013)
12. Cats, O., Krishnakumari, P.: Metropolitan rail network robustness. Physica A **549**, 124317 (2020)

13. Yang, Y., Liu, Y., Zhou, M., Li, F., Sun, C.: Robustness assessment of urban rail transit based on complex network theory: a case study of the Beijing Subway. Saf. Sci. **79**, 149–162 (2015)
14. Kopsidas, A., Kepaptsoglou, K.: Identification of critical stations in a metro system: a substitute complex network analysis. Physica A **596**, 127123 (2022)
15. Luo, D., Cats, O., van Lint, H.: Can passenger flow distribution be estimated solely based on network properties in public transport systems? Transportation **47**(6), 2757–2776 (2019). https://doi.org/10.1007/s11116-019-09990-w
16. Zeng, A.Z., Durach, C.F., Fang, Y.: Collaboration decisions on disruption recovery service in urban public tram systems. Transp. Res. Part E: Logistics Transp. Rev. **48**(3), 578–590 (2012)
17. Su, X., Yan, X., Tsai, C.L.: Linear regression. Wiley Interdisc. Rev. Comput. Stat. **4**(3), 275–294 (2012)
18. Gumus, M., Kiran, M.S.: Crude oil price forecasting using XGBoost. In: 2017 International Conference on Computer Science and Engineering (UBMK), pp. 1100–1103. IEEE (2017)
19. Chen, M., Liu, Q., Chen, S., Liu, Y., Zhang, C.H., Liu, R.: XGBoost-based algorithm interpretation and application on post-fault transient stability status prediction of power system. IEEE Access **7**, 13149–13158 (2019)
20. Spadon, G., de Carvalho, A.C., Rodrigues-Jr, J.F., Alves, L.G.: Reconstructing commuters network using machine learning and urban indicators. Sci. Rep. **9**(1), 1–13 (2019)

Estimating Peak-Hour Urban Traffic Congestion

Marco Cogoni(✉), Giovanni Busonera, and Francesco Versaci

CRS4: Center for Advanced Studies, Research and Development in Sardinia, 09100 Cagliari, Italy

marco.cogoni@crs4.it, giovanni.busonera@crs4.it, francesco.versaci@crs4.it

https://www.crs4.it

Abstract. We study the emergence of congestion patterns in urban networks by modeling vehicular interaction by means of a simple traffic rule and by using a set of measures inspired by the standard Betweenness Centrality (BC). We consider a topologically heterogeneous group of cities and simulate the network loading during the morning peak-hour by increasing the number of circulating vehicles. At departure, vehicles are aware of the network state and choose paths with optimal traversal time. Each added path modifies the vehicular density and travel times for the following vehicles. Starting from an empty network and adding traffic until transportation collapses, provides a framework to study network's transition to congestion and how connectivity is progressively disrupted as the fraction of impossible paths becomes abruptly dominant. We use standard BC to probe into the instantaneous out-of-equilibrium network state for a range of traffic levels and show how this measure may be improved to build a better proxy for cumulative road usage during peak-hours. We define a novel dynamical measure to estimate cumulative road usage and the associated total time spent over the edges by the population of drivers. We also study how congestion starts with dysfunctional edges scattered over the network, then organizes itself into relatively small, but disruptive clusters.

Keywords: Traffic · Urban networks · Betweenness centrality · Congestion

1 Introduction

Urban networks have been widely studied in recent years [1–3], both their growth over time and their complex dynamics under different traffic levels. Network science has considerably helped to improve our understanding of cities and to analyze and predict the reaction of the different parts of the network under stress [4,5]. Such predictive analyses may be performed by considering just the geographical and topological features of a city, without having to recur to experimental traffic data, but these datasets constitute nonetheless the reference

H. Cherifi et al. (Eds.): COMPLEX NETWORKS 2016 2022, SCI 1077, pp. 541–552, 2023.
https://doi.org/10.1007/978-3-031-21127-0_44

against which every theoretical model should be evaluated [6,7]. Urban networks belong to the special class of (almost) planar graphs and in this context there have been some notable results recently [8,9]. The geographical embedding leads to several constraints to topology such as limitations to the number of long-range connections and to the maximum connectivity observed at the single edge level [10]. The study of edge degree distributions did not improve much our understanding of cities, but non-local higher-order metrics such as network centralities have been widely used both for theoretical studies and for practical applications with notable success [11,12]. One of the metrics that have been used the most in recent years is the Betweenness Centrality (BC), which is defined as the total flow passing over each node of the network when enumerating all Origin-Destination (OD) pairs and connecting them via shortest-path routing.

This definition is easily extended to Edge BC (EBC) by counting edge usage instead. We will write BC instead of EBC in the rest of paper since our focus will be on edges.

BC and other measures have been used to predict which edges are subject to the highest traffic demand that leads to the breakup of the network into functionally independent pieces. This phase transition has been well studied via percolation theory and it is known that the size distribution of the resulting sub networks follows a power law with a critical exponent that depends on traffic intensity and on the time of the day. This transition has been observed for real world datasets in large cities such as London, Beijing and New York City [4,13,14]. During the most extreme conditions, total network breakdown has been observed to last for several hours or even days [5].

The percolation phase transition is not specific of high congestion levels, but an edge may be classified as dysfunctional even with vehicles traveling just below the speed limit [2,4,13]: this transition simply signals a change in the network behavior that happens at all traffic levels, but for different critical speeds. On the other hand, in this work we will follow the more practical definition of calling an edge dysfunctional only when it cannot receive traffic anymore, due to road density nearing the maximum value and speed approaching zero [5]. The transition happens when there is suddenly a large part of the desired travel paths that are no longer usable.

Standard BC approaches can only grasp a limited picture of the network under stress because BC relies on several assumptions [15,16]: traffic origins and destinations uniformly spread over the nodes; shortest paths with fixed cost function; non-interacting multiple paths sharing the same edge; the amount of traffic density contributed by each vehicle is the same regardless of its driving time. Edge usage obtained from a BC computation approximates well a network with very small (or extremely fast) agents, each using shortest path navigation with no congestion awareness [17]. A urban transportation network is characterized by vehicle travel times comparable to the typical timescale of congestion buildup during peak-hours.

In this work we want to extend the standard BC as a measure of road congestion, especially by taking into account the strong interaction effects observed in real traffic. The interaction among vehicles depends on the duration and vehicular volume of the network loading: the longer the edge travel times, the higher the probability of finding a vehicle in a given road segment during the observation period. The present approach stems directly from this fact and aims at focusing on the peak-hour periods, that usually last for one hour, to estimate the cumulative traffic seen on the roads during that finite time window. Thus, we propose a dynamical model to compute the contribution to traffic at a road-segment level due to each vehicle added to the network, while iteratively recomputing travel times after each addition. From the vast literature on transportation we choose one of the simplest models to describe vehicular behavior depending on geography (edge properties) and on the dynamical network state: the single regime Greenshields model [9,18], for which speed starts at the free flow value to decrease linearly to zero when maximal road density is attained. To complement the traffic model, we assume that vehicles know exactly the network state before their departure, in order to plan an optimal path. The proposed scenario mimics a network with a mixture of self-driving cars and human drivers generating a shortest-time route at the start of their trips. It is likely that this will become increasingly relevant in the near future.

In this paper we employ our dynamical model and the associated metrics to theoretically predict the behavior of five large cities under growing traffic. Results will be validated against real traffic data in future works.

2 Methods

2.1 Interaction Model

We aim at modeling the network evolution, as observed by travelers, while the traffic increases from zero up to complete gridlock. The desired network traffic over simulation time (τ) will be added incrementally, activating one new path $\pi(i)$ at each simulation step i. Thus, i can be interpreted both as the current number of added paths and as a temporal marker to define the sequence of OD pairs randomly generated for each simulation.

For simplicity, and to be able to compare results with the standard BC, traffic will be added uniformly to the network. It is however straightforward to adapt our procedure to any OD matrix.

We model the traffic network as a directed, weighted graph $G = (V, E)$ with $N = |V|$ nodes and $M = |E|$ edges. Each edge e represents a road segment between two intersections and it is characterized by three constant features: its physical length l_e, maximum speed v_e^* and number of lanes c_e.

The state of the network at each time step will be represented by the occupancy of vehicles added so far in each edge $s_e(i)$, where $s_e(i) = \sum_{j=1}^{i} \sigma_e(j)$. We will define the occupancy σ of a single vehicle so that it will sum to one only when its total traveling time is equal to or longer than the simulation time (in general: $\sum_{e\in\pi(i)} \sigma_e(i) \leq 1$). To each edge we also associate a normalized vehicle density $\rho_e(i)$, that increases monotonically as we add vehicles:

$$\rho_e(i) = \frac{s_e(i)L}{l_e c_e} \in [0,1] \ , \tag{1}$$

where L is the average space occupied by one vehicle.

We define the occupancy $\sigma_e(i)$ of a new vehicle added to the network using a simple approximation of the time it might spend on edge e:

$$\sigma_e(i) = \min\left(\frac{T_e(i)}{\tau}, 1\right) , \qquad \text{where } T_e(i) = \frac{l_e}{v_e(i)} , \tag{2}$$

and with speed $v_e(i)$ following a Greenshields linear law [19]:

$$v_e(i) = v_e^*(1 - \rho_e(i-1)) , \tag{3}$$

thus the approximate time for the complete path $\pi(i)$ will be $T_{\pi(i)} = \sum_{e\in\pi(i)} T_e(i)$. A non-interacting system is obtained in the limit $L \to 0$. In more detail, the state $s(i)$ of the network is obtained iteratively ($s(i) = f(s(i-1))$), starting with an empty network ($s(0) = 0$) and according to the following dynamic process:

- a pair of OD nodes is chosen, independently and uniformly at random, and the fastest path $\pi(i)$ connecting the nodes is computed;
- starting from O, for each edge $e \in \pi(i)$, we accumulate the average occupancy $\sigma(i)$ induced by $\pi(i)$ during τ:

$$s_e(i) = s_e(i-1) + \sigma_e(i) , \tag{4}$$

  and the number of vehicles on the edge:

$$n_e(i) = n_e(i-1) + 1 . \tag{5}$$

  Note that, as soon as the sum along $\pi(i)$ of the added $T_e(i)/\tau$ factors reaches 1, we choose to skip the remaining edges up to D, to avoid adding more than a unit factor to vehicle occupancy along $\pi(i)$;
- $v_e(i+1)$ and $T_e(i+1)$ are updated according to Eqs. 2 and 3, using the new $\rho_e(i)$ value;
- this process is iterated until the desired total traffic is reached.

Intuitively, the occupancy factor induced by a vehicle over an edge e is proportional to the time the vehicle is supposed to spend on it (T_e/τ), as forecast at the moment of its generation, and the sum over the whole path will be equal to unity (certainty of finding the vehicle within π during τ) only when $T_\pi \geq \tau$. The initial OD pairs find a nearly empty network, so their fastest paths and travel times are very similar to the $L = 0$ case. With rising traffic, however, edges slow down and subsequent fastest paths will be slower and, in general, different. Some edges will eventually reach maximum density and become dysfunctional. If the fastest route from origin to destination comprises a dysfunctional edge (i.e., the network is disconnected) we still choose to add the initial part of the path, but we will skip the edges starting from the first dysfunctional one. This allows us to model the backward propagation of traffic jams observed at high traffic volumes [14,20]. Since the order in which OD pairs are added to the network can lead to different final results, we replicate the system and perform several simulations to analyze the stability of the results.

The main limitations of this model are: the Greenshields model is very rough; no explicit time evolution as in the Nagel-Schreckenberg CA model, but just a sequence of path addition in a *first come, better served* approach; no true time evolution means that long paths added initially will generally experience faster times (uncongested) for edges that will likely become congested later by newer paths; OD pairs are uniformly extracted at random over the network: it is known, e.g., that morning and evening peak hours show an opposite average traffic direction [1].

2.2 Cumulative BC Definition

BC implicitly assumes that the different paths do not interact, i.e., that the cost (typically the shortest length between two points) does not change when adding agents in the network. If, however, we assume that the active network traffic does alter the cost function, as it is the case when choosing fastest vs shortest paths, then the standard BC gives a biased picture of bottlenecks and hotspots. A good approximation of the BC can be obtained by sampling the OD pairs uniformly [21], and it is natural to think about it as a dynamical process of subsequent path additions. In particular, if the exact paths, their order of appearance and their travel times are known, the timescale τ of the dynamical phenomenon becomes a crucial parameter to better estimate the edge visiting frequencies.

Several studies extending the original BC concept have been presented in the past [11,12,22], to deviate from the idea of shortest paths by introducing routing randomness. Here, on the other hand, we extend the BC idea by taking into account the evolution of the network state. To this aim, we define a Cumulative BC (CBC), based on the average occupancy $s_e(i)$, defined in Eq. 4 as:

$$\gamma_e(i) = \frac{s_e(i)}{i} \ . \tag{6}$$

This quantity is normalized and tends to the original BC for $L \to 0$.

3 Simulation Details

We select the vehicular transportation layer of the urban networks relative to five large cities and their surroundings from OpenStreetMap. The radius R of the circle inscribed in each region varies from 12 km for Rome to 20 km for Boston, and the number of edges of the corresponding graphs ranges from about 65k of Rome to about 200k of London. For every city, we perform 10 simulations, each with a reshuffled order of the OD additions to the network to estimate the sensitivity of the evolving configurations to external conditions. The total number of added ODs was 10^6, sufficient to bring all cities to a deeply congested state. The total computation time for simulation and analysis is less than one day for each city, running with shared-memory parallelism on Intel E5-2680 nodes (12 cores, 2 threads/core) with 250 GB RAM.

4 Results and Discussions

Transition to Fragmented Network We first focus on the behavior of the network while it receives an increasing number of paths. Since edges become progressively dysfunctional as $\rho \to 1$, at some traffic level (set by i) the graph becomes disconnected, which means that some OD pairs will belong to functionally separated subgraphs. When this occurs, all paths connecting those OD pairs will include a dysfunctional edge and we say that the OD pair is (at least partially) rejected. This is a coarse indicator of a transition in the network behavior. We present rejection ratio curves in Fig. 1: Nairobi (Fig. 1a) and Rio (Fig. 1b) are the first to collapse at $i \approx 210$k and $i \approx 320$k, respectively, with moderately steep curves. The different transition types depend upon the topological features of each city (see maps on Fig. 4): London (Fig. 1c) and Rome (Fig. 1d), being traversed by rivers, abruptly break at $i \approx 335$k and $i \approx 490$k, respectively, leaving both cities split in two halves (50% of rejected ODs). The collapsed bridges of both cities, due to heavy usage, are clearly visible in Fig. 4c. Once broken in two parts, London is able to better support local traffic better than Rome that collapses quickly even for shorter ODs, as visible from the different slopes after the transition. Boston (Fig. 1e) is apparently the most resilient to breakup, with the first hints of congestion appearing at $i \approx 560$k, and a global behavior similar to Nairobi and Rio.

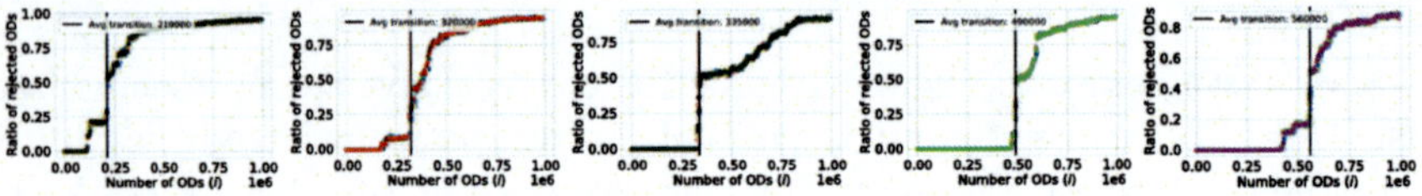

(a) Nairobi (b) Rio (c) London (d) Rome (e) Boston

Fig. 1. Fraction of (partially) rejected paths for increasing traffic (i)

We now shift our focus to the spatial configuration of dysfunctional edges: to display their typical growth over the graph, we choose Rio in Fig. 2 (left): the number of dysfunctional edges (red curve) and that of edges belonging to the Largest strongly connected dysfunctional Cluster (LC) grow almost monotonically with i (black curve). It is worth noting that, although scattered dysfunctional edges appear early and grow monotonically, the LC lags and shows a staircase-like behavior, indicating that dysfunctional edges take time to coalesce into a set of clusters. We measure only the LC size and the plateaus are explained by new dysfunctional edges appearing in distant parts of the city. Remarkably, the table on the right of Fig. 2 shows that, for all cities, a small number of dysfunctional edges of order 0.1% of the total edges, is enough to severely hinder transportation efficiency.

In order to further analyze the transition to congestion, we define two additional measures: the average path length ratio: $\mathbf{D}(i) = \frac{1}{iR} \sum_e n_e(i) \cdot l_e$ and the

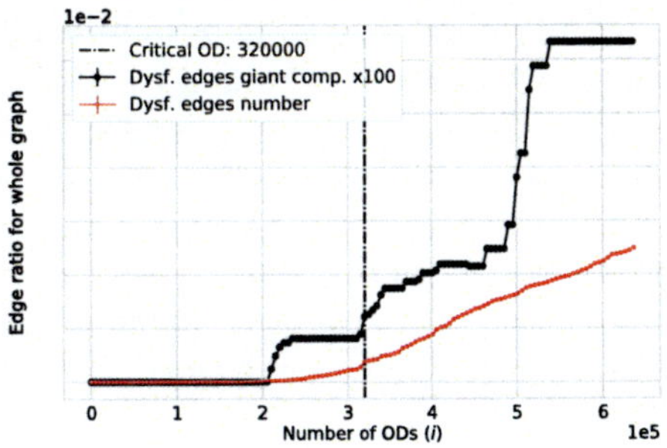

	# dysf.	LC size
Boston	$2 \cdot 10^{-3}$	$4 \cdot 10^{-5}$
London	$5 \cdot 10^{-4}$	$2 \cdot 10^{-5}$
Rome	$3 \cdot 10^{-3}$	$6 \cdot 10^{-5}$
Rio	$1 \cdot 10^{-3}$	$3 \cdot 10^{-5}$
Nairobi	$1 \cdot 10^{-3}$	$9 \cdot 10^{-5}$

Fig. 2. Spatial correlation of congested edges. Left: Fraction of dysfunctional edges (w.r.t. the entire graph) found at the transition in Rio as traffic increases in red, and in black (magnified 100 times), the associated largest connected component cluster size fraction. Right: Values of red (# dysf.) and black (LC size) curves at the transition for each city.

average path time per vehicle: $\mathbf{T}(i) = \frac{1}{i}\sum_e n_e(i) \cdot t_e$. Since $n_e(i)$ is the number of vehicles that chose edge e in τ, by summing its products with edge lengths (traversal times) we obtain the global distance run (time spent) by all vehicles in τ. $\mathbf{D}(i)$, with respect to the traffic level i, shows different regimes up to the transition: a constant or gentle slope for London and Rio, respectively, a bit steeper for Boston; a marked peak at the transition for all cities except Nairobi and Boston (Fig. 3); a final decrease, for all cities after the transition, mainly due to partial truncation of paths added after the transition. $\mathbf{T}(i)$ per vehicle are similar for all cities: Fig. 3 (right) shows a gentle slope for low traffic regimes, which suddenly steepens nearing the transition with the dysfunctional edges restricting the choice of optimal paths. For brevity, $\mathbf{D}(i)$ and $\mathbf{T}(i)$ are shown in Fig. 3 only for Boston: as traffic rises, avoidance of congested roads leads to an increase in the average travel length, but in general by less than $\approx 10\%$.

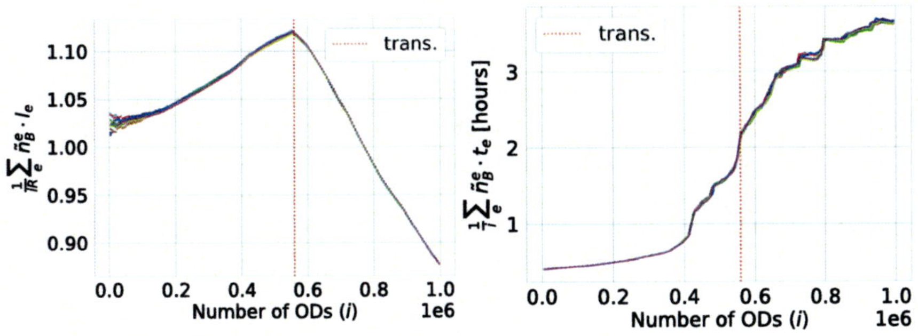

Fig. 3. (Average path length ratio $\mathbf{D}(i)$ (with respect to R) (left) and path travel time $\mathbf{T}(i)$ (right) per vehicle for Boston. Each color represents a different replica with reshuffled OD order.

BC at Increasing Levels of Traffic Standard BC may be used to probe the network state, after the addition of a number of vehicles, to obtain instantaneous information on how congestion affects path choices for increasing traffic levels. Figure 4 a, b show how the BC maps of our five cities change from the empty state to the transition. Specifically, in Fig. 4b, edges with either larger or lower BC with respect to the empty state, are shown in red and blue respectively. It is worth to note that the roads with the highest BC with little traffic become dysfunctional long before the transition as entirely different path choices emerge: blue roads in Fig. 4b are those that lost most of their flow due to early saturation while the red ones became important at higher congestion levels.

Cumulative BC The "loaded" standard BC discussed above, does not describe the network behavior during the whole simulation time, but just what would happen to a small batch of vehicles added on top of a system with a specific traffic load. The Cumulative BC (CBC), on the other hand, takes into account the total contributions of all vehicles added within τ. The CBC maps are shown in Fig. 4c exactly at the transition i_c specific for each city, so each edge carries the contribution to the global traffic experienced during the whole simulation time. Several edges show high CBC values not detected as important by a BC in Fig. 4a. This is especially true for peripheral roads that become important only when the fastest options are already saturated: notably, Boston shows a somewhat busy eastern region and Rome has several radial roads that show a relatively large usage.

Total Time Spent in Traffic (TS) After defining the CBC as a proxy for the number of vehicles to be observed during the network loading, we can use this information to obtain some interesting features, such as the total time spent by all vehicles on each edge: $\mathbf{TS}_e(i) = \frac{1}{l_e}\sum_e n_e(i)T_e(i)$. This measure provides an intuitive way of estimating the total wait experienced, per unit length, during τ by all vehicles sharing an edge. As visible in Fig. 4d, the maps show the relative importance of each road: low values belong to seldom used roads while high values emerge for roads that are frequently chosen whether with an empty network (coinciding with BC) or at near-congestion. In general, urban highways, with several lanes and higher speed limits, are the primary candidates to display top scores in this metric, but as congestion grows, other roads, often not planned for heavy use, start to attract a large fraction of vehicles and may compete with highways. It should be noted how most of the top-ranked roads in London are bridges over the Thames that in fact are the first to become dysfunctional (as seen in Fig. 1c), leading to a city split in two roughly equal parts.

CBC and Total Time Spent in Traffic We study how the distributions of CBC (Fig. 5a) and total time spent in traffic (TS) (Fig. 5b) change for all cities when traffic grows from light to severely congested. As expected, both distributions shift to the right and a double peak structure starts to emerge already for

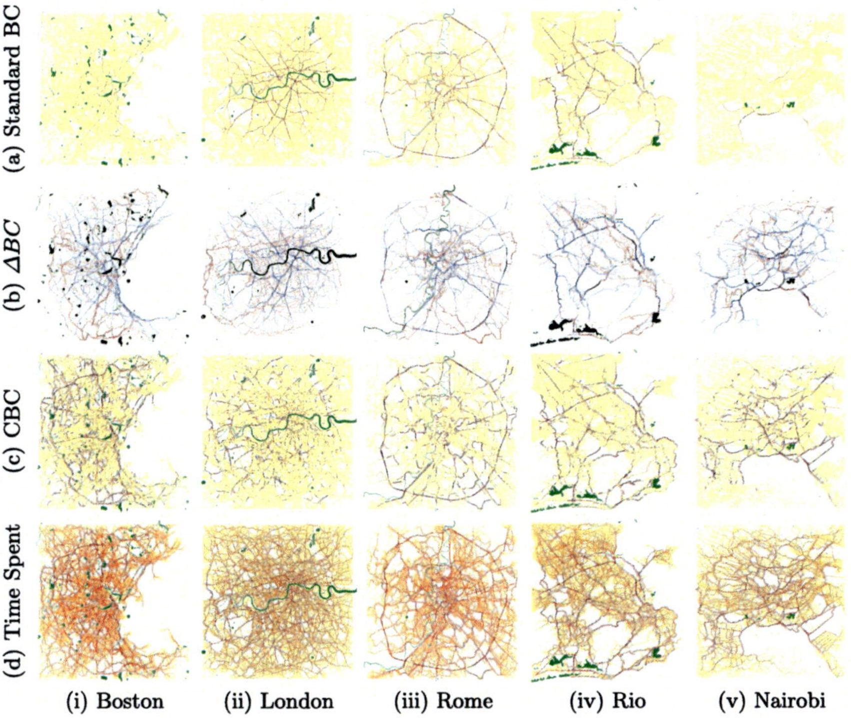

Fig. 4. (a): Standard BC maps for all cities: red edges are within the 99-th percentile of the BC distribution for the empty network. (b): ΔBC at the transition: deep red and dark blue edges increase or decrease their BC by $\approx$ 100%, respectively. (c): red edges show the 99-th percentile of the CBC distribution at criticality. (d): red edges represent the 99-th percentile of the Time Spent in Traffic distribution at criticality (per unit length, in order not to visually overestimate longer edges).

medium-low traffic levels. The indications of behavior change obtainable from the standard BC in the same conditions are much weaker as seen in [3] as the BC distribution is almost invariant for totally different cities.

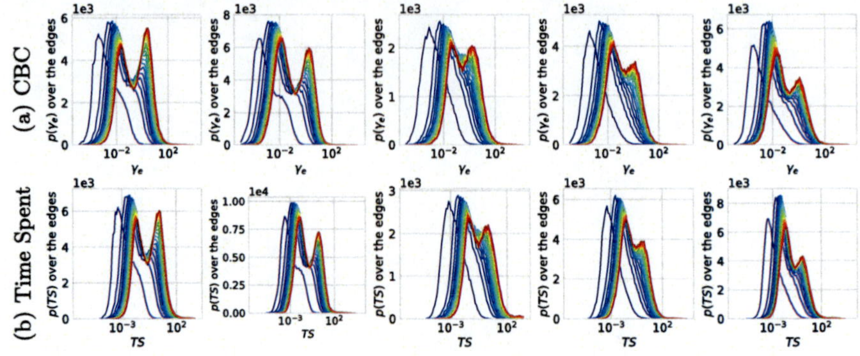

(i) Boston (ii) London (iii) Rome (iv) Rio (v) Nairobi

Fig. 5. Top: CBC [average occupancy per edge in 1h] for all cities at increasing levels of traffic (blue: light traffic, green: near transition, red: congested). Bottom: Time Spent in Traffic (TS) [vehicles·hour]. Colors as for top row.

5 Conclusion

By extending the idea of BC, we proposed a novel approach based on a dynamical model to take into account interactions among vehicles, to specifically characterize the peak-hour network loading, typically the most demanding time in terms of infrastructural stress. Our results show that the Cumulative BC is able to identify the bottlenecks of the network, when subjected to a persistent load, and to study the back-propagation of traffic jams. Interpreting the CBC as an expected road occupancy, we can also identify the edges responsible for the largest contribution to the total time spent on average by all vehicles. The results appear to be independent of OD order of addition for all the considered metrics, but this may change with different dynamical models. In the future we plan to validate the model against real traffic data (e.g., UBER Traffic Movement) and to improve it theoretically (e.g., fine temporal evolution, different dynamical systems), while also extending it to wider and more general networking contexts.

6 Authors' Contributions

M. Cogoni and G. Busonera contributed to the study conception and the model design. All authors contributed to the writing and optimization of the simulation code and to the analysis and interpretation of results. All authors read and approved the final manuscript.

Acknowledgments. This work has received funding from Sardinian Regional Authorities under projects SVDC (art 9 L.R. 20/2015) and TDM (POR FESR 2014-2020 Action 1.2.2). Map data copyrighted OpenStreetMap contributors: www.openstreetmap.org

References

1. Çolak, S., Lima, A., González, M.C.: Understanding congested travel in urban areas. Nat. Commun. **7**(1), 10793 (2016)
2. Li, D., Fu, B., Wang, Y., Lu, G., Berezin, Y., Eugene Stanley, H., Havlin, S.: Percolation transition in dynamical traffic network with evolving critical bottlenecks. Proc. Natl. Acad. Sci. USA **112**(3), 669–672 (2015)
3. Kirkley, A., Barbosa, H., Barthelemy, M., Ghoshal, G.: From the betweenness centrality in street networks to structural invariants in random planar graphs. Nat. Commun. 9(1), 2501 (2018) (Number: 1 Publisher: Nature Publishing Group)
4. Zeng, G., Li, D., Guo, S., Gao, L., Gao, Z., Eugene Stanley, H., Havlin, S.: Switch between critical percolation modes in city traffic dynamics. Proc. Natl. Acad. Sci. USA **116**(1), 23–28 (2019)
5. Hamedmoghadam, H., Jalili, M., Vu, H.L., Stone, L.: Percolation of heterogeneous flows uncovers the bottlenecks of infrastructure networks. Nat. Commun. **12**(1), 1254 (2021)
6. Bongiorno, C., Zhou, Y., Kryven, M., Theurel, D., Rizzo, A., Santi, P., Tenenbaum, J., Ratti, C.: Vector-based pedestrian navigation in cities. Nat. Comput. Sci. 1(10), 678–685 (2021) (Number: 10 Publisher: Nature Publishing Group)
7. Lee, M., Barbosa, H., Youn, H., Holme, P., Ghoshal, G.: Morphology of travel routes and the organization of cities. Nat. Commun. 8(1), 2229 (2017) (Number: 1 Publisher: Nature Publishing Group)
8. Diet, A., Barthelemy, M.: Towards a classification of planar maps. Phys. Rev. E 98(6), 062304 (2018) (Publisher: American Physical Society)
9. Helbing, D.: Traffic and related self-driven many-particle systems. Rev. Mod. Phys. 73(4), 1067–1141 (2001) (Publisher: American Physical Society)
10. Aldous, D., Ganesan, K.: True scale-invariant random spatial networks. Proc. Natl. Acad. Sci. **110**(22), 8782–8785 (2013)
11. White, D.R., Borgatti, S.P.: Betweenness centrality measures for directed graphs. Soc. Netw. **16**(4), 335–346 (1994)
12. Newman, M.E.J.: A measure of betweenness centrality based on random walks. Soc. Netw. **27**(1), 39–54 (2005)
13. Cogoni, M., Busonera, G.: Stability of traffic breakup patterns in urban networks. Phys. Rev. E **104**(1), L012301 (2021)
14. Taillanter. E., Barthelemy, M.: Empirical evidence for a jamming transition in urban traffic. arXiv:2109.00233 [cond-mat, physics:physics]. arXiv: 2109.00233 (2021)
15. Taras, A., Tortosa, L., Vicent, J.F.: A variant of the current flow betweenness centrality and its application in urban networks. Appl. Math. Comput. **347**, 600–615 (2019)
16. Gao, S., Wang, Y., Gao, Y., Liu, Y.: Understanding urban traffic-flow characteristics: a rethinking of betweenness centrality. Environ. Plann. B. Plann. Des. **40**(1), 135–153 (2013)

17. Kazerani, A., Winter, S.: Can betweenness centrality explain traffic flow. In: 12th AGILE International Conference on Geographic Information Science, pp. 1–9 (2009)
18. Rakha, H., Crowther, B.: Comparison of greenshields, pipes, and van aerde car-following and traffic stream models. Transp. Res. Rec. **1802**(1), 248–262 (2002)
19. Jin, W.-L.: Introduction to Network Traffic Flow Theory: Principles Models, and Methods. Elsevier, Concepts (2021)
20. Olmos, L.E. Çolak, S., Shafiei, S., Saberi, M., González, M.C.: Macroscopic dynamics and the collapse of urban traffic. Proc. Nat. Acad. Sci. 115(50), 12654–12661 (2018) (Publisher: Proceedings of the National Academy of Sciences)
21. Riondato, M., Kornaropoulos, E.M.: Fast approximation of betweenness centrality through sampling. Data Min. Knowl. Discov. **30**(2), 438–475 (2016)
22. Bavaud, F., Guex, G.: Interpolating Between Random Walks and Shortest Paths: A Path Functional Approach. arXiv:1207.1253 [physics]. arXiv: 1207.1253 (2012)

Adaptive Routing Potential in Road Networks

Michael Logan(✉) and Allison Goodwell

University of Colorado Denver, Denver, United States
michael.n.logan@ucdenver.edu

Abstract. Generally, in biological and ecological disciplines, the concept of adaptation refers to ecosystems evolutionarily changing as system perturbations. This paper extends these concepts to road transportation networks and focuses on the potential for adaptability of commuters to traffic disruptions. Specifically, we apply an Ecosystem Network Analysis (ENA) based upon an information-theoretic framework to demonstrate the potential, calculated as the number of alternate options available throughout a given network, for adaptive routing optimization on road networks. An initial assessment of balanced metrics of resilience and efficiency, calculated from 13 Metropolitan Statistical Area (MSA) road networks, is performed. These metrics are then compared with their respective commuter delay levels, which indicates a correlation between the balance of resilient and efficient networks and the annual commuter delay of each MSA. Whereas road network topologies that demonstrate either a highly efficient or highly resilient network structure show a tendency for higher commuter delay levels, road networks that balance efficiency and resilience suggest a tendency for lower commuter delay levels. This study's novel implication explicitly considers the road network topology as a driver of traffic delay patterns, isolated from commuter decisions.

1 Introduction

Roads, rail lines, and plane routes can be readily viewed as complex networks, where nodes are viewed as intersections, stations, or airports and edges as roads, routes, or flight paths between these nodes. Navigating these networks to optimize travel, both in terms of distance and time, constitutes an ever-evolving area of research. Adaptive routing in transportation networks is a well-researched topic that proposes methods of traffic congestion alleviation [1]. Numerous algorithms have been developed and integrated into various technologies to assess the efficiency of commuter travel. In this study, we explore a new measure to characterize adaptive routing in road transportation networks in cities across the United States.

Primary research has been conducted to optimize congestion avoidance and determine minimum time routing for individual commuters using algorithmic methods such as Dynamic and Stochastic Routing [2] and Dynamic Shortest Path Routing [3]. While some research views routing as a systems-level network assessment [4] most research focuses on traffic equilibrium as the aggregation of individual commuters [5]. Attempting to eliminate the issue of scaling from individual commuters to systemic traffic patterns, real-time monitoring, and probabilistic assessment of routing strategies have also been proposed methods of congestion alleviation [6]. Further, some methods have hybridized

H. Cherifi et al. (Eds.): COMPLEX NETWORKS 2016 2022, SCI 1077, pp. 553–562, 2023.
https://doi.org/10.1007/978-3-031-21127-0_45

these algorithmic approaches to integrate historical [7] and real-time traffic data [1], to formulate a probabilistically estimated congestion state. From this, the research aims to provide real-time adaptive routing coupled with intersection lighting schemes [8] to better scale local congestion alleviation through system-wide routing. Though less researched and diverging from infrastructure-controlled routing optimization, human decision-making systems, commuter preference, and choice have demonstrated that information conveyed to commuters [9] and choice models [10] produce a highly varied selection determinant. These varied decisions generally create uncharacterizable, subjective driver choices, further complicating congestion alleviation attempts [10].

As cross-discipline research attempts to optimize travel, the effect of system-level road network topology on commuter adaptation to traffic perturbations for routing optimization has seen little discussion. Though commuter-level traffic research is necessary and most applicable to individual commuters, the systems-level properties of road networks, how these properties constrain traffic routing options, and their effect on routing optimization have been less studied. These routing strategies do little to inform and guide road network planning on city or regional road network structures.

As metropolitan areas continue to expand, insight into the integration of existing topologies and how this expansion may affect traffic congestion through system-level topological changes may aid in integrating decades of divergent road structure philosophies.

Here we propose preliminary work on the metric of Adaptive Routing Potential (ARP) to assess the degree of commuter adaptability inherent within the topology of road networks. Our objective is to determine optimal network topologies that allow for the greatest adaptability in commuter actions while balancing the need for efficient travel throughout a road network. The proposed term, adapted from Ecological Network Analysis (ENA) [11] attempts to quantify road networks in order to quantify their balance of resilience and efficiency and assess how this balance may affect traffic delay times in small to medium Metropolitan Statistical Areas (MSA).

Generally, ecosystems, when viewed as a complex network and measured using information-theoretic methods, exist in a narrow window along a spectrum within the opposing structures of resilience and efficiency [21]. In the face of external system perturbations, such as climate change, and internal structural changes, such as species loss, current ecosystems have, by their existence, [17] demonstrated persistence in the face of change. Our hypothesis is that, given that these ecosystems exist, they may have evolved a network topology with the ability to persist through time and that network topology may present properties of optimality that may be useful in varied disciplines characterized as complex networks.

2 Method

Ecological Network Analysis (ENA) is used to analyze the dynamics of the whole ecosystem, assessing food web dynamics and trophic relationships [21] that demonstrate coherent topological properties that have formed over millennia [16]. This analysis method has been applied to various research areas such as economics [13], infrastructure [14], and industrial systems [15].

Here, we consider traffic throughput, or the edge distance between two intersections, as a random variable that indicates the usage level of a length of roadway. Meanwhile, we consider intersections as nodes and roadways as edges in the physical transportation network. In this way, information theoretic measures based on throughput between all nodes in a transportation network reveal aspects of its variability, resilience, and efficiency. We compare measures between cities and against traffic delay statistics for each city to show how they are useful constructs with which to characterize different types of road networks.

We consider that the two system properties of efficiency and resilience are equally important in route travel and adaptation in road networks. Commuters seek the efficiency of a route in that they want to travel from point A to B with a minimal distance, and potentially to a greater degree, minimal time. However, should the route from A to B become unnavigable, alternative routes need to be available to bypass congestion, making the commuters' route resilient in that they are able to adapt to a new route that conforms to their distance and time expectations of travel.

Ecosystem network investigations have led to a number of findings that may inform properties of network topology in general and network adaptability specifically [20]. To better understand the concept of adaptability, with the potential use in road networks, ecosystem definitions may provide a conceptual and analytical method of assessing road network relationships. In ENA, use of the concept of 'Reserve' provides an information-theoretic derived calculation that describes an ecosystem's ability to adapt, or more specifically, the allowable options or food web pathways available for adaptability. The concept and calculation of 'Reserve' accounts for this adaptability, the potential for adaptive routes in a road network.

3 Measures

From ecosystem dynamics research [18–20], the terms and calculations for Capacity, Reserve, and Ascendancy (Eqs. 1–3 [20] respectively) are used and applied to define the properties of adaptability inherent within a network system [20]. Our primary interest is in system throughput ($T_{..} = \sum_{i,j} T_{i,j}$) and how the scaled properties may quantitatively assess the balance of resilience and efficiency. In ecosystems, system throughput is the amount of nutrients transferred from species to species, with respect to traffic, the number of potential commuters traveling from one intersection to another. In these definitions, T is a 2-dimensional, asymmetric matrix of throughput, where $T_{i,j}$ indicates the throughput from intersection i to intersection j, creating the distance, the weighted edge from the ith to the jth intersection. In this context, the symbol '.' is used as a summation over all i's or j's. So that the term T.. Can be understood as the sum of all edges, potential commuters traveling throughout the road network system. The term T.. Calculates the total distance of all edges within the system, and $T_{.j}$ and $T_{i.}$ Calculate the total distances connecting the respective nodes. These correspond to summation over rows and columns calculated in an adjacency matrix.

Capacity - In information theoretic terms, the capacity is the equivalent to the entropy of throughput for all links in the system. A high capacity indicates that all routes are near "equally probable" since their throughput is similar to a uniform distribution. Whereas a network with a few links with very high throughput relative to others would have a lower capacity.

$$Capacity = -\sum_{i,j} T_{i,j} \log\left(\frac{T_{i,j}}{T_{..}}\right) \tag{1}$$

Ascendancy - In information theoretic terms, ascendancy is the scaled mutual constraint. A high value of ascendancy indicates that the throughput $T_{i,j}$ tends to be higher than the product of the throughputs that originate at i and end at j ($T_{i.} * T_{.j}$).

$$Ascendancy = \sum_{i,j} T_{i,j} \log\left(\frac{T_{i,j}T_{..}}{T_{i.}T_{.j}}\right) \tag{2}$$

Reserve - While Capacity and Ascendency are similar to information theory measures of entropy and mutual information, Reserve is defined as scaled conditional entropy, measure of potential actions characterized by a lack of directed operation.

$$Reserve = -\sum_{i,j} T_{i,j} \log\left(\frac{T_{i,j}^2}{T_{i.}T_{.j}}\right) \tag{3}$$

The ratio of Reserve to Capacity is calculated and introduced, for road network applications, as the Adaptive Routing Potential, ARP = R/C. Conceptually, this metric assesses the potential adaptability inherent in the structure of a network. Though terms referenced from ENA are suitable for demonstrating network properties, a parallel is useful to demonstrate applicability with more general terms used in varied disciplines. Here a representation of efficiency and resilient networks is undertaken graphically and then mathematically related to ENA terminology. Efficient networks, consisting of long uninterrupted roads, provide direct, unencumbered traffic but are susceptible to perturbations, such as traffic accidents, that can lead to congestion. Resilient networks, consisting of short roads with more intersections, those with numerous alternative routes, require increased network complexity and traffic devices. However, should a road become unnavigable alternative routes are generally available.

The properties of calculated ARP are demonstrated with three example networks analyzed (Fig. 1). We can take a perfectly efficient network (A) to demonstrate a route from the origin (1) to destination (10) consisting of 9 edges of equal length 10. This network is highly efficient, containing no intersections, and demonstrates a Reserve of 0. Should the road become unnavigable at any point in the network from 1 to 10, the route is not traversable, and the potential for adaptation is not present. Though this efficient network ensures a minimal distance, the network is fragile and offers no alternative routes.

The remaining networks would be considered resilient to varying degrees. Network B consists of 9 additional nodes and 18 additional edges with a weight of 8. Should any single node or edge be removed from the network, navigation from 1 to 10 is still possible. This is reflected in the increase in the calculated value of Reserve. In Network C increase resilience, and additional 8 edges are added such that, should any 2 edges be removed, a route from 1 to 10 is still possible. Though Network C provides numerous

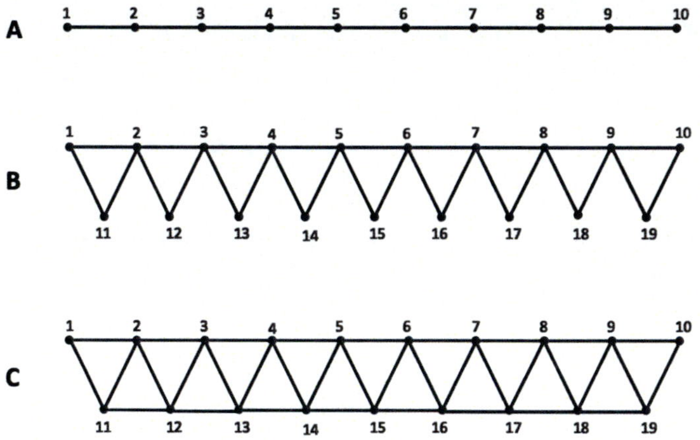

Fig. 1. Example networks with varying resilience

route alternatives, the difference in Reserve and, as a result, ARP does not increase linearly, demonstrating that adding additional resilience measures leads to a diminished return in ARP (Table 1).

Table 1. Example network properties

Property	Network A	Network B	Network C
Edge count	9	27	35
Node count	10	19	19
Capacity	285.29	1110.69	1526.34
Ascendance	285.29	789.58	949.23
Reserve	0	321.11	577.11
ARP	0	0.29	0.38

While efficiency is conceptually simple, resilience is more difficult to demonstrate. To illustrate the system properties of added resilience, a cascading node removal strategy is employed where the added resilient nodes (11-19) and their adjoining edges are removed, and ARP is calculated. The resulting curve demonstrates that added resilience demonstrates initially large payoffs but diminishing returns with added nodes and edges. This over-resilience adds increased traffic signals and other flow-regulation mechanisms that slow traffic, making it far less efficient (Fig. 2).

4 Procedure

Road networks and resultant MSA's were chosen based on reporting indicating the best and worst cities for traffic delays and a random selection of cities for additional

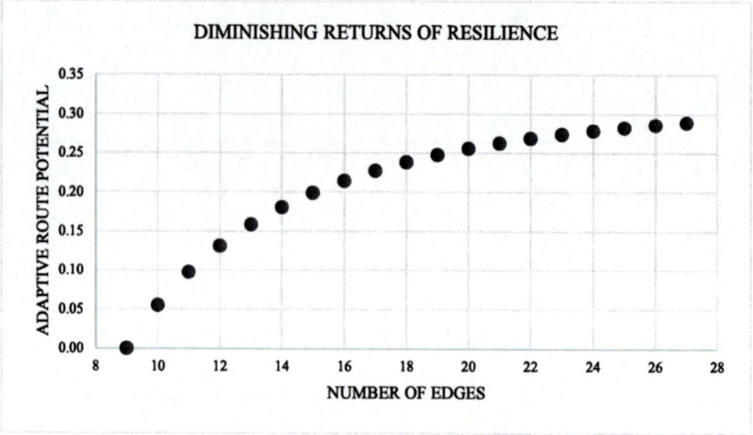

Fig. 2. Edge removal of results of resilient networks

metrics. Road network data were obtained from OpenStreet Maps [21], which produced a street intersection (node, edge) list weighted by node-to-node distance, characterized as road intersections. This list was then converted to an adjacency matrix where Capacity, Ascendency, and Reserve were calculated for each MSA. The ratio of Reserve to Capacity was further calculated to yield the ARP. To assess the relationships between the ARP and traffic delay, MSA level data was retrieved [22] and assessed against 'Annual Hours of Delay per Auto Commuter.'

5 Results

Adaptive Routing Potential was calculated for 13 small to medium MSA's of varied US locations. This calculation was plotted against Annual Hours of Delay per Auto Commuter (Fig. 3). The figure will be broken into 3 subsets: High-Efficiency Networks, High Resilience Networks, and Balanced Networks.

High-Efficiency Networks, with a lower Reserve to Capacity ratio relating to an ARP between 0.14 and 0.165 consists of cities, such as Little Rock, AR and Colorado Springs, CO, demonstrated higher delay hours, with a mean value of 40.20 Annual Hours of Delay PAC. The High Resilience Network consisting of Boulder, CO, demonstrated 35.00 Annual Hours of Delay PAC. The Balanced Networks that fell near the middle of the Efficient/Resilient spectrum demonstrated mean Annual Hours of Delay PAC of 26.14.

The balanced efficiency and resilience networks demonstrate a lower delay rate than the other sections, though the data is inconsistent with outliers such as Salem, OR, and Spokane, WA. Road delays have many causes, and a potential explanation for these cases may be that weather plays a large part in road delays. The Boulder, CO MSA is noticeably on the resilient side of the ARP calculation signifying an earlier assessment that networks with no uninterrupted roads, nearly all roads in Boulder have intersections, will be calculated as high resilience. As a comparison, the MSA with the highest efficiency calculation, Little Rock, contains 6 freeways in its road network.

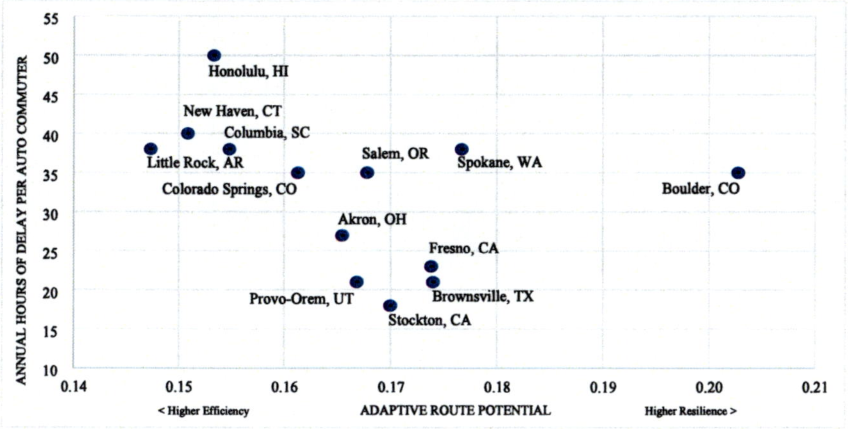

Fig. 3. Adaptive route potential of MSA's vs annual hours of delay per auto commuter.

Fig. 4. Stockton, CA road network

Local, surface streets (Black), Highways (Blue) and Red Roads and Controlled Access Highways (Red).

Mapping road networks (Table 2) of the most demonstrative variables of the road networks (Fig. 3) gives an indication of how network topology may affect commuter delays. The road network maps in Table 2 delineate 3 types of roads: 1) Local (Black), 2) Highways (Blue), and 3) Controlled Access Highways or Freeways (Red). In the context of resilience/efficiency discussion, a general attribution can be assessed:

1) Local roads consist of 2 lanes whose intersections (nodes) may be regulated by stop signs or stop lights and whose distance (edge length) between traffic regulation devices is relatively short. With respect to traffic congestion, these streets may be defined as more resilient. The short distances allow for more options for travel and congestion avoidance.

Fig. 5. Boulder, CO road network

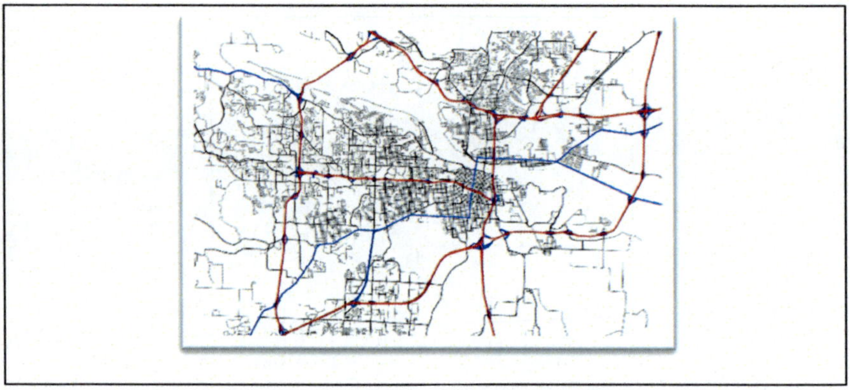

Fig. 6. Little Rock, AR road network

Table 2. Selected representative resilience/efficiency edge road networks of Fig. 3

See Fig. 4	See Fig. 5
See Fig. 6	

2) Highways are generally regulated by stop lights at greater distances and are more difficult to enter or exit. Highways generally balance resilience and efficiency by offering commuters longer edge lengths for uninterrupted travel but more difficulty in avoiding congestion and fewer route options during congestion.
3) Controlled access highways, or freeways, do not generally have in-route traffic control devices. Freeways are highly efficient; they offer uninterrupted travel for

the length of the highway. However, during congestion, there are few options for congestion avoidance unless a commuter is near an off-ramp when congestion is noticed.

These three maps demonstrate how the topology of road networks at the MSA level may affect commuter delay times. As can be seen, the Boulder, CO (Fig. 5) road network has no freeways and is composed primarily of local streets with some highways and thus, ranks highly resilient. Little Rock, AR (Fig. 6) conversely shows multiple freeways that are highly efficient but interdependent and, should an accident or congestion occur, leaves commuters with few options for adaptation. Stockton, CA (Fig. 4) ranks in the middle of the efficient, resilient rankings and has the lowest commuter delays of the 13 MASs measured. As can be seen, Stockton has a mix of highways, local roads, and 2 freeways balancing resilience and efficiency.

6 Conclusion

Research on adaptive routing techniques is an active area of research; however the system-level attributes of the networked environment that those methods and algorithms operate in have seen little attention. Research in road networks at the systems level may demonstrate that this environment, the efficiency/resilience balance of road networks, may be an important measure in commuter delay. Due to the availability of road network and commuter congestion data comparisons, localized traffic congestion assessments could not be obtained however, localized congestion assessments are a possible avenue of research.

While initial results allude to an optimal balance of network resilience and efficiency, clearly, a greater number of MSAs require calculation. Further, a more diverse network topology needs to be assessed, especially with respect to high resilience networks. Though outlying balanced networks may be explained, there are surely an innumerable number of variables that may account for delay times, and the effect of potential variables of delay require inclusion.

Adaptability, while it can be viewed and acted upon from a commuter perspective, must account for the system, the networked environment within which it operates. To view that environment, conceptualizing the system network through disciplines where adaptability has been tested and refined through millennia of evolutionary processes may demonstrate optimal network properties.

The metric proposed, Adaptive Routing Potential, could demonstrate that the system-level network topology of roads plays an integral part in traffic congestion. This finding could aid in the development, planning, and integration of future road networks in urban areas. Potentially, the development of road networks may take into account the current state of resilience/efficiency balance and model network additions that move the system-wide network towards more balance, alleviating and preventing a further commuter delay.

References

1. Fu, L.: An adaptive routing algorithm for in-vehicle route guidance systems with real-time information (2021). https://doi.org/10.1016/s0191-2615(00)00019-9
2. Ritzinger, U., Puchinger, J., Hartl, R.F.: A survey on dynamic and stochastic vehicle routing problems. Int. J. Prod. Res. **54**(1), 215–231 (2016). https://doi.org/10.1080/00207543.2015.1043403
3. Chabini, I.: Discrete dynamic shortest path problems in transportation applications: complexity and algorithms with optimal run time. Transp. Res. Rec. J. Transp. Res. Board **1645**(1), 170–175 (2007). https://doi.org/10.3141/1645-21
4. Schulz, A.S., Stier-Moses, N.E., Jahn, O., Mohring, R.H.: System-optimal routing of traffic flows with user constraints in networks with congestion. Oper. Res. **53**(4) (2005)
5. Hoang, N.H., Vu, H.L., Lo, H.K.: An informed user equilibrium dynamic traffic assignment problem in a multiple origin-destination stochastic network. https://doi.org/10.1016/j.trb.2018.07.007
6. Isaac Engel, J., Martín, J., Barco, R.: A low-complexity vision-based system for real-time traffic monitoring. https://doi.org/10.1109/tits.2016.2603069
7. Afrin, T., Yodo, N.: probabilistic estimation of traffic congestion using Bayesian network. https://doi.org/10.1016/j.measurement.2021.109051
8. Zaidi, A.A., Kulcsár, B., Wymeersch, H.: Back-Pressure traffic signal control with fixed and adaptive routing for urban vehicular networks. https://doi.org/10.1109/tits.2016.2521424
9. Alizadeh, H., Bourbonnais, P.-L., Morency, C., Farooq, B., Saunier, N.: An online survey to enhance the understanding of car drivers route choices. https://doi.org/10.1016/j.trpro.2018.10.042
10. Peeta, S., Whon, J.Y.: A hybrid model for driver route choice incorporating en-route attributes and real-time information effects. Springer Science + Business Media, Inc. Manufactured in The Netherlands (2005)
11. Ulanowicz R.E.: Quantitative methods for ecological network analysis. https://doi.org/10.1016/j.compbiolchem.2004.09.001
12. Shannon, C.E.: A Mathematical Theory of Communication (1948)
13. Goerner, S.J., Lietaer, B., Ulanowicz, R.E.: Quantifying economic sustainability: implications for free-enterprise theory, policy and practice. https://doi.org/10.1016/j.ecolecon.2009.07.018
14. Cui, D., Zeng, W., Ma, B., Zhuo, Y., Xie, Y.: Ecological network analysis of an urban water metabolic system: Integrated metabolic processes of physical and virtual water. Sci. Total Environ. **787** (2021). https://doi.org/10.1016/j.scitotenv.2021.147432
15. Morris, Z.B., Weissburg, M., Bras, B.: Ecological network analysis of urban–industrial ecosystems. https://doi.org/10.1111/jiec.13043
16. Dunne, J.A.: Food-web structure and network theory: the role of connectance and size
17. Dunne, J., Yeakel, J.: Modern lessons from ancient food webs. Am. Sci. **103**(3), 188 (2015). https://doi.org/10.1511/2015.114.188
18. Ulanowicz, R.E.: Quantifying the complexity of flow networks: how many roles are there?
19. Ulanowicz, R.E.: The balance between adaptability and adaptation. https://doi.org/10.1016/s0303-2647(01)00170-8
20. Ulanowicz, R.E., Goerner, S.J., Lietaer, B., Gomez, R.: Quantifying sustainability: resilience, efficiency and the return of information theory. https://doi.org/10.1016/j.ecocom.2008.10.005
21. Boeing, G.: OSMnx: new methods for acquiring, constructing, analyzing, and visualizing complex street networks. https://doi.org/10.2139/ssrn.2865501
22. Report, U.A., et al.: Performance Measure Summary—101 Area Sum The Mobility Data for 101 Area Sum, vol. 2014 (2014)

Diffusion and Epidemics

Detecting Global Community Structure in a COVID-19 Activity Correlation Network

Hiroki Sayama[1,2](✉)

[1] Binghamton University, State University of New York, Binghamton, NY 13902-6000, USA
sayama@binghamton.edu
[2] Waseda University, Shinjuku, Tokyo 169-8050, Japan
https://bingweb.binghamton.edu/~sayama/

Abstract. The global pandemic of COVID-19 over the last 2.5 years have produced an enormous amount of epidemic/public health datasets, which may also be useful for studying the underlying structure of our globally connected world. Here we used the Johns Hopkins University COVID-19 dataset to construct a correlation network of countries/regions and studied its global community structure. Specifically, we selected countries/regions that had at least 100,000 cumulative positive cases from the dataset and generated a 7-day moving average time series of new positive cases reported for each country/region. We then calculated a time series of daily change exponents by taking the day-to-day difference in log of the number of new positive cases. We constructed a correlation network by connecting countries/regions that had positive correlations in their daily change exponent time series using their Pearson correlation coefficient as the edge weight. Applying the modularity maximization method revealed that there were three major communities: (1) Mainly Europe + North America + Southeast Asia that showed similar six-peak patterns during the pandemic, (2) mainly Near/Middle East + Central/South Asia + Central/South America that loosely followed Community 1 but had a notable increase of activities because of the Delta variant and was later impacted significantly by the Omicron variant, and (3) mainly Africa + Central/East Canada + Australia that did not have much activities until a huge spike was caused by the Omicron variant. These three communities were robustly detected under varied settings. Constructing a 3D "phase space" by using the median curves in those three communities for *x*-*y*-*z* coordinates generated an effective summary trajectory of how the global pandemic progressed.

1 Introduction

The global pandemic of COVID-19 over the last 2.5 years have produced an enormous amount of epidemic/public health datasets, including the numbers of positive cases, deaths, hospitalizations and vaccine administrations [1–6], individual-level case data [7,8], and other forms of publicly available datasets [9,10]. These datasets have been utilized extensively for modeling and visualization of the COVID-19 pandemic [1,3–5,11–13] (including some of the author's own work as well [14,15]), which have contributed significantly to both advancing scientific understanding of the disease spreading dynamics and also raising public awareness about the pandemic.

H. Cherifi et al. (Eds.): COMPLEX NETWORKS 2016 2022, SCI 1077, pp. 565–575, 2023.
https://doi.org/10.1007/978-3-031-21127-0_46

Although traditional pandemic data modeling, analysis and visualization typically consider the dynamics of epidemic variables in a population or a region in isolation, complex systems and network sciences have firmly established the need for taking the underlying network structure into account when considering epidemic dynamics [16–20]. The majority of network epidemic research mostly focuses on the networks of social contacts and mobility connections among human individuals, which would have only limited connection to larger-scale epidemic/public health data described above. Meanwhile, network-oriented data modeling and analysis can be useful for understanding and prediction of disease spreading at such larger spatial scales. Examples include epidemic studies using global air traffic and tourism networks [21, 22] and regional networks within a country [23, 24].

Very little research has been done on network modeling and analysis of the COVID-19 data at a global scale. Earlier attempts include Zhu et al.'s work [25] that conducted basic network modeling and analysis using correlations among countries in the first several months of the COVID-19 activities, although their network analysis remained rather elementary using only a simple thresholding method (i.e., correlations were not utilized quantitatively as edge weights) with no robustness check. Now that we have more than two years of detailed global pandemic data, this direction of research can be useful for detecting the underlying structure of our globally connected world in a more robust, quantitative manner. Over the last few years, we witnessed different parts of the world underwent different epidemic patterns, while some geographically distant regions showed similar activity patterns. Such differences and similarities of COVID-19 activities may help reveal the effective structure of international connectivities. In this paper, we constructed a correlation network of countries/regions for the entire globe and studied its global community structure. We further used the detected major communities as a means to generate summary variables of the global epidemic state of the world, which allowed for low-dimensional "phase space" visualization of the COVID-19 epidemic dynamics.

2 Dataset

We used the COVID-19 time series dataset maintained and publicly released by the Center for Systems Science and Engineering (CSSE) at Johns Hopkins University [1, 2]. We obtained the time series data of the cumulative numbers of COVID-19 positive cases (written as n_t^i, where i is the country/region and t the date) for 285 countries/regions over the time period from January 22, 2020 to May 29, 2022. We selected only such countries/regions that had at least 100,000 cumulative positive cases as of May 29, 2022, in order to prevent the large stochasticity in smaller countries/regions from affecting our analysis. As a result, we had time series data for 139 countries/regions (listed in Table 1), which was used for the network construction and analysis in this study.

Note that the original data contained many reporting anomalies and other kinds of erratic values. To reduce the effects of such data anomalies on the results, the data was first smoothed over time and clipped to a finite value range before network modeling and analysis, as described in detail in the next section.

Table 1. List of 139 countries/regions that had at least 100,000 cumulative positive cases as of May 29, 2022. These countries/regions were used for network construction in this study. Note that some countries such as Australia and Canada are divided into multiple regions in the original Johns Hopkins University CSSE dataset, which we kept separated without aggregation in this study.

Afghanistan	Canada: Saskatchewan	Italy	Philippines
Albania	Chile	Jamaica	Poland
Algeria	China: Hong Kong	Japan	Portugal
Argentina	China: Unknown	Jordan	Qatar
Armenia	Colombia	Kazakhstan	Romania
Australia: Australian Capital Territory	Costa Rica	Kenya	Russia
Australia: New South Wales	Croatia	Kosovo	Rwanda
Australia: Queensland	Cuba	Kuwait	Saudi Arabia
Australia: South Australia	Cyprus	Kyrgyzstan	Serbia
Australia: Tasmania	Czechia	Laos	Singapore
Australia: Victoria	Denmark	Latvia	Slovakia
Australia: Western Australia	Dominican Republic	Lebanon	Slovenia
Austria	Ecuador	Libya	South Africa
Azerbaijan	Egypt	Lithuania	South Korea
Bahrain	El Salvador	Luxembourg	Spain
Bangladesh	Estonia	Malaysia	Sri Lanka
Belarus	Ethiopia	Maldives	Sweden
Belgium	Finland	Mauritius	Switzerland
Bolivia	France	Mexico	Taiwan
Bosnia and Herzegovina	France: Guadeloupe	Moldova	Thailand
Botswana	France: Martinique	Mongolia	Trinidad and Tobago
Brazil	France: Reunion	Montenegro	Tunisia
Brunei	Georgia	Morocco	Turkey
Bulgaria	Germany	Mozambique	Uganda
Burma	Ghana	Namibia	Ukraine
Cambodia	Greece	Nepal	United Arab Emirates
Cameroon	Guatemala	Netherlands	United Kingdom
Canada: Alberta	Honduras	New Zealand	Uruguay
Canada:	Hungary	Nigeria	US
British Columbia	Iceland	North Macedonia	Uzbekistan
Canada:	India	Norway	Venezuela
Manitoba	Indonesia	Oman	Vietnam
Canada:	Iran	Pakistan	West Bank and Gaza
Ontario	Iraq	Panama	Zambia
Canada	Ireland	Paraguay	Zimbabwe
Quebec	Israel	Peru	

3 Methods

To construct a correlation network of countries/regions with regard to their COVID-19 activity patterns, we first converted the time series for each country/region into another time series about the numbers of daily new positive cases and then smoothed it out with 7-day moving averaging to reduce day-to-day fluctuations and weekly cyclical effects. We then calculated yet another time series of *daily change exponents* by taking the day-to-day difference of log(number of new positive cases, 7-day averaged). These can be summarized mathematically as follows:

$$\text{Cumulative positive cases:} \quad n_t^i \tag{1}$$

$$\text{Daily new positive cases:} \quad d_t^i = n_t^i - n_{t-1}^i \tag{2}$$

$$\text{7-day average:} \quad a_t^i = \frac{1}{7}\sum_{j=0}^{6} d_{t-j}^i \tag{3}$$

$$\text{Daily change exponent:} \quad c_t^i = \log(a_t^i) - \log(a_{t-1}^i) = \log\left(\frac{a_t^i}{a_{t-1}^i}\right) \tag{4}$$

We took this change exponent approach so as to avoid the results of correlation analysis being affected by the difference in the absolute volume of positive cases that varied greatly over space and time. The daily change exponent values were clipped to the $[-\alpha, \alpha]$ range with $\alpha = 7$ (this corresponds to $[e^{-7}, e^7] \sim [10^{-3}, 10^3]$ in linear daily change ratios) to prevent data reporting anomalies from influencing the correlation analysis too much.

We constructed a correlation network of countries/regions by measuring the Pearson correlation coefficients between their daily change exponent sequences (for the entire time period) and connecting such countries/regions that had a correlation coefficient greater than threshold ρ (we used $\rho = 0$ in the main results), using the correlation coefficient as the edge weight. Any correlations with coefficients equal to or less than ρ were ignored and not included as edges in the network.[1] This made the countries/regions that exhibited similar COVID-19 activity patterns connected with each other with edge weights proportional to their correlation-based similarities. To study the global community structure of this network, we applied the modularity maximization method [26, 27].[2] All the network analyses and visualizations were conducted using Wolfram Research Mathematica 13.0.0.

1 In the main result presented in this paper, the following countries/regions did not have any positive correlation with others so they do not show up in the correlation network and subsequent plots: Cambodia, Egypt, France: Martinique, Georgia, Kyrgyzstan, Luxembourg, and Oman.

2 The modularity maximization method was chosen for community detection because we wanted to detect clusters of countries/regions that showed higher levels of correlation within themselves than across them. Meanwhile, other community detection methods can certainly be considered as well.

4 Results

The modularity maximization method revealed three major communities in the global COVID-19 activity correlation network. Figure 1 visualizes the detected community structure in a network diagram, and Fig. 2 shows time series plots of daily change exponents for the three major communities. The first community (Community 1) consists mostly of Europe, North America, and Southeast Asia, which showed similar six-peak patterns during the pandemic (Fig. 2 top). The second community (Community 2) contains about half of the rest of the countries/regions, including Near/Middle East, Central and South Asia, and Central and South America, that loosely followed Community 1 but had a notable increase of activities because of the Delta variant in late 2020/early 2021 and was later impacted significantly by the Omicron variant in late 2021/early 2022 (Fig. 2 middle). The third community (Community 3) was largely made of Africa, Central and East Canada, and the majority of Australia, which did not have major activities after the initial spike until another huge spike was caused by the Omicron variant in late 2021 (Fig. 2 bottom). In this particular result, there were also two much smaller

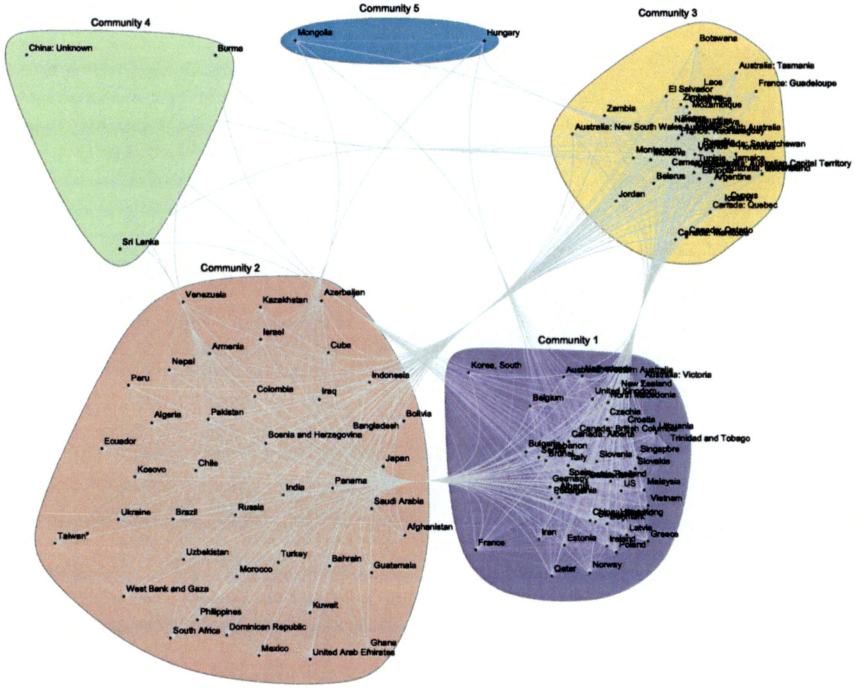

Fig. 1. Global community structure detected in the COVID-19 activity correlation network using the modularity maximization method [26,27]. Nodes represent countries/regions that had at least 100,000 cumulative positive cases as of May 29, 2022 (Table 1) in the Johns Hopkins University COVID-19 data repository [1,2]. Edges represent positive correlations of the COVID-19 activities in terms of the daily change exponent time series. Three major communities (1, 2, 3) were detected, together with two tiny ones (4, 5) in this example.

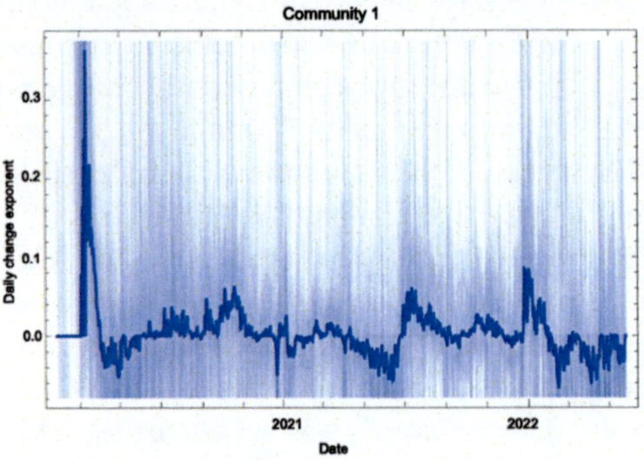

Community 1: Albania, Australia: Victoria, Australia: Western Australia, Austria, Belgium, Brunei, Bulgaria, Canada: Alberta, Canada: British Columbia, China: Hong Kong, Croatia, Czechia, Denmark, Estonia, Finland, France, Germany, Greece, Iran, Ireland, Italy, Latvia, Lebanon, Lithuania, Malaysia, Netherlands, New Zealand, North Macedonia, Norway, Poland, Portugal, Qatar, Romania, Serbia, Singapore, Slovakia, Slovenia, South Korea, Spain, Sweden, Switzerland, Thailand, Trinidad and Tobago, United Kingdom, US, Vietnam

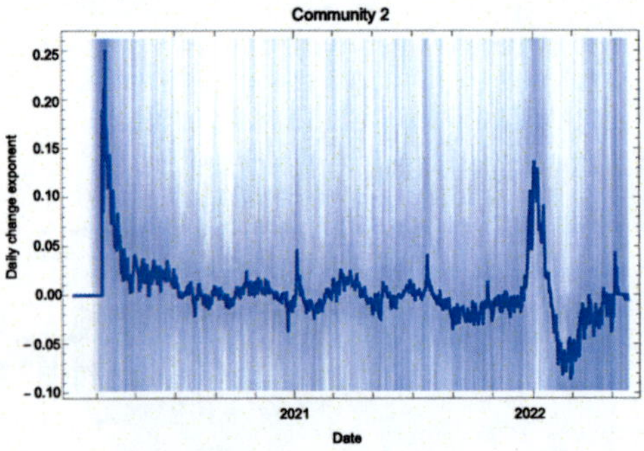

Community 2: Afghanistan, Algeria, Armenia, Azerbaijan, Bahrain, Bangladesh, Bolivia, Bosnia and Herzegovina, Brazil, Chile, Colombia, Cuba, Dominican Republic, Ecuador, Ghana, Guatemala, India, Indonesia, Iraq, Israel, Japan, Kazakhstan, Kosovo, Kuwait, Mexico, Morocco, Nepal, Pakistan, Panama, Peru, Philippines, Russia, Saudi Arabia, South Africa, Taiwan, Turkey, Ukraine, United Arab Emirates, Uzbekistan, Venezuela, West Bank and Gaza

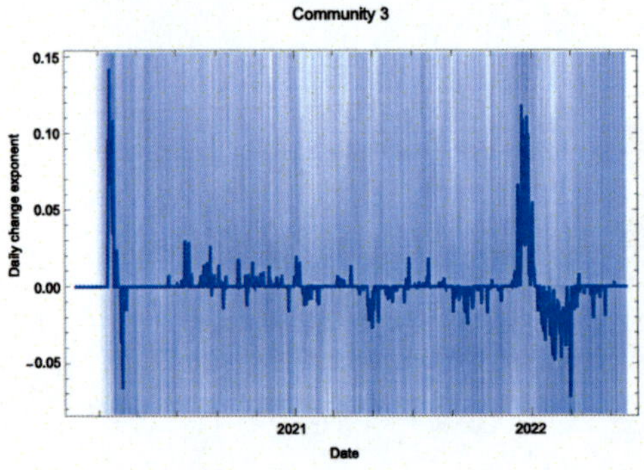

Community 3: Argentina, Australia: Australian Capital Territory, Australia: New South Wales, Australia: Queensland, Australia: South Australia, Australia: Tasmania, Belarus, Botswana, Cameroon, Canada: Manitoba, Canada: Ontario, Canada: Quebec, Canada: Saskatchewan, Costa Rica, Cyprus, El Salvador, Ethiopia, France: Guadeloupe, France: Reunion, Honduras, Iceland, Jamaica, Jordan, Kenya, Laos, Libya, Maldives, Mauritius, Moldova, Montenegro, Mozambique, Namibia, Nigeria, Paraguay, Rwanda, Tunisia, Uganda, Uruguay, Zambia, Zimbabwe

Fig. 2. Time series plots of the daily change exponents for the three major communities. Thin blue lines show the exponent time series of individual countries/regions (listed on the right), whereas thick blue lines show the median trend curve for each community. When the exponent was positive (negative) the number of daily new positive cases was growing (shrinking). Note that the actual peaks of the COVID-19 activities were where the median trend curve *switched its sign* from positive to negative, and *not* where the curve peaked in these plots.

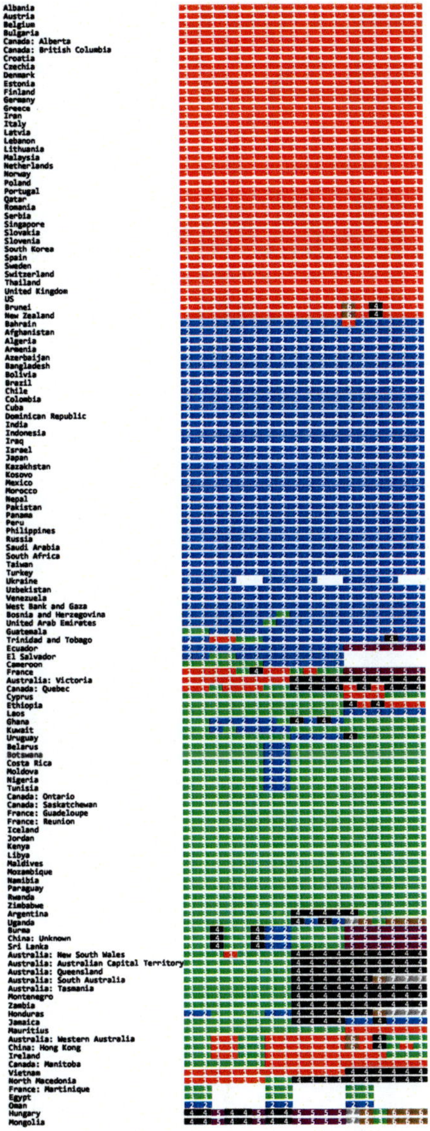

Fig. 3. Results of robustness testing experiments. Each row represents a country/region, whereas each column shows the result of network construction and community detection under a specific setting. A total of 18 setting variations were tested ($\rho : \{0, 0.05, 0.1\} \times \alpha : \{5, 7, 9\} \times$ correlation : $\{$Pearson, cosine$\}$). The third column of the results corresponds to the main result described mostly in this paper ($\rho = 0$, $\alpha = 7$, correlation: Pearson). Members of Communities 1, 2, and 3 are shown in red, blue, and green, respectively. Members of other communities are shown in other colors. Blank space means that the country/region was not included in the correlation network. Countries/regions (rows) were ordered heuristically to maximize the visual consistency of community patterns.

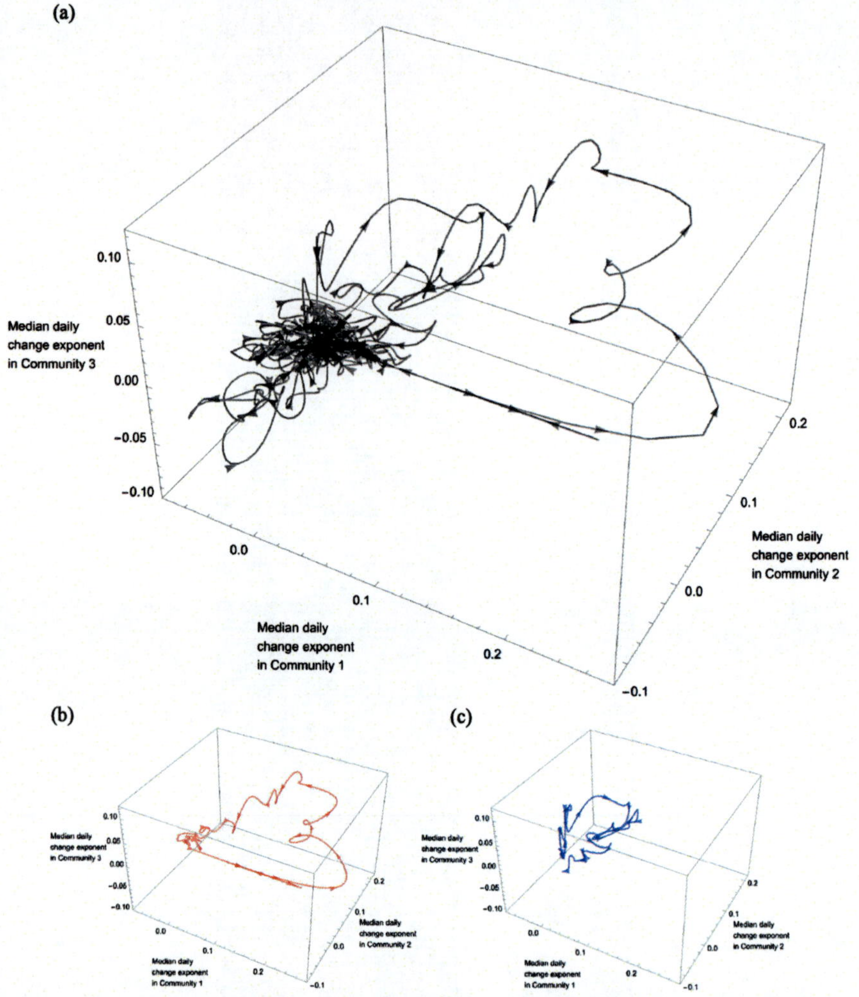

Fig. 4. Three-dimensional "phase space" visualization of the global pandemic state. The trajectory was constructed using the median trend curves of the daily change exponents for the three major communities (Communities 1, 2, and 3 as shown in Fig. 2) for x, y, and z coordinates, respectively. The trajectory was smoothed using the B-spline algorithm for visualization purposes. Arrowheads indicate the direction of time. (a) Trajectory for the entire period of collected data. (b) Initial outbreak in January–April 2020, in which the activities spread in a sequence from Community 1 to Community 2, and then to Community 3. (c) Outbreak of the Omicron variant in December 2021 and January 2022, in which the activities spread in a different direction from Community 3 to Communities 1 and 2.

communities detected (Communities 4 and 5) but these smaller communities were not robust, as described below.

To test the robustness of observations, we also conducted the same community detection task with varied settings, including the correlation threshold for edge creation ρ ($\rho \in \{0, 0.05, 0.1\}$), the daily change exponent's clipping range α ($\alpha \in \{5, 7, 9\}$), and the type of correlation measurements (Pearson correlation or cosine similarity). Figure 3 shows the results. The three major communities were robustly detected under those varied settings, as seen in the figure as the three major color regions (red, blue, and green) that appeared consistently across settings (columns). Some countries/regions switched their community memberships (especially in Community 3; e.g., the regions in Australia), but the overall structure of those three major communities remained fairly consistent across the varied settings.

The fact that the three major communities were robustly detected in the COVID-19 activity correlation network suggests that we may be able to use the aggregated activity levels within each of those communities as reasonable summary variables to capture the global pandemic dynamics. We therefore used the median curves of the daily change exponents for Communities 1, 2 and 3 (as shown in Fig. 2) for *x*-*y*-*z* coordinates, respectively, to construct a three-dimensional "phase space" visualization of how the global COVID-19 pandemic progressed over time (Fig. 4). We applied the B-spline algorithm to smooth the trajectory to improve visual readability of the trajectory. Figure 4(a) presents the overall trajectory of COVID-19 activities for the entire period of collected data, showing that the trajectory usually fluctuated around the origin ($(x, y, z) = (0, 0, 0)$) for the most of the time, but it also deviated significantly at several occasions when a major outbreak occurred. The largest loop that extends toward the right hand side (Fig. 4(b)) corresponds to the initial outbreak in early 2020, in which the COVID-19 activities spread in a sequence from Community 1 to Community 2, and then to Community 3, which is seen as a counter-clockwise rotation in Fig. 4(b). Meanwhile, the spread of the Omicron variant in late 2021/early 2022 is seen as a smaller extended loop at the center (Fig. 4(c)), in which the activities spread in a different direction from Community 3 to Communities 1 and 2. These visualizations offer an effective low-dimensional summary of complex global pandemic dynamics, which was made possible by detecting the global community structure in the COVID-19 activities.

5 Conclusions

In this study, we used the Johns Hopkins University's COVID-19 data to measure similarities of activity patterns between countries/regions over the last 2.5 years and constructed a global correlation network. We chose to characterize epidemic dynamics with the daily change exponents, instead of actual numbers of positive cases, so that the activity patterns could be captured and compared for a wide range of activity amplitudes that greatly varied from place to place and from time to time. Our network analysis robustly detected three major communities, each of which contained geographically distant countries/regions that exhibited similar COVID-19 activity history. This allowed for generating a trajectory of the global pandemic in a low-dimensional "phase space" by using the three median trend curves obtained from the three major communities as

3-D coordinates. This produced an effective summary and visualization of how the pandemic unfolded differently at different times, such as the initial outbreak and the more recent outbreak of the Omicron variant.

This study remains entirely empirical and observational, and it still leaves many open research questions. First, we have not gained any insight into what makes each of the three major communities a "community," or in other words, whether there were sociopolitical, cultural and/or economic factors that were shared among the members of each community. Second, the network was constructed merely based on simultaneous similarity of the COVID-19 activity patterns, and it was not clear whether detecting more causal linkages with time delay (e.g., Granger causality, transfer entropy) would produce a different picture of the global community structure. Third, consideration of the possible temporal dynamics of network structures, such as the difference of global community structures between different seasons/years, were completely omitted. Fourth, we studied only the dataset of the recent COVID-19 pandemic, and thus it was unclear whether the detected global community structure would apply only to this particular pandemic or to other global infectious diseases (e.g., influenza) more generally. Finally, we have not yet explored how this global network modeling and analysis could inform epidemic prediction and mitigation efforts at local/regional levels. We hope this study inspires further future research on these and other questions about the global network of epidemic dynamics.

Acknowledgments. We thank three anonymous reviewers for their constructive comments that were very helpful in improving the clarity of this manuscript.

References

1. Dong, E., Du, H., Gardner, L.: An interactive web-based dashboard to track COVID-19 in real time. Lancet Infect. Dis. **20**(5), 533–534 (2020)
2. Center for Systems Science and Engineering (CSSE) at Johns Hopkins University. COVID-19 Data Repository. https://github.com/CSSEGISandData/COVID-19. Accessed on 5 June 2022
3. US Centers for Disease Control and Prevention (CDC). COVID Data Tracker. https://covid.cdc.gov/covid-data-tracker/. Accessed on 5 June 2022
4. World Health Organization. Coronavirus (COVID-19) Dashboard. https://covid19.who.int/. Accessed on 5 June 2022
5. New York Times. Coronavirus World Map: Tracking the Global Outbreak. https://www.nytimes.com/interactive/2021/world/covid-cases.html. Accessed on 5 June 2022
6. Worldometer. COVID-19 Coronavirus Pandemic. https://www.worldometers.info/coronavirus/. Accessed on 5 June 2022
7. Xu, B., Gutierrez, B., Mekaru, S., et al.: Epidemiological data from the COVID-19 outbreak, real-time case information. Sci. Data **7**(1), 1–6 (2020)
8. Global.health: A Data Science Initiative. https://global.health/. Accessed on 5 June 2022
9. Cheng, C., Barceló, J., Hartnett, A. S., Kubinec, R., Messerschmidt, L.: COVID-19 government response event dataset (CoronaNet v. 1.0). Nature Human Beh. **4**(7), 756–768 (2020)
10. Shuja, J., Alanazi, E., Alasmary, W., Alashaikh, A.: COVID-19 open source data sets: a comprehensive survey. Appl. Intel. **51**(3), 1296–1325 (2021)
11. Binti Hamzah, F.A., Lau, C., Nazri, H., et al.: CoronaTracker: worldwide COVID-19 outbreak data analysis and prediction. Bulletin World Health Org. Preprint, 32 pages (2020)

12. CoronaTracker. https://www.coronatracker.com/. Accessed on 5 June 2022
13. Mamoon, N., Rasskin, G.: COVID Visualizer. https://www.covidvisualizer.com/. Accessed on 5 June 2022
14. Sayama, H.: COVID-19 Geographical Animation Generators. GitHub. https://github.com/hsayama/COVID-19-geographical-animations. Accessed on 5 June 2022
15. Sayama, H.: How artificial life researchers can help address complex societal challenges. In: ALIFE 2021: The 2021 Conference on Artificial Life, pp. 39–41. MIT Press (2021)
16. Pastor-Satorras, R., Vespignani, A.: Epidemic spreading in scale-free networks. Phys. Rev. Lett. **86**(14), 3200 (2001)
17. Moreno, Y., Pastor-Satorras, R., Vespignani, A.: Epidemic outbreaks in complex heterogeneous networks. Eur. Phys. J. B-Condensed Matter Complex Syst. **26**(4), 521–529 (2002)
18. Pastor-Satorras, R., Castellano, C., Van Mieghem, P., Vespignani, A.: Epidemic processes in complex networks. Rev. Modern Phys. **87**(3), 925 (2015)
19. Masuda, N., Holme, P., eds.: Temporal Network Epidemiology. Springer (2017)
20. Thurner, S., Klimek, P., Hanel, R.: A network-based explanation of why most COVID-19 infection curves are linear. Proc. Nat. Acad. Sci. **117**(37), 22684–22689 (2020)
21. Brockmann, D., Helbing, D.: The hidden geometry of complex, network-driven contagion phenomena. Science **342**(6164), 1337–1342 (2013)
22. Tsiotas, D., Tselios, V.: Understanding the uneven spread of COVID-19 in the context of the global interconnected economy. Sci. Rep. **12**(1), 1–15 (2022)
23. Della Rossa, F., Salzano, D., Di Meglio, A., et al.: A network model of Italy shows that intermittent regional strategies can alleviate the COVID-19 epidemic. Nature Commun. **11**(1), 1–9 (2020)
24. Amico, E., Bulai, I.M.: How political choices shaped Covid connectivity: the Italian case study. PLOS ONE **16**(12), e0261041 (2021)
25. Zhu, S., Kou, M., Lai, F., Feng, Q., Du, G.: The connectedness of the coronavirus disease pandemic in the world: a study based on complex network analysis. Front. Phys. **8**, 602075 (2021)
26. Blondel, V.D., Guillaume, J.L., Lambiotte, R., Lefebvre, E.: Fast unfolding of communities in large networks. J. Stat. Mech. Theory Exp. **2008**(10), P10008 (2008)
27. Wolfram Language & System Documentation. FindGraphCommunities. https://reference.wolfram.com/language/ref/FindGraphCommunities.html. Accessed on 5 June 2022

Overcoming Vaccine Hesitancy by Multiplex Social Network Targeting

Marzena Fügenschuh[1(✉)] and Feng Fu[2,3]

[1] Berliner Hochschule für Technik, Berlin, Germany
marzena.fuegenschuh@bht-berlin.de
[2] Department of Mathematics, Dartmouth College, Hanover 03755, USA
[3] Department of Biomedical Data Science, Geisel School of Medicine at Dartmouth, Lebanon 03756, USA

Abstract. Understanding the impact of social factors on disease prevention and control is one of the key questions in behavioral epidemiology. The interactions of disease spreading and human health behavior such as vaccine uptake give rise to rich dueling dynamics of biological and social contagions. In light of this, it remains largely an open problem for optimal network targeting in order to harness the power of social contagion for behavior and attitude changes. Here we address this question explicitly in a multiplex network setting. Individuals are situated on two layers of networks. On the disease transmission network layer, they are exposed to infection risks. In the meantime, their opinions and vaccine uptake behavior are driven by the social discourse of their peer influence network layer. While the disease transmits through direct close contacts, vaccine views and uptake behaviors spread interpersonally within a long-range potentially virtual network. Our comprehensive simulation results demonstrate that network-based targeting with initial seeds of pro-vaccine supporters significantly influences the ultimate adoption rates of vaccination and thus the extent of the epidemic outbreak.

Keywords: Multilayer networks · Network-based interventions · Influence maximization · Opinion formation · Epidemic models

1 Introduction

According to WHO, vaccine hesitancy—the reluctance or refusal to vaccinate despite the availability of vaccines—jeopardizes the progress made in combating vaccine-preventable diseases and was listed among the ten threats to global health in 2019 along with air pollution, climate change and global influenza pandemic [35]. Paradoxically, vaccine hesitancy has became a particularly big problem in high-income countries that manifests itself in the growing decline of trust in experts, the state, corporations, and media [21]. Unsurprisingly, vaccine hesitancy should not be underestimated in the current COVID-19 pandemic as

H. Cherifi et al. (Eds.): COMPLEX NETWORKS 2016 2022, SCI 1077, pp. 576–587, 2023.
https://doi.org/10.1007/978-3-031-21127-0_47

well [6,11,18]. Overcoming vaccine hesitancy has became a serious issue in the era of worldwide use of social media [23,27]. In particular, negative experiences like side-effects of vaccination spread quickly and undermine trust in medical care. As observed in [33], individual's vaccination decision depends in part on that of those surrounding the social network. Modeling and studying the interaction between the social environment and vaccination behavior, and thus the dynamics of a disease spread remains therefore an active research topic. Approaches from evolutionary dynamics [5,8], game theory [36], decision-making [3] come into use; we refer to [1,2] for an overview. Regardless of the methods, there are two further distinctions that can be made. On the one hand, the surrounding infrastructure—either both social and epidemic aspects are considered within one network or each one has its own environment. On the other hand, the vaccination strategy—the individuals are motivated to vaccine either by information campaign or by the apprehension of getting infected. The separation between the social and epidemic environment allows capturing the different dynamics of the processes driven thereon as well as the interaction between them. Prior work has made a number of advances on studying such disease-behavior interactions, including: a two-stage game with vaccination decision followed by the health outcome on an epidemiological process [8,13], an immunization strategy based on vaccination sentiments shared on Twitter and applied to a network with a spreading infection [32], a 2-layer network where diffusion of an infectious disease on one layer interacts with diffusion of preventive behavior on the other [25], or—supplemented by an information diffusion layer—a 3-layer network [24], and a stochastic agent-based model for influenza transmission associated with vaccination uptake considered as a socially contagious process [20]. Here, we consider a two-layer multiplex, one with a social communication network and one with a disease transmission network. The novelty of our approach is twofold. Unlike [4], which modeled public opinion formation and the spread of infectious disease as two serial processes that occur on the same contact network, we assign a separate layer of contacts to each of the processes and thus consider a multiplex. Additionally, in some previous multiplex approaches, the spreading processes taking place on the layers are of the same kind, e.g. opinion formation on different social networks [29] or diffusion or epidemic spread in complex networks [7,10,14,25,30,37]. In contrast, we consider transmission dynamics of different sorts. There is active ongoing research on how to use social media influencers for advertisement or health campaigns [17,22]. Over the years, various strategies have been developed to identify those individuals in a network who contribute to a maximal spread of information. This includes tracking the values of network centralities like degree and betweenness centrality [31], eigenvalue centrality for community structured networks [26], other community-based strategies [15] or following random acquaintances [9,32]. Generally, intrinsic motivation is more fruitful than that caused by fear. The same applies for vaccination. In addition, prophylactic vaccinations can prevent the outbreak of an epidemic if the pro-vaccine campaign has started sufficiently in advance [12]. Therefore we opt for a word-of-mouth vaccination campaign and trigger on the layer of social communication network with an opinion formation process. We propose a multiplex network framework to investigate the impact of an ongoing vaccine sentiments on epidemic spreading. Our contribution is structured as follows. First, we specify the structure of the layers of the multiplex as well as parameterize the dueling contagion processes on them. Next, we set up modeling scenarios and discuss their implications, followed by conclusions with an outlook.

2 Model

The multiplex we consider consists of two layers having equal sets of individuals but differing in the connections between the individuals. One layer represents the local community environment of an individual, where the person physically encounters family members, coworkers, friends and other people at leisure times or hobbies. We assume a uniform distribution of the number of these contacts for all individuals and thus represent these relationships with a two-dimensional lattice graph. In this environment, a disease dissemination process is emulated. On the other layer, we use a social network with non-local links, which provides a platform for opinions exchange. On this opinion network layer, the connections between the individuals are built on the basis of the Barabási-Albert graph. Its structure provides an environment in which an agent's network position affects the dynamics of the dissemination of opinions. It is known that scale-free networks enhance both the vaccination behavior and thus the effective immunization [5].

Algorithm 1: The simulation

Input: All parameters listed in Table 1.

Initialize the multiplex with a grid graph and the Barabási-Albert graph.
Set up the opinion layer on the Barabási-Albert graph and the disease layer on the grid graph.
while *more infected individuals than at start* **do**
    Select randomly an individual.
    if *rand*() $< \omega$ **then**
        Update the state of the individual on the disease layer.
    else
        Update the state of the individual on the opinion layer.
    end if
end while

Output: The state of each individual per iteration and layer.

The spread of the disease on the one layer and the opinion formation on the other take place alternately during a simulation run. With every iteration, one individual is chosen uniformly at random. The next step is to select the layer on which the respective process is to be continued. Since the processes that run separately on the two layers have different dynamics, the probability of choosing a layer does not need to be the same. For that we introduce the first parameter ω and refer to Table 1 for the overview of all model parameters for the agent-based simulation. Once the layer is chosen, the corresponding process is continued according to the state of the individual in this layer as is explained below in detail.

2.1 The Opinion Layer

Our interest is to qualify the effects of the willingness of the population to prevent an epidemic through vaccinations. Therefore, we simulate an exchange of views on the subject of vaccines. Based on the way the agents represent their opinions, we consider a voter model with single discrete variable with more—in our case three—states, according to the classification given in [19, 34]. In our setting, an individual can have one of the three discrete opinions: counter, neutral or pro, indicated by the values $-1, 0$ and 1 respectively. The number of supporters and opponents at start is given by the parameters υ^{+} and υ^{-} respectively. To compare the impact of the supporters' position in the network on the opinion formation process and thus on the epidemic, we consider several strategies for the initial assignment of the pro-opinions.

1. *hubs*: The supporters are assigned to hubs—vertices with the highest degree.
2. *betweenness*: The betweenness centralities of the nodes underlay the probability distribution, from which the vertices are drawn. The higher the value, the more likely a vertex will be chosen.
3. *high-degree*: Pro-opinions are assigned with respect to the degree distribution of the vertices. Again, the higher the value, the more likely a vertex will be chosen.
4. *random*: The vertices are selected uniformly at random from the entire population.
5. *hub-community*: A heuristic community search. Initially, the supporter set is filled with a vertex with the highest degree along with its neighbors. Next, repeatedly as long as there are places for pro-opinions available, new vertices are added to the supporter set. These are neighbors of a supporter that has not been considered yet and that has the highest degree with respect to the entire opinion layer.
6. *low-degree*: The least connected vertices are among the candidates for vaccination supporters. Pro-opinions are assigned with respect to the degree distribution of the vertices. The lower the value, the higher probability to be chosen.
7. *hub-neighbors*: The candidates are adjacent to but not themselves hubs.
8. *mod-community*: The supporters are members of communities calculated using Clauset-Newman-Moore greedy modularity maximization algorithm.[1] The communities are sorted by the number of members and taken one by one as long as the number of available supporters is not exhausted.

[1] This method is designed to find communities in, among others, scale-free networks.

9. *coworkers*: The pro-opinion is assigned to vertices that build a connected rectangular shaped sub-grid on the disease layer.

Once the pro-opinions are assigned, the opponents are chosen randomly from the remaining nodes, and all others are set to neutral. Which strategy is used in the simulation is set in the parameter $\mathcal{O}_s$.

After the initialization, while the opinion formation process unfolds, the view of a selected individual is updated according to one of the following four adoption methods. The individual assumes the opinion of a randomly chosen neighbor (*random*) or of the majority of the neighbors (*max*) or indicated by the sign of the sum of the opinions of the neighbors (*sum*) or the closest to the average of the neighboring opinions (*mean*). Which method is used in the simulation is specified by the parameter $\mathcal{O}_a$. It is possible to apply all methods by randomly choosing one at each iteration step. As a matter of fact, accepting the contrary opinion does not usually come easy. Therefore the parameter ρ represents the probability that the calculated opposite opinion will ultimately be accepted.

2.2 The Disease Layer

The epidemic dissemination process we consider on the disease layer is based on the Susceptible-Infected-Recovered (SIR) model extended by the state of immunization [16,28]. Individuals who consent to the vaccination, in particular the initial supporters in the opinion layer, enter the epidemic simulation in the immunized state. An infection begins in a predefined number of individuals in the population—given by the parameter ι—who, of course, are not vaccinated. Once the simulation is triggered and it is the turn of the disease layer, the state of the individual being considered is updated according to one of the following rules. A susceptible supporter is vaccinated without restrictions whereas a susceptible neutral with a predefined very low probability η. If not eligible for immunization, a susceptible can get infected provided there are infections in the direct neighborhood. As a matter of fact, the more infected neighbors the higher risk of infection. For that, the parameter β that represents the probability of getting infected, increases by the percentage of the infections in the neighborhood: $\beta\left(1+\frac{I_u}{N_u}\right)$, where N_u is the number of all neighbors of the considered susceptible u and I_u is the number of the infected. An infected individual recovers with a predefined probability γ. A vaccinated or recovered one remains in the assumed state till the end of the simulation. At one iteration, an individual may record exactly one state.

To give an impression of how the process on the opinion layer influences the disease spread, we show the results of one simulation run initialized with the default values as given Table 1. In Fig. 2, the curves capture the two simultaneously started and alternately altered processes on the opinion layer (far left) and the disease layer (second from the left). Both plots on the right side display the epidemic spread launched in parallel with the same initial parameters as on the left side but with the disengaged opinion layer. Here, the individuals are

Table 1. Parameters controlling the simulation.

Layer	Par.	Def. value	Parameter definition
	ω	0.75	Probability of selection of the disease layer
	n	50	Range of the square lattice graph, n^2 is the number of nodes in each layer
	m	1	# edges to be attached from a new node in the Barabási-Albert graph
Opinion	υ^-	10%	Initial anti-opinions
	υ^+	10%	Initial pro-opinions
	$\mathcal{O}_s$	*rand*	Strategy to assign supporters
	$\mathcal{O}_a$	*all*	Opinion adoption method
	ρ	0.25	Probability that the contrary opinion is adopted
Disease	ι	1%	Initial infections
	β	0.95	Probability that a susceptible gets infected by infectious neighbors
	γ	0.25	Probability that an infected recovers
	η	0.02	Probability that a susceptible neutral gets vaccinated

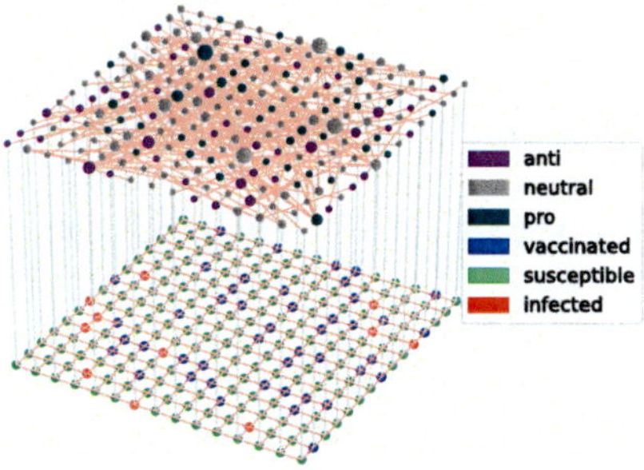

Fig. 1. An example of a multiplex with an opinion (top) and disease (bottom) layer initialized on a 15 × 15-lattice and with respect to default values of parameters $m, \upsilon^-, \upsilon^+, \mathcal{O}_s$, and ι as displayed in Table 1.

vaccinated only in the initial phase (third from the left) or not at all (far right). To measure the extent of the epidemic and thus to compare different scenarios, we count all individuals who became infected within a simulation run. Considering the three disease courses in the given example, 58% of the population gets infected when the disease spread is decelerated by increasing vaccinations upon opinion formation, while 88% in the case of a fixed number of immunized individuals and 99% if no one is vaccinated at all.

3 Results

In our numerical experiments, we focus on exploring how the pro-opinions should be scattered on the opinion layer in order to achieve the most efficient diffusion of supporters' views in terms of flattening the infection curve on the disease layer. Thus we mainly concentrate on variation of the parameters that deploy the opinion layer: methods for the supporter placement and view adoption as well as the number of initial pro and anti-opinions.

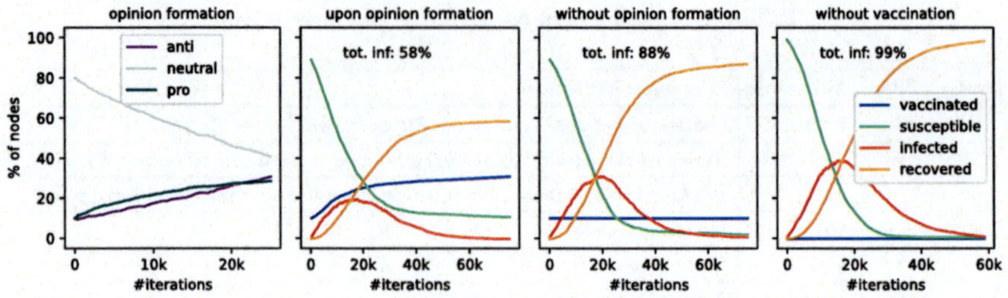

Fig. 2. An example of a disease spread course (second from the left) linked to an opinion formation process (far left) and thus affected by a growing rate of vaccination supporters, compared to the constant (third from the left) or absent rate of immunized individuals (far right). All other parameters are set to default.

3.1 Varying the Opinion Assignment and Adoption Method

In the first test scenario, we investigate the impact of the initial opinion assignment and the opinion adoption method applied to the opinion layer on the extent of the outbreak of the disease. To do this, we vary the parameters $\mathcal{O}_s$ and $\mathcal{O}_a$ and set all other to defaults given in Table 1. The results are displayed in Fig. 3. The boxplots compile percentage of the total number of infected individuals within a simulation. Per method of the initial assignment of the vaccine supporters, ordered on the x-axis, all four different adoption methods and its random mixing, distinguished by different colors, are considered.

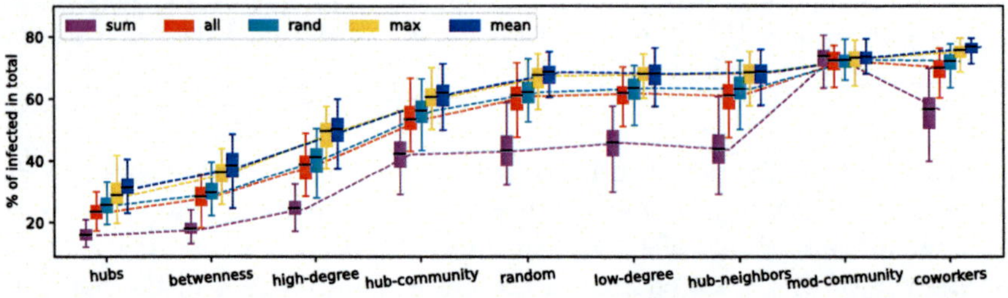

Fig. 3. Comparing the percentage of all infections while running 100 simulations per initial opinion assignment (x-axis) and adoption method (the color map in the legend).

According to the plot, both, the initial placement of the vaccine supporters on the opinion layer as well as the opinion adoption method decide on the extent of the epidemic. Hubs and generally vertices with the highest network centralities values are the most effective influencers. It may be well assumed that the potential number of nodes a supporter can influence plays a crucial role. The performance drops when the best connected nodes are stuck in a like-minded community, as the numbers for *hub-community* and *mod-community* show. In these cases, the results are close to and even worse than the fully random approach. A highly connected node may have only a few connections to like-minded neighbors so that there are enough candidates left to be persuaded. Such a pattern is certainly missing in the case of *coworkers*. Here, we notice that seeding a pro-opinion to a group of individuals connected physically, e. g. by a common working place, does not contribute to a fruitful transfer of views. The expectation that vertices of low-degree usually being simultaneously hub-neighbors would be able to affect the potent neighbors and transmit a spark that set off the entire network, turned out to be gratuitous.

Regarding the opinion adoption methods, the rank order of the colored boxplots per initial assignment method, x-tick, is the same except for *mod-community*. Taking the sign of the sum of the neighboring opinions, the purple color, appears to be the most efficient adoption method, followed by the random approach after a visible gap, and then *max* and *mean* close behind, both performing almost equally with a slender advantage of the majority approach. The outstanding case *mod-community* confirms that in a closed like-minded community there is little room for an exchange or spread of opinions. Regardless of the adoption method the flow of information is limited.

3.2 Varying the Initial Number of Pro and Anti-opinions

Beside the selection of the supporters of the pro-vaccine opinion, it is the number of supporters and opponents present when the epidemic breaks out that shapes the infection curve. In the following test scenario we vary the parameters υ^+ and υ^- which stand for the percentage of the pro and anti-opinions that are assigned on the opinion layer at the beginning of a simulation. All other parameters remain fixed to defaults as given in Table 1 and every simulation begins with a generation of a new pair of graphs the layers are based on. The results are shown in Fig. 4. In the outer frame of the chart, a boxplot corresponds to a pair (υ^-, υ^+) being an element of the cross product on the set $\{5, 20, 15, 20, 25\}\%$ and arranged along the x-axis. On the y-axis, we have the percentage of infected individuals in total. The boxplots form a noticeably regular pattern. For a fixed υ^+ and increasing υ^-, the number of total infections ascend, predictably. The growth is moderate and close to linear with the smallest slop when $\upsilon^+ = 25\%$, and it generally gets flatter the lower the y-value, except for the group of $\upsilon^+ = 5\%$ in the lower chart. On the other side, the boxplots for a fixed υ^- and increasing υ^+ form a steep descent.

A similar pattern can be assumed for higher rates. Hence for the values 30% to 50%, again with a 5% step, only the cases when $\upsilon^+ = \upsilon^-$ are considered and the results shown in the inner frame. As one reads from the plot, from 30% on minor disease outbreaks can be observed and they practically vanish by 50%, regardless of the initial placement of the pro-vaccine individuals. For comparison, we mark the outcome of the epidemic when both υ^+ and υ^- are set to zero, i. e. the opinion layer is disabled and thus no vaccination takes place on the disease layer. In this case, close to 90% of the population gets infected. As

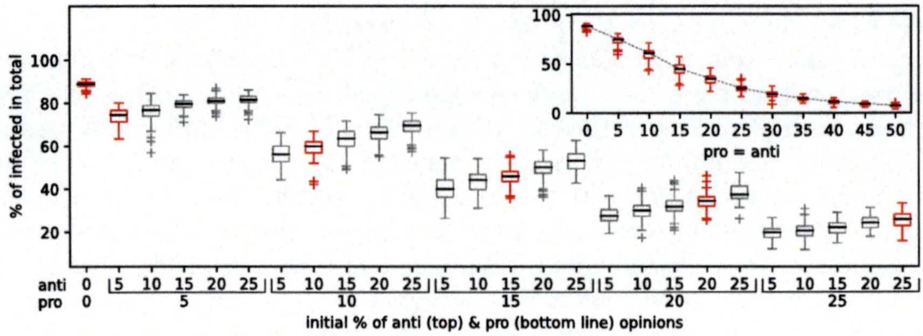

Fig. 4. Comparing the percentage of all infections when varying υ^+ and υ^-, with $\mathcal{O}_s$ set to *random*. Each boxplot compiles results from 100 simulations.

we determined in further tests, the arrangement of the boxplots as exposed in Fig. 4 is independent from the initial pro-opinion seed $\mathcal{O}_s$.

3.3 Extending the Lattice

The experiments discussed so far were carried out on a fixed population size of 2500 individuals. Figure 5 gives an insight into how sensitively our simulation reacts to the growing number of participants. Beginning with the default 50×50 lattice, the remaining ticks on the x-axis correspond to 10,000, 22,500, 40,000 and 62,500 number of nodes on each layer. To assess the impact of the initial supporter assignment in this test scenario, we consider a selection of three $\mathcal{O}_s$-methods: *high-degree*, *random*, and *coworkers*; each of them belongs to one of the three distinguishable y-value ranges in Fig. 3. Each boxplot compiles results of 10 simulations. As the number of nodes in the layers increases, the infection rates decrease and become less volatile, as shown in the left plot. The trend towards a lower infection rate as the network grows is maintained with every initial setting of the opinion layer. The relative position between the boxplots of different colors, i.e. corresponding to different $\mathcal{O}_s$-methods remains preserved. For each graph size, *high-degree* performs best, followed by *random* and then *coworkers*. Again, we observe the robustness of the simulation with regard to the initial pro-opinion assignment. It is to be expected that the relative position of the boxplots obtained for different $\mathcal{O}_s$-methods as well as (υ^-, υ^+)-pairs will lean on the patterns formed in Figs. 3 and 4 respectively.

According to the right plot in Fig. 5, the simulation times[2] grow steadily, approximately quadratic in the number of nodes. Starting with about one minute for a 50×50-lattice and reaching to over three hours on 250×250. The computation time is, however, less sensitive to the parameter $\mathcal{O}_s$. Generally and intuitively, the less infections in total the sooner an epidemic is over.

[2] The computations were conducted on 3.2 GHz 16-Core Intel Xeon W with Turbo Boost up to 4.4 GHz and 768 GB RAM.

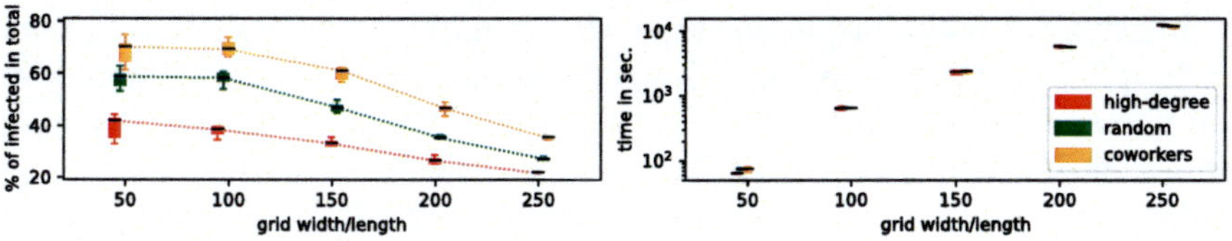

Fig. 5. Comparing the percentage of all infections (left) and the running times of the simulations (right) when increasing the number of nodes in the layers. Each boxplot compiles 10 simulations.

It should be mentioned that according to additional tests, our simulation is not sensitive to the density of the network where the disease spreads.

3.4 Further Remarks

Our scope is to observe how the dynamic process on the opinion layer influences the epidemic spread. Therefore, in our tests, we focused on varying the parameters controlling the opinion layer by keeping fixed the remaining parameters. In our tests, we assume that a disease spread process is more dynamic than the exchange and acceptance of views and thus set $\omega = 0.75$. The default values assigned to ι, η, β and γ were chosen intuitively with a reliable disease spread course on a lattice in mind. The values $\iota = 1\%$ and $\beta = 0.95$ ensure that the infection circulates despite sparse connections between the individuals. Considering the SIR model, we assume that individuals who recover from the infection become immune and thus they are not in need for a vaccination. The remaining parameters ρ and η are similarly of negligible importance. Summing up, varying the values of any of the parameters mentioned in this paragraph leads to a slightly different shape of the infection curve. However, the impact of the opinion layer on the disease spread is noticeable in any case and we assume it is analogous to the results presented above.

4 Conclusions

We presented an approach of coupling an opinion formation with a diffusion process. For this we consider a multiplex social network consisting of two layers spanned on the same set of individuals. Both processes are initially started independently on a different layer, but influence each other as they progress. The opinion formation process bases on vaccine sentiments and the diffusion process is an epidemic spread. Our main concern was to observe how the spread of the disease is sensitive to individuals' immunization readiness. As one would expect, the selection of the pro-opinion seed determines the extent of the epidemic outbreak. The notable choice are individuals with a central position in the opinion layer. What is striking, however, that individuals related through employment—referred to as neighbors on the lattice—poorly exchange their views. Comparing opinion adoption methods, we made another remarkable observation that taking the signum of neighboring opinions leads to a superior influence of the supporters. Regardless of the technical setting of the considered multiplex and its layers, our simulation shows that a social contagion such as epidemic spread is conditioned by the population's views on the immunization. It points out that spreading a positive attitude towards vaccination is a powerful tool in the fight against the virus and encourages further reflection of complex systems of this kind. The approach from the multiplexity, which we pursued in this contribution, allows investigating a coupling of two dynamic processes in an equalized complexity.

References

1. Bedson, J., et al.: A review and agenda for integrated disease models including social and behavioural factors. Nature Human Beh. **5**(7), 834–846 (2021)
2. Bhattacharyya, S., Bauch, C.: Mathematical models of the interplay between individual vaccinating decisions etc. Hum. Vac. Immun. **8**(6), 842–844 (2012)
3. Bhattacharyya, S. et al.: The impact of rare but severe vaccine adverse events on behaviour-disease dynamics: a network model. Sci. Rep. **9**(1) (2019)
4. Campbell, E., Galvani, A.P.: Complex social contagion makes networks more vulnerable to disease outbreaks. Sci. Rep. **3**(1) (2013)
5. Cardillo, A. et al.: Evolutionary vaccination dilemma in complex networks. Phys. Rev. E **88** (2013)
6. Cascini, F. et al.: Attitudes, acceptance and hesitancy among the general population worldwide to receive the covid-19 vaccines. EClin. Med. **40** (2021)
7. Chang, H.H., Fu, F.: Co-contagion diffusion on multilayer networks. Appl. Netw. Sci. **4**(1) (2019)
8. Chen, X., Feng, F.: Imperfect vaccine and hysteresis. Proc. R. Soc. B **286**(1894), 20182406 (2019)
9. Cohen, R., Havlin, S., ben Avraham, D.: Efficient immunization strategies for computer networks and populations. Phys. Rev. Lett. **91**(24) (2003)
10. Cohen, R., Havlin, S., ben Avraham, D.: Epidemic processes in complex networks. Phys. Rev. Lett. **91**(24) (2003)
11. Coustasse, A., Kimble, C., Maxik, K.: Covid-19 and vaccine hesitancy. J. Ambulatory Care Manage. **44**(1) (2021)
12. Determann, D. et al.: Acceptance of vaccinations in pandemic outbreaks: a discrete choice experiment. PLOS ONE **9**(7) (2014)
13. Fu, F. et al.: Imitation dynamics of vaccination behaviour on social networks. Proc. R. Soc. B **278**(1702) (2011)
14. Gómez, S. et al.: Diffusion dynamics on multiplex networks. Phys. Rev. Lett. **110**(2) (2013)
15. Gupta, N., Singh, A., Cherifi, H.: Centrality measures for networks with community structure. Phys. A Stat. Mech. Appl. **452** (2016)
16. Hens, N. et al.: The SIR Model, pp. 25–58. Springer New York (2012)
17. Himelboim, I., Golan, G.J.: A social networks approach to viral advertising: the role of primary, contextual, and low influencers. Soc. Med. + Soc. **5**(3) (2019)
18. Hudson, A., Montelpare, W.J.: Predictors of vaccine hesitancy: implications for covid-19 public health messaging. Int. J. Env. Res. Pub. Health **18**(15) (2021)
19. Jedrzejewski, A., Sznajd-Weron, K.: Statistical physics of opinion formation: is it a spoof? Comptes Rendus Physique **20**(4) (2019)

20. Kahana, D., Yamin, D.: Accounting for the spread of vaccination behavior to optimize influenza vaccination programs. PLOS ONE **16**, 1–15 (2021)
21. Kennedy, J.: Vaccine hesitancy: a growing concern. Pediatric Drugs **22**(2) (2020)
22. Kostygina, G. et al.: Boosting health campaign reach and engagement through use of social media influencers and memes. Social Media + Society **6**(2) (2020)
23. Larson, H.J.: Negotiating vaccine acceptance in an era of reluctance. Human Vaccines Immunotherapeutics **9**(8) (2013)
24. Mao, L.: Modeling triple-diffusions of infectious diseases, information, and preventive behaviors etc. Appl. Geography **50** (2014)
25. Mao, L., Yang, Y.: Coupling infectious diseases, human preventive behavior, and networks. Soc. Sc. Med. **74**(2) (2012)
26. Masuda, N.: Immunization of networks with community structure. New J. Phys. **11**(12) (2009)
27. Nayar, R.K. et al.: Methods to overcome vaccine hesitancy. Lancet **393**(10177) (2019)
28. Newman, M.E.J.: Spread of epidemic disease on networks. Phys. Rev. E **66** (2002)
29. Nguyen, V.X. et al.: Opinion formation on multiplex scale-free networks. EPL (Europhys. Lett.) **121**(2) (2018)
30. Pastor-Satorras, R., Castellano, C., Van Mieghem, P., Vespignani, A.: Epidemic processes in complex networks. Rev. Mod. Phys. **87** (2015)
31. Pastor-Satorras, R., Vespignani, A.: Immunization of complex networks. Phys. Rev. E **65**(3) (2002)
32. Salathé, M., Khandelwal, S.: Assessing vaccination sentiments with online social media. PLOS Comp. Biol. **7**, 1–7 (2011)
33. Shaham, A. et al.: Personal and social patterns predict influenza vaccination decision. BMC Public Health **20** (2020)
34. Sîrbu, A. et al.: Opinion Dynamics: Models, Extensions and External Effects, pp. 363–401. Springer International Publishing (2017)
35. WHO: Ten threats to global health in 2019 (2019). Accessed on 20 Jan 2021
36. Zhang, H.-F. et al.: Effects of behavioral response and vaccination policy on epidemic spreading. Sci. Rep. **4**(1) (2014)
37. Zhao, D. et al.: Immunization of epidemics in multiplex networks. PLOS ONE **9**(11) (2014)

Community-Aware Centrality Measures Under the Independent Cascade Model

Hawraa Zein[1], Ali Yassin[2], Stephany Rajeh[2(✉)], Ali Jaber[1], and Hocine Cherifi[2]

[1] Computer Science Department, Lebanese University, Beirut, Lebanon
[2] Laboratoire d'Informatique de Bourgogne - University of Burgundy, Dijon, France
stephany.rajeh@u-bourgogne.fr

Abstract. Network topology, diffusion process, and node centrality are the key elements driving the diffusion dynamics in networks. Classical centrality measures do not exploit the community structure, although it is a ubiquitous property of natural and man-made real-world networks. Recent works have shown that community-aware centrality measures can be more effective. However, in their investigation, these works generally focus on popular diffusion processes such as the Susceptible-Infected-Recovered (SIR) and the Linear Threshold (LT) models on real-world networks. This work performs an extensive comparative analysis of eight popular community-aware centrality measures using the Independent Cascade (IC) model. Besides real-world networks, we also consider artificial networks with controlled properties to better understand the influence of the network topology in the diffusion process. Results show that targeting the nodes bridging the communities or highly inter-linked nodes results in a higher diffusion when a low fraction of nodes are initially involved in the diffusion process. In contrast, when the initial budget of nodes is high, it is more effective to target distant hubs as initial spreaders. Moreover, setting a uniform threshold and a weak community structure strength hinders the diffusive power of the community-aware centrality measures.

Keywords: Centrality · Community structure · Influential nodes · Diffusion model · Independent cascade model

1 Introduction

Complex networks may be used to represent a broad variety of real-world complex systems such as ecological, social, and biological systems. Given a limited budget, choosing a set of nodes is critical whether the objective is to boost or hinder any diffusion process. Accordingly, targeting influential nodes is of great interest. Classical centrality measures are one of the main methods used for identifying key nodes [1,2]. Generally divided into local and global centrality measures, to compute a node's centrality, local centrality measures solely examine its neighborhood. They are considered efficient. However, global centrality

H. Cherifi et al. (Eds.): COMPLEX NETWORKS 2016 2022, SCI 1077, pp. 588–599, 2023.
https://doi.org/10.1007/978-3-031-21127-0_48

measures assess a node's centrality by examining its location in the whole network. They are more accurate than local ones but are computationally expensive. The problem with classical centrality measures is that they ignore the network community structure, which is of high importance in real-world networks [3].

Recently developed community-aware centrality measures take the network's community structure into account when identifying influential nodes [4–13]. They discern intra-community links from inter-community links. Intra-community links join nodes within the same community. Inter-community links join nodes from various communities. Intra-community and inter-community links are tied to the node's influence at the local and global levels, respectively. Each community-aware centrality measure calculates the node's centrality differently based on the node's inter-community and intra-community links. For instance, Comm centrality [7] identifies hubs or bridges as prominent nodes based on the network's community structure. However, it gives more importance to bridges. Modularity Vitality [8] calculates a node's centrality based on the variation of modularity after a node's removal.

Previous research has used the Susceptible-Infected-Recovered (SIR) and the Linear Threshold (LT) models to analyze the behavior of seven community-aware centrality measures [14–16]. This work extends the previous work by studying the performance of the community-aware centrality measures using the Independent Cascade (IC) model. Moreover, previous work did not examine the behavior of the community-aware centrality measures with synthetic networks. We depart from prior studies to systematically investigate eight popular community-aware centrality measures. We use three synthetic networks generated with the LFR model [17] that allows controlling the community structure strength. We also use three real-world networks originating from different domains.

The rest of the paper is organized as follows. Section 2 introduces the community-aware centrality measures that are being evaluated. The Independent Cascade model is described in Sect. 3. Section 4 presents the data and strategies used in the evaluation process. In Sect. 5, experimental results are given. A discussion of the findings and the conclusion are presented in Sect. 6.

2 Community-Aware Centrality Measures

In this section, we recall briefly the definitions of the community-aware centrality measures under evaluation. One can refer to [18] for more information. Assume that *G(V,E)* is connected, simple, unweighted, and undirected graph where V is the set of nodes of size $N = |V|$ and $E \subseteq V \times V$. In G, we have N_c non-overlapping communities where c_k is the kth community. For each node i, the total degree is defined as $k_i^{total} = k_i^{intra} + k_i^{inter}$ where k_i^{intra} and k_i^{inter} are the intra-community and inter-community links respectively. Consider $k_{i,c}$ as the number of links node i has in a given community c.

Community Hub-Bridge [4] weights the intra-community links by the node's community size and the inter-community links by the node's number of neighboring communities:

$$\alpha_{CHB}(i) = Card(c_k) \times k_i^{intra} + NNC_i \times k_i^{inter} \quad (1)$$

where $Card(c_k)$ is the community size of node i and NNC_i is the number of neighboring communities node i has.

Participation Coefficient [5] answers the question *"How well-distributed are the links of node i among various communities?"* If the links are uniformly distributed among all communities, the Participation Coefficient is close to 1, and 0 if the node is only linked to other nodes within its community:

$$\alpha_{PC}(i) = 1 - \sum_{c=1}^{N_c} \left(\frac{k_{i,c}}{k_i^{total}} \right)^2 \quad (2)$$

Community-Based Mediator [6] is based on the concept of entropy of the intra-community and inter-community links of node i. The value of the centrality gets higher as the links of a node i are more mixed:

$$\alpha_{CBM}(i) = H_i \times \frac{k_i^{total}}{\sum_{i=1}^{N} k_i^{total}} \quad (3)$$

where $H_i = [-\sum \rho_i^{intra} \log(\rho_i^{intra})] + [-\sum \rho_i^{inter} \log(\rho_i^{inter})]$, ρ_i^{intra} and ρ_i^{inter} represent the node's ratio of intra-community and inter-community links respectively, and the total degrees in the network is represented by $\sum_{i=1}^{N} k_i^{total}$.

Comm Centrality [7] identifies hubs or bridges as prominent nodes based on the network's community structure strength while giving more importance to bridges:

$$\alpha_{Comm}(i) = (1 + \mu_{c_k}) \times \left(\frac{k_i^{intra}}{max(j \in c) k_j^{intra}} \times R \right) + (1 - \mu_{c_k}) \times \left(\frac{k_i^{inter}}{max(j \in c) k_j^{inter}} \times R \right)^2 \quad (4)$$

where μ_{c_k} is the number of inter-community links divided by the number of total community links in community c_k, and R is a user-defined parameter to standardize the intra-community and inter-community links.

Modularity Vitality [8] is a signed measure indicating both how important a node is and what way the node is a key node by differentiating a hub and a bridge based on the variation of modularity after the node's removal:

$$\alpha_{MV}(i) = M(G) - M(G_i) \quad (5)$$

where $M(G)$ is the network's modularity and $M(G_i)$ is the modularity after removing node i. This study investigates bridges-first, hubs-first, and bridges-and-hubs-first ranking strategies.

Community-Based Centrality [9] quantifies a node's importance based on the node's links in each community and the size of these communities:

$$\alpha_{CBC}(i) = \sum_{c=1}^{N} k_{i,c} \left(\frac{n_c}{N} \right) \quad (6)$$

K-shell with Community [10] divides network G into two networks. The first contains the nodes and their intra-community links, while the second comprises the nodes and their inter-community links. The influence of each node is then assessed using a linear combination of the k-shell hierarchical decomposition of these networks:

$$\alpha_{ks}(i) = \delta \times \alpha^{intra}(i) + (1 - \delta) \times \alpha^{inter}(i) \tag{7}$$

where k-shell values of node i on the graphs having intra-community and inter-community links are represented respectively by $\alpha^{intra}(i)$ and $\alpha^{inter}(i)$. In this study, δ is adjusted to 0.5 to ensure equal preference for hubs and bridges.
Map Equation Centrality [11] follows the vitality principle by assessing the difference between two descriptions of the same network based on the map equation:

$$\alpha_{MapEq}(i) = L^i - L^{i*} \tag{8}$$

where L^i and L^{i*} represent the inefficient code and the efficient code, respectively.

3 Independent Cascade Model

The Independent Cascade (IC) model is a diffusion model proposed by [19]. It starts with an initial fraction of nodes f_o set in the active state. The probability $P_{u,v}$ denotes the likelihood of node u activating node v. Once activated, each node has the potential to activate its neighbor once in the following time step. The activation is based on the probability associated with that edge. Following that time step, the previously activated node transitions to the inactive state and cannot activate other nodes. Since we use the generalized version of the IC model, a threshold on the edges (u, v), denoted as $\theta_{u,v}$, can be set to hinder the diffusion. Thresholds can assume a fixed value or be uniformly distributed in the interval [0, 1]. Accordingly, node v is activated by node u if and only the the $P_{u,v} \geq \theta_{u,v}$. The diffusion iterates in discrete steps until activation is no longer possible. Since the IC model is stochastic, simulations are averaged over 100 separate iterations for each network.

The dynamics of the IC model differ from the Susceptible-Infected-Recovered (SIR) and the Linear Threshold (LT) models [20,21]. Compared to the SIR, the infection rate is not constant but changes from one edge to another. Moreover, one can set a threshold value on each edge. Compared to the LT model, an activated node has only one chance to activate its neighbor(s). However, this is not true in the LT model. The IC model can represent many real-world cases such as the propagation of opinions in a group of people. A person with opinion X can influence his/her neighbors once. Moreover, the influence varies with the neighbors' proximity.

4 Datasets and Evaluation Measure

This section briefly describes the data and the evaluation measure of the community-aware centrality measures. Table 1 reports the basic topological properties of the networks.

Table 1. Macroscopic topological properties of the synthetic and real-world networks. N is the total number of nodes. $|E|$ is the number of edges. μ is the mixing parameter. γ is the estimated exponent of the degree distribution. * indicates the largest connected component if the network is disconnected.

Network	N	$\|E\|$	μ	γ
LFR strong	2500	10,412	0.05	2.7
LFR medium	2500	10,295	0.20	2.7
LFR weak	2500	10,347	0.70	2.7
EU Airlines	417	2953	0.073	2.061
GrQc	4158	13,422	0.20	2.042
New Zealand collaboration*	1463	4246	0.564	2.112

4.1 Synthetic Networks

One can control various topological properties in synthetic networks. We generate three synthetic networks with a non-overlapping community structure using the Lancichinetti, Fortunato and Radicchi algorithm (LFR) [17]. One has a strong community structure strength ($\mu = 0.05$), the second has a medium community structure strength ($\mu = 0.2$), and the last one has a weak community structure strength ($\mu = 0.7$). Their degree distribution follows a power law. That is, $P(k) \sim k^{-\alpha}$ where $P(k)$ is the probability of a node having degree k and α is a constant such that $2 < \alpha < 3$. To mimic real-world networks, the degree and the community size distributions' exponents are equal to 2.7.

4.2 Real Networks

We use two collaboration networks (GrQc and New Zealand Collaboration) and one infrastructural network (EU Airlines)[1]. They are chosen in order to have comparable community structure strength with the synthetic networks. In the GrQc network, nodes are researchers co-authoring in General Relativity and Quantum Cosmology, and links represent co-authorship of scientific papers. In the New Zealand Collaboration network, academic institutions represent the nodes, and they are connected if Scopus lists a minimum of one common publication between authors in these institutions. In the EU Airlines network, nodes are European airports, and links represent airline routes. We use Infomap to uncover their community structure [22].

[1] Networks can be obtained from: http://networkrepository.com/index.php.

4.3 Evaluation Measure

We compare the activation size after the diffusion process of the IC model stops when the fraction of initial nodes is activated based on the community-aware centrality measures. The baseline is the degree centrality. The relative difference is defined as:

$$\Delta A = \frac{A_c - A_b}{A_b} \tag{9}$$

where A_c is the total number of activated nodes using a community-aware centrality measure c, and A_b is the total number of activated nodes with the baseline centrality. A positive ΔA value shows that the community-aware centrality measure outperforms the baseline centrality.

5 Empirical Analysis

In the IC model, one can consider identical threshold values for all the edges or randomly distributed threshold values. In this study, we investigate both types of thresholds. Let's call "fixed threshold" the case where all the edges of the nodes share a similar constraint on whether to accept or not an opinion. In the following, the so-called "random threshold" depicts the case when it is uniformly distributed. It allows incorporating a deviation between the individuals characterizing their different sensitivity to their neighbors' opinions.

5.1 Synthetic Networks

We first analyze the behavior of the community-aware centrality measures in the IC model using fixed thresholds set on the edges. Figure 1 reports the relative activation size as a function of the fraction of initially activated nodes on the three synthetic networks.

Comparing the Activation Size with a Fixed Threshold: In the network with a strong community structure (μ = 0.05), when the fraction of initially activated nodes is between 0.01 and 0.06, Community-Based Mediator (α_{CBM}) is the winner alongside Modularity Vitality targeting hubs and bridges ($|\alpha_{MV}|$), then Comm Centrality (α_{Comm}) exhibits a high activation difference for a range between 0.06 and 0.15 of initially activated nodes. The behavior of Modularity Vitality targeting hubs (α^{+}_{MV}) is very similar to Comm Centrality (α_{Comm}) when the fraction of initially activated nodes is low. So, one can say that Community-Based Mediator, Comm Centrality, Participation Coefficient, and Modularity Vitality targeting hubs perform better than others with a low fraction of initially activated nodes. Then, Participation Coefficient (α_{PC}) takes the lead in the medium range, peaking at ΔA = 15%. Finally, Modularity Vitality targeting hubs performs best with a high range of initially activated nodes.

In the network with a medium community structure (μ=0.20), when the fraction of initially activated nodes is between 0.01 and 0.05, all community-aware centrality measures have a negative value except for Community-Based Mediator

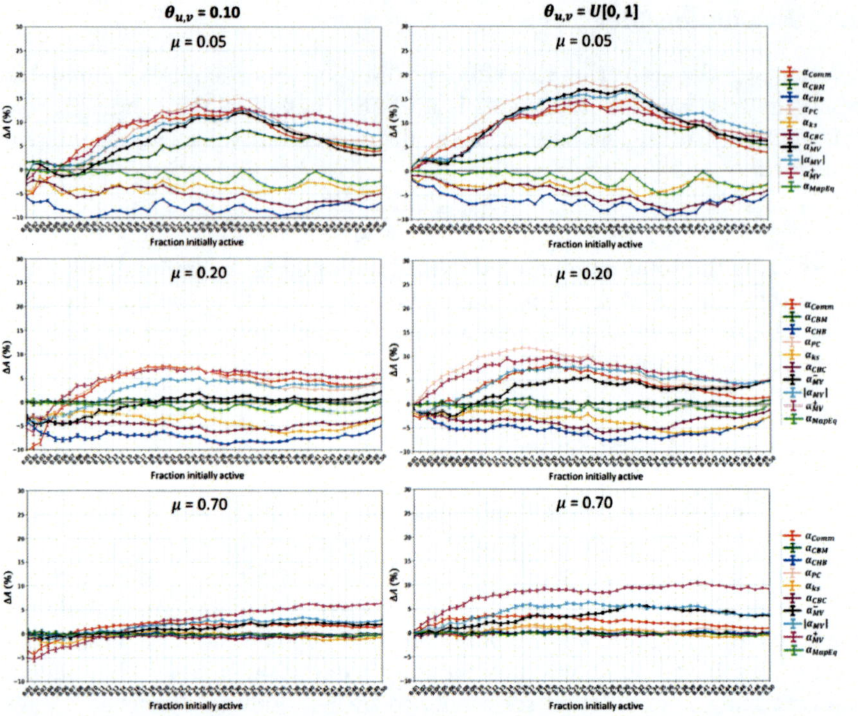

Fig. 1. Relative difference of the activation size (ΔA) as a function of the fraction of initially activated nodes for three synthetic networks. The initial spreaders set is built according to the ranks associated with a given community-aware centrality measure. On the left, a fixed threshold is set. On the right, a random threshold is set where each $\theta_{u,v}$ is distributed uniformly among edges.

(α_{CBM}) and the Map Equation Centrality (α_{MapEq}). Then, Modularity Vitality targeting hubs (α^{+}_{MV}) outperforms others with a fraction between 0.05 and 0.11. So, one can say that Community-Based Mediator, Map Equation Centrality, and Modularity Vitality targeting hubs perform better than others with a low fraction of initially activated nodes (i.e., $\leq$0.11). One can distinguish that there is a very similar behavior for Modularity Vitality targeting hubs (α^{+}_{MV}), Comm (α_{Comm}), and Participation Coefficient (α_{PC}) when the fraction of initially activated nodes is in the medium range. Then, Modularity Vitality targeting hubs (α^{+}_{MV}) takes the lead with a high fraction of initially activated nodes.

In the network with a weak community structure, overall, the Map Equation Centrality (α_{MapEq}) and Community-Based Mediator (α_{CBM}) have positive ΔA with a fraction of initially activated nodes between 0.01 and 0.08. Comm (α_{Comm}) performs better than others when this fraction is between 0.08 and 0.19. So, one can say that Community-Based Mediator, the Map Equation Centrality, and Comm Centrality perform better than others with a low fraction of initially activated nodes. Then, as this fraction increases above 0.19, Modularity Vitality targeting hubs (α_{MV}^{+}) takes over. Thus, with a medium to a high fraction of initially activated nodes, Modularity Vitality targeting hubs (α_{MV}^{+}) is the winner. The outperformance of α_{MV}^{+} persists till the fraction of initially infected nodes reaches 0.50. However, compared with strong networks, the relative activation size's magnitude is less than 5%. Nonetheless, the magnitude of α_{MV}^{+} is the same when the fraction of initially activated nodes is 0.50.

Comparing the Activation Size with a Uniform Threshold: In order to investigate effect of the threshold set on edges compared to the fixed thresholds, we investigate the behavior of the centralities with uniform threshold. The results are given in Fig. 1 on the right.

In the network with a strong community structure, Participation Coefficient (α_{PC}) performs the best when the fraction of initially infected nodes is $\leq$0.36. Then, Modularity Vitality targeting hubs and bridges ($|\alpha_{MV}|$) takes the lead but with a smaller magnitude.

In the network with a medium community structure, Participation Coefficient (α_{PC}) performs the best when the fraction of initially infected nodes is $\leq$0.27. Modularity Vitality targeting hubs (α_{MV}^{+}) takes the lead as the fraction of initially activated nodes increases. We can distinguish that all the ranking schemes of Modularity Vitality (i.e., α_{MV}^{+}, α_{MV}^{-}, ($|\alpha_{MV}|$) alongside the Participation Coefficient (α_{PC}) compete when the fraction of initially activated nodes increases above 0.45.

In the network with a weak community structure, Modularity Vitality targeting hubs (α_{MV}^{+}) outperforms the rest of the centrality measures at all fractions of initially activated nodes. Its gain over the baseline reaches 9%, with a 5% difference with other centralities that rank second.

5.2 Real Networks

In order to study the consistency of the results in the previous section, the same experiments are done on real-world networks. Here, we are using three networks. One with strong community structure strength (EU Airlines), the other with medium community structure strength (GrQc), and the last one with a weak community structure strength (New Zealand Collaboration). The results are provided in Fig. 2.

Comparing the Activation Size with a Fixed Threshold: In the EU Airlines network, a network with a strong community structure, with a low fraction of initially activated nodes, Comm centrality (α_{Comm}), Participation Coefficient (α_{PC}), Modularity Vitality targeting hubs (α_{MV}^{+}) perform well. Then, with a medium and high fraction of initially activated nodes, Modularity Vitality targeting hubs (α_{MV}^{+}) takes the lead.

In the GrQc network, a network with a medium community structure, results convey that when the fraction of initially activated nodes is $\leq$0.28, Comm (α_{Comm}) performs the best. Then, Modularity Vitality targeting hubs (α_{MV}^{+}) takes the lead.

In the New Zealand Collaboration network, a network with a weak community structure, Modularity Vitality targeting hubs (α_{MV}^{+}) outperforms the rest

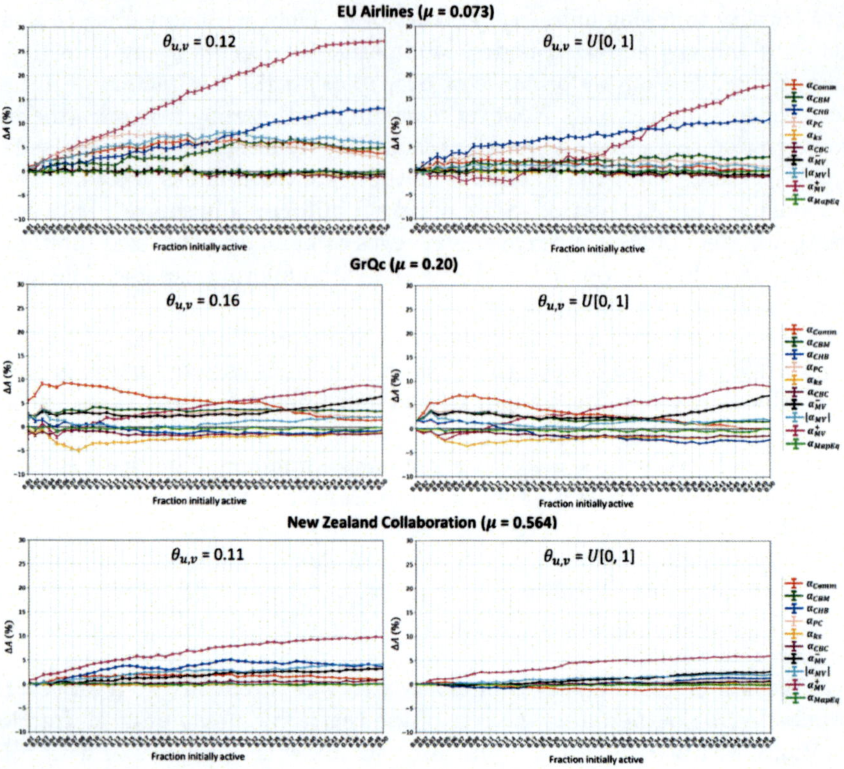

Fig. 2. Relative difference of the activation size ($\varDelta A$) as a function of the fraction of initially activated nodes for three real-world networks. The initial spreaders set is built according to the ranks associated with a given community-aware centrality measure. On the left, a fixed threshold is set. On the right, a random threshold is set where each $\theta_{u,v}$ is distributed uniformly among edges.

of the centrality measures at all fractions of initially infected nodes. Its gain is approximately 5% more than the other centralities that rank second.

Comparing the Activation Size with a Uniform Threshold: In order to investigate whether the results are stable compared to those seen with fixed thresholds, we set the also thresholds based on the uniform distribution.

It can be seen that with the EU Airlines network, the behavior of the centrality measures can be divided into two categories. A similar behavior characterizes the first compared to the results observed when the threshold is fixed but with a lower relative activation size ($\varDelta A$). For instance, when the fraction of initially infected nodes is 0.50, Modularity Vitality targeting hubs (α^{+}_{MV}) reaches $\varDelta A$ = 18% when the thresholds are uniform compared to $\varDelta A$ = 27% when the thresholds are fixed at $\theta_{(u,v)} = 0.12$. However, in both cases, it is still the best performing centrality at a high fraction of initially infected nodes. The second category is characterized by having different trends which either result in a higher or lower magnitude in terms of ΔA. For example, when the fraction of initially infected nodes is less than 0.31, the outperforming centrality measure with a uniform threshold is Community-Hub Bridge (α_{CHB}). However, when the thresholds are fixed, α_{CHB} is ranked fifth in terms of performance.

In the GrQc network, a network with a medium community structure strength, one can note that with thresholds based on the uniform distribution, Comm Centrality (α_{Comm}) performs slightly worse compared to when the threshold is fixed. The maximum ΔA reaches 6.5% while with fixed thresholds, it reaches 9.5%. Modularity Vitality targeting hubs (α^{+}_{MV}) maintains its performance when the fraction of initially active nodes crosses 0.24 with a magnitude similar to when the threshold is fixed.

In the New Zealand Collaboration network, a network with a weak community structure, Modularity Vitality targeting hubs (α^{+}_{MV}) outperforms the rest of the centrality measures at all fractions of initially infected nodes using both fixed thresholds and uniform thresholds. However, its outperformance is slightly lower with uniform thresholds.

6 Discussion and Conclusion

Identifying key nodes in networks is critical since they significantly boost or hinder the diffusion process. The importance of communities in networks cannot be ignored. Thus, using traditional centrality measures becomes inefficient in this case, and using recently developed community-aware measures becomes crucial. Despite the importance of the diffusion process and network topological properties, earlier studies have usually relied on the Susceptible-Infected-Recovered and Linear Threshold models to investigate the performance of these community-aware centrality measures only on real-world networks. In this work, we use the Independent Cascade model to study the behavior of eight community-aware centrality measures on real-world and synthetic networks.

In synthetic networks, when the threshold is fixed among the nodes, results suggest that, in general, the Map Equation Centrality and Community-based Mediator perform better than others when resources are minimal. In contrast, Comm Centrality and Modularity Vitality targeting hubs function better with a medium to a large number of initially activated nodes. When thresholds are uniformly distributed, the results are generally similar except for Participation Coefficient. It performs the best within the low to medium range of initially activated nodes in networks with strong and medium community structure strengths. Thus, allowing a disparity in thresholds among the nodes in such conditions reveals that targeting nodes inter-linked to many communities is more efficient. The behavior of the centrality measures in real-world networks is generally consistent with their behavior in synthetic networks. Indeed, it is more effective to target hubs with high budget availability in all of these networks. If the budget is limited, it is better to use community-aware centrality measures prioritizing bridges or well-mixed nodes, namely, Community-based Mediator, Comm Centrality, or Community Hub-Bridge.

Results in both real-world and synthetic networks also show that when thresholds on edges are set based on the uniform distribution, the diffusive power of the community-aware centrality measures weakens. This is expected compared to the fixed thresholds set in the range $\theta_{(u,v)} = [0.10, 0.16]$. Indeed, a variance in the thresholds hinders the diffusive power of the nodes initially activated based on any community-aware centrality measure. Results also show that as the community structure gets weaker, the diffusive power of the community-aware centrality measures decreases.

In previous work, Rajeh et al. [15] performed a similar analysis on real-world networks using the Susceptible-Infected-Recovered (SIR) diffusion model. Results showed when resources are available, it is more beneficial to target distant hubs, whereas if resources are limited, bridges are better to target. These results are consistent with the results of the IC model. Nevertheless, results with the Linear Threshold model contradict that of the SIR and the IC models [14]. Indeed, results suggest that bridges and highly inter-connected nodes are better to target compared to hubs, regardless of the availability of resources in the Linear Threshold model. In the Linear Threshold model, the change in a node's decision is highly dependent on the fraction of its neighbors that have adopted the same opinion. Therefore, hubs may inhibit the activation of other nodes, especially in dense communities where nodes have many neighbors imposing a higher threshold. On the contrary, selecting bridges and inter-linked nodes rather than hubs naturally spread out across the network, igniting a larger activation size.

These results demonstrate that three main parameters affect the performance of the community-aware centrality measures. The first is the community structure strength. The diffusive power of the community-aware centrality measures decreases as the community structure strength weakens. The second is the availability of resources. Bridges are more vital to target when resources are limited. The third is the dynamics of the diffusion model. It is more effective to target bridges than hubs in the Linear Threshold model. In contrast, hubs are more critical when resources are highly available in the Susceptible-Infected-Recovered model and the Independent Cascade model.

Acknowledgements. This work is supported by Ektidar, a Lebanese project for empowering youth.

References

1. Lü, L., Chen, D., Ren, X.-L., Zhang, Q.-M., Zhang, Y.-C., Zhou, T.: Vital nodes identification in complex networks. Phys. Rep. **650**, 1–63 (2016)
2. Ibnoulouafi, A., El Haziti, M., Cherifi, H.: M-centrality: identifying key nodes based on global position and local degree variation. J. Stat. Mech. Theory Exp. **2018**(7), 073407 (2018)
3. Rajeh, S., Savonnet, M., Leclercq, E., Cherifi, H.: Characterizing the interactions between classical and community-aware centrality measures in complex networks. Sci. Rep. **11**(1), 1–15 (2021)

4. Ghalmane, Z., El Hassouni, M., Cherifi, H.: Immunization of networks with non-overlapping community structure. SNAM **9**(1), 1–22 (2019)
5. Guimera, R., Nunes Amaral, L.A.: Functional cartography of complex metabolic networks. Nature **433**(7028), 895–900 (2005)
6. Tulu, M.M., Hou, R., Younas, T.: Identifying influential nodes based on community structure to speed up the dissemination of information in complex network. IEEE Access **6**, 7390–7401 (2018)
7. Gupta, N., Singh, A., Cherifi, H.: Centrality measures for networks with community structure. Phys. A Stat. Mech. Appl. **452**, 46–59 (2016)
8. Magelinski, T., Bartulovic, M., Carley, K.M.: Measuring node contribution to community structure with modularity vitality. IEEE Trans. Netw. Sci. **8**(1), 707–723 (2021)
9. Zhao, Z., Wang, X., Zhang, W., Zhu, Z.: A community-based approach to identifying influential spreaders. Entropy **17**(4), 2228–2252 (2015)
10. Luo, S.-L., Gong, K., Kang, L.: Identifying influential spreaders of epidemics on community networks (2016). arXiv:1601.07700
11. Blöcker, C., Nieves, J.C., Rosvall, M.: Map equation centrality: community-aware centrality based on the map equation. Appl. Netw. Sci. **7**(1), 1–24 Springer (2022)
12. Chakraborty, D., Singh, A., Cherifi, H.: Immunization strategies based on the overlapping nodes in networks with community structure. In: International Conference on Computational Social Networks, pp. 62–73. Springer, Cham (2016)
13. Kumar, M., Singh, A., Cherifi, H.: An efficient immunization strategy using overlapping nodes and its neighborhoods. Companion Proc. Web Conf. **2018**, 1269–1275 (2018)
14. Rajeh, S., Yassin, A., Jaber, A., Cherifi, H.: Analyzing community-aware centrality measures using the linear threshold model. In: International Conference on Complex Networks and Their Applications, pp. 342–353. Springer (2021)
15. Rajeh, S., Savonnet, M., Leclercq, E., Cherifi, H.: Comparative evaluation of community-aware centrality measures. Qual. Quan. 1–30 (2022)
16. Rajeh, S., Savonnet, M., Leclercq, E., Cherifi, H.: Comparing community-aware centrality measures in online social networks. In: Mohaisen, D., Jin, R. (eds.), Computational Data and Social Networks, pp. 279–290. Springer International Publishing, Cham (2021)
17. Lancichinetti, A., Fortunato, S., Radicchi, F.: Benchmark graphs for testing community detection algorithms. Phys. Rev. E **78**(4), 046110 (2008)
18. Cherifi, H., Palla, G., Szymanski, B.K., Lu, X.: On community structure in complex networks: challenges and opportunities. Appl. Netw. Sci. **4**(1), 1–35 (2019)
19. Kempe, D., Kleinberg, J., Tardos, E.: Maximizing the spread of influence through a social network. In: Proceedings of ACM SIGKDD, pp. 137–146 (2003)
20. Shakarian, P., Bhatnagar, A., Aleali, A., Shaabani, E., Guo, R.: The independent cascade and linear threshold models. In: Diffusion in Social Networks, pp. 35–48. Springer (2015)
21. Li, M., Wang, X., Gao, K., Zhang, S.: A survey on information diffusion in online social networks: Models and methods. Information **8**(4) (2017)
22. Rosvall, M., Bergstrom, C.T.: Maps of random walks on complex networks reveal community structure. PNAS **105**(4), 1118–1123 (2008)

Paths for Emergence of Superspreaders in Dengue Fever Spreading Network

L. L. Lima[1] and A. P. F. Atman[1,2,3](✉)

[1] Programa de Pos-Graduação em Modelagam Matemática e Computacional, Centro Federal de Educação Tecnológica de Minas Gerais, Belo Horizonte, Minas Gerais, Brazil
atman@cefetmg.br

[2] Departamento de Física, Centro Federal de Educação Tecnológica de Minas Gerais-CEFET-MG, Belo Horizonte, Minas Gerais, Brazil

[3] National Institute of Science and Technology for Complex Systems-CEFET-MG, Belo Horizonte, Minas Gerais, Brazil

Abstract. The identification of superspreaders is essential to contain an epidemic, especially when there is not enough information about the disease to develop precautionary measures. Unlike infections caused directly between individuals of the same species, epidemics caused by vectors have well-explored peculiarities. In this direction, we intend to study the networks obtained from the dissemination of dengue to verify, from the results of a simulation of agent based models, if the transmission of this disease follows the 20/80 rule for the proportion of spreaders and infected. We built different transmission networks considering the spread between vectors and humans up to the second generation and we observed that, despite the human-to-human transmission network follow the 20/80 rule, the other networks (human–mosquito, mosquito–mosquito and mosquito–human) did not follow this rule. Varying the density of agents, we show that the phenomenon of superspreading is accentuated with high density of mosquitoes. These characteristics of vector-borne disease networks need to be further explored, as these vectors are highly vulnerable to climate change, and a better understanding of disease spread can help better target dengue epidemic control strategies.

Keywords: Dengue · Network analysis · Agent-based model

1 Introduction

Numerous infectious diseases spread from human to human or human to animal and vice versa, creating infection networks [18]. Research in graph theory has provided a wide variety of tools over time for describing networks, many of them with applications in epidemiological studies [10]. Many epidemic control methods can be modeled with networks, such as contact tracing and vaccination, allowing the study of the effect of different strategies in an artificial environment before

H. Cherifi et al. (Eds.): COMPLEX NETWORKS 2016 2022, SCI 1077, pp. 600–609, 2023.
https://doi.org/10.1007/978-3-031-21127-0_49

their real application [10]. The introduction of many actual elements and better knowledge about the temporal pattern of the network is essential for a better understanding of the epidemic propagation [14].

In the spread of infectious diseases, heterogeneity in the behavior of a population often plays a crucial role in determining whether an epidemic occurs [2]. The social network structure is essential in the dynamics of disease spread [13], and the network topology influences the general spreading behavior of an epidemic and the fluctuations of network connectivity, which can increase the incidence of infection [14].

In the study of disease transmission, including vector-borne parasites and sexually transmitted pathogens, it is estimated that about 20% of those infected are responsible for 80% of infections (20/80 rule) [19]. This rule can be verified through the study of networks. For COVID-19, for example, a study in Hong Kong constructed the disease transmission networks and found that 69% of infected individuals did not spread the disease to anyone, while 17–19% of infectious individuals were responsible for 80% of all transmission events [1].

Regarding dengue, a mosquito-borne viral disease transmitted mainly by *Aedes aegypti*, a study showed that the disease epidemic in Singapore organized itself in a scale-free network of transmission during outbreaks between 2000 and 2005 [12].

Dengue epidemics have recently increased due to the rapid spread of the disease in many tropical countries, Europe, and the USA [7]. It has already been considered a minor public health problem but has become an important reemerging disease and was considered the most important vector-borne viral disease in the last decade [6].

The disease transmission occurs from the interaction between humans, mosquitoes, viruses, and the environment factors [7], and its epidemics usually show periodic patterns related to mosquito population dynamics, which are driven by natural climate variations [15] (Fig. 1). Given the importance of re-emerging diseases and climate change, including mosquito-borne diseases, we aim to investigate whether dengue transmission follows the 20/80 rule using network analysis applied to the results of an agent-based model simulation.

2 Methods

We use an agent-based model (ABM) to simulate the dynamics of dengue in an urban environment. Mosquitoes and humans are agents in the model, and the environment is built using a satellite image of the study area (Belo Horizonte, Brazil), which corresponds to an area of $100\,\mathrm{m} \times 100\,\mathrm{m}$. Regular square network cellular automata (each site corresponding to $\sim 1\,\mathrm{m}^2$) were used to build the environment. Thus, agents interact not only with each other but also with the environment. The total simulation time was two years, and the time step is equivalent to one minute.

Only female mosquitoes were used in the simulation, as they transmit the disease to humans [4,8]. We use the SIR compartment model dynamics

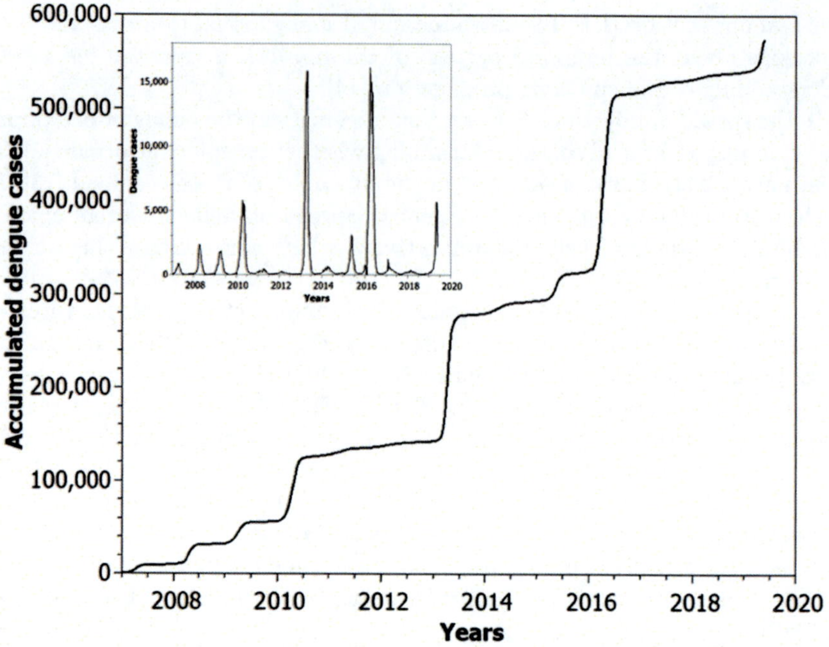

Fig. 1. Accumulated dengue cases recorded over time in Belo Horizonte, Minas Gerais, Brazil. Inset of dengue cases over time.

(Susceptible–Infected–Recovered) in our ABM. The statuses of mosquitoes and agents are: "susceptible", "infected" or "incubated". Humans can also be "recovered". Mosquitoes cannot be recovered because once infected, the mosquito remains so until the end of its life, even after repeated meals of human blood [4].

We also evaluated local precipitation and temperature as attributes of the environment. If it rained and if the maximum temperature was less than 30 °C and the minimum temperature was greater than 20 °C,which are ideal conditions for eggs to hatch [17], we increased the site attribute by 50%. To introduce this meteorological data, we used the average daily temperature and precipitation data from 2000 to 2016 from the Pampulha Meteorological Station, in Belo Horizonte, Minas Gerais, Brazil.

While each mosquito lives approximately 45 d [3], being replaced as soon as it dies, human renovation only occurs with a probability of 5% per year in the model. Mosquitoes and humans are randomly distributed across the sites. Humans can walk (random walk), and mosquitoes can fly, both in two dimensions. Mosquitoes can also bite humans or reproduce. A random walk guides the mosquito's flight with weight (quenched noise) according to the value of the environment. Mosquitoes and humans walk with a step size of one site. A mosquito decides on its activities at each time step: it can fly (25% probability), reproduce (25% probability, but only happens if weather conditions are favorable; otherwise, it flies), or bite (50% probability, but only happens if there is at least one human in the same location; otherwise, it flies).

The simulation starts with two mosquitoes infected. It was considered that there is a possibility that an egg infected with a randomly selected serotype will hatch during reproduction to simulate the introduction of new dengue serotypes over time.

The model was previously calibrated using data from mosquito traps [16]. We found that the best calibration occurred with approximately 100 mosquitoes and 500 humans per area. However, other simulations were tested with different amounts of mosquitoes and humans to build the infection network in this work. The simulations combined 250, 500, 750 and 1000 humans with 100, 180, and 360 mosquitoes. Each combination refers to a scenario. We ran 30 simulations.

We built the infection networks from the simulation results. The networks are directed, and if a node i infects a node j, there is a connection from i to j. We notice that the greater the number of mosquitoes, the larger and more connected the networks. The greater the number of humans, the smaller the networks since the simulations showed a lower percentage of infected humans.

We measure the degree of network nodes to assess superspreaders, following the same procedure as Lima and Atman [11]. However, as dengue involves a vector in its transmission between humans, we analyzed four types of networks: human-human, mosquito-mosquito, human-mosquito, and mosquito-human.

3 Results

The percentage infected is showed in Fig. 2. We can see that the higher the number of humans and the lower the number of mosquitoes, the lower the percentage infected. This indicates that, at the simulation level, even with many humans in an area, the percentage of infection will be small if the number of mosquitoes is small, highlighting the importance of vector control. In contrast, this number is high when the number of mosquitoes is high (Fig. 2L).

On the other hand, with few humans (Figs. 2A, 2B, and 2C), population saturation occurs rapidly, with almost the entire population infected within the first few days of simulation. This saturation occurs for all cases with high numbers of mosquitoes over time (Figs. 2C, 2F, 2I, and 2L). An example of a network generated at the end of each simulation scenario is shown in Fig. 3.

Figure 2 shows that the growth of dengue cases over the simulation time (2 years) represents some recorded epidemic years, especially epidemic ones (such as 2010, 2013, and 2016) (Fig. 1). Notice that this simulation result should be compared with Fig. 1 considering a time window of two years in the real data.

Network analysis shows that the human-human network recorded a high degree for some humans in all simulations with 360 mosquitoes (higher degrees ranging from approximately 130–180) (Fig. 4). This indicates that some humans indirectly infected many other humans in the simulation. The same was recorded in the human-mosquito networks, in which some mosquitoes infect a high number of humans (higher degrees ranging from approximately 140–180).

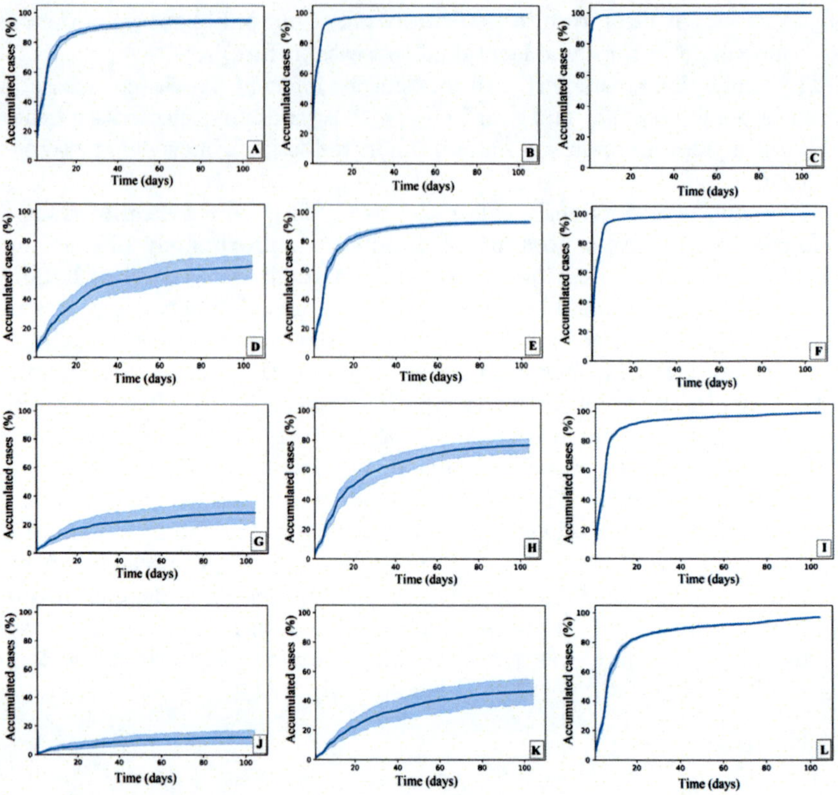

Fig. 2. Percentage of dengue infected in each scenario. Mean of 30 samples and 95% confidence interval. The percentage was calculated considering the number of humans in each simulation. Scenarios are A: 250 humans and 100 mosquitoes; B: 250 humans and 180 mosquitoes; C: 250 humans and 360 mosquitoes; D: 500 humans and 100 mosquitoes; E: 500 humans and 180 mosquitoes; F: 500 humans and 360 mosquitoes; G: 750 humans and 100 mosquitoes; H: 750 humans and 180 mosquitoes; I: 750 humans and 360 mosquitoes; J: 1000 humans and 100 mosquitoes; K: 1000 humans and 180 mosquitoes; L: 1000 humans and 360 mosquitoes.

For the human-mosquito network, the maximum degree values were 26 (simulations with 500 humans and 360 mosquitoes) and 28 (simulations with 250 humans and 360 mosquitoes). This indicates that the same human transmitted (indirectly) the disease to 26 and 28 humans, respectively. In mosquito-mosquito networks, the highest degree values are for the network of 250 humans and 360 mosquitoes (maximum degree of 37). It means that 37 mosquitoes were infected by humans who received the infection from the same mosquito.

In the case of the human-mosquito networks, mosquito control can break hubs more easily than in lower-grade networks, where it is necessary to remove many vectors to break a hub. This shows that, at the simulation level, it is easier

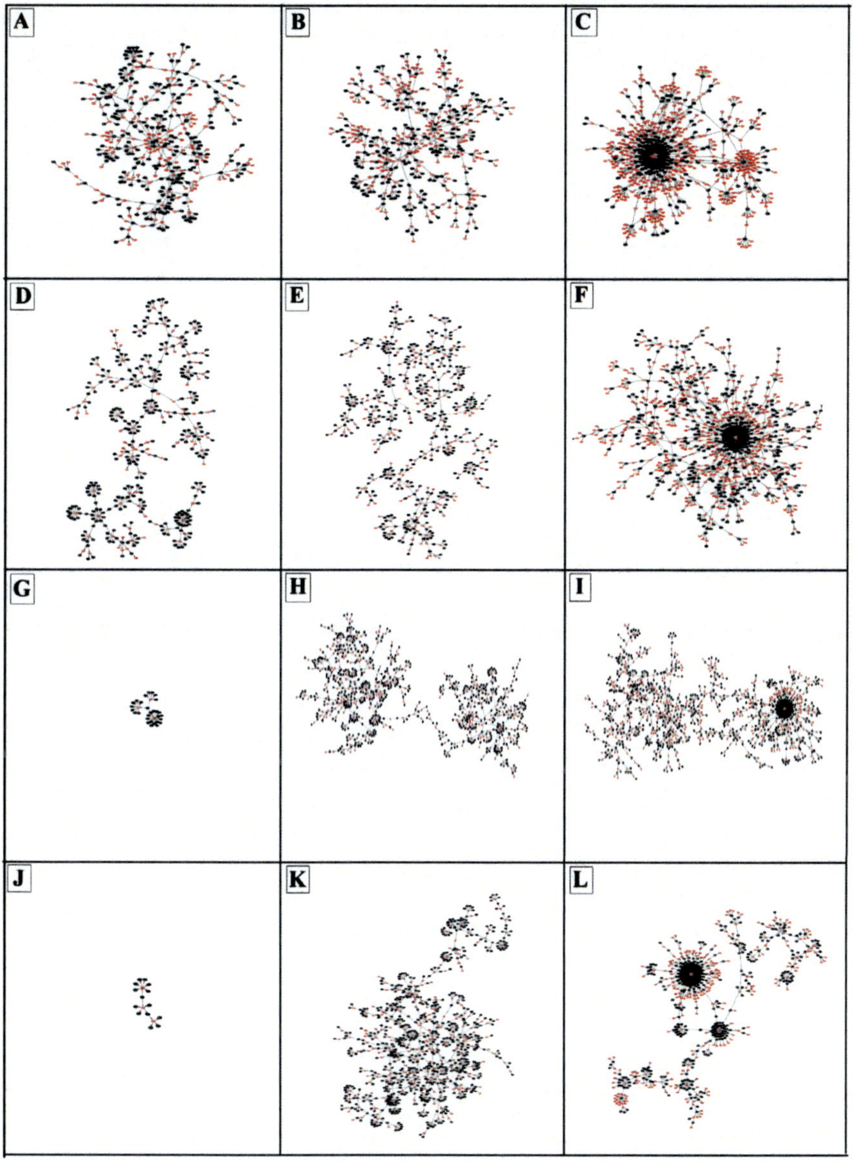

Fig. 3. Examples of directed infection networks in each scenario. Black nodes represent humans and the red ones represent mosquitoes. Two nodes are directly connected from the one who transmits the infection towards the new infected. The scenarios are A: 250 humans and 100 mosquitoes; B: 250 humans and 180 mosquitoes; C: 250 humans and 360 mosquitoes; D: 500 humans and 100 mosquitoes; E: 500 humans and 180 mosquitoes; F: 500 humans and 360 mosquitoes; G: 750 humans and 100 mosquitoes; H: 750 humans and 180 mosquitoes; I: 750 humans and 360 mosquitoes; J: 1000 humans and 100 mosquitoes; K: 1000 humans and 180 mosquitoes; L: 1000 humans and 360 mosquitoes.

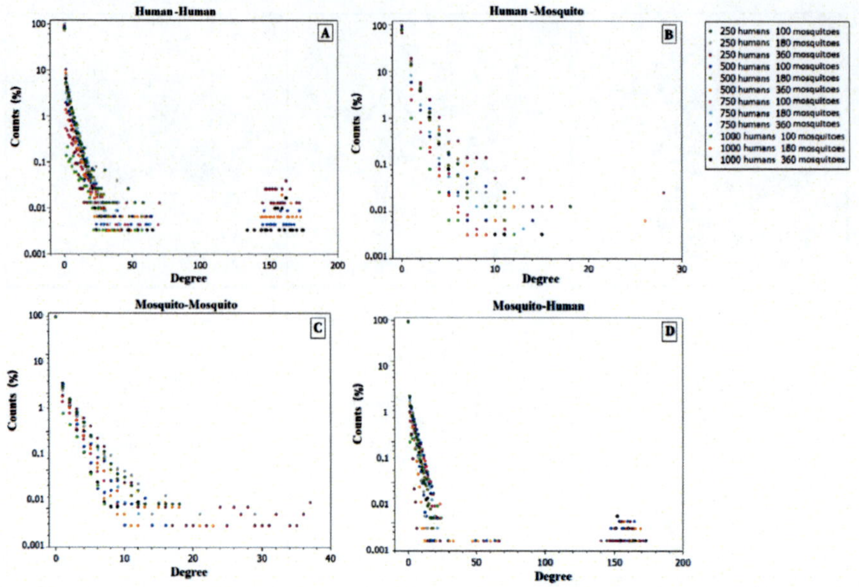

Fig. 4. Degree of (A) human-human, (B) human-mosquito, (C) mosquito-mosquito and (D) mosquito-human networks. Percentage of nodes calculated based on the 30 model runs.

to control the infection in areas with fewer vectors and humans, highlighting the need for more outstanding efforts by the community and health authorities to contain transmission in overcrowded areas and mosquito-friendly environments. Also, a study that analyzed small dengue networks showed that they allow for in-depth analysis of dengue transmission, which could be used to inform the more effective spatial application of disease control strategies [9].

The percentage of mosquitoes and humans responsible for infections in each scenario (among the infected) is presented in Table 1. It is worth noting that human-human networks followed the 20/80 rule, then approximately 20% of humans are responsible for about 80% of human infections in all scenarios. In the mosquito networks, the values ranged from 16.75 to 39.46% of mosquitoes infecting about 80% of mosquitoes. In human-mosquito and mosquito-human networks, approximately 40–70% of the respective infecting agents were responsible for about 80% of infections. This value is more than double the expected number in the 20/80 rule.

The Shapiro-Wilk normality test showed that the data from Table 1 have a normal distribution. Thus, to compare the means between the groups, the ANOVA test was used ($\alpha = 0.05$), which showed no significant difference between the percentages of infected agents when comparing the networks. However, there is a significant difference between all the networks for the percentages of agents transmitting the infections.

Table 1. Percentage of mosquitoes and humans responsible for dengue infections in each scenario and the respective percentage of infections for which they are responsible. The percentage is calculated concerning the total infected. The scenarios are A: 250 humans and 100 mosquitoes; B: 250 humans and 180 mosquitoes; C: 250 humans and 360 mosquitoes; D: 500 humans and 100 mosquitoes; E: 500 humans and 180 mosquitoes; F: 500 humans and 360 mosquitoes; G: 750 humans and 100 mosquitoes; H: 750 humans and 180 mosquitoes; I: 750 humans and 360 mosquitoes; J: 1000 humans and 100 mosquitoes; K: 1000 humans and 180 mosquitoes; L: 1000 humans and 360 mosquitoes.

Scenario	Human–human (%)	Human–mosquito (%)	Mosquito–mosquito (%)	Mosquito–human (%)
A	21.82→81.54	62.53→84.71	34.68→82.97	46.39→78.2
B	25.11→80.22	61.05→81.64	39.46→83.86	43.61→78.26
C	18.82→80.14	53.75→80.09	27.94→80.35	35.77→78.03
D	17.22→81.14	66.01→79.28	24.89→81.23	48.99→75.92
E	21.51→79.46	64.76→79.8	32.98→80.75	47.61→80.77
F	23.1→70.33	61.14→81.92	34.35→80.58	53.31→79.18
G	13.21→80.15	64.39→82.28	19.78→79.32	49.06→74.72
H	18.83→80.47	68.75→81.61	27.11→79.25	49.25→83.01
I	23.53→80.65	61.45→76.19	34.05→80.16	61.76→76.83
J	11.8→80.31	62.93→74.46	16.78→80.44	51.85→81.13
K	16.64→81.58	71.16→82.83	22.54→81.18	47.84→74.94
L	22.21→80.3	65.44→79.74	31.41→80.03	68.26→82.48

4 Discussion

Although the degree of the nodes of the human-mosquito network are high, this transmission occurred in areas with lower population density and high numbers of mosquitoes. Consequently, more mosquitoes tend to bite the same human. This did not happen in areas with many humans and mosquitoes, as there are more humans for the mosquito to bite. For mosquito networks, it is worth noting that as the mosquito population renews itself more frequently (although the population size remains constant), there are always new susceptible individuals. This does not happen in human networks, so the population can quickly become immune, saturating the infection network.

In this work, network analyzes were crucial to investigate the role of agents as superspreaders of dengue. The analysis of superspreaders individuals and super-spreading events is essential for emergency public health responses, as one or a few superspreaders can ignite epidemic outbreaks [5]. Large epidemic outbreaks can also occur when the spread of a disease starts in superspreaders [21].

Our findings show that, at the simulation level, although human-human networks have superspreaders agents and obey the 20/80 rule, this does not happen in the others. It may be a characteristic of networks involving vectors and hosts. However, a more extensive study, perhaps involving other vector diseases, is needed to understand how infections occur in these networks.

Although the initial results allow inferences about the structure of infection networks, this model has several limitations, as well as practically all models. The first limitation is that the population is constant and with low renovation. This dynamic does not reflect the reality of urban centers, where there is no frontier for the disease, which spreads temporally and spatially, hardly infecting the entire population. This population variation is one of the reasons why dengue has cyclical outbreaks every three or five years. The same limitation occurs in the mosquito population, which, although renewed more frequently, is restricted to a space and a population.

Despite being programmed to run with the four dengue serotypes circulating in Brazil, the model was used only for one serotype for simplification and because of the short simulation time (two years). If the model is run for a longer time, it is recommended to include all serotypes. The results can even be studied in networks with a layer for each serotype. A multiplex approach can also be adopted for one or more serotypes.

Even though its limitations, this initial study can be used as a starting point for a more extensive and in-depth analysis, as it is impossible to effectively track a mosquito-borne disease. It is worth noting that, unlike an airborne or sexually transmitted disease, it is not feasible to measure the transmissibility of each mosquito. In this context, one of the best ways to understand whether networks involving the vector follow the 20/80 rule is via computer simulations.

The structure of the networks of a vector-borne disease involves agents other than humans and a significant influence of environmental factors. Understanding dengue transmission networks can help in proposing measures to reduce infection by the disease, such as vector control [20].

References

1. Adam, D.C., Wu, P., Wong, J.Y., Lau, E.H., Tsang, T.K., Cauchemez, S., Leung, G.M., Cowling, B.J.: Clustering and superspreading potential of sars-cov-2 infections in Hong Kong. Nature Med. **26**(11), 1714–1719 (2020)
2. Ball, F., Mollison, D., Scalia-Tomba, G.: Epidemics with two levels of mixing. Ann. Appl. Probab. 46–89 (1997)
3. Crovello, T.J., Hacker, C.S.: Evolutionary strategies in life table characteristics among feral and urban strains of aedes aegypti (l.). In: Evolution, pp. 185–196 (1972)
4. Donalísio, M.R., Glasser, C.M.: Vigilância entomológica e controle de vetores do dengue. Revista Brasileira de Epidemiologia **5**(3), 259–279 (2002)
5. Duan, W., Qiu, X., Cao, Z., Zheng, X., Cui, K.: Heterogeneous and stochastic agent-based models for analyzing infectious diseases' super spreaders. IEEE Intel. Syst. **28**(4), 18–25 (2013)
6. Gubler, D.J.: The economic burden of dengue. Am. J. Tropical Med. Hygiene **86**(5), 743 (2012)
7. Guzman, M.G., Harris, E.: Dengue. Lancet **385**(9966), 453–465 (2015)
8. Halstead, S.B.: Dengue virus-mosquito interactions. Annu. Rev. Entomol. **53**, 273–291 (2008)

9. Hendron, R.W.S., Bonsall, M.B.: The interplay of vaccination and vector control on small dengue networks. J. Theoret. Biol. **407**, 349–361 (2016)
10. Keeling, M.J., Eames, K.T.: Networks and epidemic models. J. R. Soc. Interface **2**(4), 295–307 (2005)
11. Lima, L., Atman, A.: Impact of mobility restriction in covid-19 superspreading events using agent-based model. Plos One **16**(3), e0248708 (2021)
12. Massad, E., Ma, S., Chen, M., Struchiner, C.J., Stollenwerk, N., Aguiar, M.: Scale-free network of a dengue epidemic. Appl. Math. Comput. **195**(2), 376–381 (2008)
13. Moore, C., Newman, M.E.: Epidemics and percolation in small-world networks. Phys. Rev. E **61**(5), 5678 (2000)
14. Moreno, Y., Pastor-Satorras, R., Vespignani, A.: Epidemic outbreaks in complex heterogeneous networks. Eur. Phys. J. B Condensed Matter Complex Syst. **26**(4), 521–529 (2002)
15. Obolski, U., Perez, P.N., Villabona-Arenas, C.J., Thézé, J., Faria, N.R., Lourenço, J.: Mvse: an r-package that estimates a climate-driven mosquito-borne viral suitability index. Methods Ecol. Evol. **10**(8), 1357–1370 (2019)
16. Resende, M.C.D., Ázara, T.M.F.D., Costa, I.O., Heringer, L.C., Andrade, M.R.D., Acebal, J.L., Eiras, Á.E.: Field optimisation of mosquitrap sampling for monitoring aedes aegypti linnaeus (diptera: Culicidae). Memórias do Instituto Oswaldo Cruz **107**(3), 294–302 (2012)
17. Tun-Lin, W., Burkot, T., Kay, B.: Effects of temperature and larval diet on development rates and survival of the dengue vector Aedes aegypti in north Queensland, australia. Med. Veterinary Entomol. **14**(1), 31–37 (2000)
18. Watts, D.J.: The "new" science of networks. Annu. Rev. Sociol. **30**, 243–270 (2004)
19. Woolhouse, M.E., Dye, C., Etard, J.F., Smith, T., Charlwood, J., Garnett, G., Hagan, P., Hii, J., Ndhlovu, P., Quinnell, R., et al.: Heterogeneities in the transmission of infectious agents: implications for the design of control programs. Proc. Nat. Acad. Sci. **94**(1), 338–342 (1997)
20. Ximenes, R., Amaku, M., Lopez, L.F., Coutinho, F.A.B., Burattini, M.N., Greenhalgh, D., Wilder-Smith, A., Struchiner, C.J., Massad, E.: The risk of dengue for non-immune foreign visitors to the 2016 Summer Olympic Games in Rio de Janeiro, Brazil. BMC Infect. Diseases **16**(1), 186 (2016)
21. Zeng, Q., Liu, Y., Tang, M., Gong, J.: Identifying super-spreaders in information-epidemic coevolving dynamics on multiplex networks. Knowl. Syst. **229**, 107365 (2021)

Multilayer Networks

Structural Cores and Problems of Vulnerability of Partially Overlapped Multilayer Networks

Olexandr Polishchuk(✉)

Laboratory of Modeling and Optimization of Complex Systems, Pidstryhach Institute for Applied Problems of Mechanics and Mathematics National Academy of Sciences of Ukraine, Lviv, Ukraine
od_polishchuk@ukr.net

Abstract. The concept of aggregate-network of multilayer network (MLN), which in many cases significantly simplifies the study of intersystem interactions is introduced, and the properties of its *k*-cores are investigated. The notion of *p*-cores is determined, with help of which the components of MLN that are directly involved in the implementation of intersystem interactions are distinguished. Methods of reducing the complexity of multilayer network models are investigated, which allow us to significantly decrease their dimensionality and better understand the processes that take place in intersystem interactions of different types. Effective scenarios of simultaneous group and system-wide targeted attacks on partially overlapped multilayer networks have been proposed, the main attention of which is focused on the transition points of MLN through which the intersystem interactions are actually implemented. It is shown that these scenarios can also be used to solve the inverse problem, namely, which elements of MLN should be blocked in the first place to prevent the acceleration of spread of dangerous infectious epidemics diseases, etc.

Keywords: Intersystem interactions · Complexity · Multilayer network · Aggregate-network · Network core · Vulnerability · Targeted attack · Epidemic spreading

1 Introduction

Any real system is open, i.e. it interacts with other systems [1]. Intersystem interactions of different kinds give rise to different types of interconnected network structures, which in the theory of complex networks (TCN) are called multilayer networks [2]. At present, a number of the most common types of MLN in the physical world, nature and human society have been identified. First of all these are multidimensional networks, each layer of which reflects a different type of intra- and inter-system interaction [3]. An example of multidimensional MLN is international cooperation, which includes the layers of political, economic, military, security, cultural, sports and other types of interactions between the nodes-countries of the Earth. Each city of the country is a node of several transport networks of different types, state and local government networks, economic

H. Cherifi et al. (Eds.): COMPLEX NETWORKS 2016 2022, SCI 1077, pp. 613–624, 2023.
https://doi.org/10.1007/978-3-031-21127-0_50

and financial networks, etc. The city's life support system includes electricity, gas and water supply networks, telephone networks, cable television, Internet services, etc.

The models of intersystem interactions structures often have a huge dimensionality and complexity, associated with both a large number of MLN nodes and amount of various intra- and inter-layer interactions between them. This requires the development of methods to reduce the complexity of models of such formations, aimed at simplifying their understanding and cognition [4], as well as overcoming the problem of complexity of systems research in general [5, 6]. One way to solve this problem is the division of multiflow MLN into monoflow multilayer networks, i.e. those, that allow the movement of flows of one type by different media or systems-operators and the ability to move from one layer to another through the so-called transition points, i.e. nodes, that are part of several system-layers [7]. Transition points can have multiple connections to nodes of other layers (e-mail or telephone communication) or only to points with the same number from the total set of nodes of the multilayer network (common transport MLN). From a functional point of view, the latter case means that the corresponding node is an element of several system-layers and performs in them different functions or one function, but in different ways. For example, using this approach, it is advisable to divide the general transport system into two monoflow four-layer MLNs, each of which provides the movement of passenger or freight flows by road, rail, air or water (sea or river) transport, respectively. The next step to overcome the problem of complexity is to identify structurally or functionally most important components of monoflow MLN, i.e. its cores of different types [8, 9]. Identification of such cores allows us to develop effective scenarios of simultaneous group and system-wide targeted attacks on the process of intersystem interactions and appropriate means of their protection.

2 Structural Model of MLN

Usually, multilayer network structures are described as

$$G^M = \left(\bigcup_{m=1}^{M} G_m, \bigcup_{\substack{m,k=1 \\ m \neq k}}^{M} E_{mk} \right), \tag{1}$$

where $G_m = (V_m, E_m)$ is a description of structure of the m-th network layer of MLN; V_m is a set of nodes of the network G_m; E_m is a set of edges of the network G_m; E_{mk} is a set of edges between the nodes of sets V_m and V_k, $m \neq k$, $m, k = \overline{1, M}$, M is a number of layers of MLN. The set

$$V^M = \bigcup_{m=1}^{M} V_m$$

will be called the total set of MLN nodes, N^M is a number of nodes of V^M.

We represent the mathematical model of MLN in the form of adjacency matrix $A^M = \{A^{km}\}_{k,m=1}^{M}$. Blocks $A^{km} = \{a_{ij}^{km}\}_{i,j=1}^{N^M}$ of this matrix are defined for the total set of MLN nodes, i.e. the problem of coordination of nodes numbers in the case of their independent numbering for each layer disappears. It is obvious that the diagonal blocks of matrix A^M describe the structure of intra-layer interactions, and the off-diagonal ones describe the structure of inter-layer interactions.Below, to simplify the presentation, we assume that the structure (1) and adjacency matrix A^M describe the undirected MLN of most general form.

From a structural point of view, the most common type of multilayer networks are the partially overlapped MLN, the intersection of sets of nodes V_m of which is not empty (Fig. 1) [10]. The boundary cases of partially overlapped (PO) multilayer networks are multiplexes, i.e. MLN, the sets of nodes V_m of which coincide [8], and multi-networks, i.e. MLN, the sets of nodes V_m of which do not intersect [7], $m = \overline{1, M}$.

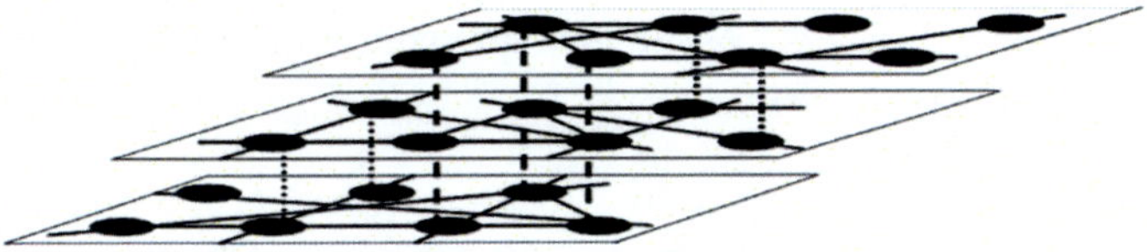

Fig. 1. Example of structure of partially overlapped multilayer network

3 Aggregate-Network of Partially Overlapped MLN

The local characteristic ε_{ij} of edge (n_i, n_j), in PO MLN, which will be called its aggregate-weight, is determined by formula.

$$\varepsilon_{ij} = \sum_{l,m=1}^{M} a_{ij}^{lm}, i, j = \overline{1, N^M}.$$

The aggregate-weight ε_{ii} of node n_i, $i = \overline{1, N^M}$, in a partially overlapped multilayer network is equal to the number of layers of which it is a part. A node that belongs to several layers of MLN and through which the flow can move from one layer to another will be called the transition point of multilayer network. For an arbitrary PO MLN the adjacency matrix $E = \{\varepsilon_{ij}\}_{i,j=1}^{N^M}$ completely defines the weighted network, which will be called the aggregate-network of partially overlapped MLN. The elements of matrix **E** determine the integral structural characteristics of nodes and edges of such multilayer network (Fig. 2).

For multiflows (multidimensional) networks, the value of aggregate-weights ε_{ij}, $i \neq j$, of weighted aggregate-network determines the number of interactions of different types between the nodes of such structures. The projection on weighted aggregate-network of multidimensional MLN loses some meaning, because the weight of each of its edge determines the total number of connections of different types. For monoflow MLN, this disadvantage is absent, because the weight of each edge reflects the number of possible carriers or systems-operators that can provide the movement of corresponding type of flow.

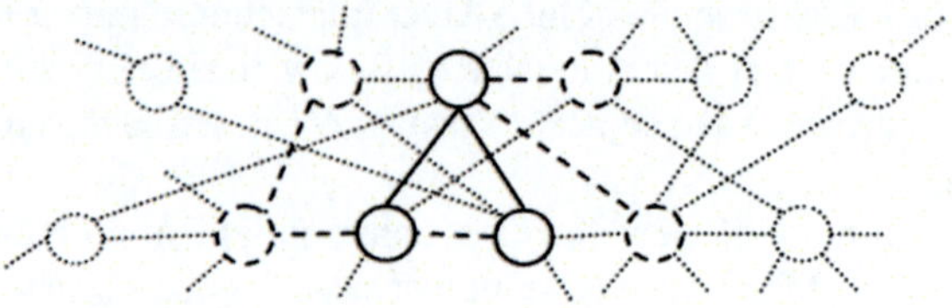

Fig. 2. Fragment of aggregate-network for the shown in Fig. 1 partially overlapped three-layer network (____—for $\varepsilon_{ij} = 3$, _ _ _—for $\varepsilon_{ij} = 2$, ...—for $\varepsilon_{ij} = 1$)

During the study of monoflow PO MLN there are many practically important problems that can be solved at least in the first approximation, using the concept of its aggregate-network. Such problems include:

1. Determination of the shortest paths through the MLN [11] (change of transport modes can significantly speed up time or reduce the cost of movement of passengers and cargos). The shortest path is built in aggregate-network, the structure of which is described by matrix $\boldsymbol{E}$. When such path is determined, it remains to choose the optimal by certain criteria carriers or system-operators on each edge of this path, the aggregate-weight of which exceeds 1.
2. Search for alternative paths of transit flows through different network layers during the isolation of certain zone in separate network layer (use of subway in large cities during traffic jams) [4].
3. Counteracting the spread of epidemics or computer viruses, which due to multi-layer interactions can spread much faster than in one layer [12]. Thus, the weighted aggregate-degree d_i^{ε}, of aggregate-network's node n_i calculated by formula

$$d_i^{\varepsilon} = \sum_{j=1}^{N^M} \varepsilon_{ji} = \sum_{j=1}^{N^M} \varepsilon_{ij}, \tag{2}$$

   determines the number of nodes adjacent to n_i from which the threat of infection of this node may come or the number of nodes adjacent to n_i in which there may be a threat of infection from this node; the aggregate-weight ε_{ii} of node n_i determines the number of layers of MLN, in which this node can contribute to the spread of infection.
4. Finding the path from arbitrary node of one layer to arbitrary node of another layer, especially if they lie outside the intersection of sets of nodes of these layers. The ability to move a flow from one network layer to another and back through transition points expands access to nodes that are unreachable in separate network layers and allows communication between unconnected components of such layers (the movement of traffic flows across the oceans or to remote regions of separate countries—the northern regions of Canada, the central regions of Australia, etc.).

Using the aggregate-network to solve the above problems means reducing the dimensionality of source MLN's model or its quantitative complexity by M times.

4 Structural *k*- and *p*-cores of Aggregate-Network of Partially Overlapped MLN

In TCN, one of the ways to determine the structurally most important components of MLN and simplify the models of multilayer networks is the concept of **k**-core [13].

$$\boldsymbol{k} = \{k_1, k_2, ..., k_M\},$$

as a combination of k_m-cores of separate layers of MLN. The values k_m, $m = \overline{1, M}$, for different layers may differ, i.e. the **k**-core determines the elements that are structurally most important more for the MLN layers than for organization of interlayer interactions in it. To determine the structurally most important components of MLN at a whole, we can use the concept of *k*-cores of its weighted aggregate-network. Then under k_{ag}^{ε}-core of weighted aggregate-network we understand its largest subnetwork, the degree of nodes d_i^{ε} of which, calculated by means of formula (2), are not less than k_{ag}^{ε}. Adjacency matrices $\boldsymbol{E}(k_{ag}^{\varepsilon})$, which fully describe the structures of weighted k_{ag}^{ε}-cores of PO MLN aggregate-network, are obviously obtained from adjacency matrix $\boldsymbol{E}$ by excluding rows and columns for nodes whose values $d_i^{\varepsilon} < k_{ag}^{\varepsilon}$, $i = \overline{1, N^M}$. Indirectly, the weighted aggregate-degrees of nodes of PO MLN's aggregate-network determine the importance of this node in multilayer network, as duplication is usually subject to elements that implement the most important functions in the system. It should be noted that k_{ag}^{ε}-core of weighted aggregate-network of PO MLN provides much more important information for the study of intersystem interactions than **k**-core.

To solve the problem of determining the structurally most important components of intersystem interactions in MLN, we introduce the notion of *p*-core of partially overlapped multilayer network

$$\tilde{G}^p = (\tilde{V}^p, \tilde{E}^p),$$

as combination of those subnetworks of separate layers with connections between nodes of them that are part of at least p, $2 \le p \le M$, layers of MLN (it is obvious that $\tilde{G}^1$ is identical to the source multilayer network G^M defined by formula (1). In other words, the *p*-core of MLN is its multilayer subnet, each of the nodes of which has an aggregate-weight $\varepsilon_{ii} \ge p$. If

$$p_{\max} = \max_{i=\overline{1,N^M}} \{\varepsilon_{ii}\} = M,$$

that is, if the structural M-core of multilayer network is nonempty, then the studied MLN we will call the kernel (Fig. 3). Such structures are generated by the general transport system of the Earth and its separate continents, systems of maintenance of vital activity of the city, postal systems, systems of telephone communication, etc. It is obviously that the kernel $\tilde{G}^M$ has a multiplex structure. If the condition

$$p_{\max} < M,$$

is met then the studied MLN will call the non-kernel. Such MLN are generated by linguistic multilayer network systems (there is hardly at least one person who speaks all existing languages), social network systems, online services, mobile operators and provider of cable TV, etc. [7].

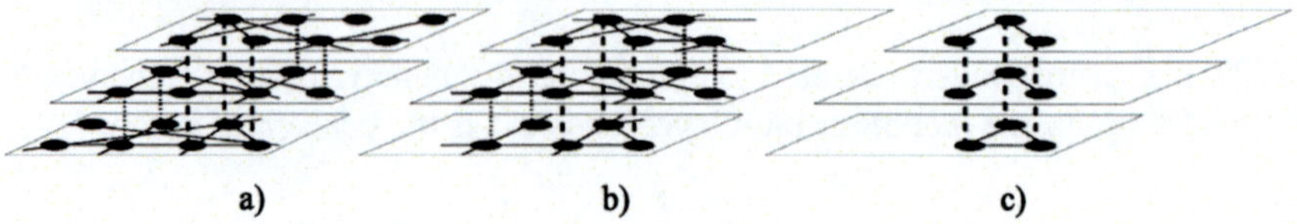

Fig. 3. Examples of reflected in Fig. 1 source three-layer partially overlapped kernel MLN (a) and its *2*- (b) and *3*-core (c)

Hereinafter, the p_{ag}-core of aggregate-network of MLN will be called such its weighted subnet, each node n_i of which has an aggregate weight ε_{ii} not less than p (Fig. 4).

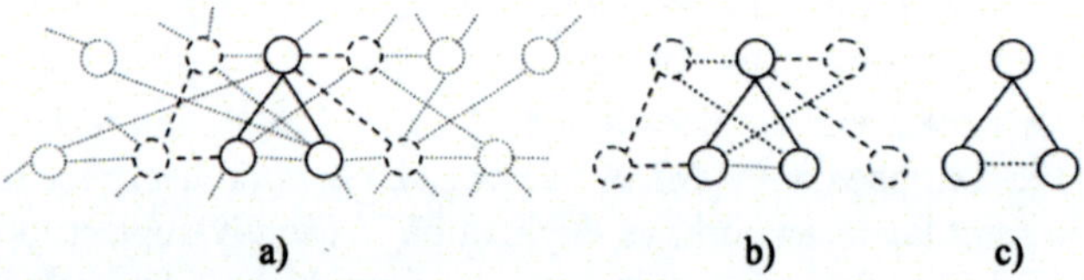

Fig. 4. Examples of weighted aggregate-network of reflected in Fig. 3 source three-layer partially overlapped kernel MLN (a) and its 2_{ag}- (b) and 3_{ag}-core (c, ____—element is part of three layers, - - -—element is part of two layers, ...—element is part of one layer)

Denote by N^p the number of elements of the total set of nodes of the *p*-core of partially overlapped multilayer network. Let us determine the specific weight η^p of *p*-core transition points in this MLN by the ratio

$$\eta^p = N^p \Big/ N^M .$$

The *p*-cores determine the set of transition points that are directly involved in the organization of intersystem interactions at least of *p* layers of MLN. Therefore, the closer the value η^p to 1 for all $p = 1,2,\ldots, M$, the stronger the interaction of MLN layers. The specific weight of transition points of the layer determines its ability to participate in intersystem interactions with other layers of the multilayer network.

The value N^M only partially reflects the complexity of multilayer network. Much more adequate indicators are the characteristics of dimensional and connectional complexity of MLN. The dimensional complexity of partially overlapped MLN can be estimated through the diagonal elements of matrix **E** by means of parameter

$$\phi = \sum_{i=1}^{N^M} \varepsilon_{ii},$$

and its connectional complexity—using parameter

$$\varphi = \sum_{\substack{i,j=1 \\ i \neq j}}^{N^M} \varepsilon_{ij}.$$

To calculate the corresponding indicators of complexity of the *p*-core of MLN, we define the matrix

$$E_p = \{\varepsilon_{ij}^p\}_{i,j=1}^{N^M}, \varepsilon_{ij}^p = \begin{cases} \varepsilon_{ij}, & \text{if } \varepsilon_{ii} \geq p, \\ 0, & \text{if } \varepsilon_{ii} < p, \end{cases}, i,j = \overline{1, N^M}, p = 2, 3, ..., M.$$

Then the dimensional complexity of *p*-core of PO MLN we determine by means of parameter

$$\phi_p = \sum_{i=1}^{N^M} \varepsilon_{ii}^p,$$

and its connectional complexity—using parameter

$$\varphi_p = \sum_{\substack{i,j=1 \\ i \neq j}}^{N^M} \varepsilon_{ij}^p, p = 2, 3, ..., M$$

Based on this, we can determine the reduction of dimensional complexity of the *p*-core model compared to the PO MLN model in general by the ratio

$$\eta_p = \phi_p / \phi,$$

and reduction of connectional complexity of the *p*-core model compared to the MF PO MLN model in general by the ratio

$$\mu_p = \varphi_p / \varphi, p = 2, 3, ..., M.$$

Note that k_{ag}^{ε}-cores are formed on the basis of values of the sum of nondiagonal elements of the rows or columns of matrix **E**. At the same time, *p*-cores are formed on the basis of values of the diagonal elements of matrix **E**. This means that there is no direct connection between the *k*- and *p*-cores. However, their properties can be combined as follows. To determine the sets of nodes with the largest values of aggregate-degree centrality, which at the same time take the greatest part in process of intersystem interactions, we introduce the notion of $p_{ag}(k_{ag}^{\varepsilon})$-core as a such subnet of source aggregate-network, nodes of which at first have aggregate-degree $d_i^{\varepsilon} \geq k_{ag}^{\varepsilon}$ and at second belong to not less than *p* layers of MLN. The adjacency matrix $E(p_{ag}(k_{ag}^{\varepsilon}))$ of corresponding subnet of source MLN is obviously obtained from the adjacency matrix $E(k_{ag}^{\varepsilon})$ by excluding rows and columns whose diagonal elements are smaller than p_{ag}. Conversely, to determine the sets of nodes that participate most in the process of intersystem interactions and among them have the greatest values of aggregate-degree centrality, we introduce the concept of $k_{ag}^{\varepsilon}(p_{ag})$-core as a subnet of source aggregate-network, nodes which at first are the part of at least p layers of MLN and at second have the aggregate-degree $d_i^{\varepsilon} \geq k_{ag}^{\varepsilon}$. The adjacency matrix $\boldsymbol{E}(k_{ag}^{\varepsilon}(p_{ag}))$ of corresponding subnet of the source MLN is obviously obtained from the adjacency matrix $\boldsymbol{E}_p$ by excluding rows and columns for which the nodes have the aggregate-degree $d_i^{\varepsilon} < k_{ag}^{\varepsilon}$. It is obvious that the structures of $p_{ag}(k_{ag}^{\varepsilon})$- and $k_{ag}^{\varepsilon}(p_{ag})$-cores, as well as the adjacency matrices $\boldsymbol{E}(p_{ag}(k_{ag}^{\varepsilon}))$ and $\boldsymbol{E}(k_{ag}^{\varepsilon}(p_{ag}))$ may differ, because the methods and purposes of construction of each of them are different. The use of k- and p-cores allows us at least partially to solve the problem of complexity of the study of real intersystem interactions, highlighting in them the most structurally important components.

5 Vulnerability of Intersystem Interactions

The study of stability of real systems and intersystem interactions of different types to targeted attacks is one of the most important problems of modern systems analysis. Let us that all layers of MLN are the free-scale networks [14], which are quite widespread and at the same time the most vulnerable to such attacks. The stability of separate system-layers can be determined using the scenarios proposed in [15, 16] etc. These scenarios are based on the defeat of network nodes with the highest degree or nodes with the highest betweenness centrality. Stability of MLN to targeted attacks is determined primarily by the vulnerability of transition points of the multilayer network, i.e. its p-cores with different values p, starting with the largest $p_{\max} \leq M$. We will build scenarios of attacks on intersystem interactions, using local and global characteristics of the elements of weighted aggregate-network, which determine the integrated indicators of their importance in PO MLN. Note that usually the removal of a particular node from the system structure leads to redistribution of traffic routes that passed through it by the network, and the establishment of new connections between the remaining nodes. This process reflects the reaction of system to changes in operating conditions and means a change of degrees and betweenness centralities of MLN nodes.

Before constructing scenarios of targeted attacks, it is necessary to determine the criteria for their success, i.e. the desired level of damage of system structure. Such criteria include the division of aggregate-network into unconnected components in which the MLS ceases to exist as a single supersystem formation, the cessation of intersystem interactions between layers, and so on. Let us build the first attack scenario on the basis of aggregate-weights and weighted aggregate-degrees of nodes of the weighted aggregate-network of MLN:

1. make a list of nodes of the aggregate-network of MLN in descending order of their aggregate-weights (values ε_{ii} of matrix $\mathbf{E}$);
2. in each group of nodes with the same values ε_{ii}, we arrange the nodes on the basis of decreasing values of their weighted aggregate degrees d_i^{ε};
3. delete the first node from the beginning of created list; if the selected criterion of attack success is fulfilled than algorithm is finished, otherwise go to the next point;

4. the removal of particular node usually leads to the establishment of new connections between the nodes that remain in MLN, i.e. the values d_i^{ε} of nodes may change; therefore, if the list of nodes with the current value ε_{ii} is not exhausted, we pass to point 2 of this scenario;
5. after processing the nodes of group with the current value $\varepsilon_{ii} > 1$ go to step 2 with the value ε_{ii}-1; if $\varepsilon_{ii} = 1, i = \overline{1, N^M}$, then the algorithm is finished.

The second scenario of targeted attack will be built on the use of aggregate-weights and betweenness centrality of nodes of the aggregate-network of MLN:

1. make a list of nodes of the aggregate-network of MLN in descending order of their aggregate weights (values ε_{ii} of matrix **E**);
2. in each group of nodes with the same values ε_{ii}, we arrange the nodes on the basis of decreasing values of their betweenness centralities in aggregate-network of MLN;
3. delete the first node from the beginning of created list; if the selected criterion of attack success is fulfilled than algorithm is finished, otherwise go to the next point;
4. the removal of particular node usually leads to the establishment of new connections between the nodes that remain in MLN, i.e. the values of betweenness centrality of nodes may change; therefore, if the list of nodes with the current value ε_{ii} is not exhausted we pass to point 2 of this scenario;
5. after processing the nodes of group with the current value $\varepsilon_{ii} > 1$ go to step 2 with the value ε_{ii}-1; if $\varepsilon_{ii} = 1, i = \overline{1, N^M}$, then the algorithm is finished.

Usually, scenarios that list the characteristics of elements after removal of the next node and are based on the use of betweenness centrality, are more effective for achieving the goal of attack than scenarios that use degree centrality [17].

Along with described above, there is a reverse problem, which is to prevent the spread of epidemics of dangerous infectious diseases, computer viruses, invasion processes, etc. This problem is especially acute during Covid-19 and is the need to block those components of MLN that most "contribute" to the pandemic spreading [18]. The greatest risk of infection in settlements arises during direct contact of people in places of their mass concentration or constant communication. The spread of infection between settlements, regions and countries is due to large volumes of passenger traffic between them. These features are quite well correlated with the elements of MLN with large values of aggregate weights and weighted aggregate-degrees of nodes of its aggregate-network. Hence the ways to prevent the pandemic spreading, which are to block the nodes where crowds are possible, and to block the paths of flows motion to/from the nodes, which have a high level of infection. Obviously, to implement these ways, we can use the proposed above scenarios, in which the criterion for success of anti-epidemiological measures is to minimize access to the nodes with the highest level of infection.

In [15] was shown that the average performance of Internet is halved if only 1% of nodes-domains with the largest degrees fail, and the Internet becomes divided into unconnected components if 4% of such nodes fail. In Ukraine, the number of state-owned banks in the country's banking system does not exceed 0.7%. At the same time, the share of their assets in this system is 55.2%, and the share of deposits of individuals is 61.6% [19]. A successful attack on this group of banks will lead to the largest losses in financial system of the state. Simultaneous *DDoS*-attacks on January 14 and February 15, 2022 on the computer networks of more than 70 major state, security and financial structures of Ukraine became a serious threat to the public administration system [20]. This means that for critical destabilization or shutdown of the system operation, it is usually necessary to simultaneously block the operation of a certain group of nodes. Indeed, successive attacks on separate, even the most important nodes, often allow us to distribute their functions among other nodes of the system. This is duly taken into account in the above scenarios. However, to counteract the simultaneous successful attack on a group of the most important elements of MLN, and the main thing to overcome the consequences of such attack, is much more difficult. To defeat intersystem interactions, such group can be a certain set of the most important transition points. The notions of p_{ag}-, $p_{ag}(k_{ag}^{\varepsilon})$- and $k_{ag}^{\varepsilon}(p_{ag})$-cores defined in the previous paragraphs allow us to determine the most important for MLN groups of nodes, simultaneous targeted attacks on which will certainly cause the most damage or even lead to the cessation of intersystem interactions.

One of scenarios of such defeat, based on the use of notion of $k_{ag}^{\varepsilon}(p_{ag})$-core of aggregate-network of MLN, consists of sequential implementation of the following steps:

1. we accept equal $p_{ag} = p_{\max} \leq M$ and $k_{ag}^{\varepsilon} = \max\limits_{i=\overline{1,N^M}} d_i^{\varepsilon}$;
2. remove from structure of MLN nodes that belong to $k_{ag}^{\varepsilon}(p_{ag})$-core and its connections; if selected criterion for the attack success is met, the algorithm is finished, otherwise go to the next point;
3. reduce the value k_{ag}^{ε} by 1; if the list of nodes of $k_{ag}^{\varepsilon}(p_{ag})$-core is not exhausted, then go to point 2, otherwise go to the next point;
4. if the value $p_{ag} > 1$, then reduce it by 1 and go to point 2 with a value k_{ag}^{ε} equal to the maximum weighted aggregate-degree of remaining nodes of aggregate-network; otherwise the algorithm is finished.

To counteract the epidemic spreading, it is also advisable to use scenarios of blocking groups of nodes (settlements) in which the highest incidence is observed. Thus, the practice of preventing the spread of Covid-19 in Ukraine has shown that sequential blocking of the most infected nodes (settlements) gives a worse result than simultaneous blocking groups of nodes (regions of the country) from which the infection can potentially spread [21].

Usually, the spread of epidemics is stopped by the introduction of quarantine, i.e. the isolation of areas where carriers of infection are found. From a functional point of view, the isolation of certain subnet of the source network means the complete cessation or significant restriction of flows motion from (in, through) it. For many reasons, measures taken to combat the deployment of Covid-19 pandemic have transformed the world into network of isolated zones, movement of flows between which (especially human ones) reduced by dozens due to the cessation or significant restriction of rail, air and road services. Moreover, as a result of introduction of such restrictions, many states have also become networks of isolated communities or separate settlements. The self-isolation of majority of citizens, caused by traffic restrictions, large fines for non-compliance with quarantine conditions and the closure of enterprises or their operation in remote access, has significantly reduced not only external but also internal flows in such isolated zones. With the constant network structure there was a kind of "granulation" of the system, which was divided into a hierarchy of successively isolated subsystems in terms of limiting the interaction between them. Thus, a new type of "granular" networks has emerged, which has not been studied so far. In this case, the losses suffered by the system are not caused by blocking its separate component, but by restricting the movement of flows between all components as a whole. The possibility of transition of network system to the "granular" state, which is characterized by complete or partial isolation of all its components, is a separate kind of system-wide targeted attack. The collapse of USSR was a successful outcome of such attack, which was carried out in response to the war in Afghanistan, and led to its division (granulation) into separate independent states. As a result, the volume of flows within the former Soviet Union has declined significantly, and the total GDP of all these states has not yet reached the level of GDP of the former USSR. Such attacks can be implemented in the form of international sanctions against countries that threaten world security and so on. It should also be noted that much more effective scenarios of targeted attacks of various directions can be built on the basis of flow models of complex network systems and intersystem interactions [4, 22–24].

6 Conclusions

The study of real complex networks and intersystem interactions of different types allows us to understand many processes that take place in the physical world, nature and human society. The main obstacle that arises in this way is the problem of complexity, caused both by dimension of such systems and the number of heterogeneous interactions in them. One way to solve this problem is to highlight the most important from structural and functional point of view components of system that determine its behavior. To simplify the study of intra- and intersystem interactions, the article introduces the notion of aggregate-network, and its k- and p-cores, which determines the most structurally important components of MLN. It is shown how the notions introduced in the paper can be used to solve a number of practically important problems of the theory of complex networks, in particular the stability of intersystem interactions to various vulnerable factors of both artificial and natural origin. As the emergence of new threats, such as Covid-19, is not excluded, the development of scenarios for such defeats and means of timely counteraction to them is an extremely important applied problem, for the theoretical and practical solution of which a lot of effort will be spent.

References

1. Scott, W.R., Davis, G.F.: Organizations and organizing: rational, natural and open systems perspectives. Routledge, London (2015)
2. Boccaletti S et al (2014) The structure and dynamics of multilayer networks. Phys Rep 544(1):1–122. https://doi.org/10.1016/j.physrep.2014.07.001
3. Berlingerio, M., Coscia, M., Giannotti, F., Monreale, A., Pedreschi, D.: Multidimensional networks: foundations of structural analysis. World Wide Web **16**(5–6), 567–593 (2012). https://doi.org/10.1007/s11280-012-0190-4

4. Polishchuk, O., Yadzhak, M.: Network structures and systems: II. Cores of networks and multiplexes. Sys. Res. Inf. Tech. **3**, 38–51 (2018). https://doi.org/10.20535/SRIT.2308-8893.2018.3.04
5. Barabasi, A.-L.: The architecture of complexity. IEEE Cont. Sys. Mag. **27**(4), 33–42 (2007). https://doi.org/10.1109/MCS.2007.384127
6. Northrop, R.B.: Introduction to complexity and complex systems. CRC Press, Boca Raton (2011)
7. Polishchuk O (2021) Aggregate-networks and *p*-cores of monoflow partially overlapped multilayer systems. arXiv:2111.10764v1
8. Lee, K.M., Min, B., Goh, K.I.: Towards real-world complexity: an introduction to multiplex networks. Eur. Phys. J. B **88**, 48 (2015). https://doi.org/10.1140/epjb/e2015-50742-1
9. Corominas-Murtra, B., Fuchs, B., Thurner, S.: Detection of the elite structure in a virtual multiplex social system by means of a generalised K-core. PLoS ONE **9**(12), e112606 (2014). https://doi.org/10.1371/journal.pone.0112606
10. Alvarez-Zuzek, L.G., Di Muro, M.A., Havlin, S., Braunstein, L.A.: Dynamic vaccination in partially overlapped multiplex network. Phys. Rev. E **99**, 012302 (2019). https://doi.org/10.1103/PhysRevE.99.012302
11. Ghariblou, S., Salehi, M., Magnani, M., Jalili, M.: Shortest paths in multiplex networks. Sci. Rep. **7**, 2142 (2017). https://doi.org/10.1038/s41598-017-01655-x
12. Ventura da Silva PC et al (2018) Epidemic spreading with awareness and different time scales in multiplex networks. arXiv:1812.01386v1
13. Osat, S., Radicchi, F., Papadopoulos, F.: k-core structure of real multiplex networks. Phys. Rev. Res. **2**(2), 023176 (2020). https://doi.org/10.1103/PhysRevResearch.2.023176
14. Albert, R., Barabasi, A.-L.: Statistical mechanics of complex networks. Rev. Mod. Phys. **74**(1), 47 (2002). https://doi.org/10.1103/RevModPhys.74.47
15. Albert, R., Barabási, A.-L.: Error and attack tolerance of complex networks. Nature **406**, 378–482 (2000). https://doi.org/10.1038/35019019
16. Al, B., et al.: Graph structure in the web. Comput. Netw. **33**, 309–320 (2000). https://doi.org/10.1515/9781400841356.183
17. Holme, P., Kim, B.J., Yoon, C.N., Han, S.K.: Attack vulnerability of complex networks. Phys. Rev. E **65**, 056109 (2002). https://doi.org/10.1103/PhysRevE.65.056109
18. Faria do Valle Í (2020) Recent advances in network medicine: from disease mechanisms to new treatment strategies. Multiple Sclerosis J 26(5):609–615. https://doi.org/10.1177/1352458519877002
19. https://index.minfin.com.ua/ua/banks/stat/
20. https://24tv.ua/kiberataka-ukrayina-2022-novini-pro-napad-na-uryadovi-sayti_n1842293
21. https://covid19.rnbo.gov.ua
22. Polishchuk, O.D.: Vulnerability of complex network structures and systems. Cybern. Syst. Anal. **56**(2), 312–321 (2020). https://doi.org/10.1007/s10559-020-00247-4
23. Nasiri, E., Berahmand, K., Li, Y.: A new link prediction in multiplex networks using topologically biased random walks. Chaos Solitons Fractals **151**, 111230 (2021). https://doi.org/10.1016/j.chaos.2021.111230
24. Nasiri E et al (2022) Impact of centrality measures on the common neighbors in link prediction for multiplex networks. Big Data 10(2):138–150. https://doi.org/10.1089/big.2021.0254

Multilayer Block Models for Exploratory Analysis of Computer Event Logs

Corentin Larroche(✉)

French National Cybersecurity Agency (ANSSI), Paris, France
corentin.larroche@ssi.gouv.fr

Abstract. We investigate a graph-based approach to exploratory data analysis in the context of network security monitoring. Given a possibly large batch of event logs describing ongoing activity, we first represent these events as a bipartite multiplex graph. We then apply a model-based biclustering algorithm to extract relevant clusters of entities and interactions between these clusters, thereby providing a simplified situational picture. We illustrate this methodology through two case studies addressing network flow records and authentication logs, respectively. In both cases, the inferred clusters reveal the functional roles of entities as well as relevant behavioral patterns. Displaying interactions between these clusters also helps uncover malicious activity. Our code is available at https://github.com/cl-anssi/MultilayerBlockModels.

Keywords: Cybersecurity · Network modelling · Multiplex networks

1 Introduction

The ever-increasing prevalence of cyber threats compels network administrators and information security specialists to constantly monitor ongoing activity inside sensitive networks. Such monitoring often boils down to collecting and analyzing event logs, which can be generally defined as structured records reflecting all sorts of actions. These logs are continuously produced in massive amounts by nearly all devices within a monitored network, which is both a blessing and a curse: while wide and systematic log collection provides a detailed situational picture, extracting actionable insights from such volumes of data is highly challenging. Data mining algorithms are thus needed to efficiently analyze and summarize all the available information.

Interestingly, various types of event logs can be naturally represented as heterogeneous graphs. Consider for instance a network flow recorded at the border of a protected network, indicating that a connection has occurred between internal host u and external host v. A collection of such events can be abstracted into a bipartite graph whose top (resp. bottom) nodes are the internal (resp. external) hosts. In addition, each network flow is further characterized by additional categorical variables, such as the transport layer protocol (e.g. TCP or UDP) or the source and destination ports. Therefore, each internal–external host pair can

H. Cherifi et al. (Eds.): COMPLEX NETWORKS 2016 2022, SCI 1077, pp. 625–637, 2023.
https://doi.org/10.1007/978-3-031-21127-0_51

be linked by various kinds of edges. This suggests that a multiplex graph [17] might be an appropriate representation. However, when considering a real-world network, such a graph could easily have tens of thousands of nodes, making it hard to interpret without any further processing.

To alleviate this issue, we propose to use a model-based multiplex graph biclustering algorithm. This algorithm extends previous work on latent block modelling [10,12] by factoring in the existence of multiple types of edges. It allows us to extract relevant clusters of entities along with their interactions across various network layers, thereby creating a simplified view of the information contained in the logs. While graph biclustering has already been used for event log analysis [20], to the best of our knowledge, the use of multilayer models to account for the heterogeneity of events is novel. In addition, we present detailed case studies on two publicly available datasets consisting of network flow records and authentication logs, respectively.

The rest of this paper is structured as follows. We first introduce the statistical model and inference procedure we use for multiplex graph biclustering in Sect. 2. The two case studies are then presented in Sects. 3 and 4, respectively. Finally, we review some related work in Sect. 5 and discuss potential areas of improvement in Sect. 6.

2 The Multilayer Latent Block Model

Before diving into the specifics of authentication logs and network flows, we first introduce the statistical tools we use to analyze them. Section 2.1 presents our generative model for multiplex bipartite graphs, and Sect. 2.2 describes the algorithms we use for model inference and selection.

2.1 Model Description

Key notations. Let $\mathcal{U} = \{u_1, \dots, u_I\}, \mathcal{V} = \{v_1, \dots, v_J\}$ be two node sets of size I and J, respectively. We consider a multiplex bipartite graph $\mathcal{G} = (\mathcal{U}, \mathcal{V}, \mathcal{E})$, where $\mathcal{E} \subset \mathcal{U} \times \mathcal{V} \times [L]$ denotes the edge set (with $[L] \overset{\text{def}}{=} \{1, \dots, L\}$). Specifically, each edge $(u, v, \ell) \in \mathcal{E}$ between a top node u and a bottom node v is further characterized by an edge type ℓ, and L denotes the number of possible edge types. The multiplex graph $\mathcal{G}$ can then be characterized by L biadjacency matrices $\mathbf{B}^{(1)}, \dots, \mathbf{B}^{(L)}$, where for each $\ell \in [L]$, the matrix $\mathbf{B}^{(\ell)} = \left(b_{ij}^{(\ell)}\right) \in \mathbb{R}^{I \times J}$ is defined by

$$\forall (i, j) \in [I] \times [J],\ b_{ij}^{(\ell)} = \begin{cases} 1 & \text{if } (u_i, v_j, \ell) \in \mathcal{E} \\ 0 & \text{otherwise.} \end{cases}$$

Finally, we aim to partition $\mathcal{U}$ (resp. $\mathcal{V}$) into a fixed number H (resp. K) of clusters, which we denote $\mathbb{U} = \{\mathcal{U}_1, \dots, \mathcal{U}_H\}$ (resp. $\mathbb{V} = \{\mathcal{V}_1, \dots, \mathcal{V}_K\}$). For each top node u_i (resp. bottom node v_j), the unique index h such that $u_i \in \mathcal{U}_h$ (resp. k such that $v_j \in \mathcal{V}_k$) is denoted U_i (resp. V_j). In order to find the optimal partition, we adopt a model-based approach relying on the generative model described in the next paragraph.

Generative model. We propose a multilayer extension of the Poisson latent block model introduced by Govaert and Nadif [12]. This model relies on the fundamental assumption that the probability of an edge between two nodes depends on the cluster assignments of these nodes. The biadjacency matrices $\mathbf{B}^{(1)}, \ldots, \mathbf{B}^{(L)}$ are then generated by the following hierarchical model:

(i) for each $i \in [I]$, sample $U_i \sim \text{Multinomial}(\boldsymbol{\pi})$;
(ii) for each $j \in [J]$, sample $V_j \sim \text{Multinomial}(\boldsymbol{\rho})$;
(iii) for each $(i, j, \ell) \in [I] \times [J] \times [L]$, sample $b_{ij}^{(\ell)} \sim \text{Poisson}\left(\mu_i \nu_j \theta_{U_i V_j}^{(\ell)}\right)$.

In words, the proportion of top (resp. bottom) nodes falling into each cluster is controlled by a parameter $\boldsymbol{\pi} \in (0,1)^H$ (resp. $\boldsymbol{\rho} \in (0,1)^K$). The probability of an edge of type ℓ linking a top node u_i and a bottom node v_j then depends on three parameters: the node-specific factors μ_i and ν_j represent the overall propensity of each node to form edges across all layers, and the rate $\theta_{U_i V_j}^{(\ell)}$ controls the number of edges between clusters $\mathcal{U}_{U_i}$ and $\mathcal{V}_{V_j}$ in the ℓ-th layer. Note that the Poisson distribution is only used here to facilitate calculations: in practice, the biadjacency matrices only contain zeros and ones.

2.2 Model Inference and Selection

Given an observed graph $\mathcal{G} = (\mathcal{U}, \mathcal{V}, \mathcal{E})$, we now aim to find the optimal partitions $\mathbb{U}$ and $\mathbb{V}$ using the model introduced above. To that end, we need to infer the parameters $\boldsymbol{\pi}$, $\boldsymbol{\rho}$, $\boldsymbol{\mu} = (\mu_i) \in \mathbb{R}_+^I$, $\boldsymbol{\nu} = (\nu_j) \in \mathbb{R}_+^J$, and $\boldsymbol{\Theta} = (\boldsymbol{\Theta}^{(1)}, \ldots, \boldsymbol{\Theta}^{(L)})$ (where $\boldsymbol{\Theta}^{(\ell)} = (\theta_{hk}^{(\ell)}) \in \mathbb{R}_+^{H \times K}$ for each $\ell \in [L]$). The number of top clusters H and bottom clusters K must also be set in a principled manner. These two problems are addressed in the next paragraphs.

Model inference. The parameters of the model and the cluster assignments are obtained by maximizing the complete data log-likelihood

$$L_{\mathrm{C}}(\mathbb{T}, \mathbb{U}, \mathbb{V}) = \sum_{i=1}^{I} \log \pi_{U_i} + \sum_{j=1}^{J} \log \rho_{V_j} + \sum_{i=1}^{I} \sum_{j=1}^{J} \sum_{\ell=1}^{L} \left\{ b_{ij}^{(\ell)} \log\left(\mu_i \nu_j \theta_{U_i V_j}^{(\ell)}\right) - \mu_i \nu_j \theta_{U_i V_j}^{(\ell)} \right\},$$

where $\mathbb{T} = \{\boldsymbol{\pi}, \boldsymbol{\rho}, \boldsymbol{\mu}, \boldsymbol{\nu}, \boldsymbol{\Theta}\}$ denotes the complete set of parameters. To that end, we use the block expectation-maximization algorithm described in [12], with minor adjustments to factor in the multilayer nature of the data. This algorithm first introduces soft cluster assignment matrices $\mathbf{U} = (\tilde{u}_{ih}) \in [0,1]^{I \times H}$ and $\mathbf{V} = (\tilde{v}_{jk}) \in [0,1]^{J \times K}$, with $\sum_{h=1}^{H} \tilde{u}_{ih} = \sum_{k=1}^{K} \tilde{v}_{jk} = 1$ for all $i \in [I]$ and $j \in [J]$. It then maximizes the fuzzy criterion

$$G(\mathbb{T}, \mathbf{U}, \mathbf{V}) = L_{\mathrm{S}}(\mathbb{T}, \mathbf{U}, \mathbf{V}) + H(\mathbf{U}) + H(\mathbf{V}),$$

where $H(\mathbf{U}) = -\sum_{i=1}^{I}\sum_{h=1}^{H}\tilde{u}_{ih}\log\tilde{u}_{ih}$ denotes the total entropy of the soft cluster assignments of the top nodes, $H(\mathbf{V})$ denotes the total entropy for the bottom nodes, and

$$L_{\mathrm{S}}(\mathbb{T},\mathbf{U},\mathbf{V}) = \sum_{i=1}^{I}\sum_{h=1}^{H}\tilde{u}_{ih}\log\pi_h + \sum_{j=1}^{J}\sum_{k=1}^{K}\tilde{v}_{jk}\log\rho_k$$
$$+\sum_{h=1}^{H}\sum_{k=1}^{K}\sum_{i=1}^{I}\sum_{j=1}^{J}\sum_{\ell=1}^{L}\tilde{u}_{ih}\tilde{v}_{jk}\left[b_{ij}^{(\ell)}\log\left(\theta_{hk}^{(\ell)}\right) - \mu_i\nu_j\theta_{hk}^{(\ell)}\right]$$

is the fuzzy likelihood function. The criterion $G(\mathbb{T},\mathbf{U},\mathbf{V})$ is maximized by alternatively performing two steps: optimizing $\mathbf{U}$ and $\mathbf{V}$ with $\mathbf{\Theta}$ fixed (E-step), and optimizing $\mathbf{\Theta}$ with $\mathbf{U}$ and $\mathbf{V}$ fixed (M-step). These two steps are iterated until G stabilizes. The node partitions $\mathbb{U}$ and $\mathbb{V}$ can then be obtained by assigning each node to the most probable cluster according to $\mathbf{U}$ and $\mathbf{V}$. Algorithm 1 describes the detailed inference procedure. Note that since the result depends on the random initialization of the parameters, we run the whole procedure 50 times with different initializations and return the model with the highest likelihood.

Model selection. In order to run the aforementioned inference procedure, we must first set the number of top clusters H and bottom clusters K. In the absence of any prior knowledge, we use the integrated completed likelihood (ICL [3]) to pick the best values out of a predefined set of candidates. More specifically, for each $(H,K)\in\{2,\ldots,16\}^2$, we run the inference procedure to obtain the optimal parameter set $\mathbb{T}$ and node partitions $\mathbb{U},\mathbb{V}$. We then compute

$$\mathrm{ICL}(\mathbb{U},\mathbb{V};H,K) = L_{\mathrm{C}}(\mathbb{T},\mathbb{U},\mathbb{V}) - \frac{H-1}{2}\log I - \frac{K-1}{2}\log J - \frac{LHK}{2}\log(LIJ)$$

for each candidate model, and the highest-scoring model is selected. The ICL penalizes the complete data log-likelihood by subtracting terms proportional to the number of free parameters, thus enabling a trade-off between goodness-of-fit and model complexity. Note that it relies on several approximations, which makes it an arguably imperfect evaluation criterion. This has motivated many subsequent contributions on model selection for stochastic and latent block models (see e.g. [4,19,27]). However, for the sake of simplicity, we leave the use of more sophisticated criteria for future work.

3 First Case Study—Network Flows

Having described our modelling tools, we now move on to our first case study. Section 3.1 describes the dataset we consider and the preprocessing steps we apply to turn it into a multiplex network. We then discuss the results obtained by applying our methodology in Sect. 3.2.

Algorithm 1: Block expectation-maximization algorithm for maximum likelihood inference of the node clusters and model parameters.

Data: Biadjacency matrices $\mathbf{B}^{(1)}, \ldots, \mathbf{B}^{(L)}$; number of top clusters H; number of bottom clusters K; stopping criterion ϵ.
Result: Set of estimated parameters $\mathbb{T}$; node partitions $\mathbb{U}$ and $\mathbb{V}$.

$M \leftarrow \sum_{\ell=1}^{L} \sum_{i=1}^{I} \sum_{j=1}^{J} b_{ij}^{(\ell)}$;
foreach $i \in [I], j \in [J]$ **do**
  $\mu_i \leftarrow \frac{1}{\sqrt{M}} \sum_{\ell=1}^{L} \sum_{j=1}^{J} b_{ij}^{(\ell)}$; $\nu_j \leftarrow \frac{1}{\sqrt{M}} \sum_{\ell=1}^{L} \sum_{i=1}^{I} b_{ij}^{(\ell)}$;
end
Randomly initialize $\boldsymbol{\Theta}, \boldsymbol{\pi}, \boldsymbol{\rho}, \mathbf{U}, \mathbf{V}$; $\Delta \leftarrow \infty$;
while $\Delta > \epsilon$ **do**
  $G_{\text{old}} \leftarrow G(\mathbb{T}, \mathbf{U}, \mathbf{V})$;
  foreach $i \in [I], h \in [H]$ **do**
    $s_{ih} \leftarrow \log \pi_h + \sum_{k=1}^{K} \sum_{\ell=1}^{L} \sum_{j=1}^{J} \tilde{v}_{jk} \left(b_{ij}^{(\ell)} \log \theta_{hk}^{(\ell)} - \mu_i \nu_j \theta_{hk}^{(\ell)} \right)$;
    $\tilde{u}_{ih} \leftarrow \frac{\exp(s_{ih})}{\sum_{h'=1}^{H} \exp(s_{ih'})}$;
  end
  foreach $j \in [J], k \in [K]$ **do**
    $t_{jk} \leftarrow \log \rho_k + \sum_{h=1}^{H} \sum_{\ell=1}^{L} \sum_{i=1}^{I} \tilde{u}_{ih} \left(b_{ij}^{(\ell)} \log \theta_{hk}^{(\ell)} - \mu_i \nu_j \theta_{hk}^{(\ell)} \right)$;
    $\tilde{v}_{jk} \leftarrow \frac{\exp(t_{jk})}{\sum_{k'=1}^{K} \exp(t_{jk'})}$;
  end
  foreach $h \in [H], k \in [K]$ **do**
    $\pi_h \leftarrow \frac{1}{I} \sum_{i=1}^{I} \tilde{u}_{ih}$; $\rho_k \leftarrow \frac{1}{J} \sum_{j=1}^{J} \tilde{v}_{jk}$;
    foreach $\ell \in [L]$ **do**
      $\theta_{hk}^{(\ell)} \leftarrow \frac{\sum_{i=1}^{I} \sum_{j=1}^{J} \tilde{u}_{ih} \tilde{v}_{jk} b_{ij}^{(\ell)}}{\left(\sum_{i=1}^{I} \tilde{u}_{ih} \mu_i\right)\left(\sum_{j=1}^{J} \tilde{v}_{jk} \nu_j\right)}$;
    end
  end
  $G_{\text{new}} \leftarrow G(\mathbb{T}, \mathbf{U}, \mathbf{V})$; $\Delta \leftarrow |1 - G_{\text{new}}/G_{\text{old}}|$;
end
return $\mathbb{T}$, ROUND($\mathbf{U}$), ROUND($\mathbf{V}$)

3.1 Data Description

The dataset we use was originally created for the Mini-Challenge 3 of the VAST 2013 competition [26]. It represents two weeks of simulated network traffic between approximately 1400 hosts, most of which belong to an enterprise network. We refer to these hosts as internal ones, while hosts that are not part of the enterprise network are called external hosts. The dataset contains benign traffic as well as various kinds of attacks. Some of these attacks are rather noisy—e.g. distributed denials of service (DDoS), port scans—while others are more subtle (e.g. data exfiltration). Overall, 53 out of 200 external hosts are involved in malicious traffic, which is a rather high ratio. This, along with the small number of hosts and synthetic nature of the data, makes this dataset a somewhat easy test case for our methodology. Moreover, a thorough description of the hosts and their respective roles is available, as well as a complete list of attack-related events. This allows us to evaluate the relevance of our results.

Note that while the dataset encompasses several data sources, we only keep the network flow records between internal hosts and external hosts. We turn these records into a bipartite multiplex graph as follows. First, each internal (resp. external) host is represented by a top (resp. bottom) node. For each flow between an internal host IH and an external host EH, we then build an edge between IH and EH, of type (*Protocol*, *Destination Port*, *Direction*). Note that due to the large number of possible destination ports, we only keep 10 distinct values corresponding to well-known protocols[1] and represent the other values as a single Other token. As for the direction of the flow, it is either inbound (the internal host is the destination) or outbound (the internal host is the source). Note that each observed (IH, EH, Type) triple is represented by one single edge, regardless of its number of occurrences. See Table 1 for some descriptive statistics about the dataset and the obtained graph.

Table 1. Description of the datasets: number of top nodes I, number of bottom nodes J, number of layers L, number of distinct edges M, and total number of events N.

Dataset	I	J	L	M	N
VAST	1220	200	18	26,597	68,793,510
LANL	74,049	16,119	44	869,547	842,282,832

3.2 Results

Our methodology yields three clusters of internal hosts and three clusters of external hosts. These clusters, as well as the aggregated edges between them, are shown in Fig. 1. Note that for the sake of readability, only edges representing at least 40 events are displayed. The size of each cluster is proportional to the logarithm of the number of hosts it contains. Similarly, the width of each edge grows logarithmically with the number of underlying events.

Starting with internal clusters, we observe that cluster 1 initiates HTTP and SSH connections, and that it has more outbound flows than inbound ones. Similarly, cluster 3 primarily interacts through outbound flows, including HTTP and FTP traffic. In contrast, cluster 2 receives many inbound connections, with HTTP, SMTP and DNS among the most represented protocols. It also sends SMTP traffic to external cluster 3. Overall, we can thus safely assume that clusters 1 and 3 contain workstations while cluster 2 contains servers, which is

[1] TCP/20 (FTP-Data), TCP/21 (FTP), TCP/22 (SSH), TCP/23 (Telnet), TCP/25 (SMTP), TCP/53 (DNS), TCP/80 (HTTP), TCP/443 (HTTPS), TCP/465 (SMTPS), and TCP/587 (SMTP message submission).

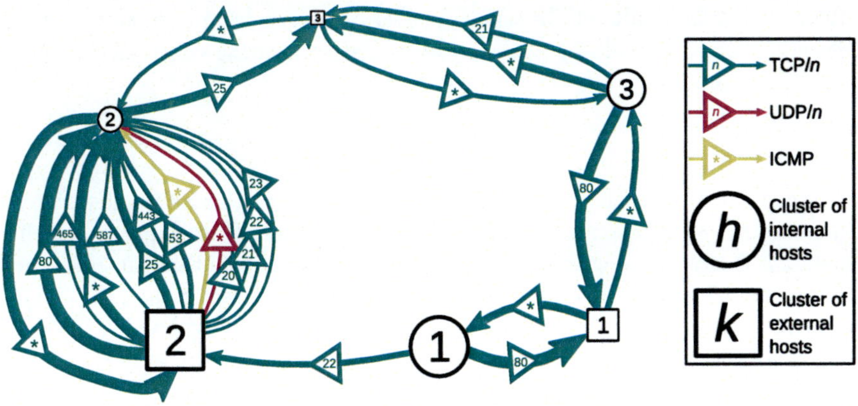

Fig. 1. Clusters found in the VAST dataset and interactions between them.

indeed confirmed by the ground truth description. As for external clusters, cluster 1 mostly receives HTTP connections from internal workstations, suggesting that it contains Web servers. This is consistent with the ground truth. Cluster 3 contains a mail server and an FTP server, which explains the inbound SMTP traffic coming from internal servers as well as the inbound FTP connections coming from internal workstations. Finally, cluster 2 contains the majority of external hosts, which are primarily observed connecting to the internal servers. In particular, attackers all fall into cluster 2, along with many benign hosts.

Even though partitioning the set of external hosts does not isolate malicious ones, studying the interactions between clusters allows us to uncover several attacks. For instance, all TCP traffic from external cluster 2 to internal cluster 2 on ports 20, 21, 22, 23, 53, 443, 465 and 587 originates from two hosts (10.9.81.5 and 10.10.11.15), which happen to be attackers. This also holds for all UDP traffic between these two clusters. These edges stand out in Fig. 1 because of the relatively small number of events they represent. Such scarce connections on many different ports suggest port scanning activity, which is confirmed by the ground truth. Similarly, ICMP traffic from external cluster 2 to internal cluster 2 also results from network scans. Finally, all SSH traffic from internal cluster 1 to external cluster 2 is directed towards one single host (10.0.3.77), which calls for further investigation. These flows actually represent beaconing activity from compromised internal hosts to a command and control server.

Overall, this first case study shows that our approach can indeed infer meaningful clusters, thereby revealing the functional roles of most hosts. It can also help identify some malicious behaviors by providing a reduced number of starting points for deeper investigation. Note, however, that some types of attacks are not easily detectable. Typically, data exfiltrations and DDoS attacks only differ from normal traffic through their volume. Since our graph-based representation does not include this characteristic, it does not allow us to distinguish exfiltrations from regular FTP traffic, or DDoS attacks from regular Web traffic. However, these high-volume attacks can be detected using other methods.

4 Second Case Study—Authentication Logs

Our second case study focuses on a different data source, namely authentication logs collected within a real-world enterprise network. The dataset, which we describe in Sect. 4.1, is significantly larger and more complex than the one studied in the previous section. It is thus more challenging to extract meaningful insights from it. However, as we show in Sect. 4.2, our methodology still allows us to infer some functional roles and uncover malicious behavior.

4.1 Data Description

We use the "Comprehensive, Multi-Source Cyber-Security Events" [14,15] dataset released by the Los Alamos National Laboratory (LANL). This dataset represents 58 days of activity within the LANL's enterprise network, recorded through several kinds of event logs. A red team exercise took place during this time span, meaning that penetration testers tried to breach the network in order to assess its security. The remote authentications performed by the red team are labelled, providing and interesting example of an advanced intrusion within an enterprise network. In particular, this attack is stealthier than those considered in the previous section, which makes it a more challenging test case for our method.

We focus on Windows authentication logs, more specifically on successful *LogOn* events. Each one of these events is described by a source user SU, a destination user DU, a source host SH, a destination host DH, a logon type LT and an authentication package AP. Note that SU and DU can be identical, and the same goes for SH and DH. As for LT and AP, they describe the type of session being created and the protocol used to authenticate the user, respectively. Examples of logon types include Interactive (local session with graphical interface) or Network (used for remote file accesses or remote procedure calls, for instance), and the most frequent authentication packages are Kerberos and NTLM. Note that these two additional fields are especially meaningful as they allow to distinguish different behaviors (e.g., remote administrative session versus simple file access). They can also help characterize users and hosts: for instance, since NTLM is considered less secure than Kerberos, NTLM authentication can sometimes be disabled for highly privileged accounts. Conversely, some legacy applications may not support Kerberos authentication, thus servers hosting such applications should frequently use NTLM.

We represent users as top nodes and hosts as bottom nodes, and we turn logon events into typed edges as follows. If SH and DH are identical, we create one edge between DU and DH with type (LT, AP, Local). Otherwise, we create one edge between SU and SH with type (LT, AP, From), and one edge between DU and DH with type (LT, AP, To). While this construction breaks the link between the source and destination of an authentication, it still highlights the difference between hosts receving many remote authentications (typically servers) and hosts from which these authentications originate (typically workstations). See Table 1 for some descriptive statistics about the dataset.

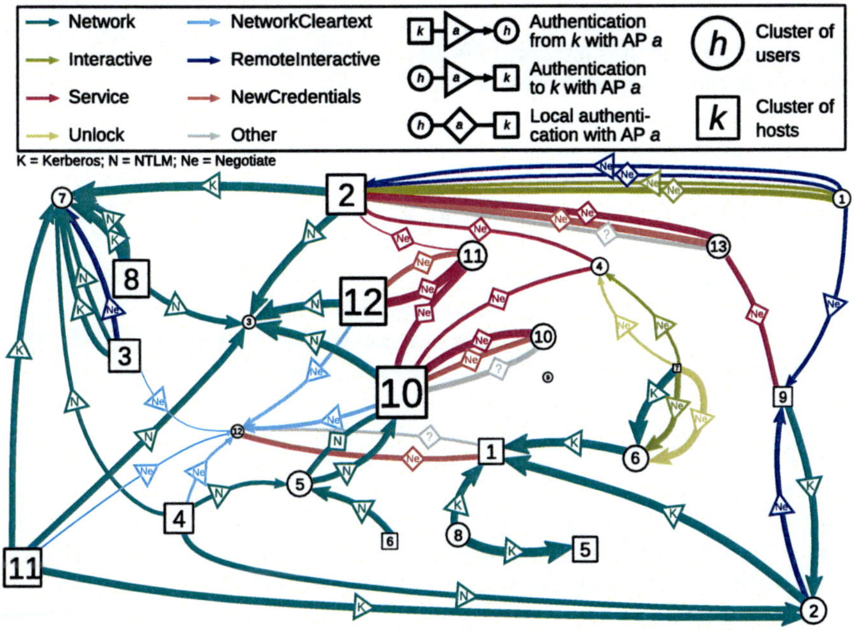

Fig. 2. Clusters found in the LANL dataset and interactions between them.

4.2 Results

The optimal model has thirteen user clusters and twelve host clusters. We display them in Fig. 2, along with the main interactions between them. Specifically, we display all edges (h, k, ℓ) such that $\theta_{hk}^{(\ell)} \geq 0.7$. Note that we select edges using the set of rate matrices $\boldsymbol{\Theta}$ rather than the number of underlying events because the behaviors we seek to detect generate relatively few events. Thus cluster–cluster edges with high rate, which represent many user–host edges occurring at least once, are more relevant than those representing many individual events.

The global picture is expectedly more complex than in the previous section. In addition, little information is available on the true functional roles of users and hosts. However, we can still infer the meaning of some clusters and confirm their relevance. For instance, user clusters 10, 11 and 13 mainly interact through local authentications with logon type Service, indicating that they contain service accounts. User cluster 1 opens remote interactive sessions on many hosts, which suggests the presence of administrator accounts. As for user clusters 3 and 5, they mostly consist of anonymous user credentials, which are typically used to access shared resources which do not require any access control (such as internal Web pages). This is consistent with the NTLM-authenticated network logons associated with these clusters. Interestingly, a significant proportion of the non-anonymous user names from these two clusters were used for red team activity. This suggests that looking for inconsistencies between the expected functional role of an entity and the cluster it falls into can help uncover malicious behaviors.

Inferring the functional roles of host clusters is more difficult, although it seems reasonable to assume that cluster 1 contains domain controllers and applicative servers. Indeed, this cluster is the one receiving the most remote authentications. Conversely, clusters 2, 3, 8, 9, and 11 are sources of remote authentications, suggesting that they mostly contain workstations. Finally, cluster 4 stands out as the source of remote NTLM authentications involving three user clusters. Further investigation reveals that most of these authentications were performed by the red team and originate from host C17693, which also happens to be the main source of red team activity. Once again, the simplified situational picture we generate thus preserves important clues that can lead to the detection of malicious behavior.

5 Related Work

Biclustering and latent block models. Biclustering, i.e., the idea of simultaneously partitioning the rows and columns of a matrix so as to form blocks of coefficients with similar values, can be traced back to the early 1970s [13]. Spectral methods [7,18] are among the most popular approaches to this task. However, model-based biclustering has gained traction in the last two decades, following the introduction of the latent block model [10]. Various inference procedures have been proposed, most of which rely on the expectation-maximization (EM) algorithm. While we apply the simple variational approximation proposed in [10], other approaches include the addition of intermediary classification steps [11] or the use of Bayesian inference [16].

Multilayer stochastic block models. Even though we are not aware of any previous work on multilayer generalizations of the latent block model, a straightforward connection can be made with multiplex stochastic block models (SBMs). More specifically, our model is analogous to the degree-corrected SBM with independent layers described in [22]. Note, however, that many other multilayer generalizations of the SBM can be found in the literature. In particular, the existence of diversely complex statistical dependencies between layers has been addressed in various ways [2,6,21,24].

Data exploration and visualization for cybersecurity. Finally, our work contributes to a vast research effort aiming to ease network security monitoring by providing insightful visualizations. In particular, displaying network flows between internal and external hosts is a recurring challenge. VISUAL [1], VisFlowConnect [28], NFlowVis [8], and FloVis [25] are some of the existing tools designed to perform this task. However, none of them focuses on statistical modelling to build a more condensed view. Note that FloVis and NFlowVis use the hierarchical nature of IP addresses to aggregate hosts by subnetwork, and NFlowVis also uses k-medoids clustering to find external hosts with similar communication patterns. Even so, the use of behavior-based biclustering makes our method more effective at reducing the number of elements to display while preserving essential information. As for authentication logs, APTHunter [23] also focuses on interactivity rather than automated data reduction, only letting the user manually apply filters to make the authentication graph legible. Finally, the work of Glatz et al. [9] adopts an approach somewhat similar to ours: they first extract frequent itemsets from network flow records, then display them as a bipartite itemset–item graph. While this does provide a simple summary of large volumes of logs, the extracted itemsets still cover a small minority of the total number of events, leading to significant information loss.

6 Conclusion and Perspectives

We propose a graph-oriented approach to event log exploration for network security monitoring. Through the use of model-based multiplex graph biclustering, we aim to extract meaningful clusters of entities, such as groups of users or hosts sharing functional roles or behavioral patterns. Our case studies demonstrate that such meaningful clusters can indeed be uncovered. In addition, displaying interactions between these clusters can facilitate malicious behavior detection.

Aside from investigating better model selection criteria (see Sect. 2.2), extending our model to factor in the temporal dimension could be an interesting lead for future work. Previous contributions on temporal latent block models [5] provide foundations for such an extension. Finally, looking for meaningful groups of edge types in addition to entity clusters, similarly to the strata multilayer SBM [24], is another promising direction. Indeed, it could lead to an even simpler situational picture, especially for large and complex datasets such as the one studied in Sect. 4.

References

1. Ball, R., Fink, G.A., North, C.: Home-centric visualization of network traffic for security administration. In: VizSec/DMSec (2004)
2. Barbillon, P., Donnet, S., Lazega, E., Bar-Hen, A.: Stochastic block models for multiplex networks: an application to a multilevel network of researchers. J. R. Stat. Soc. A Stat. **180**(1), 295–314 (2017)
3. Biernacki, C., Celeux, G., Govaert, G.: Assessing a mixture model for clustering with the integrated completed likelihood. IEEE Trans. Pattern Anal. Mach. Intell. **22**(7), 719–725 (2000)

4. Côme, E., Latouche, P.: Model selection and clustering in stochastic block models based on the exact integrated complete data likelihood. Stat. Model. **15**(6), 564–589 (2015)
5. Corneli, M., Latouche, P., Rossi, F.: Exact ICL maximization in a non-stationary time extension of the latent block model for dynamic networks. In: ESANN (2015)
6. De Bacco, C., Power, E.A., Larremore, D.B., Moore, C.: Community detection, link prediction, and layer interdependence in multilayer networks. Phys. Rev. E **95**(4), 042317 (2017)
7. Dhillon, I.S.: Co-clustering documents and words using bipartite spectral graph partitioning. In: KDD (2001)
8. Fischer, F., Mansmann, F., Keim, D.A., Pietzko, S., Waldvogel, M.: Large-scale network monitoring for visual analysis of attacks. In: VizSec (2008)
9. Glatz, E., Mavromatidis, S., Ager, B., Dimitropoulos, X.: Visualizing big network traffic data using frequent pattern mining and hypergraphs. Computing **96**(1), 27–38 (2014)
10. Govaert, G., Nadif, M.: Clustering with block mixture models. Pattern Recognit. **36**(2), 463–473 (2003)
11. Govaert, G., Nadif, M.: Block clustering with Bernoulli mixture models: comparison of different approaches. Comput. Stat. Data Anal. **52**(6), 3233–3245 (2008)
12. Govaert, G., Nadif, M.: Latent block model for contingency table. Commun. Stat. Theory Methods **39**(3), 416–425 (2010)
13. Hartigan, J.A.: Direct clustering of a data matrix. J. Am. Stat. Assoc. **67**(337), 123–129 (1972)
14. Kent, A.D.: Comprehensive, multi-source cyber-security events. Los Alamos National Laboratory (2015)
15. Kent, A.D.: Cybersecurity data sources for dynamic network research. In: Dynamic Networks in Cybersecurity. Imperial College Press (2015)
16. Keribin, C., Brault, V., Celeux, G., Govaert, G.: Estimation and selection for the latent block model on categorical data. Stat. Comput. **25**(6), 1201–1216 (2015)
17. Kivelä, M., Arenas, A., Barthelemy, M., Gleeson, J.P., Moreno, Y., Porter, M.A.: Multilayer networks J. Complex Netw. **2**(3), 203–271 (2014)
18. Kluger, Y., Basri, R., Chang, J.T., Gerstein, M.: Spectral biclustering of microarray data: coclustering genes and conditions. Genome Res. **13**(4), 703–716 (2003)
19. Lomet, A., Govaert, G., Grandvalet, Y.: Model selection in block clustering by the integrated classification likelihood. In: COMPSTAT (2012)
20. Metelli, S., Heard, N.: On Bayesian new edge prediction and anomaly detection in computer networks. Ann. Appl. Stat. **13**(4), 2586–2610 (2019)
21. Paul, S., Chen, Y.: Consistent community detection in multi-relational data through restricted multi-layer stochastic blockmodel. Electron. J. Stat. **10**(2), 3807–3870 (2016)
22. Peixoto, T.P.: Inferring the mesoscale structure of layered, edge-valued, and time-varying networks. Phys. Rev. E **92**(4), 042807 (2015)
23. Siadati, H., Saket, B., Memon, N.: Detecting malicious logins in enterprise networks using visualization. In: VizSec (2016)
24. Stanley, N., Shai, S., Taylor, D., Mucha, P.J.: Clustering network layers with the strata multilayer stochastic block model. IEEE Trans. Netw. Sci. Eng. **3**(2), 95–105 (2016)
25. Taylor, T., Paterson, D., Glanfield, J., Gates, C., Brooks, S., McHugh, J.: Flovis: Flow visualization system. In: CATCH (2009)

26. Whiting, M., Cook, K., Paul, C., Whitley, K., Grinstein, G., Nebesh, B., Liggett, K., Cooper, M., Fallon, J.: Vast challenge 2013: Situation awareness and prospective analysis. In: VAST (2013)
27. Wyse, J., Friel, N., Latouche, P.: Inferring structure in bipartite networks using the latent blockmodel and exact ICL. Netw. Sci. **5**(1), 45–69 (2017)
28. Yin, X., Yurcik, W., Treaster, M., Li, Y., Lakkaraju, K.: Visflowconnect: netflow visualizations of link relationships for security situational awareness. In: VizSec/DMSec (2004)

Check for updates

On the Effectiveness of Using Link Weights and Link Direction for Community Detection in Multilayer Networks

Daiki Suzuki(✉) and Sho Tsugawa

University of Tsukuba, Tsukuba, Ibaraki 305-8573, Japan
d.suzu@mibel.cs.tsukuba.ac.jp s-tugawa@cs.tsukuba.ac.jp

Abstract. Multilayer networks are useful representations of real-world complex networks, and community detection in multilayer networks has been an area of active research. There are several options for the graph representation of a multilayer network, for example, directed or undirected and weighted or unweighted. Although these options may affect the results of community detection in a multilayer network, the representations that are effective for community detection have been not yet been clarified. In this paper, we experimentally investigate how the graph representation of a multilayer network affects the results of community detection. Through experiments using multilayer networks of Twitter users, we show that using a directed graph for each layer of a multilayer network contributes to improved accuracy in estimating the communities of Twitter users. We also show that when there is a clear oppositional structure among nodes in a network, manipulating link weights to emphasize the oppositional structure improves the accuracy of community detection.

Keywords: Multilayer network · Community detection · Social network

1 Introduction

Community detection is a central issue in network science research. Many networks have a community structure that consists of densely connected communities of nodes and sparse links between communities [14]. The goal in community detection is to extract communities from a given network and to determine the community to which each node belongs [6].

There are many proposed methods for detecting communities in a single given network, but in recent years there have been increasing research efforts toward detecting communities in multiple layers of networks [12]. In a multilayer network, different layers represent different types of relationships between nodes. For example, a society of individuals can be represented as a multilayer network

H. Cherifi et al. (Eds.): COMPLEX NETWORKS 2016 2022, SCI 1077, pp. 638–649, 2023.
https://doi.org/10.1007/978-3-031-21127-0_52

with different layers of person-to-person relationships through different means of communication, such as telephone, e-mail, and face-to-face conversations. Multilayer networks are considered to be effective representations of complex real-life objects, and there is a growing focus on developing techniques for community detection in multilayer networks [12].

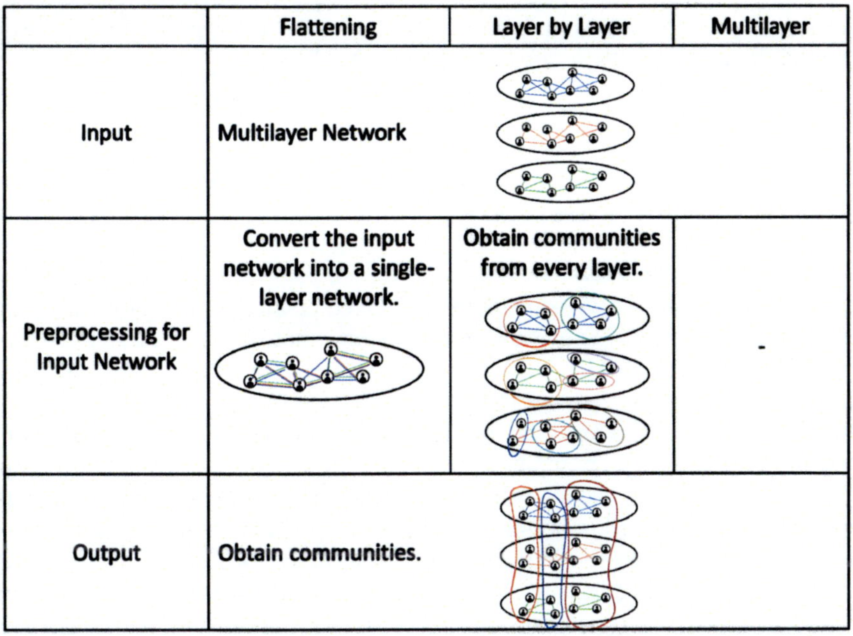

Fig. 1. Three types of community detection algorithms for multilayer networks [12]

Three major approaches are used for community detection in multilayer networks and various community detection algorithms have been proposed for each approach [12]. The following provides an overview of each of the three approaches:

1. Flattening
   This type of community detection algorithm obtains communities using an algorithm that extracts communities from a single-layer network constructed by merging all links at all layers of a multilayer network [1]. Any existing algorithm developed for single-layer networks can be used, but the downside of this method is that information about the layers of the input multilayer network is completely lost.
2. Layer-by-layer
   In this type of algorithm, communities are extracted from each layer by using a single-layer community detection algorithm, and the extracted communities are merged to obtain the communities of the multilayer network [2,18]. As in

the "flattening" approach, any community detection algorithm for single-layer networks can be used. While this approach uses information from the layers of a multilayer network, each layer is independently treated, and relations between layers are not considered.

3. Multilayer
   This approach takes a multilayer network as input and detects communities as they are in the multilayer network. Because the input to the community detection algorithm is a multilayer network, all information in the multilayer network can be used to detect communities [10,11,13], and because all the information is used, between-layer relations can be considered when detecting communities.

These methods differ in their characteristics, but it remains unclear which is most effective for which types of data.

There are also several options for representing multilayer networks to be input into a community detection algorithm. For example, we should decide whether to represent the multilayer network as a directed or undirected network, how to define the link weights, and whether to use information from all or only some layers. It is also unclear how differences in representations of multilayer networks affect the results of community detection.

In this paper, therefore, we evaluate how representations of input multilayer networks affect community detection algorithms for a typical multilayer network. In particular, we analyze how community detection results differ with and without link direction and with and without link weights. We also investigate how manipulating the link weights of certain layers affects the community detection results. As representative algorithms for community detection in multilayer networks, WF_{EC} [1] is used as a flattening approach, hybrid bipartite graph formulation (HBGF) [5,7] is used as a layer-by-layer approach, and generalized Louvain (GLouvain) [10,13] is used as a multilayer approach. We apply these algorithms to three types of multilayer networks representing relationships among Twitter users, and evaluate how varying the presence or absence of link direction and link weights in the input multilayer networks changes the community detection results.

The remainder of this paper is organized as follows. Section 2 presents the problem formulation, dataset, and experimental methodology for detecting communities in a multilayer network. In Sect. 3, we present and discuss the results of our analysis. Finally, Sect. 4 summarizes the paper and presents topics for future study.

2 Methodology

2.1 Problem Formulation

A multilayer network is defined as $M = \{G_1, G_2, \ldots, G_l\}$, where $G_i = (V_i, E_i, W_i)$ $(1 \leq i \leq l)$ is the network representing layer i in the multilayer network, and l is the number of layers. Here, V_i is the set of nodes in layer i, E_i

is the set of links, and W_i is the set of link weights. $w_i(u,v) \in W_i$ is the weight on link $(u,v) \in E_i$. We use $V = \bigcup V_i$ to denote the set of all nodes in M. Note that $(u,v) = (v,u)$ if the network is undirected at each layer, but $(u,v) \neq (v,u)$ for a directed network. If G_i is an unweighted network, then $w_i(u,v) \in E_i$ for all links $(u,v) \in E_i$, $w_i(u,v) = 1$. By this definition, $V_i \subseteq V$, and not all nodes exist in all layers. Each node $v \in V$ in the multilayer network belongs to one of the ground truth communities $C = \{c_1, c_2, \ldots, c_k\}$. The goal in community detection is thus to determine the ground truth community $c(v) \in C$ to which node $v \in V$ in multilayer network M belongs.

Table 1. Basic statistics of the DKPol dataset

	RT layer	Reply layer	Total
Num. nodes	212	137	228
Num. links	722	284	1006
Num. ground truth communities	10		

Table 2. Basic statistics of the Higgs dataset

	RT layer	Reply layer	Mention layer	Total
Num. nodes	27,234	4444	15,160	34,877
Num. links	30,375	3695	18,851	52,921
Num. ground truth communities	113			

Table 3. Basic statistics of the Mancer dataset

	RT layer	Reply layer	Quote layer	Total
Num. nodes	11,561	4921	3717	12,774
Num. links	106,700	13,707	9057	129,164
Nm. ground truth communities	77			

2.2 Datasets

In this study, we use multilayer networks from three datasets representing different types of relationships among Twitter users: DKPol [8], Higgs [4], and Mancer [16]. Each dataset is described below:

1. DKPol
   This is a multilayer network of Twitter users among Danish parliamentarians as nodes. Layers represent follow, retweet, and reply relationships between nodes. The dataset provides data on the political party to which each node belongs, which we use as the ground truth community.
2. Higgs
   The Higgs dataset was constructed by monitoring the diffusion process on Twitter before, during, and after the July 4, 2012 announcement of the discovery of a new particle with the predicted characteristics of the Higgs boson. Its layers represent the follow, retweet, reply, and mention relations among nodes representing Twitter users. As target data for the experiments in this study, we used the network as it was between July 1, 2012 and July 3, 2012.
3. Mancer
   This is a multilayer network of Greek Twitter users. Layers represent follow, retweet, reply, and quote relationships among the Greek Twitter users. We used the network as it was on February 1, 2018 as the target data for the experiment in this study.

The Higgs and Mancer datasets do not give explicit information on the community to which each node belongs, so for the ground truth community, following [17], we used the communities obtained only from the follow layer by using the Leiden method [19]. To obtain the ground truth communities, the follow layer is represented as a directed unweighted graph.

For the Higgs and Mancer datasets, we used the community obtained using the follow layer as the ground truth community, so for the Higgs dataset, we used a multilayer network comprising the retweets, replies, and mentions layers as input in the community detection method, while for the Mancer dataset, we used the retweets, replies, and quotes layers. We did not use the follow layer in the DKPol dataset, as in the other datasets, but we used the retweet and reply layers for community detection. Tables 1, 2, and 3 summarize the multilayer networks constructed from these three datasets.

2.3 Representations of Multilayer Networks

We construct a multilayer network M from each dataset to be used as input in the community detection algorithm. Link direction can be naturally defined for each layer, and numbers of retweets and replies provide natural definitions for link weights. It is also possible to represent M as an undirected or unweighted network, ignoring weights and link directions. In this paper, for each dataset, we use four types of graph representations: weighted–undirected (undirected), weighted–directed (directed), undirected–unweighted, and directed–unweighted. For a directed network, a link (u, v) in each layer represents u sending a reply or mention to v, or u retweeting or quoting a tweet by v. For a weighted network, a link weights $w(u, v)$ in each layer represents the number of replies, mentions, retweets, and quotes between u and v. We analyze how the results of community detection differ depending on how the input network is represented.

Using the characteristics of each layer in the multilayer network, we also construct a multilayer network M with manipulated link weights. As discussed in [12], the reply layer has different characteristics than the retweet and follow layers in multilayer networks that represent relationships among Twitter users with a clear oppositional structure, such the political party affiliations in the DKPol dataset. Specifically, while there are many retweet- and follow-layer links between nodes belonging to the same community, there are few reply-layer links. From these results, it is expected that the accuracy of community detection might be improved by assigning link weights in the reply layer differently from in the retweet and follow layers, instead of simply using the number of replies as the link weight.

Specifically, we manipulate the link weights as follows. First, the community C_{RT} is obtained from the unweighted undirected network G_{RT}, which represents the retweet layer, by using the Louvain method. Node $v \in V_{\mathrm{RT}}$ of retweet layer G_{RT} belongs to community $c_{\mathrm{RT}}(v) \in C_{\mathrm{RT}}$. We replace link weight $w_{\mathrm{RE}}(u,v)$ in network G_{RE} representing the reply layer with $\tilde{w}_{\mathrm{RE}}(u,v)$, defined as

$$\tilde{w}_{\mathrm{RE}}(u,v) = \begin{cases} \alpha \times w_{\mathrm{RE}}(u,v) & \text{if } c_{\mathrm{RT}}(u) = c_{\mathrm{RT}}(v), \\ w_{\mathrm{RE}}(u,v) & \text{if } u, v \notin V_{\mathrm{RT}}, \\ \beta \times w_{\mathrm{RE}}(u,v) & \text{if } c_{\mathrm{RT}}(u) \neq c_{\mathrm{RT}}(v). \end{cases} \tag{1}$$

In this paper, to determine the link weights, we set parameters α and β to one of three settings; weaken, strengthen, or cut. "Weaken" reduces link weights in the reply layer if the nodes at both end of the reply layer link do not belong to the same community in the retweet layer, with $\alpha = 1$ and $\beta = 0.5$. "Strengthen" increases reply layer link weights if the nodes at both end of the reply layer link belong to the same community in the retweet layer, with $\alpha = 2$ and $\beta = 1$. "Cut" disconnects nodes at both ends of a reply layer link if they do not belong to the same community in the retweet layer, with $\alpha = 1$ and $\beta = 0$. To construct a multilayer network M for use as input in the community detection method, we substitute the network obtained by using the weaken, strengthen, and cut settings to manipulate link weights.

2.4 Community Detection Algorithms and Evaluation Metrics

We use WF_{EC} [1], HBGF [5,7], and GLouvain [10,13] as methods for detecting communities in multilayer networks. WF_{EC} is a typical method based on the flattening approach, HBGF is a typical method based on the layer-by-layer approach, and GLouvain is a typical method based on the multilayer approach. WF_{EC} constructs a single-layer network with $\sum w_i(u,v)$ as the weight of link (u,v) from the input multilayer network, and then, the Louvain method is applied to the single-layer network to obtain communities. Since the Louvain method does not consider link directions, we used the Leiden method [19] when using a directed network as input. The Leiden method is an extension of the Louvain method and extracts communities from an input single-layer network of directed graphs. HBGF extracts communities at each layer using a community detection algorithm for single-layer networks, and constructs a bipartite graph from the set of those communities, using nodes and communities as vertices. The bipartite graph is then partitioned into communities to obtain the final communities. As the community detection algorithm for each network layer under HBGF, we used the Louvain method to extract the input undirected network and the Leiden method to extract the input directed network. GLouvain is an extension of the Louvain method and extracts communities from an input multilayer network. GLouvain can extract communities from both undirected and directed networks.

Following a previous study [15], we use normalized mutual information (NMI) [3] and the adjusted Rand index (ARI) [9] as evaluation measures. NMI and ARI are widely used for evaluating the accuracy of community detection when the ground truth communities are explicitly known.

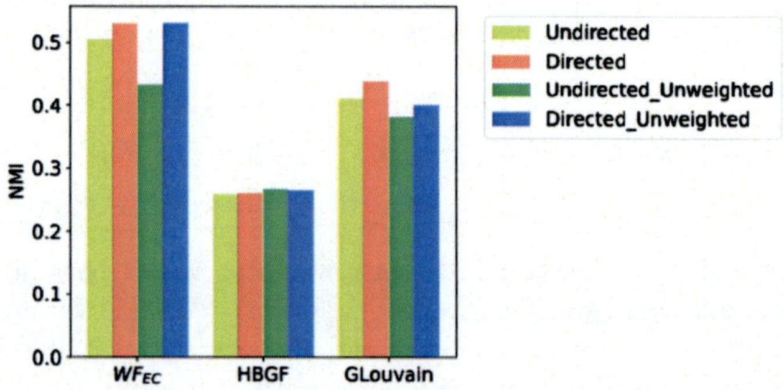

(a) DKPol

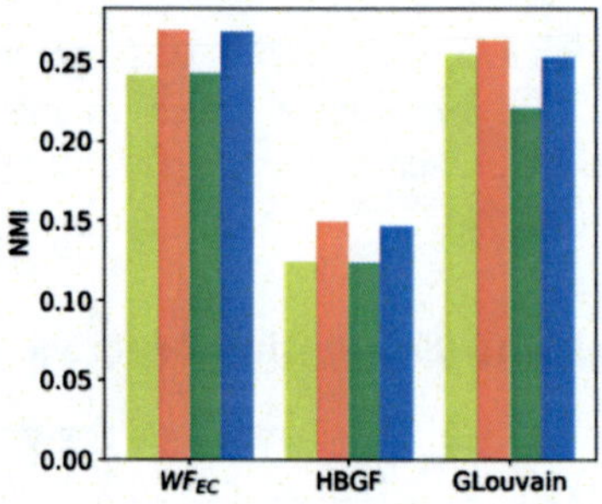

(b) Higgs

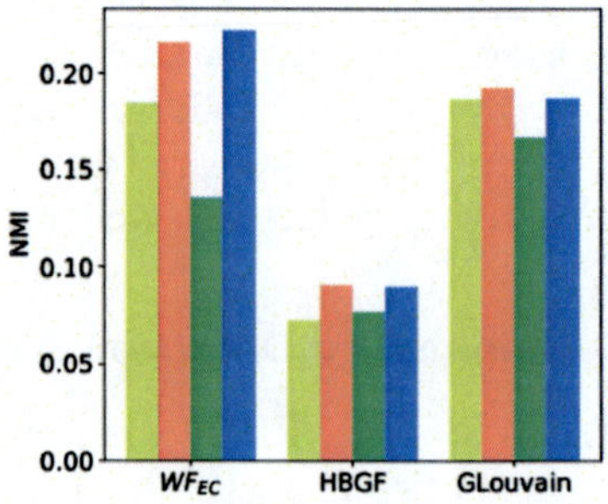

(c) Mancer

Fig. 2. Effects of link weight and link direction on NMI

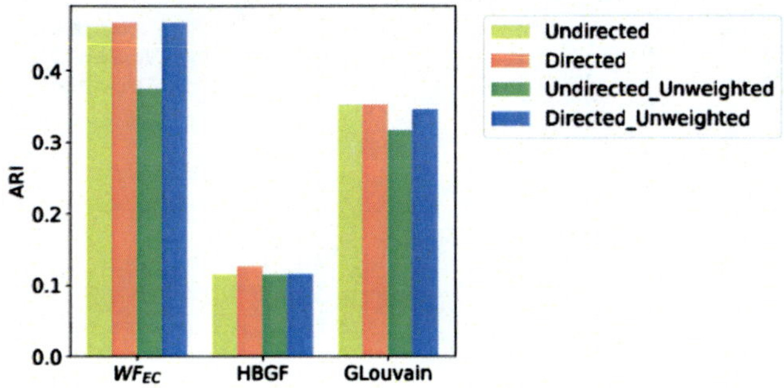

(a) DKPol

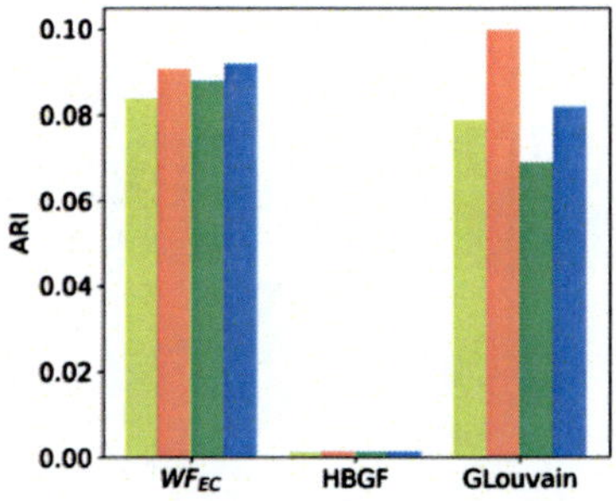

(b) Higgs

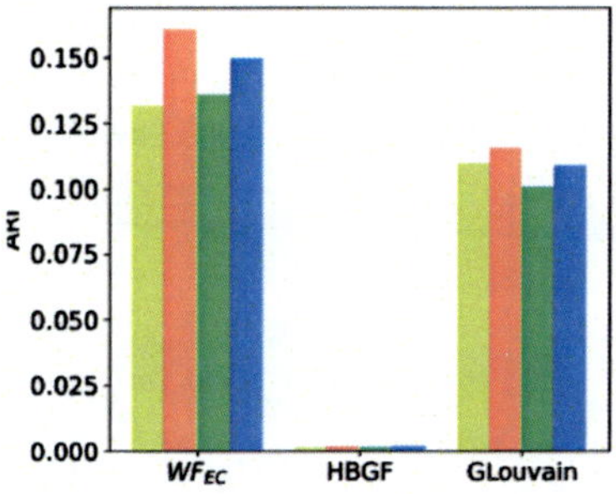

(c) Mancer

Fig. 3. Effects of link weight and link direction on ARI

3 Results and Discussion

First, we obtain NMI and ARI by detecting communities using the four different networks, with or without weights and with or without link direction as inputs. Figures 2 and 3 show the results for NMI and ARI, respectively.

Figures 2 and 3 show that when using weighted networks and directed networks, we can obtain high NMI and ARI values. In particular, the NMI and ARI values are larger when the multilayer network is represented as a directed network than when it is represented as an undirected network. This suggests that considering link direction when detecting communities in multilayer networks is important for improving the accuracy of community detection. Comparing WF_{EC} based on the flattening approach, HBGF based on the layer-by-layer approach, and GLouvain based on the multilayer approach, we can see that WF_{EC} in particular achieves higher NMI and ARI values. It seems that the layer-by-layer method extracted communities with distinctive features from each layer, and that merging communities from each layer did not result in a good merger. The multilayer method goes so far as to consider relations between the layers,

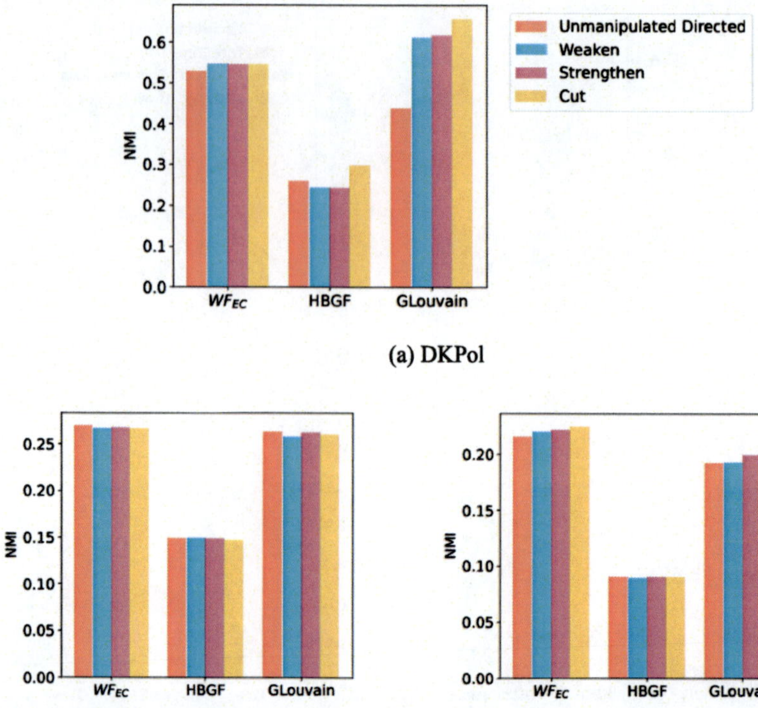

(a) DKPol

(b) Higgs

(c) Mancer

Fig. 4. Effects of link weight manipulation on NMI

but this information becomes noise in the community extraction process, which appears to results in the flattening results being better than multilayer results.

We next investigated the impact of the link weight manipulations (strengthen, weaken, and cut) on community detection results. We used directed weighted networks for the link weight manipulation. Figures 4 and 5 show the results for NMI and ARI, respectively.

First, we focus on the DKPol dataset, in which political parties create a clear oppositional structure. For this dataset, we expected that manipulating link weights would be effective in improving the accuracy of community detection. In line with this expectation, in most cases we observed improved accuracy on the DKPol dataset. In particular, the results of GLouvain are greatly improved in comparison with the case without manipulation of link weights. In a multilayer network with an oppositional structure, it can be effective to manipulate link weights and link disconnections that emphasize those differences.

We next focus on the Higgs and Mancer datasets. Because these datasets lack a clear oppositional structure, we expected manipulations of link weights to be less effective in improving the accuracy of community detection than in the case of the DKPol dataset. The results show that link manipulation does

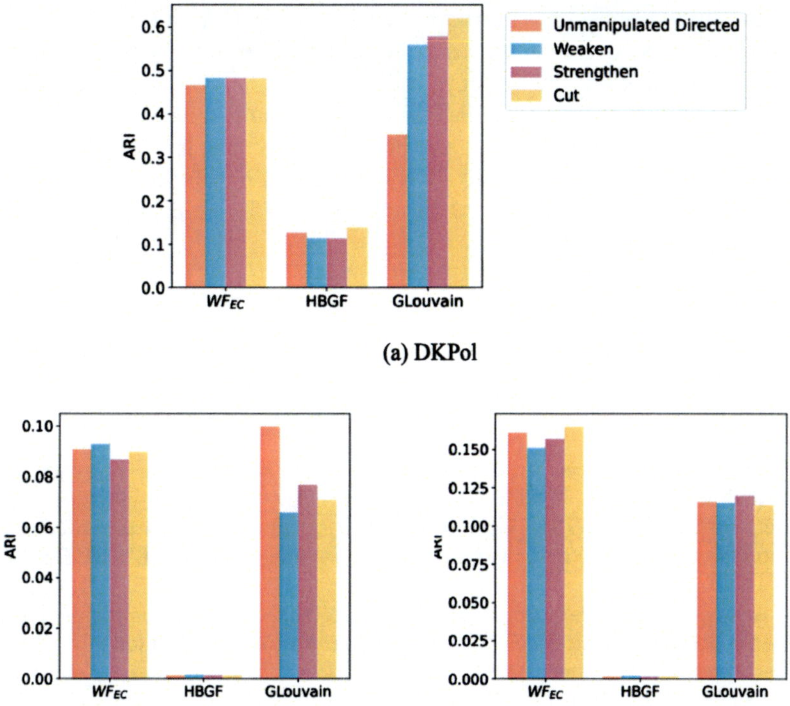

(a) DKPol

(b) Higgs

(c) Mancer

Fig. 5. Effects of link weight manipulation on ARI

not significantly change the ARI and NMI values, regardless of whether WF_{EC}, HBGF, or GLouvain is used. There are no clear oppositional structures in the Higgs and Mancer datasets, and it appears that link manipulations in the reply layer did not have much impact on the community detection results. In future work, we will continue to investigate methods of link manipulation better suited to community detection.

4 Conclusion and Future Work

We evaluated the impact of methods for representing multilayer networks applied to community detection algorithms for typical multilayer networks. We showed that representing each layer in a multilayer network as a directed network that explicitly represents relational directions among users improves the accuracy of detecting communities of Twitter users. We also showed that considering link direction is useful for improving the accuracy of several types of community detection algorithms, which include WF_{EC} based on the flattening approach, HBGF based on the layer-by-layer approach, and GLouvain based on the multi-layer approach. In addition, We showed that when there is a clear oppositional structure among nodes in a network, manupulating link weights to emphasize the oppositional structure improves the accuracy of community detection. Future work will include clarifying appropriate multilayer network representations for community detection. In particular, we will examine definitions of link weights, link manipulation methods, and methods for selecting layers for community detection. Moreover, the generalizability of the findings in this paper should be evaluated using different multilayer networks and different community detection algorithms.

Acknowledgments. This work was partly supported by JSPS KAKENHI Grant Number 22K11990.

References

1. Berlingerio, M., Coscia, M., Giannotti, F.: Finding and characterizing communities in multidimensional networks. In: Proc. ASONAM'11, pp. 490–494 (2011)
2. Berlingerio, M., Pinelli, F., Calabrese, F.: ABACUS: frequent pattern mining-based community discovery in multidimensional networks. Data Min. Knowl. Disc. **27**(3), 294–320 (2013)
3. Danon, L., Diaz-Guilera, A., Duch, J., Arenas, A.: Comparing community structure identification. J. Stat. Mech. Theor. Exper. **2005**(09), P09008 (2005)
4. De Domenico, M., Lima, A., Mougel, P., Musolesi, M.: The anatomy of a scientific rumor. Sci. Rep. **3**(1), 1–9 (2013)
5. Fern, X.Z., Brodley, C.E.: Solving cluster ensemble problems by bipartite graph partitioning. In: Proc. ICML'04, p. 36 (2004)
6. Fortunato, S.: Community detection in graphs. Phys. Rep. **486**(3–5), 75–174 (2010)
7. Ghosh, J., Acharya, A.: Cluster ensembles. Wiley Interdisc. Rev. Data Min. Knowl. Disc. **1**(4), 305–315 (2011)
8. Hanteer, O., Rossi, L., D'aurelio, D.V., Magnani, M.: From interaction to participation: The role of the imagined audience in social media community detection and an application to political communication on twitter. In: Proc. ASONAM'18, pp. 531–534 (2018)
9. Hubert, L., Arabie, P.: Comparing partitions. J. Classification **2**(1), 193–218 (1985)
10. Jutla, I.S., Jeub, L.G., Mucha, P.J., et al.: A generalized Louvain method for community detection implemented in MATLAB. http://netwikiamath.unc.edu/GenLouvain (2011)
11. Kuncheva, Z., Montana, G.: Community detection in multiplex networks using locally adaptive random walks. In: Proc. ASONAM'15, pp. 1308–1315 (2015)
12. Magnani, M., Hanteer, O., Interdonato, R., Rossi, L., Tagarelli, A.: Community detection in multiplex networks. ACM Comput. Surv. **54**(3) (2021)
13. Mucha, P.J., Richardson, T., Macon, K., Porter, M.A., Onnela, J.P.: Community structure in time-dependent, multiscale, and multiplex networks. Science **328**(5980), 876–878 (2010)
14. Newman, M.E., Girvan, M.: Finding and evaluating community structure in networks. Phys. Rev. E **69**(2), 026113 (2004)
15. Prokhorenkova, L., Tikhonov, A., Litvak, N.: When less is more: systematic analysis of cascade-based community detection (2020). arXiv:2002.00840
16. Sotiropoulos, K., Byers, J.W., Pratikakis, P., Tsourakakis, C.E.: TwitterMancer: predicting interactions on twitter accurately. arXiv:1904.11119 (2019)

17. Suzuki, D., Tsugawa, S.: Effects of hidden users on cascade-based community detection. In: Proc. CompleNet'21, pp. 1–11. Springer (2021)
18. Tagarelli, A., Amelio, A., Gullo, F.: Ensemble-based community detection in multilayer networks. Data Min. Knowl. Disc. **31**(5), 1506–1543 (2017)
19. Traag, V.A., Waltman, L., Van Eck, N.J.: From Louvain to Leiden: guaranteeing well-connected communities. Sci. Rep. **9**(1), 1–12 (2019)

Author Index

H. Cherifi et al. (Eds.): COMPLEX NETWORKS 2016 2022, SCI 1077, pp. 651–653, 2023.
https://doi.org/10.1007/978-3-031-21127-0